The Amboseli Elephants

The Amboseli Elephants

A Long-Term Perspective on a Long-Lived Mammal

Edited by Cynthia J. Moss,
Harvey Croze, and Phyllis C. Lee

The University of Chicago Press
Chicago and London

CYNTHIA J. MOSS is the director of the Amboseli Trust for Elephants and the author of *Elephant Memories: Thirteen Years in the Life of an Elephant Family*. HARVEY CROZE is a trustee for the Amboseli Trust for Elephants and coauthor of *Pyramids of Life: An Investigation of Nature's Fearful Symmetry*. PHYLLIS C. LEE is professor of psychology at the University of Stirling.

Section photographs by Martyn Colbeck

The University of Chicago Press, Chicago 60637
The University of Chicago Press, Ltd., London
© 2011 by The University of Chicago
All rights reserved. Published 2011.
Printed in the United States of America

18 17 16 15 14 13 12 11   1 2 3 4 5

ISBN-13: 978-0-226-54223-2 (cloth)
ISBN-10: 0-226-54223-8 (cloth)

Library of Congress Cataloging-in-Publication Data
The Amboseli elephants : a long-term perspective on a long-lived mammal / edited by Cynthia J. Moss, Harvey Croze, and Phyllis C. Lee.
    p. cm.
  Includes bibliographical references and index.
ISBN-13: 978-0-226-54223-2 (cloth : alk. paper)
ISBN10: 0-226-54223-8 (cloth : alk. paper) 1. African elephant—Kenya—Amboseli National Park. 2. African elephant—Ecology—Kenya—Amboseli National Park. 3. African elephant—Habitat—Kenya—Amboseli National Park. 4. African elephant—Behavior—Kenya—Amboseli National Park. 5. African elephant—Kenya—Amboseli National Park—Reproduction. I. Moss, Cynthia. II. Croze, Harvey. III. Lee, Phyllis C.
  QL737.P98A475 2011
  599.67'4—dc22
                                                                        2010011941

♾ The paper used in this publication meets the minimum requirements of the American National Standard for Information Sciences—Permanence of Paper for Printed Library Materials, ANSI Z39.48-1992.

Contents

Foreword
Robert A. Hinde ix

Acknowledgments xi

1 The Amboseli Elephants: Introduction
   Cynthia J. Moss, Harvey Croze, and Phyllis C. Lee 1

   Box 1.1 Definitions of Terms Used throughout the Book 2

   Annex 1.1 Glossary of Maa Place-Names and Synonyms 6

Part 1: The Amboseli Context: Ecology, People, and Genetics
Section editors: *W. Keith Lindsay and Harvey Croze*

2 Amboseli Ecosystem Context: Past and Present
   Harvey Croze and W. Keith Lindsay 11

   Box 2.1 Amboseli, a Non-equilibrium Ecosystem 19

   Annex 2.1 Population Size over Time 25

3 The Human Context of the Amboseli Elephants
   Kadzo Kangwana and Christine Browne-Nuñez 29

4 The Population Genetics of the Amboseli and Kilimanjaro Elephants
   Elizabeth A. Archie, Courtney L. Fitzpatrick, Cynthia J. Moss, and Susan C. Alberts 37

   Box 4.1 Population Genetic Methods 40

Part 2: Habitat Use, Population Dynamics, and Ranging
Section editors: *W. Keith Lindsay and Harvey Croze*

5 Habitat Use, Diet Choice, and Nutritional Status in Female and Male Amboseli Elephants
   W. Keith Lindsay 51

   Box 5.1 Size and Energetics of Elephants *Phyllis C. Lee* 52

6 Ecological Patterns of Variability in Demographic Rates
   Phyllis C. Lee, W. Keith Lindsay, and Cynthia J. Moss 74

   Box 6.1 Life Tables 81

7 Patterns of Occupancy in Time and Space
   Harvey Croze and Cynthia J. Moss 89

   Box 7.1 Ranging of Bulls outside the Park: Data from Radio and Satellite Tracking *Iain Douglas-Hamilton* 92

Part 3: Behavior, Communication, and Cognition
Section editor: *Joyce H. Poole*

8 Signals, Gestures, and Behavior of African Elephants
  Joyce H. Poole and Petter Granli 109

9 Behavioral Contexts of Elephant Acoustic Communication
  Joyce H. Poole 125
  - Box 9.1 Methodology 128
  - Annex 9.1 Behavioral Contexts and Associated Call and Context Types 160

10 Vocal Communication and Social Knowledge in African Elephants
  Karen McComb, David Reby, and Cynthia J. Moss 162
  - Box 10.1 Illustrations 164
  - Box 10.2 Structure of Elephant Contact Calls 166
  - Box 10.3 Contact Calls: Individual Identity in Call Characteristics 167
  - Box 10.4 Spectrograms of Contact Calls 171

11 Elephant Cognition: What We Know about What Elephants Know
  Richard W. Byrne and Lucy A. Bates 174

Part 4: Reproductive Strategies and Social Relationships
Section editors: *Phyllis C. Lee and Joyce H. Poole*

12 Female Reproductive Strategies: Individual Life Histories
  Cynthia J. Moss and Phyllis C. Lee 187
  - Box 12.1 Comparative Life Histories *Phyllis C. Lee* 188
  - Box 12.2 Dominance in Female Elephants *Phyllis C. Lee* 190

13 Female Social Dynamics: Fidelity and Flexibility
  Cynthia J. Moss and Phyllis C. Lee 205
  - Box 13.1 Some Family Histories of Fission and Fusion *Cynthia J. Moss* 209
  - Box 13.2 Personality in Elephants *Phyllis C. Lee* 214

14 Calf Development and Maternal Rearing Strategies
  Phyllis C. Lee and Cynthia J. Moss 224

15 Friends and Relations: Kinship and the Nature of Female Elephant Social Relationships
  Elizabeth A. Archie, Cynthia J. Moss, and Susan C. Alberts 238
  - Box 15.1 Non-invasive Genetic Sampling 239

16 Decision Making and Leadership in Using the Ecosystem
  Hamisi Mutinda, Joyce H. Poole, and Cynthia J. Moss 246
  - Box 16.1 Terminology and Methodology 254

17 Male Social Dynamics: Independence and Beyond
  Phyllis C. Lee, Joyce H. Poole, Norah Njiraini, Catherine N. Sayialel, and Cynthia J. Moss 260

18 Longevity, Competition, and Musth: A Long-Term Perspective on Male Reproductive Strategies
  Joyce H. Poole, Phyllis C. Lee, Norah Njiraini, and Cynthia J. Moss 272
  - Box 18.1 Genetic Paternity Analysis of the Amboseli Elephant Population *Julie A. Hollister-Smith, Joyce H. Poole, Cynthia J. Moss, and Susan C. Alberts* 274

Part 5: Elephants in the Human World
Section editors: *Cynthia J. Moss and Kadzo Kangwana*

19 The Maasai-Elephant Relationship: The Evolution and Influence of Culture, Land Use, and Attitudes
    Christine Browne-Nuñez   291

    Box 19.1 Consolation for Livestock Loss: A Case Study in Mitigation between Elephants and People *Soila Sayialel and Cynthia J. Moss*   294

20 The Behavioral Responses of Elephants to the Maasai in Amboseli
    Kadzo Kangwana   307

21 Ethical Approaches to Elephant Conservation
    Joyce H. Poole, W. Keith Lindsay, Phyllis C. Lee, and Cynthia J. Moss   318

22 The Future of the Amboseli Elephants
    Harvey Croze, Cynthia J. Moss, and W. Keith Lindsay   327

Appendix 1: Methods   337

Appendix 2: Large Animal Species Referred to in the Book   346

References   347
List of Contributors   375
Index   377

Foreword

Robert A. Hinde

IF YOU were choosing a species on which to do a long-term study of all aspects of its biology, you certainly would not choose the African Elephant. Most obviously, it lives too long. No individual research worker could hope to follow an individual from birth to death. Much of its communication is outside the range of human sense organs and can be detected only with special equipment. Individuals travel long distances, so the research worker may have to spend much of her time just looking for the animals. Furthermore, the elephants in Amboseli have a complicated relationship with the Maasai and have suffered from poachers: this might have prevented any hope for a long-term study. Anyway, elephants are known to be so clever: generalizations are bound to be elusive. Yet this book contains one of the most comprehensive studies of any non-human species. How has this been achieved?

A major issue must have been the fascination that elephants inspire. I am sure that every one of the contributors to this book was inspired to find out more about them by their very nature and in doing so, came to love their research subjects. I can empathize with this because I have a vivid memory of my first sight of elephants in 1968 when I was being driven into the Lake Manyara National Park on a visit to Iain Douglas-Hamilton and Cynthia Moss by my son, who was working with them as a research assistant. For a second, I did not see the elephants just because they were so huge. And then, there they were: a family group making their way slowly and magnificently among the trees. The next few days were special as Iain introduced me to some of the families. I confess that it was not only the elephants but also Iain's incredible knowledge of the individuals and the habitat: he knew, for instance, just how high one had to climb a tree to get out of the reach of Victoria, a powerful matriarch.

This awareness of the idiosyncrasies of individuals was preserved when Cynthia moved to Amboseli and was made possible by the common discipline that many of the contributors exercised in their record keeping, something certainly unusual when the study first got under way. It is hard enough for one research worker to maintain consistency, let alone for the records made by one individual to be interpretable by another. I suspect this reliability was facilitated by the camaraderie induced by their common fascination with the elephants. This is made evident also in the cooperative authorship of most of the chapters.

The difficulties posed by elephant communication, both chemical and auditory, have yielded to patient study, supplemented by playback experiments to show that elephants could communicate with each other over incredible distances. And the possible distortions accruing from elephant-human interactions were made the subject of special study.

But there is no need to enumerate further the difficulties that have been overcome. This book is a model for long-term species studies. Behavior and the complicated social structure are seen against a background of the ecological constraints. And the cognitive capacities that make the social structure and the social behavior possible are investigated in their own right, with appropriate emphasis on the idiosyncrasies of individuals. For reasons such as these, this book should be read not only for the detailed picture it provides of the biology of this extraordinary species but also as a model of what a comprehensive species study should be.

Robert A. Hinde FRS
Professor Emeritus
Sub-department of Animal Behaviour,
  Department of Zoology
St. John's College, University of Cambridge

Acknowledgments

OVER THE nearly 40 years of the Amboseli Elephant Research Project (AERP), there are obviously many, many individuals, organizations, and bodies that deserve acknowledgment. The institutional and monetary support of these many entities made it possible to start the research and, most importantly, to keep it going over a very long period. Without them there would have been no project, and we would know far less about elephants than we do today.

We would like to start by thanking those who have supported the overall project. The individual authors' acknowledgments will follow.

First and foremost, we thank the Kenya government for their hospitality and their permission to study the elephants, which are their precious natural resource. In particular we thank the Government of Kenya Office of the President and the National Council of Science and Technology for research clearance; the former Kenya National Parks and former Wildlife Conservation and Management Department, and the present-day Kenya Wildlife Service for permission to work and reside in Amboseli National Park. The Kenyan Wildlife Service, DRSRS (formerly KREMU), and the National Museums of Kenya provided local sponsorship. We are grateful to all these institutions for permissions and support.

We would also like to thank our hosts in the greater Amboseli ecosystem—the Ilkisongo Maasai, and particularly the leaders of the seven group ranches surrounding the Park. They have been consistently hospitable and always willing to work hard to try to find ways to accommodate elephants on their land.

Financial support for AERP has come from many sources. So many people and organizations have been generous over all these years, and their backing has been crucial. Some have given private money, others, through their own or family foundations. Many stand out for supporting AERP almost from the beginning and year after year thereafter. We would like to thank the following major donors (in alphabetical order): the African Wildlife Foundation, Bob Barker, Beginning with Children, the Born Free Foundation, the Leonard X. Bosack and Bette M. Kruger Charitable Foundation, the Lynne Chase Wildlife Foundation, Angela and Graham Chidgey, Dr. Charles Colao, the Cedar Hill Foundation, Laura and Jack Dangermond, the Disney Wildlife Conservation Fund, the Donner Foundation, the Detroit Zoological Society, the East Bay Zoological Society, ESRI, the Fair Play Foundation, the Howard Gilman Foundation, Steven Gold (through the Wildlife Conservation Network), Deanna Gursky, the International Fund for Animal Welfare, the National Geographic Society, Jan and Vic Overman, the Pumpkin Foundation, the Rogers Family Foundation, Willard and Susannah Rouse, Jane and Paul Schosberg, the Shifting Foundation, the Synchronicity Foundation, the Tapeats Fund, the African Elephant Conservation Fund of the U.S. Fish and Wildlife Service, the Wallis Foundation, and Wildlife Conservation International (formerly the New York Zoological Society). We thank Stephen Woo, who in addition to being a consistent donor, created and hosted the Elephant Trust Web site.

Special thanks go to the organizations that helped us at the very beginning back in 1972 when we were trying to get started. The African Wildlife Foundation helped to raise funds, administratively housed the project, and then continued to so for the next 28 years. Their support was crucial and much appreciated. Cash and in-kind support in those early days were also provided by the University of Nairobi Biology of Conservation program within the Zoology Department, the Ford Foundation, and the East African Wild Life Society.

Many individuals have been loyal supporters and enthusiasts of the project over the years and have shown their concern in many ways, including simply writing or sending a small gift to boost our morale in hard times. There are too many to acknowledge individually but we thank you for your caring and your steadfastness to the project and the elephants.

The Amboseli Elephant Research Project became a part of the Amboseli Trust for Elephants when ATE was created in 2000. The ATE exists as a not-for-profit trust both in the United States and in Kenya. The writing of this book has occurred under its auspices. We would like to thank the Board of Trustees for their support, help, and encouragement throughout: (in alphabetical order) David Breskin, Nan Buzard, Lynn Chase, Wilton W. Cogswell IV, Catherine Grellet, Neill Heath, Kathryn Heminway, Richard Leakey, Bruce Ludwig, Lia Reed, Isabella Rossellini, Susannah Rouse, and Don Young. (The ATE-Kenya trustees—Cynthia Moss, Harvey Croze, Joyce Poole, and Soila Sayialel—are all contributors to this volume.) The ATE's Executive Director in the United States, Betsy Swart, deserves a huge thanks for all that she did to create and develop ATE-US, all that she does to keep the organization running, and for the special help she gives all of us in big and small ways.

A number of donors and organizations helped with the writing of this book. We particularly want to thank Robert and Joan Donner and Joseph and Carol Reich, who supported writing workshops and stipends through their foundations: the Donner Foundation and the Jewish Communal Fund. The Barbara Delano Foundation paid for the creation of the Access database needed to carry out the analysis of the long-term data. Other organizations and individuals provided critical sponsorship in the form of workshops and writing retreats. We especially thank the White Oak Conservation Center (with particular thanks to John Lucas and Becky Thompson), where the editors and sub-editors held two important workshops; the National Center for Ecological Analysis and Synthesis (with particular thanks to Sandy Andelman, Jim Reichman, and Kirsten Parris), where on three different occasions the authors and editors worked in a very stimulating atmosphere analyzing data; and Susannah Rouse, who hosted our group at the Lewa Conservancy, where we held a very productive three-week writing retreat.

Some individuals were instrumental in helping with the 30-plus years of data analysis. First Cambridge University, back in 1981, with the help of Duncan MacKinder, allowed us to use their mainframe computer for storing and analyzing the data up to November of that year. Almost another 20 years went by before we were able to enter the new data. Paul Krystall designed a database and a user-friendly data entry system for the long-term sightings; all the data were entered by the end of 2000. Two years later when we wanted to amalgamate the Cambridge database with the new one, we ran into serious problems. Truly a knight in shining armor came along in the form of Hans-George Michna, a computer wizard and friend of AERP. He worked his magic and by 2003 we finally had a working database. Hans continues to sort out problems with the database as well as hosting our website and e-mail. He does all of this on a volunteer basis and we cannot thank him enough. We also thank Justine Cordingley for extracting behavioral data from the field notes.

Many of the analyses and ideas on differences in spatial distributions originated from the ideas and insights of John Calkins of ESRI, the Environmental Systems Research Institute, who, together with Liz Sarrow, set up our GIS, programmed the analysis tools, and produced the originals of many of the figures and maps used in this volume. ESRI generously supported the GIS research.

A number of other individuals on behalf of their organizations have provided data or data acquisition support, and particular thanks are due to: Jenifer Austin and Tanya Keen for an on-going license for GoogleEarth-Pro; Russ Kruska and Fred Atieno of ILRI for GIS advice and data; Paul Manson of the Trimble Corporation for support to early attempts at automated data-capture; David Murray for free access to Propel Accelerator software; Peter Ndunda (a former ATE-supported postgrad) of the Greenbelt Movement for being a 24/7 GIS helpdesk; Willy Simons and the talented staff of ESRI-East Africa for ArcGIS updates and support; Asbindu Singh, Paul Akiwumi, and Michael Mwangi of UNEP, Mark Ernste of the USGS EROS Data Center, and Rose Mayienda of AWF for providing satellite and GIS Amboseli data.

All research was carried out under the guidelines of the Association for the Study of Animal Behaviour (ASAB) for the treatment of wild animals during field work, and ethical clearance for research was obtained from all relevant institutions and departments.

Each of the authors in this book would like to thank the all-important field staff—Soila Sayialel, Norah Wamaitha Njiraini, and Catherine Katito Sayialel—for their participation in every element of the project. Without their dedication and hard work, this research would not be half as successful or productive. They have collected fecal and tissue samples, measured footprints, did focal samples, recorded sightings, carried out censuses, maintained the AERP long-term records, and rescued elephants and many people. In addition, they have contributed to the writing of a number of papers and chapters in this volume. Their commitment to the elephants is unparalleled and unwavering. We also want to thank Robert Ntuawasa Sayialel for his work on the sightings, long-term records, and tracking data, as well as general IT support in the field. Last but not by any means least, we acknowledge the hard work and dedication of Purity Waweru, our ATE administrator, who provides a large array of backstopping services for the project.

The staff at the research camp has kept us comfortable and well-fed over the years. We thank the past staff: the incomparable Masaku Sila, Wambua Kativa, Saibulu ole Kalama, Saruni ole Seleka; and the present staff: Peter Ngandi (who has now been with us for 25 years), Josephat Kiminza, Daniel Somoire, and Nkoshopu ole Kiluku. They have made working and living in Amboseli a pleasure.

The Amboseli tourist lodge managers have provided support and encouragement to the project in various ways, including delicious meals, cold drinks, meeting rooms, water, and access to freezers and mechanics. We thank the many managers at Serena, Amboseli, Ol Tukai, and Tortilis lodges. The guides from Ker & Downey Ltd. have been supportive in many ways and always hospitable.

Over the years many wildlife researchers have worked in Amboseli creating a stimulating and cooperative atmosphere. We have appreciated the kindness and friendship of these scientists. Getting together for drinks, dinner, maybe even a swim at one of the lodge pools was always a pleasure and often resulted in new ideas and new ways of looking at something our respective study animals were doing. We start by thanking Jeanne and Stuart Altmann, Susan Alberts, and the late Amy Samuels of the Amboseli Baboon Project, which has been operating even longer than the elephant project. We thank the non-elephant researchers who shared the camp: Dorothy Cheney, Marc Hauser, the late Wes Henry, Cynthia Jensen, Lynne Isbell, Stuart Semple, and Robert Seyfarth. Of the other researchers with their own housing, we thank Kay Behrensmeyer, David Klein, David Maitumo, David Western, and Richard Wrangham for their input in so many ways. The following are the editors' and authors' acknowledgments (in alphabetical order):

Susan Alberts acknowledges the generous support of the National Science Foundation (IBN0091612), which covered all aspects of the genetics work on the Amboseli elephants, and thanks the Amboseli Elephant Research Project for financial, logistical, and scientific support on all the field phases of the research.

Beth Archie thanks Tom Morrison, who helped collect elephant observations, and Jasmine Powell, Ebony Scales, and Courtney Fitzpatrick, who extracted DNA from elephant samples. The genetics work was funded by the National Science Foundation (IBN-0091612 to S. Alberts), and further supported by the Amboseli Trust for Elephants, the Amboseli Elephant Research Project, and Duke University.

Christine Browne-Nuñez thanks Cynthia Moss, Susan Jacobson, and Jerry Vaske for advice; and for funding, the U.S. Fulbright Program, the Disney Wildlife Conservation Fund, the Sea World and Busch Gardens Conservation Fund, and the Lincoln Park Zoo Africa/Asia Field Conservation Fund. At the University of Florida, she thanks the Alumni Graduate Fellowship Fund, the Department of Wildlife Ecology and Conservation, the Tropical Conservation and Development Program, and the African Studies Program for funding.

Dick Byrne and Lucy Bates were supported by a project grant from the Leverhulme Trust (F/00 268/W) and acknowledge the importance of the continual collaboration of the Amboseli Trust for Elephants in their study of the cognition of the African elephant.

Harvey Croze would like to acknowledge the inestimable contribution of his Oxford professor, the late Niko Tinbergen, who together with co–Nobel Laureates Konrad Lorenz and Otto von Frisch wrested the study of animal behavior from the confines of the laboratory to the field, where it evolves and belongs. He reiterates the project's thanks for the support of ESRI and Jack and Laura Dangermond and their staff in helping us establish AmboGIS. He also wants to acknowledge and thank the following individuals: Edwin Bulte, Richard Damania, and Randy Stringer for getting us started on thinking about payments for ecosystem services; David Campbell and Jenny Olsen for a constant and stimulating source of ideas and data on the human-wildlife interface; Iain Douglas-Hamilton and Save the Elephants for continuing encouragement and support over the years and providing some of the ranging data and operational expenses associated with conventional and satellite collaring operations; Norman Owen-Smith for ecosystem insights; and David Western for introducing us to the Amboseli elephants in 1972. Thanks also to Cynthia Moss, Phyllis Lee, Joyce Poole, and Soila Sayialel for generously welcoming me back to the project after nearly 17 years of wandering elsewhere like an Amboseli bull. And, at the end of the day, heartfelt thanks to Cristina Boelcke for encouragement, ideas, management advice, and time.

Iain Douglas-Hamilton thanks Discovery Communications, Inc., who paid for the GPS and standard radio-collars, and Save the Elephants, who paid for the operational expenses.

Julie Hollister-Smith thanks the Amboseli Elephant Research Project for scientific and logistical support, N. Georgiadis for help collecting tissue samples, and M. Lavine and E. Vance of the Institute of Statistics and Decision Sciences at Duke University for statistical advice. The genetics work was supported by the National Science Foundation (IBN0091612) to S. C. Alberts.

Kadzo Kangwana thanks Tim Clutton-Brock, Phyllis Lee, Cynthia Moss, and Joyce Poole for advice, and the African Wildlife Foundation and the Cambridge Commonwealth Trust for funding.

Phyllis Lee thanks Robert Hinde for his early encouragement and guidance, Pat Bateson at Madingley, Cambridge, for supporting her first elephant project, and all the members of the AERP for inspiration and friendship over 30-plus years. She would also like to thank the many members of the Amboseli Baboon Project, who have been friends, colleagues, and a sounding board for ideas about a variety of species. A host of PhD students have become collaborators and colleagues, and all of them have contributed to her sanity and science—thanks to you all! She thanks Vicki Fishlock and Lizzie Webber for reading chapters, Michelle Klailova for help with figure production, and Leslie Smith for growth analyses. Financial support has come from the National Geographic Society, the Association for the Study of Animal Behaviour, Downing College (University of Cambridge), the Department of Biological Anthropology (University of Cambridge), the Department of Psychology (University of Stirling), the Carnegie Trust for Scottish Universities, and the AERP.

Keith Lindsay thanks Keith Eltringham (in memoriam) for his tireless supervision, David Western, David Maitumo, and the Olgulului-Lolarashi Group Ranch members for local knowledge and assistance, KWS wardens and local staff for help and cooperation, AERP and Cambridge colleagues for advice and good friendship, Sally and the home team for their support and patience, and the Natural Sciences and Engineering Research Council (NSERC) of Canada, the New York Zoological Society, the A. J. Keith Fund, and the East African Wild Life Society for funding.

Karen McComb thanks the many people who have helped with field observations, playbacks, and analysis. The Biotechnology and Biological Sciences Research Council provided the major funding for her research (grant no. 85/S07659). Additional financial support or equipment came from the African Wildlife Foundation, the Association for the Study of Animal Behaviour, Newnham College (University of Cambridge), the Natural Environment Research Council, the Nuffield Foundation, the Royal Society, and the Tusk Trust.

Cynthia Moss wishes to thank Iain Douglas-Hamilton for introducing her to the joy of studying elephants; Robert Hinde for teaching her how to study elephants; and David Western for inviting her to study the Amboseli elephants. She is grateful to Harvey Croze for his collaboration in the first several years of the project and again in the last 14 years. She thanks all the other collaborators who have joined the project over the years and have shared the highs and lows of the long-term research. In addition to all the funders already mentioned, she is thankful to the MacArthur Foundation for her 5-year fellowship from 2002 to 2006. Andy Dobson and Phyllis Lee provided her with assistance and a temporary base for writing and analysis for this book at Princeton and Cambridge University.

Hamisi Mutinda thanks the Amboseli Elephant Research Project and especially Cynthia Moss for support and advice during data collection and analysis, and the African Wildlife Foundation and AERP for funding.

Joyce Poole is grateful for support from Smith College, Kings College (University of Cambridge), the New York Zoological Society, the Guggenheim Foundation, the National Institute for Mental Health, the National Geographic Society, Care for the Wild, the African Elephant Conservation Fund of the U.S. Fish and Wildlife Service, Born Free, Crystal Springs Foundation, Klingenstein Family Foundation, Winnick Family Foundation, International Fund for Animal Welfare, and many, many generous individuals over many years. She thanks Betty Horner (in memoriam) at Smith College for her boundless enthusiasm and mentoring; and Robert Hinde at Cambridge University and Dan Rubenstein at Princeton University for their direction, advice, and encouragement. She is grateful to all those in AERP for their friendship, support, and collaboration over so many years. She wishes to acknowledge Sarah Benson-Amram for her painstaking measurements of elephant calls. She is thankful to Petter Granli for his assistance and his understanding over the protracted years of analysis and writing; Cynthia Jensen and Virginia Poole for taking time out to look after Selengei during the writing workshops, and Selengei for her tolerance.

The editors would like to thank the following scientists for commenting on chapters in draft form or advising on analyses: Fred Bercovitch, Tim Clutton-Brock, Robin Dunbar, Kate Evans, Vicki Fishlock, Sarah Hrdy, Phil Kahl, Dan Levitis, Miryam Niamir-Fuller, Norman Owen-Smith, Henrik Rasmussen, Robin Reid, Joanna Setchell, Angela Stoeger-Horwath, Charles Vanpraet, and Cathleen Wilson.

Our editors at the University of Chicago Press have from the beginning been gentle and encouraging. We thank the late Susan Abrams for her faith in us at the start, Christie Henry for her patience and enthusiasm, and Abby Collier for seeing us through the last stages. We thank Martyn Colbeck for providing the photos for the book and for his positive spirit, sense of humor, and support when he was filming in Amboseli.

Finally, we would like to thank the elephants of Amboseli. They have given us hours, days, and years of wonder and joy.

Chapter 1   The Amboseli Elephants: Introduction

Cynthia J. Moss, Harvey Croze, and Phyllis C. Lee

Historical Perspectives on Elephant Studies

Elephants have fascinated humans for millennia. Aristotle wrote of them with awe; people in Asia have tamed, trained, and revered them for centuries; Hannibal used them in warfare; and John Donne called the elephant "Nature's great masterpiece . . . the only harmless great thing." Ivory has been sought and treasured in most societies at great cost to elephant populations. Eventually elephants were put on display in zoos and made to perform in circuses where, sadly, they still draw the laughter of children and the jibes of adults. It was only in the second half of the 20th century that people started to take an interest in elephants as elephants, that is, as free-living biological and ecologically functioning beings. This volume aims to provide a portrait of what the Amboseli Elephant Research Project (AERP) has contributed to our current knowledge of elephants as integrated components of their ecosystems and as individuals with potentially long and intricately involved social lives.

People who live in contact with elephants invariably are interested in them, whether they try to hunt them or avoid being killed by them. Knowledge of elephants among central African rainforest pygmies or savannah pastoralists is traditionally colored by myth, magic, and ethno-ecological reality derived from their experience and close contact. The Europeans who arrived in Africa to hunt "big game" had somewhat different attitudes. Those foreign hunters wrote numerous books and articles with titles such as *On Killing Elephant* (Bell 1931). Most such books were narratives of the writer's fearlessness in the face of massive charging bulls. Very little natural history appeared in their accounts, and the few descriptions of elephant life were often erroneous. Seen through the eyes of 19th- and 20th-century European men, the social life of wild animals was a clear reflection of the beholder's worldview—or perhaps more a reflection of what he wished the world to be. Almost all of them described elephants as living in a harem system with a single herd bull and his many females. Hunters' tales dominated in descriptions of the behavior of Asian elephants; the more knowledgeable mahouts or keepers of captive animals had relatively few opportunities in the colonial context to correct these misapprehensions.

In the late 1950s and early 1960s, a new group descended on Africa to pursue wildlife. These were biologists, ecologists, physiologists, parasitologists, and other scientists, whose technique for studying elephants was to shoot them and examine the carcasses. Much was learned in this way, but the elephants' death limited opportunities to study behavior and to gather longitudinal data. In the 1960s, ethologists who had been taught by Niko Tinbergen at Oxford came out to Africa to study living animals in the wild. Tinbergen, along with Konrad Lorenz and Karl von Frisch, jointly received the Nobel Prize in 1973 for developing ethology—the experimental study of animal behavior under natural conditions whenever possible—as a scientific discipline. One of Tinbergen's students was Iain Douglas-Hamilton, who began a pioneering study of wild elephants in Lake Manyara National Park, Tanzania, in 1965.

Douglas-Hamilton's study (on which Cynthia Moss worked for a year) was based on the ability to recognize individual elephants over time. Each animal in the study population was named or numbered, and its movements, activities, associations, and relationships were recorded over a four-year period. Using this novel approach, a new understanding of elephant social life, behavior, and ecology began to emerge. At the same time, Harvey Croze was working in the Serengeti National Park as a Tinbergenian post-doctoral researcher studying the interactions of elephants and their habitat. In 1972, at the suggestion of David Western, Moss and Croze started the AERP, which stands today as the longest continuously running elephant research project in the world. At the time this introduction is being written, the project is in its 38th year. Neither of the founders thought at the time that they would be personally involved in the project for so many years, but they had planned from the beginning for AERP to be a long-term research commitment. The guiding philosophy was that elephants are long-lived animals and that to gather the data necessary to understand and describe their ecology and behavior would need a lifetime—both elephant and human.

Background to the Study Population

Forty years ago, the Amboseli elephant population was known to be relatively small, to co-exist with Maasai pastoralists in a semi-arid savannah (the dynamics of Amboseli's ecology are detailed in chapter 2), and to be both approachable and observable (Western 1973). The big males were being sport hunted, and there were low levels of poaching and conflicts with people, but inside the protected area, they were tolerant of vehicles. They appeared to be good candidates for study.

During the 1970s and 1980s, the African continent saw a catastrophic decline in its elephant populations. In almost every country where elephants ranged, ivory poaching drastically reduced numbers. Throughout that period of changing fortunes for Africa's elephants, only a few populations, such as that in Amboseli, were relatively well protected. Three circumstances in the Amboseli ecosystem contribute to its relatively undisturbed state. The land surrounding the National Park belongs to the Maasai people, who

---

BOX 1.1 DEFINITIONS OF TERMS USED THROUGHOUT THE BOOK

Group

An elephant group is defined as any number of individuals of any age or sex moving, feeding, or resting together in a coordinated manner with no single member or sub-group at a distance from its nearest neighbor that is greater than the diameter of the main body of the group at its widest point. An *aggregation* is a group that consists of at least two families or parts of families and is commonly used when many families and individual bulls are present.

Group types

Bull group. One or more males of any age in the same group.

Cow-calf group. A group consisting only of females and calves; any males present are family members who are not yet independent.

Mixed group. A group of cows and calves with one or more independent non-family males.

Sub-unit, fragment or party. A section of a family separated from its original family by more than the diameter of the group at its widest point and usually a kilometer or more away from the rest of the family.

Sub-group refers to a grouping within a larger aggregation in which there is spatial delineation between groupings but the sub-groups are not separated by more than the diameter of the main body of the group at its widest point.

Age-sex classes

Adult female. A female aged 10 years or over based on the youngest age of first birth (8.9 years). Less than 1.3 percent of females have conceived resulting in a live birth under the age of 10 years (N = 304).

Adult male. Males are considered to be reproductively adult at 10 years but are not socially adult at this age. Males become independent of their natal family unit at a mean age of 14 years (range 8–18). There is typically a period of several months to several years when males come and go from their families. During this period, they are termed *transitional males*. Once they have left their families permanently, they are considered to be adult.

Age classes of elephants. In Amboseli, the majority of the individuals are of known age but early in the study, most adult ages were estimated. For some analysis purposes, we placed animals in age classes based on both known age and estimates (see appendix 1). These age classes correspond generally to unweaned infants, juveniles, adolescents, early adults/late adolescents, and several adult classes until the oldest age group of 50+. Classes are: 0A = 0–4.9 years; 0B = 5–9.9 years; 1A = 10–14.9 years; 1B = 15–19.9 years; 2 = 20–24.9 years; 3 = 25–34.9 years; 4 = 35–49.9 years; 5 = 50+.

traditionally have been tolerant of wildlife and at the same time inhospitable to outsiders hunting or poaching on their land. In addition, the Amboseli elephants have never been culled as part of a park management program nor have they been fenced into a protected area. Finally, the research project itself has provided wildlife and anti-poaching authorities with additional eyes and ears and a nearly full-time vigilante presence in the field.

In the comparative absence of poaching and culling (but see box 19.1), the Amboseli elephants have increased slowly in numbers since the late 1970s. The population in 2008 at just over 1,500 remains relatively small but is nonetheless very important for Kenya and the rest of Africa. Amboseli is one of few places in Africa where the elephant age structure has not been drastically skewed by past poaching. Known individuals span the range from newborn calves to old matriarchs in their 60s and—exceptional these days—large adult bulls in their 50s and early 60s. With a natural age structure and intact social organization, the Amboseli elephant population has become increasingly important as a source of baseline data on elephant social and reproductive behavior and population dynamics. As such, it is used as a model for determining the status of other elephant populations in Africa and Asia and as a yardstick for assessing the extremes that captive elephants experience.

Threads and Themes

Four underlying themes run throughout the chapters in this volume: longevity, size, intelligence, and the future of the elephants in a changing and threatened ecosystem. The book is divided into five parts, which are explained in detail below (See annex 1.1 for a glossary of Maa place-names and box 1.1 for definitions of common terms used throughout the book).

Longevity in the Amboseli context has two facets: the elephants and the study itself. The commitment of a group of cooperating but highly individualistic researchers, some

---

Social terms

Family unit. A single or several adult females with their immature offspring who show a high frequency of association over time relative to any other females in the population, act in a coordinated manner, exhibit affiliative behavior toward one another, and are known to be related or are putatively related (Moss 1988; Moss and Poole 1983; Poole and Moss 1989).

Matriarch. The leader of a family unit, typically the oldest female in the family. The others orient toward her and follow her decisions and leadership. She exercises a higher degree of control on the behavior of the family unit than any of the other adult females within the family (Moss 1988).

Estrus. In the field, recognition of estrus as a period of receptivity and fertility is based on female behavior. Females exhibit characteristic postures and behavior such as the head-held-high walk, which can be recognized at a considerable distance (Moss 1983).

Musth. Musth is an annual period of heightened male sexual activity and aggression characterized by a distinct posture and walk with the head held stiffly high above the body; swollen and visibly secreting temporal glands; the dribbling of strong-smelling urine, which often marks the inner back legs (Poole 1987a; Poole and Moss 1981); and distinctive vocalizations (Poole 1987a, 1999a; Poole et al. 1988).

General terms

Sighting. Visual contact with an individual elephant or group of elephants at a precise time on one day. Associated with each sighting is an individual or family identification (or age-sex class when recognition is not possible), count of size of group, composition of group, habitat type, activity, and a location (either GPS or grid square). Presence of estrous females and musth males is recorded.

Activities. We used consistent and mutually exclusive terms for different activities and behavior. These are coded as follows: Feeding while Moving; Feeding (stationary); Resting (standing or lying); Standing (alert); Comfort (dusting, mud wallowing, splashing, rubbing, or other related behavior); Interacting (playing, greeting, aggression, copulation, etc.); Drinking; Moving (walking/running); Other (including mixed activities within the group); and Unknown.

Interactions and vocalizations. Elephant behavior is defined in detail in chapters 8 and 9.

Human society

Dorobo. The term Dorobo has numerous meanings for the Maasai: non-Maasai hunter-gatherers, hunters with cattle, pastoralists who hunt or gather honey, or Maasai who lost their cattle and became Dorobo (Bernsten 1976; Distefano 1990).

Moran. A man in the Maasai warrior age-set is called a *moran*, plural *ilmoran*. These are the common spellings of those terms; the more phonetically correct spellings are *olmuranni* and *ilmurran* (Mol 1996). We have used the common spellings.

of whom have collaborated for four decades, is almost as remarkable as the unfolding of an animal's lifespan over nearly seven decades.

The Amboseli study presents an important opportunity to explore longevity in a natural evolutionary and functional context. Longevity is not easily studied in natural populations; only studies of a few large mammals such as lions (Mosser and Packer 2009), red deer (e.g., Kruuk et al. 1999b), and sheep (e.g., Festa-Bianchet, Jorgenson, and Réale 2000) have provided longitudinal individual life history data. Behavioral aging—changes in behavior or functions with age—is a poorly understood attribute common among all long-lived animals (Finch 1994). Some individuals age rapidly while others resist the effects of age. Most species that have been used as models for evolutionary and behavioral aging are small and short-lived, so they can be reasonably observed and manipulated over the duration of a research grant. In this volume, we add elephants to these classic studies, but we should stress that despite the unique dataset analyzed and presented in the chapters that follow, we have still witnessed only half of an elephant's potential lifespan. While we have been careful not to presume that we can as yet accurately assess genetic fitness over a full lifespan, we do explore the individual fitness consequences of much of the behavior.

Size, of both the animal and its spatial requirements, is the second theme. The elephant is at the extreme end of the scale in terms of terrestrial mammalian body mass and energetic relationships. It also has to cope both individually and socially with the heterogeneity and dimensions of a habitat that ranges from swamp to dense bushland to bare soil over thousands of square kilometers. Large size and a long growth period have consequences for reproduction and sociality as well as for the retention and manipulation of information about companions, enemies, and the environment.

Thus, the third theme is the exceptional intelligence that provides elephants an evident adaptive edge in habitats that vary over time in unpredictable ways. We address this issue by providing descriptive but data-rich portraits of elephant behavior, gestures, and signals, which are contextualized by experimental manipulations of communication and explorations of social knowledge. We also emphasize throughout these chapters the dynamic nature of elephant sociality: fission-fusion, where each individual is embedded in a knowledge-based network of others. While fission-fusion as a system of aggregation may not require any particular cognitive abilities, as evidenced by fish schools or flocks of roosting starlings (see, for example, Aureli et al. 2008), elephants have to rely on memory, knowledge, and information exchange to maintain their networks, and thus sociality both drives and enables elephant intelligence.

The final theme, which looms like a dreadful portent, is the uncertainty of the elephant population's future in the face of changing social and economic circumstances in the ecosystem surrounding Amboseli National Park. That uncertainty has over the years caused the project to stretch beyond its original research mandate and reach out to the local community, to demonstrate benefits from the elephants and their ecosystem, and to form partnerships with the Maasai. These partnerships provide the project with valuable information about ecosystem occupancy and community concerns and aim to maintain an atmosphere of goodwill toward elephants that is fundamental to their future.

All four themes situate the study and the volume in a single place: Amboseli. This volume is special in that it approaches a single species in one place from many perspectives, exploring how it makes a living in a specific but constantly changing habitat over a 60-plus-year lifespan from birth to death. Previous studies of single ecosystems (e.g., Serengeti [Sinclair and Arcese 1995]) have perforce skimmed the surface of behavior and ecology of many of their constituent species over time, while studies of a single species (e.g., roe deer [Anderson et al. 1998]) or species group (e.g., cetaceans [Mann and Smuts 1998; Mann et al. 2000]) have had to compare populations rather than examine them over time. In the pages that follow, we intentionally keep the perspective focused on the elephants of Amboseli and do not belabor comparisons with studies of all other elephant populations. Such comparisons will emerge in time, we hope, as a result of this volume.

The focus on a single population over time provides a stable reference point for theoretical understanding; being large, long-lived, and distributed across habitats from dense forests to deserts, hemmed in by a variety of human landscapes, means that elephant populations exhibit variation at all levels and scales. The ability to examine variation in context for one population should enhance our ability to compare, contrast, and make sense of the sweep of elephant variation across the species' range.

What the Book Is About

This volume has had the equivalent gestation length of five—almost six—elephant calves. As more researchers joined the project, bringing their individual theoretical and practical backgrounds and expertise, enthusiasm grew for compiling all our knowledge in a single publication. In September 1998, the nominated section editors, hosted by the White Oak Conservation Center in northern Florida, met to discuss content and a workplan. By October 1999, when we met again as a working group to discuss the analysis of long-term datasets at the National Center for Ecological Analysis and Synthesis in Santa Barbara, California, we confirmed the importance of the work and had mapped out a clear way forward for tackling the long-term dataset.

Many readers of this volume will not recall those days at the start of the project in the early 1970s, when we were in the computational Stone Age: we had only just shelved our slide rules and begun trying out the new handheld programmable calculators. Our original data were designed to be compatible with 80-column IBM Hollerith punch cards, and this format was retained consistently across a variety of now-extinct data-entry modes and analysis protocols. Another fortunate decision was to cast the project's location data collection into a 1 km^2 map grid, which, although well before the days of affordable commercial satellite data and global positioning systems and at the dawn of geographic information systems, proved most useful for subsequent mapping and spatial analysis (see appendix 1).

The construction, cleaning, and verification of a long-term amalgamated sightings database started in 1996 and continued until 2003 and for this reason, much of the analysis in the volume uses sightings data for only the first 30 years of the project, that is, up to the end of 2002. New data have been constantly added, but the analysis had to have an end point, and 30 years seemed a reasonable cutoff. As chapters have evolved and been revised over time, we have incorporated additional data where necessary; some chapters have included political changes and novel observations right up to the moment of going to press. Analyses of demographic and population data extend until 2006 in order to use recent events to provide further insights.

Thus, much of the data we present have not appeared in previous publications because they are based on a long-term dataset that was constructed and made manageable only during the course of writing the book. Many chapters build upon published peer-reviewed research but take advantage of the whole of the dataset to revisit and reanalyze earlier research questions. The chapters herein are thus a blend of previously published but augmented research merged with entirely novel analyses resulting from research questions mooted over the 10-year gestation of the volume. We have subjected the new analyses to peer review, commentary, and revision.

Throughout the book, we often use statistics to argue a point. But when statistical tests are based on huge sample sizes such as the sightings database (N circa 30,000), statistical power is high but explanatory effect may actually be rather small. We have thus tried to explore biologically sensible effects rather than simply using P values. Despite having nearly 40 years of almost daily observations of slowly reproducing, very long-lived individuals, analysis is further bedeviled by having relatively few records of those rare events that make up the unique life history of any individual. As Dick Byrne observes (see chapter 11), the value of very long-term studies is that as time goes on, the unexpected and, indeed, the previously unknown are eventually seen. We have tried to maintain the "individual" perspective over a lifespan as a focus within each of the themes in the book.

We ask our readers' indulgence regarding the inclusion of different datasets, approaches, and perspectives. In editing the book, we have not demanded a single style or voice for the chapters; some make qualitative arguments while others are rich in analysis. This blend of perspectives is, we feel, what makes the project itself so strong and unique. More science and philosophy will indeed emerge as the project continues and as new questions are asked and problems pursued.

How the Book Is Organized

The book is structured in five parts. Part 1 establishes the physical, human, and temporal context for the elephant population. It provides an overview of the Amboseli ecosystem and the human pressures impinging on the elephants' lives. By both human and elephant standards, 35 to 40 years is a relatively long period—sufficiently long to see changes in climate, woodland cover, species numbers and distributions, and rapid urbanization at the ecosystem edges. It is, however, only a brief interlude in the geological processes that formed Amboseli and continue to act on the habitat and in the population mixing, movements, and dispersals that underlie a deep-time perspective. The first two chapters trace the dynamics of recent rapid changes occurring for both the environment and the people, illustrating the inherent lack of ecological and societal equilibrium in postcolonial East Africa. The origins and population history of the Amboseli elephants within the context of the larger regional elephant population are drawn over genetic time in the final chapter in this part.

This ecosystem and population context is followed by part 2, which details how elephants make a living in the Amboseli habitat and the consequences for reproductive rates and population dynamics of living in an unpredictable, dynamic ecosystem. In chapter 5, Lindsay presents an analysis of season- and sex-specific feeding behavior while in chapter 6, Lee, Lindsay, and Moss link these foraging dynamics with reproductive rates and population dynamics. Croze and Moss then explore in chapter 7 how the Amboseli elephants range throughout their habitats, balancing the ecological context with their social and reproductive needs as detailed in later chapters.

Part 3 consists of an exploration of elephant behavior and communication, rounded out with a perspective on comparative cognition. The ethogram of elephant behavior, dating originally from the 1960s (e.g., Kühme 1961), is comprehensively updated by Poole and Granli in chapter 8. This new ethogram establishes the context for much of the social and reproductive strategies explored in part 4. In the 20-plus years since the early reports of infrasound (Payne, Langbauer, and Thomas 1986), elephant communication has been the focus of considerable interest and research. In chapter 9, Poole specifies the behavioral contexts of vocalizations and provides entirely new portraits of many of the calls. McComb, Reby, and Moss (chapter 10) give an overview of their recent studies, which have experimentally expanded our understanding of the social contexts of female elephant communication. The part concludes with Byrne and Bates's overview of elephant cognition (chapter 11), placed in the comparative context of the capacities of other large-brained, long-lived species, with examples from the results of their recent experimental work in Amboseli and elsewhere.

Part 4 meshes the complex skein of elephant fission-fusion sociality with its consequences for female and male reproductive success. Moss and Lee (chapter 12) describe individual female reproductive histories and general reproductive strategies. These reproductive strategies are then placed in the context of female social dynamics (chapter 13), exploring the causes and consequences of the elephant's archetypical fission-fusion model of society. The next chapter is an examination by Lee and Moss (chapter 14) of how maternal social and behavioral rearing strategies impact calf survival and physical and social development. Having established the general social systems and reproductive strategies among female Amboseli elephants, Archie, Moss, and Alberts in chapter 15 combine genetic and behavioral analysis of the population to tease apart the layers of female elephant society. Significantly, the Amboseli population has been less disturbed historically by genetic upheaval through deaths than have many other populations, suggesting, as we note above, the value of this population in helping to interpret and understand events in other populations throughout Africa (e.g., Gobush and Wasser 2009; Wittemyer et al. 2009). In chapter 16, Mutinda, Poole, and Moss follow females along their habitual trails to deduce leadership and decision making within their complex social dynamics. The female worldview in this part is balanced with Lee et al.'s exploration of the social nature of males and how it emerges and changes over time (chapter 17). Finally, Poole et al. show how male sociality relates to competition over mating and to enduring friendships (chapter 18). Despite the very different developmental and social trajectories of males and females throughout their lives, their shared trait of longevity determines how they cope with social and environmental hazards that affect their survival and contribution to subsequent generations.

From Central African forests to Southern African reserves, from Tanzanian woodlands to Amboseli swamps, elephants exist within human-dominated landscapes. Part 5 looks in detail at the particular trials and tribulations of elephants in Amboseli's human world. In chapter 19, Browne-Nuñez recounts the changing social and economic context for Maasai culture, land use, and attitudes, and Kangwana in chapter 20 outlines the behavioral responses of the Amboseli elephants to their human-dominated surroundings.

In the last two chapters, we turn to the future. In chapter 21, Poole et al. provide ethical arguments for accommodating elephants more generally in our physical world and moral space. And in chapter 22, drawing on our extensive knowledge of the animals and their ecosystem, Croze, Moss, and Lindsay speculate on the future of the Amboseli elephants.

Elephants echo the human condition in so many ways that it is on occasion difficult not to become engaged in a manner that reaches beyond the confines of scientific enquiry. If, however, the project and the elephants are to survive long enough to shed light on the full 60-plus-year lifespan, then we have to safeguard the space required for sustaining members of this unique population and the capacity for future research. We trust that the results of the study contained in this volume will stand as a milestone contribution to elephant science. At the same time, we hope that our work will ignite a spark of passion within future conservationists and scientists to keep and hold in trust the Amboseli elephants and their ecosystem.

ANNEX 1.1 GLOSSARY OF MAA PLACE-NAMES AND SYNONYMS

The written conventions for place-names in Maa, the Maasai language, are not fully canonized. In this book, we have used transliterations of place-names in common usage but from different sources: topographic maps, the literature, and native Maa speakers associated with the project. The topographical map names, which themselves were imperfectly rendered by colonial cartographers, have tended to become the most common forms. Rather than claiming to be a source for Maa nomenclature, we provide in this glossary the two or three alternative spellings for the same places commonly referred to in the text. No categorization is implied by the three columns.

Dorobo	Wandorobo	Wadorobo
Emparinkoi	Embaringoi	Baringoi
Endoinyo Oontawua	Ontawua	[Lemongo; error on topo map]
Enkong'u Narok	Enkongo Narok	
Eselenkei	Selenkay	Selengei (Eselengei)
Ilkisongo	Kisongo	
Ilmarba	Marba	Ilmarn
Kitirua	Kiterua	O'lengia (region near Kitirua)
Lemeipoiti	Lemeiboti	
Lenkalong	Longolong	Lengalong
Lonkinya	Lonkinye	Longinye
Namelok	Namelog	
Olgulului/Ololarashi	Olgulului	Olgulului/Lolarashi
Olkejuado	Kajiado	
Olkelyunyet	Ol Kelyunyet	
Olkulului	Olgulului	
Oloitokitok	Loitokitok	
Ositet	Ositeti	

Part 1
The Amboseli Context: Ecology, People, and Genetics

Section editors: W. Keith Lindsay and Harvey Croze

Chapter 2 Amboseli Ecosystem Context: Past and Present

Harvey Croze and W. Keith Lindsay

THE AMBOSELI ecosystem is unique. No other place in Africa combines the special hydrology, topography, geology, and cultural history of Amboseli. Despite modest rainfall, a gently rolling bushland surrounding a system of swamps fed by the Kilimanjaro mountain forest catchment supports a stunning array of birds and mammals, dominated in terms of biomass and visibility by a population of some 1,500 African elephants. Overlaid on this landscape, against the backdrop of Kilimanjaro looming to the south (figure 2.1), is a traditional system of nomadic pastoralism practiced by the Maasai people, whose faith and pride in their own culture is impressively steadfast in the midst of rapid social and economic development (see Kangwana and Browne-Nuñez, chapter 3). At the heart of the ecosystem nestles Amboseli National Park, which for decades has been a major attractor for wildlife and tourism.

In order to place the ecology, ranging, population dynamics, and sociality of the Amboseli elephants into the perspective of their habitat over the course of the study, in this chapter, we describe

- the variable and dynamic Amboseli ecosystem in general and
- the broad habitat changes that have taken place over the past five decades (1957–2002), including assessments of changes in wildlife species numbers.

Mutable is the best way to characterize the core of the Amboseli ecosystem. Since the 1950s and 1960s, when dedicated wardens and keen researchers began taking note of events in the Amboseli basin, the swamps have grown in size, standing water has increased, some *Acacia* woodlands (*Acacia xanthophloea*, the yellow-barked "fever tree") have changed dramatically, and salt-loving plant species have spread (Western and van Praet 1973; Struhsaker 1976; Altmann 1998). (The designation Amboseli basin is commonly used to refer to the area containing the [usually] dry lakebed, Amboseli National Park, and the immediate surrounds.)

Western (1997) saw change as the central Amboseli theme, "from the long, slow, concatenated events triggered by the eruption of Kilimanjaro to the apparently whimsical migrations and grazing succession," and asked, "Could it not be that ecological change rather than stability [is] the rule? . . . [T]he whole string of disruptions set off by geological upheavals, climatic change, even human activity, [makes] it impossible for any ecosystem to stabilize for more than a fleeting moment. . . . [C]ould it possibly be that change rather than stability [explains] Amboseli's wealth of species?" (1997, 116).

The changes under way today have precedent in the past. They are driven by large-scale fluctuation in water flow from the Kilimanjaro watershed, which is driven in turn by regional rainfall as well as human alteration to the water capture, holding, and delivery characteristics of the catchment zone. Annual rainfall data from local meteorological stations show considerable interannual variation but no clear pattern (e.g., Altmann et al. 2002; see below), suggesting that basic ecosystem drivers are in constant flux. A more clearly directional change is occurring at the scale

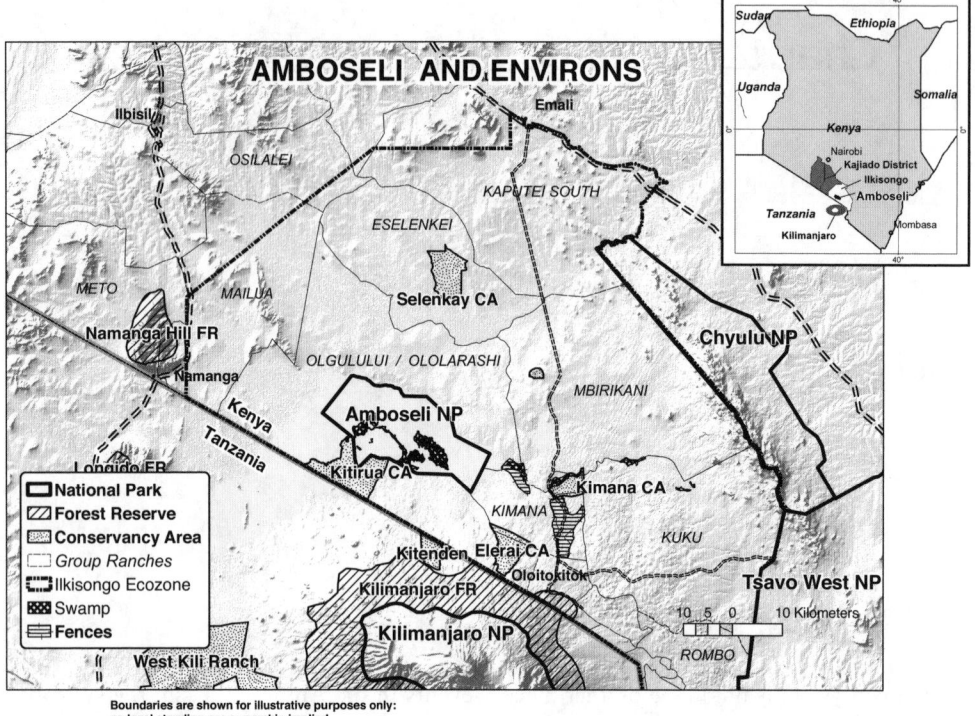

Figure 2.1 Amboseli ecosystem in the Ilkisongo secton of Kenya's Kajiado District showing major topographic features, main swamps, and protected areas. CA = conservation areas, including community conservancies and commercial concession wildlife areas; FR = forest reserve; NP = National Park.

of the global climate (e.g., Folland et al. 2001), evidenced locally by the rapid shrinking of the Kilimanjaro glaciers (Thompson et al. 2002).

The Amboseli Ecosystem

Location

The Amboseli ecosystem is located in southern Kajiado District, a 22,000 km² administrative unit that stretches from just outside the capital of Nairobi south to the Tanzanian border. Kajiado District is the administrative name. The district county council, Olkejuado County Council, has opted to use the original Maa name for the region, which means "the river that is long." The written conventions for place-names in Maa, the Maasai language, are not fully canonized. Throughout the rest of this book, we have used transliterations (provided by our Maasai colleagues [see the glossary in the annex to chapter 1]) that are as close as possible to topographical map names, which themselves were imperfectly rendered by colonial cartographers.

Kajiado District has marked geophysical features—some truly spectacular, such as the eastern step-faults of the Great Rift Valley—that create four distinct ecozones as defined by geomorphology, topography, and vegetation, namely the Athi-Kapiti Plains, the Rift Valley, the Central Hills, and Ilkisongo (UNDP/FAO 1980; figure 2.1)

The Ilkisongo ecozone is congruent with what is generally known as the Kenyan portion of the Amboseli ecosystem, as defined by a commonality of soil and vegetation types, a local rainfall regime, a distinct drainage system, and the occupancy of large herbivore populations, both residents and locally seasonal migrants. When migratory species make up a large proportion of the animals in an area, the limits of their annual movements may be taken as operational ecosystem boundaries (Pennycuick et al. 1977). The name Ilkisongo is in fact that of the dominant Maasai social group in the area. The ecosystem has been well described elsewhere (e.g., Western 1973; Lindsay 1982, 1994; Behrensmeyer, Stayton, and Chapman 1993; Reid et al. 2004), and the general account that follows will paraphrase freely from those sources.

The Amboseli ecosystem is a roughly 8,000 km² area that straddles the Kenya-Tanzania boundary, reposing at 1,100 m as a broad basin between the northern slopes of Kilimanjaro, the late (post-Pleistocene) volcanic Chyulu Hills (2,200 m) to the east, a motley range of broken basement hills to the north, and scattered granitic outcrops and earlier volcanic cones to the west and southwest, the largest of which, Oldonyo Orok, is 2,400 m. The most recent eruption of Kilimanjaro about 1.5 million years ago blocked the ancient Pangani River that flowed northwest to southeast and thus created a closed central basin and a lake with no outlet: Lake Amboseli. Today, the lake holds water for only a few weeks after heavy rains. More commonly, the lake is dry; the lacustrine silts that have accumulated over the years reflect starkly white on satellite imagery.

Geology and Soils

Quaternary volcanic soils dominate on the northeastern Kilimanjaro slope, encouraging rain-fed agriculture around the town of Oloitokitok; basement rock soils cover most of the rest of Ilkisongo, making only pastoralism possible. These dark red to reddish brown sandy clay soils are low in fertility despite the rapid growth of grass on them in the early rains. Darker brown-to-black ("black cotton") alluvial clays accumulate in seasonal runoff lines and low-lying areas of impeded drainage, where they trap nutrients and support grass growth for a while after the rains. In general, even where volcanic soils are present, soil fertility in the ecosystem is a tenuous matter, underlain as it is with nutrient-impoverished basement quartzites, crystalline limestones, schists, and gneisses. The soils in and around the Pleistocene lake bed are an unfriendly mix of saline accumulations that form calcrete pavements, support only a meager seasonal grass growth, and produce a ferocious albedo, the vertical energy of which is believed to repel clouds and delay the onset of the rains compared with surrounding areas. The soil chemistry in the immediate vicinity of the springs and swamps (see below) is less saline due to dilution by groundwater and percolation of salts to the margins of the groundwater zone.

Water

Scarcity of permanent water is the salient feature of Amboseli's surface hydrology; water is obviously a key limiting factor in the ecosystem. Apart from a handful of spring-fed rivulets that bubble from the northeastern piedmont of Kilimanjaro (such as Nolturesh), there are no perennial rivers in the ecosystem, only seasonal streams that flow for short periods during the rains (figure 2.2). The Eselenkei-Kiboko river drainage in the north and northeast portion of the

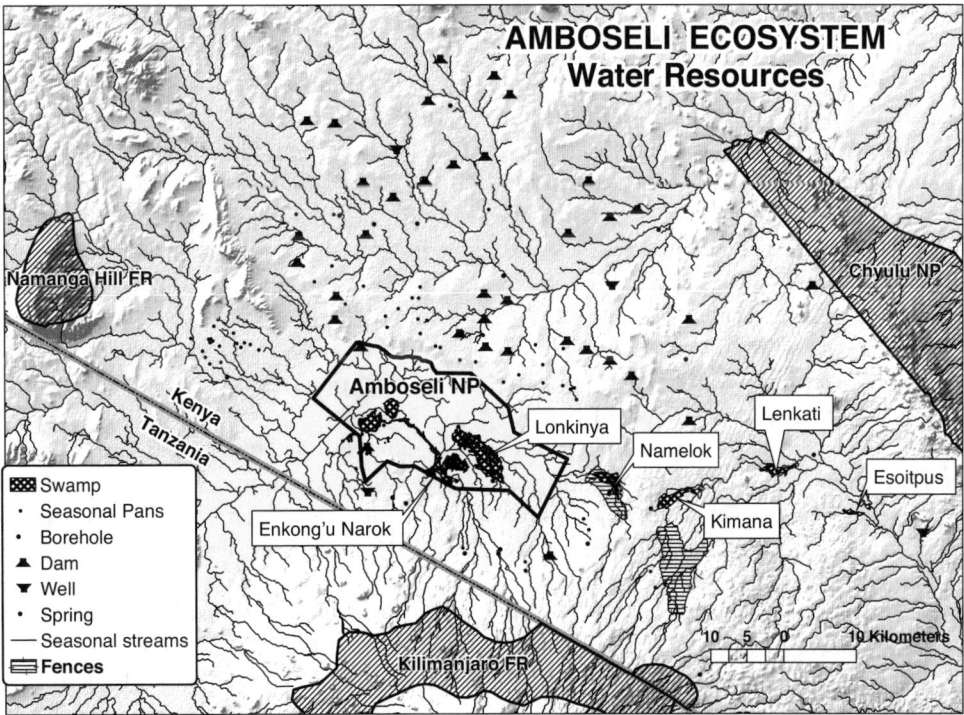

Figure 2.2 Water resources in the Amboseli ecosystem. All indicated water courses are seasonal, as are the water points indicated as dams and seasonal pans. Springs and boreholes are perennial but often secured from wildlife access. Data sources include topographic maps from aerial photography, satellite imagery, and GPS points taken on the ground or from light aircraft. The three major forest catchments—Namanga Hill Forest Reserve, Chyulu Hills National Park, and the Kilimanjaro Forest Reserve (in Tanzania)—are crosshatched. The fences enclose irrigated agriculture.

ecosystem is highly seasonal. There is no surface runoff from the Chyulus: rainfall soaks almost on impact into the porous volcanic soils. There are also no permanent streams coming from the Kilimanjaro slopes or the catchment of Oldonyo Orok, the "black mountain") to the west.

Springs and Swamps

Water, which falls as rain onto the forested catchments and volcanic soils of Kilimanjaro and the Chyulu Hills, feeds through a little-understood underground drainage system and emerges at the southern margin of the basin in a number of springs that meander in channels northward across the flat Amboseli plains. The volume of outflow determines the extent of surface water and height of the underground water table in the basin, which in turn affect the salinity of water in the rooting zones of trees (see below).

The springs feed an important series of west-east–oriented swamps that are the lifeblood of the ecosystem: Enkong'u Narok and Lonkinya within the Amboseli National Park boundaries, then eastward to Namelok, Kimana, Lenkati, and near the Chyulu Hills, Esoitpus (figure 2.2). Thanks to the swamps, the ecosystem is today a haven for biodiversity, able to sustain the impressive populations of large herbivores, small mammals, and birds as well as the Maasai and their livestock and high-intensity agriculture, especially in and around the Namelok and Kimana swamps.

The extent of outflow from the springs seems to depend on variations in rainfall amount and runoff from Kilimanjaro's forest zone. Rainfall variation may be random or cyclic—anything but constant (see below). The relationship between rainfall events and the recharging of catchments is not a simple correlation with annual amounts. There is some evidence from the Lake Victoria basin that pulses of high rainfall, such as in the late 1950s and early 1960s, which followed an extended period of lower precipitation, may have saturated catchment zones across Kenya and provided increased downstream flow for a number of years to follow (Lamb 1966; Mifflin 1991). Other possible drivers of water flow into the basin include runoff from outlying areas subject to intensive grazing and shifts in underground watercourses due to tectonic activity around Kilimanjaro. The evidence to distinguish between these factors is equivocal at best (Meijerink and van Wijngaarden 1997). It is clear, nevertheless, that prior to the late 1950s, the outflow from the basin-edge springs was relatively modest (Lovatt Smith 1997) and by 1957, both outflow and water table had begun to rise (Western and van Praet 1973) and have remained high to date (Mifflin 1991; Meijerink and van Wijngaarden 1997).

Rainfall

Over two discernible seasons—the "short rains" in November and December (sometimes starting in late October) and the "long rains" from March to the end of May—only some 340 mm fall on average annually in the Amboseli basin (figure 2.3). The short dry season between January and February is relatively hot (up to 35°C in February); the longer dry season between June and October can be cold (down to 5°C in July and August). For purposes of analysis, we have defined an Amboseli Year as beginning in October and ending the following September. The Amboseli Year captures more accurately than the calendar year the annual cycle of growth and maturation of vegetation. Since it may take a month for plant growth to respond to rainfall, and since rain may begin to fall some years in October, we have summed monthly rainfall from October through September as the total precipitation for one Amboseli Year.

The long-term rainfall data series is a composite of measurements at three different meteorological measuring points within 5 km of each other. For the 1972–82 period, data were taken from Amboseli National Park's Ol Tukai meteorological station (see also Western 1973), after which the Amboseli Elephant Research Project (AERP) camp rain gauge was then the main data source. Records from the Amboseli Baboon Research Project camp rain gauge (Altmann et al. 2002) were used to fill a small number ($n = 6$) of daily measurement gaps between 1983 and 2002 when the AERP rain gauge was temporarily out of order. Data in 1970–71 and 1972–73 were completely missing for seven and three months, respectively.

It is clear from the long-term rainfall record that there is a great deal of inter-annual variability in both rainfall amount and seasonal distribution. In our analyses of foraging in relation to primary production (Lindsay, chapter 5) and the subsequent impacts on elephant behavior and demography (Lee, Lindsay, and Moss, chapter 6) and spatial distribution (Croze and Moss, chapter 7), we use a number of precipitation measures. These include the following:

- Annual Rainfall, the total for an Amboseli Year.
- Wet Months, the number of whole months with rainfall ≥20 mm. Inspection of rainfall and vegetation (Lindsay 1994) records suggests that little plant production occurs in months with less than 20mm rainfall.
- Dry Months, the number of consecutive dry season months with rainfall <20mm.
- Dry Season Intensity (DSI). This index, proposed by Lindsay (1994), estimates the severity of the long dry season in a given year. Calculated as the number of Dry Months (as above) divided by Annual Rainfall, it is a proxy measure of the extent of plant biomass decline

Amboseli Ecosystem Context: Past and Present 15

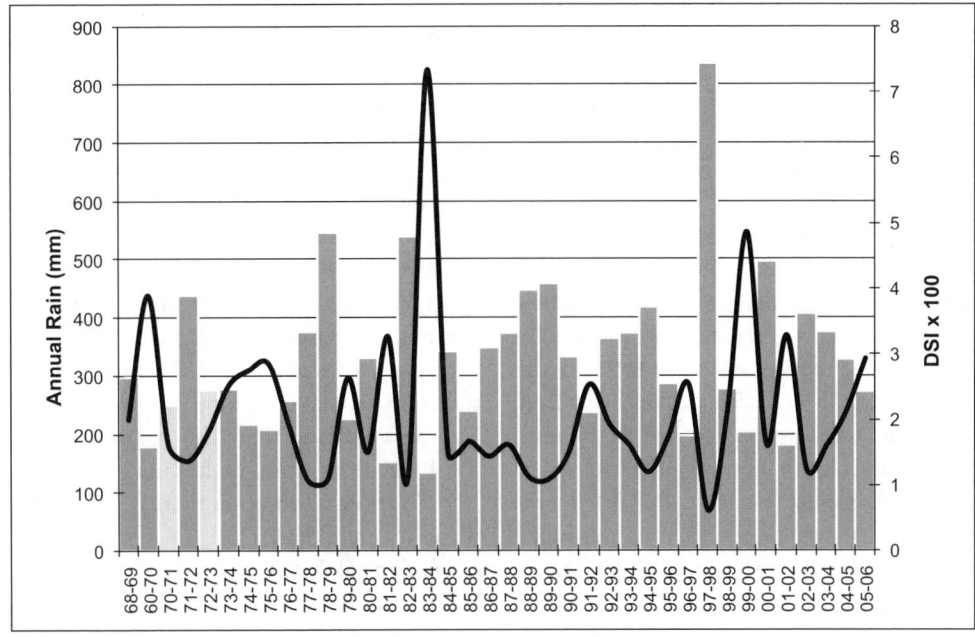

Figure 2.3 Annual rain total (bars) and Dry Season Intensity (heavy black line; see text) for Amboseli Years (Oct–Sep) from 1968–69 to 2005–6. There were incomplete data for 1970–71 and 1972–73 for 7 and 3 months, respectively, so the data were estimated using the 30-year monthly means.

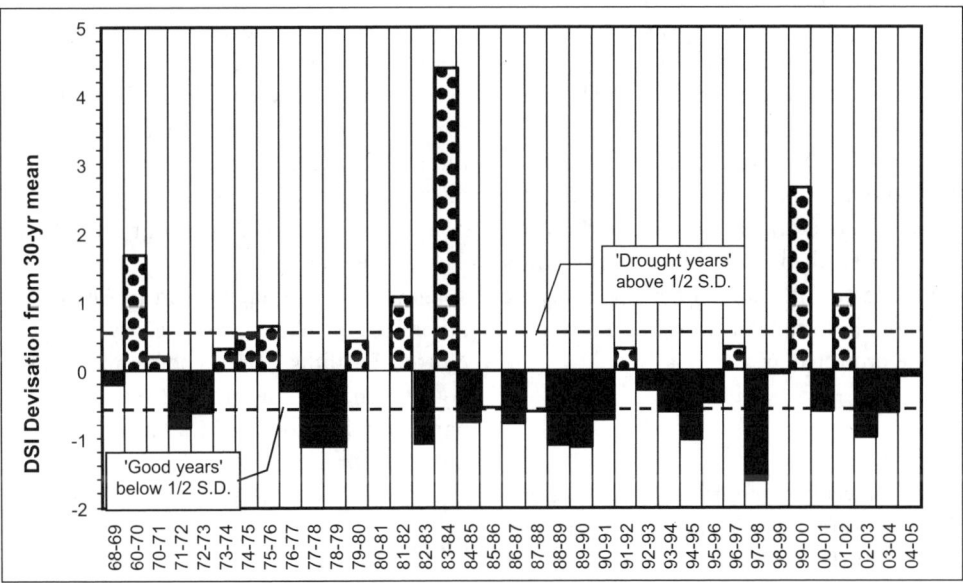

Figure 2.4 Annual DSI (Dry Season Intensity) deviation from 30-year mean. There are some 6 "drought years" in which the deviation was greater than 1/2 S.D. from the mean and 11 "good years" in which the deviation was less than 1/2 S.D.

through consumption by herbivores (length of the dry season) relative to plant biomass growth and production (rainfall).

In order to distinguish "good" from "bad" years, we determined the frequency distribution of DSIs. Six years with a DSI greater than half a standard deviation of the mean were categorized as drought, 15 years with a DSI within one-half S.D. as average, and 11 years with less than half a S.D. as good (figure 2.4). The distinctly unpatterned distribution of good and drought years over the study period underlies the inherent variability and unpredictability of the controlling factors of the ecosystem. There is compelling evidence that extremes in rainfall over eastern Africa are driven by global and regional ocean-atmosphere events such as El Niño and the Indian Ocean Dipole (e.g., UNEP 1992; Behera et al. 2005). These phenomena are markedly aperiodic, which certainly contributes to the variability of the ecosystem.

Boreholes and Pipeline

Early in the conservation history of Amboseli (Kangwana and Browne-Nuñez, chapter 3), it was recognized that if the core of today's National Park were to be sustained, then competition between the Maasai herders and wildlife for water and grazing should be reduced (Western 1997). There would either have to be reliable new water points away from the Park's edge or agreed access to a portion of the swamps within the Park, or both. A pipeline was constructed in 1977 by the New York Zoological Society (NYZS) from the perennial Enkong'u Narok spring north to a booster pump on the ridge northeast of the Park to supply storage tanks and troughs at several sites along the ridge (figure 2.2). A number of water pumps with diesel engines had been developed at borehole sites in the bushland around the Amboseli basin during colonial agricultural development programs during the 1940s and 1950s, but many of these had fallen into disrepair. The 1970s World Bank and NYZS-funded efforts to maintain the pipeline and refurbish and maintain borehole pumps at strategic points well outside the Park were only partly successful (Lovatt Smith 1997). The boreholes and waterpoints fell into disuse by the mid-1980s for two ostensible reasons. The Olkejuado County Council for its part did little or nothing to pick up the maintenance bill when the donor funding dried up. On top of that, the wildlife management authority (then the Wildlife Conservation and Management Department) had neither the expertise nor mandate to take on the burden of managing a system of boreholes outside the protected area.

Dams and Wells

The ecosystem is dotted with numerous seasonal dams and shallow wells to capture rainfall or tap into shallow groundwater sources (figure 2.2). The dams, like the boreholes, are accessible to wildlife as well as Maasai livestock. Such accessibility is beneficial on the one hand as it lures the grazing load away from the central swamps. It is potentially deleterious on the other hand since it reduces negative feedback to population growth and exacerbates the frequency of human-wildlife conflict over water.

Vegetation

The vegetation consists of trees, bushes, and grasses. Much of the grassland is *Acacia*-dominated savannah dotted with scattered trees and thickets (see below; see Lindsay, chapter 5, and Western [1973] for more thorough descriptions). (As an aside, we have noted recently proposed changes in the nomenclature of the genus *Acacia* in Africa to *Senegalia* or *Valechia*, but until the changes are generally accepted, we will continue to use the commonly accepted name *Acacia*.)

Using the National Park as a point of reference, the dominant vegetation types (Pratt, Greenway, and Gwynne 1966) are open-bushed grasslands toward the north and along the Chyulu Hills at the eastern edge of the ecosystem and *Acacia*-dominated bushland to the south up to the forest belt of Kilimanjaro. Throughout these main types, there are patches of swamp and swamp-edge grassland and *Acacia* woodland that stretch northwest-southeast along the Park's long axis, with wooded and bushed grassland found variously wherever there is seasonal accumulation of water (Mumiukha 1976). On an agro-ecological zone map, the region is labeled arid to semi-arid (Zone IV; Sombroek, Braun, and Pouw 1982).

The appearance and composition of the central Amboseli plant communities are strongly determined by the state or level of soil chemistry, water availability, and herbivore-use intensity, all of which are clearly subject to marked variation, both across the spatial extent of the basin and through time. There have indeed been changes in *Acacia xanthophloea* woodlands across the central basin and to a lesser extent, the stands of the classically umbrella-shaped *Acacia tortilis* in and just to the southeast of the National Park. But such changes—though visually striking—must be seen in the context of spatial heterogeneity and the long-term, non-equilibrium behavior of arid ecosystems (see Habitat Change below).

Grass fires are uncommon in Amboseli compared with other East African savanna ecosystems (see, for example, Norton-Griffiths 1978). The biomass and cover of the herb layer is generally low around Amboseli by virtue of its comparatively low average annual rainfall, and thus there is little fuel for fires.

People

The area is predominantly occupied by Maasai, who are the most geographically widespread and numerically largest single ethnic group in the region. There are also non-Maasai peoples permanently resident in the towns and agricultural areas: Kikuyu, Kambas, Luos, Luhyas, Chaggas, and Somalis. The total population of southern Kajiado District is roughly 113,000 of which some 36,000 are Maasai (Central Bureau of Statistics 2005). There are roughly 6,000 to 7,000 people living in the ecosystem in an area immediately surrounding Amboseli National Park at a distance of ca. 20 km (Kenya Wildlife Service 1991; Ntiati 2002).

There are also about 180,000 cattle and 230,000 sheep and goats found in the ecosystem, mostly owned by the Maasai. Other land uses in the areas surrounding the Park include semi-pastoralism, farming (rain-fed and irrigated), and wildlife-viewing and cultural tourism (Kenya Wildlife Service 1991). (See Kangwana and Browne-Nuñez, chapter 3, for a full discussion of the ethnographic context of the people, elephants, and their shared habitats.)

Wildlife

The panoply of large and small mammals (species names given in appendix 2), birdlife, and vegetative diversity has not changed markedly over the past three decades. Some striking changes in species numbers are evident. For example, under the passive protection of Maasai landowners and the watchful eye of researchers, the population of known elephants has grown during the lifetime of the study from around 600 to over 1,500 (see annex 2.1). Lions have decreased in numbers; hyenas have increased (see below). Other large mammals have remained more or less constant at the ecosystem level, with the exception of some localized diminishing—not disappearance—of numbers of browsing species such as lesser kudu and impala (Western and Manzolillo-Nightingale 2003) and the anthropogenic, total demise of rhinos in the basin (see below).

Large herbivores occupy the ecosystem in three distinct ways (Western 1975): some water-independent species such as gerenuk, oryx, eland, giraffe, and Grant's gazelle are predominantly resident in the arid bushlands. Others—impala, waterbuck, reedbuck, and buffalo, for example—favor the basin woodlands and swamp margins throughout the year. And finally, elephants, wildebeests, zebras, and Thomson's gazelles are seasonal migrants. They alternate between bushlands and basin, moving to relatively high-quality bushed grassland pastures in the wet seasons and returning to the basin swamps and surrounding woodland as water availability diminishes in the dispersal areas in the dry season (see Lindsay, chapter 5, and Croze and Moss, chapter 7). The Maasai and their livestock, the latter a majority member of the herbivore community (see below), also opt for the migratory strategy and for several hundred years have more or less mimicked the seasonal movements of the wild migrant herbivores in the Amboseli area (Jacobs 1975).

Although there is no published record of the seasonal distribution and abundance of wild herbivores over the ecosystem, these data do exist, for example, in the long-term aerial survey records of the Amboseli Research and Conservation Project (Western 1997) and within the Government of Kenya's Department of Resource Surveys and Remote Sensing (Ojwang', Waragute, and Njiro 2006). Western (2005, cited in Croze, Sayialel, and Sitonic 2006) has summarized the average occupancy of large herbivores, both seasonally migratory (wildebeest, zebra, elephant) and resident (buffalo, impala, kongoni) species, over 30 years of ecological monitoring by systematic reconnaissance flights (figure 2.5).

Several points are clear from the distribution of wildlife vis-à-vis the ecosystem features:

- The core of Amboseli National Park maps well to the core of wildlife occupancy (dark crosshatch on the map).
- Settlements, increasingly occupied on a year-round basis, have been encroaching over the years into the southern portion of the core occupancy.
- Wildlife also makes significant use of the bushlands surrounding the National Park; its survival in the long run is absolutely dependent on the goodwill of the owners and occupiers of the ecosystem dispersal areas (see Croze, Lindsay, and Moss, chapter 22).

Estimates of total numbers of wildlife in the ecosystem are few and of questionable comparative value due to differences in survey techniques and areas covered. But in round terms, the ecosystem currently is supporting on the order of 60,000 wild herbivores and 400,000 domestic stock in the form of zebu cattle, sheep, and goats (Western and Manzolillo-Nightingale 2003). Given those numbers, it appears that the 1,500-plus elephants make up roughly 4 percent of the wildlife metabiomass and, adding Maasai livestock to the sums, 2 percent of the total large herbivore metabiomass. (Metabiomass refers to the metabolic weight [body mass raised to the 3/4 power] of an animal [Croze et al. 1978]. It is more accurate than mass or biomass as a measure of the relative resource demand that species of different body size impose on the ecosystem.)

Over the past 30 years, the wildlife populations of Amboseli have responded to changes in both ecological factors and management actions. The major ecological change has been the increased outflow from the Kilimanjaro springs, leading to an elevated water table and changes in extent of both woodlands and swamp courses (see below). The major management action to affect wildlife populations in the past decades has been the planning and creation of the National Park, from which Maasai livestock and people were excluded in 1977. This change has led to reduced competition between people and wildlife over habitat resources but also increased social tension and some conflict. Other important human influences have been the increasing sedentarization of traditionally mobile human and livestock populations, increasing settlement and cultivation around water sources and swamps, and killing of wildlife by people for

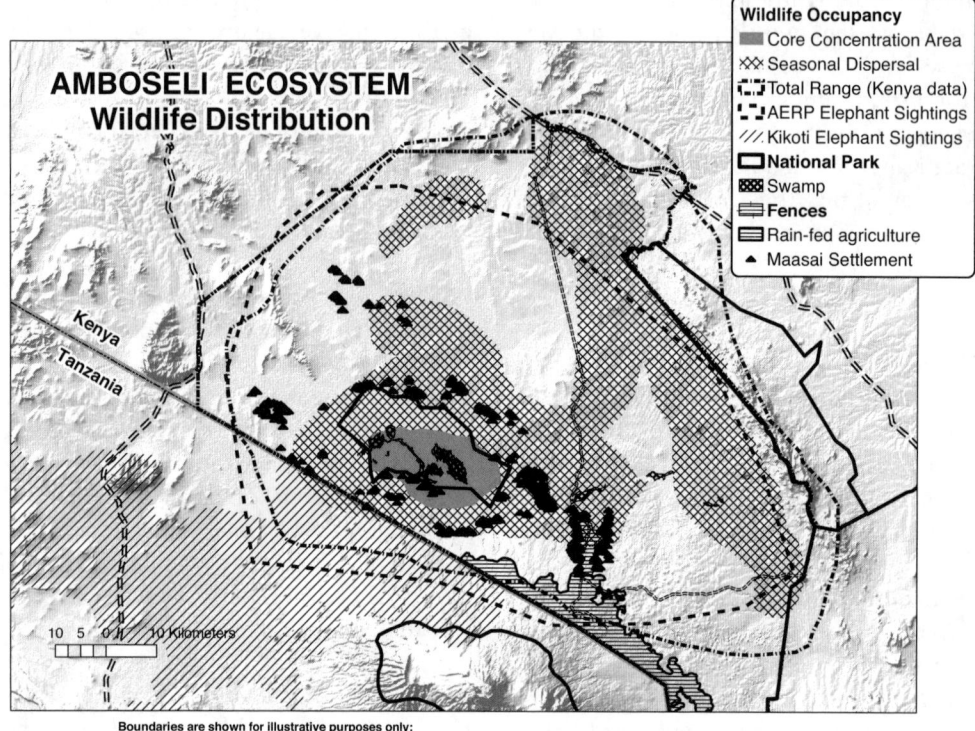

Figure 2.5 Wildlife occupancy of the Amboseli ecosystem. All wildlife species, 30-year average 95 percent kernel (grey solid), 50 percent: (crosshatch), 5 percent (dotted outline) (Western 2005; redrawn in Croze et al. 2006); elephants, 30-year Minimum Convex Polygon (dashed; AERP data); elephants Tanzania range of sightings, 1998–2001 (hatched; A. Kikoti, unpublished data, 2007). Map also shows the Park being surrounded by Maasai settlements and rain-fed agriculture invading from the southeast.

political and commercial reasons. These factors have had far-reaching and interacting effects on plant, herbivore, and carnivore communities.

From personal communication with contemporary wildlife researchers and our knowledge of recent events, we can summarize the status of the main wildlife components:

Elephants. One of the best-studied populations of large, free-ranging mammals in the world, deserving of World Heritage status. As discussed in chapter 1, the population has been relatively free from heavy poaching, largely because of the goodwill of the Maasai following the Park agreement and the continuous presence of AERP researchers from 1972 onward. Numbers have grown from some 850 (600 of which had been identified) in the mid-1970s to just over 1,500 at the time of writing (see annex 2.1). There have been changes in distribution and habitat use over this period as well. Although elephants concentrated in the sanctuary of the Park from 1977 to the early 1990s, they now spend some 80 percent of time outside the Park (see Croze and Moss, chapter 7), so that density within the Park has actually decreased in the past decade. Popular press views of elephant "over population" are unfounded given contemporary non-equilibrium ecological thinking (see below and box 2.1) and as long as dispersal areas outside the Park are available.

Black Rhinos. The rhino population declined from ca. 120 in the 1950s to some 10 to 15 in the late 1970s (Western 1982b. The killing of rhinos—either as a social protest or in poaching for their valuable horns—resulted in the last animals being nearly exterminated by the mid-1980s. The two remaining rhinos were translocated to a sanctuary. There is a nascent program to rehabilitate a remnant population in the Chyulu Hills.

Other Very Large Herbivores (hippos, buffalos, giraffe). No dramatic changes in population sizes have occurred at the ecosystem level. Giraffes have decreased within the basin and are found in woodlands elsewhere but are under threat from bushmeat trade. Buffaloes appear to have increased within the Amboseli basin (Western and Manzolillo-Nightingale 2003), likely in response to reduced grazing competition from livestock. Hippos occur in small groups in open-water patches in Enkong'u Narok swamp.

Medium-Sized Herbivores. Among grazing species, there has been significant increase in zebras (likely due to reduced dry

BOX 2.1 AMBOSELI, A NON-EQUILIBRIUM ECOSYSTEM

Much has been said about the nature of habitat change in Amboseli National Park, the bulk of it couched in negative terms such as "destruction of habitats," "woodland decline," and "loss of wildlife habitat." Conventional wisdom and published assertions hold that in response to human pressure, Amboseli elephants no longer wander freely outside the Park and are unnaturally compressed within its boundaries and as a result are negatively impacting biological diversity therein (e.g., Mitchell 2005; Western 1997; Western and Manzolillo-Nightingale 2003; see Croze and Moss in chapter 7 for AERP's picture of the changes in ranging patterns). This assertion appears to be based on the view that the normal state of nature is fixed equilibrium—the Balance of Nature paradigm. Woodlands remain woodlands unless they are "damaged" by a disruptive agent, and change is a deviation from the "correct" state. It is now clear that observed elephant impacts on habitats are a normal consequence of a large animal feeding on vegetation and that localized, short-term changes in plant and animal community structure are common rather than rare (Owen-Smith et al. 2006).

Ecologists have come to recognize that ecosystems and their components are in constant change, either continually variable or shifting between temporarily persistent "stable states" at different scales, from the local to the regional. As noted by Illius and O'Connor (1999), the characterization of such variable systems as "non-equilibrium" may be more accurately expressed as "not-at-equilibrium"; even highly variable ecosystems may have some key components that are subject to stabilizing feedback while other parts are free to fluctuate. The importance of time frame is recognized: what we observe at any one time is just a snapshot, a single frame of a moving picture show played out in ecosystems around the world, as different as coral reefs, tropical forests, and savannas shift from one set of conditions to another. Paleoecological studies find that semi-arid bush savannahs such as Tsavo National Park in Kenya have shifted back and forth between open and more densely wooded habitats over the course of hundreds of years (Gillson 2004). With the benefit of nearly half a century of research, it is evident that these "non-equilibrium" ecosystems are highly variable, devilishly unpredictable, and very resilient (Ellis and Swift 1988; Niamir-Fuller 2002; Walker and Salt 2006).

Conservation management, which carries social and political attitudes along with applied ecology, has lagged behind its scientific counterpart in accepting the new ecological paradigm and in many places still focuses on maintaining fixed habitat conditions and wildlife populations at what is termed "carrying capacity" (Rogers 1997). Human interventions to suppress change through time or to homogenize habitats on a spatial scale have been attempted in many intensively managed parks. Slowly but steadily, however, conservation agencies around the world are adopting the non-equilibrium view, recognizing, allowing, and even encouraging heterogeneity in space and through time. Examples include Yellowstone National Park (Keiter and Boyce 1991) and Kruger National Park in South Africa (Rogers 2003).

In Kenya, the prevailing policy has been to "let nature take its course," with occasional recent calls for more interventive management (e.g., Western and Maitumo 2004), ironically at a time when conservation best practice is tending toward a lighter finger on the trigger. Amboseli provides a classic example of the need for caution when interpreting change as necessarily a negative process, pointing toward the appreciation that change is the rule rather than the exception in ecology.

The complex juxtaposition of contrasting soil chemistries along with the spatial and temporal vicissitudes of the main driving variables—water flowing from the mountain-fed springs and rain falling with stochastic perversity—set the stage for the Amboseli ecosystem to be highly variable. Viewed against this unpredictable backdrop, it is unreasonable to expect constancy in the composition of vegetation and fauna in the local scale of the Park; indeed, this chapter has documented a history of change over the past five decades and postulated similar changes in the longer term past. The not-at-equilibrium ecological perspective is clearly appropriate in the Amboseli context.

---

season grazing competition) and no significant change in wildebeests; there has been an apparent gradual decrease in Thomson's gazelles and the less numerous oryx and waterbuck (Western and Manzolillo-Nightingale 2003). Ostrich numbers appear unchanged. Other observations are the evident spread of local range of Bohor's reedbuck into new swamp patches; among browsers, apparent decrease of impala, lesser kudu (never very numerous), and dik-dik within Park boundaries due to changes in *Acacia* cover; apparent decrease in gerenuk and eland in the surrounding bushlands; and no statistically robust trend in browsers at the ecosystem level.

Lions. The lion population should be healthy and perhaps even increasing, as the important prey populations, zebras and buffaloes, have increased. However, due to spearing (and poisoning) by Maasai, they are under serious threat. More than 100 were confirmed killed in the ecosystem between 2001 and 2006 (S. Maclennan, personal communication). There may be only 800 left in all of Kenyan Maasailand (Kajiado and Narok Districts).

Hyenas. Significant increase in spotted hyena numbers over the past 15 years (K. Holekamp, personal communication to Faith and Behrensmeyer 2006), with reduced numbers of lions in the Park and increased food supply in the form of zebras and buffaloes. Hyenas have largely replaced lions as predators of livestock, as lions have been reduced by spearing.

Other Large Predators. There are insufficient data to draw conclusions. Cheetahs have returned to the Park after being hounded out by off-road driving in the 1980s. Some 10 individuals are now seen regularly. Their future is uncertain. In Tanzania, Laurenson (1995) has shown that the Serengeti and Ngorongoro cheetah populations are strongly affected by lion and hyena predation on cubs. Thus, the falling Amboseli lion population could benefit cheetahs on the one hand, and the increasing hyenas could prove deleterious on the other. Wild dogs were seen on rare occasions in the late 1970s and early 1980s, but their numbers and range are unknown. Striped hyenas have also been sighted, but rarely. Many other species of small carnivores are present, but numbers and distributions are unknown.

Livestock. No significant trends have been discernible since the 1960s (Ojwang', Waragute, and Njiro 2006). Conventional opinion is that the ratio of "shoats" (sheep and goats taken together) to cattle has increased.

Habitat Change, 1957–2000

Changes in woodland vegetation are frequently attributed to "predation" by elephants, and Amboseli has been a prime candidate for the assertion. Over the past 50 years, *Acacia xanthophloea* trees have died progressively from the central Amboseli basin outward, leading to claims of a critical loss of species diversity (Western 2006). However, *Acacia* woodlands cannot be considered an immutable "climax" vegetation type. Maasai oral tradition associated with recognized age-group genealogy recalls a dearth of *Acacia xanthophloea* around the swamps at the turn of the previous century (Western and van Praet 1973), and old safari guides recall that in the 1950s, there were no or very few *Acacia tortilis* near Iltalal airstrip, which is in the eastern part of the ecosystem 60 km from the Park boundary, where they are now abundant (L. Belpietro quoting J. Fletcher, personal communication).

Young and Lindsay (1988) note that relatively even-aged stands of trees, such as dry-site *Acacia xanthoploea* and *Acacia tortilis* woodlands, tend to become established under ephemeral favorable conditions that produce tree stands of roughly even age and size. Woodland stands of mature trees in extensive areas away from groundwater sources are inherently unstable and prone to sudden change in factors affecting single, particularly the older, size classes. Since there is no emerging understory to offset lost canopy-level trees, factors affecting older plants can result in rapid disappearance of mature woodlands. The dieback of *Acacia xanthophloea* woodlands in response to dissolved salts in a rising water table has been recorded recently in Ngorongoro, Tanzania (Mills 2006). The retreat and dramatic regeneration of *Acacia tortilis* woodlands in Lake Manyara National Park, Tanzania, is another example of the episodic nature of *Acacia* population dynamics. Mature *Acacia tortilis* trees in stands with no understory died off due to feeding by the Manyara elephants. The consequent opening up of the canopy coupled with a disease-induced drop in impala numbers released tree seedlings from light competition and browsing pressure. The result was a vigorous burst of regeneration; the woodlands were restored (Prins and van der Jeugd 1993).

In order to understand better the time course and real reasons for habitat changes in the Amboseli basin, we have examined the distribution of the main vegetation communities at six points in time over the past five decades: 1957, 1967, 1978, 1984, 1993, and 2000. We based the analysis on available vegetation maps produced during ecological studies, ground observations, and evaluation of aerial photographs and satellite images.

The 1967 vegetation map, a mosaic of 1:50,000 aerial photographs taken in the mid-1960s supplemented with ground-based observations, was produced by Western (1973). It depicted 28 vegetation communities: 23 zones within the basin and 5 zones in the surrounding bushlands. For the current analysis, the basin types were pooled into 10 broader categories based on the dominant woody or herb layer cover. For 1978, the original 1967 map was updated by using ground observations (Lindsay 1982) and 1976 Landsat MSS images. Similarly, for the 1984 coverage, the 1978 map was modified using ground observations (Lindsay 1994) and a 1984 Landsat TM image. The 2000 map was produced using a Landsat TM panchromatic image and ground-truthing. Finally, a vegetation map for 1957 was extrapolated backward from the 1967 original, modified with reference to habitat descriptions and sketch maps provided by David Lovatt Smith, Warden for Amboseli Game Reserve during the 1950s (Lovatt Smith 1997). All maps were digitized and vegetation community areas were estimated with ArcGIS 9 (Environmental Systems Research Institute [ESRI]).

Amboseli Ecosystem Context: Past and Present 21

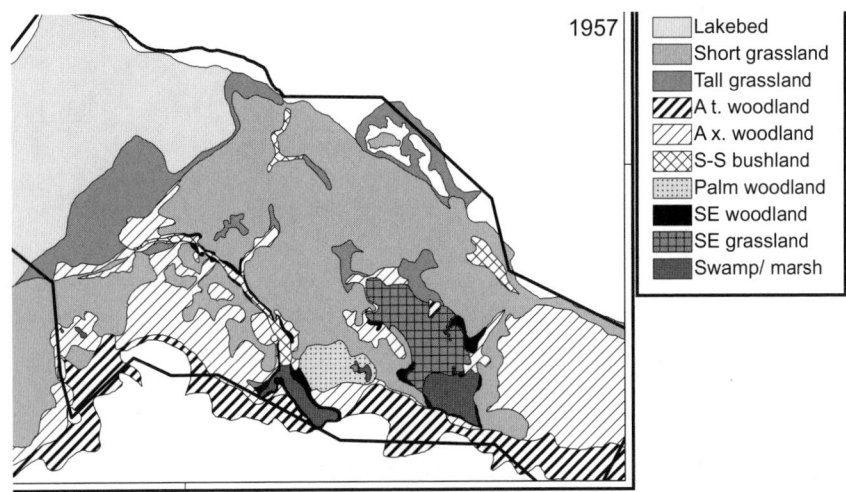

Figure 2.6 Vegetation communities of Amboseli National Park in 1957.

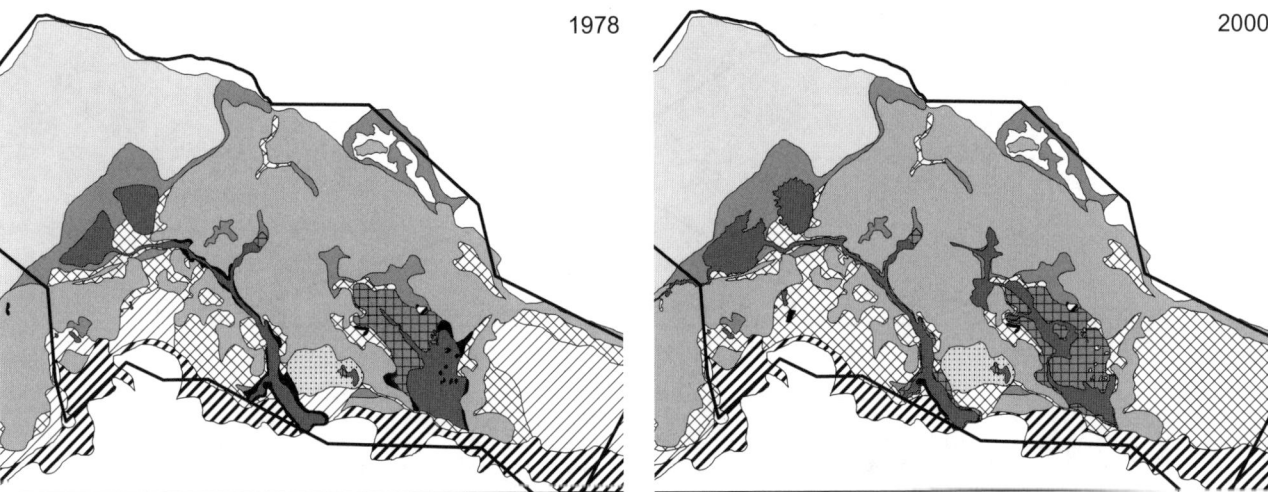

Figure 2.7 Vegetation communities of Amboseli National Park in 1978.

Figure 2.8 Vegetation communities of Amboseli National Park in 2000.

The sequence from 1957 to 1978 to 2000 (figures 2.6–2.8) shows a steady decrease in the extent of *Acacia xanthophloea* woodlands across the basin, where they have changed to bushland comprising the salt-tolerant shrubs *Salvadora persica* and *Suaeda monoica*. The changes in spatial extent of the woodlands and swamps are summarized in Table 2.1 and figures 2.9 and 2.10. The most extensive conversion in the dry-site woodlands occurred from the 1950s to 1980s, from the central areas outward, with a slower rate of change in the remaining patches at the periphery of the basin. Since the mid-1970s, on the margins of the watercourses there has been more gradual change in the mixed-age *Acacia xanthophloea* woodlands, which in any case were limited in extent to under 5 km² in 1957. There has been a slight decrease in the *Acacia tortilis* woodlands over the 1968–2000 period, and the age structure is that of a markedly senescent population with little sapling recruitment. At the same time, there has been steady growth in the extent of wet swamps and swamp-edge grasslands: their courses have altered and their areas have increased. The rate of swamp increase was greatest from the 1950s to 1970s, with a conspicuous expansion starting from a low point in late 1956 (Lovatt Smith 1997), and was accompanied by a rise in water table at wells in the central basin area and at Sinya Mine some 25 km to the west of Ol Tukai (Meijerink and van Wijngaarden 1997).

In the 1960s and 1970s, popular opinion attributed the woodland changes to overgrazing and bush cutting by Maasai herdsmen or to destructive feeding by elephants—popular interpretations that homed in on the most conspicuous agents (Western and van Praet 1973). The reality was not that simple. Changes in the two different fever tree *Acacia*

Table 2.1 Area covered by woodland, bushland, and swamp/swamp-edge vegetation communities within the central Amboseli area, by year

Vegetation type	Area (km²), by year					
	1957	1967	1978	1984	1993	2000
Woodland (*Acacia tortilis*)	42.4	42.4	40.5	40.1	39.9	39.5
Woodland (*Acacia xanthophloea*)	82.9	75.8	45.3	26.7	12.8	1.7
Bushland (*Salvadora/Suaeda*)	5.3	9.4	37.5	56.6	70.6	81.0
Swamp-edge woodland (young *Acacia*)	4.9	4.6	4.5	3.3	1.2	0.7
Swamp-edge grassland	11.2	7.2	9.0	12.1	12.3	13.6
Wet swamp, marsh	7.0	17.7	20.4	19.2	23.3	23.4

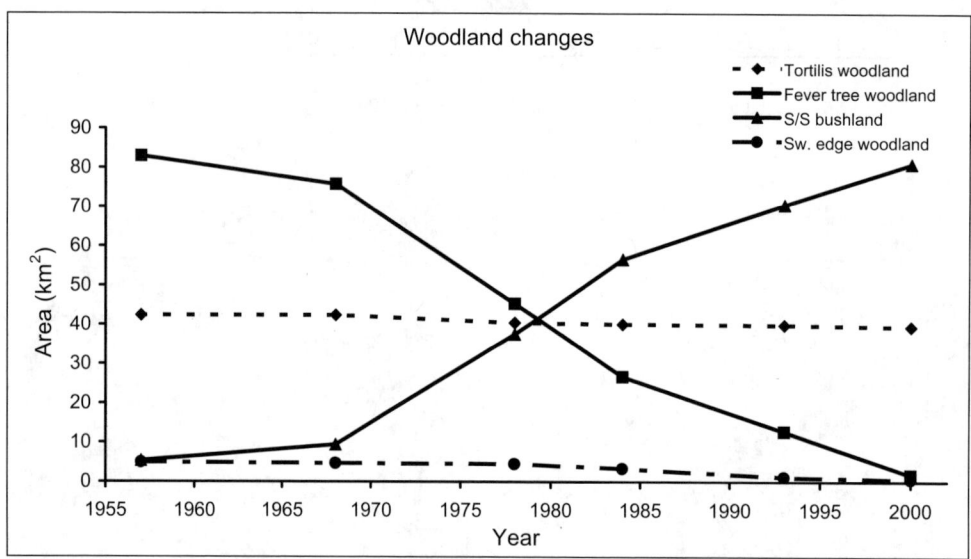

Figure 2.9 Changes in area of woodland vegetation communities, 1957–2000. Note the decrease in *Acacia xanthophloea* (fever tree) woodland and corresponding increase in (conversion to) *Salvadora/Suaeda* bushland, starting in the 1950s and accelerating in the 1960s–1970s. Swamp-edge *Acacia xanthophloea* woodland changes are also shown.

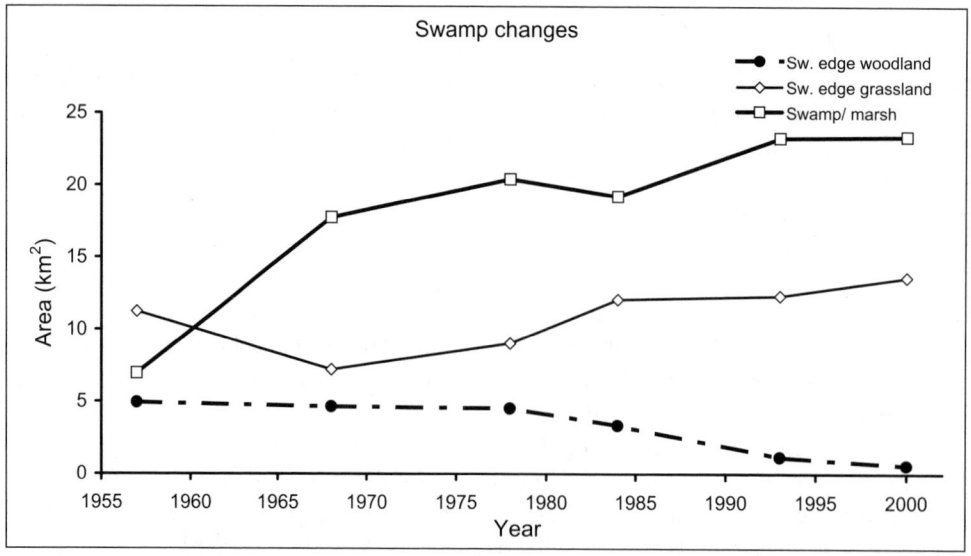

Figure 2.10 Changes in area of swamp/marsh and swamp-edge vegetation communities, 1957–2000. Note the steady increase in the marshes from the 1950s, leveling in the mid-1990s. Swamp-edge *Acacia xanthophloea* woodland is shown again in this graph; with the relatively smaller range of areas depicted, its stability throughout the 1950s to 1970s and subsequent decrease from the mid-1970s to 2000 is more clearly visible than in the previous figure.

xanthophloea woodland types—dry-site mature woodlands and swamp-edge thickets—were in fact due to two different processes. Western and Van Praet (1973), in an elegant analysis combining data from field sites, historical records by European explorers, and Maasai oral accounts, demonstrated that the main driver of woodland decline in the dry sites covering much of the Amboseli basin was directly attributable to the increase in the level of the water table that elevated a concentrated salt solution within the saline basin soils into the root zone of the mature *Acacias*, thereby altering their osmotic balance. The trees were effectively suffering "physiological drought." Elephant browsing was more likely to be a side effect rather than a cause; a catalyst to the inevitable decline of already dying trees. As noted above, since there were few understory fever trees in the dry sites, the death of the mature trees left shrub layer *Salvadora persica* and salt-tolerant *Suaeda monoica* bushes as the dominant woody species to characterize the plant community.

In recent years, AERP researchers have documented a significant die-off of *Suaeda monoica* shrubs within the *Salvadora/Suaeda* bushland areas. Discernable *Suaeda* mortality was first noted in 1998, a year of exceptionally high rainfall, and again in the high rainfall period of late 2006. It is possible that heavy rains, such as those of the two El Niño episodes of 1997–98 and 2006–7, have a "rinsing" action on the salts in near-surface rooting zones (Mifflin 1991), rendering the soils less saline and conducive to salt-loving plant species. Evaporative draw from a high water table in average to drier years may bring salts back to the surface again, reversing the process.

The swamp-edge *Acacia xanthophloea* woodlands, in contrast to the dryland sites, were relatively unaffected by the salinity changes associated with the rising water table, as the permanent groundwater in the swamp courses diluted and transported salts away from the swamp margins. With trees at all stages of growth, they were structurally different from the dry-site *Acacia xanthophloea* woodlands. However, they were very limited in area—less than 5 km² at their maximum in 1957—and located along the watercourses where elephants have always concentrated in dry seasons. As noted above, the creation in 1977 of a sanctuary from contact with Maasai and their livestock encouraged elephants to remain in the vicinity of the swamps on a year-round basis, a situation that persisted until the early 1990s. With the disappearance of *Acacia xanthophloea* woodlands across the dryland areas of the Park by the 1970s, elephants concentrated their browsing along the swamp edges, which resulted in the gradual removal of canopy-level trees and bushes. Nonetheless, an abundant seedling layer of fever trees has remained in the swamp edges.

The relatively stable state of grassland–low bushland could persist in the face of relatively light herbivory on the remaining tree seedlings, either by a small number of elephants foraging seasonally or by other browsers and roughage grazers. Similar situations have been noted in the Mara ecosystem of southwestern Kenya (Dublin, Sinclair, and McGlade 1990), where fire now acts in combination with browsers to maintain open grasslands, and in the Lake Manyara National Park *Acacia* woodlands, where browsing impalas suppressed seedling recruitment, as described above. Such short-term stable states are reversible when ecological factors favor a swing toward alternative conditions. Factors that would reduce elephant browsing in Amboseli, such as a series of wetter than average years providing alternative food supplies or competition from people in key swamp-edge areas, could result in rapid growth of swamp-edge seedlings to shrubs and trees. Recolonization of woodlands across dryland wider areas could require a longer, sustained period of favorable conditions, perhaps in combination with dispersal of *Acacia xanthophloea* seeds by elephants and other browsers.

It is instructive to look more closely at woodland distribution in the year 2000 (figure 2.11). By then, dry-site *Acacia xanthophloea* woodlands across the Park had changed to *Salvadora/Suaeda* bushland, and most of the swamp-edge *Acacia xanthophloea* woodlands had also changed to grassland. However, woodlands persisted in a few key areas for two reasons: competitive exclusion and fencing.

In the Enkong'u Narok area at the southern edge of the Amboseli National Park, a number of Maasai settlements—normally seasonally occupied—became permanent in the early 1990s. Since then, the swamp edges immediately north of the settlements have been used continuously by people and livestock. Elephants have been deterred from the area by the proximity of human activity (Croze and Moss, chap-

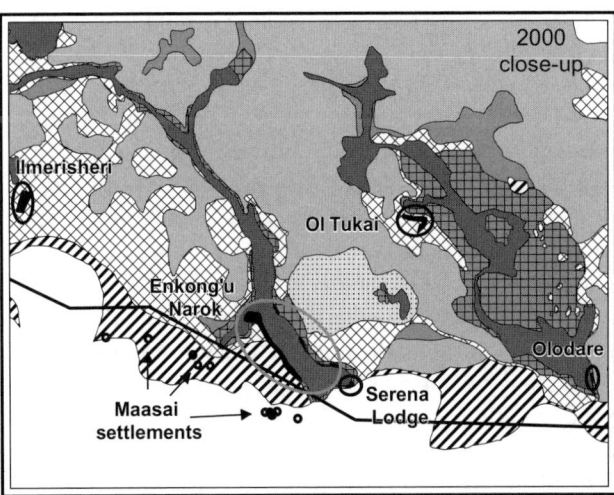

Figure 2.11 Key swamp-edge woodland areas in 2000. Areas enclosed by fences are within dark ellipses, while the woodland adjacent to the Enkong'u Narok settlements is within the light ellipse.

ter 7; Kangwana, chapter 20), and a robust stand of young *Acacia xanthophloea* has emerged.

Exclosure fences erected for ecological experiments or to protect landscaping around tourist hotels have resulted in dramatic growth of *Acacia xanthophloea* trees along the swamp edges from the high density of rootstocks to canopy-level trees within a period of 5 to 10 years (Western and Maitumo 2004). Both of these examples show that a very brief period of reduced browsing pressure can allow swift regrowth of seedlings to canopy-level trees in the swamp-edge woodlands.

The changes in Amboseli vegetation communities should be viewed in their historical context. Changes in the water table, as induced by changes in the flow regime of the swamps, in turn determined by flow from the watershed of the north slope of Kilimanjaro, result in variation in the area and distribution of the swamps. The level of water table also determines the extent of *Acacia xanthophloea* woodlands.

The result is that large swamps are accompanied by small woodlands and vice versa. It is clear from remote sensing images that watercourses cross the short grasslands to the northern edge of the basin, hinting at a wetter, swampier time in the past. Indeed, Behrensmeyer (1993) found fossil hippo and waterbuck remains at the northern edge of the lakebed where it joins with a discernible but now dry watercourse, indicating that swamps extended to the northern lake margin clearly at a time—some 70,000 years ago—when swamps were large and woodlands were likely to be small (figure 2.12).

Figure 2.13 shows the alternative scenario, with limited outflow from the swamps and extensive woodlands. This situation was found in Amboseli in the middle of the 20th century (Lovatt Smith 1997). It now appears that the system is moving back toward the large swamp/small woodland condition.

At an ecosystem scale, extensive *Acacia xanthophloea* woodlands thrive outside the Park, particularly to the north-northwest, west, and southwest. Statements that the habitat changes of the past five decades have resulted in dramatic loss of species diversity and even species extinctions (Western 2006) are accurate only for localized area within the Park. Intervention to protect trees (for example, with exclosure plots) within the Amboseli basin when water table and soil salinity are high is likely to succeed only in areas along swamp margins. However, if browsing pressure were reduced and dispersed, for example, through negotiated agreements to allow elephants to maintain their current dispersion across the ecosystem, combined with a period of several good rainfall years and herbivore deterrents, the resultant reduced impact of elephants in the National Park would surely lead to rapid regrowth of woodlands in the

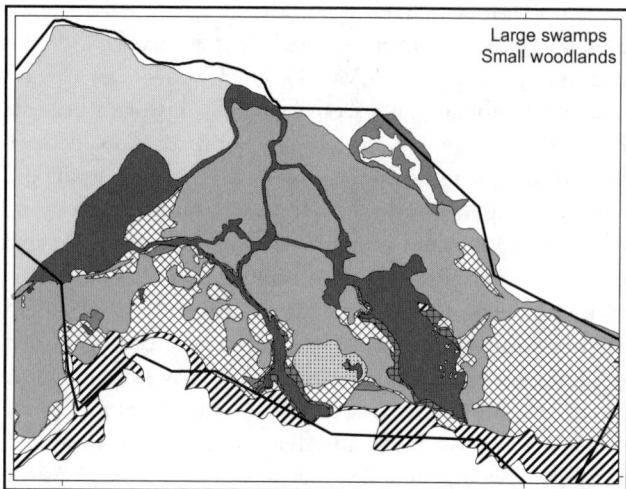

Figure 2.12 Scenario of large swamps and small *Acacia xanthophloea* woodlands.

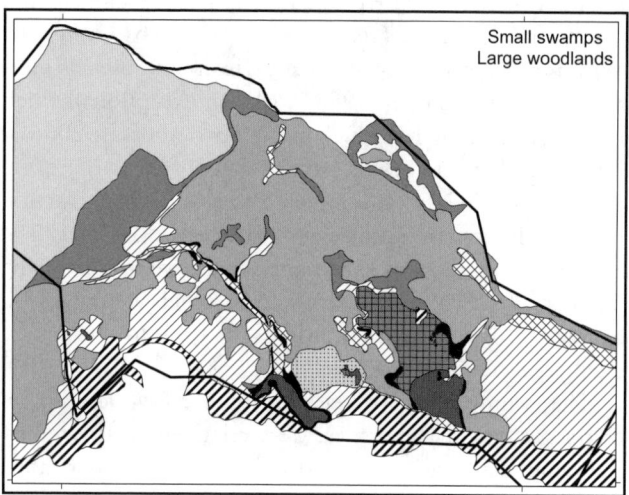

Figure 2.13 Scenario of small swamps and large *Acacia xanthophloea* woodlands.

swamp edges and perhaps elsewhere in the Park. Since none of the wildlife species that inhabit *Acacia* woodlands have been lost from the broader ecosystem, they would also return to the Park habitats.

Conclusions

Amboseli is a non-equilibrium ecosystem typical of arid and semi-arid regions that are highly variable (in physiognomy and species mix, in time and space), unpredictable, and resilient (Ellis and Swift 1988; Niamir-Fuller 2002). From time-series analyses of gross changes in proportional cover of major habitat types over three decades as well as evidence from paleoecological records and oral traditions, it is evident that the inherent nature of the central Amboseli basin is that of a system fluctuating between two extreme conditions in what appears to be a non-periodic cycle: large swamps–small woodlands and small swamps–large woodlands.

The ultimate causes have yet to be determined; absolute rainfall amount and factors affecting runoff (such as slope, soil type, and vegetation cover) are likely to be involved. At any point in time, the system is in transition from one state to another, subject to the vicissitudes of unpredictable (and little understood) water delivery from the Kilimanjaro mountain catchment combined with low and erratic rainfall that is quite likely itself changing on a continental scale from global processes (UNEP 2000).

Management actions in the National Park can affect neither the outflow rate from the Kilimanjaro springs nor the level of the associated water table, both the results of mountain hydrology and regional climate well beyond the reach of animals and people. *Acacia xanthophloea* colonization of the dry areas across most of the Amboseli basin is thus likely to occur only under certain conditions, when the water table has dropped after a series of average to low rainfall years.

The creation of the National Park and resultant removal of dry season grazing competition has had cascading effects on the Amboseli wildlife populations and their habitats, allowing population increases in grass-dependent species such as buffaloes and zebras and seasonal grazers such as elephants. Predator populations have also benefited from the increased prey base. Maasai spearings have eliminated rhinos from Amboseli and if unchecked, are likely to exterminate the remaining lions. The removal of lions together with increased food supply has allowed the hyena population to grow. Elephants discovered that the Park was a sanctuary from human contact and concentrated in its confines, thereby accelerating woodland change and displacing browsing ungulates and other tree-dependent wildlife to the areas outlying the central Amboseli basin.

Elephants play a catalytic or modifying, rather than determining or controlling, role in the habitat changes of Amboseli. At high woodland density and extent, elephants are unlikely to have a significant impact. As woodlands decline with rising water table and salinity or simply through natural senescence and disease, a population of elephants will have a greater per capita impact on a decreasing number of trees, thereby accelerating the decline. Trees may be able to persist along the margins of swamps, even with the high water table and general salinity across the rest of the Amboseli basin, but with such a limited distribution, even a small number of elephants would gradually reduce the woodlands to grassland and maintain them in this state.

The erratic presentation of "good years" and "bad years" in terms of food and water abundance is likely to have a profound impact on the demography and distribution of the Amboseli elephants, as discussed in part 2, as well as on their social dynamics and reproductive strategies (part 4).

ANNEX 2.1 POPULATION SIZE OVER TIME

Reliable estimates of the number of elephants in the ecosystem are obviously important in order to, for example, understand changes in the population size and status, identify factors causing the changes, and plan for the management support that is needed for a sustainable dispersal area. In this annex, we explore how the size of the Amboseli elephant population has changed over time.

Numerical Changes over Time

There is a perennial undertone in casual discussions and the popular press about elephants in Amboseli that hints at overcrowding and therefore of a situation that may need to be managed. Western (1997, 227), for example, in his personal narrative account warns, "elephants would eventually destroy their own habitat and die of starvation if the population continued to grow unchecked. Fewer than a couple of hundred elephants could be sustained within the park indefinitely." He goes on to say, "The only solution lay in winning back space outside the park—if the area could be made safe." True enough, and in fact, he probably overestimates: it is unlikely that the Park at its present size could support more than a few tens of elephants, let alone hundreds, on a continuous basis. Lovatt Smith (1997) observes, "[I]t would be a brave person who would stick their [sic] neck out and tell the warden of Amboseli, with any degree of confidence, to reduce the numbers of elephants to a specific number per square kilometer in order to keep the Park with some degree of acacia woodland attractiveness."

The implication seems to be that historically, Amboseli was not an "elephant area," so the presence of elephants there today is somehow unnatural, discordant, and disruptive and thus something that might require human intervention. We conclude in this chapter that in the non-equilibrium setting of Amboseli, significant state changes are the norm, and attempts to manage and modulate the state of the ecosystem (for example, by husbanding numbers) would be futile. In this light, we recount below a brief history of elephants in Amboseli by way of background rather than for purposes of setting management guidelines on numbers.

Pre-Colonial Period

There is evidence that elephants existed along the shores of Lake Amboseli as far back as the Pleistocene (Behrensmeyer and Boaz 1981). From Behrensmeyer and Boaz's examination of fossil bone distributions and frequencies, it is possible to deduce that ca. 70,000 years ago, the large mammal species assemblage was not unlike it is today.

The genetic analysis by Archie et al. (see chapter 4) provides evidence that most of today's Amboseli elephants have their origins in the south, possibly as far as southern Africa, via Kilimanjaro. The evidence from other haplotypes is less clear but hints at some arrivals from the north a couple hundred years ago. These incomers may have nurtured their genetic legacy elsewhere (in the Aberdares and Mount Kenya montane forests, for example, 250 km north) and then moved into Amboseli within the last several hundred years.

An alternative scenario to the supposed historical absence of elephants is one of long-term fluctuations in occupancy, with a substantial elephant population in the region hitting a low from the turn of the century until the 1950s, reduced perhaps by the rinderpest pandemic of the early 1890s (antibodies for rinderpest have been found in Asian elephants [Behrensmeyer, Stayton, and Chapman 1993]) or from ivory poaching. The largest pair of tusks ever recorded—226.5 pounds (102.7 kilos) and 214 pounds (97 kilos)—came from an elephant shot on the lower northwestern slopes of Kilimanjaro, well within the Amboseli ecosystem, in the 1890s. Pre-20th-century Maasai were alleged to be so hostile to invaders of their land that the ivory and slave caravans that penetrated the hinterland from the coast steered to the south of Maasailand through central Tanzania (Lamouse-Smith and School 1998; Browne-Nuñez, chapter 19), but obviously some ivory hunters were able to operate in the area. It has also been suggested that the rinderpest epidemic so reduced livestock populations that Maasai in both the Loitokitok and Serengeti-Mara areas were forced to diversify their livelihoods by involvement in the ivory trade in order to survive (Baumann 1894; Waller 1976; Dublin 1991; Browne-Nuñez, chapter 19).

Colonial Period

Notwithstanding the legendary ferocity of the Maasai (e.g., Lamouse-Smith and School 1998), colonial adventurers passed through or explored the southeastern edges of the Amboseli ecosystem, including Joseph Thomson (1883), Ludvig von Hoehnel and Count Samuel Teleki von Szet (1887), and Mary French-Sheldon (1891). Except for French-Sheldon, these explorers shot impressive numbers of buffalos, rhinos, and other animals: hartebeest, impala, wildebeest, zebra, gazelles, and so on. It seems unlikely that elephantine targets would have escaped their attention.

From mid-July to mid-August 1883, Scottish hunter-explorer Joseph Thomson passed through the heart of Amboseli from the Tsavo region on his way west to Lake Victoria. In his book *Through Maasailand*, he states, "In spite of the barren and desolate aspect of the country, game is to be seen in marvelous abundance" (Thomson 1885, 276). Despite that abundance, he mentions no elephants. As he entered and camped in the southeastern part of the ecosystem, he encountered members of the Dorobo people living in the area and "occupying themselves with elephant-hunting" (Thomson 1887, 242). Obviously, there must have been elephants within the area in sufficient numbers to make the hunting profitable.

Count Samuel Teleki von Szel and Ludwig von Hoehnel followed a similar route (von Hoehnel 1894). In July and August 1887, their party trekked from Taveta on today's Kenya-Tanzania border north and northwest into the Amboseli ecosystem, noting spoor of elephants in the region of Rombo. In late July, they camped one night on the top of a flat stony hill (possibly Ontawua hill in Elerai; see figure 2.1):

On the west rose a few low hills covered with black volcanic rocks, whilst on the east the land sunk, in one long terrace, to the plain which stretched far away to the foot of the Julu [Chyulu] chain. There was very little grass, and that little was sere and dry; even the reeds in the swamps were dead or trodden down by wild animals. In the distance we could make out a few thriving steppe plants such as euphorbia, various kinds of succulent bush, aloes and two kinds of *Sansiveria*, but the ground was everywhere sandy and bare. This dreary wilderness was, however, tenanted by a great variety of birds, including two kinds of doves, starlings with gleaming steel-green plumage, beautiful nutcrackers with turquoise-blue fathers, several kinds of fowls, hawks, and vultures, marabout [sic] storks, and bustards, whilst a little farther away roamed herds of gazelles, antelopes, rhinoceros, zebras, gnus, giraffes, ostriches, and wild boars. One night, too, we heard elephants in the swamp. (von Hoehnel 1894, 225)

That passage could almost have been written today, apart from a curious reference to a grey tiger (probably a striped hyena) that the Count later shot. A few days later, von Szel and von Hoehnel were camping on the shores of "Lake Nyiri," a "sheet of water being fed by springs only" (1894, 228). From von Hoehnel's narrative and sketch maps, the location must have been on the edge of one of the large swamps in the central Amboseli basin, probably in the vicinity of Enkongu Narok or Lonkinye, with more open water than there is today, an echo of the high water table seen in Amboseli from the mid-1960s to the present day. The name the men attributed to the lake was too close to today's designation of Njiri plains, the southeast part of the basin, to be a coincidence. There was evidence that another caravan had camped in the same place some weeks before, but history draws a veil over the identity of its principals. Although von Hoehnel was "fairly successful" with his hunting (buffalos getting the brunt), he makes no further mention of elephants in the area, and the party passed across the dry Amboseli lakebed on to the north to "Kikuyuland."

Even today, July and August generally are not good months to see elephants in Amboseli since the bulk of them leave the basin perhaps to seek particular vegetation types elsewhere or because of reduced water requirements in the cold season. Those that do use the central basin do so in small groups and are found deep in swamps and thus are hardly visible when on foot. Thus, Thomson's and von Hoehnel's accounts do not necessarily mean that the ecosystem was devoid of elephants in their time. Throughout the period of our study, it is often possible to drive across the Amboseli basin and not see a single elephant, even today when the elephant population tops 1,500. It is possible that with habitat conditions at that time similar to those today (i.e., high water table, widespread swamps, and limited *Acacia* woodlands), any elephants that were present may have preferred to use areas outside the immediate basin area as they have tended increasingly to do in the later years of the AERP study (see Croze and Moss, chapter 7).

Whether or not Thomson and the Teleki party had "bad elephant weeks" on the Njiri plains south of Ol Tukai, it is unlikely that no elephants were using the area at that time, since fossil (Behrensmeyer and Boaz 1981; see above) and genetic (Archie et al., chapter 4) evidence confirms that elephants have occupied the area since the end of the Pleistocene. But from this evidence, it is not possible to deduce if occupancy was constant, seasonal, or sporadic.

Irrespective of the date of the arrival of Amboseli's first elephant and the population's numerical starting point, Amboseli is now clearly an important elephant area. And in keeping with the complexity and flexibility of elephant sociality, the way that elephants occupy the Amboseli ecosystem is neither simple nor altogether predictable in the manner of the annual Serengeti wildebeest migration or the seasonal expansion and contraction of a resident population of impalas (Fritz, De Garine-Wichatitsky, and Letessier 1996). Along with an increase in population size and the changing ecological and human social circumstances over the years, the elephants have apparently altered their strategy of deployment over the ecosystem, probably several times.

Modern Times

The history of conservation in Amboseli has been discussed elsewhere (Kangwana and Browne-Nuñez, chapter 3). Like Joseph Thomson 70 years earlier, David Lovatt Smith found a paucity of elephants when he joined the Game Department as an assistant warden in 1952. He reports, "It was not until the late 1950s that we began to recognize individual bulls which remained month after month and eventually become resident." As we note above, it is unlikely that there were *never* elephants resident in the basin. The question is, why were there apparently so few in the early part of the 20th century?

The appearance of bulls in the 1950s smacks of the vanguard of an immigration, or perhaps more likely, re-immigration, with groups of non-sexually active adult males feeding, socializing, and gaining condition in "bull areas" (Croze 1974; Poole 1982). Bulls elsewhere, such as the Okavango, are known to be wanderers, the first to move into new areas or return to areas after local extirpation (Evans 2006). Whether the population in and around Amboseli was recovering from a climate-habitat–related demise over the past 5,000 years, from disease-related mortality over the past 100 years, or an ivory hunting surge in the late 1800s (as also seen in Serengeti-Mara and over much of sub-Saharan Africa; Dublin 1991) we can only surmise. Suffice it to say that elephants (re)established themselves in 20 years between the 1960s and 1980s as an important herbivore in the core of the Amboseli basin, if rather less important in the whole ecosystem (see Croze and Moss, chapter 7).

Figure A shows the change in Amboseli elephant population numbers from 1969 to 2005. The graph is a patchwork of sources, with data from individual estimates (Western 1973), systematic reconnaissance flights, total aerial counts, and numbers of known individuals. How these data were collected and analyzed is described in Appendix 1.

Thus, the data presented here have a highly variable quality and different levels of precision and accuracy. However, combining data from several different sources provides validation when there is concurrence among the sources. To comment on each of the five sources used in turn:

1. The 1968–70 number of 220 is certainly an underestimate since it comes from Western's (1973) counts restricted to the National Park and immediate surroundings. Western (personal communication) reckoned there to be as many as 1,200 elephants in the ecosystem in the 1960s.
2. The combined SRF estimates from the Wildlife Conservation and Management Department of 830 ± 30 percent (Croze 1978) is quite likely reasonable for 1973–76 given the statistical rigor of pooled estimates of variance and the fact that the surveys covered the entire ecosystem.
3. The numbers from this study (e.g., Moss 2001) on known individuals begin with an estimated population

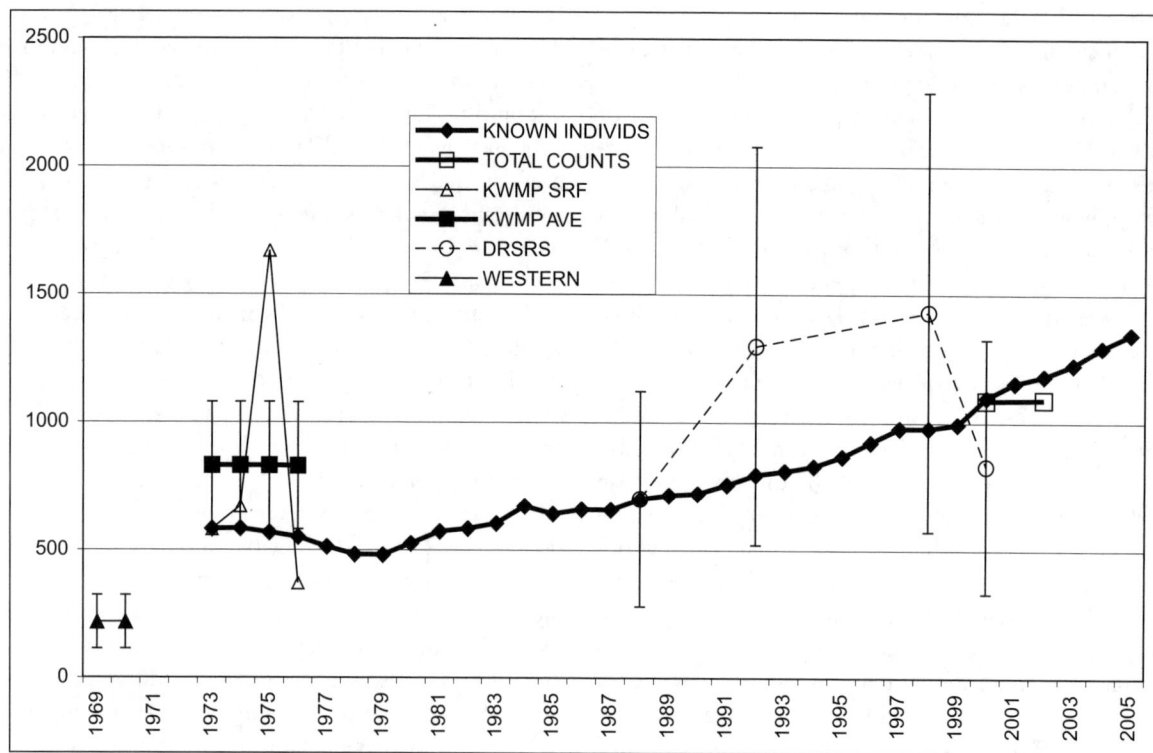

Figure A Change in elephant numbers over nearly four decades. Data are drawn from a number of sources (see text).

of just under 600 animals in 1973. By the end of 1976, when 90 percent of the families had been registered, there were some 512 known individuals. At that time, the number of registered animals would have been an underestimate of the population: the study area was concentrated mainly in the National Park, and there were still some five families left to identify, not to mention a number of transitory or occasional males.

4. Kenya's Department of Resource Surveys and Remote Sensing provided four SRF estimates from 1988 to 2000 (Ojwang', Waragute, and Njiro 2006). They are within range of our known population at that time but have the very high standard errors of ca. 60 percent.

5. The Kenya Wildlife Service total counts dating from 2000 and 2002 are uncannily close to each other (1,087 and 1,090, respectively) but also encouragingly close enough (on the low side, typical of total counts) to the known number of individuals at the end of each of those years (1,154 and 1,225) to suggest 1,200 as a solid number for the beginning of 2002.

Although the Amboseli elephants were largely spared the poaching slaughter of the 1970s and 1980s until the international accord by the Convention on International Trade in Endangered Species of Flora and Fauna (CITES) to ban the trade in ivory in 1989, they suffered some mortality, with the numbers of known individuals being reduced from ca. 580 in 1974 to ca. 480 by the end of 1978 (Moss 2001; figure A). The then irascibility of Maasai toward other peoples hunting on their land combined with the near-constant presence of watchers and alarm-raisers in the form of project researchers was a positive deterrent given the relatively easier pickings in neighboring Tsavo National Park and the game-controlled areas of northern Tanzania.

Conclusions

In sum, best estimates for elephant population sizes throughout the ecosystem appear to be around 600 at the beginning of the study in 1972, growing to 1,225 by the end of 2002, the nominal cut-off for the 30-year data analyses in this book, and rising to 1,500 by 2008. That is an average rate of growth of about 3 percent per annum over the period, a moderate rate of population growth for elephants (Moss 2001; see also Lee, Lindsay, and Moss, chapter 6). Recent Leslie matrix analyses of the rate of population increase based on known females give a population multiplication rate of 1.034 (3.4 percent annual growth; Clubb et al. 2009).

Chapter 3 The Human Context of the Amboseli Elephants

Kadzo Kangwana and Christine Browne-Nuñez

Introduction

The Amboseli elephants live in a complex mosaic of human settlement and land use. At the center of this mosaic is the Amboseli National Park, which comprises 392 km² of the 8,000 km² greater Amboseli ecosystem and serves as the dry season concentration area for the ecosystem's migratory herbivore community. During the wet season, migratory species such as zebra and wildebeest disperse beyond the Park's boundaries, while elephants move in and out of the Park year round (see Croze and Moss, chapter 7). The land outside the Park is divided into group ranches occupied predominantly by the pastoral Maasai, whose cultural practices and attitudes toward elephants have contributed to maintaining the Amboseli elephant population. Changes in land tenure and land use in the Amboseli ecosystem during the last century have impacted the Amboseli elephants. These human dimensions, therefore, not only form an integral part of the story of the Amboseli elephants but to a large extent determine the future of the elephant population.

We present here the human context within which to understand the Amboseli elephants. We describe the people of Amboseli, the Maasai, focusing on aspects of Maasai history and culture that have contributed to the coexistence of the Maasai and wildlife. We also describe the conservation and development initiatives that have defined the Maasai's use of the Amboseli ecosystem over time and have influenced Maasai attitudes toward wildlife and conservation. The challenges of conserving the Amboseli elephants within a landscape of changing land use and land tenure in the Amboseli ecosystem are discussed.

The aims of the chapter are as follows:

- To present the human context of the Amboseli elephants and explore how this sets the stage for understanding their ecology and population dynamics in the context of pressure from people and livestock (part 2) and the patterns of individual and social use of the space that elephants share with people (part 4).
- To provide the background for the analysis of human-elephant relationships in chapter 19 and the context for the behavioral responses of elephants to the Maasai in chapter 20 and to introduce the key elements that impact on the future of the Amboseli elephants as discussed in chapter 22.

The People of Amboseli—The Maasai

Wildlife conservation continues in Maasai areas such as Amboseli largely as a result of Maasai culture. Although the Maasai identity varies across space and time, with some Maasai practicing various degrees of agriculture, the "true" Maasai is typically viewed as a pure pastoralist, herding livestock, moving seasonally, and subsisting on milk and blood.

The agro-pastoralist ancestors of the modern Maasai came from southern Sudan (Spear 1993). These eastern Nilotes moved south through the Rift Valley, interacted with other ethnic groups, expanded, and subdivided. Through interaction with the eastern Cushites c. 500 years ago near the border of Ethiopia and Kenya, the eastern Nilotes were exposed to the cultural taboos against hunting and eating game, which they may have adopted at that time (Galaty 1993a). This early exposure may explain similar taboos in modern Maasai culture. In their migrations, these agro-pastoralists also made contact with farming groups migrating into the East African highlands and hunter-gatherers living in the forested areas bordering the highlands. The different economic activities of the pastoralists, the farmers, and the hunter-gatherers came to be viewed as ethnic divisions between Maa-speaking pastoralists, Bantu farmers, and Okiek hunter-gatherers (Spear 1993). Trade between these groups may have made it possible for the early Maasai to become more specialized pastoralists.

While pastoralists have been a feature of the East African savannahs for thousands of years, Maasai pastoralists only settled in the region in the last few hundred years (Kituyi 1990; Lamprey and Waller 1990; Galaty 1993b; Sutton 1993). As the Maasai adapted to the East African savannahs, they became more specialized pastoralists and developed and maintained cultural values, institutions, and practices suited to that habitat, enabling them to coexist with the abundant wildlife in the region.

The Maasai who inhabit the region surrounding Amboseli National Park belong to the Ilkisongo section of the proper Maasai. The Ilkisongo are one of the two largest sections of the Maasai. They range from southern Kenya into Tanzania, with the majority living in Tanzania. Until recent years, land was predominantly occupied and utilized by the Maasai on a communal basis in order to optimize its use in the variable, semi-arid environment. The Maasai's traditional system of using land required social and economic cooperation. The main features of Maasai society—a system of sections, clans, and age-sets—facilitated this cooperation by widening social relations and creating channels for exchange (Homewood and Rodgers 1991). A section can be thought of as a subsection of the entire population: it is a territorial group with its own political structure based on the age-set system. The section is divided into localities (*enkutoto*), which may be self-contained ecological units in which member households within a locality have access to the same permanent water source and dry and wet season grazing lands (Kituyi 1990). Sections and localities facilitate the communal land tenure system and allow for shifting grazing areas in times of environmental stress (Homewood and Rodgers 1991).

The age-set has been defined as a group of men who are initiated in youth during a definite span of time and share certain constraints and expectations for the remainder of their lives (Spencer 1988). During a man's transition through the age-set, stages from herdboy to warrior or *moran* to elder, social values, and networks are established. One's age-set also determines the responsibility over the management of livestock and how much livestock is accumulated by an individual (Evangelou 1984). A Maasai herd boy is educated about livestock and geography, and young boys begin by looking after small stock near the settlement. Boys remain herders until around the age of puberty, when they are initiated with ceremony into the warrior age-set: *moran*hood. *Moran* are exempt from herding and instead have the responsibility of protecting their homesteads and livestock. Newly initiated *moran* traditionally set up a *manyatta*, or warrior camp, and during their time in the *manyatta*, the young men compete in feats of strength and prowess, including the spearing of wildlife such as lions and elephants (Homewood and Rogers 1991). Whenever a *moran* succeeds in killing a dangerous animal, the other *moran* and women celebrate his strength and sing songs of praise for him. During this highly regarded period in a man's life, lifelong friendships are formed. These friendships can be important later in life, when it may be necessary to ask for grazing permission. From *moran*hood, men transition, with ceremony, into elderhood. As elders, Maasai men are expected to marry, settle, and concern themselves with herd management and the affairs of the community. Elders supervise the herding, plan the grazing strategy, and often go out with children to help with the herding (Evangelou 1984; Kituyi 1990). Women are not grouped into age-sets as are men, but they are identified with the age-set for whom they sang when young and unmarried (Evangelou 1984). Women in Maasai society fulfill domestic roles, including fetching water and firewood, milking livestock, caring for children, and building and maintaining huts (Saitoti and Beckwith 1980).

The traditional Maasai settlement is the *enkang*, also referred to by the Swahili term *boma*. These are more or less circular thorn bush enclosures with individual homes made of cattle dung and thin poles located on the inside periphery of the thorn fence. Households within a *boma* tend to be unrelated, and the social identity of a *boma* is associated with the head of the family that started the construction (*olmarei oitera enkang*) (Evangelou 1984). Members of a *boma* are often those with complementary resources for herding cooperation, usually in the form of labor, but this can extend to food support (Kituyi 1990). The livestock of one *boma* are usually herded together and brought into the center of the *boma* at night as protection against predators.

Each *boma* is in turn part of a local settlement. The Maasai household identifies as its home a perennial or permanent *boma* in which most of the members of the household live year-round. Some members may move away during the dry season and reside in smaller camps (*elatia*) as part of the practice of transhumant herd management (Kituyi 1990). Transhumant pastoralism is a strategy to cope with seasonal extremes of water availability and primary production by moving between dry season and wet season grazing areas (Berntsen 1976). In the dry season, the Maasai gather with their herds around a permanent source of water and graze their cattle in an ever-widening circle around the water source. In the rainy season, they move to more distant regions of their locality in order to preserve grazing near the water source for dry season use. This seasonal movement of the Maasai follows a similar pattern to the dispersal of wildlife in the Amboseli ecosystem.

The Maasai have demonstrated that pastoralism in savannah rangelands can be highly compatible with wildlife conservation (Thompson and Homewood 2002; Homewood and Brockington 1999; McCabe, Perkin, and Schofield 1992). For some four hundred years, the Maasai have occupied the extensive rangelands of the Amboseli ecosystem, living alongside the elephants and other migratory herbivores, but the last hundred years has brought many changes to the Maasai social structures and economy. Traditionally, the Maasai have been dependent on their livestock for food, turning to other food sources only in times of severe drought or other catastrophe (Galaty 1980; Evangelou 1984).

In recent times, however, they have adopted a more diverse diet (Kituyi 1990; Homewood and Rogers 1991). The Maasai have also made a shift from cattle-based subsistence toward a profit-oriented diversified economy that provides cash for purchasing goods, paying school fees, and meeting other requirements (Berger 1996). Maasai now grow and sell crops, own businesses, operate wildlife sanctuaries and cultural villages for tourists to visit, and lease land to tourist operators—all developments that are related to changes in land tenure (see below; also Browne-Nuñez, chapter 19; Croze, Moss, and Lindsay, chapter 22). The introduction of agriculture to the Amboseli ecosystem has resulted in conflict between wildlife and the Maasai, as wild animals, including elephants, raid the crops (see Kangwana, chapter 20).

Group Ranches in Amboseli

During colonial rule, communal areas were registered as trustland managed by local authorities. After independence, the Kenya Government initiated group ranches in the late 1960s. They were legislated by the Group Representatives Act of 1978 to bring pastoral areas into the commercial sector and assign the rights and responsibilities of landowners to specific groups of Maasai pastoralists (Galaty 1980; Migot-Adholla and Little 1981). Fifty-two group ranches were incorporated, comprising 75 percent of the Kajiado District (Rutten 1992). Although not fully understood, they were accepted by the Maasai due to a desire to maintain land tenure (Lembuya 1990; Galaty 1980; Migot-Adholla and Little 1981) and to protect themselves against further land appropriation from the government, incursion of non-Maasai, and landgrabbing by the elite Maasai (Fratkin 1994; Campbell 1991; Mwangi 2005). The government expected the Maasai to become sedentary and adopt a commercial livestock production system, but most Maasai continued to move with their cattle across the Group Ranch lands.

Group ranches became the major organizational structures for the development of pastoral areas (Migot-Adholla and Little 1981). Negotiations between government bodies, including the National Parks authorities, and the Maasai were held with Group Ranch leaders (Kenya Wildlife Service 1990), and the Group Ranch Committee called meetings to decide on matters affecting all members of the Group Ranch.

Group ranches have been heavily criticized for being demarcated as territorial bases that were ecologically unfeasible and not economically viable and for having poor management (Helland 1980). It has been suggested that group ranches failed to meet their objectives (Mwangi 2005). The leadership of the group ranches was to fall on an elected Group Ranch Committee. Committee members tended to be young educated men who did not necessarily hold tribal authority in the age-set system and came under many local political pressures that made it difficult to carry out their duties (Lembuya 1990). With respect to increasing the commercialization of livestock production, Group Ranch leaders were meant to enforce stock quotas and destocking by members who had too many cattle in order to maintain a carrying capacity of cattle for the Group Ranch as a whole. This enforcement was against Maasai cultural practice under which no Maasai could interfere with another's livestock (Galaty 1981). In adopting the group ranch system of land tenure, individuals could obtain collateral for loans for developments, but land held in trust for a group could not be sold without serious consequences for all, and payment of loans could not be enforced (Galaty 1981).

In the late 1970s, many Maasai were demanding the dissolution and subdivision of the group ranches. There were a number of reasons for the call for subdivision. Younger members had concerns about future access to resources (Thornton et al. 2006). Other reasons included problems with the management of the group ranches by the elected committees, a desire for title deeds in order to obtain individual loans, and a perceived threat of further land loss to immigrants (Kimani and Pickard 1998; Mwangi 2005).

The 1990s saw a political move to allocate individual titles in Group Ranch and trust land areas (KWS 1991). Subdivision, it was hoped, would give people individual assets that they could sell or raise loans against and hence bring them more into the Kenyan private-enterprise economy (Lembuya 1990; KWS 1991). By 2000, 31 of the original 52 group ranches had been subdivided and issued titles (Mwangi 2005; see Browne-Nuñez, chapter 19, for additional details).

The group ranches immediately bordering Amboseli were the last to subdivide. Kimana Group Ranch to the east of the Park completed subdivision in 2005. The subdivison of Olgulului Group Ranch to the south, west, and north of the Park was under way in 2006. The implications of individual title to both livestock husbandry and wildlife conservation were much debated at group ranch meetings during 1990s and continue to cause concern. Although the Maasai are aware that small plots of land are not viable for livestock husbandry and that agriculture is not possible in most of the Amboseli ecosystem, they are keen to get title deeds to land (Croze, Sayialel, and Sitonic 2006).

The History of Conservation in Maasailand

Pre-Colonial and Early Colonial Administration

The current range of the Maasai is largely the result of colonial interventions or colonial government–induced movements. Prior to colonial intervention, the Maasai controlled a large area of Kenya and Northern Tanzania. This region extended roughly 1,000 km in a belt stretching from near Mount Kenya in the north to Dodoma in central Tanzania in the south (Kituyi 1990; Homewood and Rogers 1991).

In 1904, the colonial administration took steps to contain the Maasai in a defined area, mainly in order to free land for European settlers. A treaty was signed between the government and Olenana, the then-recognized chief and Maasai spiritual leader (*oloiboni*). The treaty confined the Maasai to two reserves with a half-mile-wide corridor connecting them (Grandin 1985; Kituyi 1990). The area of the reserves was estimated as being just 10 percent of the pre-1880 Maasai territory (Leys 1925, quoted in Kituyi 1990). The government promised to protect these reserves at all times in the future. The first breach of contract occurred, however, when disease forced the colonial government to impose a quarantine and sever the contact between the two reserves (Simon 1963; Kituyi 1990). Subsequently, in 1911 the Maasai were moved to a single reserve as a result of further pressure to release the fertile land to the north for European settlers (Kituyi 1990).

It was recognized as early as 1904 that the Maasai regions of Kenya had the highest concentrations of wildlife. Meinertzagen's (1983) diary entry for April 18, 1904, reads as follows:

In view of the likelihood of a vast invasion by European settlers it seems that the large game must disappear. I have suggested . . . a structure for a very large area in country unsuitable for white settlement where game can exist forever. I think that the area might be some three or four thousand square miles and possibly in Maasai country. The Maasai are good game preservers but are very wasteful of grazing lands.

Simon (1962, 115) reiterated the suitability of Maasai country for wildlife conservation due to the Maasai tolerance of wildlife: "[T]he traditionally tolerant attitude of the Maasai towards wild animals appeared to indicate that wildlife would stand a far better chance of survival under the custodianship of the Maasai than anywhere in East Africa. . . ."

Increased publicity for East African wildlife and a strong conservation lobby resulted in the creation of game reserves and parks at the recommendation of the Game Policy Committee in the 1940s (Lovatt Smith 1997). Several of these conservation areas had direct bearing on the Maasai. Tsavo was designated a game reserve in 1946, and two years later, its western half was declared a national park, with boundaries enclosing water sources that had long been crucial to the Maasai in the area (Campbell 1991). Amboseli and West Chyulu were declared game reserves in November 1948 (Lovatt Smith 1997).

Amboseli Game Reserve: The Beginning of Negotiation

The Amboseli Game Reserve spanned an area of 1,260 square miles (3,260 km^2). A game reserve was meant to afford wildlife protection in areas that could not be designated national parks since they were already allocated either exclusively or partially to other interests (Cowie 1951). Essentially, game reserves were multi-use areas for both people and wild animals, although the destruction of wildlife in these areas was prohibited. At the time of declaring Amboseli a game reserve, the Director of the National Parks Trust hoped that Amboseli would be given national park status (Cowie 1951).

The early 1950s saw an increase in the number of tourists visiting Amboseli. Developments and improvements were made to the access roads to Amboseli and to the tourist accommodation at the central palm and *Acacia*-wooded area known as Ol Tukai (Lovatt Smith 1997). Tourism grew in value and with it, concern for the future of wildlife in the central part of the reserve, where the presence of the Maasai and their livestock had begun to be perceived as a problem. It was against this background of concern over the Maasai's impact on wildlife in Amboseli that negotiations and steps toward cooperation between the Maasai and the reserve administrators took seed.

In 1952, a scheme of sharing revenue with the Maasai was started. Under this scheme, the local native councils were allocated land and were paid rent for the use of the lodge site. The reserve administration hoped that if the Maasai were to obtain direct financial benefit from the development of Amboseli as a tourist area, they would come to recognize the value of preserving wildlife (Cowie 1952).

The Amboseli Game Reserve authorities were also quick to recognize that at the center of the conflict between the authorities and the Maasai was the issue of water and that if alternative sources of water were available to the Maasai in the dry season, livestock would not concentrate in the central swamps of the reserve or the Ol Tukai region. Ideas for developing water resources outside the central basin soon followed, including the digging of boreholes and the deepening of a channel to redirect water away from the central swamps (Lovatt Smith 1997). These projects were reported to have been well received by the Maasai, and wardens at the time prided themselves on the cooperation they had elicited from the local Maasai (Lovatt Smith 1997). Following these water projects, limits were imposed on the number of Maasai *bomas* and cattle allowed in the central Ol Tukai areas. No more than two Maasai *bomas* and 7,000 stock units were allowed into the region (Cowie 1952).

At first it seemed that the Maasai were cooperative with regard to the limits, but this cooperation was an artifact of favorable climatic conditions. When the agreement was made, water and grazing were available in the more peripheral regions of the Amboseli Game Reserve. However, as soon as the dry season set in and water became scarce elsewhere, livestock were moved into the central swamps. This scenario typified the situation in Amboseli. During wet seasons, the Maasai moved to more distant parts of the reserve, returning to the central swamps during the dry season.

The impact of livestock on the central swamps began to gain bad publicity and at times, the Maasai were blamed. This news filtered back to the Maasai, whose suspicion of the authorities' intentions to exclude them from the central swamps grew. They were reassured, however, by the governor, Sir Evelyn Baring, who addressed a meeting in Ol Tukai in August 1958, stating that the government had no intention of excluding the Maasai from Ol Tukai (Cowie 1958). This dispelled some of the fears of the more politically astute Maasai (Lovatt Smith 1997).

The recommendation of the Game Policy Committee to abolish the Amboseli Game Reserve was met with much disappointment on the part of the National Parks Trustees. The Director felt reluctant to give up efforts to build a relationship with the Amboseli Maasai, which he felt would lead to the greater delegation of authority to the Maasai themselves (Cowie 1958). Legal notice 374 of June 23, 1961, surrendered the Amboseli Game Reserve (Casebeer 1975), and the area was declared a County Council Game Reserve by the African District Council of Kajiado on June 14, 1961. After negotiations between the County Council and local herdsmen, an 80 km² section of the central reserve area was declared stock free. All revenue from the reserve was collected by the County Council. However, its headquarters in Kajiado were far removed from the conservation area. It was during this phase of being administered by the County Council administration that the Masaai began deliberately spearing large animals such as rhinoceros and elephants as a protest against the threatened loss of their dry-season grazing lands (Lindsay 1987).

Amboseli National Park

A presidential decree issued in 1971 declared that an area of Amboseli be set aside exclusively for the use of wildlife (Lindsay 1987). Land was therefore set aside for National Park purposes in 1972, and the Park was established in October 1974 under the National Parks of Kenya Act and placed under the control of the National Parks Trustees. Administration of Amboseli changed in 1976 to the Wildlife Conservation and Management Department with the merger of the National Parks body and the Game Department (Casebeer 1975)

After the declaration of the Park, a series of management and planning initiatives were undertaken, the first being the 1973 Development Plan (Western and Thresher 1973). An economic analysis carried out as part of the plan suggested that the best use of the Amboseli area would be a combination of tourism and commercial livestock ranching (Western 1982). The plan closely followed the declaration of the Park and recommended that land adjudication in the surrounding region be changed to the group ranch system; it included proposals to compensate the Amboseli Maasai for their exclusion from the Park. The group ranches were to receive water supplies outside the Park because it was recognized that the Maasai could not be displaced from the basin without alternative water and forage (Western and Thresher 1973). Maintenance of the existing boreholes and construction of a pipeline to carry water to outlying water tanks from springs in the central basin were undertaken (see Croze and Lindsay, chapter 2). Park authorities were to be responsible for operating the pipeline and boreholes (Lindsay 1987).

The Maasai were to be paid a grazing compensation fee equal to the theoretical market value of cattle that would have been raised instead of for the equivalent biomass of wild herbivores on group ranch land. Direct economic benefits were also to be received by the Maasai group ranches through the development of wildlife-viewing circuits and tourist campsites. Additional benefits were to be received in the form of services such as schools, a dispensary, and a community center, which would be included in the new Park headquarters (Western 1982a). Negotiations between group ranch representatives, the County Council, and the central government took place, and an agreement was reached in 1977 when the Maasai agreed to vacate the Park. The reported successes of the plan were short lived (Western 1982c). By 1980, the water system was operating poorly due to administrative and technical reasons (see Croze and Lindsay, chapter 2). The grazing compensation fee stopped being paid to the Maasai in 1981, and income from tourism was limited to small amounts collected from the sale of wood and gravel and campsite fees (Lindsay 1987). The Maasai started entering the Park in order to water their cattle.

Soon after its creation in 1989, the Kenya Wildlife Service (KWS), the parastatal body in charge of national parks and wildlife conservation in Kenya, declared its intention to share park entry fees with adjacent communities in recognition that some of Kenya's protected areas, including the cornerstones of the tourist industry, depend for their survival on compatible management of adjacent areas of land (KWS 1991). The main aim of the revenue-sharing scheme was to encourage landowners to maintain wildlife and its habitat. The primary purpose of the revenue-sharing scheme was "to pay people who live in wildlife areas adjacent to national parks and who in the national interest tolerate wildlife and consequent costs" (KWS 1991, 34).

Revenue-sharing mechanisms were not determined aåt the outset, and the initial methods were regarded as interim and experimental. However, in trying to gain credibility with the people, the KWS was keen to distribute the money quickly. Much of the debate about revenue sharing focused on choosing the most appropriate body for channeling benefits to the community and whether to distribute cash to individuals or invest in community projects. The Group Ranch Committees and the County Council were the only existing institutions that were considered possible interim channels for revenue sharing (KWS 1991).

Initially, the KWS intended to provide revenue for Group Ranch projects. Although it recognized that revenue distributed as cash to individuals was probably more effective in influencing people's land-use preferences and meeting opportunity costs, the KWS chose to fund projects because of the administrative difficulty involved in distributing cash benefits to individuals. However, the KWS did not wish to use wildlife-generated money to fund projects that by rights should have been funded by other government sectors, "[f]or communities will not benefit as fully as they might from wildlife and the resulting dissatisfaction would have a negative effect on community relations and the whole conservation effort" (KWS 1991, 40). At a meeting with Group Ranch officials on August 9, 1991, the Director of the KWS announced the allocations to be made to each of the group ranches in the Amboseli dispersal area. Each committee was to identify its priority projects within the scope of its budget and present them to the KWS for financing.

Despite the KWS's efforts, the series of land alienations and history of broken promises have led to a negative attitude among the Maasai toward the Park administration (see also box 19.1). Evidence of Maasai hostility and anger is illustrated in a memorandum sent to the Minister of Tourism and Wildlife from the Olgulului Group Ranch Committee in March 1989:

The New York Zoological Society contributed in 1972 KShs 2,000,000 for watering points outside the gazetted area to enable livestock to water without necessarily entering the park. The volume of water was inadequate and WCMD [Wildlife Conservation and Management Department] have [sic] not maintained the poor piping and troughs . . . and we have not enjoyed the benefit of the water project for the last eight years (KWS 1991, 35).

The committee went on to complain that the County Council was earning revenue in its area and had made no contribution to the community, that the annual grazing compensation that had been promised had been discontinued, and that it took exception to the attitude of the National Park authorities toward group ranch members.

Elephants and the Maasai

Elephants are an important part of Maasai life and culture. They feature in a number of Maasai stories and proverbs (Kipury 1983), and they are held in higher regard than other wild animals (see Browne-Nuñez, chapter 19, for more details). Finding an elephant's placenta or witnessing a birth is believed to bring Maasai great wealth. One Maasai story attributes the origins of the elephant to humans: Maasai girls are usually betrothed to men much older than themselves, and they tend not to be happy about their marriages and do not want to leave home. Nevertheless, a girl is not supposed to turn to face her home as she walks away to her new husband's home on the day of their marriage. One tale relates how a young Maasai bride who was particularly unhappy at the prospect of her marriage did look back toward her home and immediately turned into an elephant. Another example of acknowledging the elephant's human origin is that on passing an elephant skull, the Maasai put grass into its orifices, as they would do with a human skull.

Although traditional taboo prevented the Maasai pastoralists from hunting or eating wild game, they are said to have had an indirect involvement with the ivory trade in the last century (discussed in detail in Browne-Nuñez, chapter 19). The Dorobo, who were hunter-gatherers, hunted elephants for ivory, which they traded to the Maasai who in turn traded it to ivory traders from the coast. Ivory traders maintained good relations with the Maasai, who knew where Dorobo villages were located, and guided the traders to the Dorobo (Berntsen 1977).

The spearing of lions by Maasai *moran*s in order to prove bravery and manhood is a well-known and well-documented tradition (Simon 1962; Saitoti 1986). Elephants were included as valid targets for *moran*s practicing courage and skill (Moss 1988). As noted above, spearing of wild animals in the Amboseli region has taken a political turn, with the Maasai using these gestures to show their dissatisfaction with the Park (Western 1982b; Lindsay 1987; KWS 1991; see Browne-Nuñez, chapter 19). In total, these activities have led to the loss of several hundred elephants over the duration of the project (see box 19.1 for recent mortalities).

Challenges and Opportunities

The current landscape of the Amboseli elephants is shaped by the ecology of Amboseli, the traditions of the Maasai, and the history of conservation in the area. The Maasai pastoralist lifestyle allowed for a wide-open range for migratory wildlife, while the Maasai taboo against eating wildlife meant that wildlife populations were not decimated by hunting as they were elsewhere in Africa. Maasai also hold a reverence for wildlife, respecting it as part of nature. Elephants in particular feature in tribal folklore. All these dimensions of Maasai culture contributed to making Maasai areas suitable for conservation, a potential that was recognized early in colonial times.

The subsequent annexing of Maasai land for conservation and the forced movement of the Maasai from their traditional grazing land by the colonial government early in the 1900s had a profound effect on their attitudes to conservation. Most significantly, these actions were the beginning of the disenfranchisement of the Maasai. The link between the people and their environment was weakened, and wildlife became someone else's property and responsibility. This sinister outcome of colonialism was inherited by the newly independent government and perpetuated by its efforts to conserve wildlife resources by declaring national reserves and parks and implementing development plans. The broken promises that punctuate the history of Amboseli Reserve and Park and the failure of development projects led to yet further disenchantment with conservation authorities among the Maasai.

It can be argued that the conservation efforts of the last four decades were in fact efforts to reestablish or redefine the relationship between the Maasai and wildlife. We see embedded in the development plans and compensation schemes of Amboseli attempts to redress the injustices of protectionist conservation strategies that displaced people and excluded them from critical water sources. The distribution of revenues by the KWS in the 1990s in order to offset some of the costs to the Maasai of living with a resource of national importance is an example.

Amboseli has also been influenced by the materialistic model of conservation that swept through Africa, which mandated that wildlife needed to pay its way. One premise of this model was that people would not conserve unless it was the most profitable form of land use. Another related premise was that people were unlikely to conserve wildlife if they did not benefit economically from it. In the face of Africa's poverty, these ideas were tantalizing. However, this materialistic approach to conservation has not lived up to its promise in much of Africa (for a review of community conservation initiatives see Hulme and Murphree 2001; see also Croze, Lindsay, and Moss, chapter 22). Experience has shown that the economic benefit from conservation rarely makes it the most economically viable use of land and that even when conservation provides the benefit of community development, people still lose out economically from the presence of wildlife (Emerton 1998). Studies and experience show us what we perhaps knew at the outset: that people conserve for broader reasons than pure economic benefit. While there is still much to learn about people and conservation, it is clear that the Maasai of Amboseli do not value wildlife only for its monetary benefit (see Browne-Nuñez, chapter 19). Thus, there exists in Amboseli a foundation on which to build a relationship with conservation wherein people live with and care for resources they value for a reason beyond pure economics: a stewardship for nature.

Against this backdrop of conservation agendas, Maasai society is itself evolving. Individual economic and political interests diverge progressively through adoption of new lifestyles, formal education, and changing aspirations. What was once a society characterized by relative homogeneity of interest (pastoralism) is becoming a heterogenous community with a wider variety of economic activities. Agricultural subdivision of land with irrigation potential adjacent to the protected area, conversion of forests to plantations on Kilimanjaro, and small-scale, rain-fed subsistence farming in surrounding bushlands have all increased in area and affected the ability of elephants and other mammals to disperse from the central swamps (see Croze, Lindsay, and Moss, chapter 22). Seno and Shaw (2002) have described the emergence of a diverse community of farmers, ranchers, and entrepreneurs in areas like Amboseli as the biggest challenge for the future of wildlife conservation.

It is too early in the process to evaluate the impact of subdivsion on the Amboseli ecosystem, but it is anticipated that wildlife will be negatively affected. The AERP (Amboseli Elephant Research Project) Biosphere Report states that "[subdivision] is seen as one of the greatest threats to wildlife in the ecosystem, by conservationists and many group ranch members alike" (Croze, Sayialel, and Sitonic 2006, 16). It is feared that without a national land-use policy, individuals with title deeds can do whatever they wish with the land. At the same time, both conservationists and the Maasai recognize that possessing a title deed for land may be a strong basis for the Maasai to negotiate agreements with neighbors to use land for communal purposes and with conservation interest groups to use land for conservation.

It is tempting to surrender to the romanticism of the wonderful past in Amboseli, especially with respect to wildlife conservation, when the pastoralist lifestyle of the Maasai allowed for an open range for wildlife populations. It is also easy to view the changes in the lifestyle of the Maasai and land use in the Amboseli ecosystem as threats to conservation goals and a tide to be stemmed. This stance may be justified when we consider that many of the changes in Maasai society were the result of top-down policy impositions that disregarded Maasai culture. However, the Amboseli elephants are best served by our taking a more pragmatic approach.

The conservation of the Amboseli elephants requires us to identify the opportunities for elephant conservation presented by the emerging scenario of a protected area surrounded by individually held plots of land. Ideally, the people living around Amboseli would participate meaningfully in conservation, while the Amboseli elephants would have corridors for migration and access to at least part of the land surrounding Amboseli. With these goals in mind, perhaps the newly subdivided landscape holds the first opportunity Amboseli has ever had to foster true stewardship for wildlife. The people who were considered mere beneficiaries of conservation in the past can now become the real protagonists of conservation as owners of the land needed for wildlife. Empowered with title to the land that wildlife needs, the Maasai may now choose conservation as a viable land-use option. Certainly the danger with the group ranch system of having opportunistic elite serving their own interests at the expense of the community will be minimized. Moreover, no longer would only a distant government reap the benefits of conservation.

It is obvious that conservation goals in Amboseli can be met by collective action in defining and allocating land to corridors and wildlife access. As the subdivision of land was under way in 2006, we witnessed the emergence of new models of cooperation in Amboseli. Neighboring Maasai were coming together to manage their land collectively, and tourist and conservation interest groups were negotiating for the use of land with both individuals and groups of landowners (Croze, Sayialel, and Sitonic 2006; see also chapter 22).

Perhaps our first task is to determine mechanisms for negotiation and decision making with so many different landowners. We must also seek new models for collective action to replace those traditional institutions that collapsed with the changes in land-use policy. There are many questions. Are new institutions to replace the old ones? What is the role of the individual? What is the role of the group or collective? What is the relationship between the individual and the group? What are the decision-making processes? What capacities do individuals need in order to participate productively in the collective process? As we consider the options, we must guard against assuming that the traditional institutions of the Maasai were faultless or that they would be viable in a Maasai society in such rapid transition.

Neither can we deny that the larger context of the Maasai has changed. Wildlife that was "theirs" one hundred years ago is now a global resource, with local, national, and international stakeholders. How will these interest groups interface? What is the relationship between local and global interests, and how are they both served equitably? How do we build global concerns into local decision making?

Through this brief analysis of the human context of the Amboseli elephants, it is clear that their history as well as their future are inextricably linked with human motivation and behavior and the mechanisms of human society. We have long recognized that humans will determine the future of wildlife. The current situation in Amboseli, where elephants and other wildlife need to be conserved in a protected area surrounded by individually held plots of land, brings us to the threshold of discovering how.

Chapter 4 The Population Genetics of the Amboseli and Kilimanjaro Elephants

Elizabeth A. Archie, Courtney L. Fitzpatrick, Cynthia J. Moss, and Susan C. Alberts

The Value of Understanding Elephant Population Genetic Structure

A species' ability to cope with the changing selective forces that result from climate change, habitat loss, disease, and other types of environmental alterations is partially determined by the amount of genetic variability in populations and by the way that variation is structured within and between populations. This genetic structure is a consequence of historical processes of migration, mutation, local adaptation, and genetic drift. On the most local scale (e.g., social groups within a population), patterns of philopatry, dispersal, and mating behavior influence how genetic variation is structured within populations and how quickly that genetic variation is lost from within a population (Chesser 1991, 1998; Sugg, Dobson, and Hoogland 1996; Pope 1998). Processes that act at larger temporal and spatial scales, such as effective population size, multi-generational patterns of gene flow, and species range expansion, will determine which alleles occur in a given population as well as the degree of genetic differentiation between populations.

Understanding how these processes operate to shape the genetic structure within and between elephant populations is important for two main reasons. First, several elephant populations are currently threatened by habitat loss and illegal hunting, which limit elephant movements and population sizes and place their populations' genetic diversity at risk. Understanding which processes maintain genetic diversity in elephant populations is important for managing their genetic potential. Second, at least two aspects of their social organization—the fission and fusion of groups and the fact that males breed in several social groups—are unusual in mammals and may result in population genetic structures that differ from typical social mammals. Despite the importance of such processes, we are just beginning to understand how mating behavior, dispersal, and range expansion, for example, shape the genetic structure within and between natural populations of African elephants (see also Archie et al., chapter 13).

Here we address these topics in two ways:

- First, we briefly review what is known about the genetic structure within populations, with a focus on the elephant population in the Amboseli ecosystem.
- Second, we investigate how the genetic variation in Amboseli relates to the phylogeography of elephant populations on Kilimanjaro and the rest of the African continent.

The Amboseli population provides one of the best opportunities to understand how mating and dispersal behavior, long-term patterns of gene flow, and effective population size work together to structure genetic variation within and between populations. Amboseli has been relatively unaffected by the anthropogenic problems of poaching and habitat destruction, and much is known about the demography and behavior of the population over the duration of the project.

Local Population Genetic Structure of African Elephants

The first studies of genetic structure in wild elephant populations have focused on (1) how male dispersal affects the degree of genetic differentiation between populations (Nyakaana and Arctander 1999), (2) how age and body size influence male reproductive success and how this relationship determines the strength of reproductive skew among males (Hollister-Smith et al. 2007), and (3) how female philopatry affects genetic relatedness within groups and whether relatedness predicts the patterns of group fission and fusion in both African and Asian elephants (Archie et al. 2006; Charif et al. 2005; Fernando and Lande 2000). The results of these studies provide a preliminary picture of how elephant social behaviors—especially patterns of mating and dispersal—structure genetic variation within elephant populations.

Like most mammals, male elephants are the dispersing sex, and because mature males breed in several social groups, paternal siblings are scattered in multiple social groups across the population (Hollister-Smith 2005; Hollister-Smith et al. 2007). In addition, when males disperse, they are capable of traveling long distances, and this creates male-biased gene flow between populations. As a result, when we compare individuals in different populations, we see that nuclear loci reflect much less population differentiation than do mitochondrial loci. (Nyakaana and Arctander 1999). Thus, patterns of male dispersal and mating behavior reduce genetic differentiation between social groups in the same population and between populations across the continent.

The strength of this homogenizing force will depend on the reproductive skew between males (Melnick, Pearl, and Richard 1984; Chesser 1991). Observations of elephant mating behavior supported the hypothesis that larger older males fathered more offspring than did younger smaller males (Moss 1983; Poole 1989a, 1989b; see box 18.1). Such reproductive skew should reduce the homogenizing force of male gene flow between populations because only a small number of males in each population will contribute genes to the next generation (Chesser 1991). Genetic evidence also indicates that many males in Amboseli father offspring, including some younger males (Hollister-Smith et al. 2007). As reproductive skew between males declines, a greater diversity of genes will be represented in subsequent generations, and the force of gene flow between social groups and populations should be stronger.

Finally, while male dispersal and mating behavior homogenize genetic variation between groups of elephants, female association patterns act in opposition to increase gene correlations within groups and genetic differentiation between populations. Long-term observations in Amboseli indicate that female elephants—like most female mammals—are matrilocal (Moss and Poole 1983; Moss 1988). However, unlike most social mammals that live in temporally stable social groups, female elephant social organization is unusually flexible, and group composition can change from one day to the next (Moss and Lee, chapter 13; Archie et al., chapter 15). This fission and fusion of groups is hierarchical (Moss 1981; Moss and Poole 1983; Wittemeyer et al. 2005); the most basic social unit is an adult female and her immature offspring, but these adult females also associate with other adult females in units known as families. Some families have unusually close associations with one or two other families; these social groupings are known as bond groups. All the families that share the same home range are known as a clan (Moss 1981). This hierarchical pattern of social association is thought to have arisen from a budding process in which families fission along matrilines.

If this hypothesis is supported, then gene correlations should arise within groups at all levels of organization, and this process will increase genetic differentiation between social groups and populations. Our results (Archie et al. 2006) and those of Fernando and Lande (2000) generally support this hypothesis for both African and Asian elephants. Female elephants are usually philopatric, as there is almost always complete uniformity of mtDNA haplotypes within families. In Amboseli, average genetic relatedness within families is similar to that of aunt-niece relationships (average $R = 0.14$). However, in Amboseli we also found that there are a few cases in which females had mtDNA haplotypes that did not match the other females in their family, indicating that some social groups contain females who are not close maternal kin.

Similarly, Charif et al. (2005) found that female elephants with coordinated ranging patterns often had mismatched mtDNA haplotypes. Their study was conducted in the heavily poached population in Sengwa Wildlife Research Area in northern Zimbabwe. In general, close social relationships among non-maternal kin are probably more common when females lose their natal families due to natural or anthropogenic causes (e.g., Gobush, Kerr, and Wasser 2009). In such cases, females may sometimes join new social groups, which is unusual among mammals with female philopatry. Behavior both reflects the unusual social flexibility of elephants and leads us to predict that poaching will have severe population genetic consequences for elephants.

When families divide temporarily, females usually remain with their closest genetic relatives: average pair-wise genetic relatedness between females who spend more than

90 percent of their time together is 0.35, indicating that these groups consist primarily of mother-daughter and maternal sister pairs. Thus, females associate most closely with their first-order maternal relatives (mothers and maternal half sisters). Furthermore, bond groups appear to have arisen from a process of matrilineal fission. In support of this observation, there is usually complete uniformity of mtDNA haplotypes among bond group members, and the oldest adult females in the families that make up a bond group are significantly more closely related to each other than would be expected by chance. To a limited degree, matrilineal fission also explains the association patterns between all the families in the population. Families that share the same mtDNA haplotype are slightly, but significantly, more likely to associate than those that do not. This result is robust even if bond group associations are removed from the analysis. These association patterns appear to persist long after many of the closest relatives in these groups have died; while distant maternal kinship (i.e., mtDNA) predicts the fission and fusion of social groups, pair-wise genetic relatedness between individuals as measured by microsatellites does not. These results indicate that social groups may continue to use the same ranges from one generation to the next, even as they enlarge and permanently fission (see also Croze and Moss, chapter 7). In addition, these findings suggest that female elephants probably learn and pass on information about which individuals and families are familiar social partners. These patterns are maintained from one generation to the next.

These few studies provide an initial picture of how patterns of mating behavior and philopatry structure genetic variation within elephant populations. Like most mammals with male-biased dispersal, males act as agents of gene flow, which reduces genetic differentiation between social groups and populations. This force is countered by matrilocal females, whose association patterns act to increase genetic differentiation between social groups and populations. However, several features of elephant mating and dispersal patterns suggest that genetic variation in elephant populations is less socially structured than in many of the best-studied social mammals. Among howler monkeys, yellow-bellied marmots, macaques, and baboons, males only breed in one or a few social groups and females are generally philopatric. In contrast, male elephants may breed in many social groups, and occasionally females disperse from their natal group—the rule rather than the exception among forest elephants. Thus, gene correlations will be weaker within social groups of elephants than in some other highly cooperative social mammals, and elephant social groups will be less genetically differentiated from each other. As a result, kin selection may be a less important selective force in shaping patterns of elephant sociality than in shaping sociality in these less fluid societies.

Goals of This Chapter

While we are beginning to understand how various processes structure genetic variation within Amboseli elephants, almost nothing is known about the origin of this population relative to populations on the rest of the African continent. Such information is important for understanding both (1) how historical processes of gene flow and genetic drift have shaped the current genetic structure of Amboseli and (2) which migration routes have been major sources of gene flow and thus have become important targets of conservation for the maintenance of genetic diversity in Amboseli. Here we focus on these processes for female elephants by comparing mitochondrial DNA haplotypes (mtDNA is inherited only from the mother and hence is inherited along matrilines) among female elephants in Amboseli with the phylogeography of elephant mtDNA across the continent (see box 4.1).

We also investigate the origin and phylogeography of the small population of elephants living on Kilimanjaro. The genetic relationship between the Kilimanjaro and Amboseli elephants and their relationship to the continent is important for understanding whether these populations are distinct genetic units with different conservation issues or whether they are important reciprocal sources of gene flow and genetic diversity for each other. It has been hypothesized that this population is genetically distinct from Amboseli elephants because the morphology of a Kilimanjaro elephant differs from that of a typical Amboseli elephant. Relative to Amboseli elephants, elephants from Kilimanjaro have smaller body sizes; smaller ears with more serrated edges; narrower, triangular, and more wrinkled heads with distinctive sloping foreheads (i.e., the distance between the eye and the top of the head is shorter); oddly shaped or no tusks; and little hair on their tails (Moss 1988; Poole 1996). What little is known about the elephants living on Kilimanjaro derives from brief, opportunistic observations made when Kilimanjaro family groups visit the Amboseli basin and from a population census made almost 20 years ago (Grimshaw and Foley 1990).

The results of that survey suggested that approximately 220 (±88) elephants lived on Kilimanjaro and that the population was stable or increasing at the time. Females from Kilimanjaro have been observed mating with bulls in Amboseli, suggesting at least some genetic mixing despite their phenotypic distinctiveness. Here we investigate the genetic relationship between these elephants and the elephants that live in the Amboseli basin (see box 4.1).

The Origin of Amboseli's mtDNA Haplotypes

Six mtDNA haplotypes occur in Amboseli elephants (table 4.1). We recovered four of these in our sample of 237 adult females (table 4.2). The other two (AM10 and AM2) were sequenced by Nyakaana, Arctander, and Siegismund (2002) and are probably haplotypes from male migrants (see below). Three of the four female haplotypes (AMB1, AMB2, and AMB3) were found in the vast majority of sampled females in Amboseli National Park (pooled frequency = 0.957), while AMB4 is rare and only found within the QB family (frequency = 0.043). Comparison with the minimum spanning haplotype network (redrawn from Eggert, Rasner, and Woodruff [2002]), who examined only African savanna elephant haplotypes) reveals that five of six total haplotypes in Amboseli elephants are from the same clade (figure 4.1), which also includes haplotypes from all sampled regions across the continent (West, Central, North, East, and South Africa). Of all the haplotypes in Amboseli elephants, AMB1 is the most likely to have founded the population. AMB1 is described as an ancestral haplotype (Nyakaana, Arctander, and Siegismund 2002) that is at the center of a star-shaped

---

BOX 4.1 POPULATION GENETIC METHODS

Here we describe the specific methods we used to genotype the Amboseli and Kilimanjaro elephants at mitochondrial (mtDNA) and microsatellite loci and analyze these genotypes in order to understand their continental and local phylogeography. For more general information on noninvasive genotyping, see box 15.1.

Sample collection. Fecal samples and some tissue samples were collected from several hundred elephants within and around Amboseli National Park. Samples were collected from known individuals of the Amboseli population. However, because Kilimanjaro elephants could not be individually recognized, we collected samples opportunistically without knowing whether we were resampling individuals. Because both populations showed very high genetic diversity, we assumed that all Kilimanjaro samples that had identical microsatellite genotypes came from the same individual.

Specific collection and DNA extraction methods are outlined in Archie, Moss, and Alberts (2003, 2006). The majority of samples were collected from known individuals. However, elephants from Kilimanjaro cannot be individually recognized, so samples from these elephants were collected opportunistically when they visited the Amboseli basin. Kilimanjaro elephants can be easily recognized by experienced field researchers because they have several distinct morphological features. Moreover, they are not habituated to research vehicles but often appear nervous in the presence of tourist and research vehicles.

mtDNA amplification. For 237 individuals from Amboseli and 61 samples from Kilimanjaro elephants, we amplified a 672 bp sequence of mitochondrial control region using the primers MDL 3 [5'-CCCACAATTAATGGGCCCGGAGCG-3'] and MDL 5 [5'-TTACATGAATTGGCAGCCAA CCAG-3'] (Fernando and Lande 2000). PCR amplification was performed in 10 µl reactions containing 1 µl of DNA extract, 0.4 µl of each 5 pmol primer, 2µl of 2mM dNTP mix (Invitrogen, Carlsbad, CA), 1 µl of 100 mg ml-1 BSA, 1 µl 10x PCR buffer with MgCl2, 0.6 µl of 1.5 mM MgCl2, 0.04 µl of Taq DNA polymerase (QIAGEN, Maryland, USA), and 3.56 µl of water. Reactions were amplified using a touchdown protocol in a MJ Research PTC-200 Thermocycler (MJ Research, Waltham, MA). Amplification was preceded by a 4 min denaturation step at 95°C, followed by 11 cycles of 1 min each at 68°C annealing, 72°C extension, and 95°C denaturation. For the next five cycles, annealing decreased 1°C until it reached 63°C. This 63°C cycle was repeated 15 times and followed by 5 min at 72°C. PCR products were purified using the QIAquick® PCR Purification Kit (QIAGEN, Valencia, CA) and eluted in 30 µl buffer EB. Sequencing was carried out using an ABI PRISM® 3700 DNA Analyzer using Dye Terminator Cycle Sequencing.

Microsatellite amplification. We genotyped 236 individuals from Amboseli and 31 individuals from Kilimanjaro at 10 tetranucleotide loci (Archie, Moss, and Alberts 2003; loci LaT05, LaT07, LaT08, LaT13, LaT16, LaT17, LaT18, LaT24, LaT25, and LaT26) and one dinucleotide locus (LaFMs02; Nyakaana and Arctander 1998). Amplification conditions and primer sequences are in Archie, Moss, and Alberts (2003). PCR products were separated using an ABI PRISM® 3700 DNA Analyzer. Allele sizes were determined using Genotyper software (Version 2.5, PE-Applied Biosystems).

Genotyping protocol and reliability. Amplification success of elephant mtDNA from feces is generally good because mtDNA is relatively common in intestinal epithelial cells. Thus, mitochondrial sequencing from fecal samples is not as prone to the problems of non-invasive genetic analysis as is analysis of autosomal loci. However, in order to ensure accurate genotyping, all mtDNA genotypes were replicated with two independent PCR reactions. Although Greenwood and Pääbo (1999) have reported nuclear copies of the elephant mitochondrial genome, we have no reason to suspect that these were present in the elephant samples because only one sequence was identified for each individual with multiple replicate PCRs.

formation, which indicates that this haplotype represents a central location from which other haplotypes radiated. Thus, it likely founded several new populations (figure 4.1). Consistent with this hypothesis, many other populations across the continent share AMB1 (table 4.1), including those in Samburu National Reserve (Kenya), Queen Elizabeth National Park (Uganda [Nyakaana and Arctander 1999]), Addo Elephant Park (South Africa [Eggert, Rasner, and Woodruff 2002]), and Botswana (Nyakaana and Arctander 1999).

In contrast to the broad distribution of haplotype AMB1, haplotypes AMB2 and AMB3 are unique to elephants in the Amboseli basin (table 4.1). Because these haplotypes have not yet been sampled in any other population, they may have originated within Amboseli elephants as mutations from AMB1. However, the fact that both AMB2 and AMB3 each differ from AMB1 by two mutations suggests that the two intervening haplotypes between AMB2/AMB3 and AMB1 have become extinct or have not yet been sampled.

AMB4 is rare among sampled females in Amboseli (table 4.2). Because AMB4 differs from AMB3 by only one base pair, AMB4 may be a recent mutation from AMB3 that

---

Complete microsatellite genotypes were assigned to 236 individual adult female elephants from Amboseli and 51 fecal samples from Kilimanjaro. Of these genotypes, 278 were derived from fecal samples and 9 were derived from tissue samples. Each genotype derived from tissue was replicated twice. Fecal genotyping required more replication than did genotyping from tissue because DNA from dung is degraded and low in quantity (e.g., Buchan et al. 2005; Taberlet et al. 1996). In particular, fecal genotyping is vulnerable to allelic dropout in which one of two alleles in a heterozygote fails to amplify, so the heterozygote appears to be a homozygote. Thus, apparent homozygotes require extra replications. We used a modified version of the multi-tubes approach (Navidi, Arnheim, and Waterman 1992; Taberlet et al. 1996). Initially, two replicate positive PCRs, each from two independent extracts for the same individual (more than one extract was available for 217 of the 227 individuals from Amboseli that were genotyped from feces), were carried out for each individual at each locus. Samples were considered independent if they had been collected from separate defecations.

Results of the two initial replicate positive PCRs allowed individuals in both studies to be placed into one of three categories: true heterozygotes, possible heterozygotes, and possible homozygotes. Further replication depended on these classifications. Animals were considered true heterozygotes if both PCRs produced identical heterozygote genotypes and no Mendelian mismatches; in this case, no further replications were performed. Animals were considered possible heterozygotes when the first two PCRs produced both a heterozygote and a homozygote genotype with a common allele between them or two genotypes, each of which were homozygous for a different allele. Possible heterozygotes were replicated until both alleles were observed at least twice. If one of the alleles observed in the initial replicates failed to appear again, it was classified as an error. Animals were considered possible homozygotes if both initial replicates revealed a single, identical allele. Possible homozygotes were replicated until the same allele was observed in a total of seven PCRs. If an additional allele appeared two or more times in those replicates, the individual was considered a heterozygote. If an additional allele appeared only once, the individual was considered a homozygote, and the unique allele was labeled an error.

We assessed genotype reliability in two ways: (1) for both Amboseli and Kilimanjaro elephants, we measured whether loci were in Hardy-Weinberg equilibrium using the software program CERVUS 2.0 (Marshall et al. 1998) and (2) for known individuals, we monitored the data for Mendelian errors (i.e., mother-offspring mismatches). Maternity was known through observations shortly after parturition for all offspring born since 1972. When allelic dropout was suspected as a result of Mendelian checks, further PCR replicates were performed until the mismatch was resolved.

Comparison of mtDNA sequences across Africa. To investigate the relationship between Amboseli, Kilimanjaro, and the other populations across Africa, we combined our sequences with other control region sequences from GenBank. These sequences (table 4.1) were generated by the work of Eggert, Rasner, and Woodruff (2002), Nyakaana and Arctander (1999), and Nyakaana, Arctander, and Siegismund (2002). Eggert et al. (2002) sequenced 593 bp of mitochondrial DNA, using the same primers as this study (MDL3 and MDL5; Fernando and Lande 2000), for between 23 and 50 samples from each of six savannah elephant populations (table 4.1). These samples were collected without regard to the sex of the elephant. Nyakaana and Arctander (1999) and Nyakaana, Arctander, and Siegismunnd (2002) sequenced 388 bp of control region individuals from 16 additional locations. The sex of these elephants is unknown except in Murchison Falls, where all samples were collected from solitary males. These sequences were aligned and compared for mismatches by using Sequencher software (version 4.1.2, Gene Codes Corporation, Ann Arbor MI). The combined data set includes 52 distinct haplotypes sampled from 10 countries in West, Central, East, and southern Africa (table 4.1).

Table 4.1 Distribution of African elephant mtDNA control region haplotypes in several sampling locations across the African continent

Sampling locations are arranged in geographical order across the top of the table from West Africa across Central Africa to East Africa and South Africa. Numbered superscripts refer to citations for each population set. Numbers in parentheses are total sample sizes. Haplotype names are arranged along the left-hand side of the table; some studies used different names to refer to the same haplotype. All published names for identical haplotypes are reported in the same cell (e.g., AMB1, KIL1, Addo5, and QE1 are identical sequences).

POPULATIONS (N)	Ghana, Mole NP[2] (>23)	Ghana, Red Volta Valley[2] (>23)	Ghana[3] (18)	Mali, Gourma region[2] (>23)	Cameroon, Benoue NP[2] (>23)	Cameroon, Waza NP[2] (>23)	Uganda, Queen Elizabeth NP[3] (68)	Uganda, Murchison Falls NP[3] (9)	Uganda, Kidepo Valley NP[3] (27)	Kenya, Samburu NR[1,3] (9)	Kenya, Amboseli NP[1,3] (264)	Kenya, Masai Mara NP[3] (20)	Tanzania, Kilimanjaro[1] (30)	Namibia[3] (24)	Botswana[3] (25)	Zimbabwe[3] (6)	South Africa, Kruger NP[3] (12)	South Africa, Addo EP[2] (>23)
HAPLOTYPES																		
AMB1, KIL1, Addo5, QE1							X				X	X			X			X
AMB2, KIL2, AM1											X	X						
AMB3, AM12											X							
AMB4, KV2									X		X							
AM 2											X							
AM10											X							
KIL3, KG2												X					X	
SA 8										X								
MM 4												X						
MM 19												X						
MM 20												X						
KV 1, B36, Mali 2				X	X		X	X	X		X							
KV 7, Waza 15						X			X									
KV 8						X		X	X		X							
KV 17									X									
KV 28									X									
MF 1								X										
MF 5								X										
QE 4							X											
QE 13							X								X			
WA11			X															
WA14			X															
Mole 3, Mali 14	X			X														
Mole 9, RVV 22, Mali 7, WA3	X	X	X	X														
Mole 13, RVV 15, WA 6	X	X	X															
Mole 33	X																	
Mali 28				X														
B1					X													
B7					X													
B8, Waza10					X													
Waza 27						X												
KH 2														X				

(continued)

Table 4.1 (continued)

POPULATIONS (N)	Ghana, Mole NP[2] (>23)	Ghana, Red Volta Valley[2] (>23)	Ghana[3] (18)	Mali, Gourma region[2] (>23)	Cameroon, Benoue NP[2] (>23)	Cameroon, Waza NP[2] (>23)	Uganda, Queen Elizabeth NP[3] (68)	Uganda, Murchison Falls NP[3] (9)	Uganda, Kidepo Valley NP[3] (27)	Kenya, Samburu NR[1,3] (9)	Kenya, Amboseli NP[1,3] (264)	Kenya, Masai Mara NP[3] (20)	Tanzania, Kilimanjaro[1] (30)	Namibia[3] (24)	Botswana[3] (25)	Zimbabwe[3] (6)	South Africa, Kruger NP[3] (12)	South Africa, Addo EP[2] (>23)
WC 2														X				
WC 4														X				
WC 6														X				
WC 13														X				
Bot 2															X			
Bot 4															X			
Bot 6															X			
Bot 9															X			
Bot 15															X			
Bot 16															X			
Bot 21															X			
Zbe 1																X		
Zbe 2																X		
Zbe 3																X		
Zbe 4																X		
Zbe 5																X		
Zbe 6																X		
KG 1																	X	
Addo 1																		X

[1] This study.
[2] Eggert, Rasner, and Woodruff 2002.
[3] Nyakaana, Arctander, and Siegismund 2002.

Table 4.2 MtDNA control region haplotypes, population frequencies, and sequence differences among the haplotypes in elephants from Amboseli and Kilimanjaro

The base pair row shows the location of the variable sites relative to the entire length of the gene.

Haplotype name	Amboseli frequency	Kilimanjaro frequency	Sequence												Genbank accession number
AMB1/KIL1	0.335	0.7	A	T	G	G	C	A	G	C	T	G	T	A	AY968043
AMB2/KIL2	0.487	0.1	.	.	.	.	.	G	.	.	.	A	.	.	AY968044
AMB3	0.135	0	.	.	.	.	.	G	.	T	.	.	.	.	AY968045
AMB4	0.043	0	.	C	.	.	.	G	.	T	.	.	.	.	AY968046
KIL3	0	0.2	G	.	A	A	T	.	A	.	C	.	C	G	
		base pair	104	120	161	165	168	179	180	181	182	190	380	562	

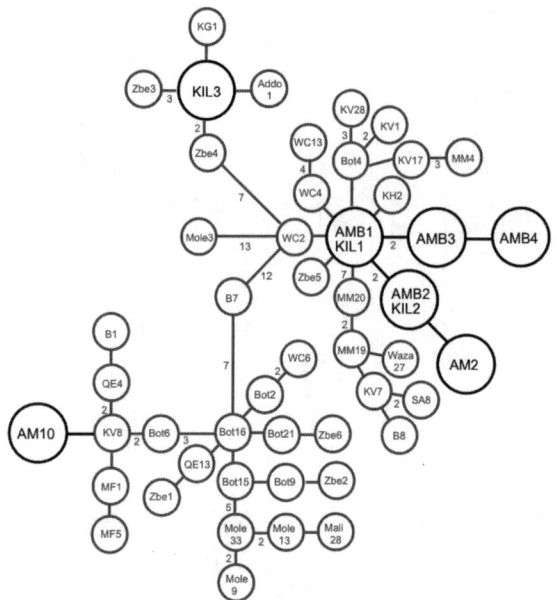

Figure 4.1 Minimum spanning network of African elephant mtDNA control region haplotypes found across the African continent, redrawn from Eggert, Rasner, and Woodruff (2002) to include Amboseli and Kilimanjaro haplotypes (in black). Large circles indicate haplotypes that occur in Amboseli and Kilimanjaro elephants; the size of the circle does not reflect haplotype frequency or importance. Connections between haplotypes are not to scale. Numbers next to connections indicate the number of changes between haplotypes; unlabeled connections are separated by one change.

occurred within one family (QB) in Amboseli. However, because AMB4 is not unique to Amboseli (table 4.1), the QB family may alternatively represent a lineage that has migrated from Northern Kenya or Uganda.

Nyakaana and Arctander (1999) sequenced two additional alleles from Amboseli elephants (AM2 and AM10) that were not found in our sample. It is possible that females in the few families that were not sampled in this study have those haplotypes. This hypothesis is supported by the fact that AM2, which is only one base pair different from AMB1 (figure 4.1), is not found anywhere else in Africa (table 4.1). However, it is also possible that one or both of these samples were collected from individuals, probably mature males, who immigrated to Amboseli from other nearby populations. In support of the hypothesis that these AM2 and AM10 were sampled from males, both haplotypes were sampled from just one individual. AM10 is considerably different from AMB1, AMB2, AMB3, and AMB4 (range of mismatches = 14–17) and was assigned to a completely different clade (figure 4.1).

The Origin of the Kilimanjaro Elephants and Their Relationship to the Amboseli Elephants

Three mtDNA d-loop haplotypes were found among 30 individual Kilimanjaro elephants (table 4.2). Two of the haplotypes, KIL1 and KIL2, are the same as AMB1 and AMB2 from Amboseli elephants. Presumably, these haplotypes represent individuals who migrated up the slopes of Kilimanjaro from Amboseli at some point in the past. The third Kilimanjaro haplotype (KIL3), like AMB1/KIL1, AMB2/KIL2, AMB3, and AMB4, also belongs to clade 3/A (figure 4.1). However, its sequence is considerably different from any other Amboseli haplotype (range of mismatches = 9–29). In addition, KIL3 has thus far only been found in one other population on the African continent besides Kilimanjaro: Kruger National Park, South Africa (KIL3 = haplotype KG2; Nyakaana, Arctander, and Siegismund 2002). KIL3 and KG2 are at the center of a star-shaped clade, the members of which also are only found in South Africa, including the other haplotype found in Kruger National Park (KG1; Nyakaana, Arctander, and Siegismund 2002), two haplotypes found in Zimbabwe (ZBE3 and ZBE4; Nyakaana, Arctander, and Siegismund 2002), and a haplotype found in Addo National Park (Addo 1; Eggert, Rasner, and Woodruff 2002). This small clade of haplotypes is separated by seven shared changes from the rest of clade 3/A that contains AMB1–4 and KIL1,2 (figure 4.1).

The complete genotypes of 30 Kilimanjaro elephants at 11 microsatellite loci provide no evidence that the Amboseli and Kilimanjaro elephants are genetically differentiated. There are few differences in the frequencies of shared microsatellite alleles (figure 4.2), and F_{ST} (a measure of population differentiation that varies from 0 = no differentiation to 1 = complete differentiation) between Amboseli and Kilimanjaro elephants is not significantly different from zero ($F_{ST} = 0.001$, $p > 0.9$, 1,023 permutations). The Kilimanjaro elephants did have a few microsatellite alleles that were not found in Amboseli elephants (allele 140, locus LafMs02; allele 208, locus LaT24; allele 290, locus LaT25), and the Amboseli population had a number of alleles not present in the Kilimanjaro population (figure 4.2). However, this finding is to be expected with highly mutable loci such as microsatellites. Overall genetic differentiation between these two groups was quite small. The lack of genetic differentiation between these populations at nuclear loci is likely a result of male-mediated gene flow. While some females may permanently migrate between the two populations, as demonstrated by two shared mtDNA haplotypes, such migration is probably uncommon relative to sexually active males from Kilimanjaro or Amboseli who travel between these populations.

Phylogeography of the Elephants in Amboseli and on Kilimanjaro

Our results indicate that the Amboseli and Kilimanjaro elephants have probably been affected by at least two historical migration events. The first of these is the range expansion that is thought to have occurred by *Loxodonta africana* at

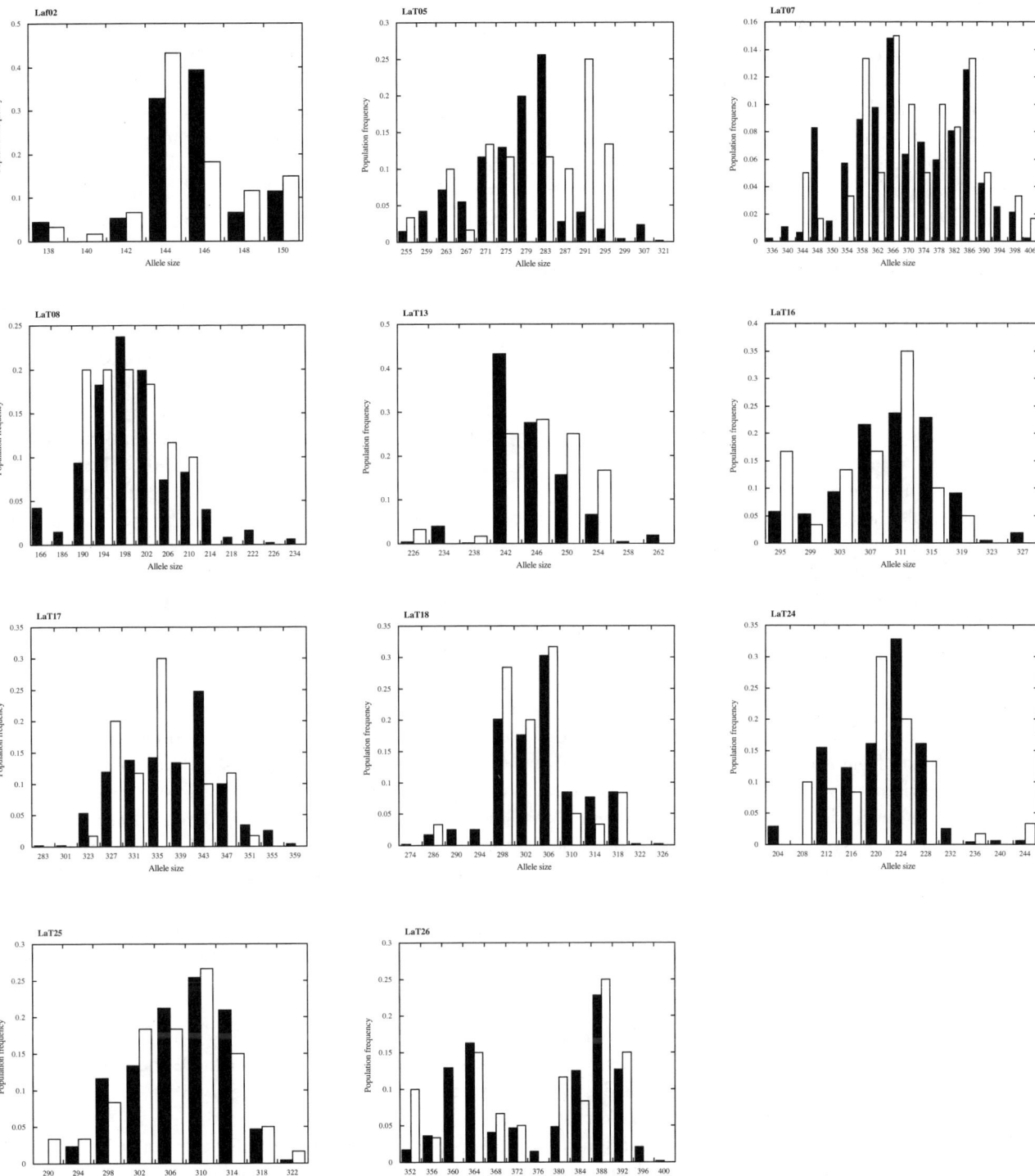

Figure 4.2 Microsatellite allele frequencies for all 11 loci from Amboseli (dark bars) and Kilimanjaro (white bars) elephants.

the end of the Pleistocene when its competitors, *Elephas iolensis* and possibly *L. adaurora*, became extinct (Roca et al. 2001; Eggert, Rasner, and Woodruff 2002; Nyakaana, Arctander, and Siegismund 2002). This range expansion left a distinctive genetic signature in the phylogeography of mtDNA: minimum spanning networks of control region sequences are star shaped, and at the center of each star is a haplotype with a broad distribution across the continent (Eggert, Rasner, and Woodruff 2002; Nyakaana, Arctander, and Siegismund 2002). One of these ancestral haplotypes is AMB1, which, when it arrived in Amboseli, probably gave rise to AMB2 and AMB3. Thus, it appears that most of the females currently living in Amboseli and on the slopes of Kilimanjaro are directly descended from the ancestors of this late Pleistocene migration into Amboseli. Fossil evidence supports the hypothesis that *L. africana* existed in the Amboseli basin at the end of the Pleistocene; *L. africana* fossils at the northeastern portion of the lake-bed edge of the Amboseli basin are probably no older than 200,000 years BP but no younger than 5,000 years BP (Behrensmeyer and Boaz 1981).

It is less clear whether the elephants bearing the haplotype AMB4 are also descendents of the late-Pleistocene migration of elephants into Amboseli. Haplotype AMB4 may be the result of a recent mutation from haplotype AMB3 that occurred within Amboseli elephants. However, this haplotype is shared with elephants in Kidepo Valley, Uganda, and is also found among populations in Samburu National Reserve in northern Kenya (E. Derryberry and I. Douglas-Hamilton, unpublished data). Thus, it is also possible that female elephants bearing AMB4 migrated to Amboseli from Uganda or Northern Kenya some time in the last few hundred years.

The origin of the female *L. africana* on Kilimanjaro that carry haplotype KIL3 is less clear. This haplotype is identical to that currently found in elephants from Kruger National Park and is similar to other haplotypes found only in southern Africa, including Addo National Park and northern Zimbabwe. Although these populations are approximately 2,000 km from Kilimanjaro, such disjunct mtDNA haplotype distributions are relatively common in elephants. Across the continent, deeply divergent clades of mtDNA haplotypes persist in sympatry across several populations of African elephants, especially in eastern and southern Africa (Georgiadis et al. 1994; Nyakaana and Arctander 1999; Roca et al. 2001; Eggert, Rasner, and Woodruff 2002; Nyakaana, Arctander, and Siegismund 2002). For example, haplotype QE13 is found only in Queen Elizabeth National Park, Uganda, and in Botswana (figure 4.1; see Nyakaana, Arctander, and Siegismund 2002). This pattern is also found in several other African species, including the great apes like chimpanzees, bonobos, and gorillas (Gagneux et al. 1999); carnivores like eastern black-backed jackals (Wayne et al. 1990); and several ungulates, including kob and African buffalo (Arctander, Johansen, and Coutellec-Vetro 1999).

Researchers have hypothesized two possible explanations for this pattern. Georgiadis et al. (1994) suggested that divergent clades might persist in sympatry as a result of protracted gene flow between populations with persistently large effective population sizes. However, there is little evidence that the effective population size of female elephants has been large enough for long enough to lead to this incomplete lineage sorting (Nyakaana, Arctander, and Siegismund 2002). Alternatively, these same patterns may be the result of repeated cycles of allopatry followed by admixture (Arctander et al. 1999; Eggert, Rasner, and Woodruff 2002; Nyakaana, Arctander, and Siegismund 2002). Specifically, the African continent is thought to have undergone a major cooling event approximately 2.8 million years ago, which was followed by several fluctuations between cool, dry conditions and warm, moist conditions (Eggert et al. 2002). These fluctuations caused the expansion and contraction of forest and savanna habitats; when the climate was harsh, species persisted in localized refugia in eastern, western, and southern Africa, and when the climate became favorable, they expanded their ranges (Arctander et al. 1999; Eggert, Rasner, and Woodruff 2002; Nyakaana, Arctander, and Siegismund 2002).

Thus, the elephants on Kilimanjaro that carry haplotype KIL3 may be descendents of savanna elephants that migrated from southern Africa when conditions were favorable. This hypothesis is supported by the fact that KIL3 is at the center of a star-shaped phylogeny (figure 4.1), indicating that it has undergone a recent range expansion. Current genotype distributions suggest that this range expansion mainly occurred in southern Africa and/or that this haplotype reached other populations in East Africa but has since become extinct.

While these results provide insight into the origins and history of the elephants living in Amboseli and on Kilimanjaro, they are preliminary and should be interpreted with caution. In order to draw more definitive conclusions, we require data from more loci and more populations, especially from elephants from Tsavo National Park in Kenya and adjacent populations in Tanzania. Such data might reveal that haplotypes AMB2 and AMB3 are not unique to Amboseli or that KIL3 is also found in other East African populations or in populations between eastern and southern Africa.

Conservation Implications for the Gene Flow between Amboseli and Kilimanjaro Elephants

Long-term observations by AERP indicate that Kilimanjaro elephants visit Amboseli National Park frequently; however, it is unclear whether these migrations have resulted in significant gene flow between these populations (Grimshaw and Foley 1990). The morphology of Kilimanjaro elephants is quite different than that of Amboseli elephants (Moss 1988; Grimshaw and Foley 1990), and Kilimanjaro males are probably too small to effectively compete for mates with larger Amboseli males. Females from Kilimanjaro do sometimes mate with Amboseli males (Poole 1982), but it is not yet known whether Amboseli females mate with

Kilimanjaro males if the females visit the forest habitats on the mountain.

Our results show that the elephants living on Kilimanjaro and in Amboseli are not genetically differentiated from each other. While some Amboseli females carry mtDNA haplotypes not found in Kilimanjaro females and vice versa, female elephants from Kilimanjaro and Amboseli both share the d-loop haplotypes AMB1 and AMB2. This suggests that females do occasionally migrate permanently between these populations. In addition, males who permanently or temporarily disperse from either Amboseli or Kilimanjaro and breed outside their natal population create a much stronger force of gene flow, as evidenced by the lack of genetic differentiation at nuclear microsatellite loci between Amboseli and Kilimanjaro individuals.

These results have two implications. First, the morphological differences between the Amboseli and Kilimanjaro elephants are more likely to be a result of environmental variation between the two habitats than genetic variation between these populations. In particular, minerals are much more available in Amboseli, which may explain why Amboseli elephants have larger bodies and tusks.

Second, and most importantly, these results suggest that conservation biologists and policy makers should not view the Kilimanjaro and Amboseli elephants as distinct populations, and conservation efforts should focus on maintaining corridors for migration between these populations. The Kilimanjaro population, because it is small, is at particular risk of "extinction vortices" caused by inbreeding depression and population decline through demographic stochasticity (Gilpin and Soulé 1986). Foley and Grimshaw (1990) found that elephants commonly use the migration corridor on the northern side of Kilimanjaro to travel between the forested mountain slopes and the savanna habitat in the Amboseli basin. Steps should be taken to protect this corridor in both Kenya and Tanzania so as to maintain genetic exchange and provide the capacity for both populations to find food in the face of a changing climate and conversion of forest to agriculture.

Part 2
Habitat Use, Population Dynamics, and Ranging

Section editors: W. Keith Lindsay and Harvey Croze

Chapter 5 Habitat Use, Diet Choice, and Nutritional Status in Female and Male Amboseli Elephants

W. Keith Lindsay

Herbivore diets and nutrient intake are determined by the time an animal spends feeding and the rate at which it feeds on different food types within different vegetation communities. Elephants, like other herbivores, select where and what to eat—from places to species to plant parts—in order to maximize their rate of nutrient intake in the context of their survival and reproductive strategies. Studying habitat use, diet choice, and feeding rates illuminates the following:

- The reasons for habitat and diet choice, in relation to variation in food abundance within and between years.
- The reasons for and consequences of differences between the sexes in diet and habitat use.
- The potential nutritional status resulting from diet choice and the implications for demographic performance at the population level.
- The interactions with, and dynamics of, plant communities.

Individual ranging patterns (Croze and Moss, chapter 7; Mutinda, Poole, and Moss, chapter 16) and demographic performance (Lee, Lindsay, and Moss, chapter 6) of elephants are both strongly influenced by foraging location and success. In this chapter, I describe a study of the time spent feeding and the use of habitats in the Amboseli basin by adult female and male elephants during two years in the mid-1980s (1982–84) in relation to changes across the seasons. Choices between the broad food classes of grass and woody browse are examined in light of the contrasting plant abundance available in a year with higher than average rainfall (1982–83: annual rainfall of 540 mm compared with the long-term mean of 332 mm) and in a drought year (1983–84: annual rainfall = 136 mm), as well as differences between the sexes. Estimated intake of two critical nutrients, digestible energy and protein, are compared with calculated requirements for maintaining body condition and investing in offspring.

The Behavioral Ecology of Food and Habitat Choice

Habitat preference by animals is usually explained in terms of feeding requirements (Partridge 1978), although other needs, such as mineral needs, safety from predators and pests, or relief from thermal stress, may be factors in some habitat choices (Belovsky 1981; Owen-Smith and Novellie 1982; Crawley 1983; Duncan 1985). Sex differences in habitat and diet choice have been described and evaluated for several sexually dimorphic mammal species, including, among many others, kangaroos (Newsome 1980), red deer (Clutton-Brock, Iason, and Guinness 1987; Gordon 1989), moose (Miquelle, Peek, and Van Ballenberghe 1992), and, indeed, elephants (Stokke and du Toit 2002; Shannon et al. 2006). Generally, sex differences in habitat use are most pronounced during periods of food scarcity: males typically occupy areas with lower quality but possibly more abundant forage, while females tend to seek out higher quality patches. The emerging consensus is that differential habitat use is driven by differing abilities to meet foraging needs relative to requirements, and that these abilities and requirements are determined by reproductive status and strategy, by allometric relationships imposed by body size and by social context, all of which differ between the sexes (Miquelle, Peek, and Van Ballenberghe 1992; Shannon et al. 2006). The overall picture painted by the combination of determinants suggests that in dimorphic species, females and males are effectively distinct ecological species (Clutton-Brock, Albon, and Guinness 1982).

Adequate body condition is a prerequisite for successful reproduction (see Lee, Lindsay, and Moss chapter 6), and thus the diet choice and nutrient intake an animal achieves

---

BOX 5.1 SIZE AND ENERGETICS OF ELEPHANTS
Phyllis C. Lee

Large size is an elephant adaptation; metabolism, energetics, mass, and body proportions are all linked in the elephant's ecology. As organisms increase in mass, cellular processes become more efficient, scaling allometrically. Allometry, or the study of size and shape and the processes operating at different sizes, has been known since Huxley's (1924) seminal work and has remained of fundamental interest in relation to the evolution of biological systems (Kleiber 1961; Peters 1983; Schmidt-Nielsen 1984; Stearns 1992). When an allometric relationship has a slope of less than one, such as the ratio of surface area to volume or weight, it is generally the case that some relative efficiency has been attained. This "savings" is illustrated by the three-fourths power efficiency of metabolism as animal size increases (Kleiber 1961). The precise values of the scaling exponents (slopes) are debated but the empirical relationships with mass remain robust. Universal scaling laws derived from particle physics are suggested as underlying the three-fourths power size-to-energy relationships observed from unicellular organisms to giant redwoods (West, Brown, and Enquist 1997; 1999). The question here is why have elephants become so large? Efficiencies of scale could be part of the answer. Less energy per unit weight is required to run the animal; when life processes cost relatively less, periods of environmental uncertainty will have a reduced impact on survival and reproduction.

Elephant energetics were first explored by Benedict in 1936 based on a small number of female Asian elephants. The standard metabolism of one female of 3,672 kg was measured as 64,800 cal/24 hr. Benedict noted that "fasting" and inactive conditions were impossible to obtain when assessing elephant respiration and also mentioned that his subjects may have been exposed to cold stress, adding a thermoregulatory cost to his values. He "adjusted" his metabolic rate values by some 30 percent, but "the elephant" still appeared to be somewhat above that predicted by Brody's (1945) regression between mass and basal metabolic rate. Using the Kleiber (1961) equations for metabolic rate, elephants of this mass are predicted to have a basal metabolic rate of 33,020 cal/24 hr, which would translate to over 60,000 cal/24 hr, assuming digestion, active locomotion, and cold stress. The measured value thus is quite close to the expected value derived from the Kleiber equation, given cold stresses, activity, and additional costs of sustaining brains and growth. Elephants may indeed be "cheap" to run as their size would suggest in the world of scaling laws, but some of their metabolic activities may be expensive (e.g., Suarez, Darveau, and Childress 2004).

Energy costs are more than just those of body mass. The size of various organs determines both the costs of running those organs and the efficiencies that can be attained with an increase in organ mass, surface area, or volume (Schmidt-Nielsen 1984). The volume of the heart determines how much blood can be pumped with each contraction, while the mass of the heart reflects the muscle fiber density needed to produce those contractions. These relationships have profound implications for the costs as well as efficiencies of an elephant's circulatory and cooling systems. The large size of elephants increases the challenge for effective heat exchange; their relatively small skin surface area in relation to body mass provides little opportunity for eliminating the heat generated by metabolism, activity, and solar absorption, so large ear size is an adaptive response (Phillips and Heath 1992). The length, surface area, and volume of the gut require similar trade-offs in the digestion and absorption of nutrients; length relates to passage rate, while the area of the digestive interface relates to absorption of nutrients as well as the elevated costs of guts in terms of enzymatic synthesis and active transport. These relationships influence energy output in relation to dietary energy intake as outlined in this chapter, while volume relates to capacity for food intake. Spermatogenesis is a function of testes volume (Dixson 1998), while water balance and blood pH are determined by kidney volume.

Metabolically costly tissues are currently thought to be brains, liver, kidneys, and guts, in that order (Passmore and Durnin 1955). Organ scaling for elephants suggests that other than the brain, these are generally of a volume or mass expected for overall size (box 5.1, table A). By contrast, elephant brains are relatively large. The absolute mass of the African elephant brain is close to 5kg (Eisenberg 1981; Shoshani 1991; Shoshani, Kupsky, and Marchant 2001, 2006), which needs 3.5 ml O^2 per 100 g/min (0.074 kj/100 g/min or about a minimum of 4000 kJ per day) to run the brain alone. The consequences of size can also be seen in stature (or standing height), which is associated with high growth costs, the need to sustain energy balance during growth, and critically for the elephant, with behavior—foraging, social, and reproductive. The brain's large size and lengthy growth period obviously add significantly to the costs of being big for elephants.

have demographic consequences. Variation in food abundance and quality between areas and between time periods would be expected to have direct influences on seasonal and annual variation in individual and population reproductive and survival variables.

The diets of elephants, at a broad level, are related to the availability within seasons, years, and between geographical locations of plants harvested in two main ways: grazing (generally grasses and forbs) or browsing (bushes and trees). Elephants eat more grass in wet periods and more woody

Table A Measurements of body size and organs for African elephants

Body mass	Male = 4,500–6,600 kg, Female = 2,500–3,200 kg
Mass at birth	Male = 120 kg, female = 100 kg
End of growth	Male >50 years, female 25–35 years
Brain mass	Male: 4.2–7.5 kg (mean = 5.0); Female: 4.0–4.8 kg (mean = 4.7)
Brain mass at birth	c. 1,800 g; 35–50% of adult brain mass
Mass of skeleton	16% total mass
Skin thickness	1.3–2.5 cm
Skin mass	10% total mass
Heart mass	12–28 kg
Heart rate	28 beats/min
Circulation	18 l blood/min
Respiration	310 l air/min
Gut capacity	150 kg/day
Gut length	35 m
Retention time in guts	33 hrs (mean); 11–54 hrs
Volume of urine	50 l/day
Penis mass (length)	27 kg (1 m)
Testes mass	6 kg
Trunk capacity	10 l
Milk production	15–25 l/day
Milk fat	17–22% fat in early lactation

Sources: Shoshani (1991); Shoshani, Kupsky, and Marchant (2006); Spinage (1994)

Energy sparing comes in the columnar structure of elephant limb bones, which is effective and economic for supporting the large body mass, but as a consequence there is little flexibility in elbow and knee joints. The feet of elephants contain a large amount of fatty and connective tissue, which cushion the shock of each step as well as the sounds of walking. Elephants seldom run—a mode of locomotion defined as lifting two limbs clear off the ground at one time. They can, however, move at up to 40 km/hr, which suggests that a rapid rate of locomotion is not limited by the constraints of limb size and shape designed to support rather than leap (McNeill Alexander 1999). Scapular articulation is flexible and unfused (Haynes 1991), allowing both for growth in the tibia and for increased rotation in locomotion. This flexibility may explain why elephants are able to climb relatively steep slopes despite the rigidity of their other joints.

Thus, there are features of elephant biology that increase the energetic efficiency attained through large mass such as limb structure and mode of locomotion and some such as brain size that add to metabolic costs. Large size as an evolved feature is the result of indeterminate growth in males or prolongation of the growth phase for at least three-quarters of the post-maturity reproductive life span for females (see also Lee and Moss [1995]; Shrader et al. [2006]). Size—and by implication, growth—seems to be a plastic trait for elephants, and they can shrink or enlarge over relatively short evolutionary periods. Elephants can be both tiny, as in the insular dwarfs of Malta that stood only 1 m at the shoulder, and very tall, such as those currently found in the Namib or historical populations (Lee and Moss 1995). Size for elephants may be both a generalized evolved response to energy availability and acquisition and a phenotypically variable trait from population to population or among individuals.

Prolonged growth as the mechanism for attaining large size has costs that could counterbalance some of the economies of scale due to being big. For most of their life, elephants need to consume sufficient additional energy to sustain growth over and above that required for maintenance and reproduction (Lee and Moss [1995]; Lindsay, this chapter). Reproductive costs for females are those expected for their size (see box 12.1), with potential "savings" due to an extended gestation and prolonged lactation. But, even low costs spread over a long period become unsustainable during short-term, intensive events of maternal energy shortage, combined with high growth rates of young calves sustained by relatively high-fat milk. Males have additional costs since they continue growing throughout their lives and due to their annual reproductive musth phase (Poole [1989a]; Poole et al., chapter 18). During the musth period, bulls reduce their energy intake, limit their time spent feeding and drinking, increase their daily movement costs while searching for females, and incur costs of contests and fights. They may only be able to sustain these high costs through an energy surplus accumulated and stored in fat during the non-reproductive periods (Poole [1989a]; Lindsay, this chapter).

The balancing of energy intake with expenditure may be made easier for elephants as a result of their large size, which allows them to exploit habitats with energy uncertainty, reduced food quality, and low availability. Their long life span, itself a correlate of their large brains and large mass, acts as a further buffer against unpredictable events in an uncertain environment. Elephants have an intrinsic ability to recover, rebound, grow, and reproduce as soon as they return to energy balance.

plants in dry periods. On an Africa-wide scale, elephants' diets are directly proportional to the relative abundance of food types (van der Merwe, Lee-Thorp, and Bell 1988), browsing more in areas dominated by woody cover and grazing more in open habitats. Seasonal food choices at a qualitative level (Field 1971; Barnes 1982; Jachmann 1989) relate to the potential nutritive value of different plant species and parts at different times of year. A few reports (Sikes 1971; Laws, Parker, and Johnstone 1975) have suggested that woody plants supply specific essential nutrients and that reduced levels of browse intake lead to lowered nutritional status in habitats that are converted to grassland, with negative consequences for reproduction and survival. While a small number of studies have estimated rates of dry matter intake by elephants on different food types, to date few researchers have related either diet choices or the demographic consequences for individuals and populations to nutrient intake rates and nutritional status.

The size of an animal's body has far-reaching implications for its physiology, ecology, and behavior (see Lee, box 5.1), and in the elephant, its extreme body size must be an important factor in determining its foraging options. As noted originally by Janis (1976) and progressively developed by Demment and van Soest (1985), Duncan et al. (1990), Illius and Gordon (1992), and Clauss et al. (2003), there is a trade-off along the trend of increasing body size in herbivores between digestive capacity (which, as a volume measure, scales at the same ratio as body weight), food requirements (metabolism scales at body weight raised to an exponent of 0.75; Kleiber [1961]), and ease of procurement (which is determined by food-gathering anatomy, having linear dimensions, and scaling to body weight at an exponent of 0.33). In the ecological abundance of plants and plant parts, there is generally a negative relationship between quality and standing biomass; high-quality food packets are rare, while low-quality, fibrous forage is more widespread.

Small animals have relatively high metabolic rates and must select food items with relatively large proportions of easily digested, soluble cell contents (e.g., in leaves, fruits, and flowers), but their needs can be met by small absolute amounts. Large animals have relatively lower metabolic rates, but high absolute demands for nutrients. Sufficient amounts of high-quality food items are not available in natural ecosystems, but since large animals have low metabolic demands relative to gut capacity, they can select greater amounts of stemmy plant material and subject the fibrous cell wall components (such as cellulose) to more thorough digestion. The range of acceptable forage quality thus becomes broader.

At intermediate body sizes, ruminants are at an advantage, since they are able to digest plant cell wall very efficiently in their foregut bacterial fermentation chambers, a mechanism that is especially advantageous when forage abundance or foraging time is limited. However, this efficiency requires long residence time in the gut for plant fiber to break down in order to release nutrients (and be passed onward for excretion). If forage quality is very low, foregut digestion cannot provide nutrients quickly enough to satisfy the needs of metabolism. In the face of fibrous plant matter on offer, limited fermentation taking place in the hindgut, near the end of the digestive tract, avoids foregut blockage and allows forage throughput to be faster. Very large-bodied, non-ruminant herbivores, such as elephants, can compensate for lower levels of food quality by higher intake and turnover. The anatomy of hindgut fermentation, with a large caecum and colon where some fermentation of plant cell wall fiber occurs (Clemens and Maloiy 1982; van Hoven, Prins, and Lankhurst 1981), combined with long feeding times, high intake rates, and short retention time of plant matter, appears to have reached its greatest development in the elephant (Owen-Smith 1988).

As the largest terrestrial herbivore, the elephant has enormous absolute metabolic needs and faces the problem of procuring sufficient nutrients from the background of varying but generally low-quality vegetation on offer in seasonal semi-arid environments. The solution is to be a dietary generalist, selecting good food items when available but consuming large amounts of fibrous forage overall. This has been a successful strategy, allowing elephants to exploit a broad range of habitat types in Africa and Asia from rain forest to arid near-deserts (Eltringham 1982; Douglas-Hamilton, Michelmore, and Inamdar 1992).

On the subject of diet choice, there has been a long and rather futile debate over whether elephants are primarily browsers or grazers (e.g., Buss 1961; Sikes 1971; Lindsay and Olivier 1984; Bell 1985; Jachmann 1987). Upon reflection, it is clear that elephants can be either or both; they can subsist on diets with a large amount of grass or of browse and generally have a mixture of both types in their overall diet (Sukumar 1989). Given that forest elephants have the same gut morphology and digestive physiology as their savannah cousins and are highly frugivorous, it is perhaps time to move beyond these not very useful distinctions. Grass, when abundant, is suitable forage: it can be harvested quickly, allowing high intake rate with relatively low toxin levels. Its disadvantages are a high cell wall fraction, particularly in dry periods, and a possible lack of some essential nutrients, such as amino acids, fatty acids, and minerals (Olivier 1982). Woody stem and leaf material, or "browse," has the advantages of a higher fraction of soluble cell contents, protein, and other readily digestible nutrients and a more diverse nutrient composition. Its disadvantages are that it may be less abundant and patchily distributed in many localities, its cell wall constituents are more lignified (which reduces digestibility), and its harvesting requires more manipulation, particularly where thorns or tough stems are present. These disadvantages result in lower dry matter intake rates at some times of year (Sukumar 1989). Woody dicotyledonous plants also have higher and more diverse toxin loadings, reducing palatability and/or digestibility, restricting the amount that can be ingested safely, and requiring diversification of diet (Westoby 1978; Bryant et al. 1992).

The social environment of an animal clearly affects habitat choice and foraging style (e.g., Fritz and De Garine-Whichatitsky 1996; Conradt and Roper 2000; Smit, Grant, and Whyte 2007a). As detailed in part 3, the composition of social groups of female and male elephants differs markedly. Females associate in families, and often in large, mobile aggregations, while groups of males are smaller, sometimes with only single animals, and more sedentary. However, when males choose temporarily to join family units or aggregations, they enter the social environment of the females and are constrained by their habitat choices, the food on offer in those plant communities, and the movement rates of these larger groups.

Observations in Amboseli

The basic pattern of seasonal habitat use by Amboseli elephants during the 1970s has been described by Lindsay (1982) and Western and Lindsay (1984). Bushlands and *Acacia* woodlands were used primarily during wet seasons, and there was a gradually increasing use of swamp edges and sedge swamps in the dry season. However, the earlier studies did not attempt to quantify food choices, so overall diet selection was not described.

I looked at habitat use and diet choice in the Amboseli basin by female and male elephants in light of intake rates of key nutrients—namely, metabolizable energy and digestible protein. In an observational study limited to daylight hours during November 1982 to October 1984, I estimated time spent feeding, habitat choice, and diet choice in scan surveys of all elephants encountered in the Amboseli basin; in 1982–83 these surveys were based on a systematic schedule, while in 1983–84, the observations were more opportunistic. During the 1980s, elephants spent much of their daylight hours in all seasons ranging within the Amboseli National Park (Croze and Moss, chapter 7). Dry matter intake rates (trunkful pluck rates multiplied by visually determined trunkful sizes) on different food types were estimated during focal sample observations of female and male elephants in each season of 1982–84, and observations were matched with measures of plant abundance. Samples of plant parts eaten by elephants were collected and subjected to chemical analysis of fiber components and protein content. Requirements for energy and protein were estimated using allometric equations extracted from large mammal nutrition literature.

Feeding Activity

The percentage of time spent feeding was calculated separately for reproductively active adult females (over 15 years old) and adult males when they associated with cow-calf groups or in bull groups. In total, 3,384 observations were made of the activity of adult females and 2,225 of adult males (1,198 with cow-calf groups, 1,027 with bull groups) during scan samples in the high rainfall year (1982–83) and as the percentage of total point samples in each of seven 24-hour follow periods during the drought year (1983–84).

Seasonal percentages of the time that elephants were observed feeding during survey scans in daylight hours for the "good" year are summarized in figure 5.1. During wet season months, less time was spent feeding and more time was spent in other activities— moving, resting, and socializing, in that order—than later in the year. Time required for feeding dropped, from early to late wet season, from 69 percent to 48 percent of daytime hours for females, hovered at about 42 percent for males with cow-calf groups, and dropped from 70 percent to 56 percent for males in bull groups. The trend was reversed in the dry months, with the percentage seen feeding steadily increasing to 79 percent for females, 54 percent for males with cow-calf groups, and 75 percent for males in bull groups by the late dry season. Across the whole year, the mean percentage of time spent

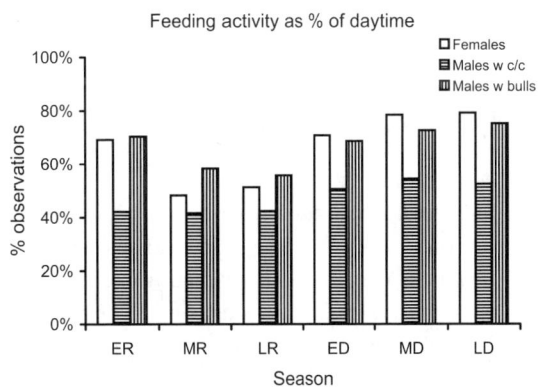

Figure 5.1 Feeding as percentage of total daytime activity. Seasonal percentages of elephants—females, males with cow/calf groups, and males in bull groups—observed feeding during survey scans in daylight hours in the "good" year of 1982–83. Seasons consisted of two-month blocks as follows: ER: Early rains, Nov–Dec; MR: Mid-rains, Jan–Feb; LR: Late rains, Mar–Apr; ED: Early dry; May–Jun; MD: Mid-dry, Jul–Aug; LD: Late dry, Sep–Oct.

in feeding was 66 percent for females, 47 percent for males with cow-calf groups, and 67 percent for males with bull groups.

In the drought year of 1983–84, the overall pattern of daytime activity was similar, although more time was spent feeding overall. A limited number of 24-hour follows recorded five adult females spending an average of 73 percent of the daytime hours in feeding, while the figure for two adult males associating with other males was some 71 percent. As with the other years, there was a trend of decreased feeding time during the rains with an increase toward the late dry season.

Habitat Use

The vegetation communities in the Amboseli basin were described in Croze and Lindsay, chapter 2. For the purposes of the diet study, the 28 communities defined by Western (1973) were pooled into six functional types:

- GR   Short Grasslands: dominated by *Sporobolus* spp, short and sparse, with no woody layer.
- TGR Tall Grasslands: patches of *Sporobolus consimilis*, a tall, fibrous grass, mainly on poorly drained sites at the edge of swamp areas.
- WD   Woodlands: *Acacia xanthophloea* and *Acacia tortilis* woodlands on dry sites away from swamp margins. Tend to be dominated by mature trees, with little seedling or sapling layer, and a seasonally productive grass/herb layer. A sub-type consists of former woodlands, now dominated by *Salvadora persica*/*Suaeda monoica* shrubs, which have the same grass layer as the *Acacia* woodlands but no *Acacias* in the woody layer.
- SEW Swamp-Edge Woodlands: *Acacia xanthophloea* stands on sites with permanent groundwater at the edge of swamp courses. Characterized by trees of all size classes and a dense seedling layer. Grasses dominated by *Cynodon dactylon* forming dense continuous mats.
- SEG Swamp-Edge Grasslands. Same as SEW, but without the tree layer. Some sparse shrubs.
- SWP Swamps: Water courses or their immediate margins, with tall *Cyperus* sedge and *Typha* spp. bulrush/reeds.

Habitat choice was examined systematically in 1982–83. During this year the majority of animals used *Acacia* woodlands and shrub bushlands during the wet season, shifting to swamp edges and swamps during the dry season (figure 5.2). The short grasslands (GR) were never used for feeding, only walking or resting (hence there is no GR bar in the figures that follow), and the tall grasslands (TGR) were used for feeding, and especially resting, at low levels throughout the year. Female elephants showed positive selection (i.e., greater use than expected on the basis of relative area) for *Acacia* woodlands in wet months and relative avoidance by mid to late dry season. Use of the swamp-edge grasslands and woodlands was greater than expected (in relation to area) at all times of year, with use becoming progressively greater as the dry season advanced. The wet swamps were also positively selected in most seasons, particularly in the mid to late dry season.

In all seasons, both males with cow-calf groups and those alone or in groups with other bulls showed an overall distribution that was different from the expected distribution based on relative area of vegetation types. Males in cow-calf groups had distributions across habitats that were not distinguishable from those of females in all seasons apart from the early and mid wet seasons (figure 5.2). At that time, bulls spent less time with females in the swamp-edge habitat types. Males in bull groups, however, had a different habitat distribution from female groups and their associating males. Males remained in the *Acacia* woodlands and in the sedge swamps and consistently avoided swamp-edge woodlands and grasslands throughout the year (figure 5.2). Within this more conservative habitat-use pattern, bulls nevertheless increased their use of the *Acacia* woodlands in the rains and of the swamps in the dry season, similar to the females' pattern. As noted above, neither females nor males ever fed in the short grasslands, while they were seen feeding in the tall grasslands at low levels throughout the year.

Annual and Seasonal Food Choice

The overall observed diet included 91 species in 33 different families, comprising 22 species of grasses or similar monocots in three families (16 species of grasses, four species of sedges and two species of reeds), 40 species of forbs and woody herbs in 21 families, and 29 species of trees and shrubs in 15 families. The bulk of the observed diet, however, was composed of a relatively small number of species. Elephants grazed on woodland grasses of several species, predominantly the stoloniferous *Cynodon plectostachyus* and a variety of *Sporobolus* spp. in the wet season, with a shift to taller *Sporobolus* spp. and *Pennisetum straminium* in the dry season. In the swamp-edge woodlands and grasslands, the dominant grass was *Cynodon dactylon*. In the wet swamps, the principal sedge eaten was *Cyperus immensus*, with additional feeding on the shorter *C. laevitagus* and *C. bulbosus* at swamp margins and the very tall *C. papyrus* in the depths of the swamps. In the swamp-edge

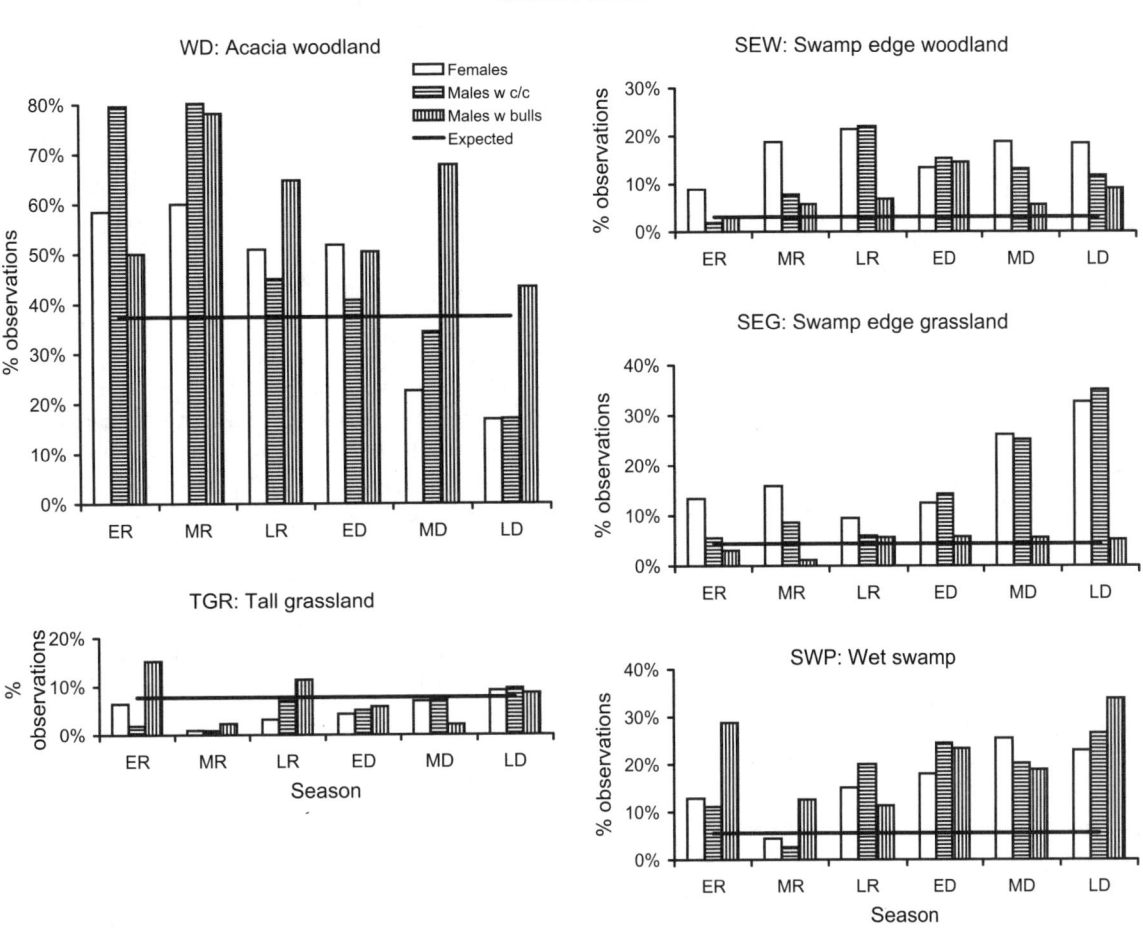

Figure 5.2 Habitat selection by female elephants, male elephants with cow-calf groups, and male elephants in bull groups in 1982–83, shown as percentage of observed feeding in different habitat types. The short grassland habitat type is not included in the figure since, while elephants were seen there, no elephants were ever observed feeding. Seasons are as in figure 5.1; habitat type definitions are in the text. The horizontal line represents the percentage area of the habitat type in the survey area, generating the expected observation level if elephants were evenly or randomly distributed across habitats—including short grasslands—without exercising choice.

woodlands (SEW) and grasslands (SEG) where much dry season grazing occurred, the herb layer was almost exclusively *Cynodon dactylon*, while the swamps (SWP) type was dominated by the four *Cyperus* spp. of sedges and two species of *Typha* (cattails or bulrushes). Many species of forbs and particularly creepers and vines formed a small part of the diet in both *Acacia* woodlands and swamps and swamp edges in the wet season, especially in years with higher rainfall, when there were bursts of diverse plant production.

The majority of elephant browsing was on twigs, branches, and the bark of branches of a variety of shrub species and of *Acacia* shrubs and trees of the two species *Acacia tortilis* and *Acacia xanthophloea*, which dominated the tree layer in the *Acacia* and swamp-edge woodlands. Some feeding on trunk bark of the two trees, particularly the latter, also occurred, but this was combined with twig browsing in the analysis. Feeding on the leaves of the African wild date palm (*Phoenix reclinata*) occurred in an area of swamp-edge woodland, but it formed a small proportion of the total diet and is not treated separately here.

Feeding records within habitat types were pooled into three food type classes: herbs/forbs, grass, and browse. The choices of food type by elephants were examined against a backdrop of seasonal patterns of food production and quality. Herb and forb biomass in woodlands and swamp edges was generally very low—on the order of 10–20 g/m^2—and in woodlands was available primarily during wet seasons, while in swamp-edge habitats, scrambling vines were available for longer periods, but still at very low biomass levels. Grass biomass varied greatly between habitat types and seasons—ranging from dry to wet seasons between 30–260 g/m^2 in *Acacia* woodlands (WD), 300–1,200 g/m^2 in swamp-edge grasslands (SEG, SEW), 900–3,000 g/m^2 in swamps (SWP), and 1,800–3,800 g/m^2 in the tall grasslands

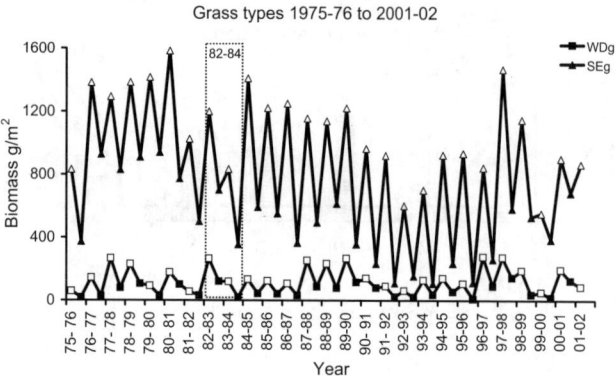

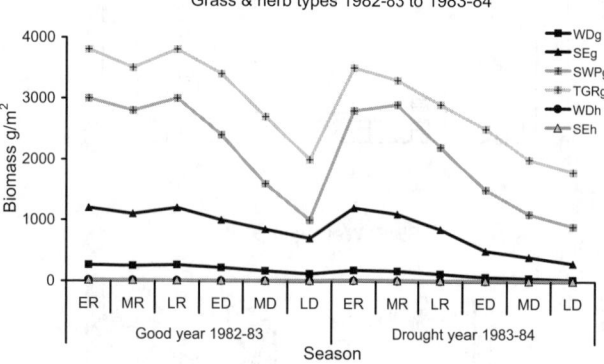

Figure 5.3 Grass/herb layer biomass measured in vegetation plots in Amboseli habitat types during 1975–76 to 2001–2. (a) Long-term records of grass biomass during late wet and late dry seasons of 1975–76 to 2001–2 in swamp edges (SEg) and *Acacia* woodlands (WDg). Records for late wet seasons are indicated with open shapes and late dry seasons with filled shapes. Records for 1982–84—the feeding study period—are shown outlined by the dashed line box. (b) Records during the feeding study (1982–84) of grass and herb biomass in tall grasslands (TGRg), swamps (SWPg), swamp edges (SEg, SEh), and *Acacia* woodlands (WDg, WDh) in 1982–83 (a good year) and 1983–84 (a drought year).

Table 5.1 Estimates of available twig and leaf biomass in habitat types.

Habitat type	Food type	Twig + leaf biomass	
		Kg/ha	g/m²
Acacia woodland (WD)	*Acacia* twigs	111	11
	Shrub twigs	857	86
	Total	968	97
Swamp-edge woodland (SEW)	*Acacia* twigs	1,123	112
	Shrub twigs	435	44
	Total	1,558	156
Swamp-edge grassland (SEG)	Shrub twigs	69	7

Note: Biomass estimates are expressed in kilograms/hectare, which is most appropriate for large trees and shrubs, and in grams/metre², which allows comparison with grass biomass figures.

(TGR)—and between the two years, as it did over the longer term (figure 5.3). Browse biomass on offer varied less dramatically, and was essentially constant, during the 1982–84 study period, on the order of 7–156 g/m² (table 5.1).

Three of the defined habitat types—tall grasslands (TGR), swamp-edge grasslands (SEG), and swamps (SWP)—were dominated by grass with no woody layer, but in two habitat types—*Acacia* woodlands (WD) and swamp-edge woodlands (SEW)—elephants could choose between grass and woody browse.

In *Acacia* (WD) and swamp-edge woodlands (SEW), seasonal choices of food type by adult females in good (1982–83) and drought (1983–84) years were skewed toward grass in the wet season, with a trend to increasing woody material by the late dry season (figure 5.4). In the wetter year, grass made up over 60 percent of observed feeding in all seasons of the year, while in the drought year a pronounced shift to 40 to 60 percent woody browse in both habitat types was in evidence by mid dry season. Females had less grass and more browse in their diets in drought than in wet years in the *Acacia* woodlands from the long rains onward and in the swamp-edge woodlands from the early dry season onward. The higher dietary browse content in the drought year than in the wetter year mirrored the general within-year pattern of higher browse content in dry seasons than in wet seasons. As noted earlier, the contribution of herbs to diet choice was low overall and generally confined to wet seasons.

Like female elephants, males shifted their seasonal pattern in food choice from a predominance of grass in the wet season to increasing woody browse by the late dry season, and again like females, there was more browsing overall in the drier year. Males with cow-calf groups (figure 5.5) in the *Acacia* woodlands also showed an overall preference for grass, which nevertheless declined in the dry season. In the swamp-edge woodlands in the wet year, there was a preference for grass during the rains and relative avoidance in the dry season, while in the drought year there was mild avoidance of grass during the rains, and this avoidance increased in the dry season. Males in bull groups (figure 5.6) made food choices that were essentially similar to those of males with cow-calf groups in the *Acacia* woodlands in both years and in the swamp-edge woodlands in the wet year. In the drought year, males in bull groups fed less on grass in the swamp-edge woodlands in the rains and continued to eat a substantial proportion of browse in the dry season.

Combining observations of habitat choice with diet choice within habitat types provides estimates of the overall proportions of different food types eaten during daytime use of the Amboseli basin (figure 5.7). Taking all animals into consideration, the overall proportion of grass in the diet was high throughout the year, ranging from 60 to

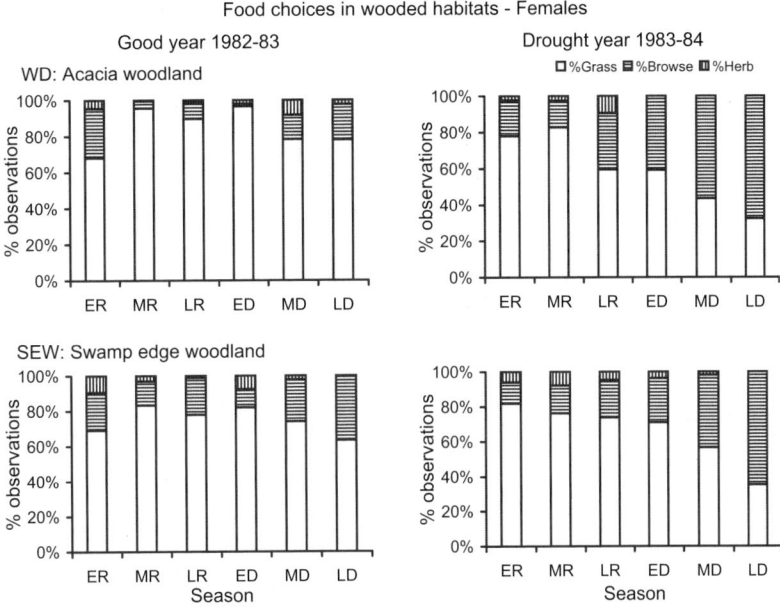

Figure 5.4 Food choices by females in wooded habitats—*Acacia* woodland (WD) and swamp-edge woodland (SEW)—in good (1982–83) and drought (1983–84) years.

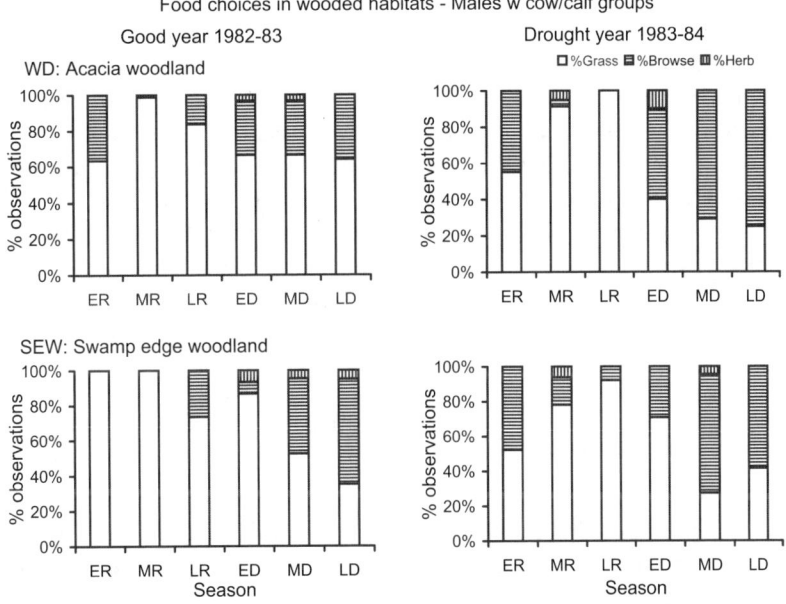

Figure 5.5 Food choices by males with cow-calf groups in wooded habitats—*Acacia* woodland (WD) and swamp-edge woodland (SEW)—in good and drought years.

99 percent of feeding observations in the good year and 40 to 98 percent in the drought year. The dry season dietary shift from grass to browse in the *Acacia* and swamp-edge woodlands was offset by the movement of greater numbers of animals into swamp-edge grasslands (SEG) and sedge swamps (SWP), where, in both cases, diets were composed almost exclusively of grasses. Browsing by females in the good year dropped from 18 percent of the overall diet in the short rains to 2 percent in the early dry season, rising to only 10 percent by the late dry season. In the drought year, the comparable figures for browse were 13 to 20 percent in the rains, rising to 23 percent by late dry season. Males in cow-calf groups were seen browsing in the good year on 29 percent of sightings in the short rains, with this figure

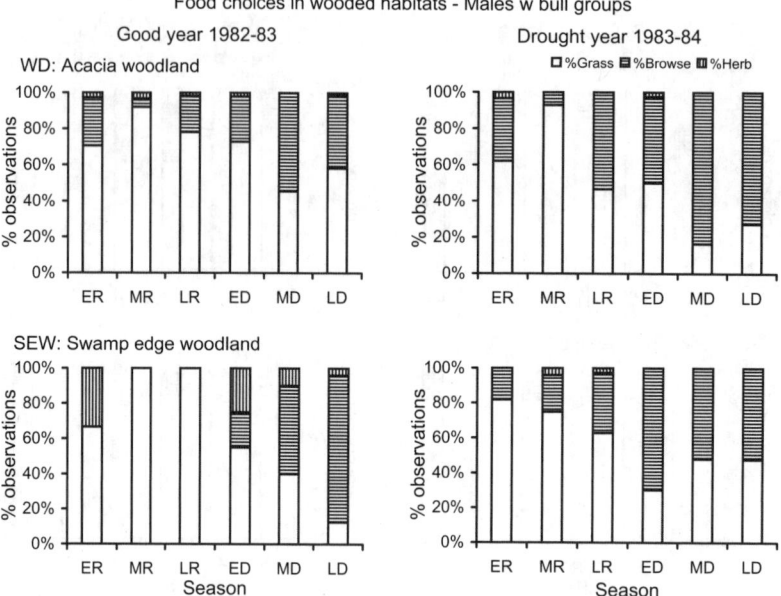

Figure 5.6 Food choices by males with bull groups in wooded habitats—*Acacia* woodland (WD) and swamp-edge woodland (SEW)—in good and drought years.

Note: There were no bulls observed in the SEW during the Late Rains in 1982–83.

dropping to nil in the mid rains and rising to 13 to 16 percent in the dry season; in the drought year, browse comprised 2 to 4 percent in the mid to late rains, rising to 20 to 33 percent by mid to late dry season. Males in bull groups generally had more browse in their overall diets, but even so, the frequency stood at only 25 to 40 percent by mid to late dry season in the good year. The figure was higher in the drought year, at 36 to 60 percent.

Nutrient Intake

Estimating Nutrient Content of Forages, Intake, and Requirements

Observations of food intake rates on herbs, grass, and browse were made during focal samples in each of the habitat types. Because elephants must pick up their food with their trunks, it was possible to identify food items, count pick rates, and estimate the size of individual trunkful picks. Trunkful sizes were cast in categories on an ascending scale of 1 to 5, using relative visual groupings that were then quantified by comparison to hand-picked vegetation samples of the respective food types; samples were bagged and dried to constant weight, and were later analyzed for chemical content. Because elephants often discarded parts of the trunkfuls as part of the food selection process, the size of the discarded trunkful was also estimated, and only the estimated trunkful size that was *ingested* was recorded.

Forage intake was calculated as the product of time spent feeding, diet choice, and food-specific intake rates to estimate overall intake of dry matter and of two key nutrients: metabolizable energy (the net energy absorbed that is available for metabolism, after the energy for digestive processing is deducted) and apparent digestible protein (the protein that is digested and available for metabolism).

The potential dry matter digestibility (percent DMD) for elephants of sampled forages was estimated from plant cell wall fiber in a modification of the summative equation for 48-hour in vitro digestion developed by van Soest (1982) for ruminants and adapted for the shorter hindgut retention time of a large nonruminant and from literature values for elephants' passage rates, estimated at roughly 24 hours (details provided in Lindsay [1994]). Fiber fractions and crude protein content of dry matter were determined using standard neutral and acid detergent treatment (van Soest 1982) and micro-Kjeldahl procedures. Digestible energy was estimated as the product of gross energy and total dry matter digestibility. Metabolizable energy comprises, on average, roughly 85 percent of the digestible energy of dry matter in large mammalian herbivores (Robbins 1983; Blaxter 1989). Estimates of the relationship between apparent digestible protein (ADP) and dietary crude protein were derived from the few published reports of feeding trials by Benedict (1936) and Foose (1982) on Asian and African elephants on diets of different quality; the quality range of forages in these studies was similar to that of the plant material available to elephants in Amboseli.

Intake rates of metabolizable energy and apparent digestible protein for different food types were calculated by

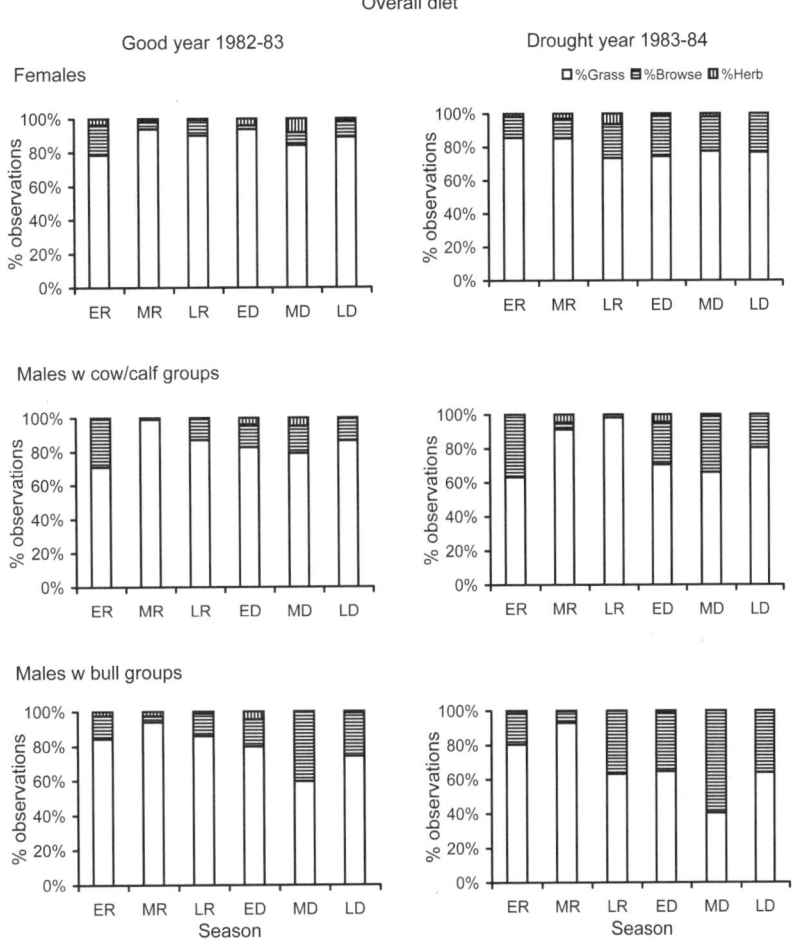

Figure 5.7 Overall daytime diets for females, males with cow-calf groups, and males with bull groups in good and drought years.

multiplying dry matter intake rates by percent DMD and the respective estimate of energy or protein content. The total amounts of daytime intake of dry matter (DM), metabolizable energy (ME), and apparent digestible protein (ADP) for the early wet season and late dry season were calculated by multiplying the percentage of animals seen feeding in each habitat type by the percentage of observations on food types within habitat types and by the dry matter and nutrient intake rates for the different food types.

The requirements for the two key nutrients, energy and protein, were derived from allometric equations relating to body mass. Amboseli elephant mass was calculated from Laws, Parker, and Johnstone (1975) equations, fit to Amboseli known shoulder heights (Lee and Moss 1995), with mean asymptotic weights of adult female and male elephants in Amboseli estimated as 2,740 kg and 4,690 kg, respectively. Active metabolism (AM), a summation of daily activity including foraging, rest, and reproduction (apart from pregnancy and lactation), was considered as a simple multiple (1.5–2 times) of basal. Metabolizable energy need during peak lactation (birth to 9 months) is a multiple or additive factor on active metabolism. A number of estimates of energy required for active metabolism and peak lactation are found in the literature (Belovsky 1978; Peters 1983; Robbins 1983; Pellew 1984; Oftedal 1985, see table 5.2). I averaged their values to estimate the energy needs of female and male Amboseli elephants.

Protein requirements in relation to mammalian body mass for active metabolism are provided by McCullagh (1969) and Peters (1983). Protein requirements at peak lactation were averaged from figures provided for mammals by Belovsky (1978), Robbins (1983), Pellew (1984), and Oftedal (1985).

The equations for the foregoing relationships and estimated values for daily energy and protein requirements are given in table 5.2. Estimates for daily metabolic requirements for nutrients were halved in order to approximate requirements for daytime (12 hours) activity, for comparison against the food intakes observed during daylight hours in my study.

Table 5.2 Estimates of daily requirements of female and male adult elephants for (a) metabolizable energy in MJ/day and (b) digestible protein in kg/day, using allometric expressions from the indicated literature sources

A. Metabolizable energy

Sex and status	Expression	MJ/day	Reference
Females (mean weight = 2,740 kg)			
Basal metabolism (BM)	$0.293 \times W^{0.75}$	111.0	1, p. 39
Active metabolism (AM)	$2 \times BM$	221.9	1, p. 39
	$1.95 \times BM$	216.4	2, p. 136
	$2 \times BM$	221.9	3, p. 112
	$1.5 \times BM$	166.4	4, p. 152
Mean AM		206.7	
Peak lactation (incl. AM)	$2.7 \times BM$	299.6	3, p. 112
	$2.3 \times BM$	255.2	4, p. 152
	Mean AM $+ .669 \times W^{0.70}$	377.2	5, p. 232
	Mean AM $+ 1.519 \times W^{0.52}$	299.8	2, p. 186
Mean peak lactation		308.0	
Males (mean weight = 4,690 kg)			
Basal metabolism (BM)	$0.293 \times W^{0.75}$	166.1	1, p. 39
Active metabolism (AM)	$2 \times BM$	332.1	1, p. 39
	$1.95 \times BM$	323.8	2, p. 136
	$2 \times BM$	332.1	3, p. 112
	$1.55 \times BM$	257.4	4, p. 152
Mean AM		311.3	

B. Digestible protein

Sex and status	Expression	Kg/day	Reference
Females (mean weight = 2,739 kg)			
Active metabolism (AM)	$0.00176 \times W^{0.75}$	0.667	1, p. 95
	$0.00156 \times W^{0.75}$	0.591	2, p. 149
Mean AM		0.629	
Peak lactation (incl. AM)	Mean AM $+ 0.01703 \times W^{0.52}$	1.673	3, p. 183
Males (mean weight = 4,688 kg)			
Active metabolism (AM)	$0.00176 \times W \; W^{0.75}$	0.997	1, p. 95
	$0.00156 \times W^{0.75}$	0.884	2, p. 149
Mean AM		0.941	

Sources: For panel (a): 1. Peters (1983); 2. Robbins (1983); 3. Belovsky (1978); 4. Pellew (1984); 5. Oftedal (1985). For panel (b): 1. McCullagh (1969); 2. Peters (1983); 3. Robbins (1983).
Note: Results were divided by two for 12-hour (daytime) requirements.

Nutrient Intake Observed by Sex and Season

The metabolizable energy (ME) and apparent digestible protein (ADP) contents of forage types sampled throughout both the good and drought years are presented in figure 5.8. The highest values for both energy and protein across all seasons were found in woodland and swamp-edge herbs, which also had the lowest biomass of all forages (figure 5.3). The most variable forage was *Acacia* woodland grass, which had high ME and ADP content in the

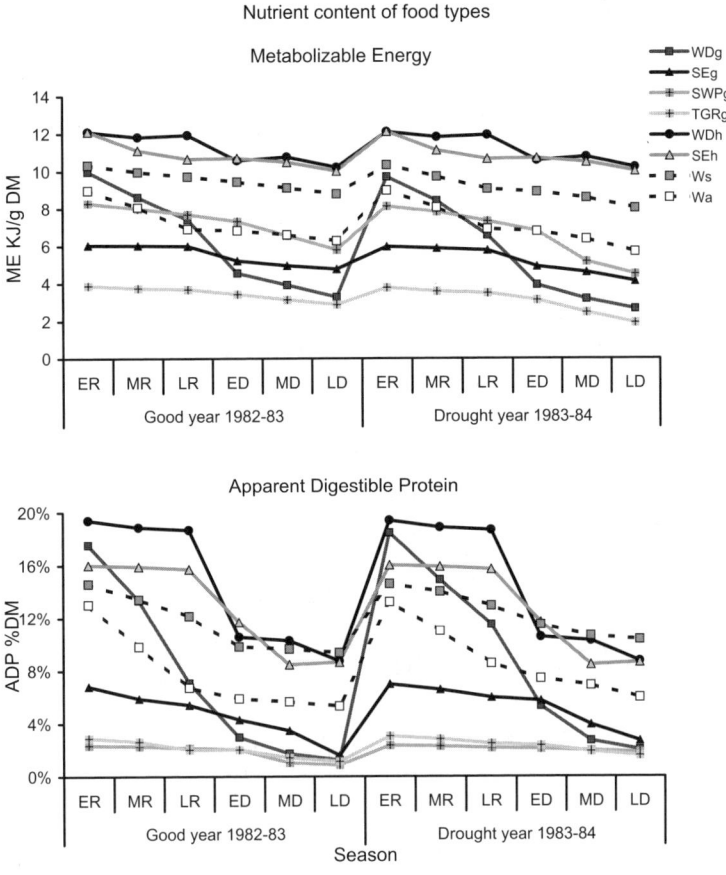

Figure 5.8 Nutrient content—metabolizable energy (ME) and apparent digestible protein (ADP) of food types within habitats in good and drought years. Habitats as in the text: g = grasses, h = herbs, Wa = *Acacias* in WD and SEW, Ws = shrubs in WD and SEW.

early wet seasons, but which dropped to very low values by the late dry seasons. Swamp sedges had moderate energy content but low protein levels, which showed a more gradual decline from wet to dry season. Swamp-edge grasses had lower energy but higher protein content, with the same gradual decline from wet to dry season. Tall grass (*Sporobolus consimilis*) had the lowest concentrations of metabolizable energy of all forage types throughout both years, although ADP content was slightly higher than swamp sedge in all seasons.

The browse (twig and leaf) food types had moderate to high energy and protein levels throughout the year, next to the herb categories. In relation to grasses, *Acacia* browse was matched in ME by swamp sedges but had higher ADP levels, while shrub browse nutrient levels were higher at all times. In relation to grasses, the browse classes had higher nutrient levels than all but woodland grasses in early to mid wet seasons, but by dry seasons both browse types were better than all the grasses. However, total cell wall and digestible protein levels in browse species may have been overestimated by the detergent analysis system because of the formation of insoluble complexes by tannins (Robbins 1983; Reed 1986; Bryant et al. 1992).

There was generally a negative relationship, seen in both energy and protein, between quality and abundance in the different habitats, although there was also a combination of both low nutrient content and low biomass of *Acacia* woodland (WD) grasses in the dry seasons. In the short rains, the highest energy levels were found in woodland and swamp-edge herbs, *Acacia* and shrub browse, and woodland grass, all at fairly low abundance levels in comparison to the lower quality (but very abundant) SWP, SE, and TGR grasses. By the late dry seasons, all biomass levels were reduced, with *Acacia* woodland grass dropping next to the bottom of the quality ranking, but the other forage types retained the same relative ranks. The picture was similar for protein, with the exception that *Acacia* woodland grass had the highest quality ranking and swamp grass the lowest in the short rains, and in the late dry seasons *Acacia* woodland grass dropped down in rank, but with TGR and SWP grass still slightly lower in quality.

Dry matter intake rates—the product of pick rate and

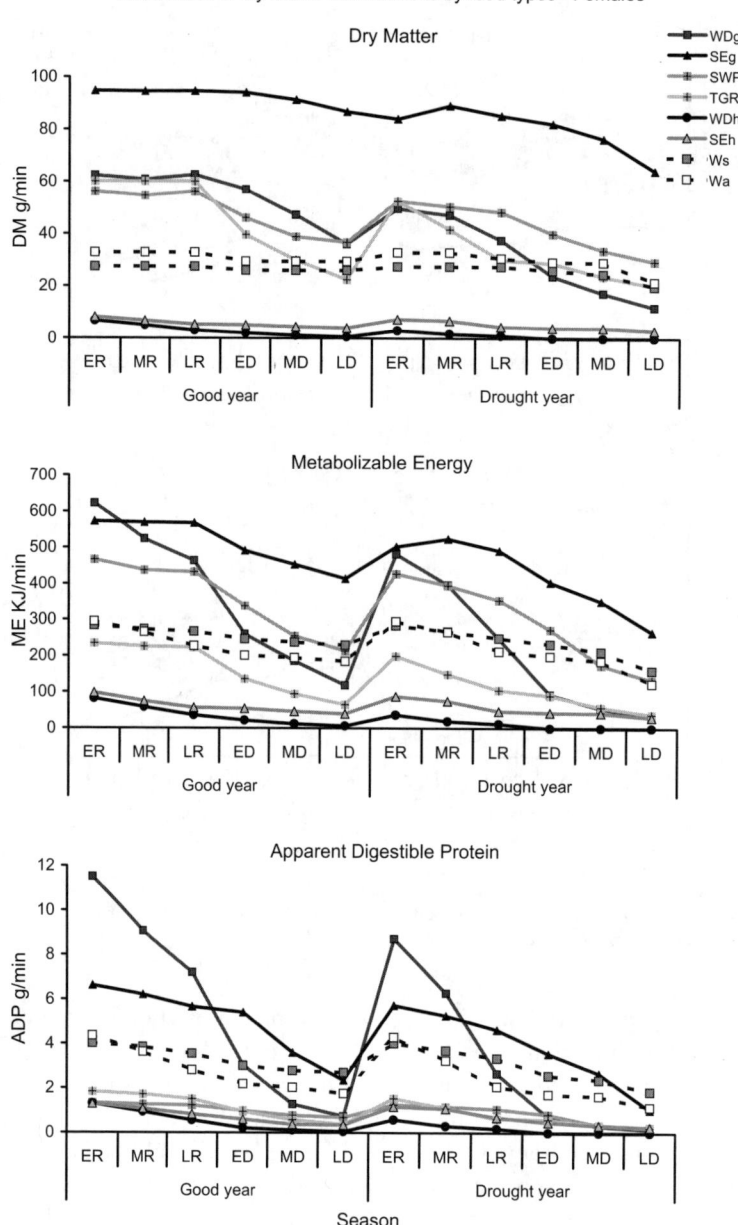

Figure 5.9 Intake rates by female elephants of dry matter (DM), energy (ME), and protein (ADP) on different food types in good and drought years.

trunkful weight—by adult females in the wet seasons were higher on swamp-edge grass than on all other forages, intermediate on the other grasses, lower on browse, and lowest on herbs in both years (figure 5.9). In the dry seasons, pick rates were highest on swamp-edge grass and swamp sedge, intermediate on tall grass and shrub browse, and lowest on *Acacia* browse, *Acacia* woodland grasses, and herbs. Trunkful weights were highest for *Acacia* browse and lowest for *Acacia* woodland grasses, shrub browse, and herbs. The relative rankings of dry matter intake remained similar to those of the wet seasons apart from *Acacia* woodland grasses, which then had the lowest intake rate next to that of herbs.

The seasonal decline in pick rates, particularly for the woodland grasses, was due to an increase in time spent searching for items and handling leaves, stems, and stolons and separating palatable items from dry plant matter. Dry matter intake rates showed the greatest seasonal decline in *Acacia* woodland grasses, with a smaller decline in tall grass and swamp-edge grass and little change in the other food

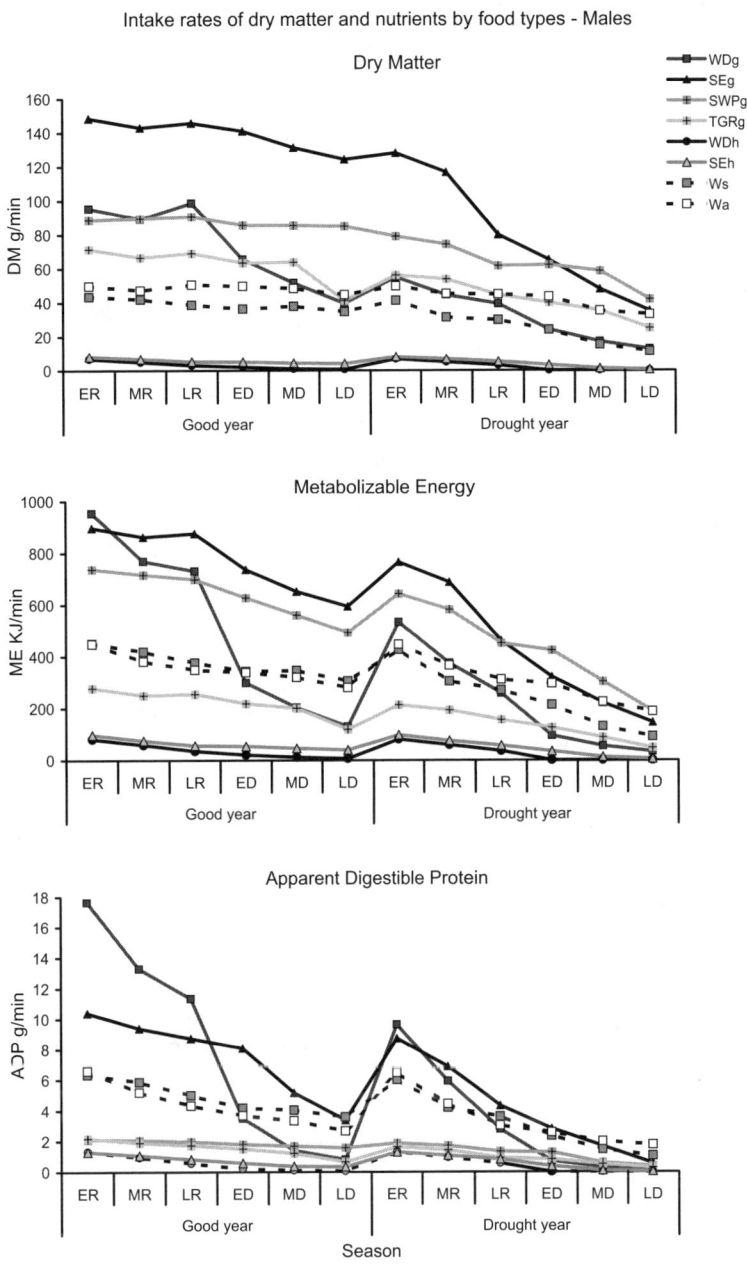

Figure 5.10 Intake rates by male elephants of dry matter, energy, and protein on different food types in good and drought years.

types. Intake measures on grass in the tall grasslands were lower at all times of year than might be expected from their high biomass abundance, apparently because of the large amount of handling time required to separate palatable material from the preponderance of fibrous stems and leaves.

Intake rates of dry matter and nutrients declined to lower levels in the dry season of the drought than in the good year for two reasons: the available biomass of all the grass types was lower in the drought, and the quality of all food types was also lowest in the drought year.

Dry matter intake rates by adult males in the wet seasons were higher for swamp-edge grass than for the other forages; intermediate for tall grass, *Acacia* woodland grass, swamp sedge, and *Acacia* browse; and lowest for shrub browse and both herb types (figure 5.10). In the dry seasons, dry matter intake rates were highest from swamp sedges, followed by swamp-edge grass; moderate from *Acacia* browse and tall grass; and lowest for woodland grass, shrub browse, and herbs. Male pick rate from shrub browse increased from wet to dry season, and dry matter intake

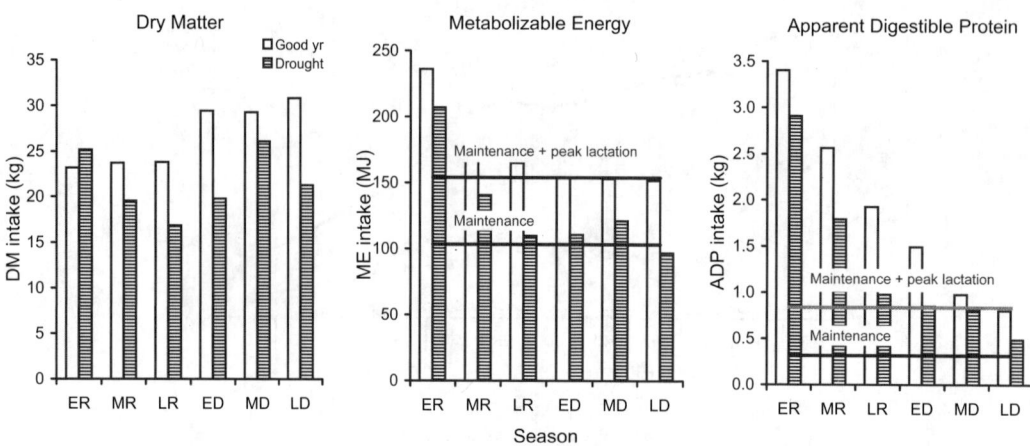

Figure 5.11 Total daytime intake of dry matter, energy, and protein by females in good and drought years, in relation to needs for basic maintenance and for pregnancy and lactation.

showed an apparent increase as well. On swamp sedges, there were clear increases in trunkful weight and dry matter intake. Intake rates of all nutrients declined further in the drought year than in the good year, as was seen with female elephants.

The rates of metabolizable energy (ME) and apparent digestible protein (ADP) intake by adult female elephants on the different forages through the seasons of both good and drought years are shown in figure 5.9. In the wet season, females could obtain the highest rates of metabolizable energy intake from *Acacia* woodland and swamp-edge grasses followed by swamp sedges, with intake rates from browse less than half as high. Rates of energy intake from TGR grass were the lowest of all forage types, apart from the sparse woodland and swamp-edge herbs. By the late dry seasons, ME intake rates were slightly higher on swamp sedges, with swamp-edge grasses a close second. Intake rates from browse were then over half the rates of the swamp and swamp-edge grasses and much higher than those for tall grass and *Acacia* woodland grasses. Protein intake rates in the short rains were highest from *Acacia* woodland grasses with swamp-edge grasses second and browse not far behind. Tall grass and swamp sedges were the poorest sources of protein at this time. By the late dry seasons, browse appeared to provide the highest ADP intake rates followed by swamp-edge grasses and swamp sedges, with those for TGR grass and *Acacia* woodland grasses very low, particularly in the drought year.

The rates of energy and protein intake by adult male elephants on the different forages through the seasons of both good and drought years are shown in figure 5.10. The ME intake levels were generally higher than with females, with a slightly different pattern across the food types. The main differences were that, during the rains, ME intake was higher on the swamp-edge grass and swamp sedges than on *Acacia* woodland grass (while females did better on woodland grass), but the other forages were in the same rank order. Browse intake rates were higher relative to those of grasses as well, when compared with females. In the late dry seasons, ME intake rates of swamp sedges were substantially higher than of swamp-edge grasses and browse, and again, intake rates of browse were relatively high relative to those for grasses when compared with the pattern for females. The overall male intake pattern for protein was similar to that of females, but again browse was relatively more important; and in the late dry seasons, browse gave a relatively higher protein intake rate than did other forages.

In general, elephants had higher rates of nutrient intake during the rains in *Acacia* woodlands than in the swamp edges and swamps, especially in terms of protein. The situation was reversed in the dry seasons, when nutrient intake in swamp-edge woodlands and swamp-edge grasslands was substantially higher than in the woodlands. In the late dry season of the drought year, browse became important, more so for males than females.

Estimates of total daytime intake by females—the summation of intake rates multiplied by time spent feeding on all the different forages—of dry matter, metabolizable energy, and apparent digestible protein are compared between the wet and drought years (figure 5.11). In the 1982–83 "good" year, dry matter intake by females increased substantially from the wet to the dry season, despite the decline in feeding *rates*, because of the increase in overall feeding *time*. However, total intake of both energy and protein decreased, with the decline in protein intake being relatively greater,

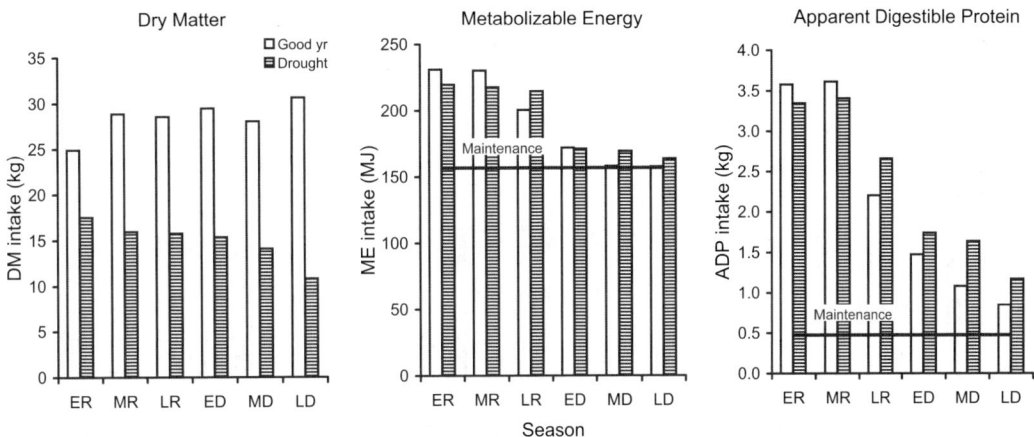

Figure 5.12 Total intake of dry matter, energy, and protein by males with cow-calf groups in good and drought years, in relation to needs for basic maintenance.

because food quality as a percentage of dry matter also declined. In the 1983–84 drought year, dry matter intake increased from wet to early dry season, then declined through the dry season, and energy and protein intake dropped to lower levels. In the wet seasons of both years, nutrient intakes were above baseline levels required for both maintenance and peak lactation demand. In the late dry season of the good year, although nutrient intake levels were above the estimated requirements for maintenance, ME and ADP intake were at, or just below, requirements for lactation. In the drought year dry season, intakes of both energy and protein were well below peak lactation requirements, and energy intake was just below maintenance level.

Estimates of total daytime intake of dry matter, energy, and protein by males in cow-calf groups are also compared between years (figure 5.12). Total dry matter intake increased in the dry season of the good year, but declined in the drought year, and ME and ADP intake decreased from wet to dry season in both years. Energy and protein intake were well above estimated male maintenance requirements in the wet season of both years, and in the dry season of the drought year. In contrast to expectations, ME and ADP intake were slightly lower in the dry season of the good year, and by the late dry season just barely at maintenance level, while in the drought year the intake of these nutrients remained above basic needs.

Total daytime intake of dry matter by males in bull groups was maintained at almost constant levels in the dry season of the good year and declined in the drought year but nutrient intakes, while declining seasonally, were above maintenance requirements in both years (figure 5.13). Energy and protein intake by bull-group males were considerably higher than those for males in cow-calf groups in the wet seasons of both years. The levels of energy and protein intake of males in bull groups remained higher in the dry season of both the good and drought years. Again, nutrient intakes were (surprisingly) higher in the dry season of the drought year compared with the good year.

Conclusions

Feeding Activity, Habitat Use, and Food Choice

Elephants spent less time feeding and more time moving and interacting in the wet seasons, with increased feeding time as forage quality declined in dry seasons, as would be expected for a non-ruminant herbivore. Males who chose to associate with cow-calf groups appeared to have lower feeding times throughout the year than either the females in such groups or males in bull groups, they evidently spent more of their time interacting with females and with other males when in these mixed-sex groups.

The patterns of habitat use by elephants in Amboseli during the mid-1980s were similar to those during the 1970s, as reported by Lindsay (1982) and Western and Lindsay (1984). Female elephants preferred *Acacia* woodlands during the rains and made increasing use of swamp edges and swamps as the dry season progressed. Males joined cow-calf groups across all habitat types, such that their habitat use patterns were not distinguishable from those of the females, in contrast to males in bull groups, who favored different habitat types, and different geographical areas (see Croze and Moss, chapter 7). Male elephants, either singly or in multi-male groups, made less use of swamp edges, particularly the swamp-edge grasslands, than females did at

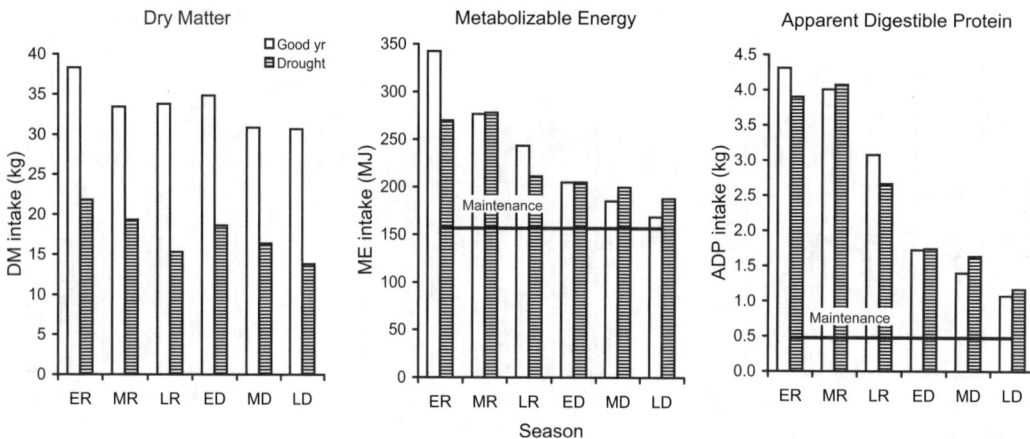

Figure 5.13 Total intake of dry matter, energy, and protein by males in bull groups in good and drought years, in relation to needs for basic maintenance.

all times of year; and males tended to remain in the *Acacia* woodlands and also to use the sedge swamps on a more continuous basis. These patterns may well have changed during the 1990s, when elephants began spending less time in the Amboseli basin throughout the year (Croze and Moss, chapter 7).

Diet choices within habitat types, for both females and males, shifted from a predominance of grass in the wet season toward increasing amounts of woody browse during the dry season, a pattern common to elephants in all parts of Africa. This shift was more pronounced during the drought year (1983–84), when grass biomass was lower than in the higher rainfall year (1982–83). However, because these elephants also chose to feed less in woodlands and more in swamp-edge grasslands and swamps in the dry seasons, where there were few or no woody plants, grasses and sedges remained highly represented in their overall diet. The presence of extensive swamps in Amboseli, and the resulting high grass and sedge content in their diet, makes them somewhat unusual in diet and habitat use by comparison to elephants elsewhere in the continent. Males in bull groups tended to choose browse more than females at all times of year, and because they also spent more time in wooded habitats throughout the year, had an overall diet with more browse in it as well.

Nutrient Intake: Observations and Implications

There were clear differences in food intake rates between food types, seasons, and years. Dry matter intake rates on all forages declined in the dry seasons, as abundance decreased and more manipulation was required to obtain better quality food parts from the background of dry leaves, stems, and soil. As grasses matured and became lignified, the relative proportion of palatable to coarse material appeared to decline, acting to decrease feeding rates, as more time was spent in "separating the wheat from the chaff." Woodland grasses, with their abundance highly dependent on rainfall, showed a much greater seasonal variation in intake rates, while swamp-edge grasses and swamp sedges, with abundance supported by an ample groundwater supply, were able to support higher intake rates throughout the year. The drought year saw greater reduction in dry season grass biomass, even in the swamp edges and swamps, and a consequent decline in intake rates.

Dry matter intake rates on browse were generally lower than those on grasses, because of the lower abundance of twig and leaf biomass and greater manipulation required, but were more consistent across seasons. Thus, by later dry season months, browse intake rates became comparable to, or even greater than, those on some grass types, particularly woodland grasses. The herbs found in woodland and swamp-edge community types had the highest quality of all forage types and were sought after by elephants, particularly early in wet seasons, but because of their very low abundance, their dry matter intake rates—and contribution to overall diets—were very low.

The rates of nutrient intake on different forages appeared to explain habitat choices. Females chose habitat types that allowed the highest intake rates of energy and protein in grasses, which were *Acacia* woodlands in the wet seasons and swamp edges and swamps during the dry periods. Protein intake appeared to be a stronger determinant of habitat use in the wet seasons, possibly because digestible energy was widely abundant in all forages. As both energy and protein intake levels declined throughout the dry season, energy intake became as important as protein, or more so. Both browse and grass intake rates influenced habitat use by males in bull groups in both seasons, which was consistent with the observation that males browsed more in the *Acacia* woodlands and swamp sedges than did females.

Nutrient intake rates also appeared to account for much of the pattern in diet choice for both sexes of Amboseli elephants. During and just after the rains, females fed primarily on woodland grasses with very high energy and protein content; in the dry season, they shifted to browse and grasses in other habitats (swamp edges and swamps). Energy intake rates were higher on grass than browse at all times of year, with swamp-edge grasslands being important throughout the year; *Acacia* woodland grasses were slightly better in the rains, and swamp-edge grasses were better in the dry season. Protein intake rates were highest on grasses in the woodlands during the rains, moderate on swamp-edge grasses throughout the year, and relatively better on browse in the dry season. Swamp sedges were a good source of energy but low in protein throughout the year.

Energy intake influenced diet choice in both seasons for males in bull groups. Males had higher rates of energy intake on grasses than on browse in the wet season, but intake on swamp sedges was higher only in the dry season. *Acacia* woodland grasses were a better source of protein in the rains, while browse became the best source in the dry season. Males were seen to feed on woodland grasses in the rains and swamp grasses in the dry season, but they fed more than females and males in female groups on browse throughout the year, particularly in the dry months. The only anomaly is the apparent avoidance by males of swamp-edge grasses, which could potentially allow for high nutrient returns. The explanation may be that the swamp-edge areas are small and geographically localized, with feeding opportunities elsewhere that would outweigh the costs of traveling to swamps and encountering the competition from herds of females in these small patches.

Social factors, such as direct competition between individuals and family groups, can complicate patterns of habitat use and thus diet (Fretwell 1972). In Amboseli, elephant numbers in habitat types with a restricted geographical area, such as the swamp edges, while relatively high, were lower than might be predicted from nutrient intake rates alone. Because densities and rates of encountering potential competitors were higher in these small patches, spacing behavior may have caused subordinate animals or family units to leave. Such interference competition may explain why elephants of both sexes did not make greater use of the productive swamp-edge community type, which would have been expected since both its grass and its woody layers allow for the highest nutrient intake rates.

The total intake of energy and protein—the product of time spent feeding, habitat choice, diet choice, dry matter intake rates, and nutrient content of forages—was compared with estimated nutritional requirements for female and male elephants. Female elephants in Amboseli were able to increase their intake of dry matter in the dry season of the good year to compensate for the decline in forage quality but not in the drought year when biomass levels were too low. They attained sufficient energy and protein for both maintenance and lactation requirements during the early to mid wet seasons of both good and drought years, with protein intake particularly high relative to estimated needs. By the late dry seasons, nutrient intake declined but was still equal to or in excess of maintenance requirements, with energy showing a greater relative decline. In the drought year, energy intake was less than estimated lactation needs after the early rains, and by late dry season, less even than maintenance needs, while protein intake was less than lactation needs by the end of the dry season. In spite of the many approximations made during the current analysis, the results do appear to confirm that droughts restrict the provision of adequate nutrients to dependent offspring or drain the reserves of lactating females (Lee, Lindsay, and Moss, chapter 6; Moss and Lee, chapter 12; Lee and Moss, chapter 14).

Males in both good and drought years achieved levels of nutrient intake higher than or equal to requirements in both wet and dry seasons. Meissner et al. (1990), working in South Africa, estimated daily energy intake by male elephants in bull groups at approximately 360 MJ/day in the wet season and 300 MJ/day in the dry season. By comparison, this study estimated 24-hour energy intake at 420–635 MJ/day in the wet season and 335–410 MJ/day in the dry season. These figures are in the same "ballpark" as, although slightly higher than, those of Meissner et al. (1990).

As noted before, males in bull groups were feeding for more time, and thus at all times of year their energy and protein intake tended to be higher than that of males with cow-calf groups. Bull-group males had substantially better overall nutrient intake in the rains, with the difference narrowing but still significantly higher in the dry seasons. Males in bull groups were thus on a higher nutritional plane than were males associating with females. This accumulation of a nutritional surplus when away from females is likely to be part of male reproductive strategy, since competition with other males for access to females—and particularly the state of musth—is energetically very expensive (see below; also Poole et al., chapter 18).

There were some apparently anomalous estimates of nutrient intakes by males in the dry season of the drought year, where both energy and protein intake appeared higher than in the preceding good year. The diets in the drier months, and more notably in the drought year, were skewed toward *Acacia* browse, which is relatively rich in nutrients but also high in tannins, which are likely to reduce the digestibility of plant fiber (Reed 1986) and proteins (Bryant et al. 1992). The extent to which tannins reduce actual nutrient intake is

Table 5.3 Pick rates, in trunkfuls/min, and mean trunkful weight, in g dry matter (DM) or wet matter (WM), in (a) the present study, (b) literature reports for Asian elephants, and (c) literature reports for African elephants

A. The present study

Sex	Season	Food type	Trunkfuls/min	Trunkful weight
F	W	G	3.8–7.5	12–15 g DM
		B	1.5–2.8	10–22 g DM
	D	G	1.8–5.1	7–13 g DM
		B	1.0–2.8	9–29 g DM
M	W	G	4.2–6.5	16–24 g DM
		B	1.4–1.8	19–41 g DM
	D	G	3.7–5.0	10–23 g DM
		B	1.3–3.8	10–40 g DM

B. Asian elephants

Sex	Season	Food type	Trunkfuls/min	Trunkful weight	Location	Source
F	—	H	1.7	69 g DM	Captivity (U.S.)	1
F	W	G	1.5	60 g DM	Ruhunu and Gal Oya NPs, Sri Lanka	2
M	W	G	1.3	"		
		B	1.6	"		
—	—	G	0.8	—	Gal Oya NP	3
—	—	G	0.7	—	Malaysia	4
—	—	B	0.4	—		
—	W	G	0.8	76.5 g DM	S. India	5
—	D	G	"	57.5 g DM		

C. African elephants

Sex	Season	Food type	Trunkfuls/min	Trunkful weight	Location	Source
F/M	W	G	5.2	—	Gounda St–Floris NP, CAR	1
		B	4.8	—		
	D	G	5.0	—		
		B	4.5	—		
—	W	G/B	4.9	27 g WM	"	2
	D	"	4.6	"		
F	W&D	—	5.9	—	QENP, Uganda	3
M		G	14.0	—		
M						
	—	G	5.3	—	Serengeti NP, Tanzania	4
	—	B	2.7	—		
F	W	G	8.0	—	Mikumi NP, Tanzanzia	5
	D	G	3.9	—		
M	W	G	8.4	—		
	D	G	3.1	—		
F	D	G	2.6–6.4	—	South Luangwa NP, Zambia	6

(continued)

Table 5.3 (continued)

Sex	Season	Food type	Trunkfuls/min	Trunkful weight	Location	Source
		B	2.0	—		
F	W	G	3.7	75 g WM	Sengwa, Zimbabwe	7
		B	2.6	"		
	D	G	2.0–2.3	"		
		B	1.5–2.1	"		
M	W	G	3.0	"		
		B	2.6	"		
	D	G	1.6–2.8	"		
		B	1.8–2.7	"		

Sources: For panel (b): 1. Benedict (1936); 2. McKay (1973); 3. Vancuylenberg (1977); 4. Olivier (1978); 5. Sukumar (1989). For panel (c): 1. Ruggiero (1989); 2. Ruggiero (1992); 3. Wyatt and Eltringham (1974); 4. Croze (1974); 5. Barnes (1979); 6. Lewis (1986); 7. Guy (1975).
Notes: Abbreviations in the columns are for sex: F = female, M = male; season: W = wet, D = dry; food type: H = hay, G = grass, B = browse. A dash (—) indicates that the relevant data were not supplied by the source.

not precisely known and the present study made no attempt to correct for these potential effects. Thus, the digestibility and nutrient intake rates were likely to be overestimated in this study during the drier periods of late dry season and drought.

The concordance of habitat and diet choice by both sexes of elephants with their nutrient intake rates supports the suggestions of Miquelle et al. (1992) and Shannon et al. (2006); that is, habitat segregation may be a result of differential foraging abilities relative to requirements in sexually dimorphic herbivores rather than interference competition between the sexes (Clutton-Brock, Iason, and Guinness 1987). In the dry season of the drought year, grass biomass levels were low in the swamp edges, and the males would not achieve acceptable intake rates relative to their requirements, while females could still do so. Illius and Gordon (1987) noted that smaller species of ungulate grazers (or the smaller females in dimorphic species) may be able to do better than larger species (or sex classes) on short grass swards.

During other seasons and in better years, however, forage abundance and intake levels were higher, making the avoidance of swamp-edge grasslands by males less easy to understand. An alternative explanation, suggested by Prins (1989) in his study of African buffalos, is that males suffer a nutritional cost when foraging with groups of females. In Amboseli, males associating with females spent less time feeding than did males in bull groups, across all seasons. This may be due to interactions among males or between males and females—most often in the context of mating strategies—in mixed-sex groups; Poole (1982) found that non-musth males spent less time foraging and more time interacting when associating with females rather than with other males. In addition, she noted that musth males also spent less time foraging overall, with or without females.

There may be differences between females and males not only in time spent feeding but also in foraging style while feeding (Barnes 1982). Females, with their companions in family units, tend to march steadily along during foraging bouts, spending limited time on one patch of ground, particularly when group sizes are large (unpublished data; see also Mutinda, Poole, and Moss, chapter 16). Bulls, in their small and socially fluid groups, by contrast, spend more time standing in one spot feeding intensively in a patch of grass or at an individual shrub or tree. This more sedentary form of feeding may allow bulls to expend less energy while in bull groups than when they are in cow-calf groups. Males appear to bear costs when associating with females— their feeding time drops, they move more, they use habitats of lower quality in terms of nutrient and energy intake, and there is the potential for high rates of interaction with other males and females. Because the swamp-edge areas are small and geographically constrained, as noted previously, males would have a high likelihood of encountering cow-calf groups and they may opt to avoid that possibility in the first place. It thus appears that both anatomical and social factors may be involved in the different patterns of habitat use by, and geographical separation of, the two sexes.

The analysis of nutrient intake outlined in this chapter provides a framework for explaining habitat and diet choice and linking habitat conditions via nutritional balance to demographic performance. Intake rates have been estimated by a small number of elephant researchers, with findings broadly similar to those of the present study (see table 5.3). Total daily intake has also been estimated by a range of authors in different parts of Africa, and these reports are also broadly in agreement with the present results (table 5.4). Despite the number of approximations made in the course of building up the overall diets and nutrient

Table 5.4 Total daily food intake, in kg dry matter (DM) or wet matter (WM) and in percentage of live weight, in (a) the present study, (b) literature reports for Asian elephants, and (c) literature reports for African elephants

A. The present study

Sex	Season	Method	Intake (kg)	% Body weight
F	W	O	43–46 DM	1.6–1.7
	D	O	45–58 DM	1.6–2.1
M	W	O	47–78 DM	1.0–1.7
	D	O	58–77 DM	1.2–1.6

B. Asian elephants

Sex	Season	Method	Intake (kg)	% Body weight	Location	Source
F	—	T	34 DM	0.9	Captivity (U.S.)	1
F	—	T	38–39 DM	1.1–1.2	Captivity (U.S.)	2
—	—	T	43 DM	1.2	Captivity (India)	3
M	—	O	60 DM	—	Sri Lanka	4
F/M	W	O	44 DM	1.9	India	5
	D	O	34 DM	1.5		

C. African elephants

Sex	Season	Method	Intake (kg)	% Body weight	Location	Source
—	—	T,F,O	135–300 WM	—	Captivity and wild	1
F	—	T	34–38 DM	1.1–1.2	Captivity (U.S.)	2
M	—	T	40–42 DM	1.2–1.3		
F/M	W	O	55 DM	—	Gounda-St Floris NP, CAR	3
	D	O	46 DM	—		
M	—	F	43 DM	1.1	QENP, Uganda	4
F	—	S	— WM	4–6	MFNP, Uganda	5
M	—	S	— WM	4		
F	W	O	200 WM	—	Sengwa, Zimbabwe	6
	D	O	115–140 WM	—		
M	W	O	225 WM	—		
	D	O	115–175 WM	—		
M	W	F	50–65 DM	1.1–1.4	Klaserie NR, South Africa	7
	D	F	64–78 DM	1.4–1.7		

Sources: For panel (b): 1. Benedict (1936); 2. Foose (1982); 3. Ananthasubramanian (1980); 4. McKay (1973); 5. Sukumar (1989). For panel (c): 1. review of several studies of captive and wild Asian and African elephants by Guy (1975); 2. Foose (1982); 3. Ruggiero (1992); 4. Petrides and Swank (1966); 5. Laws, Parker, and Johnstone (1975); 6. Guy (1975); 7. Meissner et al. (1990).
Notes: Abbreviations in the columns are for sex: F = female, M = male; season: W = wet, D = dry; method: T = feeding trials, F = fecal collection, S = stomach contents, O = observation of feeding time and intake rates. A dash (—) indicates that the relevant data were not supplied.

intakes, as well as the estimates of nutrient requirements, the results of this study appear robust. Lagendijk, de Boer, and van Wieren (2005) performed a similar analysis, comparing estimates of intake rates and metabolic requirements to assess the value of an elephant diet composed entirely of one tree species. The authors concluded that, even with its recognized limitations, the approach was useful for examining the relationship between elephant survival and their habitat resources. The present study was a first effort at assessing habitat and diet choices, in relation to nutrient intake, and used data pooled across the broad classes of sex and social grouping. In the Amboseli research project, with its detailed knowledge of individual animals, the foraging success of individual females and males could be evaluated by exploring individual patterns of habitat use, diet, and reproductive success. It is likely that such an individual-based study would parallel the results of the isotope studies of Koch et al. (1995), where the proportions of grass and browse in the diet were found to vary greatly between individuals and through time.

The analysis of foraging outcomes in relation to resource abundance is also of importance for modeling the relationship between elephants and the plant communities they inhabit. Data on elephant demography (Lee, Lindsay, and Moss, chapter 6) and intake rates of plants and plant parts could be coupled with data on plant population structure and dynamics (Dublin, Sinclair, and McGlade 1990; Western and Maitumo 2004) to produce predictive scenarios of the effects of environmental factors on the structure of the plant and animal communities in the Park and ecosystem. Such factors could include biophysical processes—water table fluctuation, population changes of wildlife species (including elephants), and climate change—or human actions such as elephant poaching, land-use competition, or different management strategies.

A pattern of continual shifting between habitats and food types to maximize nutrient intake characterized both sexes of adult Amboseli elephants. Females generally met their needs of protein and energy for basic maintenance but were unable to meet the additional costs of reproduction during dry seasons and drought years. Males, in contrast, appeared to be able to meet their nutritional and energetic needs more readily but potentially were constrained in their habitat and food use choices in some key areas by interactions with females in cow-calf groups. Social dynamics, physiological needs, and food availability were all factors underlying habitat and food choices and had consequences for elephant survival and reproduction.

Chapter 6 Ecological Patterns of Variability in Demographic Rates

Phyllis C. Lee, W. Keith Lindsay, and Cynthia J. Moss

DENSITY-DEPENDENT effects on reproduction and survival have been well documented for food-limited large mammal populations (Clutton-Brock, Iason, and Guinness 1987; Coltman et al. 1999b; Milner et al. 1999; Festa-Bianchet, Jorgenson, and Réale 2000). Environmental factors independent of density may have equally strong influences on reproductive rates, mortality rates, and population sizes (Bayliss 1985; Fryxell 1987; Sæther 1997; Mduma, Sinclair, and Hilborn 1999; Ogutu and Owen-Smith 2003; Festa-Bianchet et al. 2004). Studies of the consequences of environmental variation acting on demographic characteristics of large mammals have focused on fertility rates and survivorship, especially mortality of infants and juveniles. Seasonal and annual limitations in food abundance caused by climatic variability can reduce fertility and survival by affecting body condition (Albon, Clutton-Brock, and Guinness 1987; Owen-Smith 1990; Fairbanks 1993; Clutton-Brock et al. 1996; Portier et al. 1998; Kruuk et al. 1999b; Rasmussen, Wittemyer, and Douglas-Hamilton 2006). Constraints on fertility and an enhanced mortality rate are significant factors affecting the lifetime breeding success of individuals as well as the dynamics and regulation of populations. While body size and scale can confound the effects of ecological variation acting on populations (Economo, Kirkhoff, and Enquist 2005), explorations within a single population over time may reveal the local environmental factors of greatest importance or impact.

Annual and seasonal variation in conception and mortality rates have been described for a number of savannah elephant populations. In general, conceptions increase in frequency during rainy periods and in years with high rainfall, while calf mortality increases in periods of low rainfall (Lindsay 1994; Dudley et al. 2001; Rasmussen, Wittemyer, and Douglas-Hamilton 2006; Foley, Petorelli, and Foley 2008). As a consequence, peaks and troughs in age structures can result, and population structure and dynamics can appear highly variable over the short term but stable over the long term. Most prior studies (e.g., Wittemyer et al. 2005), with the exception of that on the Addo elephants (Whitehouse and Harley 2001; Gough and Kerley 2006), have been relatively short term and are unable to explore environmental patterns over a large number of annual cycles or assess how populations respond over the very long term. The Amboseli population provides a unique opportunity for such an exploration.

In this chapter, we aim to examine some of the ecological factors affecting the birth and death rates of elephants.

- We investigate seasonal and annual variation in rates of conception and calf mortality using 30 years of demographic data for the Amboseli population.
- We attempt to link these demographic rates to the underlying factors of food abundance and quality, using a proxy measure of rainfall and its distribution over an annual cycle and between years.
- We assess the long-term consequences of ecological impacts on individuals and the population.

Measures of Food Availability

For a mixed feeder such as an elephant that continually switches among food types and habitats (Lindsay 1994), it is difficult to develop a single simple and reliable measure of seasonal or annual food availability. Other researchers have used a greenness indicator from remote sensing data as a pooled, integrated measure of overall plant productivity (e.g., Normalized Difference Vegetation Index [NDVI]; Pettorelli et al. 2005; Rasmussen, Wittemyer, and Douglas-Hamilton 2006). The NDVI data are, however, limited to the period from 1985 to the present since they depend on satellite data; therefore, we have used rainfall for our 30-year analysis.

During the rainy months, Amboseli elephants exploit growing grass in woodlands; during dry seasons, they forage on sedges and grasses in the extensive swamps and their margins and on browse in both swamp-edge and dry bush habitats (Lindsay 1994). In general, the abundance of the different food types responds positively to rainfall (Western and Lindsay 1984; Lindsay 1994). As detailed in Croze and Lindsay (chapter 2), we use a variety of meteorological measures as indices of annual and seasonal habitat conditions. The key measures were monthly rainfall, rainfall within one annual cycle as the summed rainfall from October through September (which is applied to an Amboseli Year [see Croze and Lindsay, chapter 2]), and the length of the dry season. This latter measure was converted into a Dry Season Intensity (DSI) index, calculated as the number of consecutive dry season months divided by total annual rain for each Amboseli Year's rainfall, times 100. The frequency distribution of DSI was used to categorized years as Drought, Average, or Good. Years with the DSI in the upper or lower quartiles of the DSI distribution over 30 years were considered to be Drought or Good years, respectively (see Croze and Lindsay, chapter 2). In some analyses, we have used summed rainfall; in others, we have used lagged rainfall. Summed data were cumulative rain over one or two previous months. Lagged data were those where precipitation in previous months was compared at different lag intervals (one, two, or three months) with the demographic events of a current month.

In our analysis of the effects of population density on demography, we have to consider both the numbers of elephants and their use of space. The size of the elephant population has increased steadily from 1978 onward (Moss 2001), although with changes in the elephants' use of the central swamps and surrounding basin (Croze and Lindsay, chapter 2; Croze and Moss, chapter 7), population density has varied. From the late 1970s, most elephants spent much of the year within the protected area and density increased. However, since the early 1990s, fewer elephants have been concentrating in the central basin, and thus, geographical density has decreased despite a continued increase in total population size. In this chapter, we have used 1991 to separate the earlier period of increasing population density from more recent times when density has reduced.

Ecological Influences on Conceptions

Monthly Conception Rates

The life history data used here are those described in appendix 1. Conception dates have been calculated by backdating from all births where the birth date was known with an accuracy of one month ($N = 1,360$; total births $= 1,551$). We used a gestation of 660 days for these calculations (see also Poole et al., chapter 18). Although, as noted elsewhere, this technique underestimates conceptions lost from miscarriages, stillbirths, or undetected neonatal deaths (see also Moss [2001]), there is no evidence of a systematic bias to monthly or annual totals because the sample size of observed matings that result in a birth is relatively large. Over the entire period, there was a mean of 4 (median 2) conceptions resulting in known births per month (figure 6.1), with a maximum of 29 in a single month and a minimum of 0. Conceptions in any month, if evenly distributed, would be expected to represent an average of 8 percent of the annual total, while the median was 5.6 percent with considerable variability between months (range 0–60 percent), illustrating seasonal peaks and troughs to Amboseli conceptions.

Seasonality of breeding has been formally defined as greater than 80 percent of conceptions (or births) occurring

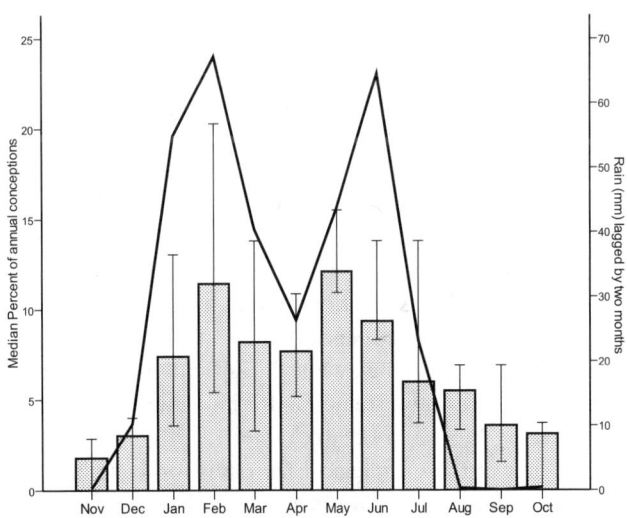

Figure 6.1 The median (±95% CI) percentage of annual conceptions by month for the 30 years of the study plotted with rainfall (mm) lagged by two months (line).

Table 6.1 Stepwise multiple regression statistics for monthly conceptions against rainfall measures

Regression model statistics	$r^2 = 0.193$		
$F_{3,328} = 26.33$		$p < 0.001$	
Rainfall lagged two months	$t = 7.97$	Beta = 0.402	$p < 0.001$
Sum three months rain, lagged one month	$t = 2.92$	Beta = 0.225	$p = 0.003$
Rain in month	$t = -2.48$	Beta = -0.136	$p = 0.008$

Notes: Model entered monthly rainfall, rainfall summed over two months, rainfall summed over three months; rainfall lagged by one month, rainfall lagged by two months, sum of two months rainfall lagged by one month, sum of three months rainfall lagged by one month. Only the final significant model is shown. All statistics on log normalized data.

within a three-month period (Sinclair, Mduma, and Arcese 2000). In this sense, Amboseli elephants are not highly seasonal, since 80 percent of conceptions occur over the nine months from February to September. Although not strictly seasonal, Amboseli elephants exhibit a bimodal temporal pattern of conceptions (figure 6.1). In an analysis of the period from 1974 to 1982, Lindsay (1994) found that monthly conception frequency was most strongly correlated with rainfall two months earlier; that is, with a time lag of two months. Analysis of the larger 30-year sample confirms that rainfall lagged over two months is the best predictor of monthly conceptions (table 6.1 and figure 6.1). In an average year, two to three months after the onset of the rains, conceptions begin to increase in frequency. The two peak months of conception (February and May) reflect the rainfall peaks in December and March. The marked trough in conceptions from October to December maps well onto a lagged effect of the driest period (June through September) when, in addition to being dry, the most nutritious standing green biomass from the April rains has been completely consumed and the elephants have to rely on the swamps or on browse outside the central basin. During this period, females consistently appear to be in poorer condition.

Conceptions resulting in a live birth are clearly a function of females attaining sufficient condition to come into estrus and carry a calf to term. Estrus itself, while less frequently observed than births, follows the same temporal pattern as that estimated for conceptions (Poole et al., chapter 18). Here, we assess the ecological factors that might limit or enhance female fecundity, while age-specific fecundity and social influences on reproduction for individual females are presented by Moss and Lee in chapter 12.

The first of these ecological factors is food availability, using rainfall as a proxy for primary plant production. Primary production in Kenya savannah ecosystems, as measured by NDVI greenness, has been found to peak around 80 days after the onset of the rains (Rasmussen, Wittemyer, and Douglas-Hamilton 2006). Between two to three months of vegetation growth after the start of the rains appear to be required before females are able to come into estrus and then to successfully gestate for the next 22 months. As the rains continue and the biomass of vegetation, particularly grasses, reaches its peak, more females achieve sufficient condition to successfully conceive. The response of vegetation to rainfall is not strictly linear; small amounts of rain early on can trigger rapid production, while a very high rainfall total may not significantly enhance existing maximal plant production, hence the often weak correlations with total annual rainfall. Once the rain begins to tail off and vegetation is consumed by elephants and other herbivores, the proportion of monthly conceptions also drops (figure 6.1).

The overall seasonal pattern found in Amboseli was not a simple function of monthly rainfall. Whether or not an individual female is able to conceive in any month is a complex function of her current physical condition, her size and age, and the age and sex of any surviving calf (Moss and Lee, chapter 12). Primiparous females are small, young (average age of first conception is 12 years), inexperienced with mating and calf rearing, and tend to be somewhat less seasonal in their conceptions. Old, experienced females generally show a more strongly seasonal pattern (figure 6.2). This difference could be a function of condition loss and recovery on the part of the multiparous females over the course of lactation, as these females are more energetically and nutritionally stressed and thus may be more seasonally entrained.

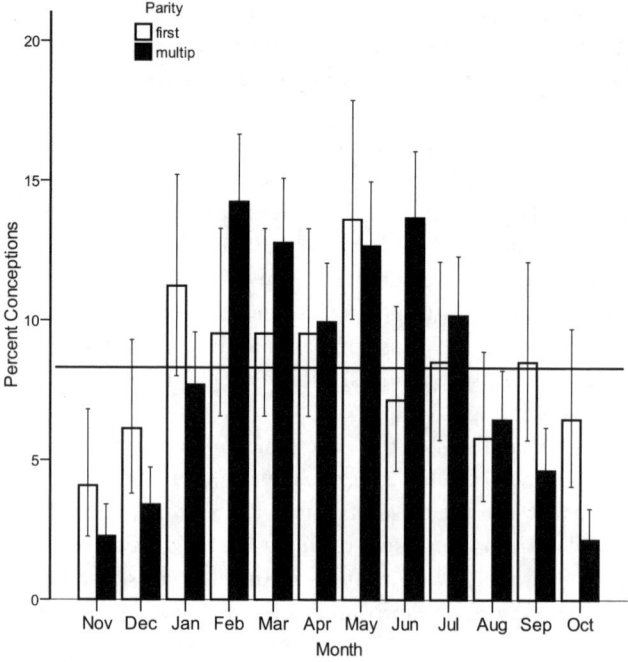

Figure 6.2 Percentage of conceptions by month for primiparous mothers (*N* = 230) by comparison to experienced, multiparous mothers (*N* = 560). The expected line for an even distribution is shown for comparison.

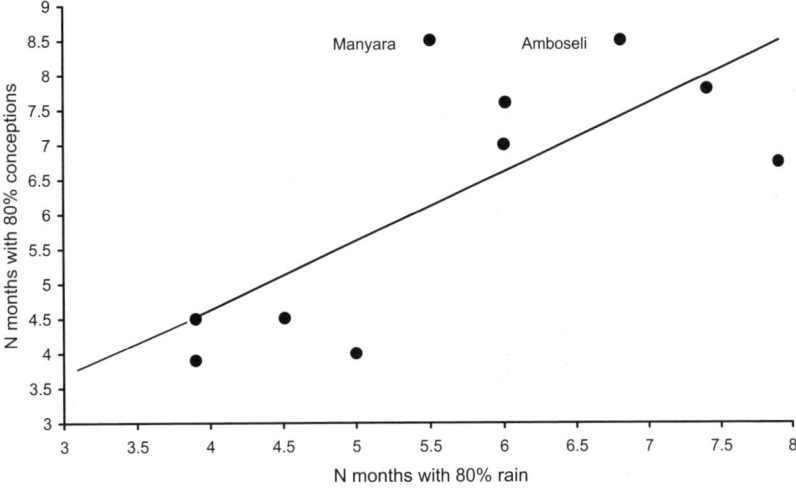

Figure 6.3 The relationship between extent of conception seasonality and rainfall for populations of elephants, using cumulative 80 percent of events.

Sources: Manyara and Amboseli with groundwater swamps or forests. Data from Lindsay 1994. Additional data from Rassmusen, Wittemyer, and Douglas-Hamilton (2006) and Foley, Papageorge, and Wasser (2001).

In contrast to expectations from female condition and observations on other elephant populations, monthly rainfall was not a powerful (although still statistically significant) linear predictor of the percent of monthly conceptions, explaining only about 20 percent of the variance. A number of other factors might also have influenced these monthly patterns. Seasonal rainfall distribution changed only slightly over the three decades of the study, yet there was a small influence of decade on the monthly conceptions (3.2 percent of variance). Between 1982 and 1991, the peak of conceptions occurred in February–March as opposed to May–June in other decades. The shift was not evidently related to gross changes in amounts of rainfall in the preceding rains. It may have been due to variation in the response of vegetation to peak rains, or as a function of changes in water tables and swamp vegetation. As noted in chapter 2, Amboseli is a dynamic ecosystem, with low predictability. Month alone explained an additional 10.5 percent of variance, suggesting—as noted by Rassmussen, Wittemyer, and Douglas-Hamilton (2006)—that females may use some internal timing cue as well as a resource–body condition cue for the onset of seasonal reproductive activity. Within-family synchrony in conceptions (e.g., Moss 2001) could also contribute to the observed "month" effect. Female age is another possible influence on variation in patterns of seasonality, but while primiparous females were less seasonal in their conceptions, there were no significant differences in the proportions of conceptions due to primiparous females over the three decades ($\chi^2 = 11.31$, df = 2, NS). The death of a suckling calf removes the lactational block on estrus (see below), which could also increase the likelihood of out-of-season conceptions.

Other elephant populations in eastern and southern Africa exhibit strong positive correlations between monthly conception frequencies and rainfall, with a lag of 0 to 4 months (see also Gough and Kerley [2007]). Populations in south-central Africa, which experience a narrow unimodal rainy season, have sharply defined conception periods, with both 80 percent of rain and conceptions falling into the same 3 to 4 month period (Lindsay 1994, figure 6.3). By comparison with other populations, Amboseli females have a broader distribution of conceptions than would be expected for the degree of rainfall seasonality. Swamp vegetation, which is available in all months regardless of rainfall, may buffer females against major seasonal declines in condition, reducing seasonality in conceptions. The conception pattern at Lake Manyara National Park, Tanzania, where there is constant access to ground-watered forests, is similarly spread across high rainfall to drier months. Condition loss as a constraint on conception has been documented for elephants in Tarangire National Park, Tanzania, where a conception peak is strongly associated with a rainfall peak from December to May. During dry periods, females have high corticosteroid levels, large daily movement ranges, low body condition, and few conceptions (Foley, Papageorge, and Wasser 2001).

Annual Conception Rates

Conceptions varied markedly between years over the period of the study, with some years having no conceptions or births

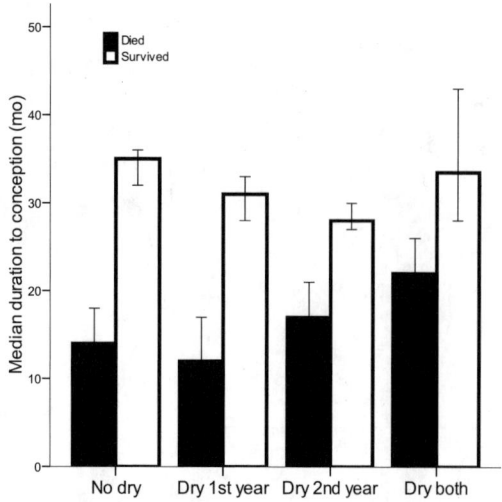

Figure 6.4 The median duration to conception as a function of environmental conditions experienced by mothers of calves that died in their first two years of life.

and others in which 35 percent of all adult females conceived (Moss 2001). During years of low rainfall, fewer females conceive, while the reverse is true for periods of higher rainfall. However, the pattern of rainfall and conception on an annual basis did not simply mirror the monthly distribution of conceptions in relation to rainfall, since, with a 22-month gestation and at least 6 to 12 months of lactational anestrus, cycles of female availability are likely to occur, with a periodicity of about four years. We also suggest that when preceding years are drier than normal, it may take longer for females to gain sufficient condition for estrus to occur, while after a succession of good years even a lactating female may only need a short period of access to good forage before she is able to conceive. When rains are high over successive years, the standing crop of biomass does not drop seasonally as much as in a single "normal" year (Western and Lindsay 1984). With a succession of very dry years and a trough in standing crop, some females simply may not be able to come into estrus even after grass growth is established.

Lactation, as noted before, acts as both a hormonal block on fertility and a prolonged drain on female condition. While females potentially can conceive again as quickly as 4 months after the birth of a surviving calf, they wait on average for 32 months when those calves survive the first 24 months of life (Moss and Lee, chapter 12). Since calf death removes the hormonal and condition constraints on resumption of estrus, and since more calves die in drought years (see below), there is an interaction between bad years and the rapidity of the next conception: the quickest time to conception was experienced by mothers when their calves died at a young age in a drought year (figure 6.4). While females can conceive again rapidly after calf loss even in a relatively poor year, in a very bad or drought year conception will be delayed or suppressed entirely. If a calf in its second year of life dies in a dry year or when there have been two successive dry years, the time to conception is longer than when the calf dies in a rainy year ($t = 3.31$, df $= 49.8$, $p = 0.002$).

There is no simple correlation between the period before the next conception and either rainfall in the year of conception or Dry Season Intensity (DSI) in the year of conception. The pattern is more complex. Mothers with surviving calves took longer to their next conception in good-average years as well as in successive drought years than did mothers who experienced a single poor year (ANOVA: $F_{2,771} = 5.68$, $p = 0.001$, post hoc test: no dry = dry both years; see figure 6.4). Mothers may lengthen the period of allocation during good times to enhance calf growth and subsequent reproductive potential (see below), while they may be forced into prolonged lactational investment when poor times are very extended, simply in order to ensure calf survival (see also Lee, Majluf, and Gordon 1991). When either of the first two years of life was dry, the time to conception was shorter (see figure 6.4), suggesting that mothers attempt to limit investment when they are energetically constrained themselves, with the potential risk of a calf death.

In order to explore conception patterns on an annual basis, we need to estimate the number of females alive and potentially available to conceive in a given year. Each female, however, experiences a different probability of conception by month, by year, and as a result of her allocation of care to any existing calf over a 4 to 5 year period. Because it is problematic to determine annual probabilities of every individual female's availability to conceive with a sample of 654 individual mothers and 1,550 calves over 35 years, and given the 22-month gestation period and long and variable inter-calf intervals, we have calculated a relatively simple index of the number of females available to conceive (e.g., Moss 2001). In the first instance, all females 9 years or older were considered "available" in a given year when they survived through an annual cycle (only 4 females have conceived under the age of 9 out of 367, and therefore we have considered available females as those 9 and above). If they conceived, they then became unavailable for the next 22 months. They were, however, considered available up to the point of conception. After a birth, females were considered unavailable until the death of their calf if it died in the first year of life, or for 12 months due to lactational anestrus. While the latter will underestimate the number of females unavailable due to lactation by 6 to 11 months, the median duration of lactational anestrus was 31 months, and only 1.3 percent ($N = 10$) of females conceived again within 12 months of their calf's birth. Rates of conception were calculated within an Amboseli Year and expressed as a percentage of the number of adult females available in that conception year.

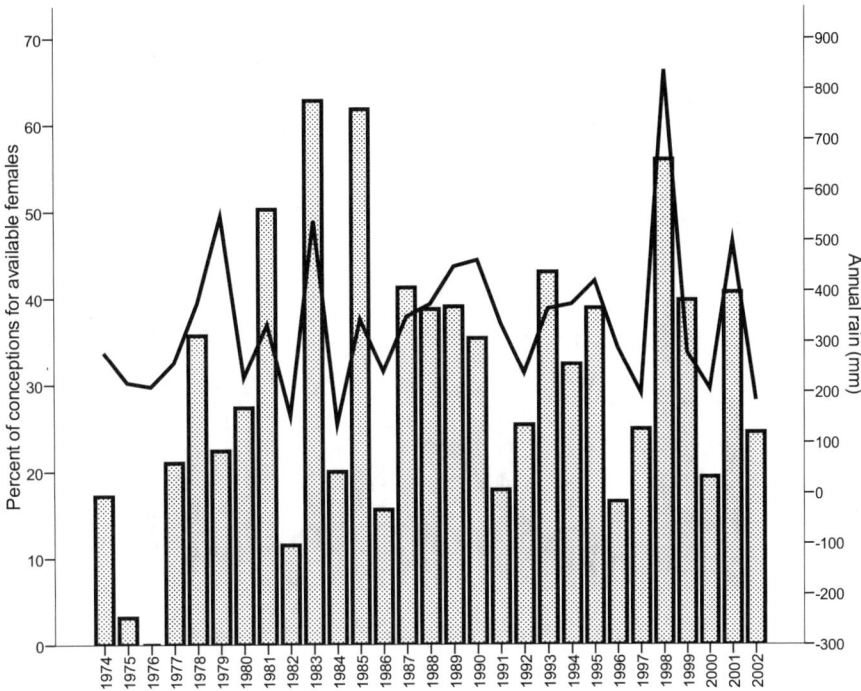

Figure 6.5 Annual conceptions as a percentage of available females by year. Total rainfall for the year (line) is also shown.

Conceptions as a percentage of available females were strongly associated with rainfall and calf mortality in the Amboseli Year (stepwise multiple regression, $r^2 = 0.536$, df = 28, $p < 0.001$). The DSI, total population size, number of adult females, and rate of female population increase were unrelated to conception rate in any year (all NS). There is little density dependence in annual conception rates, but strong environmental control on calf survival, female condition, and estrus (figure 6.5).

Seasonality of Death

Over the period of 1972 to end of 2003, 1,551 calves were born, and of these, 229 died in their first 12 months of life. An additional 64 calves died in their second year. Over 50 percent of these calf deaths occurred in the first 6 months after birth, and the major predictor of whether or not an individual calf died in its first year of life was DSI (logistic regression, natural mortality only, Wald = 10.39, df = 1, p = 0.001). Calves that experienced a drought or prolonged dry season in their first year of life were 70 percent more likely to die (20.3 percent) than were those who experienced moderate or good years (14.2 percent). The DSI also predicted mortality for individual calves within their first 24 months of life (logistic regression, natural mortality only, DSI; Wald = 7.75, df = 1, p = 0.005; figure 6.6). The effect of a high DSI on early mortality over successive years was particularly striking.

Most calf mortality in the first 12 months was due to natural causes, with only a small proportion due to human actions either affecting the family or the mother (N = 17). Droughts may, however, have exacerbated human pressure on elephants, increasing contacts that lead to deaths of adults and calves (see Browne-Nunez, chapter 19). The logistic environmental model with DSI correctly predicted 80 to 86 percent of deaths (figure 6.7), but a number of other factors also influenced the probability of calf death. For example, calves born to primiparous females and into families with few allomothers had a higher risk of death (Moss and Lee, chapter 12). Population density was a further potential influence on early mortality, explored following using annual mortality rates.

Annual rates of early mortality (see also box 6.1), expressed as the percentage of individuals born in each Amboseli Year who died of natural causes within their first 12 or subsequent 12 months until 24 months of age, demonstrated a similar trend to the individual risk of death demonstrated before. Annual rates of first- and second-year mortality due to natural causes were related to DSI (stepwise multiple regression, $r^2 = 0.171$, $F_{1,28} = 6.8$, $p = 0.015$). Annual mortality rates for calves born to experienced mothers were correlated with those for calves of primiparous mothers (r = 0.712, N = 28, $p < 0.001$), suggesting that

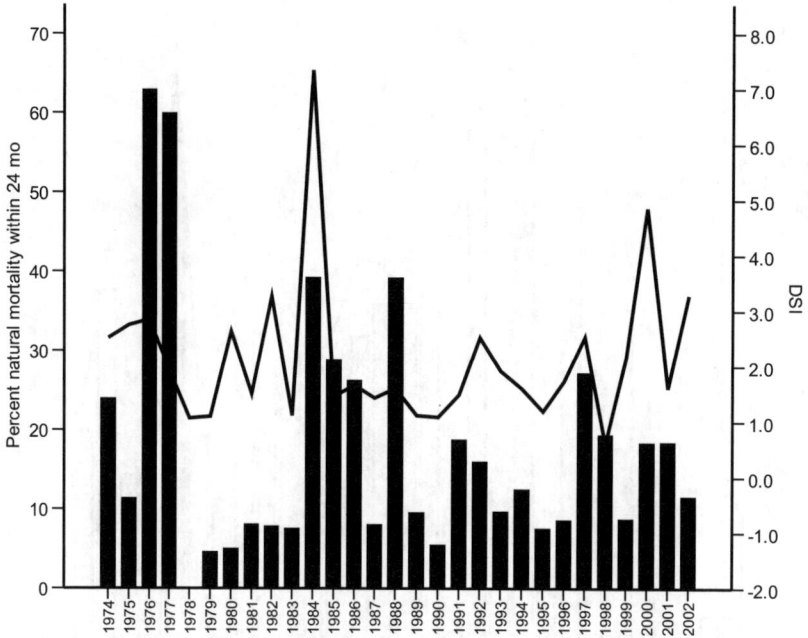

Figure 6.6 Annual rates of natural calf mortality in the first 24 months of life plotted with Dry Season Intensity Index (DSI).

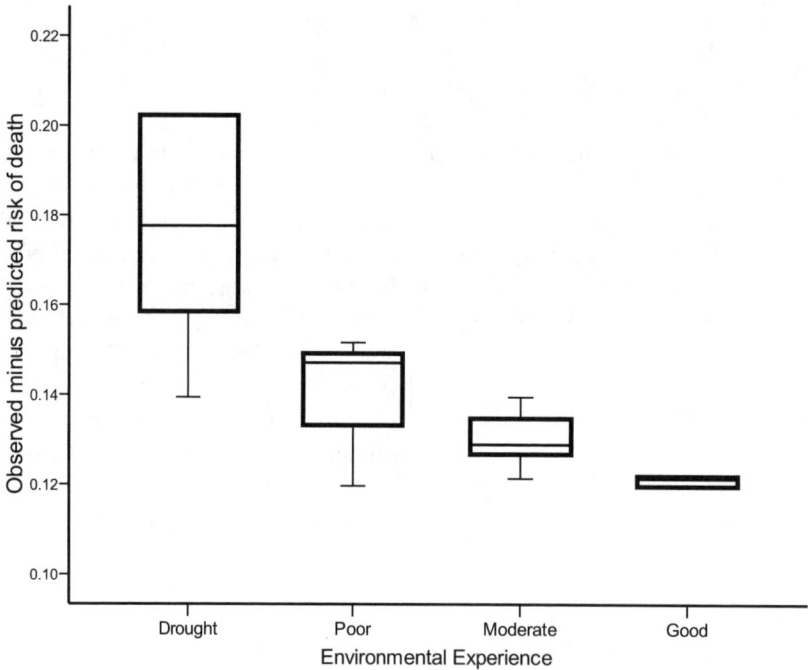

Figure 6.7 Median, range, and 95% CI of risk of mortality predicted from the logistic regression of DSI against death for calves in their first two years of life, plotted as a function of their ecological experience in the first year after birth.

environmental factors acted similarly on calf mortality irrespective of maternal experience.

Although early mortality was most strongly associated with environmental factors, there was some effect of population size. Annual rates of early mortality were related to population size of females during the period from 1972 to 1991, when elephants concentrated within the habitats of the central basin (controlling for DSI: $r = 0.52$, $N = 9$, $p = 0.051$). After the elephants dispersed to use larger areas of the ecosystem, effectively reducing density per unit area,

BOX 6.1 LIFE TABLES

A. Male life table (natural mortality only) for age at end of year (time)

Time	N Enter	N Withdrawn	N Exposed to risk	N Deaths	Probability of death q_x	Proportion surviving l_x	Cum survival	Probability density	Hazard M_x	5-year average l_x	5-year average M_x	Average life expectancy
0	847	37	828.5	106	0.128	0.872	0.872	0.128	0.137	0.872	0.137	37.4
1	704	42	683	32	0.047	0.953	0.831	0.041	0.048			
2	630	28	616	16	0.026	0.974	0.810	0.022	0.026			
3	586	18	577	7	0.012	0.988	0.800	0.010	0.012			
4	561	41	540.5	9	0.017	0.983	0.787	0.013	0.017			
5	511	45	488.5	1	0.002	0.998	0.785	0.002	0.002	0.979	0.021	41.0
6	465	16	457	8	0.018	0.983	0.771	0.014	0.018			
7	441	13	434.5	3	0.007	0.993	0.766	0.005	0.007			
8	425	18	416	5	0.012	0.988	0.757	0.009	0.012			
9	402	29	387.5	4	0.010	0.990	0.749	0.008	0.010			
10	369	25	356.5	0	0.000	1.000	0.749	0.000	0.000	0.991	0.009	37.8
11	344	9	339.5	1	0.003	0.997	0.747	0.002	0.003			
12	334	15	326.5	2	0.006	0.994	0.742	0.005	0.006			
13	317	22	306	1	0.003	0.997	0.740	0.002	0.003			
14	294	21	283.5	2	0.007	0.993	0.734	0.005	0.007			
15	271	12	265	4	0.015	0.985	0.723	0.011	0.015	0.993	0.007	33.5
16	255	13	248.5	3	0.012	0.988	0.715	0.009	0.012			
17	239	24	227	3	0.013	0.987	0.705	0.009	0.013			
18	212	6	209	3	0.014	0.986	0.695	0.010	0.015			
19	203	17	194.5	3	0.015	0.985	0.684	0.011	0.016			
20	183	15	175.5	2	0.011	0.989	0.677	0.008	0.012	0.987	0.013	31.1
21	166	17	157.5	0	0.000	1.000	0.677	0.000	0.000			
22	149	11	143.5	1	0.007	0.993	0.672	0.005	0.007			
23	137	4	135	3	0.022	0.978	0.657	0.015	0.023			
24	130	10	125	4	0.032	0.968	0.636	0.021	0.033			
25	116	17	107.5	1	0.009	0.991	0.630	0.006	0.009	0.986	0.014	28.1
26	98	2	97	1	0.010	0.990	0.623	0.007	0.010			
27	95	0	95	0	0.000	1.000	0.623	0.000	0.000			
28	95	4	93	2	0.022	0.979	0.610	0.013	0.022			
29	89	1	88.5	2	0.023	0.977	0.596	0.014	0.023			
30	86	3	84.5	1	0.012	0.988	0.589	0.007	0.012	0.987	0.013	24.5
31	82	3	80.5	1	0.012	0.988	0.582	0.007	0.013			
32	78	3	76.5	1	0.013	0.987	0.574	0.008	0.013			
33	74	3	72.5	1	0.014	0.986	0.566	0.008	0.014			
34	70	5	67.5	1	0.015	0.985	0.558	0.008	0.015			
35	64	5	61.5	0	0.000	1.000	0.558	0.000	0.000	0.989	0.011	21.0

(continued)

BOX 6.1 (continued)

A. Male life table (continued)

Time	N Enter	N Withdrawn	N Exposed to risk	N Deaths	Probability of death q_x	Proportion surviving l_x	Cum survival	Probability density	Hazard M_x	5-year average l_x	5-year average M_x	Average life expectancy
36	59	2	58	0	0.000	1.000	0.558	0.000	0.000			
37	57	7	53.5	1	0.019	0.981	0.548	0.010	0.019			
38	49	4	47	0	0.000	1.000	0.548	0.000	0.000			
39	45	7	41.5	2	0.048	0.952	0.521	0.026	0.049			
40	36	3	34.5	0	0.000	1.000	0.521	0.000	0.000	0.987	0.014	17.1
41	33	1	32.5	0	0.000	1.000	0.521	0.000	0.000			
42	32	4	30	2	0.067	0.933	0.486	0.035	0.069			
43	26	6	23	0	0.000	1.000	0.486	0.000	0.000			
44	20	4	18	0	0.000	1.000	0.486	0.000	0.000			
45	16	2	15	0	0.000	1.000	0.486	0.000	0.000	0.987	0.014	13.1
46	14	2	13	1	0.077	0.923	0.449	0.037	0.080			
47	11	1	10.5	0	0.000	1.000	0.449	0.000	0.000			
48	10	1	9.5	1	0.105	0.895	0.402	0.047	0.111			
49	8	0	8	1	0.125	0.875	0.352	0.050	0.133			
50	7	0	7	1	0.143	0.857	0.301	0.050	0.154	0.910	0.096	11.6
51	6	0	6	0	0.000	1.000	0.301	0.000	0.000			
52	6	0	6	0	0.000	1.000	0.301	0.000	0.000			
53	6	0	6	0	0.000	1.000	0.301	0.000	0.000			
54	6	1	5.5	0	0.000	1.000	0.301	0.000	0.000			
55	5	1	4.5	1	0.222	0.778	0.234	0.067	0.250	0.956	0.050	8.4
56	3	0	3	0	0.000	1.000	0.234	0.000	0.000			
57	3	0	3	0	0.000	1.000	0.234	0.000	0.000			
58	3	0	3	0	0.000	1.000	0.234	0.000	0.000			
59	3	1	2.5	0	0.000	1.000	0.234	0.000	0.000			
60	2	0	2	0	0.000	1.000	0.234	0.000	0.000	1.000	0.000	5.4
61	2	0	2	0	0.000	1.000	0.234	0.000	0.000			
62	2	0	2	0	0.000	1.000	0.234	0.000	0.000			
63	2	0	2	0	0.000	1.000	0.234	0.000	0.000			
64	2	0	2	1	0.500	0.500	0.117	0.117	0.667			
65	1	1	0.5	0	0.000	1.000	0.117	0.000	0.000	0.900	0.133	unknown

B. Female life table (natural mortality only) for age at end of year (time)

Time	N Enter	N Withdrawn	N Exposed to risk	N Deaths	Probability of death qx	Proportion surviving lx	Cum survival	Probability density	Hazard Mx	5-year average lx	5-year average Mx	Average life expectancy
0	947	30	932	70	0.075	0.925	0.925	0.075	0.078	0.925	0.078	46.7
1	847	39	827.5	35	0.042	0.958	0.886	0.039	0.043			

(continued)

B. Female life table (continued)

Time	N Enter	N Withdrawn	N Exposed to risk	N Deaths	Probability of death q_x	Proportion surviving l_x	Cum survival	Probability density	Hazard M_x	5-year average l_x	5-year average M_x	Average life expectancy
2	773	27	759.5	15	0.020	0.980	0.868	0.018	0.020			
3	731	19	721.5	14	0.019	0.981	0.851	0.017	0.020			
4	698	45	675.5	9	0.013	0.987	0.840	0.011	0.013			
5	644	42	623	3	0.005	0.995	0.836	0.004	0.005	0.971	0.030	50.3
6	599	12	593	0	0.000	1.000	0.836	0.000	0.000			
7	587	26	574	0	0.000	1.000	0.836	0.000	0.000			
8	561	28	547	1	0.002	0.998	0.835	0.002	0.002			
9	532	27	518.5	4	0.008	0.992	0.828	0.006	0.008			
10	501	26	488	0	0.000	1.000	0.828	0.000	0.000	0.998	0.002	46.0
11	475	18	466	3	0.006	0.994	0.823	0.005	0.007			
12	454	18	445	2	0.005	0.996	0.819	0.004	0.005			
13	434	31	418.5	0	0.000	1.000	0.819	0.000	0.000			
14	403	16	395	0	0.000	1.000	0.819	0.000	0.000			
15	387	6	384	1	0.003	0.997	0.817	0.002	0.003	0.997	0.003	41.5
16	380	17	371.5	2	0.005	0.995	0.813	0.004	0.005			
17	361	24	349	6	0.017	0.983	0.799	0.014	0.017			
18	331	7	327.5	1	0.003	0.997	0.796	0.002	0.003			
19	323	23	311.5	2	0.006	0.994	0.791	0.005	0.006			
20	298	9	293.5	0	0.000	1.000	0.791	0.000	0.000	0.994	0.006	37.9
21	289	26	276	0	0.000	1.000	0.791	0.000	0.000			
22	263	12	257	0	0.000	1.000	0.791	0.000	0.000			
23	251	11	245.5	1	0.004	0.996	0.788	0.003	0.004			
24	239	23	227.5	0	0.000	1.000	0.788	0.000	0.000			
25	216	22	205	1	0.005	0.995	0.784	0.004	0.005	0.998	0.002	33.1
26	193	1	192.5	1	0.005	0.995	0.780	0.004	0.005			
27	191	1	190.5	2	0.011	0.990	0.772	0.008	0.011			
28	188	4	186	1	0.005	0.995	0.768	0.004	0.005			
29	183	3	181.5	0	0.000	1.000	0.768	0.000	0.000			
30	180	4	178	0	0.000	1.000	0.768	0.000	0.000	0.996	0.004	28.2
31	176	3	174.5	0	0.000	1.000	0.768	0.000	0.000			
32	173	9	168.5	1	0.006	0.994	0.763	0.005	0.006			
33	163	4	161	2	0.012	0.988	0.754	0.010	0.013			
34	157	14	150	4	0.027	0.973	0.733	0.020	0.027			
35	139	13	132.5	2	0.015	0.985	0.722	0.011	0.015	0.988	0.012	24.9
36	124	17	115.5	0	0.000	1.000	0.722	0.000	0.000			
37	107	9	102.5	1	0.010	0.990	0.715	0.007	0.010			
38	97	9	92.5	0	0.000	1.000	0.715	0.000	0.000			
39	88	5	85.5	2	0.023	0.977	0.699	0.017	0.024			
40	81	11	75.5	0	0.000	1.000	0.699	0.000	0.000	0.993	0.007	20.9
41	70	9	65.5	3	0.046	0.954	0.667	0.032	0.047			

(continued)

BOX 6.1 (continued)

B. Female life table (continued)

Time	N Enter	N Withdrawn	N Exposed to risk	N Deaths	Probability of death q_x	Proportion surviving l_x	Cum survival	Probability density	Hazard M_x	5-year average l_x	5-year average M_x	Average life expectancy
42	58	4	56	2	0.036	0.964	0.643	0.024	0.036			
43	52	3	50.5	0	0.000	1.000	0.643	0.000	0.000			
44	49	3	47.5	1	0.021	0.979	0.629	0.014	0.021			
45	45	0	45	1	0.022	0.978	0.615	0.014	0.023	0.975	0.025	17.9
46	44	0	44	0	0.000	1.000	0.615	0.000	0.000			
47	44	1	43.5	0	0.000	1.000	0.615	0.000	0.000			
48	43	1	42.5	0	0.000	1.000	0.615	0.000	0.000			
49	42	3	40.5	1	0.025	0.975	0.600	0.015	0.025			
50	38	0	38	3	0.079	0.921	0.553	0.047	0.082	0.979	0.021	13.6
51	35	1	34.5	1	0.029	0.971	0.537	0.016	0.029			
52	33	1	32.5	0	0.000	1.000	0.537	0.000	0.000			
53	32	0	32	1	0.031	0.969	0.520	0.017	0.032			
54	31	1	30.5	2	0.066	0.934	0.486	0.034	0.068			
55	28	0	28	0	0.000	1.000	0.486	0.000	0.000	0.975	0.026	11.3
56	28	3	26.5	3	0.113	0.887	0.431	0.055	0.120			
57	22	0	22	1	0.046	0.955	0.411	0.020	0.047			
58	21	1	20.5	1	0.049	0.951	0.391	0.020	0.050			
59	19	2	18	0	0.000	1.000	0.391	0.000	0.000			
60	17	0	17	0	0.000	1.000	0.391	0.000	0.000	0.959	0.043	8.4
61	17	2	16	0	0.000	1.000	0.391	0.000	0.000			
62	15	0	15	1	0.067	0.933	0.365	0.026	0.069			
63	14	4	12	0	0.000	1.000	0.365	0.000	0.000			
64	10	2	9	1	0.111	0.889	0.325	0.041	0.118			
65	7	0	7	0	0.000	1.000	0.325	0.000	0.000	0.964	0.037	4.2
66	7	3	5.5	0	0.000	1.000	0.325	0.000	0.000			
67	4	0	4	0	0.000	1.000	0.325	0.000	0.000			
68	4	1	3.5	2	0.571	0.429	0.139	0.185	0.800			
69	1	1	0.5	0	0.000	1.000	0.139	0.000	0.000	0.886	0.160	unknown

early mortality was no longer related to female population size (controlling for DSI: $r = 0.086$, $N = 9$, $p = 0.79$; figure 6.8). These results suggest that density dependence was operating through food limitation in very dry seasons, when there was high female density. In addition, while annual mortality rates were generally unrelated to density, the absolute number of calves dying in their first year of life was weakly associated with female density ($F_{1,29} = 3.80$, $p = 0.06$) but not with DSI or likelihood of female dispersal.

The power of such density-dependent effects remains relatively small, with female population size explaining only 12.6 percent of the variance in number of calf deaths. There was no apparent increase in calf deaths over that predicted by rainfall and drought between the years of concentration in the Park and years of dispersal (as determined by inspection of regression residual values). Density and drought interact in unpredictable ways, and either habitat buffering (swamp foods or access to water) or range

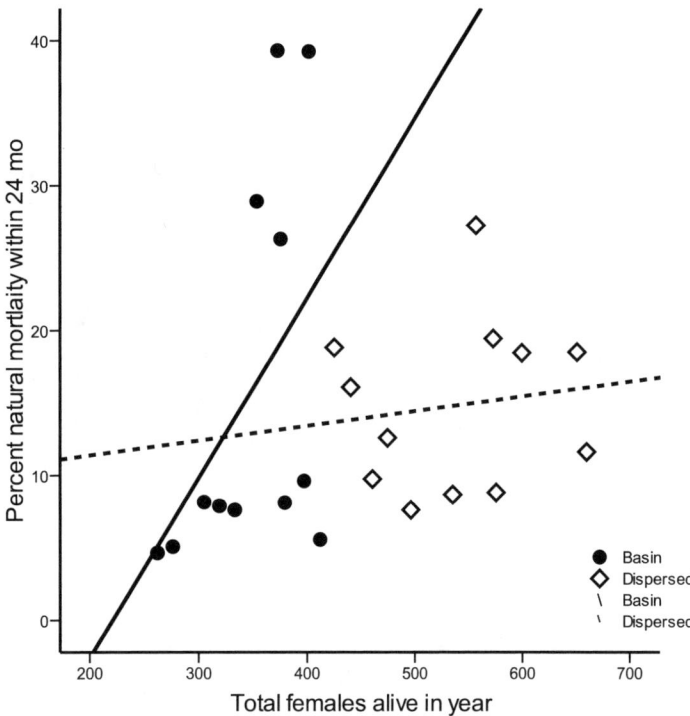

Figure 6.8 Total female population size and rates of natural calf mortality in the first 24 months for the period after population growth began (1979) when elephants focused on using the Park compared with after elephants dispersed throughout the ecosystem.

buffering can play major roles in minimizing environmental and density-dependent mortality (see also Gough and Kerley [2006] for Addo National Park).

Environmental factors impact on timing as well as rates of death (see box 6.1 for lifetables). Calves tend to die in specific months, with most natural mortality in September at the end of the dry season and October prior to grass growth (χ^2 = 85.9, df = 33, p < 0.001), a pattern that was marked in normal years (figure 6.9). There was also a discernable pulse of mortality after the less intense January–February dry period. By contrast, during drought years, calf deaths tend to be more evenly spread, with slight peaks at the start and end of the dry periods. There was no consistent effect of month of birth on number of calves dying in years of differing rainfall and habitat quality, although there was some slight monthly variation in mortality (χ^2 = 18.25, df = 11, P = 0.076, NS). Being born in a particular month did not appear to increase calf vulnerability, despite an expectation of greater mortality among calves born close to the start of a particularly harsh dry season. The late dry season and the first month of rains prior to significant grass growth appeared to be risk periods for calves. The slight mortality peak in March in not dry years was associated with poor short rains (November–December) in years that went on to have reasonable long rains.

Mortality from natural causes in elephants over the age of two was unrelated to the harshness of the year of death: 38.7 percent of deaths occurred in good years, in comparison to 26 percent in poor or drought years (N = 218). By contrast, 51.8 percent of non-calf elephant mortality due to humans (N = 313) occurred in poor or drought years (χ^2 = 14.1, df = 3, p = 0.003), again suggesting that ecological pressures may have increased the intensity of competition and contact between elephants and humans, leading to higher risks for elephants in poor times.

Long-Term Effects of Ecological Variation

Early Experience and Growth

What are the consequences of being "born in a bad year"? Having shown that early drought experience is a risk factor for death, we now ask whether it has persisting effects on the growth, maturation, and reproduction of elephants.

Calves experiencing dry conditions in their first two years of life were smaller for their age (sex-specific size-for-age residuals from the asymptotic growth curve) in comparison to non-drought calves (relative height, $F_{1,53}$ = 3.3, p = 0.045; relative foot $F_{1,74}$ = 3.7, p = 0.058), and small calves were also more likely to die in their first two years of life (logistic regression died versus relative foot: Wald = 6.29,

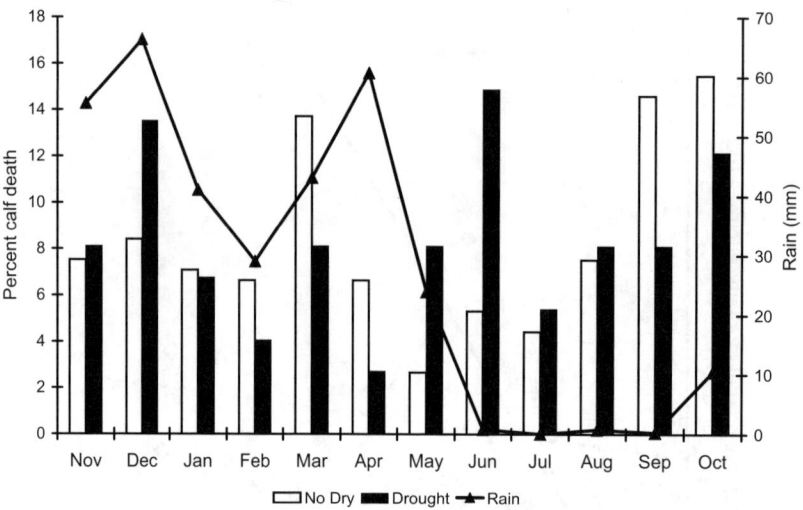

Figure 6.9 Percentage of deaths due to natural causes in each month for calves dying under the age of 24 months, in drought and not dry years (bars) with average monthly rainfall (line).

df = 1, p = 0.012; relative height: Wald = 4.87, df = 1, p = 0.027). Individuals who experienced an early dry period but survived appeared to incur growth faltering, possibly due to food shortages that limited maternal milk production or to increased energy expenditure through additional travel and foraging costs. As a result, for at least the first 10 years of life, drought-impacted individuals tended to remain smaller for their age relative to the size predicted by sex-specific growth curves (relative foot length $F_{3,231}$ = 3.46, p = 0.064; relative height $F_{3,231}$ = 8.4, p = 0.004). Small survivors of bad years also tended to have a delayed age at first reproduction if female (foot length and first reproduction $F_{1,131}$ = 4.34, p = 0.039) and a later age at first musth if male (foot length and first musth: $F_{1,45}$ = 9.88, p = 0.003; figure 6.10).

Even though individuals escaped the initial mortality risk due to drought early in life, some of these "survivors" may have been less able to cope in the long term, especially when they were smaller than expected for their age. At this stage, we cannot directly test the association between growth faltering and lifetime mortality risks, but we do find tendencies for females who experienced two dry years at the start of life and survived to have a reduced mean life expectancy (28.8±0.90 years) by comparison to females who experienced no dry in their first two years (32.9±0.77; Cox regression: L = 3.25, p = 0.07). Differences between males were less marked (29.95±1.23 no dry year vs 27.71±1.07 dry both years). It is particularly striking that the cohorts born into the "worst" years for food availability (1975–76, 1983–84, and 1999–2000) suffered a reduction of mean longevity of 20 years for females and 18.7 years for males. Evidently, early experience of particularly bad years influences hazard rates of mortality and thus long-term survival prospects for drought calves (figure 6.11).

Early stunting or temporary growth faltering could increase the susceptibility of animals to diseases, as is the case for humans (Barker 1994). Alternatively, the costs of catch-up growth over a period that can last for 20 years may be energetically unsustainable if a succession of bad years is encountered. Smaller individuals are socially more subordinate and thus may suffer higher foraging costs (Archie et al. 2006). However, whether long-term mortality risks are mediated socially or by physical factors such as growth, condition, or disease remains to be determined.

Age-Structure Pulses in the Population

Another long-term consequence of environmental influences on conceptions and mortality may be the appearance of exaggerated population pulses (figure 6.12), with brief, exceptional rates of population increase if females become synchronized in their conceptions and births. Instantaneous samples of a population and calculations of reproductive rates can thus be very problematic. Short-term analysis of population changes may provide a misleading picture of age-structures and population dynamics during both increasing and declining phases, reporting trends that may not, in fact, be sustained over the longer term (see also Gough and Kerley [2006]). Cyclical peaks and troughs in population growth rates resulting from an ecologically entrained periodicity to births and deaths tend to make elephant populations appear to be declining or increasing at unsustainable rates. These pulses may be artifacts of a short-term "boom-bust" cycle when there is ecologically mediated synchrony to demographic events. An understanding of the sensitivity and responses of elephant populations to ecological variation, therefore, is vital in the interpretation of their future viability.

Ecological Patterns of Variability in Demographic Rates 87

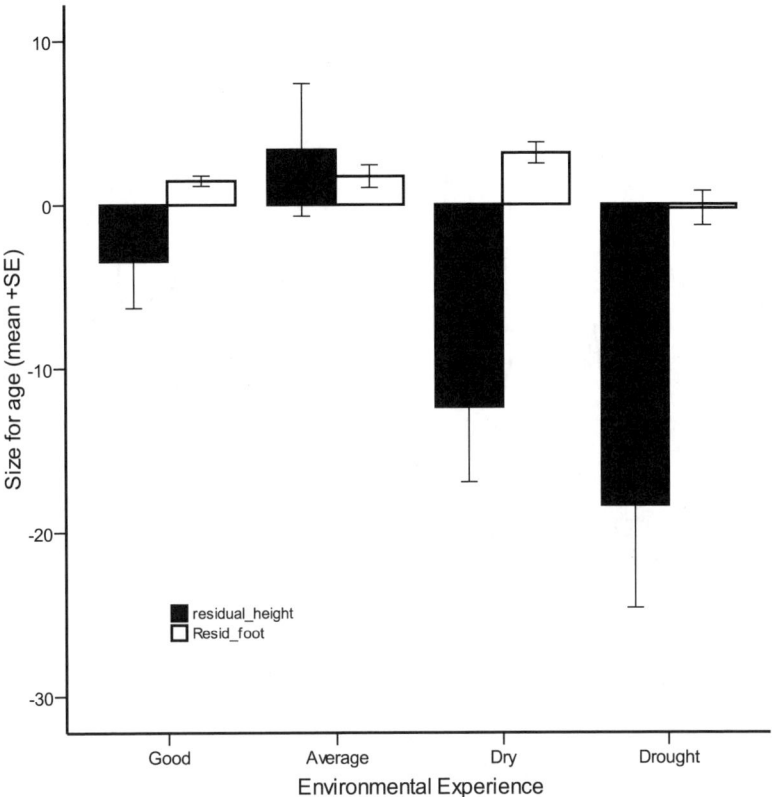

Figure 6.10 Effects of early environmental conditions on growth up to 10 years of age. Foot length and shoulder height residuals from the age- and sex-specific asymptotic growth curve (*N* males = 119 height, 134 foot; females = 112 height, 102 foot).

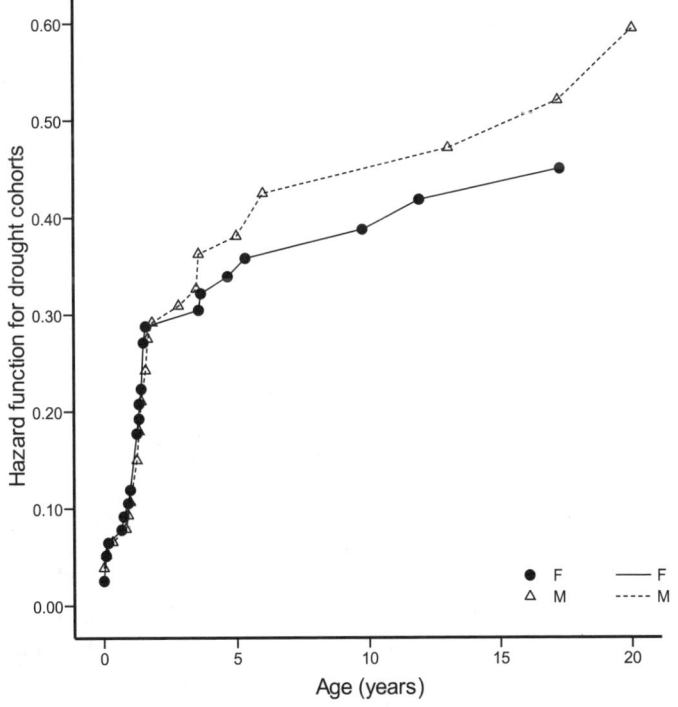

Figure 6.11 Hazard rates for male and female mortality at each age in three extreme drought cohorts (*N* males = 79, females = 80). Lowest curve fit.

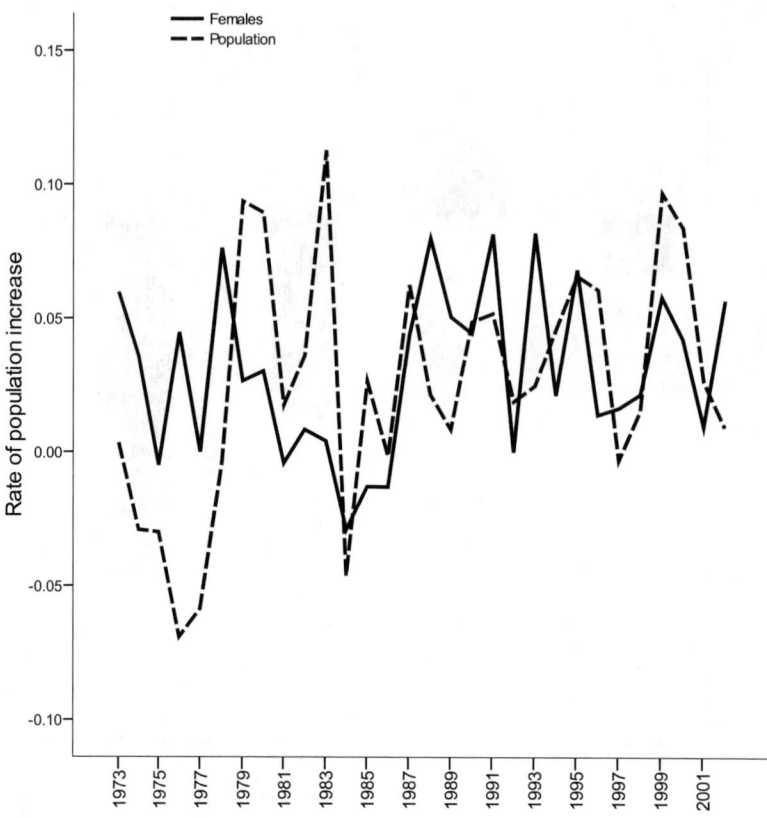

Figure 6.12 Instantaneous annual rate of increase of the total population and the adult female population over the 30 years of the study.

Conclusions

In this chapter, we have described marked seasonal and annual variation in the rates of conception and calf mortality. As expected from previous studies of ungulates and elephants, two key factors affected the birth and death rates of Amboseli elephants. These factors were the available resources and associated individual energy balance—mediated by the effect of rainfall on food abundance and quality—and the ecological and social effects of population size and composition. When resources (food and water) were scarce, widely distributed, and hard to process during dry seasons and drought years, few females gained sufficient condition to enter estrus; they may have expended their existing and stored energy locating and processing nutrients, leaving insufficient energy to devote to successful reproduction. When times were tough, few females were able to conceive, and mortality was high for the calves born into bad years.

The severity or richness of the year, however, did not explain all early mortality, and there were some effects of population density on mortality. When density was low, calf survival was determined by local environmental and social conditions, such as the age of the mother and the availability of allomothers. However, after controlling for resource limitations, more calves were likely to die during periods of high density. Despite density effects on mortality, fecundity appeared unaffected by population density. As has been documented for many large ungulates (Clutton-Brock, Iason, and Guinness. 1987; Coltman et al. 1999b; Milner et al. 1999; Festa-Bianchet, Jorgenson, and Réale 2000), environmental factors influence demography via rate of reproduction. Density dependent effects operate most strongly to increase mortality when populations are restricted from dispersal. Elephant populations thus are highly responsive to annual variability in resource availability and the competitive pressures of other elephants in the context of constraints on dry season food supplies.

The long-term consequences of environmental variation on individual survival and success and on population dynamics are only beginning to be understood, despite over 30 years of data. At this stage, we suggest that, at an individual level, the persisting effects of the experience of droughts in early life—acting potentially to reduce life expectancy, delay reproductive potential, and limit growth—could ultimately feed into population dynamics. Such effects require modeling over the very long term of an elephant's 40-plus-year reproductive life span in order to assess the consequences of such factors for populations.

Chapter 7 Patterns of Occupancy in Time and Space

Harvey Croze and Cynthia J. Moss

Introduction

Knowing where elephants go is at least as important as understanding what they do, particularly in the Amboseli ecosystem. The National Park is the core of the range of the current populations of wildlife species—elephants, wildebeest, and zebra, in particular—that regularly move in and out of the Park. As shown in chapter 2, the ecosystem the Amboseli elephants use is roughly 20 times bigger than the Park. The elephants' future depends on us knowing where and when areas outside the Park are used and needed so that we can make "arrangements" to assure them right of access.

In this chapter, we examine the ways that various social groupings have occupied the Amboseli ecosystem over time. We look at inter-annual changes in occupancy of the central basin (defined in chapter 2) and the differential, coordinated occupancy by various families and bond groups within the basin. We also record what is currently known about the spatial requirements for elephant corridors to and from the central basin using several direct and indirect techniques (see appendix 1).

Specifically, we look at the following:

- Ecosystem occupancy and changes in population use of the Park over time.
- Differences among families in occupancy of the Amboseli basin.
- Shared occupancy by social groups.
- Difference in occupancy between males and families.
- Requirements for "corridors" outside the Park.

The behavioral responses of families moving in and out of the basin and the decision making involved in such movements are dealt with by Mutinda, Poole, and Moss (chapter 16). The history of changes in the population size from around 600 to over 1,500 over three decades is presented as an annex to chapter 2.

Occupancy at the Ecosystem Level

The Amboseli elephants occupy a range that far outreaches the 392 km^2 confines of the gazetted National Park (see chapter 2). This crucial fact weighs heavily on the future of the population: if access to the extended range were denied, the population would be certainly doomed, as would most other large herbivores. For this reason, it is important to try to map the extent and relative occupancy with some precision in order to serve as a basis for discussion and negotiation for living space and access corridors for the elephants.

Elephant Range in the Ecosystem

From aerial surveys and radio tracking (see appendix 1) of six individual females in the 1970s, Western and Lindsay (1984) concluded with laudable precision that the total range of the population by 1976 was 3,588 km^2. That figure might have been a slight underestimate for the time. The aerial survey spotted elephants some 15 km outside of the putative range boundary that were not included in the range estimation. The elephants were deduced to have come from Tsavo West National Park on the basis of their

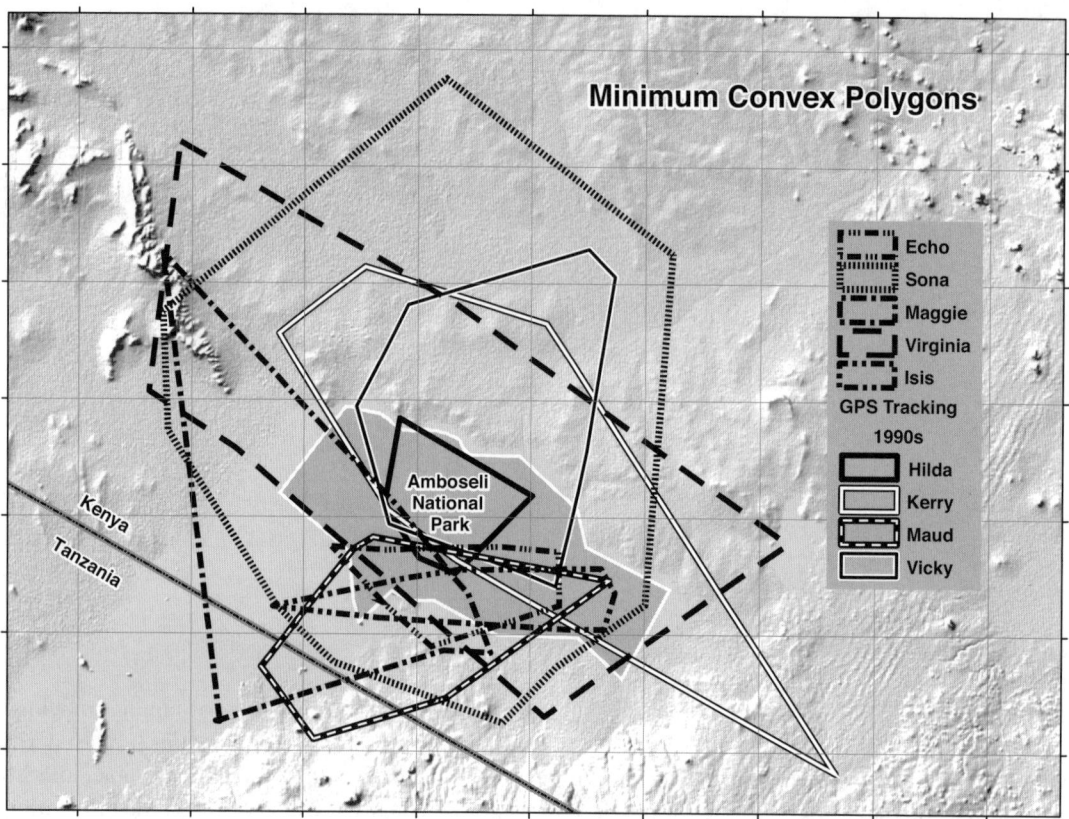

Figure 7.1 Minimum convex polygons of seven families in the 1970s (dotted) and 1990s (solid)—EB (Echo), SB (Sona), MB (Maggie, Maud), VA (Virginia, Vicky), IB (Isis), HA (Hilda), and KA (Kerry).

bunching alarm reaction to the passing aircraft, atypical of the relatively calm Amboseli elephants. However, Amboseli elephants also behave nervously when far from or recently returned to the safety of the National Park, so it is difficult to be certain of which population those elephants belonged.

With regard to putting tracking collars on elephants, the project had set a deliberate policy to limit the use of remote telemetry monitoring (see appendix 1) due to the potential trauma to the elephants and the breach of trust inherent in an immobilization exercise. Thus, over the years, only seven families had been "marked" by having a matriarch or one of the large females fitted with a radio collar.

The various radio and satellite collaring exercises over the years (1974–79, 1996–99) also provided a composite picture of range extent by means of drawing the minimum convex polygons (MCP) (Mohr 1947) of the tracking fixes (figure 7.1). The mean MCP of the seven families with collars was 570 km^2 (rounded to 5 km^2; range 95 to 1,690). The mean includes two polygons each from different females from two families in the two periods in the 1970s and 1990s, the VAs (Virginia and Vicky) and the MBs (Maggie and Maude). The composite MCP of all seven families was 2,100 km^2.

Conventional radio-tracking data points are not dense enough to analyze changes over time. In 1996–97, two adult bulls, Ganesh and Mr. Nick, fitted with the new generation of GPS satellite collars, showed extensive movements into the ecosystem (see box 7.1). As shown in chapters 5, 17, and 18, males march to their own tune, staying within the population range as dictated by females and food, but moving at different rhythms and timings that take them frequently to the edge of the range, often into harm's way. In 1994, so-called sport hunters, outfitted by two Tanzanian companies, had shot several Amboseli bulls just across the international boundary. They denied that the bulls were from Kenya, but the Amboseli Elephant Research Project (AERP) research team knew quite well that they were cross-border elephants. One of us (CJM) was quoted by the press at the time to observe that "shooting an Amboseli bull is like shooting your neighbor's poodle." Accordingly, AERP invited colleagues from Save the Elephants to test the emerging GPS tracking technology in Amboseli and at the same time to confirm the transborder elephant movements (see box 7.1).

The compilation of sightings from the Maasai scouts from 1999 to 2003 (see appendix 1) gives a composite

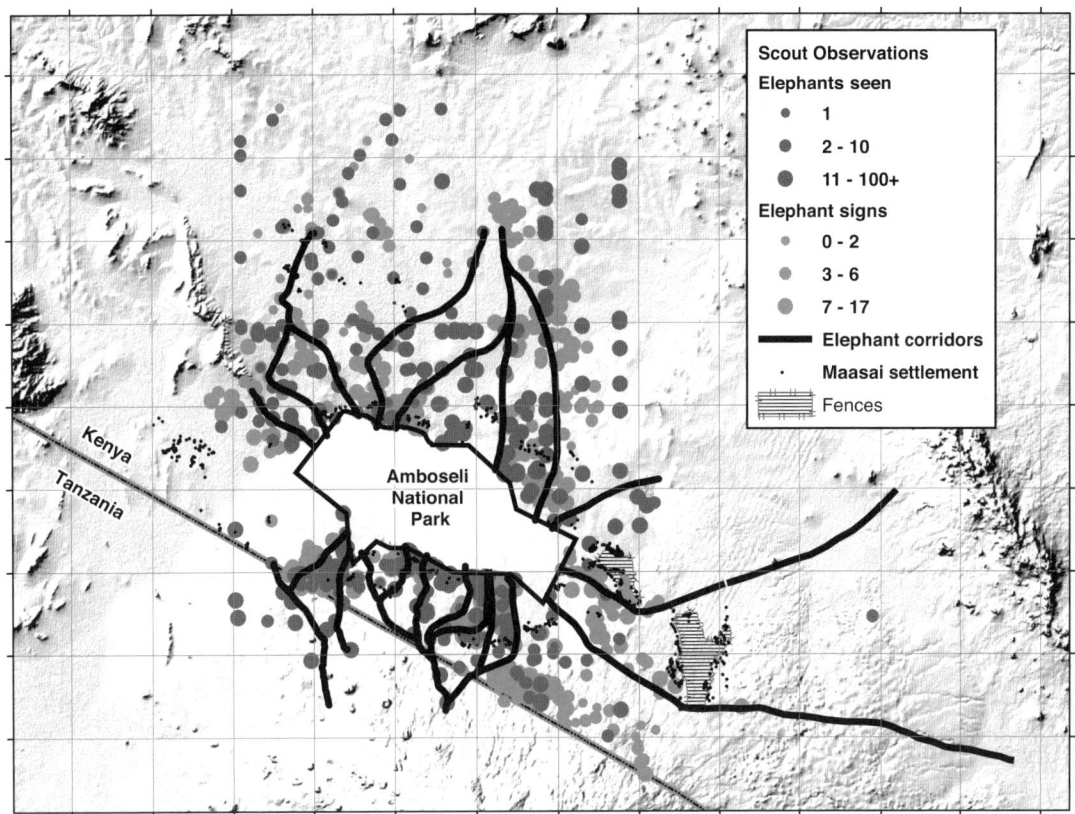

Figure 7.2 Combined ground observations made by 10 Maasai scouts from 1999 to 2003, including actual elephants seen as well as elephant signs (tracks and dung piles).

Note: Also shown are the "familiar paths" taken by Amboseli elephants moving in and out of the Park. The shorter lines can represent either daily or season movements, or both.

picture of average elephant distribution over that period (figure 7.2). The data were not gathered within a systematic sampling frame and thus suffer from lack of knowledge about null observations: does an area contain no elephants because there are no elephants or because there was no observation? Nonetheless, the picture of the extent of the range is considered to be accurate given the general paucity of sightings from other sources beyond the range.

Today's total range of the Amboseli elephant population is illustrated in chapter 2 (figure 2.5), along with that of all other wildlife, as an amalgamation of data from several sources (see appendix 1). Since the view cuts across time and season and combines different sample frames, it should not be used to comment on occupancy between seasons. It can, however, be used to estimate total range, which, when the minimum convex polygon is drawn (also shown on figure 2.5), comes to a range for known Amboseli elephants of some 7,500 km^2 (including approximately 700 km^2 contributed to the MCP by AERP and African Wildlife Foundation [AWF] observations made on the Tanzanian side of the border). Density can thus be estimated at 0.22 elephants per km^2. Adding range information on known individuals from the Tanzanian portion of the ecosystem (A. Kikoti, personal communication, 2006; AWF [2006]), nearly doubles the range to some 11,500 km^2. Overall, then, the Amboseli population has a density of 0.13 elephants per km^2, which is unlikely to amount to more than some 4 percent of the ecosystem large herbivore biomass (see chapter 2). Locally, for example in the National Park or in seasonal aggregations in and around the Park the density can approach one elephant per km^2 for short periods, usually measured in a couple of weeks.

Although there was really no reliable way to tell if the elephants sighted by air southeast of Oloitokitok by Western and Lindsay (1984) were or were not Amboseli elephants, the relatively small range (ca. 3,600 km^2) they calculated is consistent with the change in population size over the years. In 1973–79, as we have seen, the population was around 600. At the start of writing (2005), it stood at 1,450, an increase of some 2.4 times. An apparent 1.8 times increase in range from Western and Lindsay's figure of 3,600 to our later estimate of 6,800 km^2 is commensurate with the growth in population size, especially given the approximate nature of range estimates.

BOX 7.1 RANGING OF BULLS OUTSIDE THE PARK: DATA FROM RADIO AND SATELLITE TRACKING

Iain Douglas-Hamilton

From 1995 to 1998, for the first time in Africa, four bulls were fitted with GPS (global positioning satellite) tracking collars. Their movements were followed by downloading a voluminous stream of data that provided a picture of the variation in individual movement patterns and an addition to knowledge of the geographical extent of the population, particularly into Tanzania. See the map for the pooled total of bull sightings (figure A).

In August 1995, a prototype collar was put on Solanga (M207), a then 30-year-old western bull known to wander across the international border. Although the GPS element failed after 10 days, the VHF radio beacon continued to transmit, providing 62 fixes over the course of 1996. The timings and locations of 10 sightings in Tanzania indicated Solanga made at least five separate forays across the border during the year.

Another circa 30-year-old bull, Parsitau (M162), was fitted with a conventional VHF radio collar in April 1996. By May 1998, he had yielded only 24 sightings. Fifteen of the records were in Tanzania. The time pattern also indicated five separate

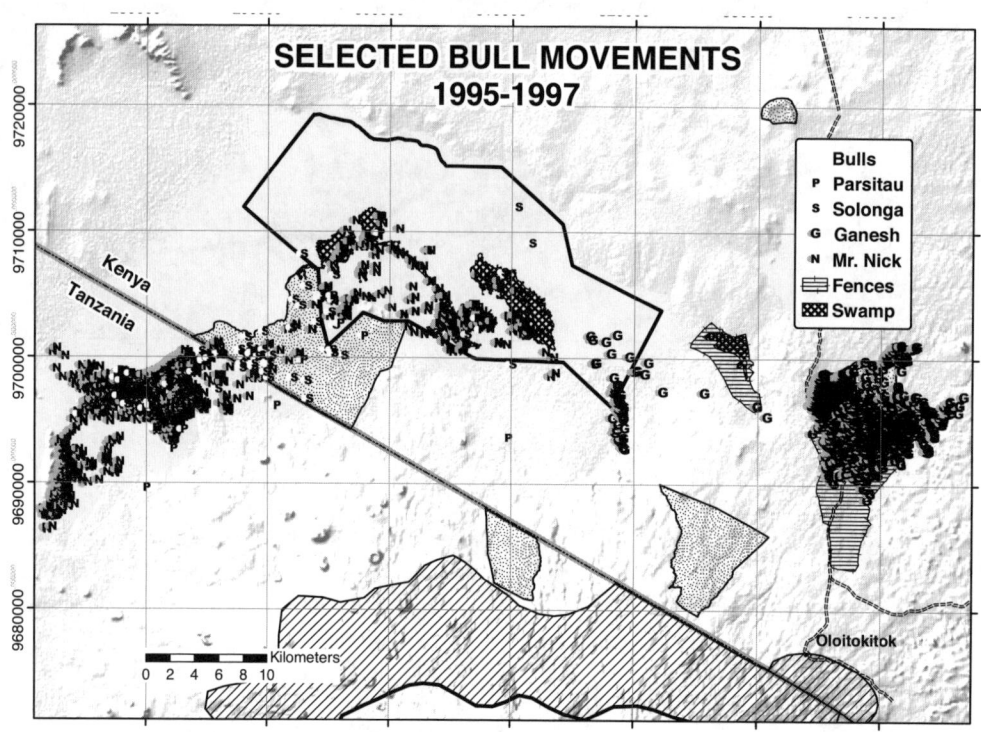

Box 7.1, figure A Total sightings of four adult bulls tracked by conventional VHF and satellite GPS technology in 1995–97 (see text).

Note: See figure 2.1 for place-names.

transboundary stays (unconnected to those of Solanga). One sojourn apparently lasted from October 1996 to March 1997, but given the low number of fixes, he may well have crossed the border more than once during the six-month period.

Both Solanga and Parsitau continued to frequent the region where the other bulls had been shot in Tanzania two years earlier. It was surprising that after such a relatively short time, bulls were still visiting a high-risk part of their range. A third bull, Discovery, was tracked in the same period in the East of the range from Kimana to the Chyulu Hills, establishing a connection that could link Amboseli Bulls with the Tsavo population.

In the first week of December 1996, we immobilized two bulls with the help of Kenya Wildlife Service (KWS) veterinarians. The individuals were selected on the basis of their tendency to disappear from the core study area for prolonged periods. Mr. Nick (M86) was about 38 years old at the time and was believed to cross the border into Tanzania to the west. The other, Ganesh (M169), was a pre-musth bull of some 33 years, suspected to be a crop raider in the Kimana area to the east, and KWS was interested to know more about his fine-grain movement patterns.

Over the course of the following months, data were gathered from light aircraft by locating the current position of the elephant and then downloading through a VHF modem while circling overhead (see Douglas-Hamilton [1998] and Douglas-Hamilton, Krink, and Vollrath [2005] for details of the technique). On average, each elephant was located 20 times a day; data were downloaded every two weeks. Mr. Nick was tracked for four-and-a-half months before the collar dropped off.

Over the whole period Mr. Nick averaged 10.3 km per day (range: 15–29) and he had a total range for the period of 210 km^2. He was in early musth when immobilized and consequently roamed long distances. He made long (ca. 10 km) sweeping arcs along the Enkong'u Narok swamp between Longolong and Sinet. He then crossed further west and looped back round northeast past Lake Conch in a circle for three days running. His fast patrolling within the Park appeared to be in search of females in estrus. On December 18, Mr. Nick made a quick three-day reconnaissance deep into Tanzania and then came back into the Park to resume patrolling.

Later, on the nights of December 24–25, he crossed quickly and directly again into Tanzania along a corridor frequently used by family groups between the two countries. He then remained in Tanzania, settling into a pattern of back and forth movements: 10–15 km in one direction one day and then back the next, evidently patrolling for females. Subsequently, he meandered about apparently feeding in a well-wooded area within the Longido Game Controlled Area. He averaged 10 km a day with occasional bursts of directional movement. He was then often seen from the air in the company of other bulls, some of whom were identified as Amboseli bulls by Katito Sayialel, who could recognize them from the air. By then Mr. Nick had presumably come out of musth (there were no ground records during the period) and was in his seasonal bull holding area. His movements remained slow for the next three months until the collar finally dropped off on April 21, 1997, after 136 days.

Ganesh, the pre-musth bull, was immobilized in the east of the National Park on December 5, 1996. After lingering for one day, he made an extended and rapid nighttime march to the Kimana swamp to the east. He covered 20 km in 12 hours. Once Ganesh reached the relative safety of the Kimana Wildlife Sanctuary, he took up a daily abode in its thick swamp-edge *Acacia xanthophloea* woodland and settled down to an average rate of 8.6 km per day. The GPS tracking allowed precise logging of Ganesh's movements: he sallied forth repeatedly southward at night to raid crops in the nearby fields and then pulled back at dawn to the safety of the dense woodland.

In total, during the 5.5-month tracking period Ganesh used a range of 140 km^2. This is unlikely to be the full extent of his range, as a full year or more is usually needed before the range of an elephant can be defined from tracking data.

Both Mr. Nick and Ganesh were well-known, often-observed occupants of Amboseli National Park, but during the tracking exercise neither spent much time there. Although Mr. Nick intensely patrolled for estrus females within the Park, he actually spent 92 percent of the period in Tanzania's Longido Game Controlled Area, 27 km to the west of the Park at his farthest reach. Ganesh spent virtually all of the tracking period 15 km from the Park in or around the Kimana Wildlife Sanctuary. Kimana is a major ecosystem attractor for elephants, both bulls and family groups, who tend to spend the day sheltering in dense vegetation of the sanctuary, thereby avoiding the front of cultivation that has expanded from the south into the ecosystem. The sanctuary is also a staging ground for nighttime crop raiding as appears in other areas where cultivation invades elephant range (Graham et al. 2009).

Although Ganesh's diurnal activity changes evinced a clear awareness of the risks of crop raiding (Browne-Nuñez, chapter 19; Kangwana, chapter 20), his movements indicated that he was quite at home in that part of the ecosystem. The "streaking" pattern (Douglas-Hamilton, Krink, and Vollrath 2005) appeared to be typical of unconstrained elephants moving from one part of their range to another. Similar localized movements, punctuated by longer, quicker forays (so-called "Lévy walks"), have been observed in many animal movement pattern studies (Mueller and Fagan 2008). The function is not always clear (Edwards 2008), but in the context of African elephants, it is usually done at night when the animals cross a known danger zone (Graham et al. 2009).

The tracking studies confirm the extent to which Amboseli elephants depend on unprotected areas well beyond the Park boundaries and reinforce the need for far-reaching management plans that must include the whole ecosystem as defined by the range of the elephants (Croze, Moss, and Lindsay, chapter 22). The data from Mr. Nick in particular support the decision of the Tanzanian Government to close elephant hunting in the Longido Game Controlled Area as an activity incompatible with the objectives of the neighboring Kenyan protected areas and wildlife concessions.

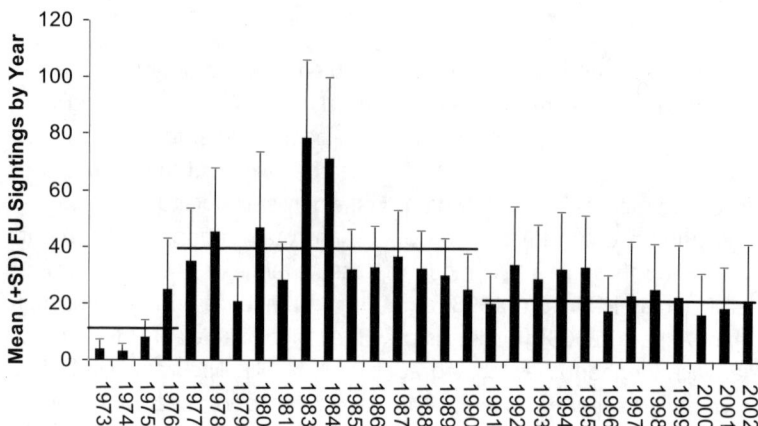

Figure 7.3 The mean (±SD) number of times that individual families were sighted each year.

Note: The overall mean for the period prior to the gazetting of the National Park is compared (horizontal lines) with those during periods of high occupancy, and then with a period in the early 1990s when family occupancy of the central basin decreased, despite an increase in observer effort (see appendix 1).

Changes over Time: Inter-annual and Seasonal

One benefit of a long-term study is to be able to monitor changes in occupancy over time and space. The study perforce concentrated effort in the basin, since the primary object was to find elephant families, often particular ones, and to observe and record their demography and social interactions. Apart from a handful of aerial surveys, limited project resources precluded a program of systematic coverage of the entire ecosystem to monitor changes in distribution, over time and season. However, the nature of the long-term data, particularly individual recognition of adults and knowledge of family structure, allows a number of indirect approaches to examine, for example, changes in seasonal occupancy of the basin over time.

As discussed above, the size of the elephant population has increased steadily from 1978 onward (Moss 2001; Lee, Lindsay, and Moss, chapter 6), although with changes in the elephants' use of the central swamps and surrounding basin (Croze and Lindsay, chapter 2), population density has not consistently increased. From the late 1970s to the early 1990s, most elephants spent much of the year within the National Park and density increased. Since the early 1990s, however, fewer elephants have been concentrating in the central basin, and thus, geographical density has decreased despite a continued increase in total population size.

The long-term sightings data can be analyzed to illustrate changes in occupancy within the National Park. The data must be first corrected in order to adjust for observer effort (see appendix 1). There is obviously some relationship between total number of elephants in monthly sightings and the number of researchers out searching for elephants (wet months, $r = 0.674$, $P < 0.001$; dry months, $r = 0.704$, $P < 0.001$). However, as the number of elephants occupying the central areas declined, increased effort was required in order to find elephants, so these observed relationships are not independent. There was no apparent relationship between number of sightings of family units and researcher effort, suggesting that the same families (and bulls) that used the central areas were being seen time and time again.

Using the total number of individual family observations in a year, which takes into account observer effort, we obtain an enhanced picture of changes in elephant occupancy in and out of the basin. Figure 7.3 shows the mean (±SD) number of times that individual families were sighted each year. The mean number of times that a family was observed each year for the period prior to the securing of the National Park (1976) was relatively low, around 10, probably due to avoidance of the Maasai, who were regularly using the core basin. From 1976, when 90 percent of the families had been identified, each family was seen 3 to 4 times every month (mean = 40 times per year). From 1990 on, family occupancy of the central basin decreased, despite an increase in observer effort.

As discussed elsewhere (appendix 1; Lee, Lindsay, and Moss, chapter 6), between 1989 and 1991, the number of sightings of each family unit per year decreased and the average family was now being seen only once every two months. This initial shift out of the central basin was associated with several relatively wet years, and water and forage may have begun to be more widely available and exploited. At the same time, the first cultural *boma*s (see Kangwana and Browne-Nuñez, chapter 3) were established along the southern margin of the Park, and livestock use of the basin increased. Thus, when observer effort is taken into consideration, the long-term sightings database confirms that

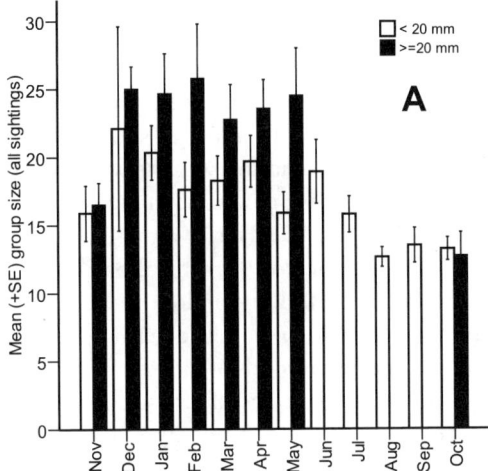

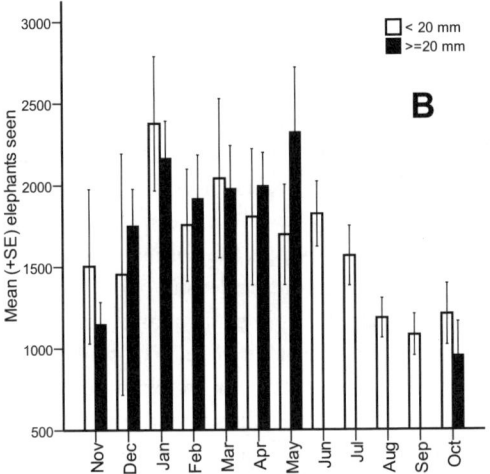

Figure 7.4 Amboseli basin monthly occupancy by elephants across all years contrasting wet and dry month occupancy within and between seasons.

Notes: In wet season months (November–May), mean group size (A) was consistently larger in months with >20 mm rain (see chapter 2). In the dry season (July–October) and in dry years, elephants used the basin swamps in small groups, and overall fewer elephants were seen (B).

elephants have been spending greater amounts of time outside the basin over the last decade than the previous 10 to 15 years.

The elephants have varied in their seasonal occupancy regime—in and out of the basin—over time. Taking the cutoff between wet and dry months as a monthly total of greater or less than 20 mm of rainfall (Croze and Lindsay, chapter 2), the number of observations in wet and dry months were compared (figure 7.4). Early in the study, the elephants concentrated in the basin during the dry season and moved off into the surrounding ecosystem as soon as the rains began. With the removal of cattle and a period of higher rainfall in the late 1970s, elephants congregated in the central areas during wet months. Across all years, in wet season months (November to April) and especially when rainfall in those months was relatively high, group size was larger as the elephants aggregated in areas with high food availability (figure 7.4a). In the dry season (May to October) and in dry years, elephants used the swamps in small groups during dry periods, and overall fewer elephants were seen (figure 7.4b). The elephants' "August Vacation" is clearly evident.

Food requirements are the major determinant of dry/wet patterns of occupancy. It is likely that the elephants aim to relieve the relative nutritional paucity (and perhaps just plain monotony) of the *Cynodon* grass sward at the swamp edge and under the *Phoenix* palms by seeking variety from foods in the surrounding woodlands and bushed grasslands during the rains (see Lindsay, chapter 5), particularly when there is also water to drink close to the feeding grounds. By contrast, after a long dry spell, hungry elephants return briefly to the basin in small groups to consume the still relatively green *Cynodon* grasses and *Acacia xanthophloea* browse in the swamp margins at high feeding rates. From the early 1990s onward, Maasai livestock incursions into the basin increased again, reverting in part to the pattern existing before the 1977 Park agreement implementation and increasing the competition for grass forage. The abundance of *Acacia* trees had also decreased to very low levels by this time (Croze and Lindsay, chapter 2), and this combination may have reduced the attractiveness of the basin habitats as foraging destinations.

Deadly encounters with humans are obviously a negative influence on elephant distribution: the elephants learn and remember the relative safety of particular areas (Kangwana 1993; Kangwana, chapter 20; Bates et al. 2007). The process of seeking refuge (or its obverse, avoiding danger zones) works both ways in the Amboseli ecosystem. On the one hand, the dangers of poaching, which persisted through the 1970s and into the 1980s, drove the elephants into the safety of the Park. The implementation of the Park agreement in 1976, whereby Maasai no longer entered the Park with their livestock, reinforced this perception of security. On the other hand, by the 1990s the Maasai had begun entering and crossing the Park with their herds in the dry season to get to the agreed-upon watering sites, and this presence tended to repel elephant families. Nonetheless, and despite growing numbers of both people and elephants in all parts of the ecosystem, the last decade has seen decreased elephant occupancy in the basin (and hence the Park) and increased time spent in the surrounding areas. Although the apparent tolerance of the surrounding human population could be seen as the outcome of positively changed attitudes and as an indication of the success of wildlife outreach programs (see box 19.1), the main challenge still remains to maintain goodwill through education and revenue sharing (see chapter 22).

Corridors to the Ecosystem

The future of the Amboseli elephant population as we know it today depends essentially on one thing: safe and secured access to the ecosystem surrounding the National Park. With expert knowledge from project observations of commonly used elephant trails, it is possible to propose a set of corridors that should be the basis for negotiation of passage easements. Figure 7.2 shows potential elephant corridors in and out of the Park. The corridors are overlaid on the sum of scouts' observations of elephant presence (sightings plus signs) discussed previously.

The map, along with figure 2.5, also shows the spatial challenges to wildlife movement: location of Maasai settlements, permanent fences around intensive swamp-edge agriculture, and the relentless advance of rain-fed agriculture from the northern slopes of Kilimanjaro. As discussed in chapter 3 and part 5, the traditional form of transhumant pastoralism is breaking down, resulting in settlements that tend to be occupied permanently or for longer periods throughout the year. This trend is gradually diminishing the degrees of freedom for wildlife movement. Crops, whether tomatoes and onions in the fenced high-intensity zones or seasonal harvests of maize and beans along the rain-fed front, provide irresistible attractors to wildlife, from elephants to porcupines, and a constant catalyst for increased human-wildlife conflict (see also Croze, Moss, and Lindsay, chapter 22).

Occupancy at the Core of the Range

Having illustrated overall trends in elephant use of Amboseli space over the long term, we now look at the ways individuals and families use their habitat. We might expect to see trends in occupancy by members of the general population to be mirrored at the level of the individual; however, this is not always the case, as we discuss below.

Family Groups

Given the gregarious fission-fusion nature of elephant society (see part 2), the instantaneous distribution of elephants throughout the population range should be highly clumped, with family members clustered around the matriarch and with each other. Moreover, the families themselves should not be continuously distributed across the landscape, but clustered from time to time around key habitat features that provide food (Lindsay, chapter 5), water, shelter, or comfort.

An obvious first question might be: do all families occupy the core of the population range in the same way? The short answer is no, and our analyses show that there are groups of families that tend to associate with each other in one part of the core range more often than other parts (see Moss and Lee, chapter 13). Here we present a map-based expression of occupancy of the central basin by elephant families, one that suggests non-random and differential spatial tenures comparable to the self-determined deployment of humans into neighborhoods.

The 30-plus-year sightings database contains a spatial reference field in the form of a one-by-one km grid square basemap that was established at the project's outset (see appendix 1). By placing the frequencies of family unit sightings into a Geographic Information System (GIS) grid framework, we can characterize the distribution pattern of each family over time, or by season, or in slices of time to examine change.

For the following analyses, we have used the Kernel Density Analysis (KDA) (Silverman 1986) tools available in ArcGIS (ESRI, version 9.2). The KDA is particularly useful for visualizing differences between distribution events across a common landscape (K. Johnson, ESRI, personal communication). Based on the density of occurrences (number of group observations in grid squares, in our case), the tool creates a smooth surface of relative probability of the event over the sample space. As an example of the output (and an illustration of the spatial limit of the analyses in this chapter), figure 7.5 shows the density kernel for all sightings of all elephants in the study area from September 1972 until October 2002. The shades indicate probability ranging from black, 90 percent probability of occurrence, to light grey, 10 percent.

Social Clusters

The association analyses of chapter 13 found 10 clusters of families within the overall association matrix. Highly associated families are likely to be positively associated in space as well as in time, so we examine here if there are any obvious differences in the average spatial clustering within and between families.

Forty-three of the current 55 family groups had more than 500 records in the sightings database between 1972 and 2002 inclusive. Their distributions in the study area for the whole 30-year period are plotted in figure 7.6. In order to avoid the visual muddle with over 40 kernel densities, we have created a visualization of the "core neighborhood" of each family by extracting and plotting as a black ring its 50 percent isopleth. Broadly speaking, each ring bounds

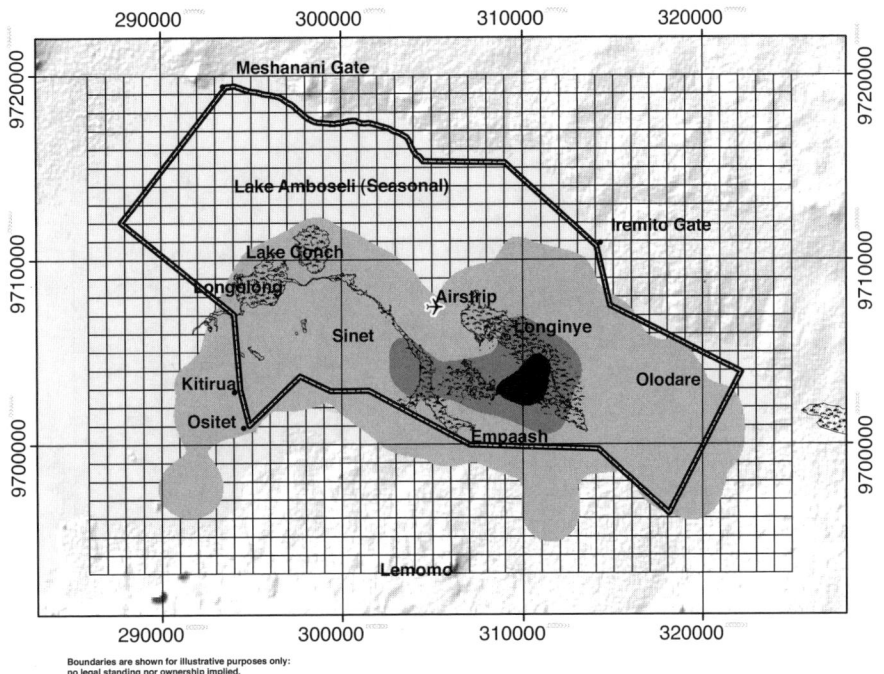

Figure 7.5 Total occupancy of the Amboseli basin by all elephants over 30 years.

Notes: Light grey shows 99 percent kernel; mid-grey, 50 percent; black, 10 percent. The one-by-one km grid squares are the UTM core sample frame for the long-term database. Several key place-names that occur throughout the book are also shown.

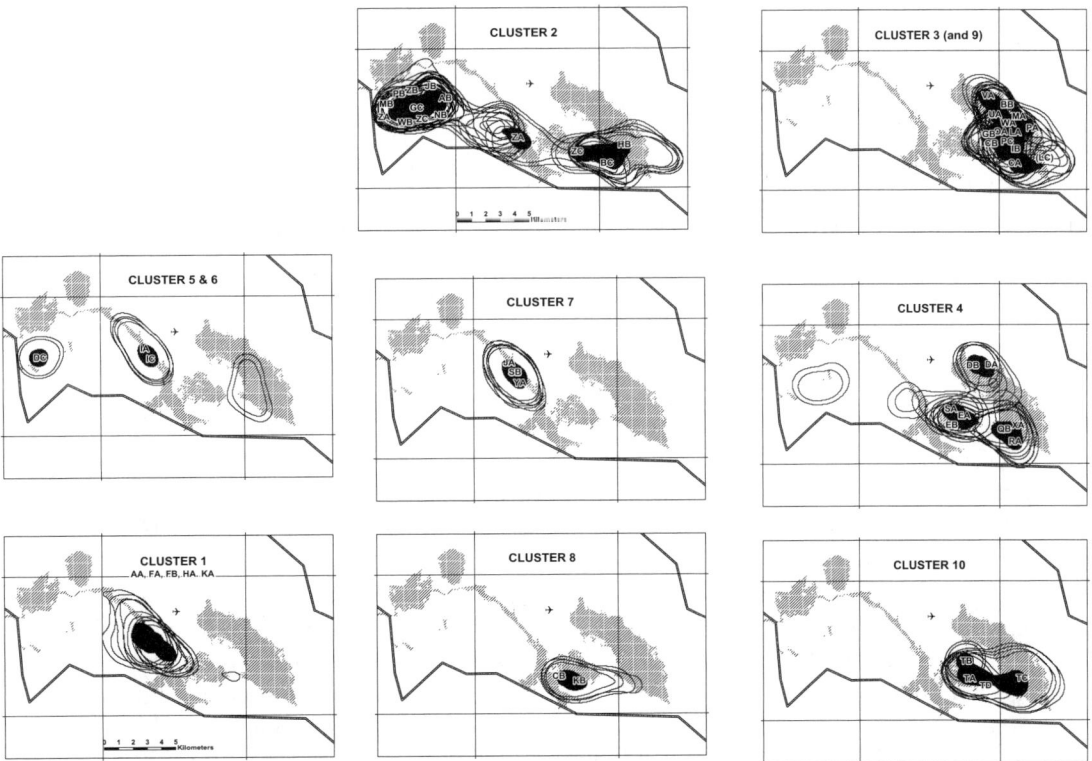

Figure 7.6 Core spatial distributions aggregated for whole 30-year study period.

Note: The isopleths (solid black, 10 percent lines, 50 percent) identify average spatial occupancy in the Park of the 10 family-group clusters shown in the association matrix results (see text and chapter 13).

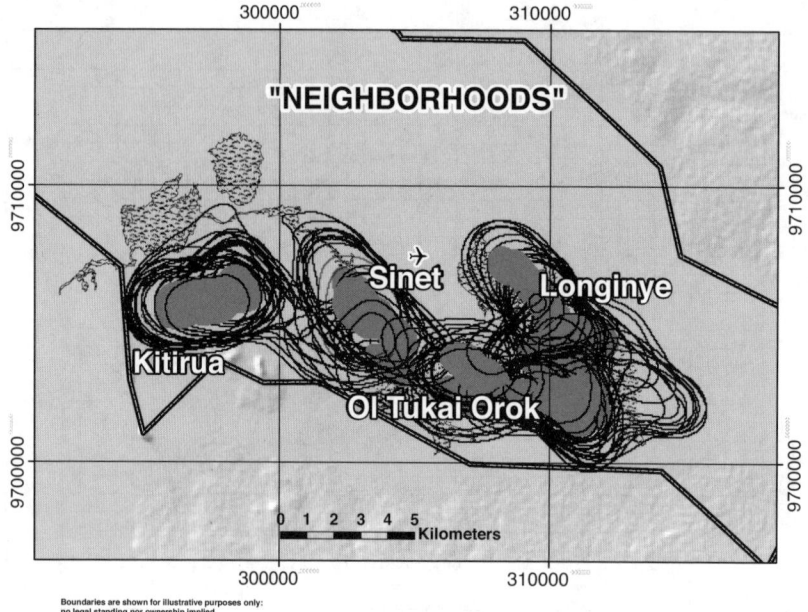

Figure 7.7 The average kernel distribution of all 10 family-group clusters (see text and chapter 13) overlaid to form recognizable "neighborhoods."

Note: The solid dark grey indicates 10 percent kernels; the black lines, 50 percent.

an area in which half of the observations of the family are likely to be found.

By inspection, four "neighborhoods," or clusters of clusters, emerge from west to east (figure 7.7): O'lengia/Kitirua (occupied by clusters 5 and part of 2); Sinet (clusters 1, 6, and 7); Ol Tukai Orok (OTO) (clusters 8, 10, the SA-EA-EB part of 4, and the eastern portion of 2); and Lonkinye (clusters 3, 4, and 9, and over time, the TC portion of 10). Not surprisingly, since all analyses derive from the same long-term database, the kernel neighborhoods are closely congruent with the social clusters and bond groups defined in chapter 13. The maps on their own cannot be taken as definitive evidence of behavioral bonding: it might be, for example, that two families tend to use the same space, but at different times of day or in different seasons. But the statistically positive association demonstrated by pair-wise matrix analyses in chapter 13 underlines the social determinants of the spatial distributions.

The occupancy of the four core neighborhoods by the cluster families is not completely consistent; for example, cluster 2 spans three neighborhoods, and the EAs and EBs use OTO more than the rest of cluster 4 members, who tend to congregate in Lonkinye. Such spatial flexibility in the long-term averaged data appears to be driven by a combination of seasonal and social determinants. A handful of families are seasonal "migrants" within the basin, whereas other families have undergone subtle changes in their core ranges over time due to group fission and fusion.

Seasonal Range Shift within the Study Area

Breaking down the average distributions into discrete time slices over the study period, we can then track core shifts over time. Forty-three families had sufficient sightings from which to generate meaningful kernels. Of those, cluster 2 in particular (figure 7.6, top center map) lacks the tight average spatial distribution of virtually all the other clusters. If we break the data set into four average seasonal distributions, it becomes clear why cluster 2 is different. Figure 7.8 shows the occupancy of the typical cluster 2, MB family, grouped by four seasons: early dry (May–July), late dry (August–October), early wet (November–January), and late wet (February–April). In the dry seasons, the MBs concentrate around the small swamps in the west of the Park. Come the early wet season, they shift eastward to stay, on average, in the Lonkinye neighborhood, just below "Kudu Corner" (UTM 37M 310102E 9703445N). Then as the wet season progresses through the long rains, they shift back to the west, "pausing," as it were, in the vicinity of the Public Camp Site before moving back to the west. Ten of the 12 families of cluster 2 show the same seasonal pattern with the exception of the WBs and the ZBs, who only go as far east as the Sinet neighborhood.

It is tempting to suggest that the MBs and other families of the cluster 2 are only able—or allowed, given possible subordinate status—to join the more resident Lonkinye families on the growing wet season swamp-edge grass

Patterns of Occupancy in Time and Space 99

swards when there is an abundance of food (see also Wittemyer et al. [2007]).

Long-Term Core Shifts

Some of the observed core range shifts over time can be explained or at least accompanied by changes in the inter- or intra-group relationships and behaviors among and between the adult females, specifically those that account for fission and fusion episodes (see Moss and Lee, chapter 13; and the following). Others cannot. Here we illustrate two such shifts with the spatial data set: a temporary (measured in years) concentration in a particular area and an apparently permanent (within the data set's life span) retreat from an area of human activity.

Over the years of the project, it was clear that the MBs were generally western elephants from the Kitirua neighborhood, the JAs were central in Sinet, and the CAs eastern in Lonkinya (figure 7.6). However, there is an apparent anomaly: in the 1977–82 period, all three families concentrated in one area (figure 7.9) which was in those days an open *Acacia* wooded grassland (see figure 2.7) south of Observation Hill west of the Sinet swamp, called Inkankeri by the Maasai. In fact, nearly all other families exhibited a similar attraction to this place during that period. Researchers observed some of the largest concentrations of elephants ever seen: in May 1981, some 550 animals, nearly the whole population at the time, was gathered there.

What was going on? There was a period of exceptionally good rains after a prolonged dry period from 1974 to 1976 (see figure 2.3), which produced abundant tree regrowth in this particularly diverse part of the basin. In addition, we can suggest that a social drive for gathering together and

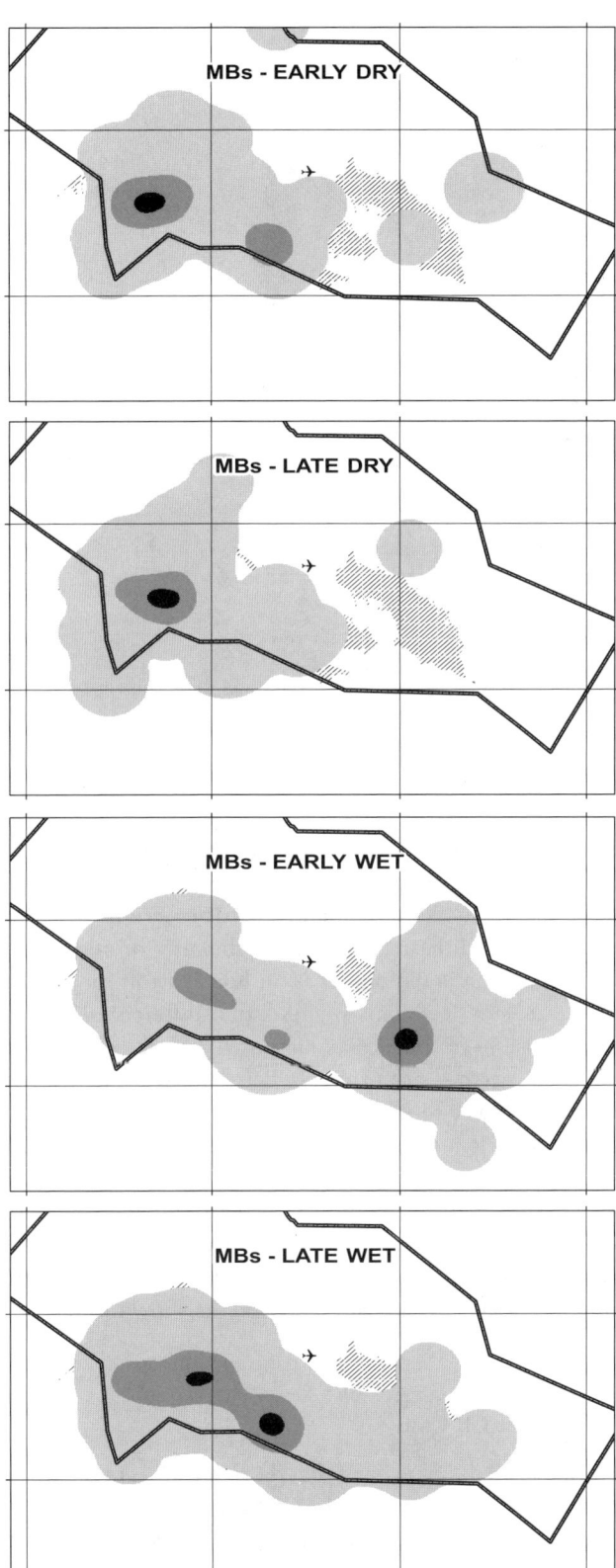

Figure 7.8 Average seasonal movement of the MB family over the 30-year study period.

Notes: Kernels shown: 10 percent (black); 50 percent (dark grey); 99 percent (light grey). Grid is 10 x 10 km.

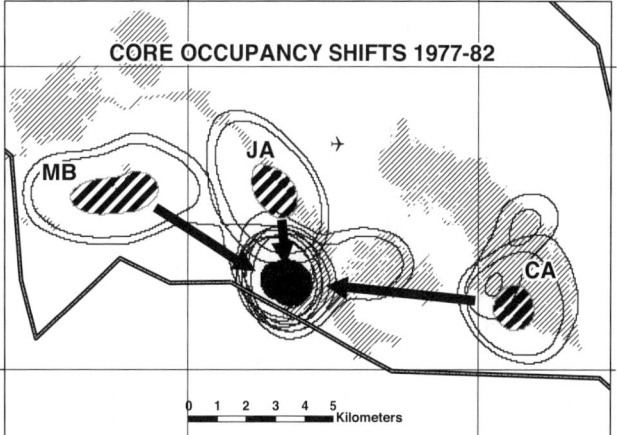

Figure 7.9 Three families—MB, JA, and CA—from three separate neighborhoods shifted their core ranges to the same area during a five-year period.

Notes: Cross-hatched shows 30-year 10 percent kernel; solid, 10 percent kernel for 1977–82. Black lines are 50 percent kernels.

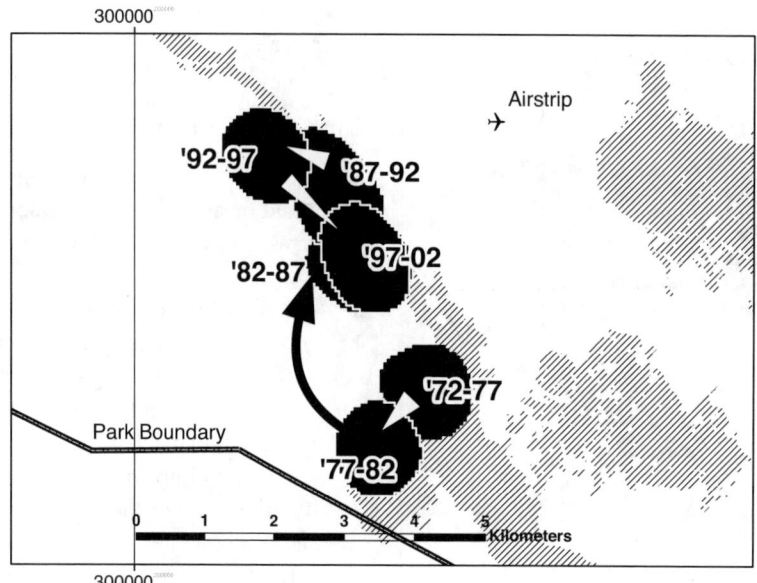

Figure 7.10 The JA family changes its core occupancy (black, 10 percent kernels) over five-year time periods between 1972 and 2002 along the Sinet swamp (see text).

forming mating aggregations had been thwarted by poor conditions for years, which might explain the observed mixed concentration of families and males.

The time series of one family, the JAs (figure 7.10), sheds additional light on the dynamics of core range shifts in the basin and illustrates the impact of a combination of events and processes on distribution. Early in the study, the JAs moved from an apparent core south of Observation Hill further south to join many other families in the Inkankeri gathering area. Then, and over the next 20 years, they moved north to settle firmly along the swamp near the Sinet causeway. There were three likely factors at work to repel the JAs (and others) from Inkankeri. One was that food resources became depleted in Inkankeri after the heavy concentration of families between 1977 and 1982, followed by a particularly difficult dry period in 1983–84. The public campsite was opened on the southern Park boundary in 1982, which brought permanent structures and occupancy by people. In addition, there were suddenly far greater numbers of Maasai herds and herdsmen impinging on the southwestern arm of Enkong'u Narok due to an agreement that gave the Maasai access to water once a day at the southernmost tip of the swamp arm (see also Kangwana, chapter 20). This situation persists in areas adjacent to the very edge of the Park boundary being occupied by people and their domestic stock virtually all the daylight hours. The elephants have habituated to the human presence to some extent, and both man and elephant seem quite relaxed, if wary, in each other's presence. But the proximity has led to a number of unfortunate incidents (see Croze, Lindsay, and Moss, chapter 22).

Fission and Fusion: Patterns within the Study Area over Space and Time

Individual families exhibit a complex variety of mechanisms and strategies of range exploitation that cannot be easily related to obvious habitat structures. From any point in the Amboseli basin, access to fresh water or one of 20-odd vegetation communities is an hour's stroll for an elephant, usually less. This suggests that differences in core occupancies—in some cases measured by only a couple of hundred meters—are probably determined by something other than basic food and water needs. In chapter 13, it is shown from long-term behavioral observations that family structures are complex and mutable over time (see box 13.1). The changing relationships between individual elephants and, consequently, between their families affect the patterns of occupancy of the core study area with subtle shifts in range over the study period, as shown in the following examples.

Fusion: The CB Case

From box 13.1, it is evident that the first decade of the study was a turbulent period for the CB family. The loss of a matriarch in mid-1976 threw the family into a disarray that expressed itself in a relatively loose spatial distribution over the first decade (1972–82), with family members equally likely to be seen in either Sinet or Lonkinye (map 1 in figure 7.11, $n = 195$ observations: the number of observations in this and the following sections refer to the size of the data

Patterns of Occupancy in Time and Space 101

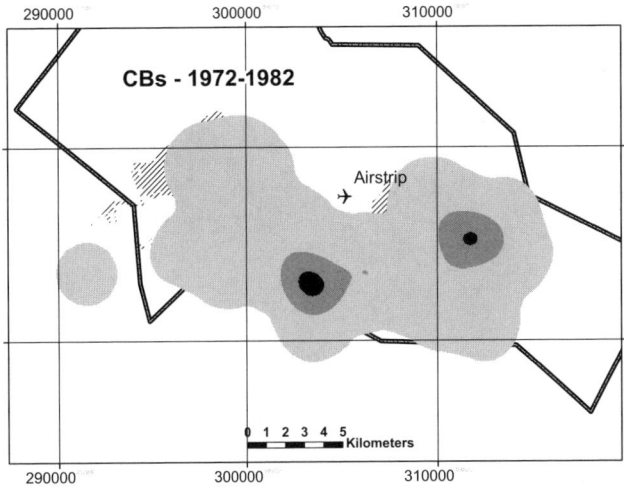

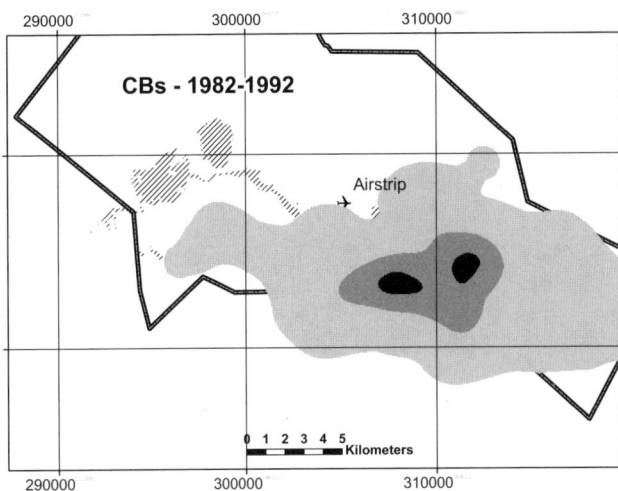

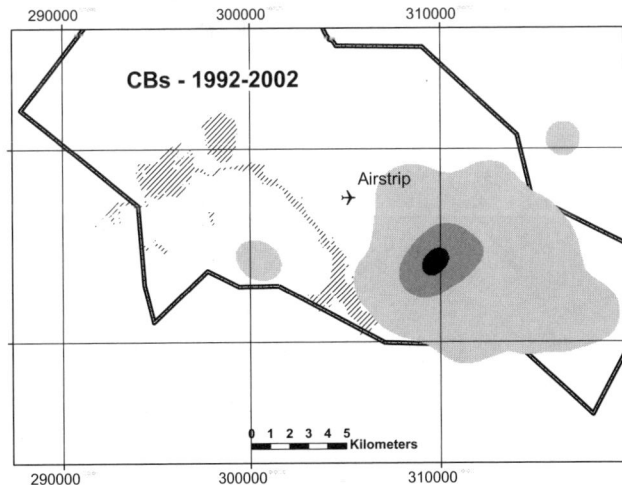

Figure 7.11 Spatial as well as social fusion: The CB family case (see text).
Note: Density distribution kernels shown: 10 percent (black); 50 percent (dark grey); 99 percent (light grey).

set of location-specific observations that were used in generating the kernels of probability). The second decade was also traumatic, with more deaths of senior family members, but as the remaining members and adult siblings formed sub-groups in the wake of yet another matriarch's death, their average spatial occupancy began to coalesce around the matriarch Chloe throughout the 1980s (maps 2 and 3, $n = 266$ and 183), consolidating on a consistent east Lonkinye occupancy for the last decade (map 4, $N = 435$).

Fission: The PA and PC Case

The largest family in the first 10 years of the study, the PAs (figure 7.12, map 1, $N = 402$) maintained a common core area, despite having fissioned into two families in 1978 (box 13.1). From 1983 onward, the two families gradually shifted their core areas (maps 2 and 3, $n = 337$ and 267), and the two matriarchs increasingly used slightly different core areas (maps 4 and 5, $n = 181$ and 183). Therefore, in the last decade, 1992–2002, the social split became a geographical reality as each family firmly stamped its distribution core on the eastern (map 6, $n = 426$) and western (map 7, $n = 465$) sides, respectively, of Lonkinye swamp.

Fission-Plus-Fusion: The DB and QB Case

In the 1972–82 decade, the QBs (figure 7.13, map 1, $n = 213$) and the DBs (map 2, $n = 340$) maintained relatively distinct western and eastern range separation, respectively (the QBs' eastern peak of high occupancy reflects the family's tendency to join the general "invasion" of the eastern Lonkinye swamp during the wet season). After Delia of the DBs and Quilla, then-matriarch of the QBs, struck up a close relationship with the birth of Delia's 1978 calf, they both split from their respective families to form a new family, the DCs, with distinct western tendencies (map 3, $n = 183$), while the DBs stayed in the east (map 4, $n = 993$). This is a rare but interesting example of social bond formation that is not based on kinship but apparently on an elusive quality that it is tempting to term "friendship." The spatial probability distribution follows the storyline of the changing social relationships (see box 13.1).

Bond Group Distributions

Bond groups are described in chapter 13 as associations of families that occur more than would be expected from random mixing of families in the core study area. From long-term observations of known individuals and families, it is likely that bond groups echo a history of relatedness and fission.

Figure 7.14 shows pair-wise map comparisons of the

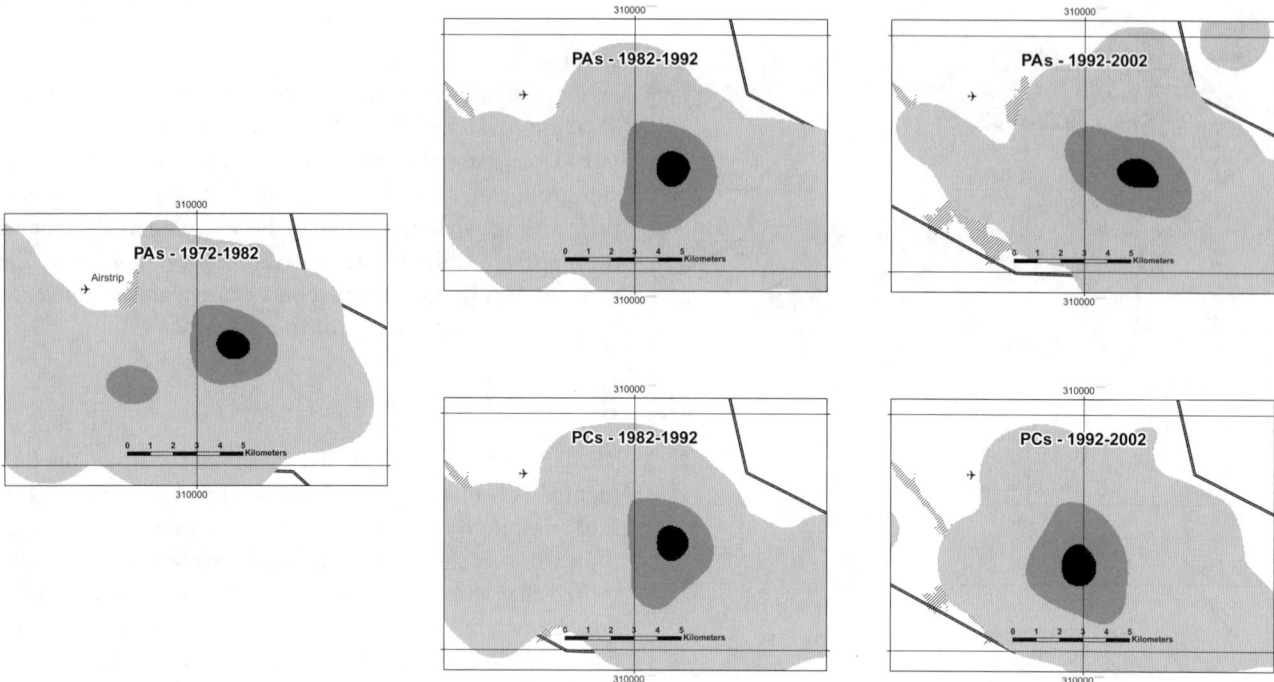

Figure 7.12 Family fission: The PA–PC case over three decades (see text).
Note: Kernels shown: 10 percent (black); 50 percent (dark grey); 99 percent (light grey).

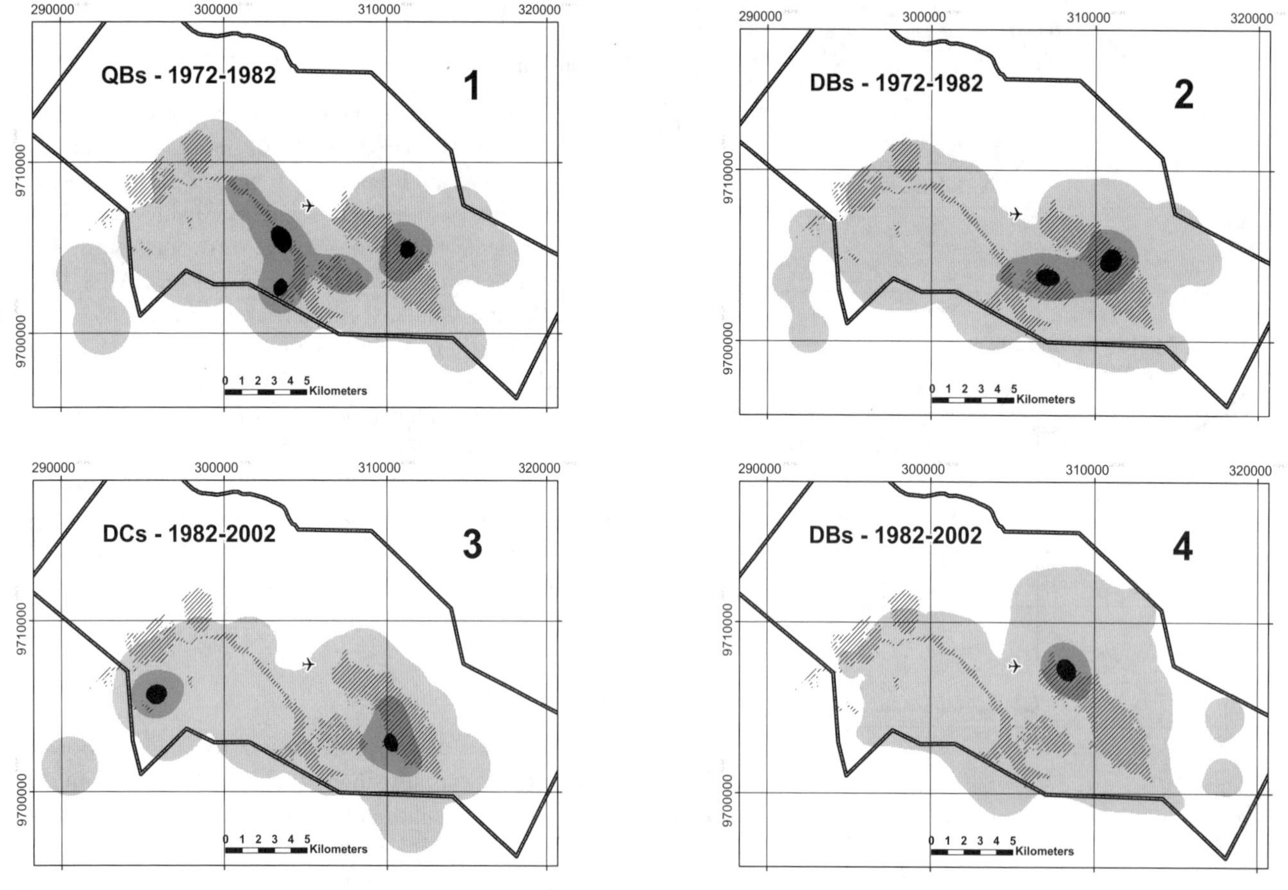

Figure 7.13 Fission and fusion: The QB–DB "friendship" case (see text).
Note: Kernels shown: 10 percent (black); 50 percent (dark grey); 99 percent (light grey).

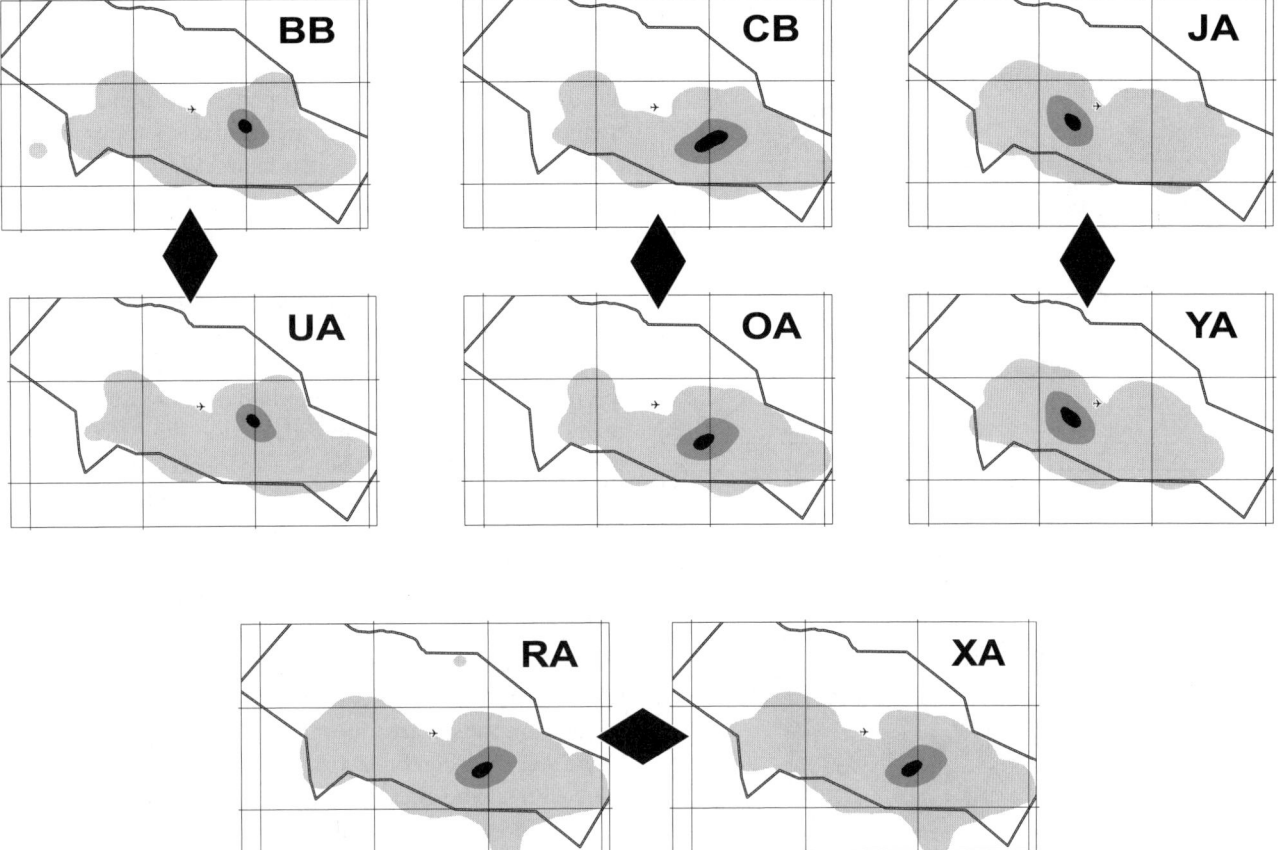

Figure 7.14 Families within bond groups (see text and chapter 13) show strong spatial concordance in their overall pattern of occupancy of the Amboseli basin.
Note: Kernels shown: 10 percent (black); 50 percent (dark grey); 99 percent (light grey).

occupancy of the basin by eight families from four bond groups: each bond group pair shows almost perfect spatial concordance. Although the genetic relatedness signal may fade quickly over time, the functional behavioral responses leading to occupancy choices appear to remain constant, probably mitigated through mother-daughter learning (see Byrne and Bates, chapter 11).

With the marked increase in mean family size over time (see chapter 13), bond groups in Amboseli may no longer exist in the same way as in the early years of the study. The families have reached such large sizes that they no longer associate with other families in the same way as they did, given that most of their close family associates are more or less constant companions. It will be interesting to observe changes over the coming years.

Bull Groups

The behavioral changes in a maturing bull, as detailed in chapters 17 and 18, involve the gradual change from being a dependent family member through transitional independence at sexual maturity to a serious competitor for breeding females some 20 years down the road. Along that journey, there are marked changes in the specific ways that adult males use space. The key elements determining male occupancy relate to their age, their sexual state, and their friendships. Young males after dispersal from their natal families are highly sociable, both with female groups and with peer males (see also Evans and Harris [2008]).

Ed (M48), who is introduced in chapter 17, can serve to compare the spatial occupancy changes associated with male elephant ontogeny with that of his natal family. Ed was estimated to be around nine years old when first seen as a young bull in the EA family in 1972. Over the first decade of the study, when Ed was between 9 and 20 years old, he was observed almost exclusively in the presence of his natal family and his range was more or less the same (figure 7.15). Over the next decade, the conservative EAs stuck to their Ol Tukai core, while Ed showed a clear shift of his core and his overall range to the west by some 5 km. By the third decade, with Ed becoming socially mature between the ages of 30 to 40, his core was now 7 km to the west of the EAs, and his range now stretched into Tanzania. In spite of his "central" roots, it is likely Ed joined the

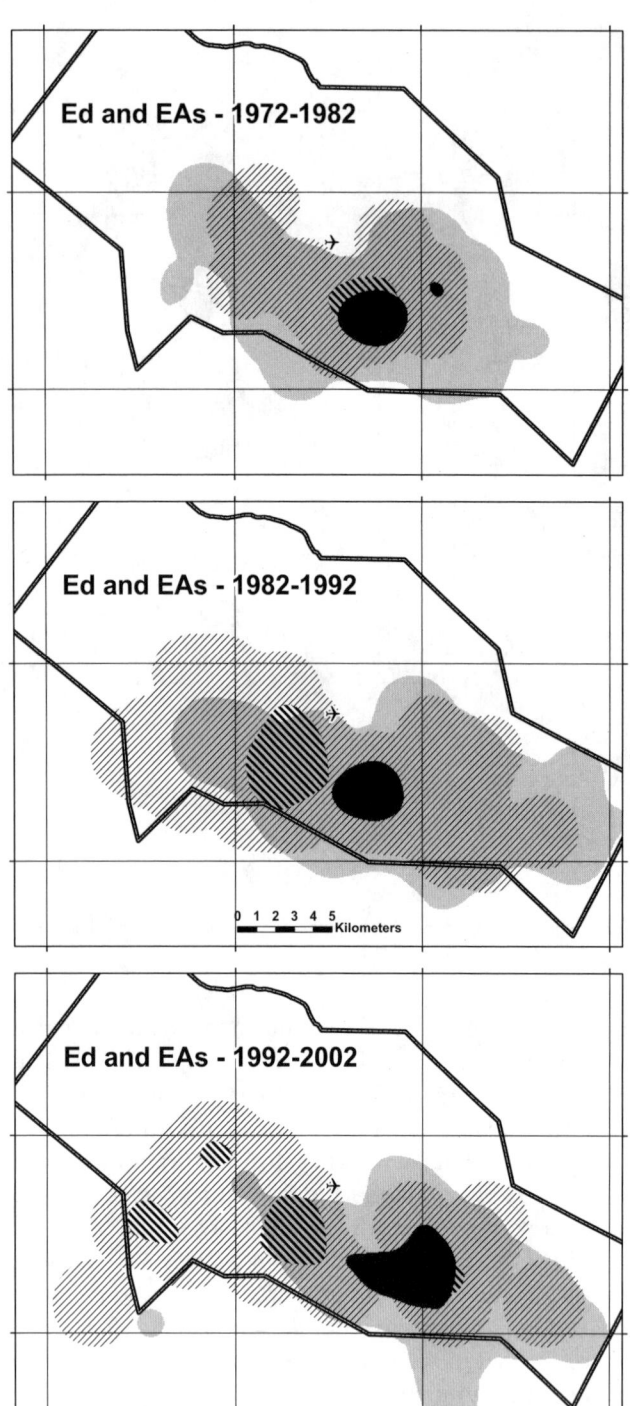

Figure 7.15 Ed grows up. Ed's pattern of occupancy gradually splits away from his matriarchal family, the EAs, over three decadal time periods.

Note: Kernels shown: EAs 50 percent (black solid) and 99 percent (grey solid); Ed 50 percent (black thick cross-hatched) and 99 percent (black thin cross-hatched).

cadre of western bulls, such as Mr. Nick and Solanga (see box 7.1), who regularly plied the corridor to West Kilimanjaro. Sadly, Ed was shot in 2004 in a tragic incident involving the death of an itinerant Maasai from outside the community.

Bull Areas

The old, large males who have been monitored throughout the study show the classical patterns of non-musth retirement to bull areas, where they feed and socialize (Moss and Poole [1983]; and see also figure 17.9). During their musth period, they can range over the entire distribution of females until they find an estrous female. The patterns of large male ranging and searching for females captured in Amboseli by the satellite tracking (box 7.1) are also seen in other populations (Douglas-Hamilton, Krink, and Vollrath 2005; Osborn and Parker 2003). As is noted in chapter 18, the precise elements that make bull areas so attractive to males have yet to be determined, but the ability to socialize with other males of all ages and experiences in the absence of female distractions may play an important role (see also Evans and Harris 2008).

The non-musth core ranges of 81 independent males with more than 50 observations each is shown in figure 7.16 (using the 30 percent probability kernels). There are clearly four distinct areas favored by non-breeding bulls. As the three examples in box 7.1 showed, males tend to use a particular non-musth area more frequently than other areas. There were a dozen bulls apart from those shown in figure 7.16 who spread their time more or less equally between two or more of the core bull areas.

It is typical of bull areas that they occur at the periphery of the females' centers of occupancy (Croze 1974; Evans 2006). It is also unlikely to be a coincidence that in many areas they occur within the relative safety of the National Park boundaries, since both elephants and designers of parks tend to seek pleasantly vegetated, well-watered areas within easy striking distance of amenities. Although the present analysis highlighted the central bull areas (given the restrictions on the data set of observations to central areas), the bull areas extend well out into the edges of the population range: southwest across the international boarder and east to Kimana (see box 7.1).

Conclusions

The distribution of elephant families and mature males is clearly no simple matter. Although gross diurnal and seasonal distribution patterns are determined by availability of forage and water, at a finer scale, elephants exhibit a spatial differentiation that can best be explained by inter-individual relationships, most likely between individual adult females, matriarchs in particular (see Moss and Lee, chapter 13; Wittemyer 2005; Wittemyer et al. 2007). The behavioral responses to a combination of resource exploitation and competition on the one hand, and social relations (attraction or repulsion) on the other, result in observed and

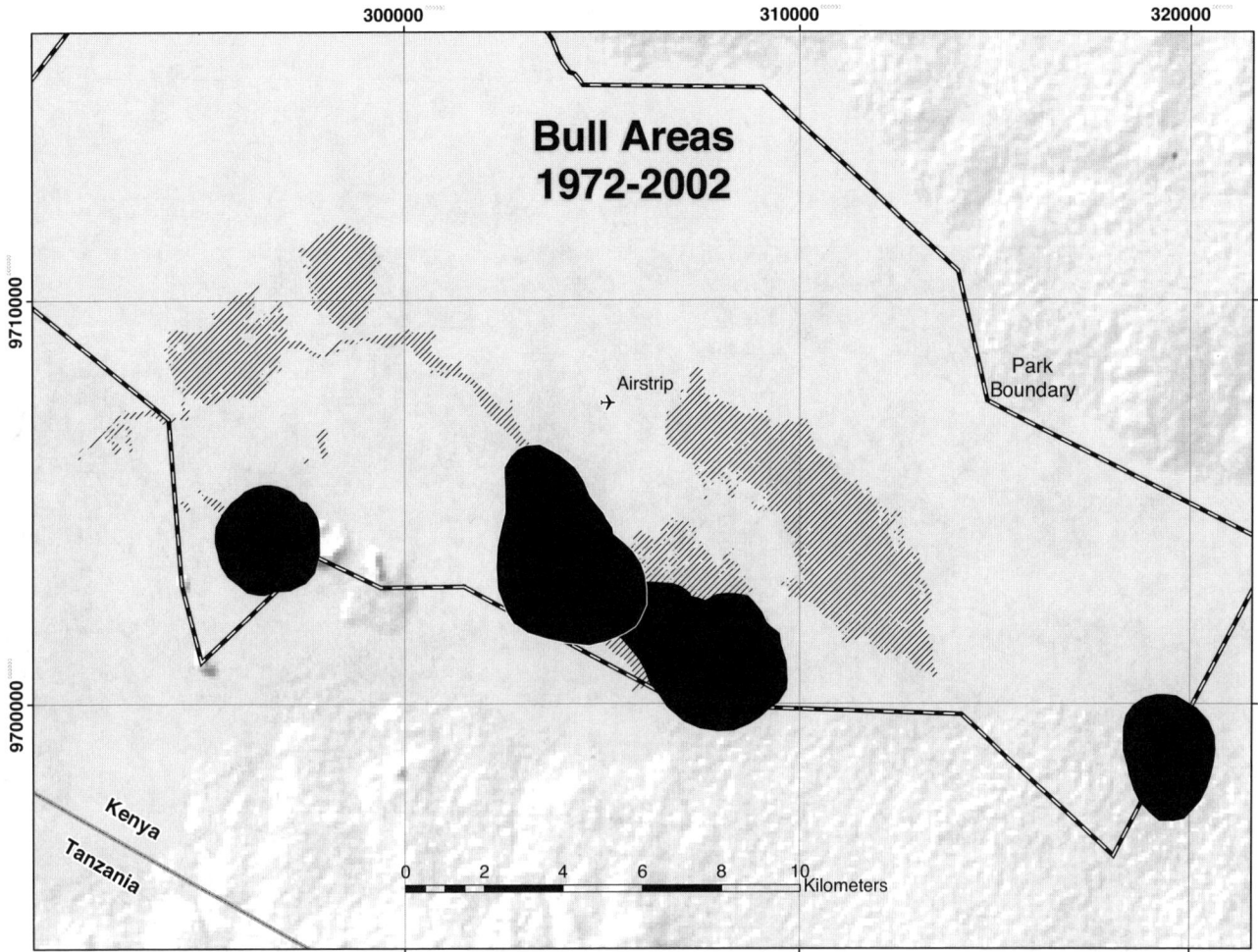

Figure 7.16 Non-musth core ranges of 81 independent males with more than 50 observations (30 percent probability kernels). There are clearly four distinct areas favored by non-breeding bulls within the Amboseli basin.

mappable "cores of subsistence," to play on the term coined by Marx (1977) referring to Middle Eastern nomadic tribes as "units of subsistence." These cores, for both elephants and nomads, are the centers of gravity of areas that provide multi-annual subsistence. And, as with the unit of subsistence, which is "articulated mainly by networks of institutionalized relationships" (Marx 1977, 343), the core of subsistence of an Amboseli elephant family is a zone of high probability of occupancy that is overlain by a network of social linkages. It results in a family center of spatial gravity that is distinct from most other families, even though there may be considerable overlap as well as social interaction with other families throughout their range. Mutinda (chapter 16) has observed and quantified those periods and places of overlap in the aggregations of groups on the swamp edges.

Fine-scale geographical cores of subsistence demonstrate one more layer of the complexity of elephant society. The evidence of distinct hubs of occupancy extracted from the long-term Amboseli elephant database, with a coincident lack of strict territoriality and high degrees of social tolerance, even among mature males during the "off season," reveals yet more subtlety to elephant social organization, perhaps only echoed today in the distribution of nomadic encampments or, until recently, in the unfenced neighborhoods of the semi-rural West.

Elephants are creatures of preferences, idiosyncrasies, and habit—up to a point. For humans, if a familiar route is blocked or a favorite restaurant shuts down, we quickly regroup, draw on our knowledge base, and go elsewhere. Elephants appear to do much the same, drawing on an impressive spatial-temporal knowledge of the ecosystem most likely stored in the brains of the matriarch and her sisters (e.g., McComb et al. 2000) or in the older males. As the Amboseli elephants face the challenges of an increasingly constrained ecosystem, we can only hope that their extraordinary capacities for flexibility will steer them through uncharted landscape to a safe but as yet unpredictable future.

Part 3

Behavior, Communication, and Cognition

Section editor: Joyce H. Poole

Chapter 8 Signals, Gestures, and Behavior of African Elephants

Joyce H. Poole and Petter Granli

African elephants, like other intelligent, long-lived, and highly social, group-living animals, use a wide range of sensory channels to communicate with one another. Visual, tactile, auditory, and olfactory signals all appear to be important to the social life of an elephant. Considerable work has focused on both auditory (see Poole, chapter 9; Berg 1983; Poole et al. 1988; Poole 1982, 1989a; Payne, Langbauer, and Thomas 1986; Langbauer et al. 1989; Langbauer et al. 1991; Soltis, Leong, and Savage 2005a, 2005b) and olfactory communication (mainly with captive Asian elephants: Rasmussen [1988]; Rasmussen et al. [1982]; Rasmussen et al. [1993]; Rasmussen, Hall-Martin, and Hess [1996]; Rasmussen and Schulte [1998]; Rasmussen and Wittemyer [2002]; Rasmussen and Krishnamurthy [2000]). Surprisingly, few systematic studies have described the visual and tactile signals of elephants except where specific behaviors explain or reveal underlying patterns of social or sexual behavior (e.g., Douglas-Hamilton 1972; Moss 1983; Poole 1987a).

In the early 1990s Kahl and Armstrong embarked on an ambitious study to compile an ethogram of the African elephant. In 2000, they named 83 ritualized visual and tactile displays based on their own observations as well as previously published descriptions. This initial inventory was simply a list without description, awaiting the detailed, systematic analysis of hours of video footage.

In 2002, with input from Kahl and Armstrong and pending their more detailed study, we developed an online searchable database, www.ElephantVoices.org, describing the appearance (written descriptions and photographs) and contextual use of these 83 displays as well as scores of additional displays and gestures of African and Asian elephants (Poole and Granli [2003], later updated as Poole and Granli [2009]). Our descriptions were based on the long-term observations of elephants and inspection of thousands of photographs and descriptions in published material, rather than the systematic analysis of video material. This chapter draws on that body of work to describe many of the displays, signals, postures, and gestures of African elephants. Its purpose is to provide a detailed description of the varied elements in the behaviors of elephants so that they can be recognized and contextually understood in the field. In some cases we have supplemented the written descriptions with a photograph, in other cases photographs of the behaviors may be viewed on the Gestures Database on www.ElephantVoices.org.

Most of the entries in this chapter are true displays or signals. In other words, they are behaviors used to transfer information, reducing the observer's uncertainty about the actor's or reactor's future behavior (Krebs and Davies 1987). In addition to signals, we have included movements, postures, and gestures that are not necessarily intentful, but may be a consequence of an elephant's state or situation. Most of these latter behaviors also convey specific information to observers (both human and elephant) and may function as true signals (e.g., Waiting; Periscope-Sniff). Finally, we include a few postures and behaviors that elephants just do in specific situations (e.g., Testing-Semen). These may or may not act as signals to other elephants, but they are interpretable by humans as measures of behavior and are, therefore, admissible in a comprehensive ethogram.

Following the convention of Kahl and Armstrong (2000),

the first letter of names given for behaviors are capitalized; if the name contains two words, the first letter of each word is capitalized and the name is hyphenated. The behaviors have been divided into several general contexts: Aggressive, Ambivalent, Socially Integrative, Mother-Offspring, Group Defense, Sexual, Play, and Generally Attentive. Each of these has been further subdivided. The behaviors are in bold the first time they are mentioned in the sections where they are described. Otherwise they are italicized.

Aggressive

Threat

An elephant may threaten another simply by a resolute reorienting of its body to gaze at an opponent (Poole, personal observation; Kahl and Armstrong 2000), or by purposeful, directed walking toward another. Sometimes just one step toward another is enough to cause a reaction (Eisenberg, McKay, and Jainudeen 1971; McKay 1973; Poole 1987a, 1989a, 1999a; Moss 1988, 1992; Ben-Shahar 1999). A **Turn-Toward** and an **Advance-Toward** may be associated with other aggressive postures such as *Ear-Spreading*, *Standing-Tall*, or *Ear-Folding*, and so forth.

In an **Ear-Spreading** threat an elephant faces its opponent head-on and fully spreads its ears (perpendicular to its body), thus increasing its apparent size to the viewer (Kühme 1961; Douglas-Hamilton 1972; McKay 1973). In a more exaggerated form, an elephant spreads its ears while **Standing-Tall**, holds its head held well above its shoulders, and with tusks lifted, directs its gaze toward its adversary, as in figure 8.1 (Douglas-Hamilton 1972; Poole 1982, 1987a, 1987b, 1987c, 1999a; Kahl and Armstrong 2000; Langbauer 2000). In an exaggerated form of *Standing-Tall*, an individual may stand on a log or an anthill to appear even taller—a tactic often used by males when they are sizing up one another.

A threatening elephant may press or "fold" the lower portion of its ears under and back so that a prominent horizontal ridge appears across the ear as depicted in figure 8.2 (Moss 1988; Poole 1982, 1987a, 1989a). **Ear-Folding** may be used in combination with a variety of other threats such as *Standing-Tall*, *Advance-Toward*, and so forth, to emphasize that an elephant "means business." *Ear-Folding* that is used in association with *Head-Raising*, *Ear-Lifting*, and/or *Rapid-Ear-Flapping* is an affiliative, not aggressive, display and is a typical greeting posture (see Social Integration, below).

An elephant expresses irritation or impatience with an individual or circumstance with an abrupt shaking of the head, or **Head-Shaking**, which causes the ears to flap sharply

Figure 8.1 *Standing-Tall*, a threatening elephant appears larger by facing its opponent head-on, with the head well above the shoulders and the ears fully spread.

Figure 8.2 An elephant threatens another by *Ear-Folding*, pressing or "folding" the lower portion of its ears under and back so that a prominent horizontal ridge appears across the ear.

and dust to fly (Eltringham 1982; Poole 1987a; Payne and Langbauer 1992; Langbauer 2000). Alternatively, an elephant may frighten a predator (e.g., lions, humans), an irritant (e.g., egrets, ground-hornbills, warthogs), or an elephant adversary with a **Forward-Trunk-Swing**, stepping or lunging forward and swinging or tossing the trunk in its direction, while simultaneously blowing out through the trunk (Kühme 1961; Poole 1987a). Similar swinging of the trunk is also observed in play (see Play, following). Lifting or uprooting objects and throwing them in the direction of an opponent, or **Throw-Debris**, is an escalated form of the *Forward-Trunk-Swing*. An elephant's aim can be very accurate even at some distance (Kühme 1961; Poole 1987a, 1987c, 1996; Kahl and Armstrong 2000).

The **Head-Nod**, a rather jerky up and down movement of the head, which at higher intensities may cause the trunk to flop about, is often associated with, or a prelude to, a *Mock-Charge* (Kühme 1961, 1963; Estes 1991; Douglas-Hamilton and Douglas-Hamilton 1992; Ben-Shahar 1999).

Two musth males involved in, or leading up to, an *Escalated-Contest* may pace back and forth in a **Parallel-Walk** to one another, typically placing a road, log, or other "barrier" between them (Poole, 1987b, 1987c). Tusking of the ground and uplifting of vegetation (**Tusk-Ground**), or tossing of the head and tusks back and forth through bushes or other vegetation (**Bush-Bashing**) creating noise, commotion, and a demonstration of strength is usually seen during the maneuvering of two musth males during an *Escalated-Contest* (Kühme 1961; Moss 1988; Poole 1987c, 1996; may also be seen in play, below). Escalated-Contests are usually terminated with the winner tracking the loser over several kilometers in an aggressive, persistent, and **Prolonged-Pursuit** (Poole, personal observation).

Escalation

Lunging and rushing at, chasing after, and physical contact all characterize more escalated elephant aggression. Elephants may be observed to run after or chase an opponent in a **Pursuit** (Poole 1982, 1987a, 1989a; Lee 1986). This increased aggression on the *Advance-Toward* takes several forms. A **Mock-Charge** is a rushing toward an adversary or predator, usually exhibiting *Head-High* (see Sexual Advertisement below) and *Ear-Spreading* that stops short of its target (Douglas-Hamilton 1972; Langbauer 2000); an elephant may aggressively *Forward-Trunk-Swing* or *Kick-Dust* (see below) as it abruptly stops (Poole, personal observation). A *Mock-Charge* is often associated with a shrill *Trumpet-Blast* (see Poole, chapter 9). In a **Real-Charge** an elephant rushes toward another elephant, predator, or other adversary with the apparent intention of following through. The head may be held high or lowered; the ears are spread and trunk curled under (so that tusks make contact first). A *Real-Charge* is usually silent (Poole 1987c; Kahl and Armstrong 2002). In the **Bow-Neck** charge, the aggressor (usually female) lowers her head by bowing her neck downward and simultaneously tilting her head upward so that her tusks are approximately horizontal. The *Bow-Neck* may be associated with *Ear-Flattening* and/or *Ear-Folding*. This posture may be held at a fast walk or during a *Mock-Charge* and/or *Real-Charge*, especially when the subject of the charge is of smaller stature. In a sense the aggressor brings her head and tusks down to her victim's level (Kahl and Armstrong 2000). It is one of the more common forms of high-level aggression between females of different families or toward young, non-family males (Poole, personal observation). Kahl (personal communication) notes that it is reminiscent of the Forward Threat Display performed by geese. In a **Coalition**, two or more individuals work together to threaten and/or pursue another individual or individuals. Generally an older female, often the matriarch, comes to the aid of a younger family member. The helping individual may approach her companion in parallel; they may vocalize in unison, sometimes clicking their tusks together (*Tusk-Clicking*; see below), and then together *Advance-Toward* and/or *Run-After* their adversary with a *Bow-Neck* display (Poole, personal observation). In a **Group-Charge** an entire family charges toward an adversary en masse in a highly coordinated manner (Poole 1996; Ben-Shahar 1999). This behavior is most often observed as an anti-predator display.

Escalated aggression may also involve a variety of physical contact. An elephant may push its head into the back, side, head, or trunk of another; such **Pushing** may lead to one animal being supplanted, and differs from the *Herding-Push* (Lee [1986]; Moss[1988]; Langbauer [2000]; Archie, unpublished data, see below). An individual may hit or slap another with its trunk, especially older individuals toward calves. Although **Slapping** is uncommonly seen in *Loxodonta*, it is apparently more common in *Elephas* (Lee 1986; Kahl and Armstrong 2000). Poking an opponent with the tip of the tusk, or **Tusking**, is an escalated form of pushing another out of the way. An elephant may tusk another in competition over resources, or in irritation (Lee 1986; Poole 1987a, 1989a; Kahl and Armstrong 2000; Archie, unpublished data).

A combined highly aggressive form of *Pushing* and *Tusking* results in **Ramming**, in which an individual lowers its head with its trunk curled under (to expose tusks) and rushes toward another, goring or poking it with its tusks, breaking the skin and potentially causing serious wounds (Kahl and Armstrong [2000]; Poole, personal observation; Archie, unpublished data).

Males, particularly those in musth, take physical contact to an extreme form, rushing at one another and attempting to gore each other with their tusks, as in figure 8.3. **Dueling** males tusk and push one another head to head, attempting to lock tusks and leverage their opponent over or maneuver him into position where he can be gored. *Dueling* is associated with the *Head-High, Tail-Raising, Ear-Spreading, Ear-Folding, Trunk-Out-Stretched* (to reduce any blow) (Poole 1982, 1987b, 1987c, 1996, 1989a).

Figure 8.3 Physical aggression takes an extreme form among males in musth. *Dueling*—males rush at one another. *Ramming*—males ram into one another. *Pushing*—males push each other head-to-head. *Tusking*—an attempt to gore the other with the tusks.

The term **Escalated-Contest** covers all displays that may be seen in the context of a serious battle between two males. These may include: *Dueling, Tusk-Ground, Parallel-Walk, Throw-Debris, Bush-Bashing, Trunk-Out-Stretched, Head-Toss, Trunk-Drag, Trunk-Curl, Run-After, Run-Away,* and others (Poole [1987a, 1987b, 1987c, 1989a]; some described below).

Submission

A submissive elephant holds its head low (**Head-Low**) so that the top is below the level of the shoulder blades, making the individual appear smaller. This posture is often associated with a *Turn-Away* and followed by a *Retreat-From* (see Retreat below). A *Head-Low* can be a persistent display by a low-ranking male just as *Head-High* can be a persistent display by a high-ranking male (Eisenberg, McKay, and Jainudeen 1971; Moss 1983, 1988, 1992; Poole 1987b, 1989a, 1996, 1997; Estes 1991; Spinage 1994); the former is often observed in young males in presence of musth male. The *Head-Low* may be an appeasement gesture, which suppresses aggression in the other individual. A submissive elephant may also hold its ears flattened (Ear-Flatten) against the body (Poole, personal observation; Kahl, personal communication). An elephant normally holds its ears in a relaxed position such that they rest approximately 10 to 30 cm from the side of the body. In this relaxed position the ears do not appear stiff or tense, lifted, spread, or pressed against the body.

In the presence of high-ranking individuals around a limited resource (at a fallen tree, or in the company of an estrous female), lower-ranking individuals (most often young males) may be observed to make a small arc or **Skirt-Around** the group to try to establish a better position (Poole, unpublished data). Similarly, around a scarce resource or object of interest such as a drinking place, mud-wallow, fallen tree, or an object of play, lower-ranking individuals may engage either in displacement activities or stand quietly as they wait their turn.

When joining a resting group and also in the context of high-level greetings (see social integration below) an elephant presents its posterior and walks **Back-Toward** another, usually higher-ranking animal (Moss 1988; Kahl and Armstrong 2000; Langbauer 2000; Archie, unpublished data). A Back-Toward another may be an appeasement gesture or sign of respect, especially within families (Poole, personal observation) or used to avoid triggering an aggressive response in a larger social context (Kahl and Armstrong 2000). As an interesting aside, Hyraxes (*Procavia*) also Back-Toward in similar circumstances (Croze, personal communication). Elephants may also be seen to Back-Toward a dead elephant or to the bones of an elephant (Moss 1992; Poole 1996; and see below).

Retreat

An elephant may avoid conflict with a threatening individual by adopting a *Head-Low* posture, orienting or walking away, or retreating from, the aggressor while looking back (**Turn-Away, Retreat-From, Look-Back**; Poole [1982, 1987a, 1989a]. Such avoidance behavior is typically associated with a wary *Look-Back* over the shoulder (Poole 1987a, 1999a; Kahl and Armstrong 2000). An individual may also *Retreat-From* an aggressor by backing away (**Back-Away**; Kühme [1961]; Poole, personal observation). During escalated aggression an individual may avoid conflict with a **Run-Away**; this is as a common response to a *Pursuit* (Poole 1982).

Post-Conflict Display

Following an aggressive interaction or more prolonged conflict, elephants may engage in a number of post-conflict behaviors. For example, in an apparent sign of bravado an individual may adopt an **Exaggerated-Walk**, an obvious bouncing swagger in which the front legs appear to bend more, and the movement of the head up and down and the trunk from side to side is overstated (Buss and Estes 1971; Hendrichs 1971; Poole 1987a; Moss 1988, 1992; Balfour and Balfour 1997). A similar walk is seen as any age/sex comes down a slope, particularly toward water. Croze (personal communication) refers to this as the "water-walk," which he believes may take its form from the inevitable down-slope toward water. While this may not be a signal, he has seen it displace subordinate males from the water's edge.

Following a *Mock-Charge*, an elephant takes a step in the direction of an offending object (elephant, other animal, human being) and at the end of the movement kicks up a cloud of dust (**Kick-Dust**) in the direction of its antagonist. The elephant typically stands tall, towering over its adversary (Kahl and Armstrong 2000). Often when an aggressor runs after another and then decides to give up the chase, he or she extends the trunk upward and outward toward his or her opponent in an apparent sign of frustration or "last ditch effort" (Jainudeen, McKay, and Eisenberg 1972; Kahl and Armstrong 2000). This **Reaching-High** is a common display between competing males, particularly in association with an estrous female (Poole, personal observation). During a *Dueling* contest a male may stretch his trunk toward his rival as they rush toward one another; the **Trunk-Out-Stretched** appears to be used by a male as a form of defensive protection against the force and sharp tusks of his rival (Poole 1987b, 1987c). It may be a form of *Distant-Frontal-Attitude*, described below.

Redirected-Aggression is aggression aimed at an individual that is irrelevant to the current situation. When a tendency to attack is thwarted, for some reason (e.g., by fear of the opponent), an elephant may redirect his or her aggression to some other individual or object, such as thrashing bushes, pushing trees (*Bush-Bashing*), throwing sticks and/or grass (*Throw-Debris*), or threatening or attacking other, lesser elephants, smaller species, or humans in the vicinity (Sanderson 1907; Spinage 1994; Deraniyagala 1955; Eisenberg and Lockhart 1972; Krishnan 1972; McKay 1973; Ben-Shahar 1999; Lahiri-Choudhury 1999).

Within elephant families, though not between adult males, escalated aggression (e.g., *Pushing*, *Tusking*) is often followed by post-conflict **Reconciliation**. Typically a third party, such as the matriarch or a close associate of the aggrieved individual, initiates the reconciliation. She approaches the conflicting elephants from the side and, while standing head to head, rumbles while *Head-Raising* and *Ear-Lifting* (see Affiliative, below) and reaching toward the other with an outstretched trunk in an affiliative gesture (Poole, personal observation).

Ambivalent

Displacement

Displacement-Grooming and **Displacement-Feeding** is behavior characterized by its apparent inappropriateness or irrelevance to the situation in which it occurs. For example, in conflict situations an individual may begin to throw dust or grass on him or herself, when such grooming is inappropriate (Kühme 1963; Eisenberg and Lockhart 1972; McKay 1973; Adams and Berg 1980; Douglas-Hamilton and Douglas-Hamilton 1992; Spinage 1994; Sukumar 1994; Kahl and Armstrong 2000). Alternatively, an individual may begin to pluck at vegetation, as if foraging, but may not actually ingest any of the material. Even if the elephant does eat, it does so in a halfhearted or distracted fashion and may slap the vegetation against foot or other part of own body and then discard it (Kühme 1963; Eisenberg, McKay, and Jainudeen 1971; Estes 1991; Spinage 1994; Sukumar 1994; Daniel 1998).

Apprehensive

In situations where an elephant is slightly apprehensive or ambivalent, unsure of what action to take, it may be observed to engage in a number of conflict behaviors. For instance, an individual may stand, listening or observing, while twisting the tip of its trunk back and forth, or **Trunk-Twisting** (Poole 1999a). Or an elephant may **Foot-Swing**, raising and tentatively swinging its foreleg (less commonly the hind leg) intermittently (Moss 1988; Poole 1999a). In contexts where an elephant feels uneasy, or unsure of the next action, he or she may also engage in **Touch-Face**, a self-directed touching of the face, mouth, ear, trunk, tusk, or temporal gland, apparently for reassurance (see figure 8.4; Kühme 1961; Poole 1999a; Kahl and Armstrong 2000). For comfort, a young individual, typically an infant

Figure 8.4 In situations in which an elephant feels uneasy or unsure of the next action, he or she may also engage in self-directed touching of the face, mouth, ear, trunk, tusk, or temporal gland, or *Touch-Face*. This behavior appears to be a form of reassurance, or displacement behavior.

or calf, may engage in **Trunk-Sucking** (Poole, personal observation; Kahl and Armstrong 2000).

As an elephant experiences rising apprehension or alarm, it may adopt several postures that may or may not be observed in combination. The elephant may stand or move with the lower potion of the **Trunk-Curled-Under** and slightly back (Kühme 1961; Poole 1999a). It may hold the head and **Jaw-Tilted-Upward** (so that the tusks become more parallel to the ground) and the ears stiffened (**Ear-Stiffening**) and slightly spread. This posture is unlike that of an aggressive elephant, who looks squarely at its adversary with ears and tusk more perpendicular to the ground (Poole 1999a). As an apprehensive elephant moves away it tilts its head to look back over its shoulder and with its ears slightly spread and jaw tilted up, gives a characteristic **Tilted-Ear-Spread** posture (Poole, personal observation). As apprehension gives way to increasing fearfulness an elephant raises its tail. **Tail-Raising** is also observed when elephants are highly excited (Moss 1988; Poole 1999a). An elephant standing or walking with its jaw tilted upward and its tail raised causes it to have an exaggerated swayback. This overstated **Back-Bowing** posture (Poole, personal observation) may also be seen in the *Estrous-Walk* adopted by receptive females (Moss 1983). A fearful elephant, or one that is highly aroused, opens its **Eyes-Wide** so that the whites may be seen (Poole 1996, 2000a). A fearful elephant may stand or move while exhibiting a combination of several displays including *Tail-Raising, Eyes-Wide, Jaw-Tilted-Upward, Tilted-Ear-Spread, Trunk-Curled-Under,* and *Back-Bowing* (Poole, personal observation). Once an elephant has decided to retreat, it may run quickly, quietly and smoothly exhibiting Ear-Flattening and Tail-Raising. This gait is referred to as **Panic-Running**. If elephants are in a group they bunch closely together (Moss 1988).

Social Integrative

Affiliative

Elephants engage in numerous forms of low intensity and more exuberant displays of affiliative behavior that function to reinforce social bonds between family and bond-group members. An elephant approaching another family member may rub (**Social-Rub**) its head or side against the other individual as in figure 8.5 (Lee [1987]; Moss [1988]; Archie, unpublished). In the context of greeting and reassuring calves an individual may reach its trunk toward or into the mouth of another (Moss [1981]; Lee [1987]; see *Caress*, below). This **Test-Mouth** behavior is also observed between non-relatives, especially males in context of greeting (Moss

Figure 8.5 In an affiliative gesture, an elephant approaching another family member may *Social-Rub* its head or side against the other individual.

Figure 8.6 During many affiliative or socially exciting interactions, elephants raise their heads and lift their ears. The *Head-Raising* and *Ear-Lifting* posture is most commonly seen when elephants greet infants or, as in this illustration, their close associates. Note the posture of the female being approached by the long-tusked matriarch who is vocalizing with her *Mouth-Open*. The young female raises her head and lifts her ears while reaching her trunk back toward the approaching older female.

1981; Poole 1982). As part of social learning calves touch the mouths of older individuals to sample food items (Lee and Moss 1999).

During many affiliative or socially exciting interactions (e.g., when greeting, reaffirming bonds, expressing solidarity, as well as following any exciting event—mating, birth, fight, reconciliation), elephants raise their heads as in figure 8.6. **Head-Raising** is different from the threatening *Head-High, Standing-Tall* posture in that the elephant's neck appears to be extended upward and outward so that the head appears separate from the rest of the body. In the *Standing-Tall* posture the elephant's neck is extended upward but not outward (Poole personal observation). In association with *Head-Raising* elephants lift their ears so that a gap appears between the lower flap and the neck, also seen in figure 8.6. This **Ear-Lifting** posture is most commonly seen when elephants greet infants (*Coo*; Poole, chapter 9) or one another (*Little-Greeting*; Poole, chapter 9). As individuals become more aroused **Head-Raising** and **Ear-Lifting** is associated with **Rapid-Ear-Flapping**, and often, *Ear-Folding* (see under Threat). This combination of postures is almost always seen in the context of greeting or bonding interactions and during other excited, social, and chorused calling (Moss 1988; Poole et al. 1988; Poole 1998a; Poole, chapter 9). As excitement levels soar elephants rumble, roar, and scream at extremely high sound pressure levels (Poole et al. 1988). Covered by the trunk, the mouths of elephants are not very obvious sources of visual displays, but during excited vocal interactions elephants open their mouths wide (**Mouth-Opening**), often holding their trunks in a curled posture, which exposes the mouth (Poole, personal observation). During these interactions of extreme social arousal, close associates may gather together (**Gathering**) in a close cluster, pressing their bodies against one another. As they vocalize loudly they often turn abruptly toward and away from each other in a rapid movement that we refer to as **Spinning** (Moss 1988). During both lower- and higher-intensity greetings, elephants may reach out to briefly seize the trunk of a close associate (**Trunk-Grasping**; Poole [1982]; Moss [1988]). Between greeting males the entwining of trunks is always slow and sensual; whereas between female relatives during a bonding or greeting ceremony the action is more sudden and dramatic, like a sudden clasping of another during an excited human greeting (Poole, personal observation). Finally, during highly social interactions such as a *Greeting-Ceremony* (Moss 1988) or during interactions in which an elephant expresses solidarity with a family member, elephants may click their tusks together (**Tusk-Clicking**; Poole, personal observation).

These more exuberant displays, known collectively as bonding ceremonies, may occur during greetings, births (Poole 1999b), calf rescues, matings (Poole 1987b), or any number of socially arousing events (see Poole, chapter 9, for more detail) and involve vocal, chemical, tactile, and visual communication. During a *Greeting-Ceremony*, for example, members of a family or bond group may run to meet one another while emitting loud, modulated, throaty rumbles, trumpets, roars, and screams (Moss 1988; Poole et al. 1988; Poole, chapter 9). The elephants raise their heads and lift and spread their ears, secreting *Temporin* profusely. As they meet they flap their lifted ears rapidly, while rumbling, trumpeting, screaming, and roaring. Rubbing against one another they stand in parallel and while holding their heads high, they click tusks together and entwine trunks. As the greeting continues they may back into one another and spin around, while urinating and defecating. The *Greeting-Ceremony*, as with other *Bonding-Ceremonies*, is a constellation of displays including: *Head-Raising*, *Mouth-Opening*, *Ear-Spreading*, *Ear-Lifting*, *Ear-Folding*, *Rapid-Ear-Flapping*, *Tusk-Clicking*, *Trunk-Twining*, *Back-Toward*, *Spinning*, *Temporin* secretion, *Tail-Raising*, *Urination*, and *Defecation* (Moss 1981, 1988; Poole 1987b; Poole et al. 1988; Poole 2000a).

A **Male-Male-Greeting** has a very different quality from a greeting between family members. Males typically approach one another slowly and reach their trunks into one another's mouths. This trunk-to-mouth greeting may be extended to sniffing one another's temporal glands and genitals, presumably to test one another's sexual state or identity (Buss, Rasmussen, and Smuts 1976; Rasmussen 1988). Greetings may lead to gentle sparring and *Trunk-Twining* (see Social Play, below). Greeting males may occasionally rumble to one another, but they do not vocalize with the excitement shown by greeting family members.

Spatial Proximity

Elephants may use several friendly methods of maintaining proximity or of monitoring how close another individual is. While they are moving together in a group one individual may give another a gentle **Herding-Push** (see figure 8.7), with the apparent purpose of keeping the group moving together (Archie, unpublished data). While standing or moving an elephant may use its tail to forcefully swat an elephant behind it to indicate that it should back off (Kahl and Armstrong 2000). A gentler **Tail-Swat** is given to feel what is behind or to ensure that a calf is still there (Kahl, personal communication; Poole, personal observation). To push another individual (especially a calf) out of the way, an elephant may **Kick-Back** with its hind legs (Lee [1987]; Kahl and Armstrong [2000]; Archie, unpublished data).

Figure 8.7 Close associates use a gentle *Herding-Push* as a friendly way of keeping the group moving along together.

Figure 8.8 An individual initiating a movement to a new habitat adopts a *Let's-Go-Stance* and rumbles, "Let's go" every minute or so. The calling female adopts a particular posture: she stands on the periphery of the group, sometimes lifting and swinging a foot (*Foot-Swing*), and purposefully faces in the direction she wishes to travel. Her persistent calling attracts the attention of others who may slowly move to join her.

Movement Initiation and Leadership

Within families the coordination and initiation of movement is a vital component of group living (Mutinda, Poole, and Moss, chapter 16). An individual may initiate a movement to a new habitat by adopting a **Let's-Go-Stance**, standing on the periphery of the group, purposefully facing in the direction she wishes to travel, waiting, and rumbling "Let's go" every minute or so (Poole et al. [1988]; Poole, chapter 9; as shown in figure 8.8). In between calling the elephant waits. **Waiting** behavior may also be observed when part of a family is lagging behind. One or more individuals purposefully wait for the others to either catch up or to initiate a procession again; contextually this posture is obvious as the waiting individual is attentive (rather than resting), often glancing back over her shoulder, and monitoring the other's activity with her trunk but otherwise showing a lack of activity (Moss 1988). The individuals begin to walk again as soon as the individuals they are waiting for approach. Elephants may indicate to one another a change of activity or direction by flapping the ear (usually one more than the other) moderately loudly against the neck and shoulder at the same time as the elephant tilts its head upward and to one side, causing the ear to slide against the shoulder. The **Ear-Flap-Slide** causes the ears (or ear) to make a rasping sound as they rub downward against the shoulders. This signal is especially used by adult females, but may sometimes be made by adult males and juveniles of either sex (Moss 1988). A loud **Ear-Slap** against the shoulders, typically by a dominant female, appears to be used to call attention to her location and activity. Dominant females also use this signal to announce "status" in slightly aggressive situations, or to initiate a coalition (Poole, personal observation).

Mother-Offspring

Suckling

In attempting to gain access to the breast, an infant or calf walks parallel to its mother, either pushing against her legs or touching the trunk onto or near to her breast. This begging or **Solicit-Suckling** behavior is often accompanied by a *Suckle-Rumble* or other begging calls (Lee and Moss [1986]; Lee [1987]; Langbauer [2000]; Poole, chapter 9). A mother can reject such solicitations, or **Reject-Suck**, by blocking access to her breast with her leg or by walking away. If the calf is already suckling, the female can terminate nipple contact by stepping forward and pulling the nipple from the calf's mouth; she can more aggressively bump her elbow into the suckling calf to force it off, and she can slap or push the calf off the nipple (Lee 1986, 1987). Rejections from suckling are often accompanied by the calf giving protest cries or loud roars (Poole, chapter 9).

A female suckling an infant, or a juvenile female "pretending" to suckle an infant often holds very still with the *Ears-Lifted* in an attentive posture and her eyes opening and closing or *Eye-Blinking* (see General Attentive, below) in what we term a **Suckle-Stance** (Poole, personal observation; figure 8.9).

Figure 8.9 A female suckling an infant (*Suckle-Stance*) typically holds very still, *Ear-Lifting* in an attentive posture.

Reassurance and Protection

Using the trunk to touch, or **Caress**, another in a protective, reassuring, or comforting manner is a common behavior in an elephant family. While adult females or juvenile females most often direct caresses toward infants and calves, adults and juveniles may also *Caress* one another. Caresses come in many forms: wrapping a trunk over the back and around the belly of a calf or over calf's shoulder and under its neck, often touching its mouth; reaching out to touch the genitals, temporal glands, face, legs, mouth, or trunk of another individual (Lee [1987]; Poole, personal observation). A gentle rumble often accompanies a *Caress*.

Adult and juvenile females are often observed to use the body, feet, trunk, or tail to shepherd, check for the presence of, gently guide, or assist an infant or calf to safety. This **Shepherding** action may also be used to gauge the proximity of an elephant behind (Lee [1987]; Kahl and Armstrong [2000]; Poole, personal observation; Archie, personal communication).

A series of more physically active contacts use the feet or trunk to retrieve young infants when they are lying down and potentially in danger of being left behind. For example, a front or hind foot lifts a calf onto its feet; trunk or trunk and tusks together pulls or pushes a calf onto its feet, over a log, or out of a mud wallow. Mothers and allomothers utilize these behaviors to maintain close contact with and assist infants (Lee [1987]; Poole, personal observation). Elephants may also use the hind feet, tusks, or trunk to attempt to lift injured, sick, dying, and dead elephants. They have even been observed carrying dying infants (Moss 1992). When adults perceive a threat, they respond by *Bunching*, thus protecting calves in the center of a tight defensive circle of elephants. Adults face outward on all sides, heads up and moving from side to side, and vocalizing, with their rumps to the center, like the spokes to a wheel (Lee, personal communication).

Group Defense

When a group of elephants is under threat, they respond first by *Freezing* (see General Attentive, below) and then by clustering or **Bunching** together (usually with the youngest animals in the center) so that the diameter of the group decreases (Douglas-Hamilton 1972; Moss 1988; Langbauer et al. 1989; McComb et al. 2000; see figure 8.10). This reaction is different from the *Gathering* together of family members seen during bonding, which also results in a smaller group diameter. In the former tight group elephants exhibit a combination of Apprehensive and/or Aggressive postures. *Bunching* may be observed in response to any situation that

Figure 8.10 When a group of elephants detects a threat, they respond by clustering or *Bunching* together (with younger individuals protected in the center). In this tight group they may exhibit a combination of fearful and/or aggressive postures as they decide how to respond. The elephants in this illustration wait apprehensively (note trunks curled under and jaws slightly raised) for the arrival of the family matriarch before departing.

is dangerous or potentially dangerous, whether caused by other elephants, predators (including people), or strange sounds, smells, or events. *Bunching* may precede a *Full-Retreat*, *Panic-Running*, or a *Group-Charge*. Following a threat to a predator, an individual or group of elephants may do an about-face and make a rapid and **Full-Retreat** from danger. A large adult female places herself several meters behind a bunched group of elephants in *Full-Retreat*. This **Rear-Guarding** female carries her head very high while looking back over first one shoulder and then the other in a *Look-Back*. While this posture indicates an alarmed elephant, the wrong move by the human observer or predator may provoke an attack (Douglas-Hamilton 1972).

Sexual

Advertisement/Attraction

Among males, the postures associated with musth function in both male-male competition and aggression as well as sexual advertisement. For example, musth males carry their heads well above their shoulders with exaggerated rigidity, tuck in their jaw, and stiffen their ears. These **Head-High, Chin-In, Ears-Tense** postures are elements of a dominant musth male's swagger, or **Musth-Walk**, and characterize all of their behavior and interactions (Poole and Moss 1981; Poole 1987a; Kahl and Armstrong 2002; Payne 2003). A feature of musth is the continual **Urine-Dribbling** from a retracted and sheathed penis. The flow may vary from an intermittent drip to a gushing stream, particularly if the male is threatening another (Poole 1987a). Males may lose up to

300 l per day, and consequently they drink more than non-musth males (Poole 1989a). The continual dribbling gives the inner hind legs a black shiny appearance and may turn the sheath of the penis a greenish color (Poole and Moss 1981; Hall-Martin 1987; Moss 1988). The urine has a very pungent odor (Poole and Moss 1981; Poole 1987a; Kahl and Armstrong 2002). Musth males also secrete from the temporal glands, creating a dark streak down the side of the face, mid-way between ear and eye. This **Musth-Temporal-Gland-Secretion** may be distinguished from *Temporin* by its congealed appearance and strong odor (Poole and Moss [1981]; Poole [1987a] defines it using categories of 1–4). Musth males rub their temporal glands against trees or the ground (when mud wallowing). **Marking** may be so vigorous that the male departs with bark and debris on the side of his face. Dissected temporal glands have been found to have pieces of bark embedded deep inside them. Non-musth elephants also mark but the behavior is more ritualized among musth males (Poole 1987a; Kahl and Armstrong 2002). *Urine-Dribbling* and *Musth-Temporal-Gland-Secretion*, while primarily olfactory signals, may also have a visual signaling effect.

Another characteristic behavioral trait of musth males is the **Ear-Wave**, as shown in figure 8.11. A musth male swings the upper portion of the ear stiffly and vigorously forward. The speed and forcefulness of the swing causes the lower, unsupported portion of the pinnae to go even further forward and flap upward. The motion creates a wave appearance across the ear (Poole 1982; Kahl and Armstrong 2002). The *Ear-Wave* may function to waft the odor of *Musth-Temporal-Gland-Secretion* toward other elephants (Poole 1987a). In addition, musth males may be seen to **Trunk-Curl**, furling and unfurling the trunk in a sinuous snakelike motion, often curling it over a tusk (Kahl and Armstrong 2002) or lifting the head up and, with open mouth, reaching with curled trunk to rub across the temporal gland and forehead (Trunk-to-Head; Poole [1987a]). Musth males may drag, or **Trunk-Drag**, the distal portion of the trunk on the ground, making a rasping sound. This behavior leaves a snake-like mark on the ground and is likely to be related to the **Trunk-Bounce** seen in Asian elephants (Kahl and Armstrong 2002). Musth males may *Trunk-Drag* as a threat at very close range (Poole, unpublished data). Occasionally in combination with a *Trunk-Curl* a musth male raises and lowers his head, or lifts and swings his head and trunk with vigor, sometimes in figure-eight movement. In the most intense form of this **Head-Toss** an elephant bends his back legs and lowers hind portion of body, causing the head and trunk to rise even higher (Poole 1987a; Kahl and Armstrong 2002). Less exaggerated forms may be observed in non-musth male and female elephants, particularly in play (Moss, personal communication).

A female or a group of females may respond to the sound, smell, or physical presence of a musth male with a **Female-Chorus** display. The female(s) steps forward with mouth open rumbling and head lifted and outstretched. Ear postures include a combination of *Ears-Lifted* and *Rapid-Ear-Flapping*. The trunk is extended down and may be swung slightly as the head is raised. *Temporin* streams down the side of the face. The female(s) typically urinates for the musth male to test and may *Rump-Present* to him (Poole et al. 1988; Poole 1999a).

When a non-musth male elephant urinates he fully extends his penis and usually thereafter obtains a semi-erection and sometimes a full erection. Infants and calves usually obtain a semi-erection when they are suckling and both young and adult males often obtain semi- and full erections when they are engaged in sparring activities, especially when these end in **Male-Male-Mounting**. In some cases erection appears to be a form of display, though this requires further investigation (Poole, personal observation).

Sexual Monitoring

The monitoring of sexual state is composed of a medley of different behaviors, which includes the testing of temporal glands, urine, feces, and genitals. During the testing of genitals an individual reaches its trunk toward the genitals of another (sniffing) for the purpose of assessment of reproductive state (Rasmussen and Wittemyer 2002). **Test-Genitals** may be Male-Female, Male-Male, Female-Male, or Female-Female. In *Male-Female-Test-Genitals* a male reaches out with his trunk and sniffs the genitals of a female (Jainudeen, Eisenberg, and Tilakeratne 1971; Eisenberg, McKay, and Jainudeen 1971; Rasmussen et al. 1982). In *Male-Male-Test-Genitals* a male reaches out his trunk

Figure 8.11 *Ear-Wave*: A musth male swings the upper portion of the ear stiffly and vigorously forward. The speed and forcefulness of the swing causes the lower, unsupported portion of the pinnae to go even further forward and flap upward.

to smell another male's genitals. This behavior may occur following a meeting between two males in association with *Male-Male-Greeting* and also by non-musth males toward musth males in relation to assessment (Poole 1982). In *Female-Male-Test-Genitals*, a female reaches her trunk in the direction of and touches or almost touches a male's penis. Females show most interest in musth males and estrous females display this behavior very prominently immediately following a mating (see also *Test-Semen*, below; Eisenberg, McKay, and Jainudeen [1971]; Poole [1982, 1987b]). *Female-Female-Test-Genitals* is also observed.

In **Test-Urine**, elephants may also touch the tip of the trunk over, on, or in urine or a urine spot (Jainudeen, Eisenberg, and Tilakeratne 1971; Rasmussen et al. 1982; Rasmussen, Schmidt, and Daves 1986; Rasmussen 1988; Rasmussen and Schulte 1998) for the purpose of assessment of reproductive state (either estrus or musth; Rasmussen and Wittemyer [2002]) and individual identification (Rasmussen and Krishnamurthy 2000). Testing of urine may be Male-Female, Male-Male, Female-Male, or Female-Female. After an individual places the tip of the trunk in or near a liquid substance (often urine) the trunk may be curled, raised vertically, and the trunk tip placed on paired orifices leading to the vomeronasal organ in the roof of the mouth in a **Flehmen** response (Rasmussen et al. 1982). The vomeronasal organ plays an important role in processing chemical signals relevant to reproduction (Rasmussen et al. 1982; Rasmussen et al. 1993; Rasmussen 1998). The *Flehmen* response may be observed in males to female urine, to female estrous urine, and to male musth urine. Sometimes an individual may **Place** the tip of the trunk over a scent of interest and immediately flatten the entire end of the trunk over the sample; sucking or blowing may ensue (Rasmussen 1998). Placing the tip of trunk over or on dung, or **Test-Dung**, is also commonly observed.

Elephants often reach out with their trunks to smell another's temporal glands. Among females and calves **Test-Temporal-Glands** may be a component of the *Caress* behavior and among adult females within a family, sniffing another's temporal glands may obtain information about a companion's emotional or physical state. Greeting males often sniff one another's temporal glands and this behavior is probably for assessing physiological and sexual state (Poole 1982; Rasmussen and Wittemyer 2002).

Musth males are often observed to persistently follow the scent of another, actively scenting with trunk by sweeping the ground and/or scanning the air. A **Tracking** musth male may pursue rivals or estrous females, sometimes for many kilometers (Poole, personal observation). Moss describes a musth male who, with aggressive intent, followed the tracks of her Land Rover for more than a kilometer (Moss 1988).

Sexual Solicitation

Musth males often adopt a non-threatening stance or **Casual-Walk** when approaching a group of females, while in the presence of females, or when moving through a group testing females. This walking (or standing) with the trunk draped over a tusk in a relaxed and "casual" manner signals friendly intent (Moss 1988, 1992).

Females may be observed to *Back-Towards* a male and **Rump-Present** for testing or to solicit mating (Poole 1987b; Poole and Moss 1989; Kahl and Armstrong 2002).

Courtship

Estrus is characterized by a series of specific behaviors that have been described in detail by Moss (1983) and Poole (1989b). A female in estrous acts with **Wariness** in the presence of males, holding her head high, her eyes opened wide, with her gaze directed at other elephants. She is intolerant of male approaches (Moss 1983) and avoids them with an **Estrous-Walk**, a characteristically long gait in an arc away from family with her head held high and turned to one side, giving her body a slight twisted motion as she walks (Moss 1983). The *Estrous-Walk* becomes a **Chase** as a male(s) pursues an estrous female at a fast walk or shuffling run (Moss 1983; Poole 1982, 1989b). Once a pursuing male has caught up with the female he reaches his trunk across her back in a **Reach-Over**, or may place his trunk along her back, and may attempt to prevent her forward movement (Kühme 1961; Eisenberg, McKay, and Jainudeen 1971; Jainudeen, Eisenberg, and Tilakeratne 1971; Moss 1983). Females are invariably lighter than males and can thus outrun them; an experienced female chooses to allow a male to *Reach-Over*. Once a male has stopped the forward movement of a female and begun to mount her, she may either choose to step forward or stand still. Within an established consort between a musth male and an estrous female, the pair may not engage in either the *Chase* or *Reach-Over* phase. Instead, the male may approach the female and, prior to *Mounting*, rest his chin on her back or begin pushing or **Driving** her, sometimes roughly, with his forehead (Eisenberg, McKay, and Jainudeen 1971; Poole, personal observation; see figure 8.12). During *Driving* the female attempts to stand her ground (Jainudeen, Eisenberg, and Tilakeratne 1971; Moss 1983). **Standing** by the female involves actively pushing back against the *Driving* male with her legs locked. **Mounting** occurs when a male has succeeded in placing his forelegs on the female's back as depicted in figure 8.13 (Moss 1983). *Mounting* may be successful (intromission and ejaculation occur) or unsuccessful (Poole 1982; Poole 1989b). A successful mating largely depends upon whether the female chooses to

Figure 8.12 A *Consorting* pair typically precedes a mating with a bout of *Driving*. The guarding musth male approaches a female and begins to push her forward with his trunk or forehead. A female at the height of estrus will attempt to stand her ground, by locking her forelegs and actively pushing back into the weight of the male (*Standing*).

Figure 8.13 *Mounting* occurs when a male has succeeded in placing his forelegs on the female's back. A successful mount (in which intromission and ejaculation occur) largely depends upon whether the female chooses to *Stand* or move forward.

Stand or move forward. **Guarding** behavior is said to occur when a male (usually in musth) attempts to prevent all rival males from obtaining access to an estrous female by staying within 5 to 15 m of the female and threatening all approaching male rivals (Poole 1982, 1989b). When both the *Guarding* male and the estrous female exhibit responsibility for maintaining proximity to one another, they are said to be **Consorting** (Moss 1983; Poole and Moss 1989; Poole 1989b; Kahl and Armstrong 2002). As a female passes the peak of her estrous period her guarding male begins to lose interest in her. During this phase the female is often observed overtly **Soliciting-Guarding** behavior from her ex-consort by following him, and when approached or pursued by other males she moves rapidly to stand next to him (Poole 1989b).

Post-Copulatory

Following a mating, the mated female adopts a **Post-Copulatory Stance**, stepping forward with streaming temporal glands and *Mouth-Opening*, she calls with a powerful series of rumbles (see Poole, chapter 9). With her head raised and outstretched, she exhibits *Rapid-Ear-Flapping*. The female alternately turns toward the male, and then rapidly away, holding her head up high and her ears spread (probably to ensure long-distance advertisement of availability) and calls powerfully and repeatedly at initially regular intervals, but then with increasing intervals and diminishing intensity (Poole 1987b; Poole et al. 1988; Poole 1989c). These extremely powerful and characteristic calls attract the attention of distant males (Poole 1987b; Poole et al. 1988; Poole and Moss 1989). Between calls, the female may reach her trunk to touch the male's penis or his semen on the ground, sometimes even flicking semen onto her body using a trunk movement similar to dusting (**Test-Penis** and **Test-Semen**; Poole [1982]; Poole [1987b]). The mated female's relatives rush to her side and join her in exuberant calling, displaying, urinating, and defecating in what we refer to as a **Mating-Pandemonium**. They, too, reach to smell and/or touch the male's penis or semen. The calling by the mated female is distinct from the pandemonium produced by her relatives.

Male-Male

Young males sometimes exhibit sexual behavior toward one another. For example, a male may chase another male with an erection much as he would a female. A male with an erection may be observed to reach over the back of another male, and even mount him, much as he would with an estrous female (**Male-Male-Chase, Male-Male-Reach-Over, Male-Male-Mounting**). The mounting male's penis is sometimes rubbed rapidly—using its own muscle-power—up and down between the other male's back legs. Ejaculation sometimes occurs (Kahl and Armstrong, personal communication). This behavior is most often seen between younger bulls during a friendly interaction (Moss 1988; Poole 1996).

Play

Solicit Play

Many of the aggressive and affiliative signals detailed previously are found in play (see also Lee and Moss, chapter 14). However, when they are used in play, these signals are (a) out of context and typically out of sequence and (b) exaggerated in their nature and expression. In playful or expectant situations, an elephant may pause with the trunk up in a periscope or S-shape waiting for its partner's next move. As two individuals approach each other with intent to spar, one (occasionally both) raises the trunk above the head and curls the tip toward the other individual in a **Distant-Frontal-Attitude** (Kühme 1961, 1963). Except for context, this display appears very similar in form to *Periscope-Sniff* (see General Attentive). Individuals may adopt several other postures in anticipation of, or when soliciting, a playful interaction. In **Solicit-Play** an elephant stretches its head down and out and looks out over its tusks at a potential playmate. The trunk may be resting folded on its tusks or placed in its mouth, the ears resting against its body (Poole, personal observation). In an invitation to play, or when an individual is considering an object of play or engaging in play it gently waggles its head from side to side (**Head-Waggle**; Poole [1996]; Kahl and Armstrong, personal communication). To solicit play with younger individuals, an older calf lies down, or gets down on its knees (Lee 1986; Poole 1998b), thereby appearing smaller, and turning off the threat latent in many play moves. This behavior appears to attract younger calves who wander over and climb on top or *Climb on* (see Social Play, below) the older individual. In similar fashion, a larger male may lower himself down on his knees, **Kneel-Down**, in order to playfully spar with a smaller male who has shown signs of being afraid to participate (Moss 1988; Poole 1996; see figure 8.14).

Lone and Object-Play

Young elephants, in particular, engage in numerous forms of object and lone-locomotor play. An elephant may sit down, raise its trunk high in the air and then allow it to flop down on its head, and repeat these actions over and over (**Sitting**; Poole, personal observation). **Trunk-Squelching** is an often repeated wrinkling up of the trunk and forcing air through it to produce a "squelching" sound (Poole, chapter 9). Or an elephant may suck water into its trunk and then spray it out while swinging its trunk like a child with a garden hose (Poole and Granli, personal observation). This appears to be done for amusement.

Elephants are very curious and are quick to investigate and play with anything new in their environment. Elephants are particularly engaged by novel man-made "toys" (e.g., a used cement bag, a cushion, paper bags, a film canister, flip-flops, tins, cans, a piece of cloth), though a palm-frond, swamp vegetation, or a stick will also do (see figure 8.15). This kind of **Object-Play**, especially with novel objects, can last 10 minutes or longer with the elephant totally absorbed. He or she may alternatively kick back at it, roll it under a foot, touch it with a hind foot, pick it up, pierce it on a tusk, bite it, swing it around, put it on its head or back, throw it away, and finally, focus attention on it in silent deliberation—only to retrieve it again and begin to toy with it once more. Similar behavior may be observed when elephants come across the bones of their own kind. They commonly pick up an elephant bone, mouth it, bite it, step on it, roll it under a foot, wave it about, and toss it. The actions while toying with bones are very gentle and done in quiet contemplation (Douglas-Hamilton 1972; Poole 1996; Moss 1992; Payne 2003). Kicking back at an object, another individual,

Figure 8.14 When soliciting play with younger individuals, an older calf may attempt to appear smaller by getting down on its knees, *Play-Kneel-Down*, or lying down, which stimulates younger animals to clamber on top in a *Play-Climb-Upon*. In this photograph the older individual also displays a *Play-Tusk-Ground*.

Figure 8.15 In *Object-Play*, an elephant's curiosity is easily stimulated by a novel object. The "toy" may be kicked or touched with a hind foot, rolled under a foot, picked up, pierced by a tusk, bitten, swung around, placed on the head or back, tossed, and then quietly contemplated.

or simply at the air is a common action during play (**Play-Kick-Back**) that may be repeated for long periods, at an object of interest. While two or more play together, an elephant may guard his "toy" from access by another. Other youngsters may queue up and appear to wait quietly for their turn with the toy (Lee 1986; Moss 1988; Poole1996, 1998b; Poole and Granli 2003, 2004).

As play becomes more exuberant an elephant may tusk the ground and lift up vegetation (Kühme 1961; Poole 1987a, 1996). *Play-Tusk-Ground* is often a form of lone play although it can also occur during social play. Exuberant play is usually associated with *Play-Trumpets* (Poole, chapter 9), which attract the interest of other elephants and leads to social play. Highly energized, playful elephants enjoy running back and forth through long grass or shrubs (**Play-Bush-Bashing**) or lifting up and throwing objects (**Play-Throw-Debris**). Many of these behaviors—*Tusk-Ground*, *Bush-Bashing*, and *Throw-Debris*—are observed in escalated contests between musth males.

However, in play elephants adopt contradictory postures. They stand or run about in **Feigned-Fear**, adopting the *Jaw-Tilted-Upwards* and *Tail-Raising* as if highly alarmed. With *Eyes-Wide* they stare out over their tusks at "imaginary enemies," alternately charging in *Bow-Neck* type posture or running away as if fearful (Moss 1988; Poole 1996). As part of this high-spirited play elephants spin or charge and turn about (**Pirouette-Run**), often *Standing-Tall* or *Head-Shaking*, and *Trumpeting*. Unlike an aggressively charging elephant, the running gait of a playful elephant is loose and floppy; the individual shakes a lowered head from side to side, allowing its ears to flap against its neck and curling its tail up high. **Floppy-Running** is associated with *Pulsated-Trumpets* (Moss 1988; Poole 1996; Poole, chapter 9).

Social Play

The best-known type of social play is **Sparring**, which occurs predominantly between males. Young males in particular test one another's strength by placing their trunks on one another's heads and pushing their opponent down and back. Unlike fighting males, the ears are relaxed. *Sparring* can range from gentle trunk wrestling and pushing in a playful or greeting context, and more boisterous shoving, to rough and aggressive tusking (Kühme 1961; Poole 1982; Lee 1986; Langbauer 2000). *Sparring* typically occurs between males, less frequently between a male and female, and rarely between two females (Poole, personal observation). *Sparring* may be initiated by a bout of **Trunk-Twining**, a gentle contact form of play where two elephants twist their trunks together in a spiral (Lee 1987). It may be interrupted by a pause in which one individual runs after another with the head, ears, and tail raised. **Play-Pursuit** partners switch between chasing and being chased (Lee 1986). One individual may grasp the tail (**Grasp-Tail**) of another, and can pull on the tail (or other body parts) in a playful situation (Moss 1988) During periods of play, and often after a play-chase, young calves and juveniles of both sexes **Play-Mount** others irrespective of the other's sex (Lee 1986; see Lee and Moss, chapter 14).

In a gentle form of play, elephants **Play-Social-Rub**, pushing at the body, head, or legs of another, who may be lying down or playing with another elephant (Lee 1986). Young calves will also climb on top of another elephant who is lying down, pushing and shoving against the lower elephant, or they will lie fully ventrally on the other and wiggle. This game of **Climb-On** play may lead to a big pile of wiggling, kicking out, and squirming elephants (Lee 1986; Moss 1988; Poole 1996). This behavior also can be a prelude to chases, *Sparring*, or gentle *Trunk-Twining* between two calves (Lee 1986; Moss 1988).

Generally Attentive

Finally, there are a number of postures and gestures that are components of an elephant's sensory input. A **Sniff-Toward** is the most commonly used form of sniffing. The trunk is held relatively straight and pointed in the direction of interest. The trunk lifted up in an s-shape, or **Periscope-Sniff** (see figure 8.16), is used to detect scents carried on the wind (Rasmussen and Schulte 1998; Poole 1999a) and is particularly used when additional information is required, such as

Figure 8.16 An elephant's nose is extraordinarily flexible and may sniff toward an object of interest in a variety of positions, which tend to be adopted in specific circumstances. The elephant in the foreground uses a *Periscope-Sniff* to detect a scent carried on the wind. This position is frequently adopted when additional information is required, such as when meeting strangers. The smallest elephant uses a *J-Sniff*, often seen when elephants are monitoring an ongoing situation. The other two use the *Sniff-Toward*.

when meeting strangers or danger (Poole, personal observation). In the **Hovering-Sniff**, the trunk is held suspended over a particular scent on the ground or object of interest (Rasmussen and Schulte 1998; Poole 1999a). The shape of the trunk may take many forms and may be surreptitious, as in the **J-Sniff**, where the trunk tip is curved slightly back as if the elephant is attempting to appear not overly interested. Younger males in the company of a musth male and an estrous female often employ this surreptitious form of sniffing, as if not to draw undue attention to themselves. When an elephant is curious about an object or scent, but dares not approach more closely, it reaches out with a full extension of the trunk in a **Horizontal-Sniff** to inspect the item (Poole, personal observation; Kahl and Armstrong 2000). During **Tracking**, an elephant's trunk moves regularly back and forth following a scent trail along the ground as the (usually) male walks purposefully and persistently in prolonged pursuit of another individual (Poole 1999a). This behavior can also be seen when an elephant tracks a person.

Elephants are capable of discriminating between the voices of elephant individuals over distances of up to 2 km (McComb et al. 2003); such auditory recognition is associated with listening behavior. Listening behavior is also observed when an elephant is aware of other socially interesting sounds or those associated with potential danger. An individual rarely stands stock-still except when *Listening* or resting; usually some part of the body, ears, trunk, or tail is in motion. A resting elephant relaxes its head and ears, allowing its head to hang below its shoulders and its ears to flop forward. A **Listening** elephant stands with its head raised and its ears lifted and slightly extended at an angle of at least 45 degrees (see figure 8.17). The body and extremities of a *Listening* elephant suddenly cease moving, as it simultaneously raises its head and stiffens its ears. Sometimes an elephant may turn its head from side to side to localize a sound (Poole et al. 1988; Heffner et al. 1882). Similarly, during **Freezing** an individual or group of individuals suddenly ceases all movement, holding stock-still, apparently *Listening* or feeling distant vibrations (Moss 1988; Langbauer et al. 1989).

African elephants normally stand or move with their eyes cast down. A direct gaze with **Eyes-Open** is a component of many postures. An elephant may look in the direction of interest with its *Eyes-Open*, visually attending to objects (Poole, personal observation). An attentive elephant may slowly open and close, or blink its eyes. This **Eye-Blinking** behavior is perhaps most noticeable during suckling, but can be combined with apprehensive head tilting, or observed when an elephant is attentive and otherwise motionless (Poole, personal observation; Lee, personal observation). When **Monitoring** an unfolding situation, particularly an interaction, an attentive individual holds its ears slightly lifted and partially extended, and gazes with *Eyes-Open* or *Eyes-Blinking* at the subject of interest. Simultaneously, the elephant sniffs by pointing its trunk tip in the direction of interest, the fingers of the trunk typically opening and closing (Poole, personal observation).

In **Explore-Touch** an elephant uses a foot (particularly the hind foot) or its trunk to investigate an elephant or object. This behavior can be seen in object play and is often associated with *Play-Kick-Back*, or as a form of *Shepherding*, but is perhaps most clear and prolonged when an elephant investigates a dead elephant or elephant bones when it may touch or stroke the body or bones of the deceased animal (Douglas-Hamilton 1972; Moss 1988; Moss 1992; Poole 1996; Payne 2003, McComb, Baker, and Moss 2006).

Death

A description of elephant behavior would be incomplete without mentioning their behavior around injured, sick, dying and dead elephants, and their remains. Similar to the Retrieving behavior observed in the care of infants, elephants may use the tusks, trunk, or feet to attempt to lift and even carry sick, dying, or dead elephants (Douglas-Hamilton 1972; Moss 1992; Poole 1996; Payne 2003; Douglas-Hamilton et al. 2006). Males may attempt *Mounting* a dead elephant. Elephants have been observed to feed those who are not able to use their own trunks to eat (Moses Kofi Sam, personal communication) and to attempt to feed elephants who have died (Croze, cited in Moss [1982]). They are also known to collect vegetation and dirt with the trunk, feet, or tusks and use it to cover a dead elephant or dead human (Moss 1992; Poole 1996). When an elephant

Figure 8.17 A *Listening* elephant stands with its head raised and its ears lifted and slightly extended at an angle of about 45 degrees or more. The body and extremities of a listening elephant suddenly cease moving, as it simultaneously raises its head and stiffens its ears. Sometimes an elephant may turn its head from side to side to localize a sound.

has been responsible for the death of a person or another animal they have been observed to vigorously flatten the ground (**Trample-Ground**) around the animal (Poole 1996). Similar behavior is seen when a baby is born (Poole 1996). Finally, elephants may engage in **Body-Guarding**, standing over the body of a dead elephant or person and protecting it from the approaches of predators or other elephants, and threatening approaching individuals (Croze, cited in Moss 1982; Douglas-Hamilton 1972; Poole 1996; Payne 2003). As mentioned before, elephants typically also investigate the remains of elephants sniffing with the trunk, using the feet and trunk to Explore-Touch, lifting, carrying, and playing with the bones or considering them in quiet reflection (Douglas-Hamilton 1972; Moss 1988, 1992; Poole 1996; Payne 2003; McComb, Baker, and Moss 2006).

Conclusions

The visual and tactile signals, gestures, and behaviors of African elephants are extremely intricate and highly varied. Some behaviors are visually dramatic and powerful, even terrifying to the casual observer. Other gestures are exquisitely understated, requiring a highly trained eye to discern the subtle shift in movement and posture. This chapter has been an attempt to introduce others to many of the behaviors we use in our quest to understand and describe the social, emotional, and cognitive world of African elephants. These behaviors are, in effect, our "tool box." Over many years, we have learned that certain relationships (e.g., mother-infant, mother-daughter, male competitors, sexual partners, friends, foes) are characterized by specific behavioral patterns made up of explicit components. Armed with this knowledge, we feel confident in interpreting elephant behavior, interactions, and relationships wherever we find them. We hope that through this chapter we have been able to impart some of our knowledge for others to use toward the furthering of scientific understanding and improving the conservation and welfare of elephants wherever they may be.

Chapter 9 Behavioral Contexts of Elephant Acoustic Communication

Joyce H. Poole

THE FISSION-FUSION society of elephants is built upon a complex network of social relationships within and between families, bond groups, and clans (see chapters 13 through 16) and between individual males (see chapters 17 and 18). Added to this multi-layered social network are fleeting interactions and temporary consortships that form between reproductively active males and cycling females. This elaborate system of associations, partnerships, coalitions, and enduring relationships is in part established, mediated, and maintained via an intricate suite of acoustic signals (see also chapter 10). The survival of females and their offspring depends upon the cohesion and coordination of the family and upon their ability to compete with other groups for access to scarce resources (McComb et al. [2001], and see also Mutinda, Poole, and Moss, chapter 16). Consequently, most calls produced by female and juvenile elephants give emphasis to the importance of the social unit. Family members call to reinforce bonds between relatives and associates, to care for calves, to reconcile differences between "friends," to defend close associates, to form coalitions against aggressors and predators, to coordinate movements, and to keep in contact with one another over long distances (Poole et al. [1988]; Poole [1994a]). Adult male elephants lead relatively more independent lives (but see chapter 17) than do females, where reproductive success and survival depend upon an individual's ability to detect sounds made by others; males tend to use calls to advertize their sexual state, identity, and rank (Poole 1999a; chapter 18).

In this chapter, I build upon an established body of work on elephant communication to examine the calls emitted by African elephants, *Loxodonta africana*, from the standpoint of social function. This chapter does the following:

- Proposes several new call types.
- Describes a broad range of behavioral contexts and examines, qualitatively and quantitatively, the acoustic signals with which they are associated.
- Makes a first attempt to discriminate between proposed contextual call sub-types based on a number of acoustic measurements.
- Uses these and contextual differences to illustrate how vocal signals form an essential and integral component of the complex dynamics of elephant society.

I begin by describing what is known about elephant acoustic communication and vocal repertoire, in general, and then examine in detail how elephants in Amboseli use acoustic signals in the context of what we know about their sociality.

Elephant Sound Production and Detection

African elephants produce a broad range of sounds from very low frequency *rumbles* to higher frequency *trumpets*, *snorts*, *barks*, *roars*, *grunts*, and *cries* (e.g., Berg 1983; Poole et al. 1988; Poole 1994a; Langbauer 2000; Leong et al. 2003; Poole and Granli 2004; Stoeger-Horwath et al. 2007) as well as a range of idiosyncratic, novel, and imitated sounds (Poole et al. 2005).

Two features make elephant vocal production unusual: their very large size and the inclusion of the trunk in the vocal tract (Soltis 2009). Elephants are capable of producing powerful, very low frequency sounds in part due to their large body size and correspondingly large vocal organs. The majority of elephant sounds are laryngeal in origin, produced by air passing from their large lungs over the vocal chords or larynx, a structure reportedly 7.5 cm long, suspended on the hyoid apparatus (Sikes 1971). The elephant's vocal tract includes the extent of the trunk, the honeycomb nasal passages in the skull, and the length from the lips to the larynx and has been estimated to be 2.5 m (Soltis 2009). This extended resonator is also unusual in that most of it (some 1.8 m of trunk) lies outside the cranium, thus providing great flexibility in both length and shape and affecting the sounds produced accordingly. Two more anatomical features may influence sound production by affecting the size and shape of the elephant's vocal tract. First, the structures of the hyoid apparatus (a series of bones at the base of the tongue) and the musculature that support the tongue and the larynx in elephants are different from other mammals. The hyoid apparatus of elephants has five rather than nine bones, and these are attached to the skull by muscles, tendons, and ligaments, rather than by bones, as in most other mammals. This rather loose arrangement allows for a greater movement and flexibility of the larynx (Shoshani 1998). Second, this looser arrangement also houses a pharyngeal pouch, a structure unique to elephants located at the base of the tongue, which in addition to providing an emergency source of water, appears to function in the production of low frequency calls (Shoshani 1998).

Elephants also produce *trumpets*, *snorts*, and some idiosyncratic sounds by blowing air through the trunk. By altering the positioning of the trunk and the speed and duration of air moving through it, elephants can produce an astonishingly versatile mixture of sounds via this "second voice," as Soltis (2009) refers to it.

Research by Heffner and Heffner (1980, 1982) demonstrated that elephants have very good low frequency hearing (frequencies below 17 Hz not tested) and very accurate localization skills. Together with Sirenians, they are unique among modern mammals in having reverted to a reptilian-like cochlear structure (Fischer 1990) that may facilitate greater sensitivity to lower frequencies (O'Connell, Hart, and Arnason 1998). Furthermore, elephants are able to detect seismic vibrations, or Rayleigh waves, through two possible means: (1) Bone conduction and the use of massive ossicles of their middle ears (Reuter, Nummela, and Hemila 1998); and (2) mechanoreceptors in the toes or feet that are sensitive to vibrations (O'Connell, Hart, and Arnason 1998). The tip of an elephant's trunk has densely packed layers of cells called Pacinian corpuscles that are extremely sensitive to vibrations (Rasmussen and Munger 1996). O'Connell-Rodwell et al. ([2006] and personal communication) and Bouley et al. (2007) suggest a high density of these cells in the front of the foot and along the edges, consistent with the notion that elephants are sensing seismic signals when they suddenly freeze.

Elephant Acoustic Communication Background

Soltis (2009) provides a detailed review of the literature on African elephant vocal communication; therefore, I provide only a summary here.

Berg (1983) made the first descriptions of African elephant call types in captivity and showed that elephants produced more high frequency calls (e.g., *trumpets*) during periods of high excitement. Soon thereafter, Payne, Langbauer, and Thomas (1986), also working in a captive setting, discovered that the low frequency rumbling sounds, or *rumbles*, made by Asian elephants (*Elephas maximus*) contain frequencies below the level of human hearing, or infrasound. Payne, Langbauer, and Thomas described fundamental frequencies of 14 to 24 Hz, with some calls produced at very high sound pressure levels. Since very low frequency sound attenuates more slowly than higher frequency sound (Dneprovskaya, Iofe, and Levitas 1963; Eyring 1946; Ingard 1953; Wiley and Richards 1978), Payne, Langbauer, and Thomas proposed that elephants might use very low frequency sounds at high sound pressure levels to communicate with one another over long distances. Their discovery prompted a flurry of studies on infrasound in free-range conditions.

In Amboseli, Poole et al. (1988) established that the *rumbles* of African savannah elephants, *Loxodonta africana*, also extend into the infrasonic range and documented frequencies as low as 14 Hz and sound pressure levels of up to 103 dB (re 20 µPa extrapolated to 5 m from source), although later, some of these calls have been shown to be as low as 8 to 9 Hz (this chapter). Subsequently, Payne, Thompson, and Kramer (2003) established that African forest elephants, *L. cyclotis*, also produce low frequency *rumbles* with infrasonic components as low as 5 Hz.

Poole et al. (1988) also described several well-known and commonly heard rumble sub-types (e.g., *greeting-rumbles*, *contact-calls* and *answers*, *"let's-go"-rumbles*, *musth-rumbles*, *female-rumbles*, *chorus-rumbles*, *post-copulatory-rumbles*, and the *mating-pandemonium*). Using specific examples in which vocal exchanges between affiliated females or mating pairs resulted in specific changes in behavior, they argued that long-term records on the behavior of elephants and the contexts of these specific *rumble* sub-types showed that elephants make use of low frequency *rumbles* in the spatial coordination of groups and as they search for mates (see also Poole and Moss [1989]). Much later, Soltis, Leong, and Savage (2005a) documented that vocal exchanges or "anti-phonal calling" such as that observed among wild elephants (e.g., Moss 1981; Poole et al. 1988; Poole and Moss 1989; Poole 1999a; McComb et al. 2000) also occurs among females in a captive setting and that, as we have argued, these constitute true communication events (Leighty et al. 2008).

Since the early work of Payne, Langbauer, and Thomas (1986) and Poole et al. (1988), theoretical and experimental studies have confirmed that elephants are able to broadcast and detect acoustic signals over several kilometers (Larom et al. 1997; Langbauer et al. 1991; Garstang et al. 1995), and playback experiments have shown that they respond appropriately to specific calls (Langbauer et al. 1991; Poole 1999a; McComb et al. 2000, 2001, 2003). Moreover, McComb et al. (2003) and chapter 10 (McComb, Reby, and Moss) showed that elephants are capable of discriminating between individual voices at distances of up 2 km. The range of acoustic communication may more than double at night, due to temperature inversions that typically form before sunset and decay with sunrise and could travel up to 10 km (Garstang et al. [1995], though even under the best conditions the distance is likely to be significantly less for individual recognition). Furthermore, O'Connell-Rodwell, Arnason, and Hart (1997) discovered that when an elephant emits a low frequency *rumble*, a corresponding seismic wave with similar characteristics is transmitted in the ground (O'Connell-Rodwell et al. [2006] were able to show via playback experiments that elephants are able to detect this seismic component and to discern subtle social cues from this information [O'Connell-Rodwell et al. 2007]).

Recently, there has been increasing focus on the vocal repertoire of the African elephant, producing a growing body of literature on the acoustic structure of different call types and sub-types, and attempts to describe the social signals contained therein. Expanding on Berg's early work, Leong et al. (2003), Stoeger-Horwath et al. (2007), and this chapter have described a variety of different call types. These are described in detail below and may be listened to online at www.elephantvoices.org/multimedia-resources/elephantvoices-calls-database-call-types.html.

Adding to the typical call types of African elephants, Poole et al. (2005) discovered vocal imitation in elephants, documenting the first case in a non-primate terrestrial mammal, and strengthening the hypothesis that a primary selection pressure for vocal learning involves the communicative demands of maintaining social relationships in fluid societies. Poole et al. suggested that the evolution of vocal imitation in humans, some birds, bats, dolphins, and elephants might result from the need to use acoustic signals to maintain individual-specific bonds when animals separate and reunite. This hypothesis predicted vocal learning in other species where long-lived social bonds are based upon individual-specific relationships, with fluid group membership, and where vocal communication is used for maintaining contact and individual or group recognition. The capacity for vocal imitation is a further indication of the elephant's highly flexible vocal tract and suggests that we should expect them to exhibit a wide range in sound production.

As noted by Soltis (2009), call types can be separated into sub-types from a structural or functional standpoint, and the relationship between the two can take many forms. Following on from Poole et al. (1988), Poole (1994a) proposed a long list of *rumble* and other call sub-types based on behavioral context and sound quality. Yet, working in captivity and based on acoustic structure, attempts by Leong et al. (2003) to subdivide *rumbles* into five sub-types was largely unsuccessful, and they and later Soltis, Leong, and Savage (2005b) concluded that *rumbles* are best viewed as a single call type with graded variation. Also working in captivity, Stoeger-Horwath et al. (2007) drew similar conclusions for calf *rumbles*. Using cluster analysis on calls recorded in the wild, however, Wood et al. (2005) were able to divide *rumbles* into four sub-types. Other researchers have separated *rumbles* into a priori categories and then, using multivariate statistical analysis, have established differences in acoustic structure across the previously defined groups. In this manner, researchers have shown differences in *rumble* structure between individual callers (wild: McComb et al. [2003]; captive: Soltis et al. [2005b]) and between broad social contexts in which calls were produced (wild: Wood et al. [2005]; captive: Soltis, Leong, and Savage [2005b]; Soltis et al. [2009]). Poole and Granli (2004), too, were able to document structural differences in *trumpet* sub-types produced during specific contexts associated with exuberant play behavior. While these studies demonstrated statistically different means across call sub-type categories, each of the studies showed overlap in acoustic structure, which is consistent with graded rather than discrete structural sub-types.

While such overlap in acoustic structure exists, playback experiments using *rumbles* and *trumpets* recorded under specific contexts show that elephants are able to discern subtle differences between these call sub-types and respond appropriately. For instance, the playback results of Poole (1999a) show that elephants distinguish between two *rumble* sub-types (*post-copulatory* or *estrus-rumbles* and *musth-rumbles*), and these responses are different, again, from those of elephants to playbacks of *contacts-calls* (Poole unpublished, McComb et al. [2001]). Playbacks also show that elephants can discriminate between different *trumpets* (Poole et al., in preparation).

More recently, researchers have started to look into the effect that context has on *rumble* structural forms. Soltis, Leong, and Savage (2005b); Soltis et al. (2009); and Li et al. (2007) provided evidence that *rumbles* produced by lower-ranking individuals during agonistic interactions with those of higher rank show increased and more variable fundamental frequencies and amplitudes and increased call durations compared with the *rumbles* they produced in periods of social tranquility. Similarly, Wood et al. (2005) documented that *rumbles* associated with "social interactions and agitation" were characterized by increased and more variable fundamental frequencies as well as decreased duration. Soltis et al. (2009) also showed that high-ranking opponents produced *rumbles* with features that signal large body size: decreased fundamental frequency and formant dispersion and increased amplitude and duration when they were interacting with one another. Furthermore, Stoeger-Horwath et al. (2007) and Wesolek et al. (2009) showed that the *rumbles* of infants who were denied access to the breast were associated with more energy in the

BOX 9.1 METHODOLOGY

Equipment and Field Techniques

Between 1984 and 1990 and from 1998 through 2006, I recorded elephant vocalizations in Amboseli. During 1998, I also recorded from elephants in Laikipia and Maasai Mara, Kenya, and from semi-captive orphan elephants in Tsavo, Kenya. I made recordings through 1990 on a Nagra IVSJ (see Poole 1999a); between 1998 and 2003, at 44.1 kHz on an HHB PDR 1000 DAT recorder (frequency response: 8 Hz: −0.43 dB; 12 Hz: −0.26 dB; 15 Hz: −0.22 dB; 20 Hz: −0.15 dB; 60–1,000 Hz no roll off); after 2003, on a modified Nagra Ares BB (frequency response: 10 Hz: −1 dB; 20 Hz: −0.4 dB; 50 Hz: −0.2 dB; 100 Hz: −0.1 dB; 200–20,000 Hz no roll off). The majority of recordings were made with an Earthworks QTC1 omni-directional microphone (frequency response: 4 Hz–40 k Hz ±1 dB).

Field data included the general area and specific location, date, tape and track number, track start and end times, elapsed time, channel settings, group size and type, and the individuals present. When a call was heard, the elapsed time was noted as well as information on the call type, caller, and distance to the source. I use the term call *type* to refer to the broad structurally differentiated categories of sounds and the term *context-type* to refer to a priori subtypes initially differentiated from a combination of sound quality and social context. Call types are in italics (e.g., *rumble, roar, trumpet*, etc.), while specific call context-types are italicized and hyphenated (e.g., *begging-rumble, musth-rumble, low-intensity-husky-cry*).

Caller, call type, and context-type were recorded with level of confidence (A: certain, B: fairly confident, C: educated guess, D: no idea). Also noted was whether the decision regarding call context-type was based primarily on the quality of the sound or on its behavioral context. A call context-type assigned confidence "A" required that both the behavioral context and sound quality matched the context-type designation. Finally, any contextual or other comments about the situation or the behavior of the calling animal were noted.

After January 1999, I focused recording sessions on members of the EB family (n = ca. 27). Once I located the family I parked near a sub-group that provided good visibility (5–20 m to the nearest elephant). As I began recording, I noted the nearest adult female and her nearest adult female neighbor. Once the elephants moved greater than about 25 m away, I moved the vehicle again. Though many elephant calls persist over long distances, the best quality calls and field data are typically from the closest individuals; accordingly, I varied the nearest elephant.

Elephant ages were categorized in two ways: by absolute age (years) and by age class (0A, 0B, 1A, 1B, 2, 3, 4, and 5; see General Methods). The terms infant (<6 months), calf (0–4.9 years), juvenile (5–9.9 years), and adult (>10 years) were used.

Sound Acquisition and Measurement

Sounds from 1998 to 2006 were systematically logged from field notes into a custom-designed MS Access Database with a Visual Basic Interface (n = 6,592). Using the sound analysis program SIGNAL RTSD, calls were acquired through a low pass anti-alias filter onto a Gateway 2000 and saved to disk as .wav files (n = 3,934). Low frequency calls were acquired at a sample rate of 2,000 Hz and displayed at a range of 0–500 Hz, while higher frequency calls were acquired at 22,050 Hz and displayed at range 0–5,000 Hz. Due to the high frequency of overlapping calls, measurement was not automated. Time-frequency measures were taken in Signal 4.0 from the spectrograph view using the cursor, and additional measurements were taken using the selection function in Raven 1.0 (see table 9.1 for a description of measurements).

Statistical Analysis

Of the measured calls, 70 percent (confidence A, n = 2,014) were recorded from 59 elephants whose identities were known. Of known callers, some contribute as few as one call to the data set, while all except one adult female member of the EB family each contribute over 30 calls. Nevertheless, once call context-types and level of confidence are considered, sample sizes per individual become considerably smaller. Many call types (e.g., *roars*) and some context-types (e.g., calls directed at predators) are only rarely produced. In addition, the use of particular context-types is unevenly distributed across individuals by age, sex, reproductive state, and even by personality. For example, confident elephants are more likely to emit *"let's go"-rumbles* (see below), male calves are more likely to protest when denied access to the breast, *begging-rumbles* are only produced by calves, *musth-rumbles* only by males in musth, *estrous-rumbles* only by females in estrus, and mothers with small calves contribute more calls to infants to the higher frequencies compared to *rumbles* by infants in other contexts.

These studies provide fertile ground with which to examine calls recorded in the wild from known individuals in known behavioral contexts. Here I attempt to present a picture of how elephants in a natural setting use vocalizations to communicate with one another and to compare the form that these calls take under a wide range of situations (see box 9.1 and table 9.1).

data set than those without. The data set is, therefore, unbalanced across individuals and contains calls from individuals at different ages across many years. The analysis is further complicated because we know that it is possible to discriminate between callers (McComb et al. 2003; Soltis, Leong, and Savage 2005b) and because we know that emotional state (Soltis, Leong, and Savage 2005b), as well as age and body size, affect call duration and fundamental frequency, two key variables (see figure 9.3). Due to the complexity of the data set, my statistical approach was tailored to the availability of specific data and is explained in further detail within the relevant sections.

In general, however, I used a combination of non-parametric descriptive statistics, and stepwise discriminant function analysis (DFA) with cross-validation to examine patterns across elephant calls. In each of a series of stepwise DFAs, I partitioned the data, using two-thirds of the data set to train the model and the remaining one-third to test which group the calls belonged to. This cross-validation method characterizes the ability of the model to recognize the group membership of the calls. In some discriminations, individual callers were largely unknown (for instance, in the case of trumpets and other high frequency calls). In these cases, the data were partitioned by random selection, two-thirds for training and one-third reserved for cross-validation. In the analysis of the more commonly heard and recorded rumbles, in which the caller was known with confidence A or B, the data were first divided into those produced by adult females and those by calves. Within each data set, calls were partitioned such that individuals who contributed to training the model were different from those used in cross-validation. Adult female calls (*cadenced*, "*let's go*," *contact*, *greeting*, *little-greeting*, and *coo*) included individuals across a broad spectrum of ages, from age 10 to 55, while calls produced by calves (*begging*, *separated*, *as-touched*, *baroo*, and *grumbling*) included between the ages of 0–9.9 years. Calls emitted by both adult females and calves (e.g., *coo*, *little-greeting*, *grumbling*, etc.) were considered within the set containing the largest sample size. The inclusion of such a broad range of ages and the partitioning of the data such that different individuals contributed to the testing and validation means that the results presented are very conservative.

Adult Female Rumbles

The adult female rumbles analyzed using DFA included *estrous*, *female-chorus*, *mating-pandemonium*, *cadenced*, "*let's go*," *contact*, *greeting*, *little-greeting*, and *coo*. Adult female reproductive calls were very unevenly distributed among individuals and due to the nature of the sample, I was not able to use cross-validation. These three calls were therefore analyzed separately (see under Mating Signals).

The overlapping nature of *greetings* and *little-greetings* prevented the inclusion of a number of predictor variables; they were, therefore, also analyzed separately (see below). *Greetings* and *little greetings* were compared using a balanced data set consisting of four calls of each type contributed by six females (Echo, Eleanor, Eliot, Elspeth, Enid, Erin, and Eudora) ranging in age from 10 to 55 years.

The data set for the remaining adult female calls (*cadenced*, "*let's go*," *contact*, and *coo-rumbles*) was unbalanced with respect to individuals, so I partitioned the data, using two-thirds of the data set with six individuals (Echo, Eudora, Enid, Eleanor, Eliot, and Elspeth) to train the model and the remaining one-third with four different individuals (Ella, Erin, Edwina, and Emma) to test call membership. The inclusion of such a broad range of ages and the partitioning of the data such that different individuals contributed to the testing and validation mean that the results presented are very conservative.

Calf Calls

Among calves, call duration and fundamental frequency change very rapidly with increasing age, thus making discriminations between proposed call types more difficult. In addition, due to the short longitudinal window, the data set was even more unbalanced than that for adult elephants. I was, therefore, forced to compare groups of similar context types, rather than across the entire data set. I first used a stepwise DFA with cross-validation to test whether discrimination between *rumbles* given in the context of begging versus those given in the context of separation from mother was possible. Next, I used a stepwise DFA with cross-validation to discriminate between three proposed call types, two given by calves in the context of being comforted and another given in context of being thwarted. In each analysis, I used two different groups of individuals, one set to train the model and another for cross-validation. Since the *rumbles* in the latter discrimination were represented by characteristic contours, I added a score for contour to the model (see table 9.1).

Table 9.1 Acoustic measures

Main acoustic feature	Parameter measured
Manner features	Feature 1: Tonal—a call or portion of a call showing distinct harmonics (frequency contours of the harmonics in tonal signals: flat (0); ascending, descending (1); bent, bent left, bent right, slightly arched, slightly wavy (2); bimodal, multi-modal, jittery, wavy (3); skewed left, skewed right (4); arched (5).
	Feature 2: Tonal with noise—A sound displaying distinct harmonics with noisy sections (frequency contours above apply).
	Feature 3: Noisy harmonics—a call or portion of a call showing noisy harmonic bands (frequency contours above apply).
	Feature 4: Noisy—a call or portion of a call with no distinct harmonics or overtones.
	Feature 5: Pulsatile
	Feature 6: Transient noisy
	Feature 7: Transient tonal
a. Duration: Cursor placement—Signal	Signal duration. Units ms.
b. Fundamental frequency: Cursor placement—Signal	F_0 start, F_0 end, F_0 minimum, F_0 maximum. Units: Hz.
c. Location of F_0 maximum: Cursor placement—Signal	The location of F_0 maximum (elapsed time from F_0 start to F_0 maximum) as a percentage of duration
d. Curve: Cursor placement—Signal	Range F_0/Duration
e. Dominant frequency: Selection "Maximum Frequency"—Raven	The frequency at which maximum power occurs. Units: Hz.
f. Bandwidth: Selection—Raven.	Upper frequency bound of the selection/ highest visible frequency of the call. Units: Hz. Not a particularly reliable measure due to variable recording distances.
g. Maximum time: Selection—Raven	In a spectrogram view, the first time in the selection at which a spectrogram point with the power equal to Maximum Power occurs. Units: ms.
h. Peak time: Selection—Raven	In a waveform view, the first time in the selection at which a sample with amplitude equal to peak amplitude occurs. Units: ms.
i. Sound pressure levels	Measurements are (re 20 µPa) extrapolated to 5 meters and from previously published work (Poole et al. 1988). Units: dB. Available for a few call types.

Range in Frequency and Duration of Elephant Calls

Elephant calls recorded in Kenya (mostly in Amboseli) contain frequencies ranging over more than 10 octaves from a minimum fundamental frequency of 8 Hz measured in some *rumbles* to maximum frequencies of around 10,000 Hz produced in some *trumpets* and *snorts*. Indeed, within a single call the fundamental frequency may range over 4 octaves, starting with a *rumble* at 27 Hz grading into a *roar* at 470 Hz. Sounds produced by African savannah elephants range in duration from less than a tenth of a second to almost 15 seconds ($n = 2,299$; see table 9.2 and figure 9.1).

Call Type Classification

The vocalizations produced by African elephants can be divided into acoustically and structurally distinct call types within which there are varying degrees of graded variation. Following from the work of Stoeger-Horwath et al. (2007), I divided calls into two main categories based on the way that the sound was produced. Laryngeal calls are those originating in the larynx and trunk calls are those produced by a blast of air through the trunk. I further categorized putative call types by ear and by manner features (those that specify the manner of articulation: in other words, how the tongue, lip, and other vocal organs are involved in making a sound)

Table 9.2 Time frequency characteristics across all measured calls

All calls	N	Median	I-Q range	Range
Duration (ms)	2219	3,248	1,671–4,818	87–14,372
F_0 start (Hz)	2108	18.0	14.3–26.2	8.6–727
F_0 max (Hz)	2099	20.5	16.4–29.6	11.7–1,016
F_0 min (Hz)	2113	16.6	13.0–24.0	8.4–658
Max Freq. (Hz)	1893	41.0	31.2–107.4	13.7–4,078
Curve	2098	0.001	.001–.002	.000–1.740
Bandwidth	2091	237	168–559	38–32,000

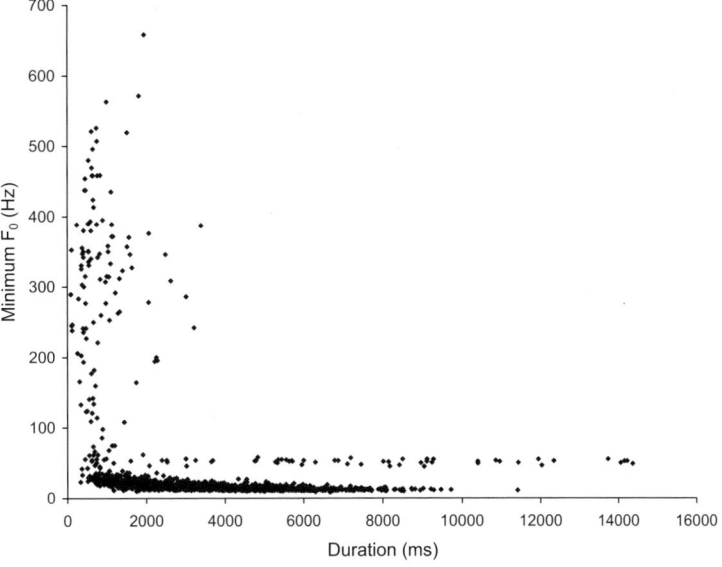

Figure 9.1 Duration and minimum fundamental frequency of all elephant calls. Note the learned truck-like calls produced by a single individual stand out as distinct from the other calls at approximately 50 Hz and multiple durations.

based on visual inspection of the spectrogram, for instance tonal versus noisy features (Fant 1960; see also table 9.1), and by structural differences in time and frequency characteristics.

I tried to fit measured calls into previously established nomenclature based on available qualitative descriptions, spectrograms, and acoustic detail as well as exchange of sound files and discussions with Angela Stoeger-Horwath, Kirsten Leong, and Joseph Soltis. Working on captive animals and following on from Berg (1983), these authors have established the mutually exclusive call types: *rumble*, *rev*, *croak*, *trumpet*, *snort*, *chuff* (Leong et al. 2003) and, additionally, *bark* and *grunt* (Stoeger-Horwath et al. 2007). Each of these call types are also produced by free-ranging elephants (though I exclude *chuff*, which I believe to be an elephant cough/sneeze), although, due to the broader and more complex contextual possibilities offered in a natural setting, call types are heard in different frequencies and circumstances than in captivity, often resulting in different interpretations. I will not attempt to discriminate between the previously described types, but will, where possible, examine how they are used in a natural social setting. To these call types, I propose three additional types: *cry*, *husky cry*, and *nasal trumpet* (see below).

The elephant's capacity for vocal production learning or imitation creates the potential for the description of call types that, while structurally unique, may not be socially relevant (e.g., *croak* [Leong et al. 2003; this chapter] and *truck-like* call [Poole et al. 2005]). I also describe these call types (*croak*, *truck-like*, and *squelch*), listing them separately as Idiosyncratic.

Accordingly, after listening to many thousands of calls and examining and measuring 2,366 spectrograms from hundreds of individuals in a free-range setting, I propose 10 primary call types used in communication: the laryngeal call types (*rumble*, *rev*, *roar* [with sub-types *noisy*, *tonal*, and *mixed*], *cry*, *bark*, *grunt*, and *husky cry*); and the trunk call types (*trumpet*, *nasal trumpet*, and *snort*). In addition to these principal call types are three idiosyncratic calls: *croak*, *truck-like*, and *squelch*. Besides these call types, elephants frequently emit composite calls that grade from one type into another. This rich range of amalgamated calls includes *snort rumbles*, *roar rumbles*, *rumble roar rumbles*, *cry rumbles*, *bark rumbles*, and *trumpet rumbles*. Composite calls are most likely to be produced when elephants are disturbed or excited.

A list of call types is presented in table 9.3 and illustrated in figure 9.2, while broad structural characteristics of each are presented in table 9.4.

Laryngeal Call Types

Examples of laryngeal call types may be heard at www.elephantvoices.org/multimedia-resources/elephantvoices-calls-database-call-types.html?catid=3.

Rumble. Rumbles (Poole et al. [1988]; table 9.4a; figure 9.2a) are the most frequently heard call type in the repertoire of all age/sex classes of African savannah elephants and are easily distinguished from other types by their very low frequencies and clear harmonic structure. During a subset of 2,055 minutes of recording with the EB family in the

Table 9.3 A comparison of nomenclature (numbers represent structurally distinct call types)

	Poole (this chapter)	Stoeger-Horwath et al. (2007)	Leong et al. (2003)[c]	Berg (1983)
	Laryngeal calls			
1	Rumble	Rumble	Rumble	Growl
	Rumble	Rumble	Rumble (noisy)	Rolling growl
2	Rev	—	Rev	—
3	Roar/Roar rumble	Roar/Roar rumble	—[a]	Roar
	Tonal roar	Tonal roar	—	Cry[b]
	Noisy roar	Noisy roar	—	—
	Mixed roar	Mixed roar	—	—
	Pulsated roar	—	—	—
4	Cry/Cry rumble	Tonal-roar	—	—
5	Bark/Bark rumble	Bark/Bark rumble	—	Bark
6	Grunt	Grunt	—(No infants)	—(No infants)
7	Husky cry	Rumble	— (No infants)	—(No infants)
	Trunk calls			
8	Trumpet	Trumpet	Trumpet	Trumpet/Trump
	Pulsated trumpet	—	—	Pulsated trumpet
9	Nasal trumpet	—	—	—
10	Snort	Snort	Snort	Snort
	Imitated and idiosyncratic			
11	Truck-like	—	—	—
12	Croak	—	Croak	—
13	Squelch	—	—	—

Note: The dashed lines indicate that the calls were not mentioned by the authors.
[a] Soltis (2009) reports *roars* from same elephant group.
[b] Soltis (2009) interprets Berg's "*Cry*" as a *Trumpet*.
[c] Based on recordings sent to me by Leong, I believe her class *chuff* is a wheeze/sneeze.

first half of 2000, I made note of 519 calls in my field notes. Of those, 474 (91.3 percent) were *rumbles*. Since many *rumbles* would have gone undetected (due to their very low frequencies), this percentage is likely to be an underestimate. Similarly, in Leong's study 87 percent of all calls were *rumbles*. A large number of specific stimuli evoke rumbling sounds and, unlike other types of calls, it is not possible to generalize in a few sentences the contexts in which *rumbles* are emitted. These are, instead, described in detail in the sections that follow.

Rumbles originate in the larynx and resonate as they filter through the pharyngeal pouch, the nasal passages of the skull, and through the trunk. *Rumbles* can be produced with the mouth open or closed, though louder, more modulated *rumbles* tend to be associated with open mouths. *Rumbles* are highly variable, graded calls. *Rumbles* in this data set contain fundamental frequencies ranging between 8.4 and 34.2 Hz (table 9.4a) depending upon the caller's age, size, and excitement level as well as the call context-type (see below). Some *rumbles* are highly modulated, powerful sounds; others rise or fall in pitch; some *rumble* contours are flat, some undulate, and still others are jittery. *Rumbles* range in duration from less than half a second to almost 12 seconds and may be emitted as a soft fluttering whisper or an explosive throaty resonance with sound pressure levels up to 103 dB extrapolated to 5 m from source (Poole et al. 1988). Very powerful *rumbles* tend to be both highly modulated and rather noisy. *Rumble* bandwidths range from 38 to 996 Hz, with some of the variance being explained by the distance of the microphone from the source. Although there is considerable variation within age classes, the *rumbles* of older individuals are significantly longer in duration and lower in frequency than those of younger individuals (Duration: Spearman $R = 0.636$; $t(n-2) = 28.6$; $n = 1,210$, $P < 0.001$; minimum F_0: Spearman $R = -0.759$; $t(n-2) = -40.52$, $P < 0.001$; figure 9.3).

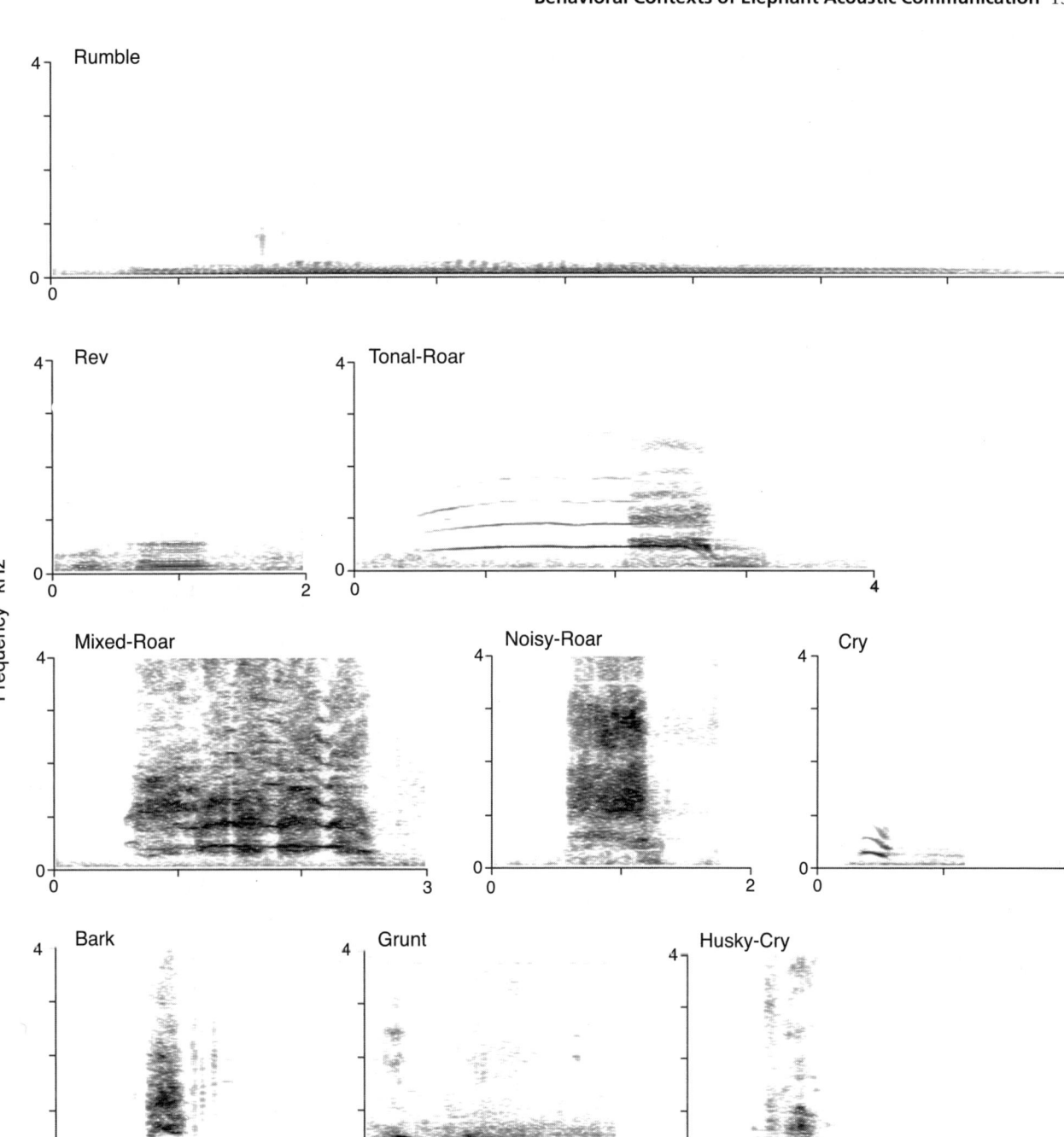

Figure 9.2a Comparative laryngeal call types.

Rev. The *rev* (table 9.4a; figure 9.2a) has been described by Leong et al. (2003) as a short tonal harmonic vocalization, less than a second in duration, and almost always followed immediately by a *rumble*. The *rev* has a harmonic structure and appearance similar to a short *rumble*, but its fundamental frequency at between 50 and 90 Hz is significantly higher than any known *rumble*. Based on recordings sent to me by Leong, which sounded buzzing or revving, I was able to locate four *revs* in my collection, although none of these occurred in a known context. Therefore, very little more than what Leong described is currently known about these infrequent call types. Their very rare occurrence in the wild suggests to me that they may be an artifact of captivity and possibly another example of a learned call.

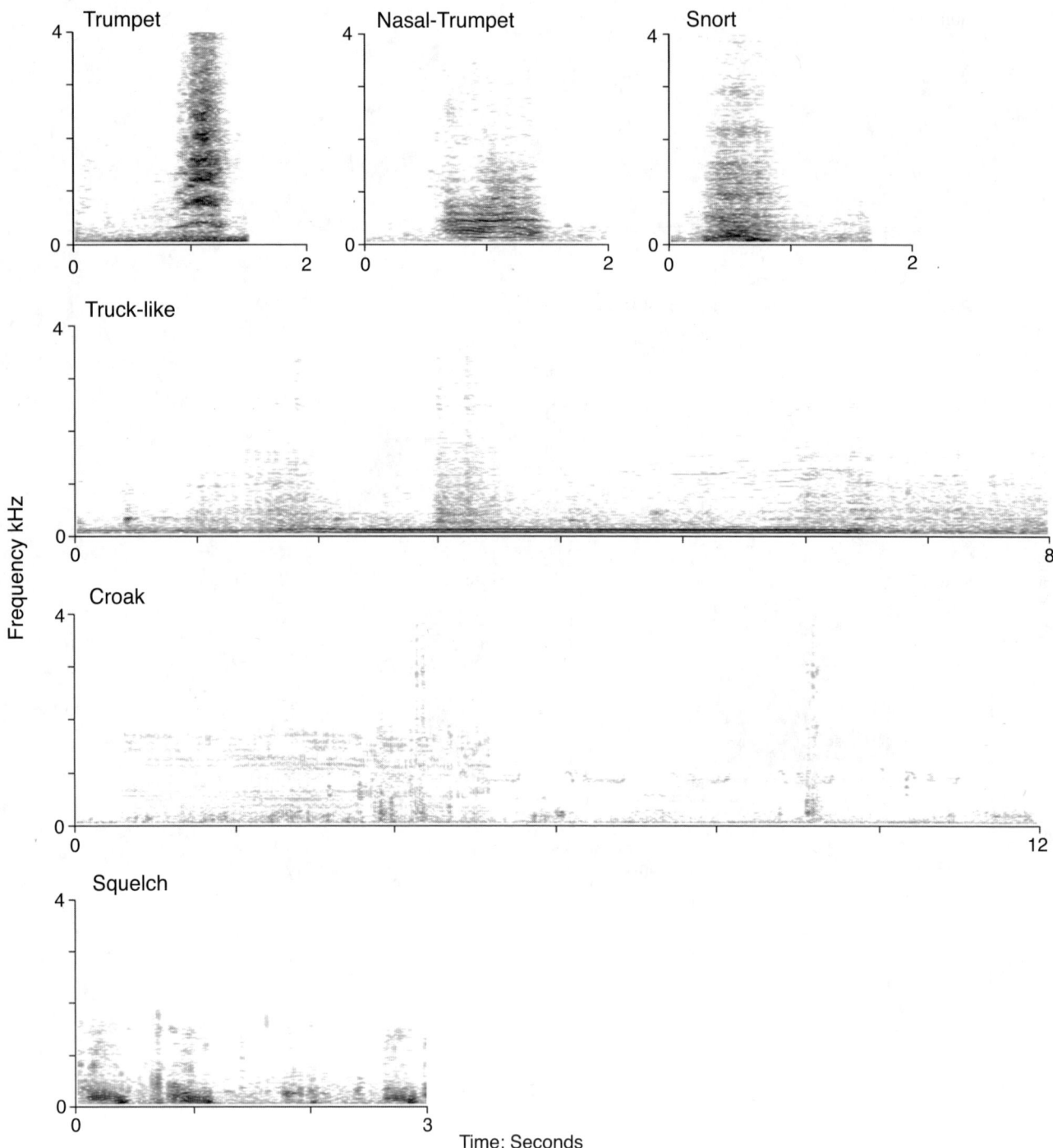

Figure 9.2b Comparative trunk and learned call types.

Roar. *Roars* (Berg [1983]; Stoeger-Horwath [2007]; table 9.4a; figure 9.2a) are powerful, highly variable bellowing, screaming, shrieking, or squealing sounds. I used the definitions of Stoeger-Horwath et al. (2007) in describing *roars*. Using manner features they divided *roars* into three main types: *noisy roars*, *tonal roars*, and *mixed roars*. *Noisy roars* are totally noisy, in other words they contain no harmonic elements. *Tonal roars* are those in which greater than 50 percent of the call is tonal (that is, containing clear harmonics). Noisy elements, if there are any, occur only at the end of the call. *Mixed roars* alternate between noisy and tonal or contain noisy components, making up more than 50 percent of the call. *Noisy roars* tend to be unmodulated, while *tonal roars* may be highly modulated, some reaching a clear

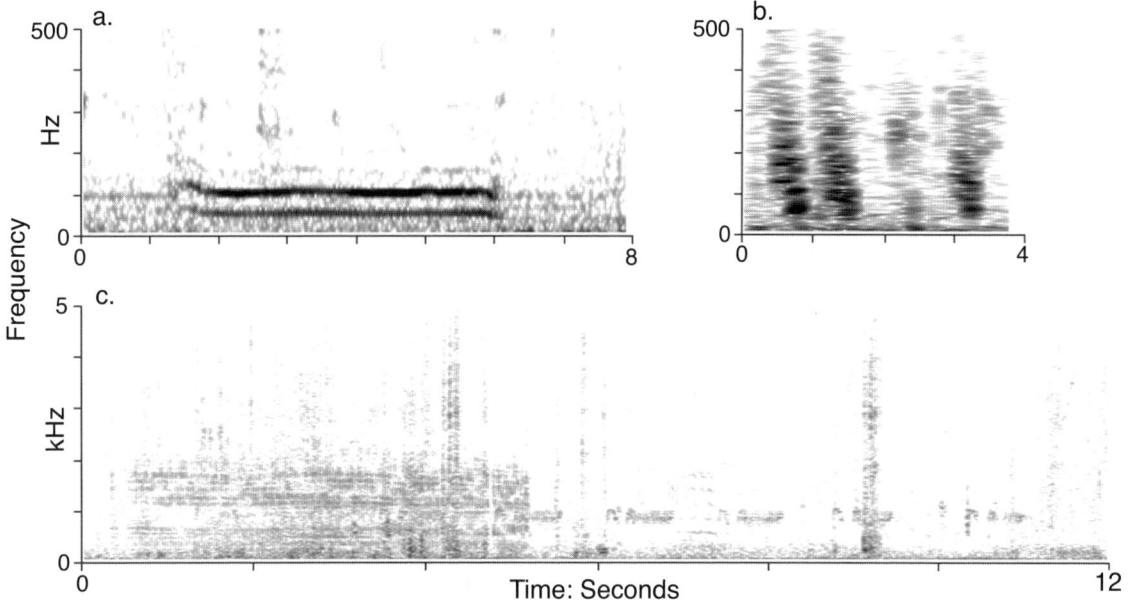

Figure 9.2c Idiosyncratic calls: (a) *Truck-like*, Malaika imitates the sound of distant trucks; (b) *Squelch*; (c) *Croak*, Gail.

crescendo before falling. All *roars* sub-types may start, end, or start and end, with a *rumble*; these composite calls are termed *roar rumbles*, *rumble roar rumbles*, or rarely, *rumble roars*. *Roars* produced by infants are generally higher pitched than *roars* produced by adults and the terms "screaming," "shrieking," "squealing," or "crowing" better describe the quality of the calls by younger individuals. As elephants become larger *roars* take on a more resonant "bellowing" or "roaring" quality.

Elephants in some form of distress emit most *roars*, with the vast majority being produced by infants, calves, and juveniles. Of the 120 *roars* that I measured, 17 were produced by calves begging for, or being denied access to, the breast or a food item; 10 were calves protesting unwanted or rough contact by another elephant; 23 were emitted by a calf or juvenile who had become separated from its mother; 30 were the recipients of agonistic behavior (generally pushed, tusked, poked, or kicked); and 1 was pounced upon by a lion. Among adults *roars* were utilized in additional contexts. Adults *roar* when chased or tusked (included above), when attacking predators (2), when chased by males during estrus (11), and during highly exciting social events (9). An additional 17 *roars* were recorded in unknown contexts.

While it is not possible to distinguish statistically between these *roars* with such small sample sizes, it is fair to say that *roars* produced in these diverse contexts do sound different. For example, the *roar* of an estrous female being chased by males pulsates, the *roar* of a female threatening a lion is a deafening scream, and the *roar* of an elephant being tusked or poked has a bellowing quality, as does that of a calf who is denied access to the nipple. *Roars* that occur during intense social excitement such as during a greeting ceremony, the birth of a calf, or in response to the arrival of a musth male are more likely to occur as a *rumble roar rumble*. Similarly, the *roars* of a lost calf may often be in the form *rumble roar rumble*, though the quality of these is different.

Roars elicit the support or attention of others. For instance, *roars* by calves draw the attention of caretakers, those by an estrous female attract males, and those by a family member who has been tusked draw the immediate response and support of close associates.

Cry. Infants and calves under the age of five years may emit a *cry* or *cry rumble* (figure 9.2a) in situations in which they are in some form of distress. Of the 20 *cries* whose spectrograms have been measured, nine were made by calves either denied access to the breast or a food item, two were protesting unwanted touching by an older elephant, one was stuck on its side unable to get up, and eight more were in unknown circumstances. *Cries* are very short whimpering sounds lasting less than half a second in duration. A *rumble* without an inhalation follows many *cries*. One composite *rumble cry rumble* given by a calf begging for a piece of palm heart was both preceded and followed by a *rumble*. The *rumble* portion in the composite calls ($n = 12$) typically lasts longer than the *cry* portion (median (ms) = 850; inter-quartile range = 547–1,628; range = 335–3,824).

Although the *cry* is structurally most like a short *tonal roar*, the two types sound very different. I used a stepwise

Table 9.4a Broad structural characteristics of laryngeal call types

Call types	Callers	Manner features	Duration (ms)	F_0 max (Hz)	F_0 min (Hz)	Curve	Maximum frequency (Hz)	Bandwidth (Hz)
Rumble	Infant–adult	Tonal/tonal with noise	3,795 (2,525–5,025) 300–11,857; 1,729	18.6 (15.7–23.2) 10.6–53.7; 1,701	14.7 (12.6–18.8) 8.4–34.2; 1,700	.007(.003–.009) .000–.060; 1661	35 (29–45) 14–312; 1347	213 (160–285) 38–996; 1,620
Rev[a]	Calf–adult	Tonal with noise	721 (347–836) 331–1,125; 6	79 (53–98) 35–117; 6	68 (43–88) 35–92; 6	.013 (.012–.018) 0–.053; 6	156 (127–282) 71–439; 6	570 (527–700) 238–4,615; 6
Roar								
Noisy roar	Infant–adult	Noisy	1,797 (1,147–2,172) 330–3,000; 59	346 (248–435) 108–568; 43	260 (183–314) 39–525; 49	.055 (.018–.085) 0–.359; 42	409 (291–632) 150–2,849; 59	3,524 (2,418–4,299) 1,364–8,791) 49
Tonal roar	Infant–juvenile	> 50% Tonal; noise occurring only at end	1,253 (1,108–2,137) 467–3,004; 12	518 (448–715) 371–740; 12	280 (241–339) 131–435; 12	.193 (.089–.356) .038–.689; 12	409 (312–561) 215–687; 12	3,407 (2,118–5,000) 1,560–8,674; 10
Mixed roar	Infant–adult	Alternating tonal and noisy elements; with noise more than 50%	1,554 (1,108–2,062) 532–3,922; 38	376 (277–496) 172–776; 38	278 (191–328) 108–658; 38	.073 (.036–.117) 0–.425; 38	557 (353–732) 183–4,078; 36	3,746 (2,968–5,143) 1,657–8,885; 37
Cry	Infant–calf	Tonal	327 (289–456) 178–528; 20	451 (334–519) 183–800; 20	274 (246–315) 159–547; 20	.314 (.246–.322) .049–1.497; 20	388 (279–642) 279–1,273; 20	3,596 (1,835–5,460) 973–6,968; 18
Bark	Infant–adult	Transient and noisy	456 (392–620) 245–787; 12	Not measurable	Not measurable	Not measurable	420 (345–566) 194–796; 12	4,234 (2,242–5,153) 941–6,094; 12
Grunt[a]	Infant	Tonal	218 (184–237) 139–255; 8	319 (269–429) 250–479; 8	259 (211–350) 205–439; 8	.24 (.19–.32) .13–.69; 8	448 (345–487) 259–753; 8	2,232 (2,549–2,072) 1,258–3,116; 8
Husky cry	Infant	Tonal with noise	976 (715–1,244) 27–1,544; 33	37 (34–45) 25–69; 32	31 (25–33) 18–62; 33	.007 (.004–.009) .000–.019; 35	66 (59–90) 35–414; 33	584 (409–685) 87–1,000; 33

Notes: Figures represent median (inter-quartile range) range and *n*. All calls that had been assigned a level of sureness "A" are included.
[a] View with caution due to low sample size and uncertain classification.

discriminant function analysis (DFA) (see box 9.1) to compare cries (*n* = 22) with *tonal roars* (*n* = 12). The two call types were significantly different (Wilks' lambda: 0.087; $F(5,17) = 35.860$; $P < 0.0001$). In the training set (see box 9.1), 100 percent of the calls were correctly classified (reduced to 91 percent in cross-validation; table 9.5).

Bark. The *bark* (Stoeger-Horwath et al. [2007]; table 9.4a; figure 9.2a) is a transient and primarily noisy call that differs from a *noisy roar* in its very short duration. Similar to the roars, *barks* may be combined with a *rumble*, producing a composite *bark rumble*. In the Amboseli population, *barks* and *bark rumbles* were heard and recorded very infrequently. *Barks* occurred in contexts similar to *roars*: three occurred in the context of begging (though these were recorded from the Tsavo orphans when calves were pushing and shoving to reach the bottle), seven occurred when elephants were tusked or pushed, one during rough play and one from a 10-year-old female as she was mated for the first time. Eight occurred in unknown contexts.

Grunt. The *grunt* (Stoeger-Horwath et al. [2007]; table 9.4a; figure 9.2a) is a soft, short, beeping or honking sound produced by infant elephants in the first days of life. Stoeger-Horwath et al. (2007) report that in captivity the infants cease producing *grunts* by two months of age. In the wild situation *grunts* are barely audible and, therefore, difficult to record. The very few *grunts* that I measured were associated with attempts to suckle.

Husky Cry. The term *husky cry* (table 9.4a; figure 9.2a) refers to gruff or husky sounds produced only by newborn and infant elephants up to about four months of age. *Husky cries* (*n* = 32 measured) range from barely audible breathy exhalations to surprisingly powerful hoarse rasping sounds. They are usually made during moments of distress and are heard frequently during the first hours and days of life as the unstable infant struggles to remain standing and as it is touched and fondled by family members (13 of the calls whose spectrograms were measured). Each call brings a vocal response and additional fondling by mother and allomothers, which in turn elicits more vocalizing by the infant. Infants may also emit a *husky cry* when separated from their mothers and frightened by something (*n* = 5), when touched and woken by their mothers (*n* = 6), when engaged in rough play (*n* = 1), or when frightened by a loud sound or

Table 9.4b Broad structural description of Trunk Call types

Trunk calls	Callers	Manner features	Duration (ms)	F_0 max (Hz)	F_0 min (Hz)	Curve	Max frequency (Hz)	Bandwidth (Hz)
Trumpet	Infant–adult	Noisy harmonics	724 (506–1,044) 198–3,385; 224	426 (372–515) 178–917; 211	318 (266–372) 108–606; 215	.142 (.087–.220) .012–.630; 211	460 (351–790) 105–2,405; 211	8,654 (1,274–21,931) 5,192–11,100; 223
Nasal Trumpet	Calf–adult	Tonal with noise	709 (578–864) 403–2,732; 23	118 (68–146) 34–329; 19	62 (56–107) 32–246; 21	.050 (.026–.076) .003–.420; 19	241 (131–306) 32–563; 23	3,729 (2,501–4,965) 786–10,425; 23
Snort	Infant–adult	Noisy	783 (488–884) 338–1,374; 18	Not measurable	Not measurable	Not measurable	86 (65–130) 43–517; 16	9,285 (4,973–11,152) 470–16,874; 18

Note: Figures represent median (interquartile range) range and n.

Table 9.4c Learned calls: Imitated and idiosyncratic calls

Production learning	Callers	Manner features	Duration (ms)	F_0 max (Hz)	F_0 min (Hz)	Curve	Max frequency (Hz)	Bandwidth (Hz)
Croak	Two individuals	Noisy harmonic	4,123 (1,651–4,643) 1,128–9,032; 6	Not measurable	Not measurable	Not measurable	611 (579–633) 526–633; 6	4,254 (4,000–4,293) 3,862–5,661; 6
Squelch	Calf–adult	Noisy harmonic	204 (117–378) 87–554; 13	306 (50–343) 30–377; 7	245 (42–290) 28–352; 7	.232 (.022–.883) .004–.896; 7	215 (65–428) 59–1920; 12	2,372 (1,078–5,940) 411–9,545; 13
Truck-like	One individual	Tonal	6,984 (4,763–9,305) 685–14,372; 60	60 (58–62) 45–597; 60	52 (50–54) 45–66; 60	.001 (.001–.002) .000–.060; 60	Not measured	168 (161–173) 95–252; 59

Note: Figures represent median (interquartile range) range and n.

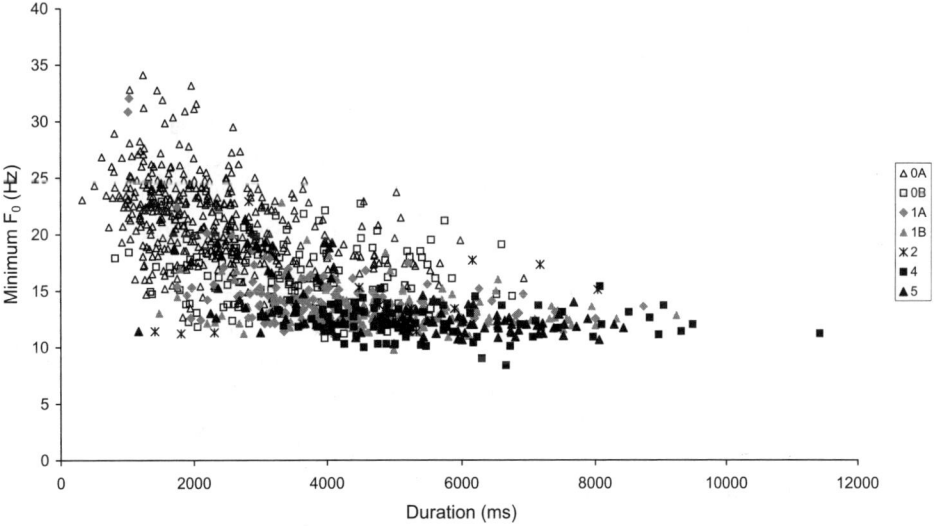

Figure 9.3 Call duration and minimum fundamental frequency of *rumbles* (males and females combined) by age class. The *rumbles* of older individuals are significantly longer in duration and lower in frequency than those of younger individuals. Duration: Spearman $R = 0.636$; $t(n-2) = 28.6$; $n = 1,210$, $P < 0.001$; Minimum F_0: Spearman $R = -0.759$; $t(n-2) = -40.52$, $P < 0.001$.

if receiving rough treatment ($n = 3$). Most of these involve some form of being touched. The remaining calls ($n = 5$) were in unknown contexts.

Husky cries contain noisy harmonics similar in structure to some *rumbles* yet their sound quality is dissimilar.

Comparing the *husky cries* ($n = 32$) and *rumbles* ($n = 110$) of calves less than a year of age reveals significant differences (Wilks' lambda 0.412, $F = 14.466$, $P < 0.0001$; table 9.5). In the training set 91 percent of calls were correctly classified (87 percent using cross-validation procedures; table 9.5).

Table 9.5 Results of stepwise DFAs for the three newly proposed call types: *Cry*, *husky cry*, and *nasal trumpet*

Call types	Cries and tonal roars	Husky cries and infant rumbles	Nasal trumpets and trumpets
n in the training set	23	79	156
Eigenvalue overall	15.9	1.43	1.073
% Variance	100	100	100
Wilks' lambda	0.059	0.412	0.482
F	34.077	14.466	19.711
P	<0.001	0.001	<0.001
Expected probabilities	.50	.50	.50
Classification % correct	100	91	96
Classification cross-validation % correct	91	87	94
Variables in the model	Duration, max time, location F_0 max, curve, F_0 max-min, bandwidth	Max frequency, duration, bandwidth, max time, curve, F_0 max-min, peak time	F_0 start, bandwidth, F_0 max, location F_0 max, max frequency, curve, max time

Notes: Each was compared with the call types they most closely resemble. *Cries* with *tonal-roars*; *husky cries* with *rumbles* produced by calves under a year; *nasal trumpets* with *trumpets*.

Trunk Call Types

Examples of trunk call types may be heard at www.elephantvoices.org/multimedia-resources/elephantvoices-calls-database-call-types.html?catid=4.

Trumpet. *Trumpets* are produced by a forceful expulsion of air through the trunk and come in several forms (Berg [1983]; Poole and Granli [2004]; table 9.4b; figure 9.2b). *Trumpets* are mainly tonal sounds, though harmonics are overlaid with noise. Most last less than a second (though extremely long *trumpets* may last over three seconds), with fundamental frequencies falling around 300 Hz and peak energy concentrated at approximately 450 Hz (table 9.4b).

An elephant can produce a wide variety of sounds by varying the speed of air she forces through her trunk, by the shape in which she holds her trunk, and by her own body posture and movement. Elephants tend to *trumpet* when they are highly stimulated—in situations in which they may be fearful, surprised, aggressive, playful, or socially excited—and the quality of *trumpeting* varies with the context (see descriptions below). *Trumpeting* is often associated with intensely social events such as a birth, mating, or greeting ceremony, where group participation is important. In these situations I postulate that *rumbling* may define the context (e.g., greeting, mating, etc.) while *trumpeting* may function as a kind of "exclamation mark," expressing the very high level of excitement and stressing the importance of the event.

Of the 224 *trumpets* whose spectrograms have been measured and contexts were known, 107 were associated with play; 85 with intense social excitement (69 with a birth, 5 with presence of a musth male, 6 with a mating, 3 with a greeting ceremony, 2 with the rescue of a calf); 16 were made by frightened calves separated from their mothers; 13 by elephants chasing predators; 1 by a male expressing irritation at noise in the research vehicle; 1 by a female in response to a member of her family being pushed; and another by an elephant responding to a distant *trumpet*.

Nasal Trumpet. *Nasal trumpets* (table 9.4b; figure 9.2b) are noisy with underlying tonal structure and sound like a large man blowing his nose. They are much noisier than *trumpets*, considerably lower in frequency, and almost invariably heard in the context of play. Of the 22 *nasal trumpets* whose spectrograms I have measured, 18 were in the context of play while 4 were associated with the intense excitement surrounding an elephant birth.

In discriminating between *nasal trumpets* and *trumpets*, it was not possible to balance data sets by individual since unidentified callers produced these calls during exuberant play or frenzied social events. Thus, to test the validity of this newly proposed call type, I performed a stepwise DFA on all measured *nasal trumpets* ($n = 23$) and *trumpets* ($n = 243$; Wilks' lambda = 0.482, $F(8,217) = 19.711$, $P < 0.0001$; table 9.5) in which 96 percent of the calls were correctly classified (94 percent in cross-validation). Fundamental frequency values mainly separate the two call types.

Snort. *Snorts* (Leong et al. [2003]; Stoeger-Horwath et al. [2007]; table 9.4b; figure 9.2b) are short, noisy, broadband sounds produced by blowing air purposefully through the trunk. Elephants may *snort* when they are surprised by something (these may simply be less explosive *trumpets*) during intense social excitement or to alert other members of their group to a new situation. *Snorts* are usually audibly distinguishable from the more common "blows" (blowing, sneezing, wheezing, and coughing sounds) that appear to be made for the purpose of clearing the nasal passages; *snorts* sound more sharp and purposeful than a blow and may also be distinguished by context.

Vocal Learning: Imitated and Idiosyncratic Sounds

Elephants produce a variety of novel sounds, some of which are learned through imitation of other elephants or things in their environment (Poole et al. 2005). Sounds reported thus far include *humming* (Randall Moore, personal communication), *squelching* (this chapter), *croaking* (Leong et al. 2003; this chapter), and the imitation of *whistling* by Asian elephants (Wemmer and Mishra 1982; Wemmer, Mishra, and Dinerstein 1985), the imitation of trucks and Asian elephant chirps by African elephants (Poole et al. 2005), and the imitation of Korean words by a 16-year-old male Asian elephant named Kosik at Everland, an amusement park in South Korea (Kwang-Seog Heo, personal communication; www.wayodd.com/south-korean-elephant-mimics-human-sounds/v/4023; www.reuters.com/news/video?videoId=1231). Many of these unusual or idiosyncratic sounds are produced only rarely, and consequently I do not have a good sample of recordings. There are two unusual sounds, *croaking* and *squelching*, for which I have some information and one, the *truck-like* call, for which I have numerous recordings from one individual. Examples of these sounds may be heard at www.elephantvoices.org/multimedia-resources/elephantvoices-calls-database-call-types.html?catid=5.

Truck-Like. In 2005, we (Poole et al. 2005) reported the imitation of a truck sound by a semi-captive orphan elephant in Tsavo (table 9.4c, figures 9.2b and 9.2c). With a fundamental frequency hovering around 50 Hz and highly variable duration (range: 685 ms to almost 15 seconds), this sound is unlike any call in the normal repertoire of African elephants. We were able to show that the elephant was imitating the noise of distant trucks on a highway 3 km away. Since the publication of the 2005 paper, additional elephants in the small Tsavo group of orphans learned to produce the same sound.

Croaking. Studying elephants in captivity, Leong et al. (2003) first reported *croaking* (table 9.4c; figures 9.2b and 9.2c) as pulsatile sounds lasting between 1 and 10 seconds. These sounds were produced by several different individuals and were often associated with the sucking of water or odors into the mouth and usually occurred in a series of two or three *croaks*. In Amboseli, as far as we know, only two elephants in the entire population of more than 1,500 individuals emit this highly unusual sound, and both are adult females from the same family. For many years only Gail was heard to *croak*, but subsequently, I heard Gwen *croaking*. Unlike the observations by Leong et al., both Amboseli elephants *croak* when they are relaxed and feeding, and based on the lack of reaction and absence of production by other elephants, it appears to have no communicative function. Gwen and Gail are sisters and close associates; I suggest that the *croak* is another example of vocal learning.

Squelching. Occasionally elephants may be heard to make a "squelching" sound (table 9.4c; figures 9.2b and 9.2c), apparently produced by forcing air through a "scrunched-up" trunk. Sometimes the individual gives the impression of having a genuine itch in his or her trunk, but at other times production of this rather odd bubbling sound seems to be an end in itself. *Squelching* is most often heard when elephants are relaxed (for example, standing by the side of a waterhole or waiting patiently for other family members to move).

Chorused Calling

There are numerous situations in which elephants call in chorus. In some cases, these are simply overlapping low to moderate level *rumbles* between two or more individuals, but very often choruses occur as a powerful series of layered calls that include a variety of different *rumbles*, *trumpets*, *snorts*, and *roars*. These powerful chorused calls typically occur during excited social events such as a birth, a greeting ceremony, the arrival of a sexually active musth male, a mating, a calf rescue, or during offensive or defensive action.

During a sub-set of 2,015 minutes of recordings from 1999, a total of 3,685 vocalizations were counted on the spectrograms, of which 42 percent occurred as single calls and 58 percent occurred within a chorus of two or more overlapping calls. The median number of calls in a chorus was 2 (inter-quartile interval = 2–4; range 2–29). In the entire data set, the maximum number of overlapping calls in a chorus was 64, following a mating recorded in the Maasai Mara population.

Behavioral Contexts and Acoustic Quality of Calls

Since many of the laryngeal and trunk call types described in the previous section are graded and, based on the measurements taken thus far, it is not possible to discern discrete call sub-types, I ask: Are *specific* behavioral contexts associated with particular acoustic-structural patterns? And if so, what clues might this give us for further understanding elephant communication?

Many years before I began examining the structural characteristics of vocalizations, members of the Amboseli Elephant Research Project (AERP) collectively gave names to calls as a way of referring to them when in the field. We assigned labels based upon a combination of sound quality and observed patterns of behavior. While the names of the call types described earlier reflect sound quality (e.g., *rumble, roar, cry, husky cry, grunt, snort, trumpet*, etc.), the names we gave to what we perceived to be different *rumbles* tended to reflect their behavioral context (e.g., *begging-rumble, separated-rumble, contact-rumble, musth-rumble, estrous-rumble*). I refer to these proposed sub-types as context-types. Although the contextual naming of calls may have led to a bias in interpretation, the very subtle differences (at least to the human ear) of many low frequency calls made it difficult to come up with unique names based purely on sound quality. Where such a distinction is possible, we have used names that describe the sound quality (e.g., *cadenced-rumble*). The remainder of this chapter gives a broad overview of the different behavioral contexts in which vocalizations occur and a description of the calls with which they are associated. Descriptive statistics of these call context-types are presented in annex 9.1; context-types include all measured calls included in the analyses that were designated a level of confidence "A" (see box 9.1; $n = 2,014$).

These results presented below include some 35 context-types, with a wide range of sample sizes. Some of the proposed context-types, such as those given during group defense, are represented by small sample sizes and cannot be used in further statistical discriminations. For those context-types with sufficient sample sizes, a standard DFA was first used to indicate whether significant differences might exist between the many proposed *rumble* context-types. For this analysis, I simply selected all context-types with classification level A and with a sample size of at least 25 measured calls across all individuals. These included 12 context-types: *anti-predator, begging, cadenced, grumbling, contact, coo, greeting, "let's-go," little-greeting, musth, separated*, and *baroo* (or "*protest*") *rumbles* ($n = 902$). I used the following predictor variables: duration, F_0 max, F_0 max-min, F_0 start, curve, high frequency, and maximum frequency. Although the results suggest that *rumbles* do vary with context (Wilks' lambda = 0.159, $F77, 6,473 = 30.148, P < 0.0001$; 35 percent classified correctly; expected 0.08), I was not able to balance the groups with respect to the contribution of calls by different individuals given the complex, non-parametric nature of the data as elaborated in box 9.1, in the Statistical Analysis section.

In order to discriminate statistically between specific context-types, I followed the procedure outlined in box 9.1, using a combination of non-parametric descriptive statistics, and stepwise DFA with cross-validation to examine patterns across elephant calls. Due to limitations presented by elephant sociality, life histories, and field collection, analyses had to be customized to suit data availability (see box 9.1) and were thus sub-divided into adult female reproductive *rumbles* (*estrous, female-chorus, mating-pandemonium*); calf *rumbles* (*begging, separated*, and, separately, *as-touched, grumbling, umbrage*); adult female social and caregiving *rumbles* (*cadenced, contact, "let's-go," coo*, and separately, *greeting, little-greeting*); finally six *trumpet* context-types were compared. Calls produced by adult male calls were excluded, since only *musth-rumbles* were suitably represented. The results of these analyses are presented in tables 9.6, 9.7, 9.10, and 9.11 and are described further in the text in the following sections. Further statistical information is presented in the methodology.

The results presented in the text below and in tables 9.6, 9.7, 9.10, and 9.11 clearly show that elephants produce a wide variety of *rumble* forms that are associated with behavioral context. While these forms, or context-types, are certainly highly variable (e.g., reflecting emotional level and individual caller in a range of measures), and in some cases may be only subtly different, the elephants show by their consistent behavior that they are able to discriminate emotion and meaning in these graded calls. For this reason, I suspect that more detailed measurements will, in the future,

Table 9.6 Results of discriminant function analysis in reproductive *rumble* types

Adult female *rumbles*	*Estrous-rumble, female-chorus, mating-pandemonium*
N	47
Eigenvalue overall	1.527
Wilks' lambda	0.291
F	F = 2.099
P	<0.01
Classification expected probabilities	0.33
Classification % correct	77
Variables in the model	Contour, duration, max time, location F_0 max, frequency modulation, curve, F_0 start, peak time, bandwidth, maximum frequency

Table 9.7 Discriminant function analysis (DFA) results for common calf *rumbles*

Calf calls	*Begging, separated*	*As-touched, grumbling, umbrage*
N (training)	148	80
Eigenvalue overall	0.647	0.828
Wilks' lambda	0.607	0.471
F	15.208	6.681
P	<0.0001	<0.0001
Classification expected probabilities	.50	.333
Classification % correct	79	70
Classification cross-validation % correct	60	67
Variables in the model	Duration, bandwidth, maximum frequency, max time, curve, F_0 max	Contour, F_0 max-min, F_0 start, duration

Notes: In the *begging, separated* DFA the training sub-set included calls by Ejac, Eldon, Elettra, Elmo, Emmet, Ewaso. The validation sub-set included Elaine, Emily Kate, Erica, Eudora 00, Explorer, and Lewa. In the *as-touched*, *baroo* and *grumbling* DFA, the training sub-set included calls by Echeri, Ejac, Elettra, Ella 03, Emmet, Erica, Eudora 00, Europa, and Explorer. The validation sub-set included Charlotte, Elaine, Eldon, Eliot 03, Elmo, Emily Kate, and Ewaso.

reveal further discriminations than I am able to show at the present.

Examples of the elephant call context-types that I describe here may be heard online at www.elephantvoices.org/multimedia-resources/elephantvoices-calls-database-contexts.html.

Group Defense

Calls associated with anti-predator behavior include those used in the context of alerting companions to the presence of a predator, intimidating, or "mobbing" a predator as well as those used while taking defensive action. Family members produce several different call types when they confront predators or when they find themselves in potentially threatening or frightening situations. These include *rumbles*, *snorts*, *trumpets*, and *roars*. Much has been written about the complex and highly coordinated defensive and offensive behavior of elephants in the presence of predators (Douglas-Hamilton 1972; McComb et al. 2000; see also chapter 8), but the variety of calls produced, and the dramatic responses of other elephants to these calls, has received little attention.

When exposed to the sound, sight, and smell of lions, hyenas, humans, or other potentially dangerous predators or situations, females and calves typically respond by first Freezing, then rapid Assembly (rapid walking or running toward one another), and then Bunching (see Poole and Granli, chapter. 8). Once bunched, the elephants vocalize, secrete from the temporal glands, and reach out with their trunks to touch one another with reassuring gestures. Through this multi-faceted exchange of information the elephants appear to assess the level of danger present and collectively decide what their response will be. A single adult female may charge forth, the group may attack en masse, or members of the group may make a hasty retreat. Their highly coordinated response is communicated via a complex combination of vocal, olfactory, and tactile signals and is an area for further research. Unfortunately, I do not have enough recordings of any of the calls given in these contexts to include them in statistical analysis.

Alert. When family groups are exposed to an unusual or disturbing situation, Listening and/or Freezing (Poole and Granli, chapter 8) behavior may follow a sharp *snort* or *snort-rumble*. Immediately following a *snort* given in this context, a number of soft, medium-length *rumbles* by one or more individuals may then be heard. The calls of several individuals may overlap. Elephants in the calling group stand alert, listening and looking.

Events that elicit such behavior may include non-elephant disturbances such as unusual commotion in a vehicle, a helicopter passing overhead, the discovery of Maasai herdsmen in the area, or the roaring sounds of lions. A disturbing or exciting event among con-specifics (such as a fight between two musth males or serious aggression directed at a family member) may also elicit this form of soft *rumbling*, as will playbacks of unexpected callers. I refer to this as *comment-rumbling* (figure 9.4, panel a), so called because elephants appear to use it to "comment on" or "call attention to" an unusual or disturbing event.

Defensive and "Mobbing." An elephant confronting a dangerous predator can produce terrifyingly powerful sounds, which may include deafening *roars*, *trumpets*, and extremely loud and noisy *rumbles*. Having bunched together, older individuals at the fore and calves occupying the center (see Bunching; chapter 8), one or more individuals may Advance-Toward or Charge (chapter 8) the predator while emitting powerful *roaring-rumbles*, *trumpet-blasts*, and *noisy roars* (figure 9.4, panels b, c, and d) that would literally put fear in the hearts of men! As the elephant charges it flings its trunk toward its adversary and may stop abruptly, kicking up dust. The primary function of these calls appears to be to intimidate.

Throughout a confrontation with a dangerous predator the bunched elephants may continue to vocalize with noisy, throaty, rolling *rumbles*, their heads raised, ears extended,

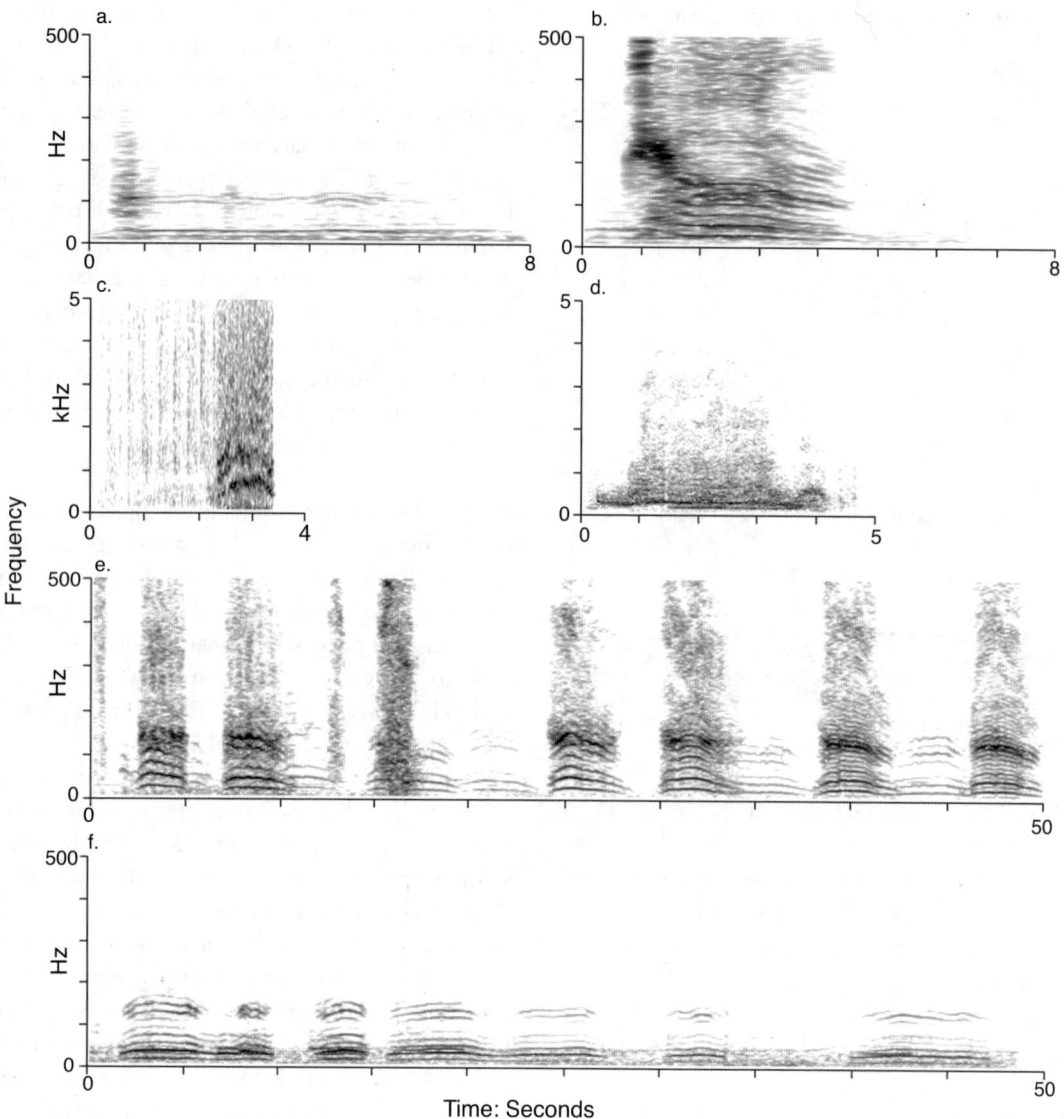

Figure 9.4 Calls produced in the context of predators (note differences in frequency scale). (a) *Snort* followed by *comment-rumble*; (b) *roaring-rumble* as Enid threatens lions; (c) *trumpet-blast* at a lion; (d) *noisy roar* in response to playback of lions roaring; (e) *roaring-rumbles* as Amy and Alison threaten a lion after it pounced on Amy's calf. Note more distant calls of approaching family members. (f) Young female charges my vehicle while *trumpeting* and *rumbling*; her behavior causes her family to bunch together and emit these *rumbles* while facing toward me.

temporal glands streaming, and trunks reaching out to touch one another. These calls have the effect of both intimidating the predator and simultaneously calling in support from family members not in the immediate vicinity. I refer to the powerful noisy *rumbles* given in this context as *roaring-rumbles* (figure 9.4, panel e). Softer calling by bunched elephants alternatively appear to be involved in decision making (figure 9.4, panel f) and may be followed by a hasty retreat.

Startled. Startled elephants make sounds associated with a sudden exhalation of air, which may take the form of a *trumpet* or a *snort* or something in between.

Food

Among many species of primates and birds, food calls are a common type of vocal signal. In Amboseli, resources such as grass, herbs, shrubs, water, and minerals are fairly evenly distributed, and, thus, calls to alert members of a group to an exciting food source may not be particularly relevant. I have occasionally heard *rumbles* given in the context of an individual locating a rare and nourishing food source, such as a fallen *Acacia xanthophloea*, yet these instances are so infrequent that I do not have a large enough sample of calls to comment on whether Amboseli's elephants have food calls. I expect that if such calls exist, they are likely to be heard in forests where food is more unevenly distributed and elephants, like primates and birds, have an interest in fruiting trees.

Sexual

Adult male and female elephants live in two essentially separate social systems and both sexually active musth males and estrous females are confronted with the problem of finding and attracting suitable mates who may be uncommon in space and time. They locate potential mates through searching behavior, conspicuous postures (see chapter 8), the secretion of strong odors, and loud and characteristic calling.

Male-Male Competition. During the heightened sexual and aggressive period of musth, males emit a distinctive rumble with a characteristically pulsated "put-put-put" or "glug-glug-glug" quality, like water gurgling through a deep tunnel. This *musth-rumble* (Poole 1987a; Poole et al. 1988; figure 9.5, panel a) is associated with an increased rate of urine dribbling and a particular ear posture known as an Ear-Wave (Poole 1987a; chapter 8).

Males emit *musth-rumbles* in many different, but rather specific contexts, including in aggressive and sexual situations, while marking, drinking, or wallowing, as well as in situations where they feel challenged. For example, the sound of another musth male, an approaching vehicle, or airplane is often enough to trigger a musth male to *musth-rumble* (Poole 1987a). Musth males also frequently *musth-rumble* before or after a bout of listening and are presumably responding to distant elephant calls or using calling to search for potential mates (Poole and Moss 1989).

Musth-rumbling in the company of females elicits a cluster of overlapping calls that we refer to as a *female-chorus* (Poole et al. [1988]; see below). Musth males are, however, most likely to *musth-rumble* when they are alone, less often in the company of females, and least often when guarding an estrous female (Poole 1999a). Since musth males often listen after *musth-rumbling*, it is likely that they use *musth-rumbles* to advertise their heightened aggressive and sexual state to rivals and potential mates and to search for female groups. Once a male has located an estrous female, he may have neither the need nor the desire to further advertise his whereabouts.

Some males have very distinctive *musth-rumbles*. Calls by different males range from rather short to very long in duration and from rolling to highly pulsated. The "typical" *musth-rumble* is a long, very low pitched, pulsated, tonal call with considerable overlaying noise. I used a stepwise DFA to test whether the measurements taken were capable of discriminating between the randomly selected calls of four males (Dionysus, RBG, Rashid, Long Left) from whom I had recorded at least six *musth-rumbles*. Despite the small sample size ($n = 24$), the model was able to classify 96 percent correctly (cross-validation was not possible due to the very small sample sizes [Wilks' lambda 0.056, $F = 2.723$, $P < 0.003$]). Although *musth-rumbles* are highly distinctive, I was not able to statistically compare them with other *rumble* sub-types, as males do not produce the other call types under consideration and females do not produce *musth-rumbles*.

Female Choice. A series of highly modulated and typically overlapping, or "chorused" calls, may be heard when a musth male joins a family group or when he moves among the group testing the genitalia and urine of females for possible receptivity. Adult and juvenile females join in the *rumbling* chorus that may also include *trumpeting* and, occasionally, *roaring*. Females may *rumble* when tested, in response to hearing a *musth-rumble*, including a playback (Poole 1999a), and may even call upon locating the scent-trail of a musth male (Poole, unpublished data). As females call in this context, they characteristically urinate, defecate, and secrete from their temporal glands. When females are very excited, their chorusing is particularly powerful. We refer to this type of calling as a *female-chorus* (Poole et al. 1988). While *female-choruses* are relatively common, it is extremely difficult to obtain representative measurements of calls due to their overlapping nature. Figure 9.5, panel c, shows a section of a *female-chorus* as Rashid joined the EB family. What is immediately obvious from examining spectrograms is that the calls are highly variable in shape but have in common a high degree of modulation and energy in the upper harmonics, especially during the peak of the chorus. The measurements presented in annex 9.1 are unlikely to reflect the calls at the peak of a chorus due to the difficulty of measurement.

Estrous females typically *roar* when pursued by a non-musth male (Moss 1983). The *roar* may begin as a series of short, pulsated growling sounds before developing (or not) into a pulsating *roar* (figure 9.5, panel d), which has the effect of attracting distant males. We refer to this as an *estrous-roar* (or *estrous-bellow*, Moss [1983]; Poole [1989b]). Females are more likely to *estrous-roar* in early and late estrus—when young, low-ranking non-musth males chase them—than during peak estrus, when they are in consort with a high-ranking musth male and, therefore, have less reason to attract the attention of additional males (Moss 1983; Poole 1989b).

When a female has been mated, and immediately after the male dismounts, she begins a series of distinctive *rumbles* that are repeated at intervals, as if in song (figure 9.5, panels b and f). These long rolling *rumbles* start at relatively

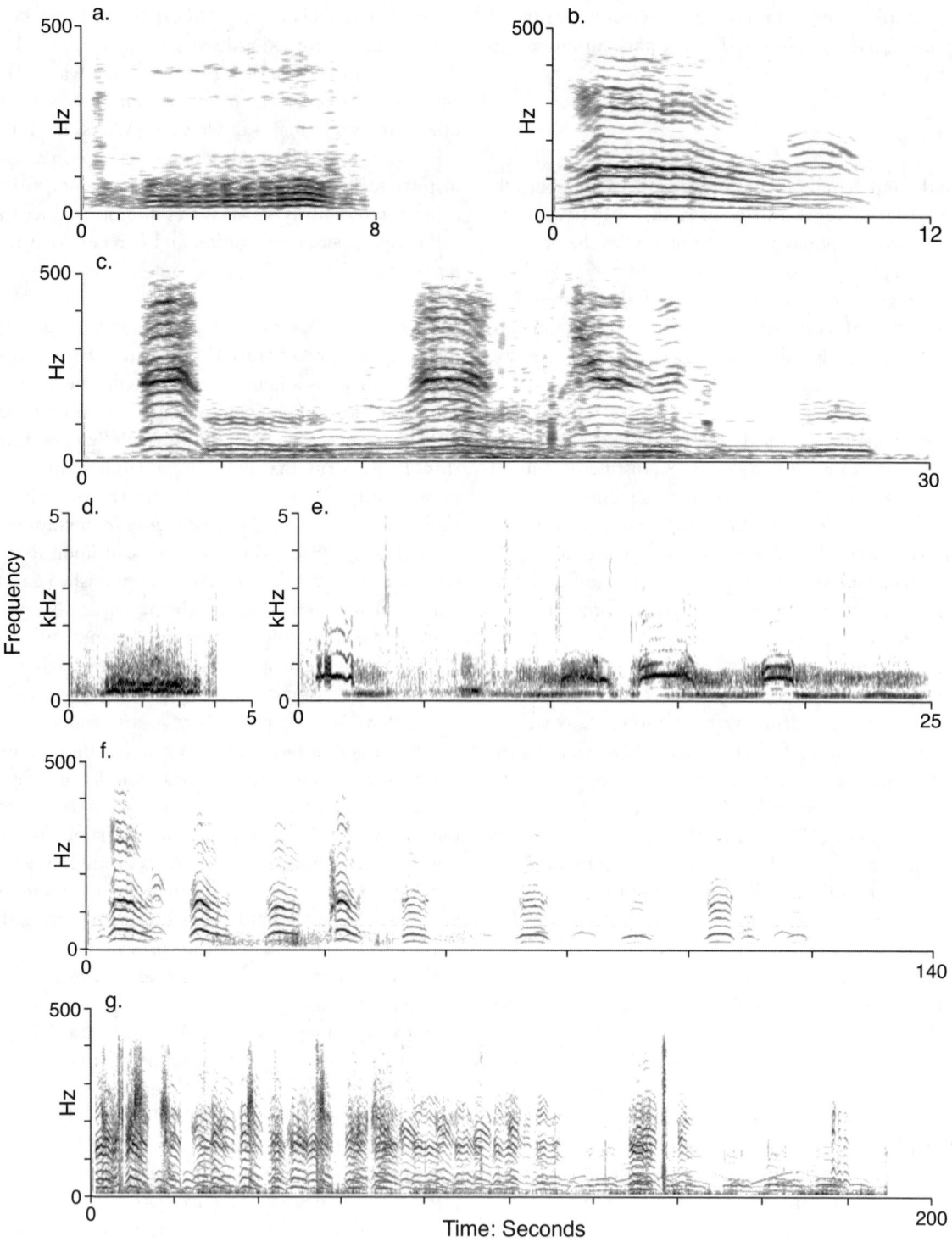

Figure 9.5 Calls produced in a reproductive context (note different frequency scale): (a) *Musth-rumble* note pulsations; (b) Single *estrous-rumble* skewed left; (c) Segment of a *female-chorus*; (d) Pulsated *estrous-roar*; (e) *Social-roars* emitted during *mating-pandemonium*; (f) Series of *estrous-rumbles* by Shirley; (g) *Mating-pandemonium*; calls by the mated female are those that reach a rapid peak in frequency. Note how the rate of calling as well as its intensity and frequency modulation decreases with time after the mating.

low frequencies, rise sharply and descend slowly, and may be repeated for up to 45 minutes (Poole, unpublished data). Though typically emitted in association with a mating, females at peak estrus may call in this manner in the absence of a mating. We refer to this pattern of calling as a *post-copulatory* or *estrous-rumble* (Poole et al. 1988). Inter-call duration is variable, starting at around 1,500 ms and increasing with time after the mating. The calling female initiates her song at very high sound pressure levels, gradually tapering the intensity of her calling. She may reach out repeatedly to touch her partner's penis or his semen on the ground, calling with more vigor at each renewed sniff. Her calling is associated with loud, rhythmic ear-flapping and copious secretion from her temporal glands. *Estrous-rumbles* have the effect of attracting (often distant) males (Langbauer et al. [1991]; Poole 1999a; see also chapter 8).

Following a mating, relatives and associates join the calling female in what we have termed a *mating-pandemonium* (Moss [1983]; Poole et al. [1988]; figure 9.5, panels e and g). The arrival and participation of the mated female's family increases her own level of excitement and calling. Adult females, juveniles, and calves all participate, overlapping the distinctive *estrous-rumbles* of the mated female with powerful *rumbles, trumpets,* and *roars* of their own. True pandemonium exists: it is difficult to tell who is calling and whether the component calls can be classified as a particular context-type (though the *estrous-rumbles* stand out as distinctive, both audibly and structurally). For now I refer to all *trumpets* and *roars* that occur during matings, greetings, births, and other highly social, excitable events as *social-trumpets* and *social-roars* (see Social Integration, below). While the *social-trumpets* do appear to be distinctive (see below), it may be the tempo of the repeated *estrous-rumble* that distinguishes the event rather than any detectable differences in the *rumbling, trumpeting,* or *roaring* of the other participants. What is notable is that many of the *rumbles* made by females other than the mated female appear to be bimodal or multi-modal.

I used a standard DFA with 11 variables (see table 9.6) to compare adult female reproductive calls: *estrous-rumble, female-chorus,* and *mating-pandemonium*. The results were significant (Wilks' lambda = 0.291, $F = 2.099$, $P < 0.01$; the model classified 77 percent correctly). The results should be viewed with caution, however, as the data set was unbalanced, and due to the nature of the sample, I was not able to use cross-validation. Three females contributed to the *estrous-calls* ($n = 15$), 15 females to the *female-chorus* ($n = 21$), and 4 females to the *mating-pandemonium* ($n = 9$).

Mother-Offspring

Begging. A broad range of call types may be heard in the context of begging, including *rumbles, cries, cry rumbles, roars, roar rumbles, barks, bark rumbles,* and *grunts*. An infant or calf typically initiates begging by approaching or moving along its mother's side with its trunk raised, often touching the mother's side or leg, and emitting one or a series of several very soft, short *rumbles* characterized by an "rrrrmmmm" sound of descending pitch. This behavior and associated calling is frequently heard in an elephant family with infants and calves, and I refer to it as a *begging-rumble* (figure 9.6, panel a; annex 9.1). *Begging-rumbles* are typically flat or slightly descending in frequency, but about a third are modulated, rising slightly and then falling. Generally only elephants under the age of five years emit *begging-rumbles*. Although juveniles as old as eight years of age may be heard to beg in this manner, it is rare and occurs primarily after the death of a younger sibling, when an older calf may attempt to access the breast again. Among free-ranging elephants, this type of *rumbling* is heard almost exclusively when calves are begging for access to the breast, although I have twice heard a calf emit this sound when it was begging for food from its mother's mouth. In captivity, these calls may be heard with great frequency and intensity at feeding time, whether food is a bottle of milk or coconut cakes.

In 1999 and 2000, I recorded a total of 166 *begging-rumbles* from the EB family with level of sureness A. Figure 9.7 illustrates the number of calves in the EB family during those years and the number of *begging-rumbles* that were recorded by age, illustrating that the occurrence of *begging-rumbles* declines around three years old, approximately the age when calves are beginning to be weaned.

A calf's mother usually responds to a *begging-rumble* by stopping and adopting a Suckling-Stance (see Poole and Granli, chapter 8) whereupon calling ceases. If she denies the calf access to her breast, begging continues or may escalate into a more modulated and noisy call with a whining tone that I refer to as a *grumbling-rumble* (see below), a higher-pitched *cry, cry rumble, roar,* or *roar rumble* (figure 9.6, panels b through f). The *roars* show a wide range of structural characteristics, ranging from noisy to tonal. Sound quality of the *roaring* calls is highly variable and might be described as squealing like a pig, screeching, roaring, shouting, yelling, crying, and even crowing like a rooster.

Eliciting Care or Support. Young elephants when distressed also produce a wide range of calls including *rumbles, cries, cry rumbles, roars, roar rumbles, husky cries,* and *trumpets*. Distress can be defined broadly to include instances of being physically hurt (e.g., pushed, poked, kicked by another elephant); becoming "stuck" somewhere (e.g., as in a mud-wallow); becoming frightened or alarmed (e.g., separation from mother or family); or being thwarted (e.g., unable to access the breast or other resource; see section specifically

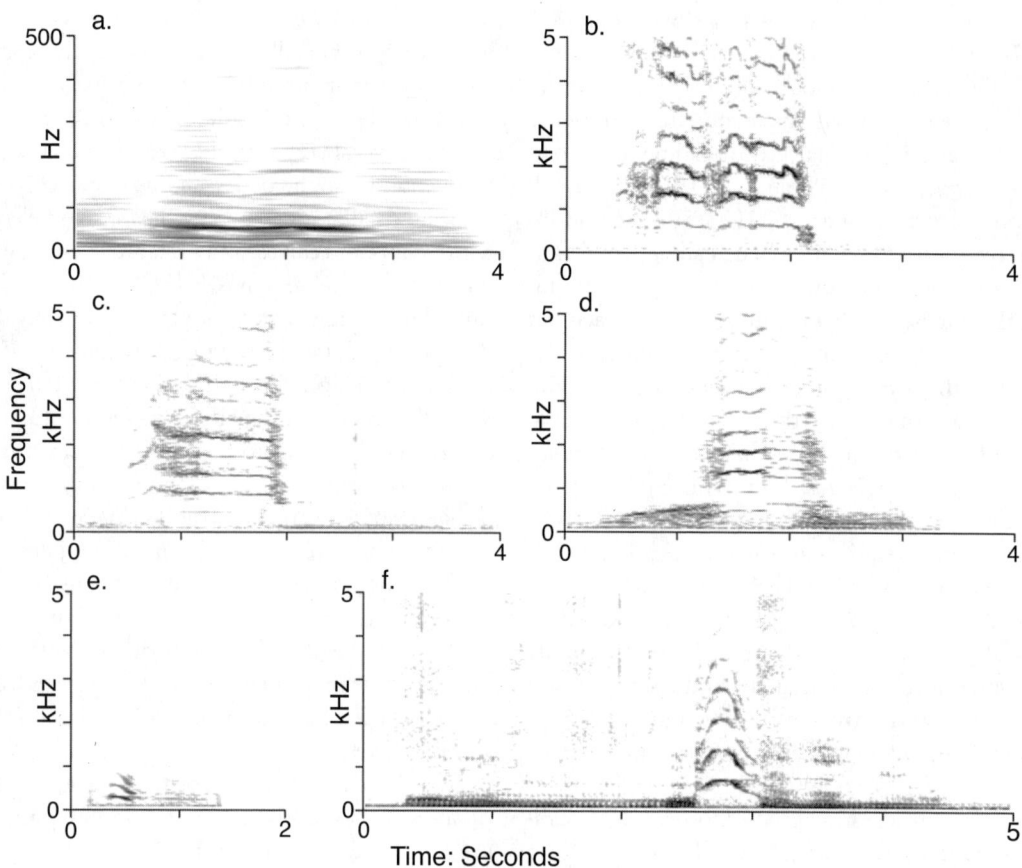

Figure 9.6 Mother–Offspring: Calls produced by calves in the context of begging. (a) *Begging-rumble* by Ella's 1999 calf; (b) *Mixed roar* in association with begging; (c) *Mixed roar rumble* by EA calf attempting to suckle; (d) *Rumble mixed roar rumble* by Elspeth's c00 while begging; (e) *Cry rumble* as begging; (f) *Rumble cry rumble* by Elaine while begging for a palm heart from mother Edwina.

about begging). Within these broad call types, there is a wide range of call patterns. To some extent the variety reflects the developmental stage of the elephant (e.g., infant, calf, juvenile, or even adult; see under Conflict section), but the level of distress is also reflected.

When moderately distressed, infants produce husky cries. These sounds are commonly heard in a group with a newborn and family members respond rapidly to aid the calling infant, touching it and *rumbling* softly. Husky cries range from barely audible, low-intensity sounds to surprisingly powerful abrupt and gravelly sounds. I refer to the softer, less intense call as a *low-intensity-husky-cry* and the louder type as a *high-intensity-husky-cry*. In the high-intensity example illustrated in figure 9.8, panel a, an adult male pushed Eudora's hours-old infant. Infants may also *roar* when highly distressed.

Older calves may *cry* or *roar* when distressed and several such sounds are illustrated in figure 9.8, panels b through e. Each of these calls elicited the support of other elephants. I do not have large enough sample sizes of these call types to know whether the *cries* and *roars* given in the context of begging are different from those emitted by infants who are distressed for other reasons. At this stage, I assume that they are not, although calls that are associated with higher levels of distress appear to be coupled with more noise.

Infants, calves, and even juveniles of either sex emit a series of characteristic calls when they are separated from and searching for their mothers. The body posture of these youngsters (Head-Raising; Tail-Raising; Trunk-Curved-Under; see Poole and Granli, chapter 8) indicates that they feel insecure and frightened. A separated calf begins by calling with a barely audible, often pulsated humming-sounding *rumble*, "mm-mm-mm-mm." The calling calf then Freezes (see Poole and Granli, chapter 8), standing with his head raised and ears spread, listening, presumably for an answer to his call, and then calls again. This pattern is repeated over and over, sometimes increasing in amplitude, until he locates his mother or other family member. The calf's typically rapid movement, as it searches for its family, is likely to be the cause of the pulsating quality of the call. I refer to calls of this quality given in this context as a *separated-rumble* or *lost-call* (figure 9.8, panel f).

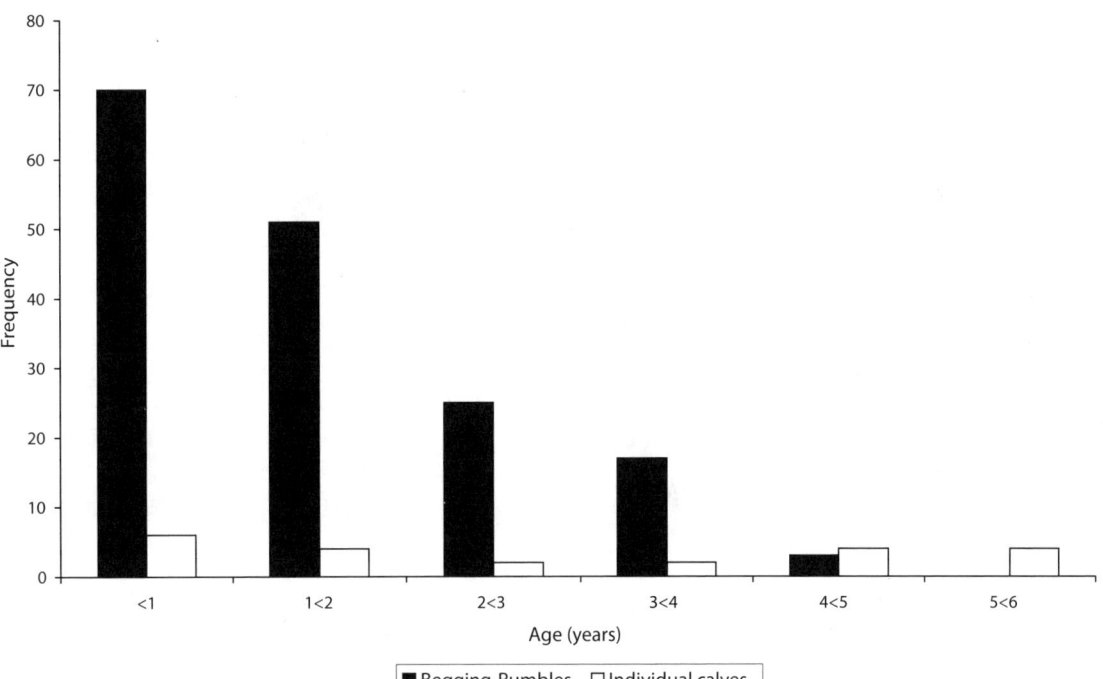

Figure 9.7 Number of *begging-rumbles* recorded and calves in age group in EB family.

While the *rumbles* emitted by separated and begging calves look structurally similar, they are audibly distinct. I used a stepwise DFA to test whether discrimination between these two proposed calls was possible (table 9.7). Although the results suggest they are significantly different, with duration and bandwidth contributing most to the discrimination, the ability of the model to correctly classify calls was not particularly good (79 percent; reduced to 60 percent in cross-validation). Several factors play a role in the poor discrimination of the model. The calls are very similar, with the primary audible difference being the pulsating nature of the call emitted by separated calves—something the measurements were not able to discern. Second, the frequency and duration of calls change rapidly with age in calves, thereby confounding the training/validation data sets, particularly considering the relatively small sample sizes and the spread in ages of the calves. I therefore ran one final test using a balanced data set from only three individuals. Once again the results are significant, but Wilks' lambda shows that the model is not particularly good at discriminating between the calls given in these two contexts (75 percent correct; see table 9.7). Additional data are necessary to conclude whether these are two different calls or not. For now we can say that since they are usually audibly distinct to my ears they probably carry specific information to the elephants, too.

As a lost calf becomes increasingly distressed, *rumbling* escalates to shrill *trumpeting* or load *roaring*. I refer to *trumpets* emitted in this context as *alarm-trumpets* (figure 9.8, panel g; see table 9.10 for statistical differences between *trumpet* context-types).

Receiving Comfort. Calves respond to reassurance and comfort given by adults and allomothers by emitting two rather different-sounding calls. When infants and calves are approached and/or touched in a caring way by an older member of their family, they may emit a soft, short, low-pitched call lasting about two seconds in duration. This *rumble* has an "aauurrrrr" quality, and the infant or calf raises its head and lifts its ears as it is calling. I refer to calls given in this context as *as-touched-rumbles* (figure 9.8, panel h). The calling calf holds herself in a posture comparable to that exhibited by older individuals in the context of a *little-greeting-rumble* (see Social Integration, and chapter 8) and the quality of the associated calls is also similar. My guess is that *rumbles* given in this context are precursors to *little-greeting-rumbles*. Similar to *little-greeting-rumbles*, calls given by calves in the context of receiving an affiliative gesture are highly variable, ranging from rather soft unmodulated tonal calls to loud, highly modulated and sometimes rather noisy calls.

Infants and calves who are comforted by a family member after having suffered some "injustice" (e.g., pushed, kicked, tusked, denied access to the breast) or who experienced something frightening or distressing typically emit a loud, noisy, and characteristically highly arched

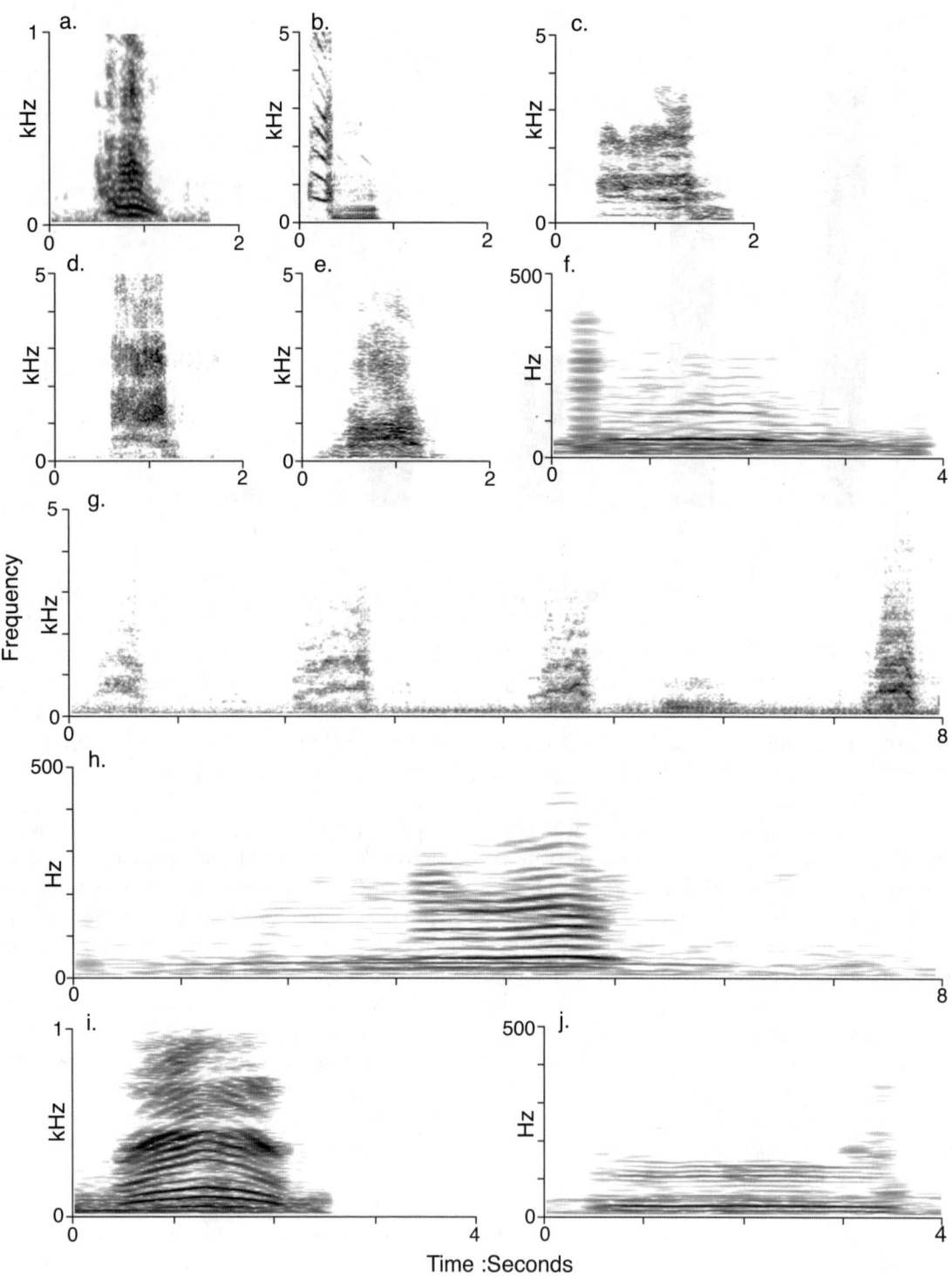

Figure 9.8 Mother–Offspring: Calls produced by calf in the context of other forms of distress. (a) *High-intensity-husky-cry*, Eudora's 2000 newborn, as he is pushed by an adult male; (b) *Cry rumble*, one-year-old infant, Elaine, stuck on her side in a depression; (c) *Noisy roar rumble*, infant Explorer, after Eudora pushed him away from her hours-old infant; (d) *Noisy roar rumble*, Amy's one-year-old calf as a lion pounces upon it; (e) *Noisy roar*, Eudora's 2000 newborn as it is tusked and kicked by an adult male; (f) *Separated-rumble* by Edwina's 1999 calf; (g) Series of trumpets, separated Ejac as he runs to his mother; (h) *As-touched-rumble*, Emmet, as he is comforted by Ella and Elettra (note: his call overlaps with a reassuring call by Elettra); (i) *Baroo-rumble*, Elspeth's 2000 calf after being pushed away from an infant by Eudora; (j) Eudora, *coo-rumble* to her infant.

"wooooaaaarrrrr" or "barooo" sound. I refer to this protesting sound as a *baroo-rumble* (figure 9.8, panel i).

In situations where a calf is in conflict with another (for example, over access to food or when one is pestering another), a calf may emit a call with a distinct "whining" tone. In such calls of longer duration, a very clear undulating or wavy pattern is distinguishable with frequencies rising and falling. I refer to this as a *grumbling-rumble* (see under conflict and figure 9.9, panels b through e). A number of *grumbling-rumbles* in my collection come from the Tsavo orphans. At night, in their stockade, without the moderating influence of adults, the calves pestered one another (e.g., repeatedly touching or sucking on one another's ears, etc).

I used a stepwise DFA with cross-validation to test whether discrimination between these three proposed call types, as produced by calves, was possible (table 9.7). Since the majority of *baroo-rumbles* are highly arched and the *grumbling-rumbles* display a wavy contour, I included values representing contour in the model (see table 9.1). Overall, the model was able to discriminate between the three calls well above the level of chance (70 percent and 67 percent with cross-validation; table 9.7), but discriminations were much higher for the *baroo* and *grumbling-rumbles* (73 percent and 78 percent, respectively, in cross-validation) than for the *as-touched-rumbles* (only 40 percent). As mentioned before, the *as-touched-rumble* is highly variable, perhaps due to the numerous situations in which a calf may be fondled, some of which may be unwelcome! At least two of these calls appear to be rather clear signals of behavioral context with emotional state reflected in the contour and in the degree of modulation, noise, and intensity.

Reassuring and Calling Calves. Calls directed toward calves are one of the more frequently heard calls in a family group. These calls are, by and large, rather similar soft, unmodulated, low-pitched *rumbles* of medium duration. Many of these calls appear to function to generally reassure and bond with calves.

I distinguish two broad behavioral contexts in which adult and juvenile females *rumble* to calves. One context is when they greet, touch, suckle, or generally "coo over" a calf in the absence of an expression of distress by the calf. Calls emitted in this context are most often directed toward infants and are especially frequent when there is a newborn in the family. Juvenile females may also call in this manner when they encourage infants to try to "suckle" from them. The other context is when adult or juvenile females are reassuring a calf following an event that has caused the calf to give some form of distress call. *Husky cries*, *begging-rumbles*, *separated-rumbles*, *as-touched-rumbles*, *baroo-rumbles*, *alarm-trumpets*, *cries*, and *roars* all elicit this type of calling by family members.

The distress of younger calves brings forth calling by more participants.

A comparison of acoustic measurements, however, suggests no significant difference between these calls given in these two reassuring contexts, suggesting that they should be viewed as one context type. I therefore refer to these calls collectively as *cooing-rumbles* (figure 9.8, panel j). A comparison of *coo-rumbles* with other common adult *rumbles* is presented in table 9.10.

Some *rumbles* given by mothers have the immediate effect of calling a wandering calf back to her side. Unfortunately, I have very few examples of calls emitted in this context, so they are not included in annex 9.1 or in any analyses.

Conflict

Aggression. During conflicts between females, especially between females from different families or between a female and a young male, the matriarch or another large adult female will come to the aid of a younger member of the family. During such coalitions, the older female rapidly approaches the female in conflict, and as she reaches her side, they may raise their heads high and *rumble* in unison as they prepare for a joint attack on a young male or on another female(s). A female seriously threatening another elephant stretches her head forward and outward (Bow-Neck, see chapter 8), and as she chases her opponent, her aggressive *rumbling* may be associated with audible ear-flapping (figure 9.9, panel a). I do not have enough examples of calls recorded in either of these contexts to include in annex 9.1 or in any analyses, but the specific behavior of the elephants when calling, particularly during coalitions, suggests an area for further research.

Protesting. The *grumbling-rumble* mentioned before (figure 9, panels b through e), with its distinct undulating or wavy contour and whining tone, is also emitted by adults. *Grumbling-rumbles* vary between one or several undulations, resulting in a rather bimodal distribution of call duration, with the longer calls sounding distinctly complaining.

When adult males are on the receiving end of a serious threat, they frequently emit a sound similar in structure to the shorter *grumbling-rumble* but of rather different sound quality. This very short, groaning *rumble* has the quality of an outboard or V-8 engine and is referred to, thus, as a *V-8-rumble* (figure 9.9, panel f; annex 9.1). Adult males may produce this sound in several specific circumstances, including when seriously threatened by a higher-ranking, often *musth*, individual at close range ($n = 3$); when chased or lunged at by a higher-ranking individual ($n = 5$); during rough sparring matches ($n = 5$); and during rough play ($n =$

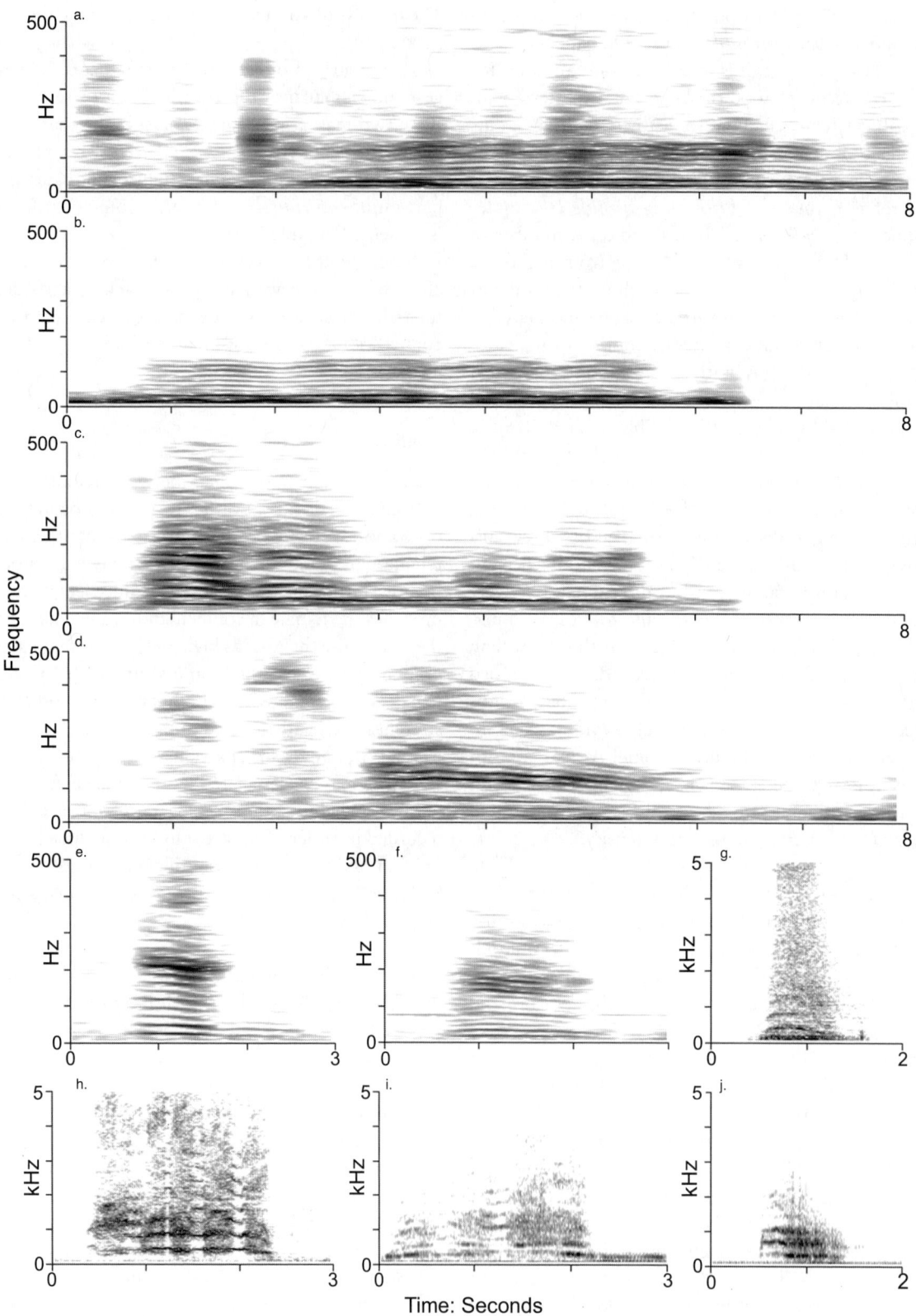

Figure 9.9 Calls produced during conflict: (a) Geraldine, as she threatens Echo's family; note loud earflaps; (b) *Grumbling-rumble*, Malaika, as keeper prevents her going where she wants; note characteristic undulating pattern; (c) *Grumbling-rumble*, Emily, when Imenti sucks on her ear, note undulation; (d) A short *grumbling-rumble*, female during scuffle as a male mounts Ella; (e) *Grumbling-rumble* by young male denied access to salt lick; (f) *V-8-rumble*, Vronsky, when sparring with Ella; (g) *Trumpet*, Eleanor, when mother, Erin, pushes her; (h) *Mixed roar*, young male as tusked by an adult female; (i) *Mixed roar*, young male when pushed at salt lick; (j) *Mixed roar*, Echeri, as access to a palm frond is denied by another elephant.

2). Two calls occurred in unknown circumstances. This call appears to be produced only by males and seems to indicate submission; the attacker usually stops when his or her (in one case) opponent has made this call.

In an aggressive and/or competitive situation, an elephant may *roar* when pushed, poked, tusked, chased, or thwarted by another individual (figure 9.9, panels h through j). I recorded 32 instances of such calls. Seventeen elephants had been pushed, 5 poked or tusked, 2 chased, and another 2 chased and then tusked, 2 lunged at, and 1 kicked; 3 occurred in unknown contexts. Such *roars* attract the attention of family members who may rush to the caller's aid either individually or as a coalition. A *roar* by an individual from another family elicits listening behavior but no assistance. *Roars* may be longer and more pulsated if the calling elephant is being chased.

Following an aggressive interaction, the elephant who has been chased or thwarted may shake his or her head vigorously in "annoyance" and blow air through the trunk, resulting in a *trumpet* or *snort* (figure 9.9, panel g).

Reconciliation. Vocalizations associated with reconciliatory behavior often follow protesting calls, particularly within a family group. Just as distressed calves are reassured by older individuals, a juvenile or adult who *roars*, or emits a protesting call in the context of received aggression, is typically approached by other members of the family and physically and vocally consoled. Vocalizing family members may reach out to touch the aggrieved elephant who may, in turn, respond with more *grumbling*- or *baroo* "woe-is-me"- type *rumbles*. Typically, such an event causes the gathering together of several family members including the victim, the mother of the victim, or another leading female(s), or the matriarch as well as the perpetrator of the aggression. As they stand together, they call with overlapping *rumbles* and reach out to touch one another and the victim. This reconciliatory behavior is likely to be a key to the maintenance of close bonds in an elephant family.

Social Integration

Bonding. Greetings between members of a close social group, including family, bond group, or, on rare occasions, clan members, take many forms. Greeting-Ceremonies as described by Moss (1981, 1988; see also chapter 8) are the most intense form and typically occur when closely bonded elephants come together after a period of separation. We know, however, that many of the visual and tactile behaviors that occur during a Greeting-Ceremony may also be seen in a wide variety of situations where elephants are not strictly greeting one another but rather giving a show of solidarity or, as with social grooming in primates, reinforcing bonds with individuals who are important to them. I will describe some of the more common of these greeting and bonding behaviors.

A relatively frequent event in an elephant family is one in which an elephant (usually older and female) approaches another (often in parallel and from behind). If the pair is closely bonded, typically the approaching elephant emits a soft and relatively short *rumble* to which the elephant being approached responds by Head-Raising, Ear-Lifting, and sometimes Backing-Toward (see chapter 8) the approaching elephant and emitting a relatively short *rumble* of medium duration and moderate intensity (although these, again, are variable). These *rumbles* are associated with the onset of temporal gland secretion, or Temporin. Examination of spectrograms indicates that in almost every case both the approaching and approached elephant *rumble* (though the softer call of the former may be drowned out by the more powerful call of the latter) with their calls overlapping. In some cases, nearby close family may join in. Elephants may also call in this manner when one approaches another face-to-face, although this pattern is less common. The Tsavo Orphan Keepers were able to elicit the same response from the orphaned calves by individually calling out their names. Members of an elephant family appear to use this common vocal exchange as a way of saying something like, "Hello, its good to be near you again," or perhaps, "You are important to me." I refer to calls given in this context as *little-greeting-rumbles* (figure 9.10, panels a and b; annex 9.1).

Little-greeting-rumbles may occur between any members of an elephant family, including between two males, but they are most common between females. Of the 45 events in which Echo participated, she was mother to the other participant in 26 events (of 6 possible pairs), grandmother in 9 (of 6 possible pairs), and great-grandmother in 2 (of 4 possible pairs). Examining just those *little-greeting-rumbles*, which occurred between EB females who were over age 10 in 2000, *little-greetings* occurred more often between mother-daughter, sister-sister, and grandmother-granddaughter pairs than expected by chance and less often between aunt-niece, great-grandmother–great-granddaughter, or pairs who were second cousin or greater ($X^2 = 107.84$, df = 6, $P < 0.0001$; table 9.8). While the number of *little-greeting* exchanges between aunt and niece is lower than expected by chance some pairs are very high. In each of these pairs, the niece was an infant prior to the aunt having her first calf, and the aunt was the most likely candidate in the family for taking the role of allomother.

Calves and their caretakers (often sisters and aunts) emit the *as-touched-rumble/coo-rumble* exchange under somewhat similar circumstances, and I believe that these

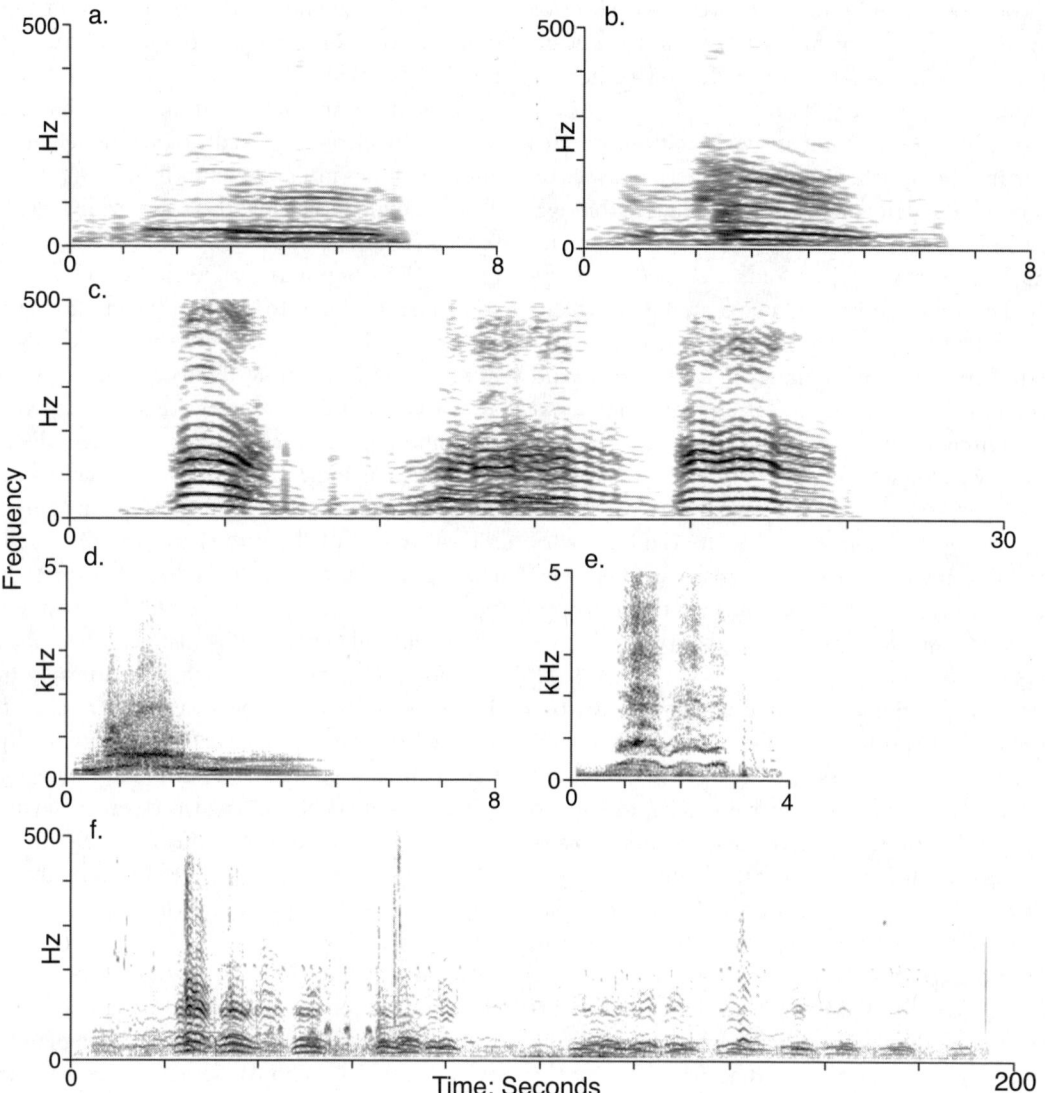

Figure 9.10 Calls produced in a bonding context: (a) *Little-greeting*, Enid then Echo, as Echo approaches Enid; (b) *Little-greeting*, Erin then Enid, as Erin approaches Enid; note typical partial overlapping of calls; (c) *Greeting-rumbles*, Emerald, several more distant elephants, Elfrida, another, and then Emerald; (d) *Rumble noisy roar rumble*, Ebony as she greets mother, Echo, after period of separation; (e) *Social-trumpet*, member of Echo's family during excitement surrounding birth of Ella's 1999 calf; (f) *Greeting-ceremony*, Echo's family, lasting three minutes. Note the softer rumbles prior to and following the intense calling that occurs between 20 and 80 seconds.

calls probably develop into the *little-greeting-rumble* exchange. I venture to propose that the relationships established through early vocal and tactile caregiving by mothers, sisters, and other allomothers form the basis of close bonds, which are then strengthened and reinforced through the customary *little-greetings* between closely bonded pairs of adults.

When members of a social group come together after a period of more prolonged separation, they approach one another face-to-face, and as calling begins, they turn, "spin," or "pirouette" to stand in parallel while Head-Raising, Ear-Lifting, and streaming with Temporin. Audible Rapid-Ear-Flapping (see chapter 8) may be heard as calling explodes with a burst of powerful, throaty, highly modulated, and overlapping *rumbles* that we refer to as *greeting-rumbles* (Moss 1988; Poole et al. 1988; figure 9.10, panel c).

The longer the separation and the closer the relationship between the calling individuals, the more intense the greeting is likely to be. A simple greeting between two individuals typically involves a few overlapping *rumbles*, while the reunion of two closely bonded families may involve

Table 9.8 *Little–greeting–rumbles*: Frequencies between pairs in the EB family

Pairs	No. of possible pairs	Observed no. of events	Expected no. of events
Mother-daughter	14	61	21.7
Sisters	10	23	15.5
Grandmother-granddaughter	4	9	6.2
Aunt-niece	18	21	27.9
Cousins	5	3	7.8
Second cousins or greater	35	16	54.3
Great-grandmother–great-granddaughter	1	2	1.6

up to 50 or more elephants and last for up to five minutes (I. Douglas-Hamilton, personal communication). During more intense Greeting-Ceremonies (figure 9.10, panel f), females may exhibit any or all of the following: Head-Raising, Ear-Lifting, Rapid-Ear-Flapping, Mouth-Opening, profuse secretion of Temporin, Urination, Defecation, Spinning, Tusk-Clicking, and Reach-Touch (see chapter 8). Associated with intense greetings and these bonding behaviors are *social-trumpets* and *social-roars* (see below).

Prior to an elephant Greeting-Ceremony, calling may be heard back and forth between two converging groups. These initial calls sound like, and may be better classified as, contact-calls (see Logistical). As the groups converge, the tone of *rumbling* explodes into a series of powerful, modulated, and overlapping *greeting-rumbles*, which give way to softer, less modulated *rumbles* that may continue for some minutes (see below). These softer *rumbles* are structurally quite different from the initial intense *greeting-rumbles*. The more forceful initial *rumbles* vary enormously, although they all show some degree of modulation in frequency and are of medium to long duration. Unfortunately, these powerful calls are difficult to measure due to their overlapping nature. Within a single Greeting-Ceremony, arched, skewed, arched with a wiggly contour, bimodal, bimodal and skewed, and multi-modal *rumbles* occur. During intense greetings, elephants reach out toward other individuals or actually touch them with their trunks (Reach-Touch, chapter 8). While the variability in calling may simply reflect the intensity of excitement, it may, alternatively, reveal additional information, such as a caller's signature or perhaps even reference to specific individuals, and is an area for further research.

While the behavioral contexts associated with *little-greeting-rumbles* and *greeting-rumbles* are very distinct, structurally both context-types are highly variable. As described previously, I used stepwise DFAs with cross-validation to compare *greeting-rumbles* and *little-greeting-rumbles*. Since duration and fundamental frequency are key predictor variables, and these change substantially with age, the model, though significant, was not very accurate in discriminating between the two call types (classified 73 percent correctly). Nevertheless, cross-validation using 11 randomly selected calls of each type showed that the model was able to discriminate between the two calls at a level well above chance (81 percent).

Elephants are highly expressive and demonstrative animals, and they vocalize loudly and in chorus under a wide variety of different circumstances. Behavior similar to the Greeting-Ceremony may occur following the birth of an elephant, a mating, the rescue of a calf, an aggressive interaction with another group, when the group has been threatened in any way, or when close associates are reunited. In all of these cases, a series of often powerful, overlapping *rumbles*—including a wide variety of contours and often interspersed with *trumpets* and *roars*—may be heard. It may be more appropriate to term these collectively as "Bonding-Ceremonies": a signal to participants and to more distant listeners that the callers are members of a supportive unit and that, together, they form a united front.

Trumpeting in this context typically overlaps with the lower frequency *rumbling* produced by other individuals. I refer to *trumpets* produced during social events as *social-trumpets* (see table 9.11). Based on the measurements I used, these *trumpets* are not significantly different from *harmonic-play-trumpets*, though they are significantly different from the other proposed *trumpet* context-types (see table 9.11).

Social-trumpets, in effect, function as a form of exclamation mark, defining the level of significance of an event. While the context-type of *rumbling* may indicate the type of event (e.g., mating, greeting, conflict), I propose that the frequency, and perhaps even the placement, of *trumpeting* may be an indication of the level of excitement and "importance" that the elephants collectively confer on an event. In a sense, the use of *social-trumpets* can be viewed as a simple form of syntax, qualifying the sequence of calls.

During moments of peak excitement female adults, juveniles, and even calves may emit powerful *roars*. Of the nine *roars* recorded during exciting social events, all except one were *noisy roars*, and all except one were combined with a *rumble*, with five of these being of the form *rumble noisy roar rumble* (figure 9.10, panel d). Similar to *social-trumpets*, *roaring* appears to signify the emotional intensity of an event. I suspect that additional data may well reveal these *roars* to be different from *roars* emitted during distress.

154 Joyce H. Poole

Logistical Calls

Maintaining Contact. Elephants use powerful, modulated *rumbles* at sound pressure levels of up to 115 dB (extrapolated to 1 m) to keep in audible contact with one another over distances of up to 1 to 2 km (Poole et al. 1988; McComb et al. 2000; chapter 10). We refer to these as *contact-rumbles* or *contact-calls*. A sequence may include several calls: a caller's initial *contact-rumble* is associated with rhythmic ear-flapping and is followed by listening behavior; the caller's head is held in an attentive lifted position with ears cocked as if waiting for a response, as if querying, "I am here, where are you?" An answering elephant typically responds with an unexpected (because the initial caller is often distant and such *rumbling* often inaudible to human listeners though discernible on spectrograms), explosive *rumble* that is preceded by an abrupt lifting of the head as if listening. Her response is seemingly stating, "I am over here." The initial caller, upon hearing an answer, may respond with another call, often associated with a more relaxed posture, as if sending confirmation that an answer has been received. Nearby family members may also add their voice to the second or third phase of the sequence, and calling back and forth may continue intermittently over hours until the callers meet again (figure 9.11, panel a).

Contact-calls are used primarily between family and bond group members, but on occasion, individuals may be heard to answer the *contact-call* of a non-family or

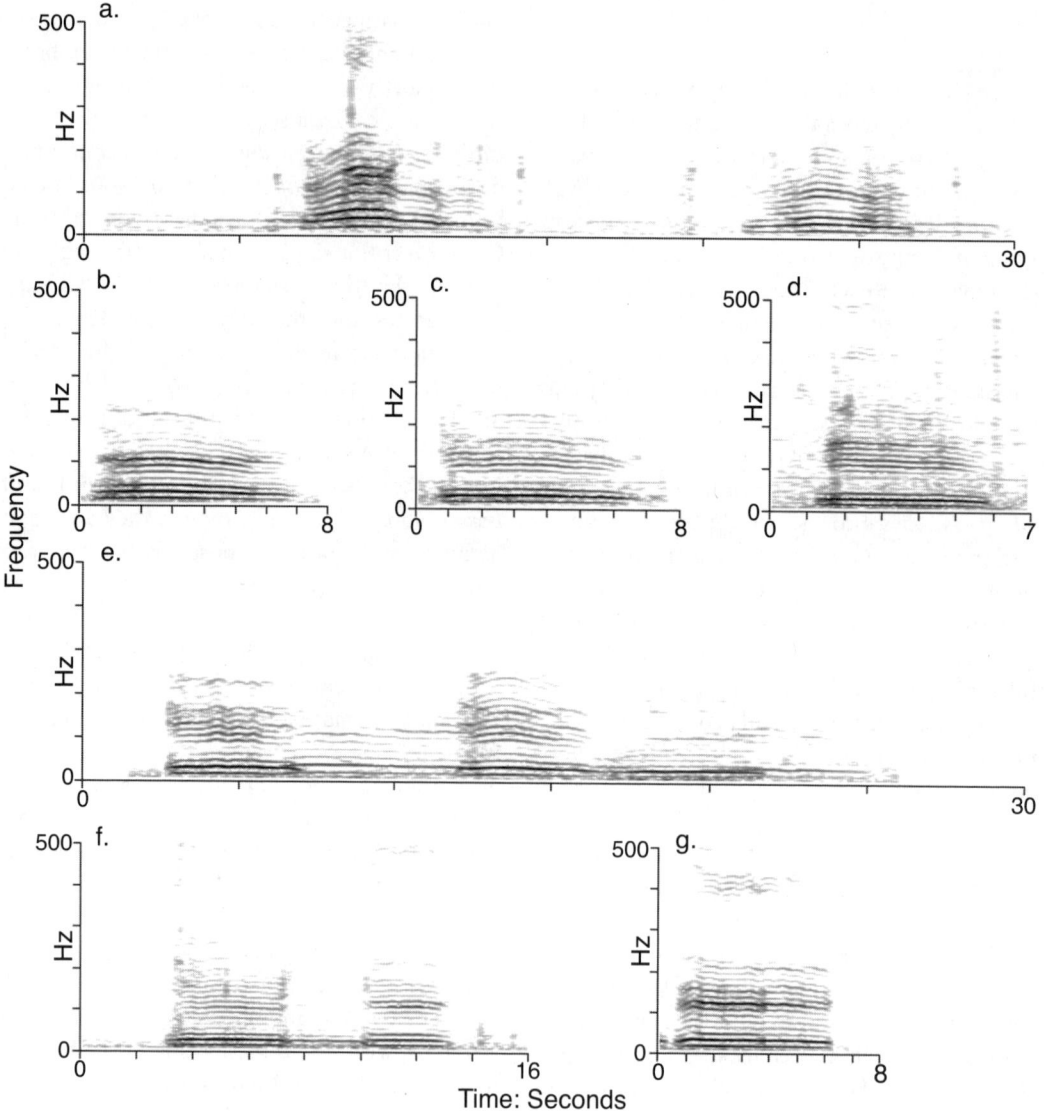

Figure 9.11 Calls produced in logistical context: (a) *Contact-call*; Eudora, Edwina, with distant elephants, note nearby and distant exchange; (b) *Contact-call*, Echo; (c) "*Let's-go*"-*rumble*, Enid; (d) "*Let's-go*"-*rumble*, Enid; (e) *Cadenced-rumble*, Enid and Eleanor, part of hour-long vocal exchange through which they work together to turn entire group around; (f) *Cadenced-rumble*, Eliot and Elettra, as they move toward swamp; (g) *Cadenced-rumble*, Ella, during exchange with daughter Emma after she abruptly moves off.

bond-group elephant. In these cases, the calling elephants have been members of the same clan. *Contact-calls* are relatively frequently heard *rumbles*, especially during drier periods when members of a family are more likely to be spread out over a larger area or split into sub-groups (see Moss and Lee, chapter 13).

Contact-calls are among the best studied of all elephant calls (see Poole et al. 1988; McComb, Moss, and Sayialel 2000; McComb, Moss, and Durant 2001; McComb et al. 2003) and are among the most powerful solo (as opposed to chorused or clustered calls) low frequency calls. Separated by long distances, elephants *contact-call* at high sound pressure levels, but as might be expected, elephants also call back and forth to one another over shorter distances at lower sound pressure levels (J. Poole, personal observation; captive: Soltis, Leong, and Savage [2005a]). Though elephants may be able to hear calls over distances of up to 10 km, information regarding individual identity probably does not travel more than around 2.5 km and usually less (McComb et al. 2002).

Adult females, juveniles, and calves all use *contact-calls*, though, as far as we are aware, adult males in Amboseli do not use the powerful long-distance *contact-call*. *Contact-calls* are typically powerful, throaty *rumbles* that are relatively long in duration. Most *contact-calls* last between 4 and 6.5 seconds and are modulated in frequency contour, typically rising sharply and falling more gradually (skewed left). There exists a wide range of variation in the contours of *contact-calls*, however, that may be related to whether the individual is the initial caller or the one answering. McComb et al. (2003; chapter 10) has shown that the *contact-calls* of individuals are structurally distinct and audibly identifiable to other elephants. In other words, *contact-calls* contain an acoustic signature. It is possible that considerable additional information is contained in the different variations of the calls, perhaps related to excitement level of the callers, their inter-individual distance, the sequential arrangement (i.e. call, answer, confirmation), or logistical or locational information. Future research may answer some of these questions.

Contact-calls are significantly different in acoustic structure from other common calls produced by adult females (table 9.10). They are distinctive both in quality and in behavioral context from other calls, though they may be confused with the onset of a *greeting-rumble* sequence. This confusion is partly due to the fact that as separated elephants approach one another, they may continue to call back and forth; once they come into close proximity, their calling shifts to *greeting-rumbles*. In these cases, it can be difficult to determine when *contact-calling* ends and *greeting-rumbles* begin, particularly as both types are loud modulated calls.

Departure Signals. An elephant indicates that she wishes to depart by calling, by exhibiting intention movements, and by pointing in the direction she wishes to travel using the axis of her body. Standing intently at the edge of her group, facing in the direction she wishes to travel, an elephant emits, with slow, rhythmic ear-flapping, a moderately loud, low-pitched, long, relatively flat *rumble* (figure 9.11, panels c and d; annex 9.1). The calling elephant repeats her "*let's-go*" appeal once every minute or so, sometimes calling for periods of up to half an hour, as she tries to persuade others to depart. She may gain the support of other individuals who join in her calling, but typically this is a solo call that we refer to as the "*let's-go*"-*rumble* (Poole et al. 1988). The "*let's-go*" is, in essence, a proposal: "I want to go this direction, let's go together." It is audibly distinguishable from most other elephant *rumbles* due to its characteristically long, drawn-out quality and the posture of the calling elephant, but it can be confused in the field with the *cadenced-rumble*.

The "*let's-go*"-*rumble* is one of the more commonly heard sounds among a family of elephants. There is a tendency for older females to produce this call more frequently than younger females (table 9.9). Less commonly, juvenile and adult males may be heard to call. Although one might expect the matriarch to be the primary producer of this call, as she gathers together her charges for a change in direction, this is not necessarily so. In the EB family, for example, Echo's second-eldest daughter, Enid, a young female with strong leadership qualities, gave the "*let's-go*"-*rumble* notably more than anyone else (table 9.9). Females with extrovert personalities (see box 13.2) are more likely to emit "*let's-go*" calls than those with introvert personalities ($R = 0.71$, $t (n - 2) = 2.83$, $n = 10$, $P = 0.022$; Ebony was excluded since she was so much younger than the others with personality scores (see box 13.2). While a matriarch may propose a plan of action using the "*let's-go*"-*rumble*, very often she simply moves off without making any obvious audible suggestion, except perhaps a Flap-Slide (see Poole and Granli, chapter 8) motion with her ears (indicating a change of activity), presumably expecting (or at least hoping) others will follow without discussion! This is certainly not always the case, and many a "discussion" and even "disagreement" takes place regarding plans of action in an elephant family (J. Poole, personal observation; C. Moss, personal communication).

Decision Making. Not uncommonly, vocal exchanges between related adult females may be heard that have the cadence of a conversation, rising and falling, as first one individual and then another contributes her voice. Other females may join the initiating individuals in a sequence of low-pitched, moderate intensity, relatively flat, long

Table 9.9 Use of "*let's-go*"-*rumble* by members >4 yrs old in the EB family 1999–2003

Elephant	Sex	Birth	A	B	Total	Personality
Echo—Matriarch	F	1945	34	13	47	+0.7
Ella—Echo's sister	F	1965	11	10	21	+0.5
Erin—Echo's daughter	F	1969	4	2	6	−2.0
Eudora	F	1972	2	4	6	−0.4
Enid—Echo's daughter	F	1982	72	21	93	+1.2
Edwina	F	1982			0	−0.6
Eleanor	F	1985	2	1	3	−1.0
Eliot—Echo's daughter	F	1985	11	7	18	+0.8
Emma—Ella's daughter	F	1987		2	2	+0.25
Elspeth—Eudora's daughter	F	1988			0	−0.5
Ely—Echo's son	M	1990			0	—
Esau—Ella's son	M	1990		2	2	—
Erwin—Erin's son	M	1991			0	—
Ebony—Echo's daughter	F	1994	1		1	+1.0
Elettra—Ella's daughter	F	1995	8		8	—
Echeri—Erin's daughter	F	1995		1	1	—
Europa—Edwina's daughter	F	1995	3		3	—
Eudora c96—Eudora's son	M	1996			0	—
Ejac—Enid's son	M	1997	1		1	—

Notes: Columns A and B refer to the number of calls with level of sureness A or B. Personality *Z* scores are from box 13.2 Component 1 "Sensitive to Insecure" and represent a measure of leadership and social integration

rumbles. My 1999 and 2000 field notes include 38 references to such bouts of calling, including a total of 239 vocalizations (including sureness level A and B). Although this pattern of *rumbling* may be heard as a solitary call (*n* = 16), it is more often a series of overlapping or closely adjacent calls interchanged over the course of several minutes and up to an hour, with the longest uninterrupted bout measured including 49 calls. The patterning of vocal exchange has such a cadence of conversation (in particular, what sounds like a higher pitch/lower pitch exchange) that I refer to this as a *cadenced-rumble* (figure 9.11, panels e through g). These calls may overlap, although typically the degree of overlap is small. Adult females appear to use *cadenced-rumbles* to "lend their voice to" a proposed plan of action, usually, it seems, regarding where to go and when to depart (e.g., related to "*let's-go*"-*rumbling*). As mentioned before, this call is similar in structure to the "*let's-go*," but unlike the "*let's-go*," I have been unable to detect any particular axis of the body that would indicate directionality. In fact, the call is unusual in that the vocalizing females continue to feed (or whatever activity they were engaged in) without looking up or showing any particularly attentive behavior. I have heard elephants from age four and upward (including juvenile though not adult males) participate in such bouts of calling.

Cadenced-rumbles are often heard following a series of "*let's-go*" calls (see "*let's-go*"-*rumble*) and a series of *cadenced-rumbles* may contain calls that are difficult to distinguish from "*let's-go*"-*rumbles*. This may be because there are "*let's-go*"-*rumbles* interspersed among the other calls or because they are one and the same call. Much additional research is needed to clarify the temporal pattern and usage of these calls, but based on the behavior of the elephants the *cadenced-rumble* appears to represent a complex level of consensus building between family members. It appears as if one individual proposes a course of action (using the "*let's-go*"-*rumble*, for example) and then a period of negotiation and consensus building follows, ending in either concordance or disagreement. Agreement is sometimes evident when individuals gather together *rumbling* with heads raised and touching or when individuals who were suggesting different directions of travel (as indicated by their Let's-Go-Stance, chapter 8) compromise on one direction. Alternatively, the elephants may continue to disagree and go separate ways.

While *cadenced-rumbles* sometimes appear to start "spontaneously" (five bouts calling), they are more often associated with another event such as: group departure (four bouts); a group member proposing departure ("*let's-go*"-*rumbles*; seven bouts); a group member making contact

Table 9.10 Discriminant function analysis (DFA) results for common adult female *rumbles*

Adult female *rumbles*	Cadenced, contact, "let's-go," coo	Greeting, little-greeting
N (training)	206	56
Eigenvalue overall	1.087	0.474 (100)
Wilks' Lamda	0.419	0.679
F	11.333	6.041
P	<0.0001	<0.0001
Classification expected probabilities	0.25	0.50
Classification % correct	57	73
Classification cross-validation % correct	54	81
Variables in the model	F_0 max-min, duration, bandwidth, max time, location F_0 max	Duration, max time, F_0 max, maximum frequency

from a distance (*contact-calling*, five bouts); a group member rejoining the group (*greeting-rumbles*; three bouts); or the reinforcing of bonds (*little-greeting-rumbles*; seven bouts). The association between *cadenced-rumbles* and *little-greeting rumbles* may be related to a reinforcement of bonds in an attempt to gain support for a proposed plan of action.

As described earlier, I used two stepwise DFAs with cross-validation to examine the commonly heard calls produced by adult females (table 9.10). In the model involving *coo-rumbles*, *"let's-go"-rumbles*, *cadenced-rumbles*, and *contact-rumbles*, 57 percent were correctly classified (54 percent in cross-validation). The results are well above what would be expected by chance. Due to the conservative nature of the model and the spread in the ages of the callers, I think it is fair to accept the differences in the calls as real.

Play

In the context of play, calves, juveniles, and adults of both sexes produce a variety of *trumpets* (Poole and Granli 2004). The majority of these are shrill harmonic sounds; others are distinctly nasal, while still others are pulsated. *Harmonic-play-trumpets* are typically short high-pitched, high-intensity sounds associated with lone locomotor and social play (Poole and Granli [2004]; figure 9.12, panel a; table 9.11). Some *harmonic-play-trumpets* have a flatter reverberating sound, almost like a loud nose-blow, and spectrographically these show more noise. Elephants of both sexes and all ages also produce a shortened form of *trumpet* during play referred to by Berg (1983) as a *trump*.

Similar to the *harmonic-play-trumpet* but exhibiting a more prolonged and resolute quality is a *trumpet* that is associated with Mock-Charging behavior (see chapter 8). Cavorting elephants often chase other species in their environment (such as hares, hyenas, wildebeests, monkeys) during which they *trumpet* loudly. We term *trumpets* associated with this form of play *mock-charge-play-trumpets* (figure 9.12, panel c; table 9.11). These *trumpets* are significantly longer in duration than *harmonic-play-trumpets*.

Highly spirited lone locomotor play is characterized by Floppy-Running (Moss 1988) behavior and *trumpeting* that is expelled in a sequence of breathy pulses as the elephant moves at a fast gait. We refer to this *trumpet* as a *pulsated-play-trumpet* (figure 9.12, panel d; table 9.11). *Pulsated-play-trumpets* are significantly longer in duration and more modulated in frequency contour than *harmonic-play-trumpets*.

As play escalates, a fourth form of *trumpet* may be heard that, as air is forced slowly through the upper part of the nasal passages and reverberates down the length of the trunk, sounds like a large man blowing his nose. These *trumpets* are noisier and significantly lower in frequency than all other *trumpets* and are referred to as *nasal-play-trumpets* (Poole and Granli [2004]; figure 9.12, panel b; table 9.11).

Both *nasal-play-trumpets* and *pulsated-play-trumpets* are associated with exuberant play. It has been my impression that elephants may be imitating the form of *trumpet* made by nearby playmates, as there is a tendency for *nasal-play-trumpets* to be temporally associated with other *nasal-play-trumpets* and *pulsated-play-trumpets* to be associated with *pulsated-play-trumpets*. Additional data will be required, however, to determine whether elephants are using vocal imitation in play.

Trumpeting by elephants is associated with high levels of excitement ranging from exuberant play, stimulating social events, and threatening, frightening, or startling situations. Just by hearing a *trumpet*, I can make a fairly accurate assessment of its behavioral context. I performed a stepwise DFA on six proposed harmonic *trumpet* sub-types (table 9.11; *nasal trumpets* have already been classified as a separate type), including the three play *trumpets* previously described and the so-called *social-trumpet*, *alarm-trumpet*, and *anti-predator-blast*. The model was able to classify 60 percent correctly (down to 52 in cross-validation). Though not particularly high, the discriminations are well above the level of chance (table 9.11).

Conclusions

Elephants produce discrete call types, but within these primary types, calls are highly graded (Leong et al. [2003];

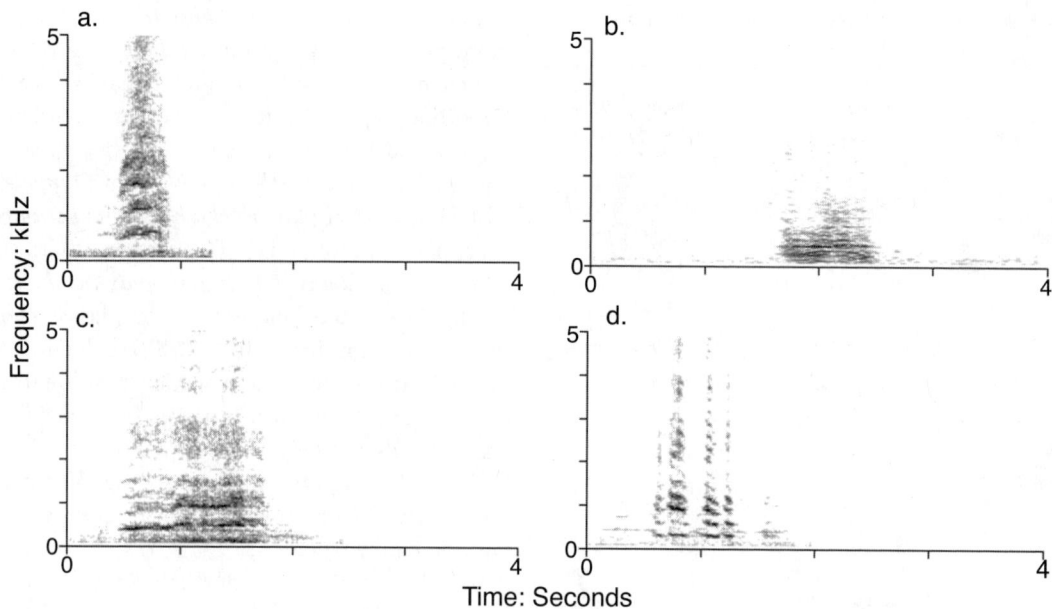

Figure 9.12 Calls produced in the context of play: (a) *Harmonic-play-trumpet*; (b) *Nasal-play-trumpet*; (c) *Mock-charge-play-trumpet*, juveniles while chasing a hare; (d) *Pulsated-play-trumpet*.

Table 9.11 Forward stepwise discriminant function analysis of six *trumpet* subtypes

Harmonic trumpet subtypes	Harmonic-play, pulsated-play, mock-charge-play, alarm, anti-predator-blast, social
N (training)	136
Eigenvalue overall	0.794
Wilks' Lamda	0.262
F	4.839
P	<0.0001
Classification expected probabilities	0.167
Classification % correct	60
Classification cross-validation % correct	52
Variables in the model	Duration, F_0 max, curve, F_0 max-min, bandwidth, location F_0 max, max time, and peak time

Note: The model did not consider the pulsated nature of the *pulsated-play-trumpet*.

Soltis, Leong, and Savage [2005b]; this chapter). Many species have repertoires that are acoustically graded yet are perceptually discrete (Hauser 1997a). Elephant calls, like human speech, exhibit gradedness on the production end but discreteness on the perceptual end. The degree to which within-call type distinctions are made by elephants will ultimately depend upon their significance to reproduction and to survival. Our ability to make the sort of acoustic discriminations that elephants are clearly able to make will depend on a better understanding of which features are most salient with regard to signal recognition.

Clearly, some elephant context-types are very distinct, while others are much more difficult for humans to discriminate. The *musth-rumble* and the *estrous-rumble*, for example, stand out from other *rumbles* even to the layman, and DFA shows the more powerful contact-calls, too, to be highly distinctive. We might expect calls, such as the *estrous-rumble* and the more powerful of the *contact-call*s, which function in long-distance advertisement and contact and, thus, rely purely on acoustics, to be discrete. We might also expect calls that define reproductive state, such as the *musth-rumble* and, again, the *estrous-rumble*, to be highly distinctive.

"*Let's-go*"-*rumbles*, *cadenced-rumbles*, and the softer *greeting-rumbles*, on the other hand, are more difficult to discriminate between. For calls that are given within the close context of the family, in which animals are in visual contact and interacting with known personalities, more subtle factors may come into play. For instance, the "*let's-go*" signal is extremely distinctive to an observer (and to the elephants) due not purely to the audible characteristics of the call but to the posture of the calling animal, the pattern of repetition, and often the emphasis and sense of purpose with which the call is given. One can literally hear and see the impatience of the calling animal. The *cadenced-rumble*, though similar in acoustic structure, does not carry the same resoluteness in its tenor. Moreover, the calling animal's behavior is significantly different. She does not stand facing away from her family, waiting in a determined and impatient manner, but is, in activity and posture, integrated with the others. The call is not repeated every minute or so but is used in a vocal exchange with another or several other elephants. Accordingly, we can define these as two discrete signals since the message is perceptively distinct. And while the structural differences between the calls may not always be obvious to us at this stage, the elephants appear able to distinguish between them. It is likely that the difference is one of tenor, a subtle shift in emphasis.

Two other calls that may be confused based on acoustic structure (both being calls short in duration) are the softer versions of *little-greeting-rumbles* and *coo-rumbles*. *Little-greeting-rumbles* are calls given in exchange between two closely bonded individuals, most commonly between mother and daughter or between sisters, and are typically composed of a softer flatter call and a louder, more modulated response. Mothers or allomothers (who are often older sisters) give the soft, flat *coo-rumble* as a small greeting or reassurance to calves who respond, similarly, with a louder, modulated call. The adult-calf/juvenile-calf pairs would be the same pairs likely to engage in *little-greetings* later in life. Thus, it seems likely that these calls, a pair of *little-greeting-rumbles*, and the *coo-rumbles* and the calf responses, *as-touched-rumbles*, are related. The bonds formed as calves through this tactile and vocal exchange with older individuals create an enduring relationship sustained through acoustic exchange and mutual cooperation.

Arousal level is clearly encoded in elephant calls, as was first noted by Berg (1983). Since then, detailed analyses have characterized the expression of affect in the structural patterns of *rumbles* (Soltis, Leong, and Savage 2005b; Soltis et al. 2009; Li et al. 2007; Wood et al. 2005; Stoeger-Horwath et al. 2007; Wesolek et al. 2009). Soltis et al. 2009 focused on dominance interactions, showing that the calls of females tended to contain higher, more variable fundamental frequencies when they were the lower ranking of a pair interacting aggressively. The fact that elephants' voices reflect affective state helps to explain the broad variability in the calls of individual females recorded in similar behavioral contexts. For example, although *coo-rumbles* and calf responses and *little-greeting-rumbles* are affiliative gestures, they are typically given between pairs of unevenly ranked individuals, and usually the younger individual's call is of higher and more variable fundamental frequency and of higher amplitude.

Estrous-rumbles, greeting-rumbles/bonding-rumbles, mating-pandemonium, and *roaring-rumbles* (heard when elephants are mobbing predators) are all marked during the peak of excitement by increased amplitude, increased noise, and increased modulation, with energy distributed in the upper harmonics (rather than in the second harmonic as in most *rumbles*), with calls becoming softer, less modulated, and less noisy with time. Comparing individual calls from the beginning of a greeting or immediately following a mating with those at the end reveal enormous differences. Are elephants able to acoustically recognize each individual call as being related to a greeting, a mating, or the arrival of a *musth* male? Or are they, like us, only able to recognize the greeting in the entirety of the sequence by its overall pattern of calling?

Irritation and anger are also clearly encoded in elephant calls, as described by Soltis et al. (2009). The calls of complaining or protesting elephants, such as *grumbling-rumbles*, have an audibly whining tone. The spectrograms of these calls show a rising and falling frequency contour and patches of noise. It is this undulating frequency and alternating noisy and tonal sound that gives these calls their distinctive complaining tone. The noisy component is also evident on other complaining calls such as the *baroo-rumble* and the *V-8-rumble*.

If the level of excitement and anger can be heard in the voices of elephants and seen in the spectrographic structure of calls, what else does the infinite variability in the frequency contours of calls encode? *Greeting-rumbles* or *bonding-rumbles*, in particular, show an extreme range in the frequency contour of calls. They may be flat, slightly arched, highly arched, bimodal, multi-modal, skewed left or right. What does all this variability signify to the elephants?

Considering that elephants are capable of vocal production learning and have unusual vocal tract flexibility and a large resonance chamber, we should not be surprised that elephants have the capacity for producing such a wide variety of sounds. Elephants are intelligent and complex social animals; the increasing understanding of elephant communication generated by this chapter, together with the very significant research of my many colleagues reflects this intricacy. As noted by Soltis (2009) and seen clearly in this chapter, the separation of elephant call types into sub-types can take many forms. If individual identity, age and body size, relative dominance, level of arousal or affect, and specific behavioral context all contribute to the shape a call takes, it is easy to understand why a *rumble* given by one individual in the context of a Little-Greeting, for example, might take many forms.

In this chapter, I have presented the broad range of contexts in which African elephants are heard to vocalize and endeavored to make a first attempt to discriminate between some of the more common calls that elephants make in different behavioral contexts. The discriminations made thus far are based on rather limited and fundamental structural measurements. In the future, more detailed observations, measurements, and analyses are likely to reveal a highly versatile repertoire of calls matching the multi-faceted interactions and relationships observed among wild elephants.

ANNEX 9.1 BEHAVIORAL CONTEXTS AND ASSOCIATED CALL AND CONTEXT TYPES

Behavioral context	Call type	Call sub-type	Manner features	Duration (s)	F_0 max (Hz)	F_0 min (Hz)	Max frequency (Hz)	Bandwidth (Hz)
Anti-predator	Rumble	Comment	Tonal; flat	3,438 (2,636–3,808) 2,501–4,129; 5	15.2 (15.1–15.4) 13.8–19.0; 5	13.0 (13.0–13.8) 12.2–16.8; 5	29.3 (29.3–29.3) 29.3–29.3; 2	166 (143–183) 160–180; 2
		Anti-predator	Tonal, upper harmonics noisy; arched	4,081 (3,273–6,233) 2,156–11,781; 32	21.8 (18.7–26.6) 14.2–33.7.4; 32	15.2 (12.4–19.0) 10.3–24.6; 32	35.2 (29.3–128.0) 15.6–154.3; 24	253 (211–560) 162–701; 30
	Roar	Anti-predator		2,538 2	290 1	235 1	215 2	4,435 1
	Trumpet	Blast	Noisy harmonics	1,269 (952–1,398) 583–2,188; 18	533 (476–658) 368–757; 18	345 (322–393) 312–498; 18	668 (457–949) 293–2,028; 17	5,004 (4,295–5,595) 2,940–6,449; 8
Mating	Rumble	Musth	Tonal with noise	4,723 (4,054–5,160) 1,739–8,744; 76	17.1 (16.4–17.9) 12.5–28.2; 77	11.7 (10.5–12.3) 8.4–14.0; 77	35.2 (33.2–35.2) 25.4–169.9; 70	124 (103.0–155) 55.0–417.0; 77
		Female-Chorus	Tonal; arched	3,797 (3,210–5,304) 1,958–7,107; 25	19.1 (17.7–21.4) 14.7–39.2; 25	14.7 (13.8–17.4) 10.8–25.8; 25	29.3 (26.4–34.2) 13.7–58.6; 20	221 (168–297) 107–562; 20
		Estrous	Tonal, upper harmonics noisy; highly arched skewed left	5,667 (4,777–6,060) 4,367–8,036; 14	24.8 (22.5–26.8) 18.3–34.9; 18	17.3 (15.1–17.4) 12.3–17.7; 18	129 (127–131) 35–203; 5	376 (337–400) 180–929; 18
		Mating Pandemonium	Tonal, upper harmonics noisy; variable	4,152 (3,367–4,933) 1,957–8,657; 22	18.0 (16.1–18.8) 14.8–22.8; 21	14.0 (13.4–16.1) 10.5–17.9; 22	32.2 (27.3–36.2) 25.4–121.1; 16	294 (235–419) 177–818; 20
	Roar	Estrous	Variable pulsated mixed noisy	2,377 (1,852–3,418) 566–3,506; 11	197 (160–362) 153–602; 10	131 (103–174) 96–458; 11	172 (140–430) 107–585; 11	1,483 (1,184–3,746) 352–4,200; 11
Parent–Offspring	Rumble	Begging	Tonal; primarily flat–descending	1,659 (1,218–2,256) 511–4,151; 139	24.4 (22.5–26.3) 16–33.5; 137	22.4 (20.7–23.5) 12.7–28.1; 139	46.9 (43.0–49.7) 29.3–242.2; 134	188 (139–283) 63–645; 139
		Separated	Tonal; flat–descending	2,480 (2,070–3,250) 621–5,163; 98	24.8 (21.4–26.2) 16.1–36.0; 97	21.8 (18.5–23.4) 13.4–27.8; 98	44.9 (39.1–49.7) 29.3–288.1; 98	162 (123–228) 63–670; 98
		As-touched	Tonal and with noise; variable	1,544 (1,138–2,185) 415–3,404; 26	28.5 (25.2–33.1) 18.9–48.5; 26	22.8 (20.5–24.5) 17.2–34.1; 26	54.7 (48.8–66.4) 43–238.7; 26	337 (294–405) 145–872; 25
		Baroo	Tonal with noise; arched	2,037 (1,514–2,727) 630–6,752; 37	32.2 (27.3–36.3) 16.1–53.7; 37	24.8 (21.5–28.0) 11.4–32.8; 37	63.9 (55.6–81.1) 27.3–303.1; 36	418 (296–700) 95–1,000; 37
		Coo	Tonal; contour primarily flat	3,388 (2,695–4,404) 1,113–6,811; 167	16.4 (14.7–18.0) 12.0–24.8; 163	14.5 (12.8–16.0) 8.7–21.8; 166	31.2 (27.3–34.7) 15.6–91.8; 164	186 (143–226) 38–684; 166
	Husky cry	Low-intensity	Tonal with noise	735 (637–940) 328–1,282; 18	35.5 (32.0–40.4) 25.1–54.1; 17	31.4 (27.5–33.2) 19.5–50.4; 18	62.5 (50.8–74.2) 35.2–101.6; 18	584 (408–693) 276–946; 16
		High-intensity	Tonal with noise	1,245 (891–1,458) 770–1,544; 13	41.6 (37.6–44.4) 33.9–51.0; 13	32.1 (28.0–33.9) 23.4–41.3; 12	88.0 (74.2–93.8) 66.4–414.1; 13	600 (571–800) 391–1,000; 13
	Roar	Suckle protest	Mixed, noisy, tonal	1,410 (994–1,845) 606–3,004; 16	459 (414–651) 172–727; 16	337 (205–443) 109–658; 16	345 (366–858) 301–2,093; 16	4,582 (3,182–7,102) 2,070–8,834; 12
		Separated	Mixed, noisy, tonal	2,055 (1,852–2,329) 467–3,564; 23	370 (279–447) 236–568; 23	278 (205–304) 163–526; 23	391 (311–632) 214–2,062; 23	2,968 (2,418–4,117) 1,816–5,143; 23
	Cry	Protest	Tonal; descending	314 (218–453) 107–528; 21	454 (337–522) 183–800; 21	276 (250–315) 159–547; 21	388 (279–642) 236–1,273; 20	3,198 (1,745–5,460) 724–6,968; 19
	Trumpet	Alarmed	Noisy harmonics, arched	620 (491–701) 435–830; 8	687 (656–737) 625–791; 8	458 (437–458) 390–469; 8	761 (691–1,044) 624–1,317; 8	9,425 (5,538–11,133) 4,811–11,235; 8
Conflict	Rumble	Grumbling	Tonal with noise; wavy	2,142 (1,520–3,390) 816–7,434; 60	24.6 (19.7–28.5) 14.8–39.2; 60	19.8 (17.1–23.0) 12.4–27.8; 60	46.9 (35.2–62.5) 15.6–246; 60	302 (208–423) 92–1,000; 60
		V-8	Tonal with noise	1,271 (1,103–1,738) 994–1,806; 6	17.8 (15.3–24.0) 15.1–28.1; 4	13.2 (19.2–14.5) 11.2–23.4; 4	32.2 (15.1–86.4) 14.6–124; 4	424 (316–450) 308–661; 6
	Roar	Agonistic and rough contact	Noisy, mixed, pulsated, tonal	1,422 (1,105–2,056) 330–5,419; 40	362 (251–448) 108–776; 31	260 (168–314) 39–507; 35	477 (332–679) 172–2,849; 39	3,688 (2,733–5,000) 1,583–8,886; 35

Behavioral context	Call type	Call sub-type	Manner features	Duration (s)	F_0 max (Hz)	F_0 min (Hz)	Max frequency (Hz)	Bandwidth (Hz)
Social integration	Rumble	Little-greeting	Tonal with noise; variable	3,370 (2,571–3,997) 1,206–7,265; 167	18.2 (15.9–22.0) 12.1–30.9; 167	15.1 (13.1–18.6) 10.4–22.9; 166	35.2 (29.3–43.0) 19.5–209.0; 157	207 (173–250) 55–712; 160
		Greeting	Tonal with noise; variable	4,801 (4,079–5,953) 1,743–10,294; 123	16.9 (15.0–20.0) 11.7–40.2; 122	12.9 (12.0–15.4) 9.9–26.8; 123	31.2 (27.3–37.1) 15.6–231; 85	250 (215–389) 135–991; 102
		Bonding	Tonal with noise; variable	2,842 (2,451–3,759) 1,479–4,446; 11	24.4 (19.7–27.1) 18.7–35.3; 12	18.3 (15.2–19.7) 13.2–21.1; 12	168 (160–178) 29.3–207; 7	301 (261–564) 156–628;9
		Contact call	Variable tonal and with noise; variable arched	5,008 (4,035–6,329) 1,844–9,478; 180	19.4 (17.1–21.7) 13.7–31.0; 174	13.4 (12.3–15.0) 10.3–25.5; 180	35.1 (29.2–39.5) 15.6–218.8; 174	253 (171–470) 57–1,015; 175
		"Let's-go"	Tonal; flat, slightly modulated	5,234 (4,308–6,521) 1,164–9,229; 123	15.6 (14.7–16.8) 12.8–21.3; 123	12.5 (12.0–13.4) 9.6–19.6; 123	27.8 (26.3–29.3) 14.6–40.9; 122	228 (178–246) 41–615; 113
		Cadenced	Tonal; flat, slightly modulated	5,106 (4,243–5,624) 3,036–7,675; 84	15.3 (14.4–16.4) 12.6–22.1; 84	12.8 (12.2–13.6) 11.0–18.0; 84	27.8 (26.3–29.3) 13.7–113.0; 70	190 (159–237) 42–645; 76
	Harmonic Trumpet	Social	Noisy harmonics	670 (469–936) 273–1,951; 70	394 (349–450) 179–841; 68	276 (255–321) 138–606; 68	453 (323–818) 132–2,019; 69	9,700 (6,450–11,000) 1,274–19,000; 70
	Roar	Social	Noisy, mixed	1,755 (1,280–1,891) 1,055–2,112 9	341 (239–419) 220–624 8	233 (148–260) 139–452 9	378 (253–483) 172–619 8	3,185 (2,152–4,640) 1,556–5,160 8
Play	Trumpet	Mock-charge	Noisy harmonics	1,116 (717–1,461) 536–2,559; 8	521 (458–581) 375–600; 7	386 (317–417) 90–464; 8	450 (424–632) 329–1,009; 8	6,794 (3,733–10,094) 3,036–11,128; 8
		Pulsated	Noisy harmonics; pulsated	1,494 (956–1,712) 215–3,385; 29	460 (407–554) 336–732; 29	333 (301–387) 124–519; 29	460 (373–878) 105–1,324; 29	6,636 (4,802–9,086) 1,326–11,256; 29
		Harmonic	Noisy harmonics; arched	546 (405–765) 198–2,486; 65	407 (371–478) 178–790; 61	342 (271–380) 108–496; 65	415 (345–604) 108–2,405; 63	7,200 (4,600–10,900) 1,600–32,500; 65
	Nasal trumpet	Nasal-play	Noisy harmonics; flat slightly modulated	676 (576–809) 445–2,426; 19	101 (67–144) 34–329; 16	61 (46.6–107) 22.7–122.9; 19	248 (132–306) 32–563; 19	3,729 (2,256–4,927) 786–10,425; 19

Notes: Manner features list sound quality and primarily observed contours. Figures represent median (interquartile range) range; *n*.

Chapter 10 Vocal Communication and Social Knowledge in African Elephants

Karen McComb, David Reby, and Cynthia J. Moss

Introduction

A multitude of studies on wild animals have investigated the functional significance of behavioral differences, but few have attempted to elucidate the nature and fitness consequences of differences in animal minds. Thus, despite widespread interest in the evolution of social intelligence, we still know relatively little about how wild animals acquire and store information about social companions or what might be the fitness consequences of differences in social knowledge. This significant gap in our understanding of the factors affecting survival and reproduction in the wild is at least partly a result of the intrinsic difficulties associated with assessing what is occurring in an animal's mind, particularly in the natural environment. While behaviors are accessible to the researcher and, by and large, open to direct observation, mental processes are not.

The study of vocal communication provides a powerful tool for investigating the mental abilities of animals. While mental processes themselves are inaccessible, we can record the vocalizations that animals give in response to other individuals and to external objects and events. These signals can then be played back in the absence of the original stimulus, and the responses of subjects to playback can reveal what animals know either about the vocal signal itself or about the individual(s) who produced it. By designing experiments carefully, it is possible to gain quite detailed insights into mental processes underlying behavior, working on animals in natural or semi-natural environments (e.g., Cheney and Seyfarth 1990; Cheney, Seyfarth, and Silk 1995; Proops, McComb, and Reby 2009). In this chapter, we will describe how we have used playback experiments, in tandem with sophisticated digital sound processing techniques, to address questions about social knowledge based on vocal signals in African elephants.

At first sight, African elephants can appear somewhat inscrutable. This may be because their sensory world differs quite radically from ours. The primary response of elephants to stimuli in the external environment is usually auditory (i.e., through hearing) or olfactory (i.e., through smell) rather than visual. Thus, when faced with a potentially threatening stimulus elephants will first tend to extend and orientate their ears and raise their trunks rather than attempt to localize the source of danger visually. Such responses provide the key to gaining insight into the knowledge that elephants have about their social companions. It is possible to record the contact calls that adult females give in order to broadcast information on their identity and location to family and bond group members (Poole et al. [1988] and chapter 9 in this volume) and play these back to other females known to have varying levels of association with the caller. Particular diagnostic reactions may then be used not only to determine whether subjects listening to the playback can classify the caller as a family or bond group member but beyond that, whether they are familiar or unfamiliar with that individual's call. Also, when detailed acoustic analysis is conducted in conjunction with playback experiments it is possible to identify the particular acoustic cues that elephants use to make these distinctions and investigate the distances over which such cues remain intact. Finally, abilities to accurately distinguish the calls of other females that are known from field measurements of association patterns to be genuinely strange or familiar to the subjects can be quantified and used in conjunction with demographic information to evaluate the causes and consequences of differences in social knowledge. Using this range of approaches in the Amboseli study population, in our chapter we do the following:

- Evaluate the extent of social knowledge based on vocal signals in female African elephants.
- Examine how information on social identity is coded in calls and the distances over which this information remains intact.
- Consider the causes and consequences of differences in social knowledge acquisition between family groups, highlighting the key role of the matriarch as a repository of social information.

Recognition of Family and Bond Group Members

It is very clear that females can distinguish family and bond group members from other individuals on the basis of their contact calls (McComb et al. 2000). When a family unit is played the contact call of an absent family or bond group member, the subjects typically contact call themselves shortly after playback and often move toward the source of playback (details in box 10.1). This very characteristic response is not obtained when the caller does not belong to one of these categories. Callers that are neither family or bond group members elicit one of two reactions—subjects either listen to the call but remain relaxed, resuming their original behavior soon after the playback has finished, or else become agitated and bunch together into defensive formation, the calves typically moving in toward their mothers and the diameter of the whole group shrinking (box 10.1). The association index between the caller and the subjects on a day-to-day basis is a good predictor of which of these reactions will occur, more frequent associates typically generating the first reaction, and those that are rarely encountered generating the last reaction (see below).

Although our results demonstrate that elephants are capable of categorically discriminating the calls of family or bond group members from those of others, they do not examine directly whether this response is based on "true" recognition of the individual who gave the call. The term "individual recognition" is reserved for instances where a signal has not only been recognized (i.e., identified as encountered before) but also perceived as belonging to a particular known individual (see Beer [1970] re: individual recognition and Barnard [1991] re: kin recognition). The specific nature of the reaction of female elephants to family and bond group members' calls, which involves distinctive behaviors that are typically reserved for family and bond group members during normal social behavior, indicates that the calls were perceived as belonging to individuals in a distinctive category. However, calls in this category would not necessarily have been perceived as originating from individuals. It could be argued that they may have been classified as part of a broader group, one that includes all highly familiar calls (whose high familiarity derives from the fact that they are given by animals that have unusually high levels of association with the subjects). Whatever the cognitive basis for the distinction made, the particular reactions shown demonstrate that this social discrimination ability would be operationally efficient, enabling females to coordinate their activities with others in their immediate social group and meet up with them again after they become separated.

While the exact nature of family-bond group recognition in elephants remains ambiguous, true individual recognition of calls has rarely been demonstrated in any animal (but see Proops, McComb, and Reby [2009]). It is often proposed that the neighbour-stranger paradigm used in birdsong research is the diagnostic test for individual recognition (reviewed in Lambrechts and Dhondt [1995]). This paradigm relies on demonstrating that male subjects respond weakly to playback of a neighbor's song from the correct territorial boundary position and strongly to a stranger's song played from the same position compared with giving a strong response to both songs when they are played from elsewhere (e.g., Catchpole and Slater [1995]). On closer inspection, however, this simply tests whether or not receivers are able to associate the call of an individual with a particular territorial location. To investigate individual recognition more effectively at a cognitive level, a test that assesses whether receivers associate an individual's call with other stored sensory information about that same individual is needed (e.g. Proops, McComb, and Reby 2009). Such a test might investigate whether animals can associate the visual image or smell of an individual with its call. While this sort of design has typically been employed in captive or semi-captive environments, elephants are one of the wide range of species to which it can potentially be applied.

Long-Term Memory of Family/Bond Group Members' Calls

Elephants are unusually long-lived, and their patterns of communication and social interactions are affected by relationships that have existed in the recent or remote past (Archie, Moss, and Alberts, chapter 15, this volume). In the Amboseli population, we had the opportunity to examine whether recognition of a family member's call persists after that individual has been dead for some time. One adult female, aged 15 years, died during the course of the study and

BOX 10.1 ILLUSTRATIONS

When exposed to playbacks of contact calls from other females, female elephants react in well-differentiated and relatively easily identifiable ways, the most important of which are illustrated here.

- Listening: elephant holds still with ears held out from the head in a stiff, extended position.

- Contact calling: contact calling is usually preceded and followed by periods of listening; here, back and forward movement of ears visible as the call is given.

- Smelling: olfactory exploration of the air.

- Approach: moving toward the loudspeaker; here, smelling the ground on the way.

- Bunching: individuals in the group become agitated and bunch together so that the diameter of the group (or constituent sub-groups) decreases; calves tend to move closer to adults; here, the hindquarters of the adults are orientated toward the center of the defensive circle—the matriarch is second from right (front row).

The strong association between certain reactions and categories of stimuli is now well established and can be used to assess how subjects classify the call they hear (see text).

we monitored the responses of members of her family unit to her call on two occasions, once 3 months after her death and again 23 months after her death. When played the call after 3 months, the family gave contact calls themselves, and after 23 months they contact called and approached the loudspeaker, with several females producing unusually large amounts of temporal gland secretion (McComb et al. 2000). These were strong reactions that are usually only witnessed when a family or bond group member's call is played. Taking advantage of another natural event, we were also able to play the call of a female who had changed family units to her original family unit, 12 years after the transfer had taken place. In this situation the rest of the family had not been totally separated from the caller but had experienced reduced contact with her since the transfer. Here again the playback elicited contact calling from the subjects (McComb et al. 2000).

In fission-fusion mammal societies, where individuals may habitually come into contact after long periods of separation, the ability to remember the calls of others may be particularly beneficial. Long-term vocal recognition has also been shown in some passerine birds, with males retaining the ability to recognize neighbors' songs between years despite separation during migration (hooded warblers *Wilsonia citrina*: Godard [1991]; see also McGregor and Avery [1986]). More recently, vocal recognition between mothers and young on the breeding grounds has been shown to persist for at least four years in northern fur seals, even though individuals are thought unlikely to associate during the period of the year that they spend at sea (Insley 2000).

How Social Identity Is Coded in Elephant Contact Calls

The acoustic structure of a female contact call is a function of the particular voice box (larynx) and vocal tract that produced it and thus would be expected, a priori, to have features that are individually distinct. In elephants as in other mammals, a vocalization is the product of a source signal (generated by vibration of the vocal folds in the larynx) that is subsequently filtered in the cavities of the vocal tract (Fant 1960). The source signal, typically an almost periodical wave with a fundamental frequency and integer multiple harmonics, determines the pitch of the vocalization. However, after being generated at the larynx, the source signal must pass through the cavities of the vocal tract before radiating out through the mouth and nostrils into the environment. Because the vocal tract is a tube of air, it has natural resonances that selectively amplify certain frequencies in the source spectrum. This filtering process thus shapes the spectral envelope of the signal, producing peaks called "formants" (Fant 1960). Since characteristics of vocalizations that arise from inherent properties of the filter can vary independently from those that arise from the source, either or both may provide receivers with important information.

Although the contact calls of female elephants have a fundamental frequency that is in the infrasonic range (typically c. 20 Hz), harmonics that are whole number multiples of the fundamental extend well into the audible range, with sound energy often present up to frequencies of at least 1 kHz. In box 10.2, the fundamental frequency (F0) and its harmonics (H2, H3, etc.) are visible as narrow evenly spaced bands present in the spectrogram and represented by regular peaks in the power spectrum display. Superimposed upon the pattern of fundamental and harmonics are the formants—wider bands of energy produced by resonances of the vocal tract that typically span several harmonics. The center frequencies of the first four formants are marked in the displays in box 10.2, with the most prominent of these being the second formant at around 115 Hz.

We examined contact calls recorded from different adult females in the Amboseli population using Digital Sound Processing techniques (full details in McComb et al. [2003]) to determine whether there were sufficient acoustic cues to identity in these vocalizations for them to be reliably assigned to individual callers. Characterizing the calls simply on the basis of seven measures of variation in the source signal and the first two formants (box 10.3), 77.4 percent of calls were correctly attributed to callers in a discriminant function analysis. We had selected a limited number of acoustic measurements for the analysis that reflected variation in the clearest source-related and filter-related characteristics. If filter-related characteristics had been classified in further detail by including bandwidth as well as center frequency, it may have been possible to assign calls to individual callers with even greater accuracy. However, it is unlikely that this characteristic would play an important role in recognition over long distances (see below).

Differences encompassed by the rate at which the vocal fold vibrate, and the modulation of this as the call progresses, potentially provide a rich source of inter-individual variation. The center frequencies of the formants might be expected to be less reliable in assigning identity because they depend on the size and posture of the caller, and different individuals may overlap in these respects (see also Reby and McComb [2003]). Moreover, while the detailed patterning of a set of formants can provide information on individual identity by reflecting idiosyncrasies in vocal tract shape (see Rendall et al. [1998]), information of this sort is more likely to be important in short- and medium-range communication. When female elephants communicate over very long distances, the complex pattern of formant frequencies—in particular, the bandwidths—is likely to be dramatically altered by attenuation effects that will not be constant across the frequency domain. As a consequence of this distortion

BOX 10.2 STRUCTURE OF ELEPHANT CONTACT CALLS

Figure A displays an oscillogram (lower panel) and a spectrogram (upper panel) of a typical elephant contact call. The oscillogram (in black) represents the sound wave itself (variation of sound pressure level with time as the call is given), whereas the spectrogram represents the distribution of energy (represented in levels of grey) across frequency (Y axis) and time (X axis).

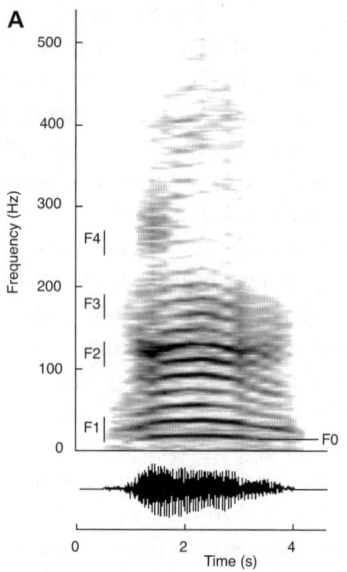

The frequency spectrum of elephant contact calls is complex and has two main components:

1. *The fundamental frequency (F0)*. The ripple pattern (narrow, evenly spaced bands) represents the fundamental frequency and its harmonic series (H2, H3, etc.). The fundamental is the lowest frequency component—marked as F0. It corresponds to the rate of vibration of the vocal folds in the larynx. The range and variation of the fundamental frequency are responsible for the "pitch" and the "intonation" of the call, respectively. In this example, the pitch is very low as it peaks at 19 Hz.

    The harmonics (H2, H3, etc.) are all multiple integers of the fundamental frequency; their position is determined by the fundamental frequency and does not provide additional information.

2. *The formants (F1 to F4)*. The broader bands of energy distributed over adjacent harmonics are called "formants" and represent an important source of information. The position of these bands is determined by the shape and length of the resonating cavities of the vocal tract, through which the sound must travel before it radiates into the environment. The longer the resonance cavities, the lower the formants, and consequently the smaller the spacing between them.

Both the fundamental frequency and the formant frequencies are likely to provide information about the identity of the caller. The fundamental frequency will reveal inter-individual variation in the larynx and its control, the formants will reveal inter-individual variation in the shape and length of the vocal tract.

In our studies of the acoustic structure of elephant contact calls, we have used different tools to extract the characteristics of the fundamental frequency and the characteristics of the formant frequencies.

The fundamental frequency contour (lowest of the dark evenly spaced bands in figure Bi) is extracted using an autocorrelation algorithm. The extracted contour (figure Bii) is then used to derive a set of variables: the minimum, the average, and the maximum fundamental frequency; the number of inflection points of the contour; and the percentage of time elapsed when the peak fundamental frequency was reached (details in McComb et al. [2003]).

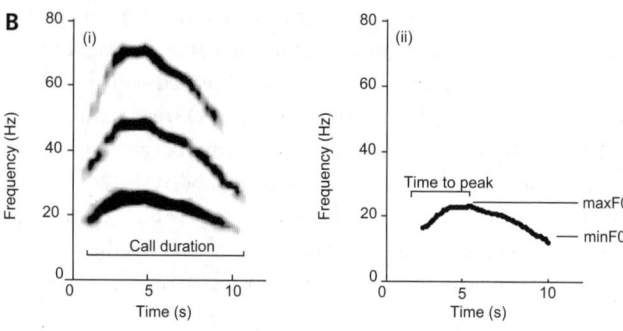

The format frequencies are extracted using an algorithm called Linear Predictive Coding (LPC), which enables us to separate the spectral envelope, characterizing the formants from the underlying harmonic structure. This process is illustrated in figure C: the frequency spectrum is "smoothed," revealing broadband peaks whose center frequencies are the formant frequencies.

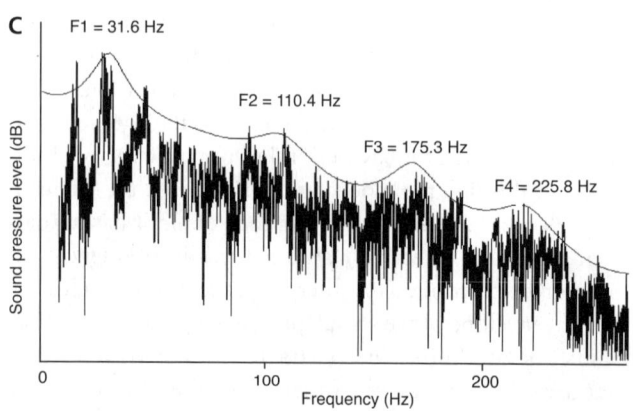

BOX 10.3 CONTACT CALLS: INDIVIDUAL IDENTITY IN CALL CHARACTERISTICS

The calls of four highly distinct individuals are plotted against the first two functions of a discriminant function analysis. The scores of each call are calculated using the discriminant functions based on 9 acoustic measurements on 86 calls from 13 individuals (McComb et al. 2003). Tilly's vocalizations are represented by the symbol +, Ysolde by ×, Pick by –, and Odette by o.

The vectors indicate how each of the acoustic variables used in the analysis is correlated with the functions. Grey arrows represent source-related variables (the variables depicting the duration, range, and contour of the fundamental frequency), whereas the green arrows indicate the filter-related variables (the first two formants: F1 and F2). Since the three variables depicting the range of F0 (mean F0, max F0, and min F0) are very strongly correlated, they are represented by a single average vector.

The calls plotted in the top right quadrant are likely to be characterized by a fundamental frequency and a second formant higher than average, whereas the calls in the bottom right quadrant are likely to be characterized by a lower fundamental, a lower first formant, a short duration, a small number of inflection points, and a fundamental peak achieved later in the call.

The spectrograms of the four "cardinal" individuals present the acoustic structure of the lower 100 Hz of the calls, illustrating inter-individual variation of the fundamental frequency contour. As discussed in the text, information on the contour could be extracted over long distances from harmonics in the second formant. These spectrograms and the means of the nine acoustic measures for all four individuals presented in the table confirm the pattern expected from the discriminant functions: Tilly's calls are relatively higher-pitched, longer, more modulated, and reach the peak F0 at the beginning of the call, whereas Odette's calls are lower-pitched, less modulated, and reach their max F0 toward the middle of the call.

	meanF0	maxF0	minF0	Duration	inflexions	sum(var)	%elaps	F1	F2
Odette	13.7	15.8	8.9	5.2	2	17.6	45	26.8	103.2
Tilly	18.1	20.5	14.7	5.8	4.4	12.8	28	46.7	110.4
Pick	14.9	16.5	12.3	5.1	3.1	9.4	45	35.5	107.3
Ysolde	21	23.6	16.3	3.6	1.5	15.9	32	36.7	144.1

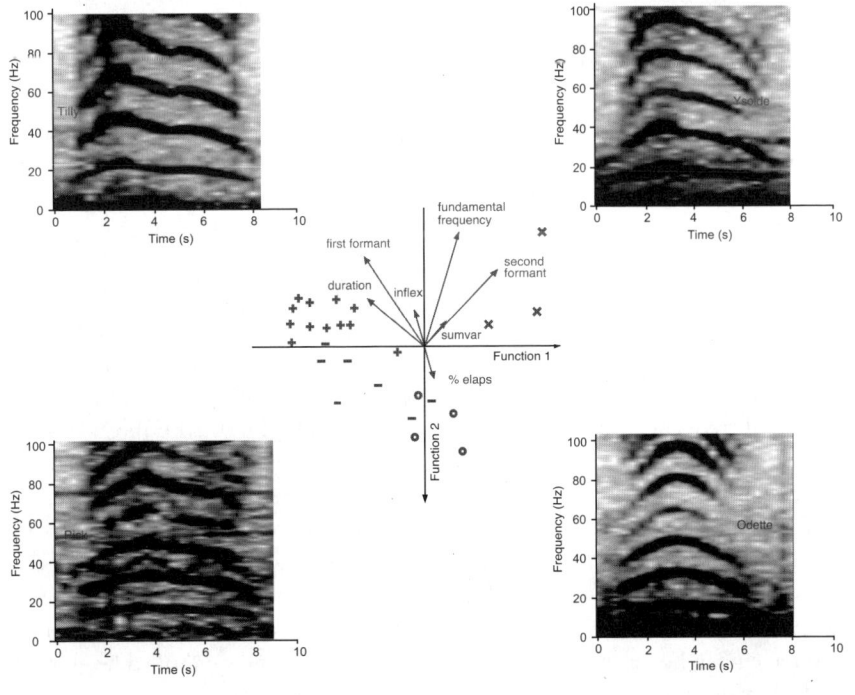

of the spectral envelope, possibly resulting in a differential reduction of formant bandwidths and ultimately in the loss of certain formants, the ability of the formant frequencies to carry information on individual identity over long distances is likely to be severely reduced (confirmed by our re-recording measurements; see below).

The unusually low fundamental frequency of elephant contact calls is likely to be a simple byproduct of extremely large vocal folds. Fundamental frequency is determined by the mass and length of the vocal folds as well as additional factors such as vocal fold tension and sub-glottal pressure (see McComb et al. [2003]). Other large herbivores such as rhinos have also been reported to produce calls with infrasonic fundamental frequencies (e.g., Budde and Klump 2003). It is clear that elephant contact calls are also produced with an unusually long vocal tract. Formant spacing can be used to provide information on the length of the vocal tract in mammals (Fitch 1997; Reby and McComb 2003). Based on the frequencies of the first four formants, and assuming a vocal tract that is a uniform tube closed at the larynx and open at the radiating end, average formant spacing in our analyses of elephant contact calls was 62.4 Hz (McComb et al. 2003). On the basis of the physical relationship between formant spacing and vocal tract length (formant spacing = sound velocity / (2 times vocal tract length): Fant [1960]; Fitch [1997]), this would predict an unusually long vocal tract length for female elephants. Assuming that sound velocity in the vocal tract is 350 ms^{-1} (Titze 1994), a formant spacing of 62.4 Hz would predict a vocal tract length of approximately 2.8 m, suggesting that the trunk and possibly a pharyngeal cavity (resulting from a mobile larynx that may be pulled downwards: see Gasc [1967] and Shoshani [1998]) interconnect to form an extended filter. The exceptionally low resonance frequencies resulting from this very long filter accentuate the lower harmonics in the spectrum of the female contact call and are undoubtedly important in facilitating long-distance communication of social identity.

Measuring the Distances over which Information on Social Identity Is Available

Considerable emphasis has been put on the potential value of the infrasonic fundamental frequency for long-distance communication in elephants. In particular, because of the predicted resilience of infrasonic frequencies to attenuation, it has often been suggested that African and Asian elephants (*Loxodonta africana* and *Elaphus maximus*) can communicate over very long distances (e.g., Payne and Webb 1971; Payne, Langbauer, and Thomas 1986; Garstang et al. 1995). It is important to remember, however, that the maximum distances over which a species can use an acoustic signal to communicate may not be equivalent to the distances over which components of that signal are physically detectable. In mammal calls, a wide range of acoustic characteristics typically carry information on individual identity, and frequency components that may be crucial in social recognition could be distorted or lost as distance from the source increases. A receiver may well be able to register that another elephant has called while that individual is a long distance away but be unable to identify the caller.

In initial attempts to quantify the distances over which African elephants could communicate using infrasonic calls, Langbauer et al. (1991) showed that females could detect a variety of infrasonic calls at 1.2 km from source, while males responded at 2 km. Then, on the basis that playback volumes were lower than the maximum sound pressure levels at which some of these calls had been recorded in the wild, they speculated that elephants could communicate over distances of at least 4 km. Other researchers (Garstang et al. 1995; Larom et al. 1997) have since used computer modeling based on this estimate to predict that under optimum atmospheric conditions elephants could communicate over distances in excess of 10 km. However, none of these studies actually considered whether information related to social identity might be extracted from calls by other elephants at the proposed transmission distances. More surprisingly, the potential importance of the non-infrasonic frequencies in coding individual identity and social meaning in elephant communication has not been given due consideration.

The lack of solid data on how the full complement of frequencies in so-called infrasonic calls are degraded over long distances, particularly components that are likely to code information on individual identity, currently falls considerably short of substantiating the popularly held belief that elephants communicate primarily using infrasound. In order to investigate the distances over which social recognition (as opposed to simple detection) is possible, it is necessary to identify a diagnostic response that can demonstrate that subjects have not only detected the call but also categorized the social identity of the caller. To do this, we made use of the distinctive calling/approach response described previously, given when subjects have identified a call as belonging to a family or bond group member (see also McComb et al. [2000, 2003]). Calls of family and bond group members were played to subjects first from distances of 2 to 3 km and then from successively closer distances until this recognition response had been obtained. Then, in order to examine how different frequency components in the calls degrade with distance, we carried out re-recordings of contact calls at distances of 3 km down to 0.5 km from the loudspeaker and quantified levels of degradation.

In practical terms, playback experiments were conducted

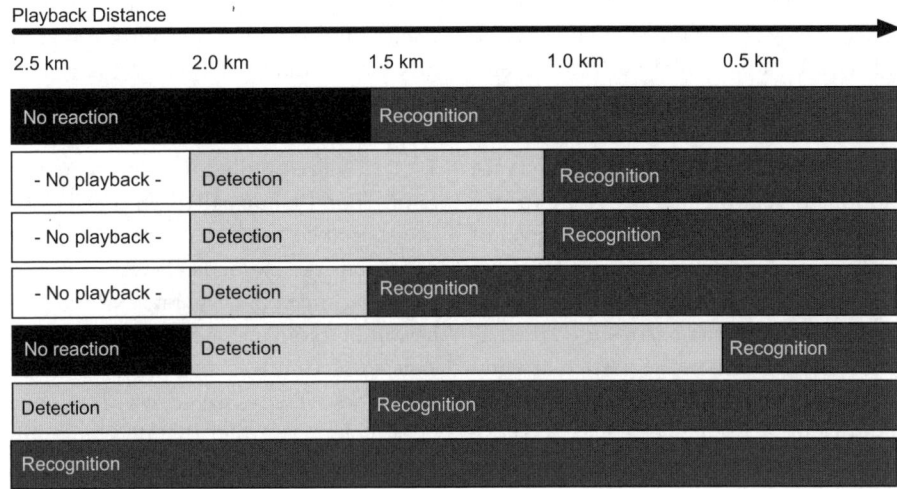

Figure 10.1 Responses of seven different families to playbacks of family or bond group members broadcast from distances of 2.5 to 0.5 km from the subjects. Playbacks were given first at the furthest distances and then at successively closer intervals until the diagnostic recognition response was obtained. See text for details.

using two vehicles that were in radio contact. One vehicle was used to play the calls, and this vehicle measured the distance from the subjects using an odometer, moving first to the maximum distance at which calls were to be played and then backtracking to successively closer positions. By playing subjects the contact call of a family or bond group member first from a long distance away and then from successively closer distances, social recognition distance could be calculated as the distance at which the subjects first give the diagnostic response of contact calling and/or approach, and the experiment terminated at this point. This experimental design, involving moving successively closer to the subjects during the experimental series rather than further away from them, avoids potential effects of habituation confounding the results and leading to an underestimate of social recognition distance.

In the playbacks of family/bond group members (figure 10.1), the distances at which the diagnostic social recognition response of contact calling and/or approaching the loudspeaker were given ranged from 2.5 km to 0.5 km (mean = 1.21 km, standard deviation = 0.64). The modal distance for the recognition response was 1 km. Typically the subjects showed signs of detecting the call (as indicated by listening, smelling, or streaming) from distances of 2 km and 2.5 km, but did not categorize it as a family member (as indicated by the diagnostic response) until playback distances had narrowed to 1 km or 1.5 km. Control trials confirmed that subjects did not give false positive response to calls that were not from members of their own family or bond group, irrespective of what distance they were played from (McComb et al. 2003, 2000). Indeed, the occurrence of bunching and avoidance reactions at distances of 1 km and 0.5 km in these trials was consistent with the calls of infrequent associates having been identified over these distances (see below; also McComb et al. [2000, 2001]).

Our long-distance playback experiments thus indicated that social recognition on the basis of call characteristics was possible over distances of up to 2.5 km but more usually achieved at around 1 km from the loudspeaker. Usually the subjects listened when the contact calls of family or bond group members were presented from the furthest distances, indicating that they had detected the call. However, it was typically only when the playback distance had narrowed to 1 km that they gave the appropriate categorization response of calling back and approaching in the direction of the loudspeaker, indicating that they had identified the caller as belonging to a family or bond group member. Observation of family members that have become separated and are using contact calls to relocate each other suggest that females put more rather than less effort into calling when the distances between them are large (Poole et al. [1988]; chapter 9, this volume; McComb, personal observation). It is therefore unlikely that subjects recognized the family or bond group member at the distance that they gave their first listening response but did not respond at this stage because they were waiting for the individual to come closer. It is also notable that in the control trials, where the calls played were not from family or bond group members, bunching and avoidance responses indicating that the caller had been categorized as an infrequent associate were obtained at distances of 1 km or less.

Our acoustic analyses of contact calls re-recorded at increasing distances from the loudspeaker reveal a possible basis for the loss of social identity cues at the furthest playback distances. As distance from the source of playback increased, the frequency spectrum became less well defined (box 10.4). However, a particular vocal tract resonance (the second formant: F2), spanning several harmonics and centred around 115 Hz, is very prominent and persistent, and decays at a lower rate with increasing distance than frequency components below and above it. In our re-recordings, this band of energy dropped to the level of background noise at between 1.5 km and 3.0 km from the source. Although the specific distance at which this happened is likely to have been a function of acoustic characteristics of the individual's call and also of the particular wind and atmospheric conditions that prevailed at the instant of rerecording, these distances are consistent with those over which loss of social recognition occurred when long-distance playbacks of the same calls were given our subjects (see earlier description). It has been established that information on the fundamental frequency contour can be extracted from a set of harmonics when the fundamental itself is actually absent (perception of the missing fundamental: Houtsma [1995]). In the case of the elephant contact call, the contour of the fundamental frequency could be derived by extrapolation from harmonics in the region of the second formant (the 115 Hz area), which appear to experience less interference from wind noise than the fundamental frequency contour itself or than harmonics in the region of the first formant.

Finally, the hearing sensitivity of African elephants has not been measured directly. However, data available on hearing in Asian elephants (Heffner and Heffner 1980, 1982) suggest that while Asian elephants have a lower low frequency hearing threshold than other mammals (measured as 17 Hz at an intensity of 60 dB: Heffner and Heffner [1982]), they are considerably less sensitive to frequencies below 100 Hz than to those between 100 Hz and 5 kHz (Heffner and Heffner 1982). Thus, even the hearing curve itself provides some indication that elephants may be better equipped to extract frequency characteristics in the 115 Hz region (second formant) than those in the lower part of the contact call spectrum. In order to establish directly the relative importance of the infrasonic and audible source and filter components in communicating individual identity over varying distances, playback experiments using filtered and resynthesized vocalizations, selectively eliminating infrasonic or audible source and filter components, are now a priority.

Social Recognition Beyond the Level of Bond Group

As well as reliably discriminating the contact calls of members of their family and bond group from those of other females in the population, female elephants do not appear to classify the calls of these other females as a single category (e.g., non-family/bond group). As outlined above, females give two distinct reactions to the calls of non-family/bond group that can be predicted on the basis of the association index between the subjects and the caller (see box 10.1). Twelve families were presented with calls from non-family/bond group callers who had either high or low levels of association with the experimental family. The families consistently remained relaxed in response to the high association index callers, but exhibited increased group cohesion or avoidance behavior to the low association index callers (McComb et al. 2000). As the order in which these playbacks were delivered was randomized, subjects could only have performed this discrimination if they were familiar with the vocal characteristics of individuals in the high association index category. Subjects interacted with an average of 14 different families at this level of association in their normal ranging patterns, containing around 100 adult females. Thus, our results indicate that females can potentially pick out the contact calls of around 100 frequent associates (McComb et al. 2000), distinguishing these from the calls of less frequent associates—an unusually large network of familiar individuals (cf., Waser 1977; Cheney and Seyfarth 1982; McComb et al. 1993; Barfield, Tangmartinez, and Trainer 1994; Rendall, Roman, and Emond 1996). The detection of the presence of less familiar females (low association index families) is potentially beneficial because this section of the population is more likely to initiate agonistic disputes or harass young calves.

Between-Group Differences in Recognition Abilities

The playback experiments described previously indicated that adult females are familiar with the contact calls of around 100 others in the population, being able to discriminate between pairs of calls on the basis of how often they associate with the caller. However, these initial results also suggested that different families differed in how good they were at this discrimination task. We were able to use such differences as a means of assessing the causes and consequences of differences between groups of elephants in social knowledge acquisition. Detailed between-group differences in discriminatory abilities could be examined in relation to factors that had the potential to affect ability to acquire and store information—including the number and ages of females in the group (this section) and the opportunities that groups had to learn to recognize the calls of others.

In the Amboseli study population, data on life histories and association patterns have been obtained for almost 2,000 individual elephants (Moss, this volume). Vocal discrimination abilities were tested by giving each of 21 family units, over the course of seven years, a series of playbacks

BOX 10.4 SPECTROGRAMS OF CONTACT CALLS

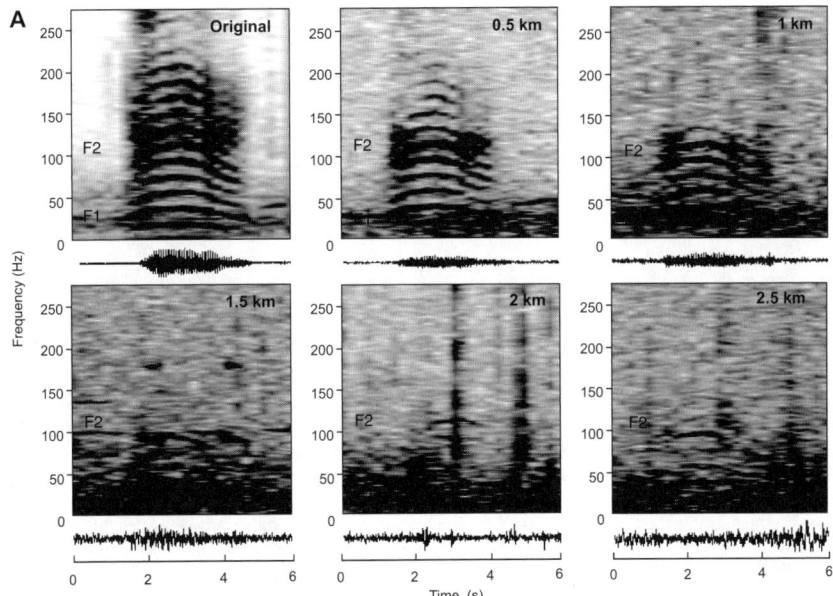

This series of spectrograms shows the degradation of the frequency spectrum of a contact call recorded from Esme, when played and re-recorded at increasing distances from the loudspeaker (original call, 0.5, 1, 1.5, 2, and 2.5 km).

As distance increases, harmonics disappear from the frequency spectrum, and the spectral components become engulfed in background noise.

At 2.5 km, the best-preserved harmonic, H5, is in the second formant (around 100Hz). Also visible are H2 (40 Hz) and H3 (60 Hz). Note that none of these harmonics lies in the infrasonic frequency range.

Figure B represents the attenuation curves for calls from one individual (Ysolde) averaged across four different re-recording sessions. The lines show how the call-to-background noise ratio varies with frequency at each re-recording distance (0.5–3 km). The call-to-background noise ratio illustrates the level at which each frequency component emerges above background noise. It is calculated by subtracting the long-term average spectrum of the current background noise (as represented in samples of recordings taken immediately before or immediately after the call) from the long-term average spectrum of the re-recording of the call (including the simultaneous background noise).

These attenuation curves provide a confirmation in the frequency domain of the phenomenon observed in the spectrograms presented above: the strongest and most robust components are found in the area of the second formant; in this case, it corresponds to fourth and fifth harmonics. These components are lost above 2.5 km.

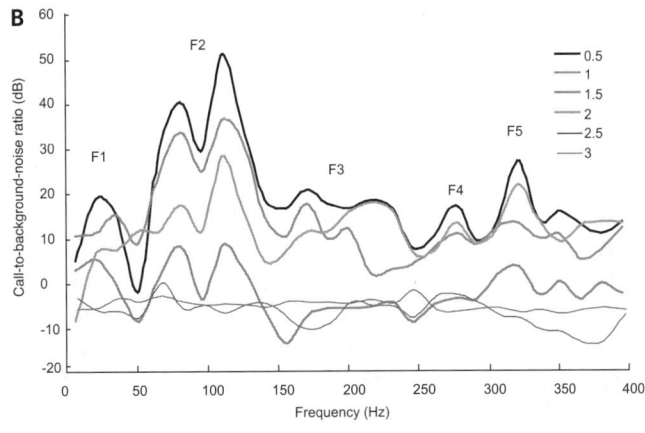

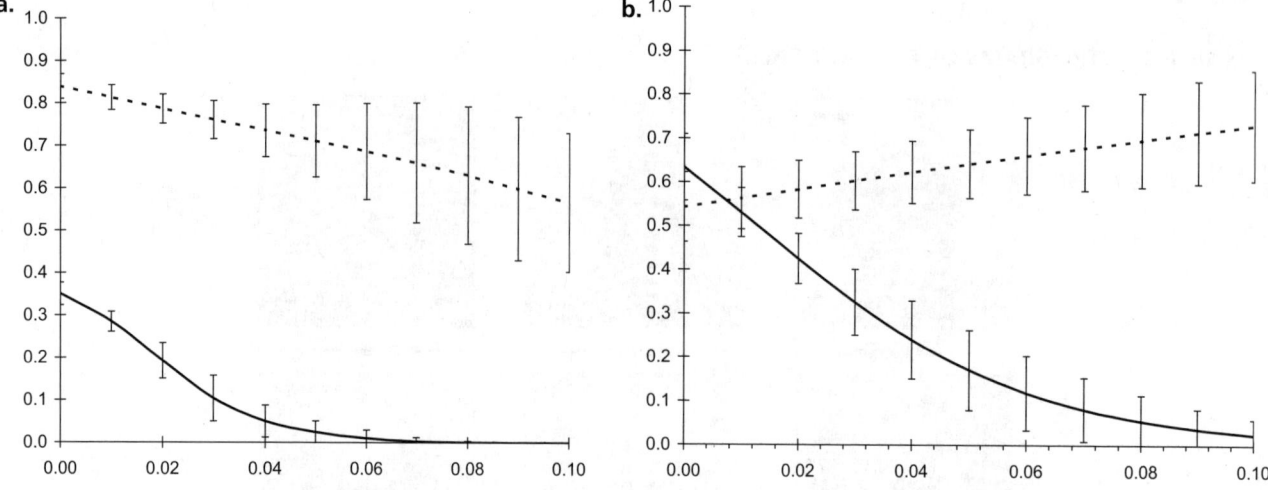

Figure 10.2 The graphs shown illustrate the probabilities of subjects (a) smelling and (b) bunching in response to playback in relation to the association index between themselves and the caller. The values depicted are those predicted from a logistic regression model (detailed in McComb et al. [2001]) for families with matriarchs of 35 years (dashed line) and matriarchs of 55 years (solid line). In the graphs, we focus on these two age groups, representative of young and old matriarchs, to illustrate how the interaction between age of matriarch and association index with the caller operated in the case of each of our key responses. The crucial differences are in the relative probability of obtaining the response (bunching or smelling) to genuine strangers versus more familiar associates for families with old matriarchs compared with those with young matriarchs. Families with older matriarchs were vastly more likely to bunch (or smell) to genuine strangers than to more familiar associates, whereas families with younger matriarchs were not.

of contact calls from adult females in other families in the population with whom they had a range of association indices. The probability of families of subjects bunching into defensive formation on hearing playbacks of calls from other families decreased with the association index with the caller, conforming to a logistic curve (figure 10.2). This curve describes the probability of bunching as association index increases, generally leveling to zero at high indices. We were able to use logistic regression to test the hypothesis that family units with older matriarchs were superior at discriminating the calls of close associates from those of distant associates, statistically excluding other potential explanatory factors. We also examined the relationship between matriarch age and reproductive success, linking reproductive success to matriarch age and response to playbacks.

Our analyses revealed that the age of the matriarch had a significant effect on the probability of bunching when controlling for the association index with the caller and family identity. More crucially, there was a significant interaction between age of matriarch and association index (McComb et al. [2001]; figure 10.2). Other potentially confounding variables, such as number of females in the group and the mean age of females other than the matriarch, could be excluded from the model because their effects were not found to be statistically significant (McComb 2001). The probability of bunching decreased with increasing matriarch age, suggesting that families with older matriarchs may either have larger networks of vocal recognition than families with younger matriarchs or greater social confidence. However, of greater importance, with respect to our hypothesis

that social knowledge increases with age, was the interaction between age of matriarch and association index with the caller. Specifically, the sensitivity of the bunching response to association index increased with the age of the matriarch so that families with older matriarchs were relatively much more reactive to females with whom they had a low association index than to those with whom they had a high association index. For example, while families with a 55-year-old matriarch are several thousand times more likely to bunch in response to calls from families with whom they have a low association index (0.01) than to those with whom they have a high association index (0.1), the probability of bunching for families with a 35-year-old matriarch increase only marginally ($\times$ 1.4) across these conditions. If families with older matriarchs were simply more confident, we would predict their lower level of defensiveness overall but not these marked differences in the slope of the logistic curves. Instead, families with older matriarchs appear considerably more adept at using auditory signals to discriminate correctly between familiar and unfamiliar females in the vicinity and respond appropriately.

The previous conclusions were strengthened by our analyses of the occurrences of smelling in response to playback. While bunching in response to playback is primarily a defensive behavior, smelling must constitute a means of further exploration; when subjects smell after playbacks, they attempt to gather olfactory information on the caller's identity. An examination of the binary variable "smelling" in response to playback also showed an interaction between age of matriarch and association index that characterized the bunching response (figure 10.2; McComb et al. [2001]). Subjects in groups with older matriarchs were much more likely to use their trunks to smell if played calls from low rather than high association index families (figure 10.2). In contrast, groups of subjects with younger matriarchs failed to show this relatively inflated probability of gathering olfactory information about infrequent associates. It is important to note that here families with older matriarchs were at least as reactive as families with younger matriarchs to callers with low association indices. However, their ability to distinguish between calls from low and high association index families was apparently much greater.

The finding that the pattern of smelling and bunching responses was not affected by the number or age of the other females in the group (McComb et al. 2001) indicates that it is the matriarch who signals to the rest of her group if defensive and exploratory behavior is necessary. Such a signal might take a number of forms, although a subtle acoustic, olfactory, or gestural cue is the most likely possibility. There are anecdotal reports in the literature of females following the lead of their matriarch in other coordinated group activities (Poole et al. 1988; Spinage 1994). Further research is needed to investigate in detail how a matriarch transmits information to other group members on her affective state or intention to change behaviors. Initial analyses also suggest that matriarch age may interact with the encounter rate that families have with others in the population to influence opportunities for learning to recognize calls, and research is in progress to evaluate this effect directly.

Reproductive Consequences of Differences in Social Knowledge

We predicted that the superior discriminatory abilities of older matriarchs should translate into reproductive benefits for the family unit, since time is more efficiently allocated by reserving defensive behavior for circumstances where it is appropriate and opportunities for cooperation with more frequent associates are provided. In support of this hypothesis, the age of the matriarch was a significant predictor of our standardized measure of recent reproductive success (the number of calves produced by the family per female reproductive year over the course of the study: McComb et al. [2000]). However, additional factors, including knowledge that older matriarchs had accumulated in a variety of other domains, might have contributed to this association. To explore the relationship between responses to playback and reproductive success more specifically, we calculated inflection points for the logistic curves of bunching and smelling on association index for each family (McComb et al. 2001).

These inflection points were used to describe differences between families in response to playback. When we entered these values into an analysis of the residual variation in reproductive success after removing the variance due to age of matriarch (and number of females, a significant variance component here) we found that families appeared to derive a marginal reproductive benefit by bunching and smelling more readily—that is, showing greater caution and exploratory behavior—when presented with another female's call (McComb et al. 2001). Thus, it is likely that families with old matriarchs benefit reproductively because their matriarchs target caution at the appropriate individuals—callers who are strange to them. The fact that exhibiting more caution and exploratory behavior (once the effects of matriarch age have been removed) has an independent reproductive benefit may explain why families that have young matriarchs, with their reduced social knowledge, should show a greater overall tendency to bunch and smell.

Until now, it has proved difficult to assess how social knowledge and experience might translate into fitness benefits for individual animals or groups. Previous researchers have focused on how individuals can derive fitness benefits from an improvement in ecological knowledge that accompanies aging (Ratcliffe, Furness, and Hamer 1998; Durant 2000). Our results indicate that aging may also influence reproductive success through its effects on the acquisition of social knowledge. More specifically, it is now clear that the possession of enhanced discriminatory abilities by the oldest individual in a group of advanced social mammals can influence the social knowledge of the group as a whole. Because tusk size in elephants is related to age and poachers are known to focus their efforts on individuals that have large tusks (Dobson and Poole 1998), these results have important implications for conservation biology. In view of our findings, it is clear that the removal of matriarchs from elephant family units could have serious consequences for the conservation of this endangered species. Indeed, in many mammal societies the oldest individuals are also the largest, and these tend to be particular targets for hunting whether legal or illegal (Stevick 1999; Dobson and Poole 1998). Species likely to be particularly affected by the removal through hunting of the largest and oldest females include killer and sperm whales, where a clear matrilineal structure is evident and the oldest females may play an important role in coordinating group activities (Baird [2000]; Whitehead and Weilgart [2000]; see also Pennisi [2001]). If groups rely on older members for their store of social knowledge in elephants and other advanced social mammals, then whole populations may be affected by the removal of a few key individuals.

Chapter 11 Elephant Cognition: What We Know about What Elephants Know

Richard W. Byrne and Lucy A. Bates

Elephants have a reputation for formidable memory and intelligence, but research specifically investigating their cognitive abilities is still a fledgling enterprise. In this chapter, we shall assess what is currently known about the cognitive skills of elephants by reviewing evidence available from a wide range of sources: we include observations from field studies of wild elephants and anecdotal reports gathered from the trainers of semi-captive, working elephants, in addition to the relatively few studies of elephant cognition published in scientific journals. To put this preliminary assessment into perspective, we will make comparisons with what is known about another, much better-studied mammalian group, the primates.

Thus, there are two main aims of this chapter:

- To describe and assess all the available scientific evidence that pertains to elephant cognition, including both experimental and observational data. In evaluation of this evidence, we shall also draw upon anecdotal reports of elephant cognitive skill.
- To compare what we know of primate cognition with what we currently understand about elephant cognitive skills. This will illustrate any differences and/or shared cognitive abilities as well as demonstrate which areas may be the most productive to pursue with further scientific investigation.

As data from Amboseli and other research sites suggest, elephant society may be one of the most elaborate of any vertebrate. There is a considerable body of theory and supportive data to suggest that living in an extensive social network often correlates with, and likely promotes, cognitive sophistication (see Byrne and Bates [2007] for a brief review). Flexible, fission-fusion sociality means it is possible that individual elephants know and differentiate among several hundred other individuals, and if so this would far exceed anything found in non-human primates. Complementing this hypothesis of "social intelligence," it has been suggested that the most advanced primate intelligence, that of the apes, derives from the technical ability that is allowed by prehensile hands and the flexible motor control of the hands and fingers (Byrne 1997; Deacon 1997; Napier 1962). Based on this theory, a case can also be made that the elephant might possess special abilities. The elephant's trunk is prehensile, and under exquisite brain controls (Shoshani, Kupsky, and Marchant 2006), such that it can pick up and put down an egg without breaking it.

Relative to their body size, the elephant brain is large, although not in the extreme range of humans or dolphins; the encephalization quotient varies between 1.3 and 2.3, depending on sex and species (Cutler 1979; Jerison 1973), comparable indices to those of apes (Eisenberg 1981). Encephalization gives a useful measure of investment in brain tissue relative to total metabolic energy available and thus indicates the cost to the animal of so large a brain (Byrne 1996). It does not, however, give the best indication of "brain-power": as with any computational system, it is the actual number of processing units available for use that is important for assessing power (Byrne 1996). Elephants have the largest absolute brain size among land animals: up to 5.5 kg in Asian elephants and up to 6.5 kg in African savannah elephants (Cozzi, Spagnoli, and Bruno 2001;

Shoshani, Kupsky, and Marchant 2006). Although neurons are less densely packed in the brains of elephants than of primates, elephant brains nevertheless contain as many cortical neurons as do human brains (Roth and Dicke 2005). Moreover, the pyramidal neurons are larger than in humans and most other species, with a large dendritic tree giving the potential for many more connections, and perhaps hinting at superior learning and memory skills (Cozzi, Spagnoli, and Bruno 2001). Furthermore, a specific class of neuron has recently been identified in the cortex of both African and Asian elephants, otherwise found only in great apes and humans, but not other primates, as well as killer, humpback, fin, and sperm whales (Hakeem et al. 2009). These "Von Economico" neurons are believed to be involved in social awareness and fast decision making in complex and rapidly changing social situations.

In structuring our review, we divide the evidence into that for physical versus social cognition, although we are aware that this may not fully reflect the organization of the underlying cognitive apparatus.

Physical Cognition

Knowledge of Environmental Spaces and Objects

African savannah elephants are known to move over very large distances in their search for food and water. Leggett (2006) used GPS collars to track the movements of elephants living in the Namib Desert. He recorded one group traveling over 600 km in five months, and Viljoen (1989) showed that elephants in the same region visited water holes, some of which were over 60 km apart, approximately every four days. Even more impressive, elephants inhabiting the deserts of both Namibia and Mali have been described traveling hundreds of kilometers to arrive at remote water sources shortly after the onset of a period of rainfall (Blake et al. 2003; Viljoen 1989), sometimes along routes that researchers believe have not been used for many years. These remarkable feats suggest exceptional cognitive mapping skills, reliant on the long-term memories of older individuals who traveled that path sometimes decades earlier. Indeed, a recent study has confirmed that family groups with older matriarchs are better able to survive periods of drought (Foley, Pettorelli, and Foley 2008). Families with older matriarchs range over larger areas during droughts, apparently drawing on the knowledge of the older females about the locations of permanent, drought-resistant sources of food and water (see also McComb, Reby, and Moss chapter 10; Mutinda, Poole, and Moss chapter 16).

We cannot yet draw firm conclusions about the cognitive mapping skills of elephants, as most data are restricted to observations where the strength of the evidence is not entirely apparent. For example, elephants can detect low frequency rumbles at distances of several kilometers (Garstang et al. 1995; Langbauer 2000), even seismic rumbles (O'Connell-Rodwell et al. 2006), so it may be that elephants can follow the sounds of distant thunder to reach fresh water sources. However, there are currently several populations of African savannah elephants where the movements of multiple individuals are being accurately mapped with GPS tracking devices, so future insight into elephant cognitive mapping skills can be expected.

Even in primates, our knowledge of the cognition involved in travel is limited. All primates whose travel has been mapped use a network of familiar routes, whether these are arboreal runways or terrestrial paths. This may not imply a limited spatial memory, however; when tamarins switched from their year-round fruit diet to feed on nectar, they traveled from tree to tree along highly direct but quite different routes to their usual ones (Garber 1988). A similar navigational efficiency was noted earlier in orangutans; when a favored fruit crop failed and individuals had to visit several trees, far apart in the rain forest, their travel was a least-effort route (Mackinnon 1978). But it is only in the last few years that strong tests have been made of cognitive mapping skills in primates (Janson and Byrne 2007). Experimental studies have shown that capuchin monkeys are able to head directly for distant locations, and choose the most valuable to visit first (Janson 2007); however, they do not compute routes that show efficiency over more than one target (i.e., they do not solve "the traveling salesman problem"). Baboons have shown greater ability in this regard: they take direct routes to distant, high-quality resources that are likely to be exploited by other groups, returning later to consume other, more reliable resources bypassed earlier (Noser and Byrne 2007b).

Moreover, primate route planning goes beyond merely spatial information: mangabeys have been shown to take into account the likely effect of warm, sunny weather on fruit ripening rates (Janmaat, Byrne, and Zuberbuhler 2006). When they pass close to a tree that held unripe fruit on their last visit, their decision whether to check the tree or not depends on the weather since their last visit. This is not due simply to more enthusiastic ranging on warm sunny days, as the effect persisted when the conditions on the day of travel were statistically controlled.

On a smaller environmental scale, we recently showed that elephants in Amboseli are able to track the relative positions of their family members (Bates et al. 2008a). Taking advantage of elephants' highly sensitive olfactory abilities (Shoshani 1997; Langbauer 2000; Rasmussen and Krishnamurthy 2000), we moved urine deposits from known individuals to positions where they would be discovered by target individuals. With samples from individuals who were at least 1 km away, urine from kin elicited significantly more interest from the target individual than did samples from unrelated individuals. We also presented test elephants with urine deposits from related individuals actually present in their group that day, which were either walking some way ahead of the target elephant or behind it. We reasoned that if elephants are able to identify specific individuals from their urine, and each is continually updating its memory of where other key individuals are, then discovering a fresh urine deposit from an individual who was walking behind should violate its expectations. Target individuals investigated samples from family members who were behind them at the time of the test significantly more than samples from individuals who were in front. From this, we concluded that elephants are able to continually track the locations of family members in relation to themselves, as either absent, present in front, or present behind (Bates et al. 2008a). These results suggest that elephants are able to hold in mind and regularly update information about the locations of at least 17 other female group members as well as implying that they recognize individual identity from scent and have some understanding of invisible displacement and person permanence. While we tested only scents of adult females, it seems likely that individuals also monitor some males and immatures in addition. That they can keep track of so many independently moving companions implies that elephants have particularly large working memory capacity.

Use of Tools and Understanding of Causality

With their opposable thumbs, primates are adept at object manipulation (Beck 1980), with great apes having the greatest manual dexterity (Byrne, Corp, and Byrne 2001; Napier 1962). Tool use is widespread among animals and is evident in all great apes and several monkey species (Beck 1980). Tool use itself is not necessarily indicative of advanced cognitive processing, however. Instead, systematic tool manufacture or modification is recognized as cognitively more demanding, especially if the species makes several different tools for different purposes. This is evident in only a handful of species: all four great apes in captivity (McGrew 1989), though only chimpanzees and orangutans in the wild (Fox, Sitompul, and van Schaik 1999; McGrew 1992), and the New Caledonian crow (Hunt 1996, 2000b).

Both Asian and African savannah elephants have been seen to use multiple tool types for up to six different functions, mostly in the context of body care such as scratching and removing ticks (Chevalier-Skolnikoff and Liska 1993). Asian elephants presented with branches that were too long or bushy to make effective fly switches (fly-swats), a commonly used tool, modified them before use by either breaking off a side branch or snapping them in half (Hart et al. 2001). Elephants may thus be added to the small number of animals that *make* tools. However, for an animal that frequently breaks branches while eating, the cognitive demands of extending this behavior to manufacture fly switches are probably not great (Bates, Poole and Byrne 2008). Elephant tool modification does not compare in either structural or logical complexity to the manufacture of ant and termite dipping tools displayed by chimpanzees (Boesch and Boesch 1990; Goodall 1986), which even includes making two different types of tools for different stages of the same task (Sanz and Morgan 2007), or with the serrated leaf probes and hook tools produced by New Caledonian crows (Hunt 1996, 2000a).

A major issue in primate studies has been to determine whether individuals who use tools do so because they understand the causal linkage between form and function and therefore appreciate which aspects of material and tool design are critical for success. Termiting probes are typically brown, flexible, c. 30 cm long, and made of plant material: do chimpanzees realize that length and pliability are important, whereas color and source may not be? Several primate species have proved able to choose tools on the basis of *task-relevant* properties like rigidity and length, rather than color—even species like tamarins, which do not naturally use tools (Hauser 1997b; Santos, Miller, and Hauser 2003). Chimpanzees also showed evidence of causal understanding when showing selectivity in what to copy (Horner and Whiten 2005). When presented with opaque puzzle boxes, chimpanzees copied two actions made by a human demonstrator to release food. When the same boxes were transparent, revealing that one action was irrelevant, they missed it out and copied only the relevant action. The judgement of relevance, however, may have been based on a simple parameter. For instance, because the "irrelevant" action made no physical contact with the food, the chimpanzees may only have attended to whether contact was made when they chose which actions to copy. Chimpanzees have not performed impressively in the "trap tube" task, which is explicitly designed to study causal understanding. In this task, poking the desired food item from the wrong side causes it to fall into a well and be lost (Limongelli, Boysen, and Visalberghi 1995); the few chimpanzees that solved the task did so only over several trials, raising concerns that they did not understand the issue. In contrast, rooks *Corvus frugileus*, a crow species that does not regularly use tools in the wild, solved the task rapidly; one individual also showed immediate transfer to a different task that relied on the same basic insight (Seed et al. 2006).

Evidence of causal understanding in elephants is sparse. Nissani (2004) reports a string-pulling experiment and a tube task conducted with two zoo-based female Asian elephants. The elephants were required to pull a retractable cord in order to obtain a food reward in the first experiment, and suck or blow through a tube to gain a reward in the second. Of the two subjects, one performed reasonably well, the other less so, and careful analysis of the pattern of results suggested that performance was dependent on trial and error learning rather than an understanding of the causal relationships between the action and the outcome.

Nissani also reported an apparent lack of causal understanding on the part of working Asian elephants in a modified discrimination task (Nissani 2006). The subjects were trained to remove food from a bucket only after they had touched the positive stimulus of the bucket lid, which during training was always placed on top of the bucket. In the test phase, the lid was placed next to the bucket instead of on top of it, so there was no need to touch it before accessing the food. In only 3 of 77 trials did any elephants ignore the lid and reach straight into the bucket for the food. In the other trials, the elephants touched or even threw the lid away, as they had previously been trained to do. Nissani argued that this showed a lack of causal understanding by the elephants, but it might simply represent a lack of understanding of the task demands. Working elephants are trained from an early age to follow precise sequences of behavior and are punished for any deviation. Thus, in trials where the lid was placed next to the bucket, the subjects may have seen that the lid was now irrelevant but nevertheless interpreted the task as one that required the previous specifically trained practice. Until the experiment is repeated on animals that are normally allowed to exercise their behavioral choices freely, we should probably not draw conclusions from it about elephants' lack of causal understanding.

Indeed, some understanding of physical causality is suggested by the results of testing elephants on a task that depends on an understanding of the concept of "support" (Irie-Sugimoto et al. 2008a). Elephants were given a series of problems in each of which food bait was out of reach, but in some cases, a bait item was supported by a tray that the elephant was able to pull. In all cases, one of the two elephants tested performed above chance in selecting the correct tray to pull, showing that it took a means-end approach to the problem and had some understanding of the notion of physical support.

Learning to Discriminate among Features and Categories

Tomasello and Call (1997) review extensive evidence that apes and monkeys are generally adept at learning feature discriminations and categories, both natural and artificial. In this, primates do not differ from pigeons and rats, the animals more typically used in animal learning laboratories. Evidence suggests that elephants are similarly capable of discrimination learning and, moreover, that they have classification abilities that may exceed those of many animals.

One of the first systematic tests of elephants' discrimination learning ability was conducted by Rensch (1957) as he explored the long-term memory of a captive Asian elephant. Rensch taught a juvenile female Asian elephant 20 different visual discrimination pairs and 6 acoustic discrimination pairs; one pattern of each pair was rewarded. It took the female 330 trials to learn the first visual discrimination, but by the fourth and subsequent pairs, it took her only 10 trials to learn the correct target, showing that she developed a learning set, as do Old World monkeys and apes (Passingham 1981). The elephant was retested with the same visual discrimination pairs after a one-year delay, and she performed with 73 to 100 percent accuracy. As well as straightforward discrimination learning, there was some evidence that the elephant could transfer what she had learned about the features of the positive stimulus and apply it correctly to novel stimuli. For example, she appeared to generalize over orientation. In the initial tests, a "+" symbol was the correct choice; when subsequently presented with an "×" symbol for the first time, she immediately recognized it as correct. Generalization over size was also shown. The elephant was initially asked to discriminate between two stimuli showing black and white stripes, with the bands placed 2 cm apart in the positive stimulus and 4 cm apart in the negative stimulus. At a later date, she was presented with novel striped stimuli, this time with the spacing either 1.5 cm or 2 cm. She chose the correct stimulus of the thinnest stripes, 1.5 cm, even though in previous trials the 2 cm stripes had been the correct choice. However, with only two such examples, it is difficult to draw firm conclusions. In contrast to Rensch's paper, a discrimination task conducted on working Asian elephants (Nissani et al. 2005) reported that some of the 20 animals tested never learned to pass the tests, although others did perform comparably to the young female tested by Rensch. There was an age effect in these results, however, with fewer individuals over the age of 20 to 30 years able to acquire the discriminations. As we noted above, older timber elephants may have been justifiably cautious about exhibiting behavioral flexibility, given a regime of punishment for "novel" behavior.

We conducted a field experiment in Amboseli that can be seen as a more ecologically valid test of elephant discrimination and categorization, presenting individually known elephants with garments that gave either visual or olfactory information about their human wearers (Bates et al. 2007). We used garments that had been worn by members of two different ethnic groups, Maasai and Kamba, which pose different levels of danger to elephants: Maasai in the Amboseli area spear elephants; the Kamba people are resident agriculturalists that rarely interact with elephants. In the first set of trials, we separately presented elephant groups with three different red cloths, using a within-subjects design, giving each group the same range of choice. Each of the cloths had been worn either by a Masaai *moran*, or a similar-aged man from the agricultural Kamba tribe, or by no one at all. The only thing that differed between the cloths was the smell, derived from the ethnicity or lifestyle of the wearers. With access only to this olfactory information, the elephants showed significantly greater flight reactions to garments worn by Masaai *moran* than similar age Kamba men. In a second experiment, elephants in the same population were presented with two cloths that had not been worn by anyone, but here one was white (a neutral stimulus) and the other was red—the color that is ritually worn by Masaai *moran*. Elephants are dichromats, with the same color vision pigments as human "color-blind" deuteranopes (Yokayama et al. 2005), so they are easily able to detect a difference between red and white but would not generally differentiate red from green hues. With access only to these visual cues, the elephants showed significantly greater reaction to red garments than white, often including signs of aggression. We concluded that elephants are able to categorize a single species (humans) into sub-classes, a classification ability that may be exceptional among animals, judging by the absence of similar reports in other species (Bates et al. 2007). As discussed by Kangwana (chapter 20), both scent and sounds associated specifically with Maasai are categorized as "risky," while neutral scents and sounds are attended to but categorized as "low risk," judging by orientation, approach, or retreat and by displays or vocalizations.

Numerous vocal, gestural, and chemical signals are used by elephants (Kahl and Armstrong 2000; Poole and Granli 2003; Langbauer 2000; Poole et al. 1988; Rasmussen and Schulte 1998; see chapters 8 and 9). Although it is not yet clear what many of these visual and vocal displays mean to other elephants, as few signals have been formally tested, the fact that so many distinct signals exist is suggestive of subtle discrimination learning and elaborate categorization. Where playback experiments have been used to test reactions to specific signals, this notion has been supported (Poole 1999a; McComb et al. 2000; chapter 10). From behavioral reactions, it is clear that elephant calls can convey information about male musth state, and some information about the caller, such as familiarity and perhaps individual identity. Thus, although primates have been tested much more extensively in this area, there is reason to suspect that abilities at least matching those of any primate will be shown for elephants.

Quantity Judgements

Irie-Sugimoto et al. (2008b) examined relative quantity judgements of Asian elephants, following earlier work by Rensch (1957), who found that his five-year-old female Asian elephant was able to distinguish three from four dots, regardless of their arrangement and spread. Irie-Sugimoto et al. presented five captive Asian elephants with two baskets containing different quantities of food; the elephant had to choose the basket with the larger amount. All five elephants chose the larger quantity significantly more often than the smaller, performing with 67 to 89 percent accuracy. Elephants were as good at picking the larger quantity when it was only slightly bigger (e.g., 6:5) than when it was considerably bigger (e.g., 5:1), and performance did not vary with the total number of items presented (up to 12). In a second experiment, four additional Asian elephants watched and listened to the baskets being baited, but they could not see the final amounts in the baskets. All elephants chose the basket containing the larger amount, significantly more often than expected by chance, performing at 72 to 82 percent accuracy. As in the first experiment, the elephants did not exhibit disparity or magnitude effects, in which performance declines with a smaller difference between quantities, or as the total quantity increases, respectively. The lack of disparity or magnitude effects contrasts strikingly with the performance of great apes and even human infants in similar relative quantity judgement tests (see Anderson et al. 2007; Beran 2001; Boysen, Bernston, and Mukobi 2001; Feigenson, Carey, and Hauser 2002; Xu and Spelke 2000).

It is not yet known what cognitive mechanisms underlie the Asian elephants' numerical ability, but we suggest that elephants are able to keep track of a larger number of items in immediate, working memory than can great apes, even including humans. We predict that the larger working memory size of elephants means that they will show the same disparity and magnitude effects as apes and human babies but only when tested with much larger numbers of items. The hypothesis of an unusually large working memory capacity would also serve to explain the remarkable ability of African savannah elephants to keep track of the movements and positions of a large number of family members, shown in our urine-moving experiments in Amboseli, discussed above. Indeed, the unusual numerical abilities of elephants may derive, in evolutionary terms, from the elephant's need to monitor and coordinate movement of their extensive families and those of friends and "strangers" (see Mutinda, Poole, and Moss, chapter 16).

Social Cognition

Knowing about Others and Their Interactions

While individual recognition is considered an important component of animal social life, experimental demonstrations of it are relatively sparse. Monkey and ape species, however, are known to understand both the direct and third-party relationships of others (Cheney and Seyfarth 1990), and alliances and coalitions with specific individuals are a prominent feature of monkey and ape sociality (Harcourt and deWaal 1992). Note that knowledge of third-party relationships is not restricted to large-brained species. Some fish and bird species have also been found able to track the relative relationships of third parties, through what has been termed "eavesdropping" on the behavior of others (e.g., Oliveira, McGregor, and Latruffe 1998).

Our urine-moving experiments, discussed above, showed that elephants have knowledge of individual identities of up to at least 17 different female family members (Bates et al. 2008a). McComb et al. (2000), using experimental playback of long-distance contact calls, also in the Amboseli population, showed that each adult female elephant was familiar with the contact-call vocalizations of individuals in an average of 14 families in the population, totaling around 100 elephants. When the calls were from a familiar family—that is, one that had previously been shown to have a high association index with the test group—the test elephants contact-called in response and approached the location of the loudspeaker. When a test group heard unfamiliar contact calls (from groups with a low association index with the test group), their spatial cohesion increased, and they retreated from the area. It is uncertain, however, whether this vocal familiarity is based on individual recognition.

Whether elephants understand and take advantage of third-party relationships has so far not been tested. Field observations of African elephants show clear coalitions and alliances (Bates, Poole and Byrne 2008; Moss and Poole 1983), but it is unclear if there is any reciprocity—exchange of help—between these temporary coalition partners. With regard to cooperation, Asian logging elephants frequently work together to roll heavy logs up ramps, but they have been trained to work like this so they may not themselves understand the cooperation (Rensch 1957). Cooperative problem solving is observed fairly regularly in long-term behavioral studies of African elephants (Moss 1988). For example, two or more individuals may work together to help individuals that are trapped by muddy riverbanks or drainage ditches (Bates et al. 2008b), to chase off vehicles when an individual is darted for veterinary purposes, or to stand on either side of a darted elephant and attempt to hold her up between them (Moss, personal observation). Related individuals in Amboseli have also been observed to form coalitions when attempting to retrieve infants that have been commandeered by other, unrelated families (Moss, personal observation), which can also be viewed as cooperative problem solving.

Communication and Social Manipulation

African savannah elephants are known to have an extensive gestural and vocal repertoire (Poole and Granli 2003; Poole and Granli, chapter 8; Poole, chapter 9). Moreover, as previously discussed, they discriminate between the contact calls of familiar and strange individuals (McComb et al. 2000). Social knowledge apparently accrues with age: old females have the best knowledge of the contact calls of other family groups (McComb et al. 2001). Monkey and ape vocalizations sometimes encode "functionally referential" information; that is, hearers react to the calls in the same way as they would to entities in the world, such as specific predators (Seyfarth, Cheney, and Marler 1980; Zuberbuhler 2000). McComb and Shannon (in progress) are currently exploring whether anything similar is latent in elephant vocalizations.

Deception and other forms of social manipulation have been hot topics in primate work since the 1980s (Byrne and Whiten 1988; de Waal 1982). Deceptive tactics have been reported in all taxa of primates, with frequency of use well predicted by the species' neocortex size (Byrne and Corp 2004), and in great ape species at least some deception appears to be done intentionally, with some understanding of others' mental states (Byrne and Whiten 1992). In elephants, there is only one report of possible deception (Morris 1986): certain captive elephants in a zoo were noted to finish their ration of hay quickly, and then move near others who were still eating, swinging their trunks in an "aimless" manner, but occasionally eating some of their hay. It is not clear, however, whether the trunk-swinging was used tactically, and whether the other elephants were deceived. Researchers of wild populations have yet to report any deception at all between elephants, so it may be that elephants' extensive social network renders deception an inappropriate way for them to manipulate others.

Social Learning

There are, as yet, no formal studies of social learning in elephants or observations of behavioral traditions. Evidence of information exchange between the young and adults in a social context is, however, well documented (Lee and Moss 1999), and it would be unwise to conclude that such a long-lived, slowly maturing, highly social species does not learn through observing group members. The impressive spatial knowledge shown in some populations, for instance, is surely acquired by young individuals from following older, more knowledgeable relatives, and knowledge of food selection is likely to be learned through observation, since young animals have neither the size nor the strength to obtain many of the foods they eat (Lee and Moss 1999).

Elephants may eventually be shown to possess social learning abilities absent in non-human primates. For instance, there is evidence of vocal imitation in African elephants (Poole et al. 2005), and even one intriguing report of a captive Asian elephant copying human speech, although this has not yet been formally verified. Non-human primates, like most mammals, lack any significant abilities in vocal imitation (Janik and Slater 1997). Also, currently unpublished data suggests that older female African elephants may teach young, naïve, nulliparous females how to behave when they come into estrus for the first time (Bates et al. 2010). Despite some compelling observations (Boesch 1991), there is no consensus that any species of non-human primate is able to teach (Caro and Hauser 1992; Thornton and Raihani 2008).

Thus, while there is little evidence from which to compare the social learning skills of elephants directly with those of the extensively studied primates, this is representative of a lack of research effort on the topic rather than a lack of social learning ability in elephants. Reports of vocal imitation and potential teaching behavior suggest this is one area where greater research effort could prove particularly fruitful and instructive.

Theory of Mind

The term "theory of mind" embraces a wide range of possibly distinct abilities: understanding the gaze and perception of others, understanding intentions and attention, understanding feelings and emotions, understanding knowledge and beliefs, and understanding the self. There has been much speculation, but little consensus, as to the cognitive underpinnings of theory of mind and self-recognition abilities (Byrne and Bates 2006; Gallup 1985; Povinelli and Vonk 2003; Tomasello, Call, and Hare 2003).

There is experimental evidence that many species of primate follow gaze, even where the gaze is directed behind obstructions (Tomasello, Hare, and Agnetta 1999). Puzzlingly, individuals of the same species often fail to give evidence that they can use gaze as a cue to object choice, but recent work suggests a resolution. Lemurs were found to use gaze successfully as a cue, by following another's gaze and then choosing objects in the direction that they were looking, a tendency that the researchers called "gaze priming" (Ruiz et al. 2009). Gaze priming produces less striking rates of object choice than would be expected from human comparisons and may have been missed in earlier work. In addition, field observations suggest that both Old World monkeys and apes are able to understand the geometric perspective of others (Byrne and Whiten 1992); this has been confirmed experimentally only in the case of the chimpanzee (Hare et al. 2000). The observational data, based on collated records of deception, also implied that primates—but in this case, only great apes—are able to take account of the knowledge of competitors (Byrne and Whiten 1992). Again, this has been confirmed experimentally in the case of the chimpanzee (Hare, Call, and Tomasello 2001; Tomasello, Call, and Hare 2003) and the orangutan (Cartmill and Byrne 2007).

Nissani (2004) modified experiments devised by Povinelli and Eddy (1996) to examine what two zoo-based Asian elephants understood about visual attention. Povinelli and Eddy had found that chimpanzees chose to beg from a person whose whole body was oriented toward them rather than oriented away, and these two elephants performed similarly. However, chimpanzees did not discriminate between people according to their head orientation alone: for instance, when offered a choice between two people standing sideways-on, one with the head turned 90 degrees to look at the animal, and one not, or between two people both with whole body oriented toward them but one with the face covered or with a bucket over their head, the chimpanzees performed at chance, choosing between the two people at random. Nissani gave eight trials of each sort to each of the two elephants, and none of the results were individually significant. However, with sideways body presentation it appeared that the elephants were able to take account of the person's face orientation (six out of eight and seven out of eight trials successful, respectively).

Recent analysis of the behavior of the Amboseli elephants suggests that they understand at least the emotions and the intended goals of others, acting empathically toward individuals who are distressed and helping them in ways that were appropriate to their predicament (Bates et al. 2008b). A description of the reactions to a dying and subsequently dead matriarch by both family members and unrelated elephants supports the notion that elephants can act empathically (Douglas-Hamilton et al. 2006). Observations that group members respond appropriately to postural and vocal signals such as the "let's-go"-stance (chapter 8) and "let's-go"-rumble (chapter 9; Poole et al. 1988) also suggest that elephants may understand the goals of others. However, Bates et al. (2008b) did not find any positive evidence for the capacity to understand others' attention, knowledge, or beliefs. The same could be said of field studies of apes, so explicit tests of elephants' understanding of others' mental states are necessary.

Straight copies of studies used with monkeys and apes may be systematically unsuitable for probing elephant cognition; new and creative experimental designs are required. Monkeys and apes predominantly use vision to learn about the social world. For elephants, audition and olfaction are demonstrably more important than vision, with comparatively much larger brain areas dedicated to these areas than visual cortex (Hakeem et al. 2005; Shoshani, Kupsky, and Marchant 2006). It therefore seems probable that tests using auditory or olfactory stimuli will be more relevant to elephants and more likely to accurately measure their mental abilities.

One area in which it will not be possible to avoid the visual domain is that of mirror self-recognition, used to explore understanding of the concept of self. Most animals plainly cannot learn to understand the operation of mirrors: after repeated exposure they continue to make social responses as if to another individual of their species, or they habituate to mirrors altogether, paying no further attention to them. Monkeys, however, do learn to understand the geometric properties of mirrors, using them to detect the presence of individuals who appear behind them or to reach food rewards more successfully by locating their reflection in a mirror (Anderson 1984). However, they consistently fail to recognize a reflection of their own face as such, despite extensive experience. In contrast, members of all species of great apes can learn to interpret a mirror reflection or CCTV image of their face as "themselves," although not all individuals ever do so (Gallup 1970, 1982; Patterson and Cohn 1994).

Accounts of two tests in Asian elephants have been published, with somewhat contradictory results. Both relied on Gallup's (1970) "mark test" paradigm developed for apes. In this, the subject is first given extensive experience with a mirror. If signs of self-recognition are observed, such as self-monitoring while making repetitive or unusual movements, then the mark test is applied. Surreptitiously, for instance while the subject is anesthetized, a conspicuous mark is applied to its body in two places, one visible directly and one not. The subject is then observed for a period without access to a mirror in order to control for the possibility that the marks can be detected by scent or tactile sensation. Provided the subject examines only the visible mark and not the concealed one, the mirror is then restored. Then, once the subject catches sight of its image in the mirror, the critical observation, which is seen in many great apes, is that it suddenly touches and explores the concealed mark while monitoring the mirror to guide its hands. Povinelli (1989) followed this procedure when testing two elephants: he observed no signs of self-recognition, and both individuals subsequently failed the mark test. He concluded that Asian elephants do not show self-recognition. However, these elephants were only given a few days' exposure to the mirror prior to testing; chimpanzees that have passed the mark test have typically had weeks or months of prior mirror exposure. In the second experiment, because their three elephants could not be separated, Plotnik, de Waal, and Reiss (2006) used sham marking, where the procedure of marking the animal is followed but no mark is actually made in order to control for the possibility that the subjects might be able to detect a mark by non-visual means. They report that one of three adult females they tested did show mirror self-recognition: she touched the visible mark several times but never the sham mark. Moreover, all three elephants showed suggestive behavior in front of the mirror, prior to any marking: most strikingly, cases where "the elephant is standing *at* the mirror and moves its head in and out of mirror view, like a kid playing with his mirror image by running in and out of view of it" (Josh Plotnik, personal communication). However, the initial response shown by the "successful" elephant after spending time in front of the mirror was to walk away for seven minutes. Then she returned and moved in and out of mirror view a couple of times, then moved away again, still having shown no mark-touching. Finally, she moved away from the mirror and only then, when away from the mirror, did she first touch the mark with her trunk. This pattern is so different from that observed in the great apes who recognize themselves in mirrors that it cannot be considered definitive, but we suspect that future data will show that elephants do indeed have the competence to recognize themselves in a mirror and thus have some concept of the self as an entity.

Summary

A recurring theme throughout this survey has been the limitations on the data available relating to elephant cognition compared with the extensive data for primates. It is not simply that less is known, for both practical and historical reasons, about elephant cognition, although that is certainly true. However, in addition, the positive things that we do know about cognition within the two taxonomic groups are often hard to compare directly. In some cases, experiments with elephants have revealed abilities that are not shown by primates; for instance, our studies with olfactory social stimuli enabled demonstration of individual identity recognition in elephants, an ability not shown experimentally in monkeys and apes. Yet the differences we have reviewed are often more likely, perhaps, to reflect differences in study methods and facilities available than to point to any profound difference in cognition. We have repeatedly seen that the ready availability, ease of manipulation, and straightforward motivation of captive primates has enabled experiments that are not easily repeated with elephants, and have yet to be attempted.

Undoubtedly, elephants perform well on laboratory tests of learning, discrimination, and memory and spontaneously engage in simple forms of tool use—no obvious differences from primates are apparent here. Similarly, it seems likely that elephants, like the great apes, have the cognitive capacity to recognize themselves in a mirror. When it comes to categorization, some intriguing findings have recently emerged that may point to capacities unusual for mammals. In number discrimination, elephants are able to distinguish quite small quantity differences, yet strangely they showed no effect of the size difference in making their judgements nor any variation in performance with the total number of items presented. One possibility is that this lack of disparity and magnitude effects is a result of unusually large working memory capacity: for elephants, even groups of five to six items can be appreciated and compared in the manner that humans and other great apes can appreciate groups of two to three. Wild elephants have been shown to subcategorize humans into groups according to the varying levels of risk to them that different groups present and to make this categorization independently on the basis of scent or color. In the social realm, there is no doubt that elephants show empathy for the problems faced by others as well as reacting to their expressed emotions, but much more work needs to be done before the initial hints of elephant abilities in cooperation, imitation, and teaching can be properly understood. In contrast to the lack of any special signs of cognitive ability in laboratory tests of memory, data from the field suggest that the elephant's vaunted reputation for memory may have a basis in fact in two ways. Faced by the need to remember spatial information over very long periods, for instance the locations of waterholes in a desert, it is thought that elephants are able to re-find, over vast distances, places not visited for many years: elephants may be specialized for cognitive mapping. And in the immediate social realm, the ability to keep track of the current locations of 17 or more family members also seems remarkable. As with the data on quantity judgements, the most obvious possibility is that elephant working memory is larger than in humans or other great apes.

We finish with two speculations. First, we doubt it is a coincidence that the tests giving the strongest positive results so far are those based on abilities elephants show naturally in the wild, whereas tests that do not lend themselves to the natural environments of elephants have often been inconclusive. Thus, in order to go beyond the somewhat limited picture we have painted here and explore elephant cognitive skills in causal reasoning, social learning, and theory of mind, we suspect that investigation using ecologically valid stimuli will be required. Second, if elephant and primate cognition do indeed prove to be similar in many ways, which still seems entirely possible, a convincing explanation will be needed. Elephants are more closely related to hyraxes, dugongs, and aardvarks than they are to primates (Murphy et al. 2001): by contrast, rats and people are close cousins. Any coincidence in cognitive skills between elephants and primates, therefore, points to convergent evolution for specific abilities, so the future study of elephant cognition offers the potential to better understand the evolutionary forces that select for particular mental skills.

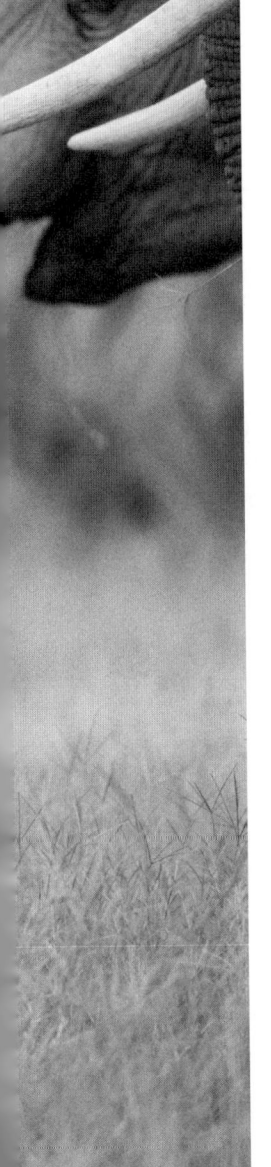

Part 4

Reproductive Strategies and Social Relationships

Section editors: Phyllis C. Lee and Joyce H. Poole

Chapter 12 Female Reproductive Strategies: Individual Life Histories

Cynthia J. Moss and Phyllis C. Lee

A FEMALE ELEPHANT has one of the longest reproductive lives of any terrestrial mammal. She may produce calves for 40 or even 50 years. During that time she faces many reproductive alternatives, which can be considered as tactical or strategic life history and behavioral "decisions." These alternatives occur in an ecological, demographic, and social context, and the decisions are influenced and constrained by life history parameters. Reproductive decisions include behaviorally optimizing the time for giving birth; acting to ensure the survival of her offspring; allocating resources to calves so as to minimize the time between reproductive events in the form of inter-birth intervals; gaining access to helpers; exercising mate choice; avoiding mortality risks so as to maximize her reproductive life span; and responding behaviorally to environmental and social conditions that can fluctuate wildly between years and over a life span. This chapter examines the breeding histories of female elephants in Amboseli on an individual basis over the 30-plus years of the study. We also present some general population-specific trends in female reproduction and life history.

Elephants mature within a milieu of diverse social and demographic circumstances in a variable habitat, and they then reproduce for 40 years in constantly changing circumstances. As such, over the very long term, a female's decisions and outcomes may change. We explore whether these changing contexts for reproduction are due to a female's experience with raising calves, successfully or unsuccessfully, over successive reproductive events within different social and environmental contexts. Changes in social circumstances relate to family size and composition and whether the female, by virtue of age and survival, becomes a matriarch. As noted in chapter 2 and part 2, the environmental context, which may range from droughts to periods of abundant food and may include predators and competitors, is also dynamic and fluctuating.

Reproductive strategies in long-lived species such as elephants are unlikely to reflect those patterns seen in relatively short-lived mammals with more invariant life histories (box 12.1). The analyses that follow suggest that the context for elephant reproduction also differs conspicuously from that of most mammals where reproductive variance is a function of relative status, and where contests among females over resources are frequent in order to ensure or enhance individual reproductive condition (e.g., ungulates: Clutton-Brock, Albon, and Guinness 1988; primates: van Schaik 1989; Isbell 1991; carnivores: Gittleman 1986). Status underlies or drives fertility, maternal investment strategies, and infant growth and survival for such species. Skewed or hierarchically determined reproductive success, with some females consistently performing well and others poorly within their social context, is a model that is unlikely to relate to elephant females. We argue that while there is an age-graded hierarchy among elephants (Archie et al. 2006; see box 12.2), the female elephant mode of reproduction is more egalitarian (e.g., Hrdy and Hrdy 1976), with reproductive cooperation within family units. This cooperation takes the form of active allomothering: sharing the time and energy costs of protecting calves from predation, ensuring that calves remain with the group, and helping calves when

BOX 12.1 COMPARATIVE LIFE HISTORIES
Phyllis C. Lee

Elephants are the classic example of a "slow" life history, unsurprising given their size, and represent the endpoint of the "mouse to elephant" relationships in life history studies. Their life history has also attracted considerable interest since their tempo of maturation, reproduction, and mortality are superficially similar to those of humans. The distinction between "live fast, die young" r-strategists and "live slow, have few infants, die old" K-strategists was historically related to size and underlying energetics but is now viewed as an adaptive consequence of age-specific mortality and fertility schedules (Promislow and Harvey 1990; Stearns 1992; Brommer 2000). Species experiencing high rates of juvenile mortality tend to be r-strategists with rapid age-specific fertility schedules, while low rates of juvenile mortality are associated with K-strategists and slower age-specific fertility.

Perhaps because female elephants have a very large mass, some of their relative or size-specific life history parameters are markedly different from those of other groups of mammals. Gestation is both absolutely long at 660 days and relatively long for their size even when compared with the anthropoid primates who are known for prolonged gestations both absolutely and relative to body size. The costs of infant production, reflected in investment in embryonic growth during gestation, stand out for elephants: relative to mass of the infant at birth, elephants again have longer than expected gestations (Eisenberg [1981]; Martin and MacLarnon [1985]; figure A), suggesting rates of growth in utero that are slower than would be expected for their size (e.g., Allen 2006).

Some elephant life history traits are independent of size.

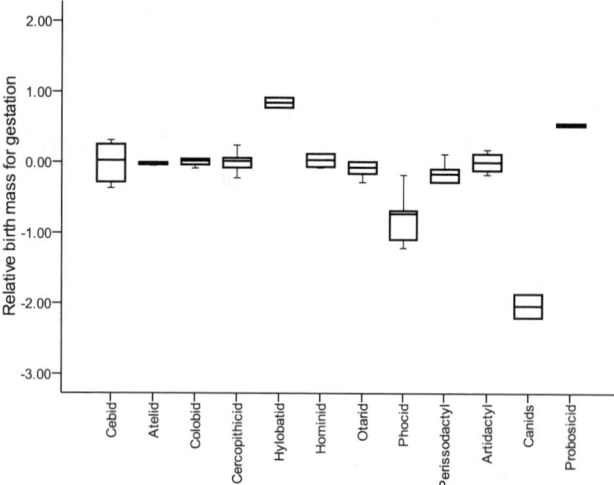

Box 12.1, figure A Mass of infants or litters at birth relative to the duration of gestation (median, 95% CI and range; phylogenetically corrected) for groups of mammals. Elephants fall outside the range, giving birth to larger infants for their gestation.

Litters are singletons, with rare events of twinning, consistent with most large ungulates, anthropoid primates, whales and dolphins, and seals, which also have median litter sizes of one. Elephant calves are weaned at sizes relative to their mother's mass as expected from inter-specific regressions (Lee, Majluf, and Gordon [1991]; figure B).

In contrast to "slow" traits, elephant reproductive rates can be surprisingly rapid. For example, a typical interval between births is three to five years (Moss 2001), compared with an interval of five years for a chimpanzee weighing only 45 kg or eight years for orangutans weighing 40 kg (Lee 1999). By comparison to anthropoid primates, elephants wean earlier and intervals between calves are far shorter than expected for their mass, as notably are those of the fasting seals (Lee, Majluf, and Gordon 1991).

Age at first reproduction, c. 8 to 9 years for first conception and a mean age of 11 to 14 years for first birth (Moss 2001; Owens and Owens 2009), is younger than expected relative to age at sexual maturation and adult size for other female mammals (Stearns 1992). In this feature of their life history, elephants may resemble whales more than other terrestrial mammals (Gaillard et al. 1989). Relatively early age at first reproduction poses energetic problems for mothers: first, in having calves with small mass at birth; second, in a small mother's ability to sustain calves during lactation; and third, in sustaining her own growth. While maximum longevity is still unknown, estimated life span for elephants (box 6.1) is somewhat extended relative to that predicted for mass, as is that of bats and primates (Carey and Judge 2001; Carey 2003). Media reports in 2005 of a known-aged Asian elephant in Thailand dying at age 80-plus in a zoo due to tooth loss makes the estimate of 70 to 75 years maximum longevity reasonable for the wild. A life span extended beyond that predicted even for their large body size suggests that, as with humans, there are significant evolutionary advantages associated with longevity for elephants. As discussed throughout this volume, social and environmental knowledge can be retained and used among the oldest females; reproductive potential increases with age for males; and experience with rearing calves enhances their survival. Elephant longevity and its correlations with reproductive success are again unusual aspects of their overall life history.

Elephants appear to be both an "average" K-strategists and an extreme example. Age at maturation is younger and reproductive rates are faster than expected for their size, while gestation is slower than expected, and growth period and life span are longer. Two factors may explain these contradictory patterns. The first is mortality, with elephants potentially experiencing high calf mortality (over 25 percent in the first year of life) combined with very low subsequent mortality until old age (less than 1 percent per year for females; box 6.1), making early and rapid reproduction advantageous. Second, the relatively large brain of elephants could impose costs on growth

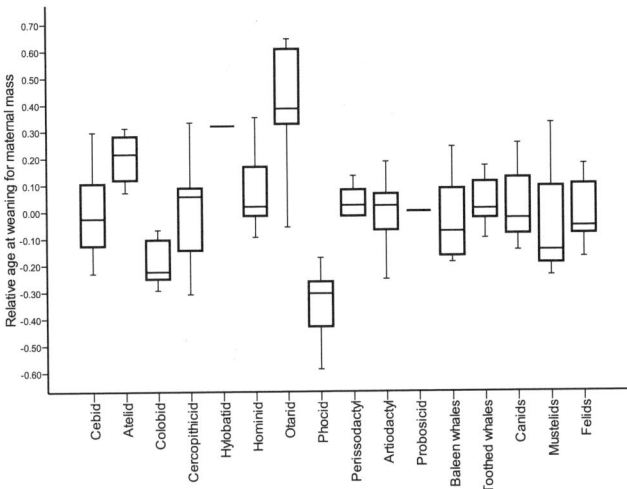

Box 12.1, figure B Age at weaning relative to maternal mass (median, 95% CI and range; phylogenetically corrected) for several groups of mammals. Elephants fall well within the range for most mammals, including the Old World primates, with seals and Neotropical primates as outliers.

and slow development generally (e.g., Martin 1996). In addition, differences between males and females in life history trajectories could contribute to a mixed strategy in relation to the tempo of elephant life history.

Reproductive Senescence and Longevity

Among mammals, only human females show universal cessation of reproductive hormone activity for all who live past a specific age. As more data become available for older female mammals, this pattern of reproductive rates slowing to complete cessation with advanced age becomes more apparent. When they live to maximum longevity, many female primates appear to cease reproductive activity some time prior to death, and the underlying hormone profiles are now being teased apart (e.g., Caro et al. 1995; Atsalis, Margulis, and Hof 2008).

Changes in mass and condition with age and multiple reproductive events as an underlying factor in reproductive senescence (e.g., Wood 1994) are poorly documented for non-human mammals. Among short-finned pilot whales (*Globicephala macrorhynchus*), old females show no signs of ovulatory activity (determined from luteal scars; Marsh and Kasuya [1984]), but at least some old females were lactating and may have been in a phase of lactational suppression of reproductive hormone activity or their body condition was too low to sustain a further reproductive event while lactating. Elephant females also have a marked decline in reproductive rates with age, but females 55 and over exhibit estrous behavior and hormonal function (Moss 2001; Freeman, Whyte, and Brown 2009).

While a decline in condition and/or ovulatory hormonal functioning leading to a reduced fertility rate with increasing age is a general feature of mammalian reproduction (Packer, Tatar, and Collins 1998), we have little conclusive evidence for a form of universal reproductive senescence equivalent to menopause (Hawkes et al. 1998). This lack of menopause combined with gradual senescence raises two questions for elephants. First, do older females invest for longer in the lactation of their last offspring and as a consequence suffer a longer period of lactational anovulation or low body condition, leading to the appearance of reproductive senescence as might be suggested for pilot whales? Second, if life span was extended beyond the "normal" range, would elephants show complete reproductive senescence similar to human menopause?

With multiple pregnancies and lactation events, females are less and less able to repair and replenish tissues as they age. Cellular damage, the accumulation of genetic mistakes, and a reduction in the efficiencies of cellular repair mechanisms are common to aging (Carey 2003). Females may simply be unable to sustain a pregnancy at the end of a life span, despite some level of hormonal competence or activity. For elephants, we have no evidence of a negative relationship between longevity and reproductive rates (Moss 2001), suggesting that such trade-offs are unlikely.

Elephant females ovulate multiply in a single estrous event, producing 6 to 8 (range 2 to 26) *corpora lutea* before a successful implantation (Laws 1967; Hodges 1998; Allen 2006). They might thus reach reproductive senescence earlier than monovular species, with long intervals between births due to a reduction in the availability of ova. However, elephants do not have a particularly extreme life span relative to their mass and thus are unlikely to "live beyond" their ova supply. As such, elephants may not be good candidates for insights into a model of ovarian senescence.

Of course, some element of all these factors could influence the decline in elephant reproductive rates with age. Maternal depletion may be associated with the need to invest more in the last offspring, with mothers delivering fewer nutrients per unit time when they are older and in poorer condition. Mothers require a longer time to regain condition at older ages while lactating, which could have an additional inhibitory effect on reproductive hormones. And finally, any decline in ova viability would decrease the chances of a subsequent conception, leading to a lower fertility per estrous event. While possibly accounting for the observed patterns of elephant reproductive decline with age, the issue of reproductive senescence remains unexplained in either adaptive or physiological terms. It also might suggest that the human form of menopause is indeed a highly derived trait, unshared in form or function with other mammals. As with so many questions about elephants, these puzzles remain to be addressed with greater knowledge over time, hopefully incorporating a lifetime perspective from individuals as attempted here.

confronted with environmental or social hazards. Due to the costs of lactation, individual females within families face conflicts of interest over access to food and, critically, water (Archie, Moss, and Alberts 2006; see also Mutinda, Poole, and Moss, chapter 16), but nevertheless act as a unit when there are vulnerable calves. Thus, the family unit is the foundation for shared infant care and knowledge between close female kin, as well as for intergenerational exchange of help, information, and support.

The questions we focus on in this chapter are the following:

- How do individuals maximize their reproductive potential? In this context, we explore age at first reproduction, sex ratios and calf sex, calf survival, and age-specific variation in inter-birth intervals.
- How does the social context affect reproduction for elephant females? We examine whether there is evidence of hierarchical or status effects on reproductive potential; how does the size of the family and the number of allomothers impact on reproductive parameters?
- Finally, how do both the individual and the social context for reproduction change over a life span?

For the most part, what we present here are outcomes; we have few insights into elephant cognitive capacity for reproductive decision making. But we use the conventional concept of a strategy as that of consistent and measurable outcomes, which vary between age-classes, between matriarchs and subordinate females, and among individuals in relation to complex individual, group, or time-specific traits. We explore reproductive strategies from the perspectives of consistency and variation in tactics—an exploration of alternatives available to females. We leave open the questions of intentional "decisions" and "choices."

BOX 12.2 DOMINANCE IN FEMALE ELEPHANTS
Phyllis C. Lee

In 1972, Iain Douglas-Hamilton observed that female elephants appeared to have a strong age-related system of status or dominance, based on kinship, age, and size, with all family females being subordinate to the matriarch. Dublin (1983) subsequently speculated that unequal social status among females might determine the extent and nature of competition and cooperation among members of a family. She suggested that one potential consequence of subordinate status would be reduced reproductive success (or at least breeding success). Archie et al. (2006) statistically demonstrated age-related hierarchies, while Moss (1988) described considerable variance in status between separate families, although the factors that explained this variance were not defined quantitatively.

Here, I briefly review evidence for differential status among females within and between families and what the outcomes might be for family cohesion, competition, and cooperative interactions. While Douglas-Hamilton (1972) suggested the existence of a linear hierarchy ranging from subordinate younger females to older dominant females, and with the matriarch at the top, Lee (1987) found clear evidence of age-related dominance, with 89 percent of older-larger females able to dominate younger-smaller females in direct avoidance or aggressive contest interactions. That study was unable to describe linear hierarchies within families since overt or escalated dominance-subordinance interactions were too infrequent in 1,750 contact hours of observation to be able to reliably assign individuals a linear hierarchical status. This finding, rather than suggesting a paucity of data, suggested a norm of non-aggressive or subtle modes of interaction, as least during that study period. Archie et al. (2006), studying females in both Amboseli and Tarangire, also found that the outcome of dyadic interactions was strongly age-size dependent but transitive—in other words, consistently structured if not significantly linear. In Tarangire, the number of dyadic competitive interactions was approximately three times (3.28 per dyad) that of Amboseli (1.16 per dyad), reflecting differences in the nature of resources over which females may compete—discrete, woody, and clumped in Tarangire versus dispersed, low quality, and abundant in Amboseli.

In Samburu, another wooded bushland habitat, although with low rates of competitive interaction, Wittemyer and Getz (2007) again found transitive age-size related dominance; in this case, these interactions occurred between matriarchs from different families. Both the estimated age and correlated shoulder height of the matriarch were significant predictors of relative rank, with a small but not significant contribution due to family size. They observed that interference competition occurred even in the context of dispersed resources and suggested that the capacity for memory of past events, past outcomes, and perceived risks were important elements determining the relative status between two matriarchs. Again, relatively few within-family overt aggressive or competitive events were recorded in Samburu, and Wittemyer and Getz (2007) proposed that fission-fusion sociality in elephants has the potential to allow for greater individual avoidance of costs imposed by despotic individuals,

Demography and Life History

Data Used in Analyses

At the beginning of the study in 1972, there were 169 mature females (aged nine and above) born before 1964. Between 1964 and the end of 2004, 918 females were born. The life histories of these 1,087 females provide a unique opportunity to examine and attempt to understand female reproductive strategies. The demographic data on which this chapter is based are explained in appendix 1.

In detailed analyses of life histories, we focused on 238 females who reached sexual maturity, mated, and gave birth to at least one calf as well as following the reproductive careers of the adult females present from the start of the study. These data were used to explore age at first birth, inter-birth intervals, calf sex ratio, survival, and the effect of maternal experience on reproduction. Additionally, the size of each female's family unit at different events was assessed, as was the relative status of the female within her family (see box 12.2). Since all the data were derived from individuals in relation to specific events or points in time, sample sizes vary among statistical tests. Analyses of age-specific changes in fertility, calf mortality, and other time-dependent events were carried out using proportional hazards analyses, while analysis of variance (ANOVA) and logistic regression were used to test between major effects and covariates (see appendix 1).

Life History Patterns of Females

For any animal, including the longest-lived, a female can only make reproductive choices within the constraints of the "typical" life history for that species (figure 12.1). Age at sexual maturity, gestation length, and life span are significant constraints, but there is potential for some variation

which may explain why so few overt competitive or aggression interactions are seen within a family. It is clear, however, that those within-family contests that do occur are generally "won" by the older, larger female. Their finding that matriarchs of similar dominance strength associate more frequently and in larger groups was also suggestive of the benefits of fission and fusion to individual females within families and aggregations.

The issue of nepotistic interactions remains one of considerable interest. Archie et al. (2006) explored whether a lower rate of interaction between first-order relatives, combined with more reversals, would illustrate some degree of nepotism and found no effect of close relatedness on either rates of interaction or on reversals in outcomes, even taking into account opportunities for interaction due to close proximity. Despite a lack of support for nepotistic status, there could potentially be some enduring effect if females are matched in size or age. Would a same-aged daughter of a matriarch be able to win in competition with the daughter of a lower status female? Such a question remains to be tested on a larger data set, and can only be suggested at this stage.

Another unexplored element in inter-family dominance is that of Dublin's (1983) conjecture of individual disposition as a contributor to relative status. Moss (1988) also proposed that some Amboseli matriarchs appeared to be aggressive and confident, others were either avoidant of engaging, while yet others were apparently disliked and therefore avoided, which affected their relative status as well as propensity to be gregarious. Andelman (1985), in a brief study in Amboseli that also included the Lake Manyara population in Tanzania, suggested that both family and bond group size were associated with inter-family status, in addition to age of the matriarch, as was noted for Samburu elephants by Wittemyer and Getz (2007). The relative group size experienced by a family might be a proxy for tolerance or aggressiveness on the part of the leaders—tolerant matriarchs might form larger and more stable aggregations and be less likely to provoke fissions so that younger or subordinate females and families are able to avoid competition.

Does elephant female status relate to her reproductive performance or lifetime success, as Dublin speculated? Amboseli families with older, experienced matriarchs have a higher per capita reproductive rate (McComb et al. 2000), but this effect may also be a consequence of longevity as a determinant of reproductive success. Reproductively successful families are large, have more allomothers and protectors, have older, dominant, and knowledgeable matriarchs, and may be more gregarious—all factors that could potentially contribute to a high age-specific birth rate and enhanced calf survival. Given that status is related to age, however, there is unlikely to be status-based variation in reproductive success within families that is independent of age and experience, nor is reproductive suppression of subordinates by dominants within families (e.g., Dublin 1983) either observed or likely to occur given elephants' cooperative rearing systems.

As a final point, the explanatory power of any single factor such as age or size, or even relatedness, is relatively low in the empirical studies of elephant dominance—much of what we observe in relation to both intra- and inter-family dominance remains unexplained and, indeed, unexplored.

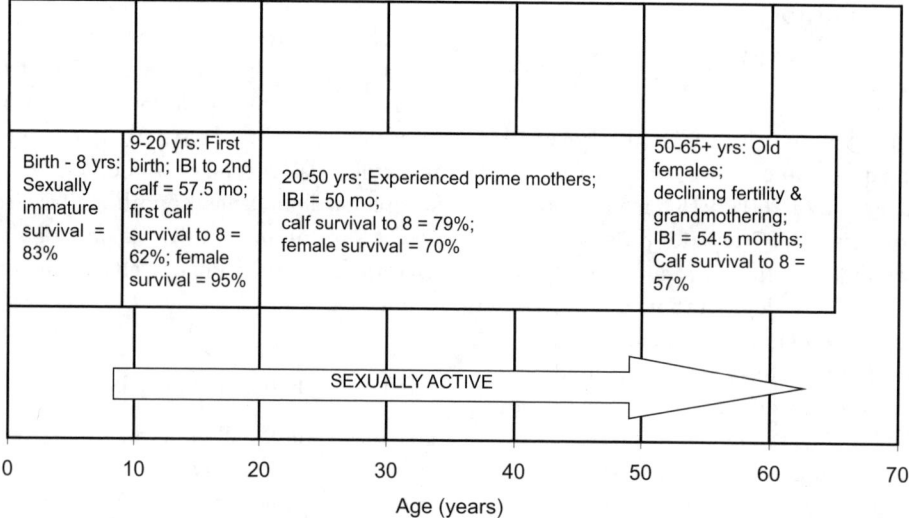

Figure 12.1 The reproductive life history of an Amboseli female.

in almost all the reproductive parameters found in African elephants. It is in this variability that we seek insights into individual female reproductive strategies.

Life Expectancy and Reproduction. Median overall life expectancy for Amboseli females was 41.67 years (±1.24 SE: mean = 38.33±1.02), across all sources of mortality (Moss 2001). Life expectancy for females exposed only to natural sources of death (illness, accident, drought, predation) was 54.42 (±2.51), while for females exposed to human-induced mortality it was 46.75 (±1.98) (see box 6.1, Life tables). Most females, therefore, survived 25 years beyond the age where they commence reproduction.

Considering only those parameters associated with a maximal rate of reproduction, a female elephant potentially could become reproductively competent and conceive at 7 years old, give birth at 9, and subsequently at three-year intervals until she has her last calf in her 60s, resulting in 18 calves born in her lifetime. As we illustrate below, an Amboseli female will have her first calf at an average age of about 14 years and give birth every 4 or 5 years until she dies at the average life expectancy of 41. Thus, she would have had 5 to 7 calves, some of which (approximately 17 percent) would not have survived.

Age at Sexual Maturity and First Birth. One key component underlying lifetime reproductive success (Clutton-Brock 1988) is a female's age at first reproduction. Starting to reproduce even one year earlier could potentially improve overall output for elephants (Croze, Hillman, and Lane 1981). Here we explore age at first reproduction for individuals (see Moss 2001). We have used the age at first birth in this analysis, since it is possible that females conceive but are unable to successfully carry infants to term; known first birth is thus a conservative measure. For 192 females, their exact age and their calf's birth date were known within one month. For a further 91 females, a reliable estimate of both their age and their first birth were made (see appendix 1). Using a proportional hazards analysis of all females potentially exposed to the "risk" of a first birth, the median age of first birth was 14.2 years. Less than 1 percent gave birth between 8 and 10 years, and only 5 percent experienced their first birth over the age of 18. Using only the 192 females whose ages and first births were known to within one month, the median age was 15 years, and again less than 1 percent gave birth under 10 years, and 6.5 percent at 18 or older (see also Moss 2001). For Amboseli females over the period of this study, first successful conception, therefore, occurred at a minimum age of 7.08 years, but at an average age of 12.2 years.

Growth and Size as Influences on First Reproduction. Individual size plays a role in the onset of reproduction in many female mammals (Birgersson 1998; Birgersson and Ekvall 1997; Setchell et al. 2001). Among elephants, physiological fertility can commence at 7 years but a female does not attain her asymptotic size or mass for a further 15 to 20 years (figure 12.2). Younger and smaller females may be less able to partition total available energy between that needed for reproduction and that needed for their own growth before their growth rates begin to slow significantly. Growth in height is most rapid in the first five years of life at 10 to 12 cm/yr and decelerates to 4.5 cm/yr between 5 and 10 but remains relatively high at 3 cm/yr up to age 15. After 15, growth velocity slows to around 1.5 cm/yr, and height increases only gradually after this age

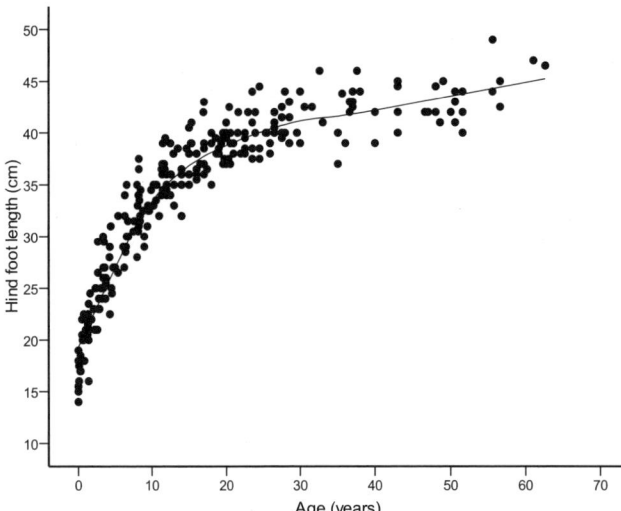

Figure 12.2 The curve for growth in female hind foot length with age ($n = 436$) (lowess curve fit).

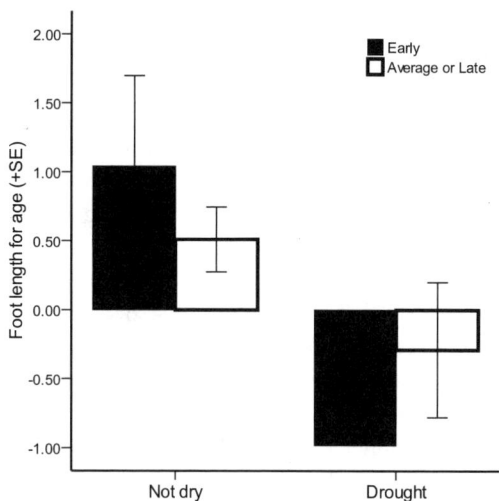

Figure 12.3 The relative foot length for age for females who reproduced early (12 years or younger) compared with that for average or late females (>12 years) as a function of whether the female experienced a dry period in her first two years of life (early $F_{1,131} = 4.34$, $p = 0.039$).

($N = 282$). Females may reach some critical size, mass, or condition level at around 8 to 10 years, such that the combined costs of growth and gestation are relatively minimal, and some females are able to successfully conceive at these young ages.

However, for most females who first give birth at 14 to 15 years, the trade-offs between sustaining the lactation necessary to maintain calf growth and ensuring individual growth may still be problematic. While females may be able to sustain the lower costs of gestation (see Prentice and Whitehead 1987), they could have difficulty supporting the costs of milk production prior to a slowing in their growth velocity. Individual growth could be maintained simultaneously with a pregnancy, but as we discuss below, the calves of young females are less likely to survive after birth, potentially reflecting higher energetic costs of providing milk, which may be difficult for smaller, younger females to sustain.

In addition to growth, condition also appears to play a role in the timing of a female's first reproduction. Although we have little data on condition at an individual level, we do have some suggestive information on reproduction and growth under varying environmental conditions. From 1974 to 1976, there was a serious drought in Amboseli (see Croze and Lindsay, chapter 2), resulting in the cessation of reproductive cycling among females of all ages. Only 2 calves were born to the population in 1977 and 5 in 1978. When good rains finally started to fall in November and December 1976, the females gradually recovered and started to cycle, giving birth to 57 calves two years later (Moss 1988; see Lee, Lindsay, and Moss, chapter 6). A delay in first reproduction was noted during this period. While the sample is small, 16 females reached nine years of age (the minimum for reproductive onset) during the 1974–76 period, and they did not have their first calves until 15.8 years of age, a delay of two years compared with those that reached nine after 1977.

Females born during drought years, defined by low annual rainfall and an extended dry season, were small for their age, especially for the critical years up to the age of 10 (Lee, Lindsay, and Moss, chapter 6). Females born into dry years ($N = 54$) had slightly later ages at first birth (14 years compared with 13 for known-age females in good years), and the average or late reproducers tended to be smaller for age (figure 12.3).

Social Roles and Social Status Affecting Age at First Birth

In some species of birds and mammals, females postpone their own reproduction to assist with raising offspring in their group (Clutton-Brock 1991). In other group-living species, only the dominant female breeds (meerkats: Clutton-Brock et al. 2001; canids: Malcolm 1985; African hunting dogs: Frame et al. 1979; tamarins: Goldizen 1987), and some females may never breed. In Amboseli, female elephants help in the rearing of younger calves in their families (Lee 1987; Lee and Moss, chapter 14), but only two females who have reached the age of sexual maturity have never produced calves. Neither has ever been recorded in estrus. To date, we find no evidence of reproductive suppression in Amboseli females where a younger or subordinate female does not breed either as a function of living in a large unit or as a result of subordinate status (Dublin 1983; Freeman, Weiss, and Brown 2004; and see box 12.2).

If females experience inter-individual competition (Archie et al. 2006) affecting their physical condition or hormonal function (e.g., Foley, Papageorge, and Wasser 2001), then females maturing within larger families would be expected to reproduce at a later age than those who matured within smaller families. We found no significant difference in the age at first reproduction among females living in large, medium, or small families (figure 12.4).

In despotic species or where there is high reproductive skew, the relative status of the mother or her dominance rank may affect her daughter's reproductive potential. For example, in primates, high-ranking mothers have daughters who start reproduction at younger ages, shorter inter-birth intervals, and higher infant survival (Altmann, Hausfater, and Altmann 1988). In pronghorn antelope (Byers 1997) and red deer (Clutton-Brock, Albon, and Guinness 1986), higher status mothers again have higher reproductive success; and among red deer, daughters of high-ranking mothers are also more successful over their reproductive life span. We examined the relationship between maternal status and a daughter's age at first reproduction for the Amboseli elephants. If having a high-ranking mother suppresses a daughter's fecundity, there should be a negative relationship between status and daughters' success. By contrast, if having a high-status mother "advantages" her daughter, then such daughters would be expected to have younger ages at first reproduction.

Status was defined for female elephants on the basis of the composition of a family unit in terms of numbers of adult females, and an individual female's age and leadership (Moss and Lee, chapter 13; Mutinda, Poole, and Moss, chapter 16). Within each family, females were categorized as either Matriarch, Split Matriarch (a female who split from her natal family with her offspring and possibly other adult females to form and lead a new family), High (below a matriarch but who provides occasional leadership and is able to win in direct contest competition with other, non-matriarch females), or Other (the remainder of the adult females in a unit). The mother's status was assigned to each calf born and was defined as Mother is Matriarch (assigned to all females born whose mothers were or at some stage became matriarchs, $N = 284$); Grandmother is (or was) a Matriarch ($N = 167$); Mother was High ($N = 141$); or Mother was Other ($N = 185$). For 213 females, maternal status was unknown, although their own status was known. A total of 127 females were Matriarchs of the 53 families at some point in their lives, 81 were High, and 782 were Other. When the mother was a matriarch, daughters were more likely to become matriarchs ($N = 24$) or high ranking ($N = 19$) themselves than were those females whose mothers were not matriarchs ($N = 2$ became matriarchs upon the death of older family females).

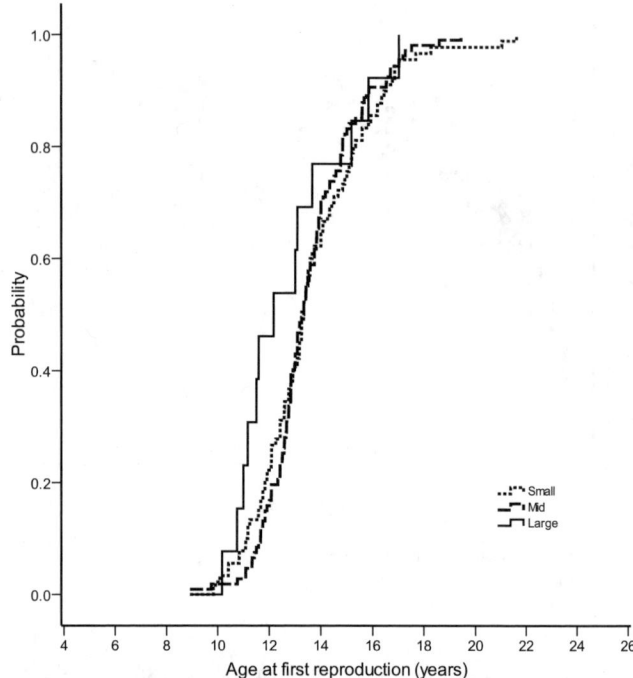

Figure 12.4 Probability of first reproduction at each age by size of family (small = <10, middle = 10–20, large = >20); family size at the age when the first event occurred (Gehan = 2.77, df = 2, NS).

The effect of the mother's status on her daughter's age at first reproduction was examined for those daughters who reproduced at least once in the study, excluding mothers of unknown status. There was no significant effect of maternal status on age at first birth for primiparous females; daughters of matriarchs did not give birth earlier or later than any other females, nor were the daughters of subordinate mothers late to reproduce.

Furthermore, we found no significant interaction between maternal age, maternal status, and age at first birth for daughters. The age at which the mother had her daughter bore no relationship to the age at which the daughter first gave birth herself ($N = 235$, $r = 0.013$, NS). Older matriarchs (>35 years), who can be considered as females with the highest relative status, did not have daughters who reproduced at a later age, which would be predicted if they were suppressing their daughters' reproduction. The lack of any significant differences in age at first reproduction between daughters who had high- or low-ranking mothers is convincing evidence that suppression of subordinate females was not acting to delay the onset of reproduction.

By contrast to predictions of suppression, and in agreement with the suggestion that elephants engage in cooperative infant care, the mere presence of a mother of any status alive within the family appeared to confer an advantage on daughters in their age of first reproduction. Females whose mothers survived until the daughter was at least nine had an earlier age at first birth (13.2 years, N = 131) than did females whose mothers died before their daughters were nine (14.9 years, N = 80) (ANOVA, maternal age at death category, $F_{2,287} = 10.31$, $p < 0.001$; post hoc tests $p < 0.001$, equal variances not assumed). This effect occurred both within and between families and across all status categories and was thus unassociated with differences in matriline rank.

Young females within a family could have lower fertility if they were suppressed in their capacity to ovulate and conceive. Alternatively their fecundity, the ability to carry a calf to term, could be suppressed. While it is possible that young females within a family may be less able to gain condition and enter estrus if there is marked female-female competition for resources, the data from Amboseli do not suggest any obvious differences in fecundity between younger and older females (see below).

Age at first reproduction for both known-age females and those with an age accuracy of ±1 year has decreased consistently throughout the period of this analysis, even as population size has increased from a low of 480 to over 1,500 (see annex 2.1). The relationship with age at first birth and the year of that female's birth was significantly negative ($r = -0.380$, $N = 283$, $p < 0.001$), suggesting that reproductive maturation has been getting progressively earlier over the period when families and the total population have both grown in size. While population size is not a good measure of density due to the elephants' ability to roam over larger areas with time (Croze and Moss, chapter 7), some density-dependence was demonstrated for calf mortality (Lee, Lindsay, and Moss, chapter 6). Nevertheless, the lack of any effect of family size on age at first reproduction suggests that intra-family competition is not suppressing female reproduction.

Fecundity Determinants

Fecundity for Amboseli elephants (taken here as the probability of giving birth) is influenced either by the environmental conditions existing at each reproductive event (Lee, Lindsay, and Moss, chapter 6), or are a function of individual characteristics. Of the individual traits, three major factors influence the frequency of births for individual females. Physical condition determines whether a female is able to come into estrus, conceive, gestate to term, and then sustain an infant during the lactation period, and her condition is both a product of the local environment (the resources available and the costs of obtaining these) and her past growth and reproductive history. The second factor is whether her previous calf survives, since the loss of an infant results in a female returning to estrus. Finally, the age and experience (parity) of a female influence her ability to conceive and rear offspring. A variety of other parameters interact with these primary determinants, such as calf sex and birth order. The number of helpers in a family is also critically important to the survival of a calf and thus to a mother's reproductive success. All of these will be discussed briefly here.

Condition

We started by examining the role of condition to asses whether females in good condition (Foley 2002) had more surviving calves. Did condition influence sex ratios, leading to more sons born to high condition mothers, as suggested by classical Trivers-Willard (1973) models? And did female condition affect her rearing strategy in relation to calf sex and its birth order?

We have insufficient data at this stage to relate sequential reproductive events to our condition measure of size-for-age (which is typically an instantaneous measure), and size-for-age does not assess changes in an individual's condition over the course of a 36 to 60 month reproductive event. We therefore use age, parity, and status as indicators of relative condition. Parity is considered in three categories: primiparous females giving birth to their first calf, second time mothers, and experienced mothers with three or more known births. Age and parity are obviously related since as females age, they increase in mass and stature (see figure 12.2). We are assuming that older, larger, and potentially dominant females were in relatively better condition and thus had greater physical reserves to devote to calf-rearing. These were the category of "experienced" mothers.

Once females began to reproduce, fecundity was relatively constant from 15 until the age of 45 (Moss 2001). Fecundity was similar across the age categories of females who had begun reproduction, although the youngest age classes had the lowest average fecundity since relatively few of them were reproductively active (table 12.1). Most females produced a calf at 4- to 6-year intervals (see below), and there were no significant differences between families in fecundity despite considerable variance in total family reproductive output. For example, over the course of the study, 73 calves were born to females within the VA family, the largest family with 20 reproductively active females. However, the average number of calves born to each female in that family was 0.18 per year, which was not significantly different from the population average of 0.165 per female ($N = 466$). There were no detectable factors that skewed fecundity within or between families, other than the reproductive lifespan of females. As we note above, the observed fecundity among the reproductively active youngest females

Table 12.1 The annual probability of a female giving birth for each age class

Age class of female (years)	Mean fecundity (probability of calves born/female/year)	Calf survivorship (proportion of calves surviving to 24 months)
0–9.9	0.020	.600 (N = 5)
10–14.9	0.142	.550 (N = 217)
15–19.9	0.214	.744 (N = 305)
20–24.9	0.223	.852 (N = 236)
25–34.9	0.230	.882 (N = 380)
35–49.9	0.209	.871 (N = 225)
50+	0.121	.804 (N = 56)

Notes: The mean for the age class has been calculated from the age-specific probabilities based on a longitudinal fertility analysis. Mean calf survivorship for calves born to females in each age class; N births given, mortality due to natural causes only.

suggests that they were not suffering from suppression of their fertility or of their ability to carry a calf to term.

Calf Survival and Maternal Investment

One major theoretical issue with regard to female reproductive decisions is whether (and when) to produce a son or a daughter. Studies of the differential production of one sex of offspring are generally inconclusive or conflicting. In some mammal species, birth sex ratios are biased (Gomendio et al. 1990; primates: Paul and Kuester 1990; deer: Kruuk et al. 1999a; elephants: Foley 2002), while in other studies no bias has been observed (primates: Hiraiwa-Hasegawa 1990; deer: Hewison et al. 1999). Birth sex ratio skew is thought to result either from variation in maternal condition (Wauters et al. 1995) or from sex-specific reproductive payoffs (e.g., van Schaik and Hrdy 1991; Hiraiwa-Hasegawa 1993; Kojola 1998), again leading to conflicting predictions. However, many observations of sex ratio biases in mammals are derived from tiny samples (Foley 2002), which can be shown to produce a low statistical power and a high skew away from 50:50 (Brown and Silk 2002).

Over all known sex births, Amboseli females gave birth to roughly equal numbers of sons (N = 638) and daughters (N = 698). Sex ratio for first births did not differ significantly from 50:50 despite a slight trend for more female births (126M:143F). Young females under 20 were statistically as likely to produce sons (N = 256) as they were daughters (N = 284), with no significant sex ratio bias across the ages (χ^2 = 5.7, df = 6, NS). Matriarchs were also equally likely to have sons (212) as daughters (216). We conclude that as yet, there is no evidence in the Amboseli population of any mechanism for biasing sex ratios at birth.

Mothers under 20 years lost on average 31.5 percent of

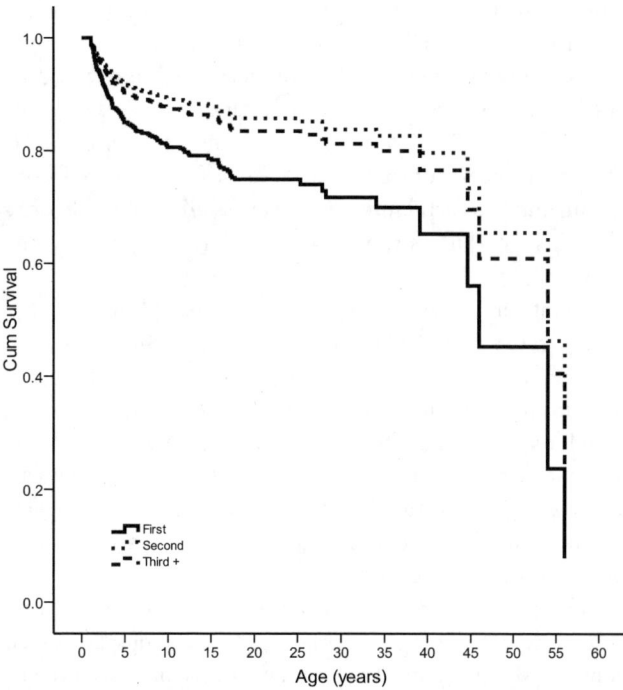

Figure 12.5 Survival for first-born calves compared with that for experienced mothers. Mortality due to natural causes only, excluding deaths in the first year of life.

their offspring to natural causes, while those over 20 lost only 12 percent (table 12.1). Over 30 percent of first-born offspring died in their first two years of life, while first-born calves of both sexes who survive their first year still had significantly reduced longevity by comparison to calves of multiparous mothers (Cox regression: Wald = 8.87, df = 2, p = 0.012; figure 12.5). Why first-born calves remained at risk after the early period of neonatal mortality was unknown. It is possible that slow or restricted growth had some permanent effect on organ development or metabolism, which placed these individuals at greater risk of death throughout the life span.

Although there was no bias in birth sex ratio, there was a difference in survival between sons and daughters, and sex-specific survival also differed as a function of the age and experience of the mother. Overall, sons were more likely to die in the first two years of life than were daughters (χ^2 = 16.0, df = 1, p < 0.001), although this effect was minimal for primiparous mothers who suffered excessive calf mortality of both sons and daughters. By the time a mother had had three or more calves, calf survival was much higher, although sons were more likely to die than were daughters (χ^2 = 6.6, df = 1, p = 0.01; figure 12.6) especially after dry periods.

Several factors could influence these sex- and age-specific trends in mortality. Females may start to reproduce before they reach the size and condition necessary to sustain offspring survival. As we note above, gestation for many large mammals is not as energetically costly as is lactation, and

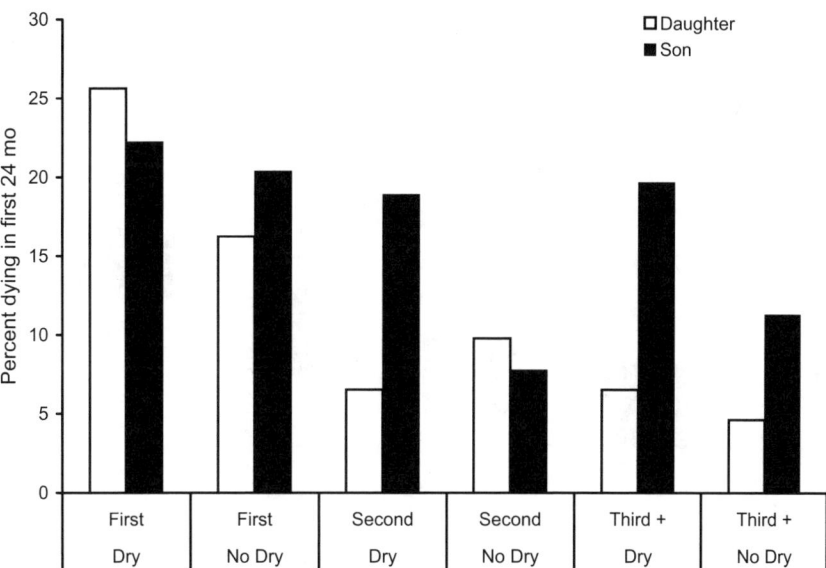

Figure 12.6 Percentage of first-born, second-born, and third-born sons and daughters that died of natural causes in their first two years of life, split by early drought experience.

thus both sexes are equally likely to be conceived and survive during a pregnancy. If all first calves born to smaller females under 15 to 20 years of age are at risk, and since males are more costly in terms of the milk energy required to sustain their more rapid growth, first-born sons may be especially vulnerable. Competition for the resources necessary for lactation potentially could put a young female at a disadvantage early in her calf's life, especially in years with low rainfall and low productivity, and lead to higher calf mortality. However, given that we have little evidence of competition affecting reproduction within a family, we suggest that a lack of experience with the care of calves may play a large role in their death.

Observations in Amboseli suggest that some primiparous females were simply inept mothers, failing to care for calves or to bond successfully. More than that, the responses of first-time mothers to their calves' suckling demands suggest a lack of meshing on the part of the mother-calf pair (Lee and Moss, chapter 14), which again could reduce the survival of first calves if mothers were not responding to genuine signals of energetic need from their calves. Considering that all females would have had some experience of calf caretaking as allomothers prior to reproduction, individual experience with suckling appears to be vital for learning how and when to allocate sufficient resources to ensure the growth and survival of calves. However, if this is so, then why were the sons of experienced females at greater risk than their daughters? At this stage, we can only suggest that even the most experienced mothers may have trouble energetically sustaining their sons' growth when they encounter drought years with low food availability.

While competition within families may not have affected fecundity, competition between families (Archie, Moss, and Alberts, chapter 15; Mutinda, Poole, and Moss, chapter 16) could have limited the energy balance of even old experienced females, reducing their condition and capacity to sustain their sons when times were tough. Comparing age and sex-specific mortality for calves born in years with average to good conditions, there was no longer a significantly increased probability of death for the sons of experienced mothers ($\chi^2 = 0.495$, df = 2, NS). In ecologically "bad" times, experienced mothers may be better able to assess the survival probabilities of their calves in relation to their own condition, and they may simply reduce investment that has potential mortality costs to themselves, resulting in higher male mortality.

In these analyses, we have explored survival in the first two years of the calf's life. This period was chosen since almost all calves that lose their mothers prior to 24 months will die. Unsustained by milk, a calf under two has only a tiny chance of survival even if cared for by other females. In addition, in statistical analyses of inter-birth intervals the only significant effect of the age at which a calf dies on the length of the interval is when the death occurs within 24 months or less (see below).

Allomothers and Calf Survival

As observed previously (Lee 1987, 1989) the presence of young, pre-reproductive females in a family appeared to enhance calf survival. Most calves irrespective of sex or birth order died when dry seasons were harsh and prolonged, and experience of a drought or prolonged dry season in the first year of life was a significant influence on calf mortality

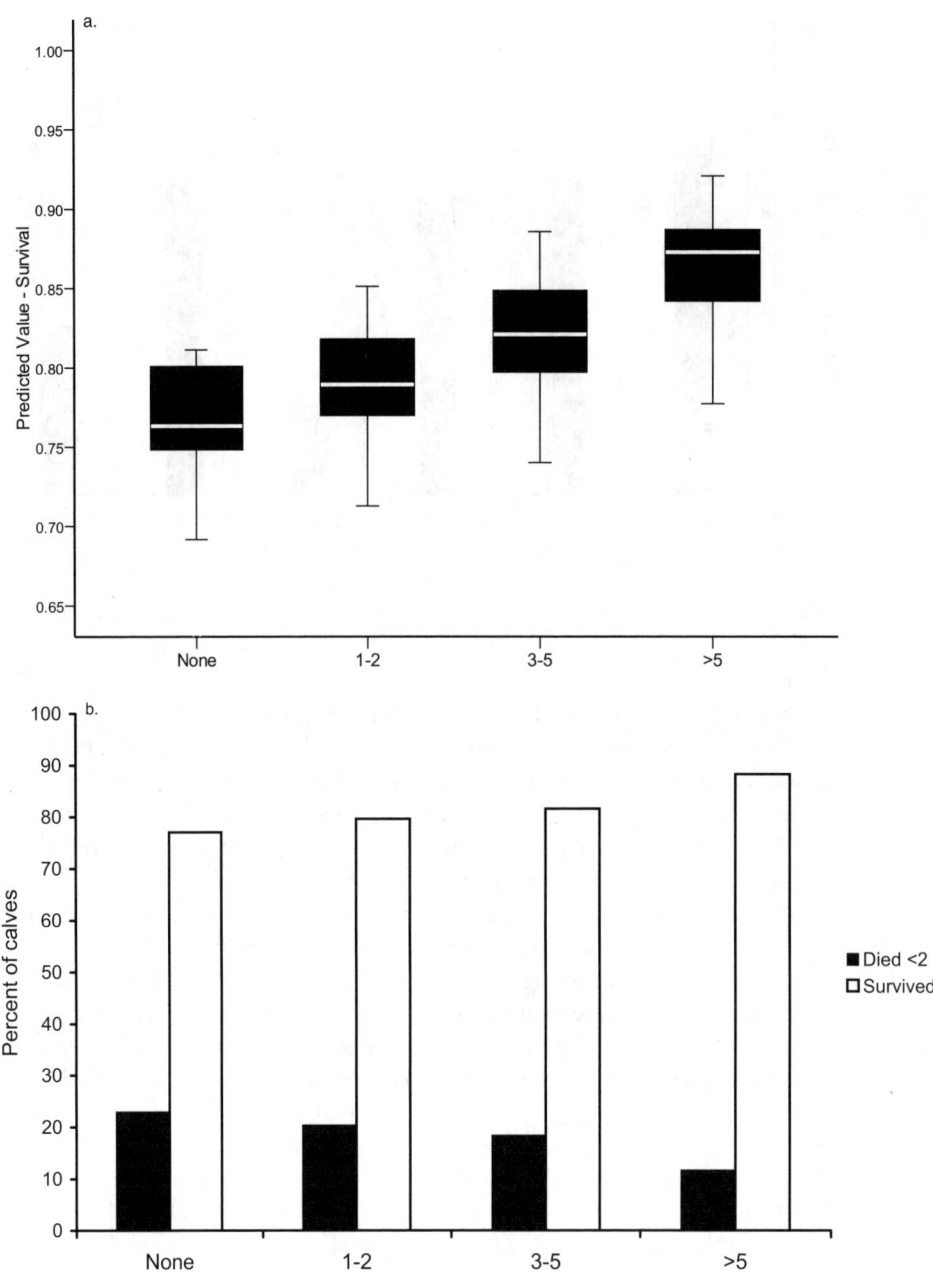

Figure 12.7 The effect of the number of allomothers on calf survival: (a) predicted survival probabilities from the stepwise logistic regression; (b) the percentage of calves that die or survive the first two years of life in relation to the number of allomothers.

(Lee, Lindsay, and Moss, chapter 6). As we noted above, older mothers had improved calf survival, as did larger families. The number of allomothers was correlated with overall family size ($r = 0.765$, $p < 0.001$, $N = 1,381$), as would be expected. However, once droughts and demographic effects of family size were taken into account, a remaining significant influence on calf survival in the first two years of life was the number of allomothers present in a family (stepwise logistic regression, removing dry season, family size, calf sex; number of allomothers: Wald = 8.7, $p = 0.03$; figure 12.7).

One of the problems faced by primiparous females could be a relatively low availability of allomothers for their calves, since allomothers are often older sisters. A female with her first calf obviously does not have an older sister of that calf available to help, while the other potential allomothers in the family may be busy tending to their closer

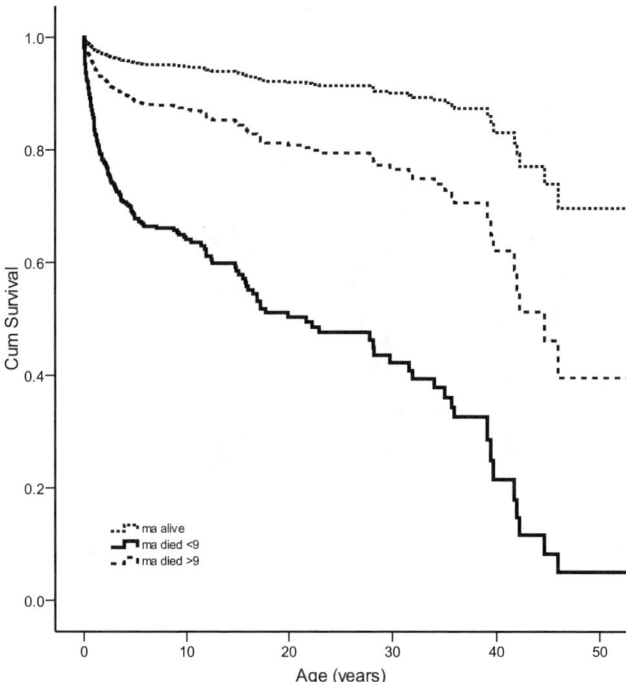

Figure 12.8 Probability of a daughter surviving when the mother has survived until the daughter is nine years old (youngest age of first known birth).

relatives. However, a primiparous female's own mother and grandmother could contribute to the survival of her calf if they are alive. When a mother survived until her daughter conceived and gestated and was well into the process of caring for her first calf (here taken as surviving until the daughter is 15 years of age), maternal presence increased the probability that the first calf will survive. Only 18 percent of first calves died with grandmothers living ($N = 207$) compared with the 31.5 percent average for primiparous females ($N = 129$, $\chi^2 = 70.5$, $p < 0.001$).

Having a matriarch for a mother somewhat improved the survival of a daughter to age nine ($\chi^2 = 6.7$, df = 3, $p = 0.08$) but did not affect the survival of her daughter's first-born calf ($\chi^2 = 0.8$, NS). Thus, maternal status may have an advantage in a cooperative sense: matriarchs improve the chances of their daughters surviving and therefore being able to reproduce, while the long-term survival of mothers, irrespective of status, significantly increases the survival of their daughters and their daughters' offspring ($\chi^2 = 28.6$, df = 2, $p < 0.001$; figure 12.8).

Causes of Calf Mortality

Overall, 19 percent of calves ($N = 1,579$) died in their first two years of life. Of mortality in the first two years, 23 percent occurred within the first month after birth and 52 percent died within six months; 89 percent was due to natural causes, while 4 percent was due to human intervention, and a further 7 percent due to human-caused maternal death. What are the main causes of this early calf mortality? There are several possible reasons for neonatal deaths: (1) the calf may have inherent defects; that is, we have recorded blind calves or calves with dysfunctional limbs; (2) the mother may not have been able to produce enough milk; or (3) the mother may not have had the experience to be able protect the calf from predators or other dangers. In Amboseli, hyenas appeared to be a significant predator of very young calves, while lions were major sources of mortality in Samburu (Douglas-Hamilton, personal communication). Calves also fall into wells or pits; furthermore, they can get stuck in mud. It takes considerable skill on the part of the mother and other family females to extricate them. Calves can become separated from their mother or other family members if the family runs from something that frightens them and the calf does not follow closely. In these latter cases, the role of an experienced mother, aided by a coalition of supportive allomothers, was a significant factor in reducing early calf mortality.

Inter-birth Intervals

In large, long-lived mammals with extended intervals between successive births, variation in maternal condition or status (Simpson et al. 1981; Gomendio 1990; Boesch 1997) and reproductive payoffs that differ depending on offspring sex (McFarland Symington 1987; van Schaik and Hrdy 1991) have been associated with variance in the length of the inter-birth interval (IBI). As we note above, the majority of models for understanding variance in IBI have been developed in the context of despotic or skewed reproductive systems. Relatively few studies have been done on species in which cooperation in infant care between reproductively active females potentially can affect IBI (Packer, Lewis, and Pusey 1991; Mann and Smuts 1998).

Maternal age and experience and calf sex are likely to be the major influences on patterns of elephant inter-birth intervals. The consequences of sex-specific care for the length of inter-birth intervals are most likely to be relevant when (1) variance in interval length is high; and (2) when a time delay to subsequent reproduction markedly reduces lifetime reproductive success. Hence, we explore three questions. First, what is the relationship between inter-birth interval and calf survival? Second, do female elephants allocate care differentially between sons and daughters, with repercussions for their subsequent fecundity? And third, does the allocation of care in a single reproductive event affect the ways in which a female allocates care over a reproductive career potentially spanning 50 years?

We analyzed 949 intervals of known length for 319

females. The median number of intervals per mother was 3 (range 1–7). The median interval length overall was 50 months, with extreme variance. The range of interval lengths was 22 to 114 months. A large part of the variance in length appears to depend on whether or not the calf died in the first two years of life (ANOVA, $F_{3,946} = 115.41$, $p < 0.001$). Post hoc comparisons of intervals following calf death or calf survival showed that mothers of calves that survived the first years of life had significantly longer intervals (53 months) than did mothers whose calves died within the first two years (37 months; figure 12.9). Conception of the subsequent calf appeared to be accelerated when the suckling stimulus was terminated by a calf death within 24 months. Average age of the first calf at the conception of its sibling was 32 months (interquartile range = 25–41 months) when the calf survived to over 24 months. By contrast, it took 17 months (IQ range = 7–24 months) to a subsequent conception when the calf died at less than 24 months. If the energetic costs of simultaneous lactation and gestation are high, then mothers may have to delay a subsequent gestation until after their calves' phase of peak growth velocity (0–22 months) and regain any condition lost during the early lactation phase.

If a calf died aged older than 24 months, the successive inter-birth interval was the same as that for calves who survived to five years or more (figure 12.9). Intervals when calves died within five years but *after* the birth of the next sibling ($N = 17$) were similar to those when calves died early in life. These intervals suggest that the mothers conceived again "too early" for their calves, and as a result the older calf died. There were occasions when the older sibling attempted to suckle alongside its younger sibling, again

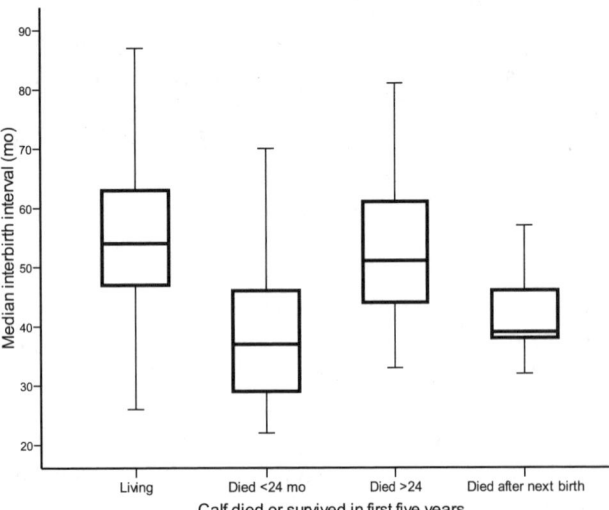

Figure 12.9 Median (IQ range and 95 percent CI) inter-birth interval for mothers of calves that survived ($N = 713$) or died as a function of the calf age at death (<24 months, $N = 176$; >24 months, $N = 40$; > next birth, $N = 17$).

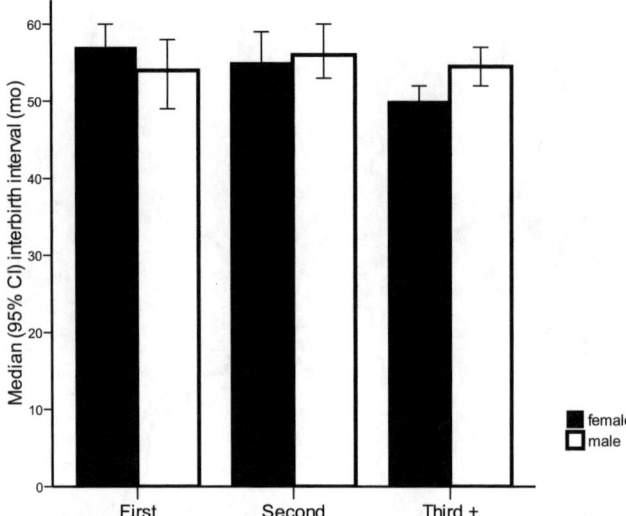

Figure 12.10 Median inter-birth intervals following the birth of male or female calves who survived >24 months by their birth order.

suggesting that the calf was not yet ready to be weaned. In some of these cases, the younger calf died as a result ($N = 3$), while in a few, the older weanling died ($N = 2$).

Inter-birth interval varied as a function of both calf sex and birth order (figure 12.10). Intervals following the birth of a male calf tended to be slightly longer (55 months) than those following a female (52.5 months) (ANOVA; hierarchical model using log-normalized data for surviving calves, sex: $F_{2,948} = 18.89$, $p < 0.001$). Inexperienced mothers of daughters had the longest intervals, while experienced mothers of sons had longer intervals than those following a daughter (ANOVA: sex and parity interaction, $F_{2,939} = 2.9$, $p = 0.056$). Post hoc analyses suggested that intervals for first-born daughters were longer than those for first-born sons ($p = 0.055$), while the opposite was true for experienced mothers of daughters and sons (daughters < sons, $p = 0.024$). Parity, rather than absolute maternal age, was the significant factor in these differences in interval length (maternal age as covariate, $F = 0.021$, NS).

Decisions for Multiparous Older Females

Mate Choice

For an animal that is going to produce relatively few calves in her lifetime, it is crucial that the fathers of these calves be the best available males. In Amboseli, females have been shown to exercise choice in mates (Moss 1983; Poole 1989b), specifically such that females chose to mate with older males and preferentially those in musth. Subsequent

DNA analysis has shown that 100 percent of fathers were over 25 years old, and 73 percent were in musth at the time of conception (see Hollister-Smith et al.; box 18.1). Moss (1983) suggested that there are both short- and long-term benefits to choosing these older males. In the short term, by choosing to mate with a large dominant male, the female will be protected from harassment and potential injury from other males. In the long term, the father of her calf will pass on any heritable traits for robustness, health, and longevity. If her calf is going to be successful itself it will need to live long enough to reproduce; therefore, longevity is a key trait for sons and daughters.

Mate choice, as is true for so much of elephant behavior, has to be learned. A young female during her first few estrous periods is usually chased by many males and frequently mated by whichever bull catches her. In Amboseli, we have recorded young females in estrus suffering a broken leg and a dislocated hip. Females soon learn to form a consort relationship with a large male, preferably one in musth who is dominant over other associating males, which helps to protect her from harassment from smaller sexually active males. A large, experienced adult female may barely get chased during estrus, thus saving herself (and often her youngest calf) the energy cost of running and the danger of injury.

Trade-Offs for Older Females

Once an elephant female has gained experience as a result of giving birth to three or more calves, what are her main strategies for achieving maximum reproductive success during the period when her fecundity is potentially at its highest? She has already survived to breeding age, and now her priorities are ensuring the production and survival of offspring. We suggest that there is a potential trade-off between producing as many offspring as possible, as quickly as possible, versus spacing these births to achieve maximum calf survival. Most elephants appear to use the latter strategy. However, since environmental and ecological cycles are unpredictable over the duration of a reproductive event that can last for three to eight years, the only "timing" issue relates to when a calf is conceived. Conceptions have a slight seasonal peak (Moss 2001; Lee, Lindsay, and Moss, chapter 6), while calf deaths occur during extended dry seasons or droughts, which are unpredictable. Thus a female, no matter how experienced she is, is unable to conceive a calf with a predictable energetic context for its birth or early development. Rather than being able to employ a consistent timing strategy, it is her own and her family's behavior after the calf's birth that will contribute to its survival and success. Old females (see table 12.2) continue to reproduce effectively until they die, since there may be little advantage to diverting investment specifically into the "last calf" (e.g., Williams 1979).

The majority of females over the age of 50, when fecundity begins to decline, have been seen in estrus; 24 females survived to 50-plus, and these females gave birth to 58 calves (figure 12.11). The oldest female to have given birth was estimated to be 64 years old (table 12.2). Older mothers have slightly, but not significantly, higher calf mortality (17.2 percent of those born to the oldest mothers die before 2 years by comparison to 12.5 percent for experienced prime mothers) and slightly longer inter-birth intervals especially after sons (49 months after a daughter [$N = 11$] and 63 months after a son [$N = 11$]). While the oldest mothers may have been extending their period of investment in their "last" sons, they were not biasing birth sex ratios (28F:29M). These old females used their social experience and ecological knowledge to respond to and care for calves rather than making "strategic decisions" in relation to their final years of reproduction. The unpredictability of the resource environment may limit a female's options for using a strategy that is age-specific, such as increased terminal investment in a son. However, behavioral experience increases with each successive reproductive event. Ultimately, a female's capacity to respond to changing contexts via behavior learned over repeated reproductive events may be the major mechanism for ensuring successful reproductive performance in old age (see also Dolhinow, McKenna, and Laws 1979).

Investment in Offspring, Adult Daughters, and Grandchildren

Females of all ages cooperate within a family to protect, aid, and rear calves. A long-lived mother increases the survival of her offspring, and especially that of daughters, to reproductive age (e.g., Hawkes et al. 1998 for human grandmothering). Elephant grandmothers can enhance calf survival by acting as allomothers specifically to their grand-calves, while at the same time continuing to reproduce themselves. In addition, there have been several cases of grandmothers sharing the suckling of their daughter's calf. We suggest that grandmothers function as repositories of caretaking knowledge, as do older matriarchs generally (McComb et al. 2001), and experienced individuals contribute to the reproductive success of the family females.

As a final note, early reproduction does not lead to shorter life span for Amboseli females in that no significant difference in longevity can yet be found between females who began reproduction early (<12 years, $N = 69$) and those who began late (>15 years, $N = 140$). Since these females in the sample are still only in their 30s, this finding is by no means conclusive. The lack of a trade-off between

Table 12.2 The characteristics and reproduction of the oldest Amboseli females over a 30-year period (1972–2002)

Name	N known offspring (1)	N known sons (2)	N known daughters (3)	Age at birth of final offspring (4)	Age at death or end of 2002 (5)	Estimated breeding life span (6)
Horatia	4	1	3	61	68	47
Freda	3	2	1	50	67	36
Slit Ear	7	0	7	58	66	44
Isis	6	2	4	63	66	49
Deborah	7	3	4	62	64	48
Phoebe	7	4	3	61	64	47
Lillian	5	3	2	56	64	42
Estella	6	1	5	51	64	37
Penelope	6	3	3	62	63	48
Teresia	3	2	1	58	62	44
Leticia	6	4	2	55	62	41
Jezebel	4	2	2	47	58	33
Wart Ear	5	2	3	48	57	34
Gloria	5	2	3	46	56	32
Wangoi	5	3	2	49	55	35
Sandy	5	3	2	50	54	36
Big T	4	3	1	52	53	38
Victoria	5	3	2	51	53	37
Karen	6	4	2	48	50	34

Notes: (1) The number of known offspring represents those either seen suckling in 1972 or born after the start of the study in 1972.
(2, 3) Sex of known offspring.
(4) Age of female at birth of final offspring.
(5) Age of female in 2002 or at death.
(6) Breeding life span is the duration from age 14 (median age of first reproduction) to age at birth of final calf.

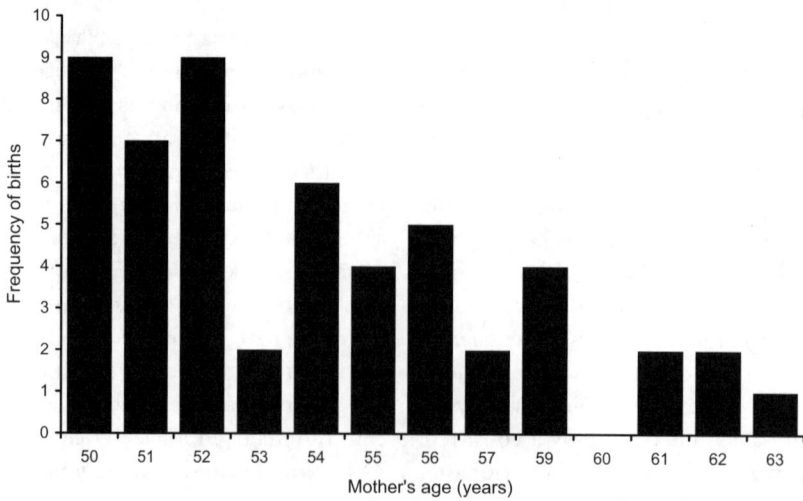

Figure 12.11 The number of calves born to the oldest females between the ages of 50 to 64 years.

Female Reproductive Strategies: Individual Life Histories 203

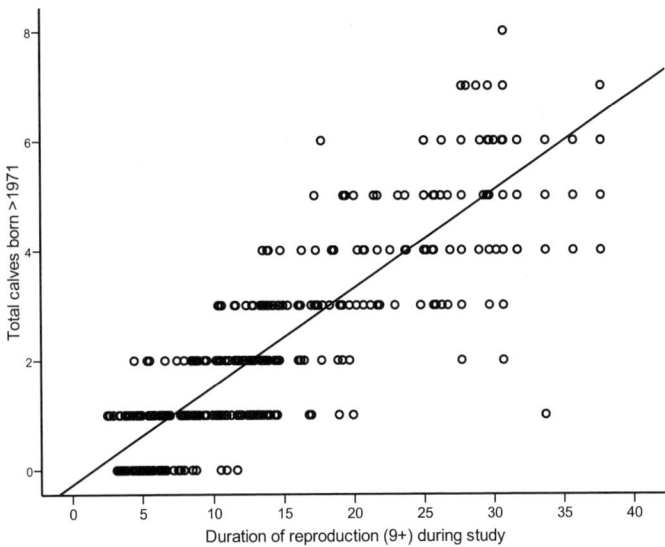

Figure 12.12 Longevity and offspring production assessed as the total number of calves born to individual females during the study period, plotted against the length of time that a female was alive and over nine years of age during the study.

early mortality and reproduction suggests that the longer a female lives, the greater the number of surviving descendants she will have (figure 12.12). As is true for so many large mammals, longevity is key to lifetime reproductive success for female elephants.

Conclusions

We have focused on three main questions in this chapter:

- How, as an individual, does an elephant female maximize her reproductive potential?
- How does the social context affect reproduction among elephant females?
- How does the individual's behavior and the social contexts for reproduction change over a life span?

An elephant female's reproductive performance is a function of her breeding life span—her age at first reproduction and her longevity. Life span is probably the most important determinant of elephant female lifetime breeding success. Age at first reproduction, the second determinant, is strongly influenced by size (a consequence of her early growth), but the social context is also important; females with surviving mothers reproduce younger and live longer than do females whose mothers are dead. Hierarchically determined status appears not to influence age at first reproduction or other elements of reproductive success (see box 12.2).

A third critical influence on reproductive success is calf survival. Calf survival is to some extent a stochastic variable strongly influenced by the ecological context of risk in the first two years of life: droughts, predators, or human conflict. However, it also depends on the mother's social context. If the family is large and successful, with many young females available to act as allomothers, calf survival during the critical dependent phase is higher. When the matriarch survives, is knowledgeable, and is able to share this knowledge across the generations of family members, again, calves survive. The context for female elephant reproduction requires cooperation within a family unit that shares the costs and care of calves. Hierarchical or status differentials among females within families (e.g., Archie et al. 2006) have a minimal influence on variance in fecundity among family females in the unpredictable environment of Amboseli. However, competition between families may play a role in the relatively higher success of some families and the near-extinction of others (see also Moss and Lee, chapter 13).

Calf sex is another component of reproductive success. The patterns predicted for sexually dimorphic polygynous mammals (e.g., Trivers and Willard 1973), with extended periods of investment in sons through longer inter-birth intervals and greater care allocation (Lee and Moss, chapter 14) are seen among elephant mothers. However, a female's experience of calf signals of needs and calf energetic requirements and, thus, her pattern of sex-specific investment varies throughout her reproductive career. Previous investment decisions and their consequences affect how a mother responds to subsequent events or challenges (Ono and Boness 1996; Lee and Moss 1999).

Inter-birth intervals, representing time allocation, were

biased in favor of first-born daughters as well as later-born sons. Since the number of allomothers available to a female influenced calf survival, investing in a first daughter who is both relatively cheap in terms of energy allocation for growth and who is more likely to survive to become an alloparent may be an effective strategy for enhancing female reproductive success, especially for smaller, young females. Experienced mothers have the physical condition and the knowledge to allocate time and energy to ensure that their sons' additional requirements for growth are met, with survival and reproductive consequences for both those sons and their mothers.

Thus, elephant reproductive events cannot be modeled without taking into account previous experience. Elephants tend to be far more egalitarian (by contrast to despotic or hierarchically competitive species; see also Hrdy and Hrdy 1976), and if a female "holds" status, this has little effect on her individual allocation of effort to reproduction. Rather than operating a "status-specific" reproductive strategy, elephant females assess contingencies based on both current and past contexts. These contexts are individual, ecological, and social: a female's body condition, offspring sex, her prior experience with sons or daughters, current food and water availability, and the family context of each event, such as the number of helpers, family size, and family dominance. As only some of these contexts are predictable, alternative behavioral decisions about investment change with each reproductive event over the extended life span. Experience accrues between the ages of 9 to over 50, and with it comes significant reproductive advantages.

Chapter 13 Female Social Dynamics: Fidelity and Flexibility

Cynthia J. Moss and Phyllis C. Lee

The largest and one of the longest-lived mammals on Earth has a fittingly large and complex social world. The multi-tiered social life of female African elephants is based on relationships that radiate from an individual and her immediate offspring out through other family members, bond group members, clan members, and sub-population members to the whole population and even further, to strangers (Moss 1981; Moss and Poole 1983; McComb et al. 2001, 2003; Wittemyer, Douglas-Hamilton and Getz 2005). Males live an almost entirely different social life once they reach independence but have an equally large network of relationships throughout the population and beyond (Poole 1982; Moss and Poole 1983; see also Lee et al., chapter 17).

In this chapter, we describe female elephant social organization and dynamics at the individual, family, and bond group levels in the Amboseli population and examine both the consistency and changes that have occurred over a period of 30 years. By analyzing association and grouping patterns among female elephants, we demonstrate the nature of the flexible fission-fusion society that allows for individual responses to environmental, demographic, and social conditions. Although many of these responses are based on costs and benefits for the individual, the social responses of elephants are far more complex than those predicted by a simple cost-benefit analysis at a single point in time. Females make social decisions over long periods of time as well as respond to social and ecological opportunities and conditions on a day-to-day and even an hour-to-hour basis.

The aims of this chapter are threefold:

- We describe the family, its nature, and socio-spatial dynamics. We then examine how families are embedded in the larger social context of the Amboseli population. With this aim, we describe the associations within and among families in the population and identify specific persisting social bonds.
- We explore the ecological and demographic factors that influence family dynamics, and we investigate the interactions and influences of environmental and social factors over the long term. We ask how resource availability affects female social decisions.
- We examine how significant demographic events such as the death of a matriarch influence family social structure and dynamics over time.

Family Units

Since much of the analysis in this chapter is done on the basis of a family unit, it is worth reiterating what the Amboseli study means by family unit. The term *family unit* was first used with regard to elephants in a study by Buss, started in 1961. Buss and Smith (1966, 376) provided this definition: "The term *family* unit refers to an adult female and its offspring, or two or more closely related females and their offspring. These units usually number 4 to 15 individuals, their activities are closely coordinated, and large bulls seen with them are generally attached loosely and temporarily." In 1972, after four years of study using individual recognition, Douglas-Hamilton (1972, iv) confirmed that "[t]he family unit [is] the basic social unit of elephant society." Other researchers around this time also described the elephant family unit (Laws et al. 1975; Sikes 1971). Studies in Samburu Game Reserve, Kenya, by Wittemyer, Douglas-Hamilton and Getz (2005) and Wittemyer and Getz (2007) rigorously analyzed individual association patterns and established statistically what the Amboseli study described originally in 1981 (Moss 1981): elephant society is structured into six levels starting with the mother-offspring unit, to family unit, bond group, clan, sub-population, and population.

The Amboseli study was started in 1972 with the premise that African savannah elephants lived in family units. We did not intend to prove that family units exist but rather undertook to understand more about elephant social dynamics. Our working definition was a *family unit* is defined as one or more adult females and calves with a high frequency of association over time relative to any other females in the population, who act in a coordinated manner and exhibit affiliative behavior toward one another (Moss and Poole 1983; Moss 1988). We use the term family and family unit interchangeably here.

We emphasize that a *group* of elephants is not the same as a family. A *group* is defined (see box 1.1) as any number of elephants of any age or sex moving together in a coordinated manner with no single member or sub-group at a distance from its nearest neighbor greater than the diameter of the main body of the group at its widest point. An *aggregation* is a group that consists of at least two families and is commonly used when the group is large (e.g., over 50) and consists of several families.

From the study's beginning, every time a group of elephants was sighted individuals were identified, photographed, and characterized by specific written descriptions. Using the individual records, it was possible to analyze who was associated with whom at the level of the individual as well as the family. In addition, group size, its composition, the behavior of the majority of group members, and its location were all recorded in routine sightings (see appendix 1).

The quantitative data on who associated with whom were the main determinants of a family unit, but there were other important qualitative attributes. A key characteristic was the general behavior of the animals sighted. Did they move and behave in a coordinated way? Did they appear to have a leader to whom they oriented or responded? It was more often than not obvious which individuals belonged to a family. These individuals coordinated their behavior so that the majority of them were engaged in the same activity (for example, feeding, walking, drinking, or resting). Affiliative behavior, such as greeting or touching one another, and cooperative behavior, such as care of infants who were not their own, added more evidence. When the same individuals were seen over and over again in the same group and exhibiting the characteristics of membership we concluded that they were a family.

By January 1979, 54 families had been identified using the above definitions (table 13.1). No doubt a few errors were made. In the majority of cases, however, the family composition was clear-cut. For example, we could say with confidence that the 13 members making up what we called the AA family (the first elephants photographed in the course of identification) were indeed a *family unit* by our definition. These 13 elephants were found with each other at a higher rate than they were with any other individuals in the population. As of January 1979, when all the families had been recognized, the AA family had been sighted 201 times. Of these sightings, 90 (44 percent) of the groups in which they were found included other females and calves. In all the other sightings, they were on their own. Out of the 201 total sightings, 83 had the highest quality of count and recognition. For those sightings that yielded partial or full censuses, the whole complement of AA adult females were present in all but eight sightings. When observed together, the individuals we considered members of the AA family coordinated their behavior and showed affiliative and cooperative interactions. The members oriented and took direction from the oldest female in the group. When they joined other elephants to form larger groups and then broke down again into smaller groups, the same individuals remained associated.

Exceptions to the AA pattern of affiliation were evident in families that were not cohesive and well coordinated. It was not so much that an individual may have been assigned to the "wrong" family but that certain individuals appeared to be more loosely associated with their putative family. One possible explanation for less evident cohesion is that Amboseli experienced poaching in the 1960s and 1970s as well as traditional and political spearing by the local Maasai people (see Browne-Nuñez, chapter 19; Kangwana, chapter 20). Through adult mortality some families ($N = 5$) ceased to exist, with some leaving a single adolescent female or a number of immature animals on their own (see box 13.1). These "floaters" either struggled on their own or attempted to join other families. When they did join another family, they tended to come and go, changing the composition of the family each time they left or associated with another. In other cases certain families were not internally cohesive and the individuals were not consistent in their associations. The DB family was one of these, as was the FB family (see box 13.1).

Some families had association levels between the adult female members that were close to 100 percent of observations, while most were over 50 percent (see below). Today, with splits and temporary fissioning, there is hardly a family that reaches 100 percent. One could say those "families"

Table 13.1 The histories of known families over the course of the study

Family	1st Record	Size	Births	Deaths	Matriarch 1st Death	Matriarch 2nd Death	Matriarch 3rd Death	Split	Fusion	Immigration	Number independent males	Demise	Size at end 2002
AA/AC	Sep–72	13	61	32	Sep–74	Mar–97		Y	0	0	6		36
AB	Sep–74	12	18	18	Feb–91	Sep–97		N	0	0	1		11
BA	Sep–72	3	1	4	Jun–75			N	0	0	0	Jun–75	N/E
BB	Dec–73	11	53	21	May–86	Mar–89		?	0	0	8		35
BC	Aug–75	11	24	14	Dec–75	Feb–88	Mar–92	N	0	0	5		16
CA	Dec–73	9	11	9	Dec–75	Feb–76		N	0	0	3		8
CB	Oct–73	13	35	19	Jun–76	Aug–76		Y	Y	0	6		23
DA	Sep–72	6	4	8	Apr–75	Sep–91		N	0	0	2	Sep–91	N/E
DB	Sep–72	5	23	8	*			Y	0	Y?	2		18
DC (F)	N/E	W/DBQB 6	11	4	*			N	Y	0	2		11
EA	Sep–72	12	43	11	*			N	0	0	11		33
EB	Nov–73	6	39	13	*			N	0	TEMP	7		25
FA	Sep–72	8	26	21	Jun–82			N	0	0	1		12
FB	Sep–72	5	33	16	*			N	0	0	3		19
FC	Aug–80	1	4	2	*			N	0	0	1		2
FD	Jan–73	3	22	13	Dec–85	Mar–95		N	Y	2	3		11
GA	Feb–73	4	0	4	by 12–74			N	0	0	4	by 12–74	N/E
GB	Mar–74	14	47	13	Dec–00			N	0	0	11		37
GC	Jan–78	5	14	6	May–81	Nov–88		N	0	0	2		11
HA	Oct–73	3	14	11	Apr–87	Sep–01		N	?	0	2		4
HB	Aug–75	4	17	5	Mar–02			N	0	0	0		16
IA/IC (F)	Sep–72	6	16	11	Apr–77			N	Y	1	1		11
IB	Mar–74	13	42	13	*			N	0	0	7		35
JA	Feb–74	11	27	14	Nov–93			N	Y	Y1	7		18
JB	Dec–73	6	14	10	Jan–88	Oct–95		N	?	0	0		10
KA	Oct–73	11	31	26	May–75	Sep–97	Oct–00	N	0	0	3		13
KB	Mar–76	8	37	14	Apr–01			N	0	0	3		28
LA	Mar–75	7	14	10	Nov–98			1979–2	0	0	2		7
LB	Jan–76	11	34	5	Apr–76			1999–3	0	0	5		32
LC (S)	Mar–79	W/LA 2	13	2	*			N	0	0	2		11

(continued)

Table 13.1 (continued)

Family	1st Record	Size	Births	Deaths	Matriarch 1st Death	Matriarch 2nd Death	Matriarch 3rd Death	Split	Fusion	Immigration	Number independent males	Demise	Size at end 2002
LD (S)	Mar–99	W/LB 3	18	2	*			N	0	0	2		17
MA	Sep–73	6	18	13	*			N	0	0	2		9
MB	Jan–77	13	53	18	Oct–85			N	0	0	2		46
NA	Nov–73	7	0	7	by 12–74			N	0	0	0	by 12–74	N/E
NB	Sep–74	4	3	5	Jan–77	Dec–96		N	0	0	2	Dec–96	N/E
OA	Oct–73	7	32	13	Dec–74			N	0	0	2		24
OB	Feb–77	5	23	15	Jan–01			N	0	0	0		13
PA	Oct–73	14 (6)	53	16	Aug–01			1982	0	1	10		42
PB	Sep–74	7	7	9	Dec–74	Jan–84		(unknown)	0	0	2		2
PC (S)	Nov–82	W/PA 6	42	13	*			1982	0	0	7		28
QA	Oct–73	2	1	3	Jun–77			N	0	0	0	Jun–77	N/E
QB	Oct–73	7(9)	32	11	Depart x 2			<2/79 1996	0	0	1		26
QC (S)	Jan–95	1	4	0	*			N	0	0	1		4
RA	Oct–73	9	30	12	*			N	0	0	7		20
SA	Sep–72	9	30	17	Dec–74	Jan–97		N	0	0	2		20
SB	May–74	5	17	8	Oct–77	Feb–85		N	0	0	2		12
TA	Sep–73	7	19	13	Dec–76	Jan–97		N	0	0	3		10
TB	Sep–73	4	2	6	May–84			N	?	0	0	May–84	N/E
TC	Sep–73	6	24	9	*			N	0	0	3		19
TD	Sep–73	7	18	5	Jun–84			N	0	0	4		16
UA	Oct–73	6	30	9	Dec–75			N	0	0	4		23
VA	Dec–73	17	69	22	Apr–88	Dec–90	Oct–00	?	0	0	11		52
WA	Mar–75	5	17	6	Dec–76	Aug–02		N	?	0	4		12
WB	Jan–77	9	52	10	Mar–98			N	0	0	2		49
XA	Dec–73	5	15	5	Oct–76	Sep–77		N	0	0	3		12
YA (F)	Oct–73	3	8	7	Aug–76	Jan–01		N	Y	0	1		3
ZA	Jan–76	12	30	19	Jun–77	Dec–86	Dec–89	N	0	0	8		15
ZB	Jan–79	10	25	10	Jul–84	Dec–91		N	0	0	5		20
ZC	Feb–78	1	7	0	*			N	0	0	0		8

Notes: 1st Record = first sightings, Size = number at first sighting; births and deaths are the total of known events over the period. Major events such as death of the Matriarch (1st, 2nd, 3rd death dates), * no matriarch death yet), splits between members of the family (Y = known to happen, N = none, date = when if known), or fusions with other families (Y = yes, N = 0, ? = possible), the number of males who went independent, date of family extinction, and family size at the end of 2002 are all presented.

BOX 13.1 SOME FAMILY HISTORIES OF FISSION AND FUSION
Cynthia J. Moss

FA: In 1982, the FA family's matriarch died, leaving two adult females, two juvenile females, and three calves. For nearly a year afterward, the family was split in three. The two adult females, Fiona and Fifi, formed separate mother-offspring units, and the juvenile females foraged with one or the other of the two adult females, and sometimes with other families. Prior to the matriarch's death, the FA family was one of the most stable. By 1985, they had coalesced and moved together as one. After Fifi died in 1998, the FA family became even more cohesive with the one large adult female, Fiona, acting as matriarch.

CB: The CB family lost its matriarch in June 1976, and the next-oldest adult female was killed in August of the same year, leaving four teenaged females—Chloe, Celeste, Raggedy, and Calandre—to lead the family. The family was in disarray for some time, but eventually Chloe and Celeste, with Chloe being marginally more matriarchal, took charge. Eight years later, during a period of severe drought and Maasai spearing, Calandre was killed, and seven years later in 1991, Raggedy disappeared. After Raggedy's death, a section of the CBs broke off to form their own family. This sub-group consisted of two females—Camilla and Cerise—and their calves plus Charlotte, the daughter of Raggedy. From their ages, I speculated that Camilla and Cerise were Raggedy's sisters. The sub-group remained independent for four years, but in 1995 Camilla died. After her death, Cerise and her calves and Charlotte rejoined the CBs and in 2002 appeared be fully integrated in the family once again. This fusion (or re-fusion) occurred when the family was relatively large, with close to 20 members (see map, figure 7.11).

PA/PC: By the early 1980s, the largest family in the population was the PAs, with 28 members. This family gradually began to split into two groups, but the composition of the two sub-groups was not always the same. Eventually the sub-groups became consistent, and by 1983 they were considered two separate family units with the two oldest females—Penelope and Phoebe—each leading a family. The two families have remained split to date and by 2002 were rarely found together. The PC family has been stable, but the PA family, as it grew, began to fission again and in 2002 was more often than not in two or three sub-groups. Not only were the PAs a large family but they also had at least one unrelated member who had immigrated into the family and was never fully assimilated (see map, figure 7.12).

DB/DC: The DB family was only moderately cohesive in the 1970s. In 1975, it consisted of seven animals: two large adult females, Deborah and Delia, two young females, one immature female, one first-year calf, and a pubertal male. In the next three years one of the young females and the calf died, and the pubertal male became independent. Deborah gave birth to a calf in 1977. Even with the reduction in numbers, the levels of association between the remaining individuals and Delia was not high. Then in late 1978, Delia gave birth to the first calf born to the Amboseli population for 16 months. From the day of the calf's birth, the QBs, who were from the western subpopulation and rarely associated with the DB family, began to spend time with Delia and the calf. A close relationship built up between Delia and the matriarch of the QBs, Quilla. At first Deborah was intolerant of Quilla and the other QB females and often threatened and chased them. They, however, initiated affiliative greetings whenever they approached the DB members, and eventually the DBs responded in turn. However, despite their apparent acceptance of Quilla, Delia left the DBs to move with Quilla and her family. After a few years, Quilla also left her own family, taking only her two daughters with her, and she and Delia formed a new family, the DCs. They remained together and moved to the opposite end of the Amboseli population range from the DBs. This case is the only one of its kind in the 30-year study (see map, figure 7.13).

JA: The JA family was relatively cohesive, but one individual, Joan, was not as consistently with the matriarch and the rest of the family as the others. We suspected that she was what we referred to as a "floater," an individual who had lost her family through deaths and was on her own before attempting to join another family. Joan joined the JAs and remained in this family until her death in 1991. She gave birth to three calves, only one of which survived. This individual, Jody, is still with the JAs and now seems to be well integrated in the family. She has given birth to three calves, two of which have survived. However, genetic analysis of this family revealed that Jody is not related to anyone else in the JA family, and in fact she is from a totally different matriline (Archie 2005). This information confirms that an individual can permanently join non-related families, although the genetic analyses also reveal that this kind of fusion appears to be a rare event.

that include dyads under 50 percent should not be considered a family unit. However, there are difficulties with this type of arbitrary cutoff, because there is history and relatedness. The DNA analysis of relatedness (Archie, Moss, and Alberts, chapter 15) has added greatly to our ability to confirm membership in a family unit and understand levels of cohesiveness. Thirty years ago the females in a particular family might have had a high association level. Today they may not. With difficulty, we have arrived at the following criteria to create new families: *Females only cease being considered as a member of their original family unit when there is a long-term split of at least two years, where previously associated females are no longer seen regularly in the same group, and when the association patterns among individuals after the split are consistent.*

Thus, families grow, change in composition (through births, deaths, and departures of young males), split, and very rarely fuse. Table 13.1 presents the demographic histories of the 61 Amboseli family units. A family that started with 2 adult females in 1974, such as the EB family, had 9 adult females by the end of 2002. They were a very cohesive family in the 1970s and 1980s. By 2002, they were less so but showed no sign of permanent fission. Another family that started with 20 members in 1973, the PAs, permanently split into two and then again since that first fission.

Bond Groups

Elephant family units join other families for periods of time ranging from less than an hour to several weeks. Families are considered associated if they are found together in the same group, as defined above and in box 1.1. Those families that frequently associate in the same group and whose members regularly and repeatedly exhibit affiliative behavior, such as greeting and touching, are considered *bond groups*. Association levels between families designated as bond groups are higher than those with any other families in the population. No particular cutoff point for association index was assigned because the definition was based on both relative association levels *and* behavior. Crucially, the members of bond group families, in particular the adult females, perform greeting ceremonies (Moss 1988) with one another, and when the families are together, they exhibit coordination in movements and activities.

Bond groups can form when a family becomes large and a portion breaks off to form a new family. The two resulting families then may spend more time with each other than with other families in the population (Moss 1988). The DNA analysis has revealed that the members assigned to a bond group are usually related (Archie et al. 2006; see also Archie, Moss, and Alberts, chapter 15), particularly the oldest females in the two families. However, the prediction that the two new families would spend more time together has not always been fulfilled (box 13.1).

Fission-Fusion Sociality

The term "fission-fusion" sociality has been applied to a number of mammal species. Initially defined by Kummer (1968), the term was used to describe the fluid structure of hamadryas baboons, which joined in huge bands at night to sleep together on cliffs and foraged during the day in small one-male units or associations of one-male units. The term has been applied to chimpanzees (Lehmann and Boesch 2004), spider monkeys (Chapman, Wrangham, and Chapman 1995), sperm whales (Whitehead 1995, 1997), dolphins (Connor et al. 2000), lions (Schaller 1972), and hyenas (Holekamp et al. 1997), among other species with flexible grouping patterns.

In the case of elephants, fission and fusion as a social and ecological strategy can be seen in short-term grouping dynamics as well as in long-term and permanent associations. Elephant females are group-living, but the size and composition of the group that each female finds herself in varies, with families temporarily splitting and coming together again, families joining other families, and families associating with adult males from time to time. In this way, there is continuous fission and fusion in the daily social life of elephants. In the longer term, a family may permanently split off as a sub-unit of individuals forming a new family. More rarely, a family might fuse permanently with another family to form a family with a new composition.

Measures and Terms Used in Analyses

The analyses in this chapter are based on two different methods for recording individual elephant presence: sightings and census data. Sightings data (box 1.1 and appendix 1) were used in analyses of group size, group type, and associations between family units. Groups containing females were categorized as cow-calf (no independent males present) or mixed groups (independent males present); these two types have been analyzed separately. Censuses of individual females present at specific sightings of families were used in the analysis of female-female associations, while associations among families are based on two or more families' co-occurrence within a single group.

As shown in table 13.1, 61 families have been recognized of which 7 no longer exist because of the death of their members. By the end of 2002, there were 55 extant families, including 5 families that were created by observed splits and two by fusions. We approached the analysis of the 61 families and the 30 years of data in several ways in this chapter.

Of the 61 recognized families, 52 had over 200 sightings in aggregations (mean N sightings = 828±358), and these 52 were used in the analysis of associations among families. Families with over 700 sightings each (N = 30) were used in analyses of gregariousness (time spent with other families or alone). For 15 families, data were available on the cohesiveness of the family (sightings when all family members were together in the same group) over the whole study period, while cohesiveness for a total of 30 families could be calculated from 1982 to 2002 since some families were either recognized later in the study or were observed only infrequently in the first decade. Each family experienced a variety of demographic events over the 30 years, such as the loss of a matriarch or other adult females with variable and complex social responses. To explore these reactions, the analyses have been divided into decades in order to avoid masking short-term consequences with the full 30 years of data.

For 10 families, individual associations between females could be examined in detail over time. Comparisons in these families were made between two periods: 1975–76 as a major drought period with few births and high mortality, and 1978–79 as a period of high rainfall and food availability, with many births. These two early Amboseli Years (see chapter 2) were chosen to minimize any potential effects of changes in family size on the female-female associations. As noted above, we use the term cohesiveness to refer to the tendency for individuals within a family to be found all together. Cohesiveness measures the propensity for all (female) members of the family to be seen together in the same group, and the data are drawn from both the sightings records (which categorized family unit presence as all or partially present) and census data (where females were individually recorded as present or absent).

Sorenson's index of association (a half-weight index; appendix 1) was used to assess the "strength" of relationships both among females and between the different family units in the population. As we were specifically interested in the choices of partners or associates when individuals are "social," this index was used rather than a simple index of association, which assumes that females are symmetrically observed. From the association indices and the percentage of sightings, we defined family gregariousness as the tendency for a family to associate with other families in groups. Solitary families were those that were seldom sighted in association with other families.

Association Patterns among Females

The Amboseli population, at just over 1,500, is relatively small compared with some of the major populations in east and southern Africa. For example, the Selous Game Reserve in southern Tanzania at one time held 110,000 elephants, while more than 42,000 were found in the Tsavo ecosystem in Kenya. Nevertheless, 1,500 individuals is a large "society." We use this term intentionally, as it is likely that an individual elephant in Amboseli knows every other adult elephant in the population (McComb et al. 2001, 2003; see also McComb, Reby, and Moss, chapter 10). The 30 years of sighting records show that each family has been found in the same group with every other family in the population at least once. While simple propinquity is not the same as sociality, the adult females are making choices about whom to associate with and whom to avoid among the population, and thus we consider the population as a social system or society.

Female Associations within Families

Female savannah elephants are highly social, living their lives in the company of other elephants: their own calves, other adult females, those females' offspring, and adult males. The stable and central core of this social life is the family unit into which male and female calves are born. Females typically stay in their natal family throughout their lives, although splits of families may occur. Males leave the family between the ages of 8 and 16 years old (Lee and Moss 1999; see Lee et al., chapter 17). Family units thus consist of closely bonded and, with very few exceptions, closely related adult females and their offspring (Archie, Moss, and Alberts 2006; see Archie, Moss, and Alberts, chapter 15). Relatedness is further enhanced by offspring within a family potentially having the same father.

Family unit size ranged from 2 to 52 and has changed considerably over the course of the study with the average family increasing from 7 in 1976, when most of the families had been registered, to 19 by the end of 2002 (figure 13.1). The number of mother-calf units (an adult female and her immediate offspring) has also increased (table 13.2).

Social Dynamics within Families

Membership in a family unit does not mean that an individual will always be present with the other members of her family or that the family will have the same composition every time it is seen. Associations among the adult females were explored to ask whether individuals were more likely to be found with those females defined as members of a family than to be found in association with females from other families. Within-family indices of association were very high. Females spent 70 to 100 percent of their time in close association with other females from their family. For 9 out of 10 families, the index of association between adult females was higher in the high rainfall and food abundance year than in the drought year, reflecting a tendency for families to be less cohesive during times of ecological stress (table 13.3 and see below).

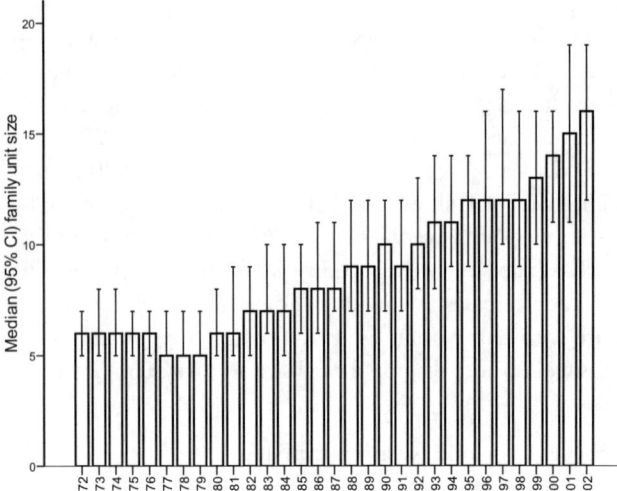

Figure 13.1 Mean family unit size (±2 SE) by year.

Table 13.2 Summary characteristics of family units in Amboseli between 1976 and 2002

	1976	2002
Number of family units	52	55
Range of family unit sizes	2–23	2–52
Mean FU size	7.22 (SD ±3.8)	19.13 (SD ±12.3)
Number of adult females in each family	1–9 (mean 2.35; median 2)	1–19 (mean 7.08; median 6)

Table 13.3 Mean Sorenson's index of association among adult females within selected families for 1976 and 1979

Family	1976 Association index-drought conditions	1976 Family size	1979 Association index-good conditions	1979 Family size
AA	.873	11	.973	10
BB	.792	11	.978	10
EA	.706	12	.984	11
EB	.914	6	1.000	8
FA	.885	6	1.000	7
KA	.766	7	.925	7
LA	.715	6	.904	6
PA	.580	23	.860	21
RA	.706	9	1.000	8
SA	.886	7	.777	7

Cohesion within Families. Although the day-to-day grouping patterns of families were flexible and dynamic, the composition of each family unit was far more consistent. Immigrations into families appeared to be extremely rare. A sub-unit of the family might temporarily split off to feed on its own for a few days, but the sub-unit very rarely split permanently and only rarely did it fuse permanently with another family. During the study up until the end of 2002, only six family units (AA, DB, LA, LB, PA, QB) experienced permanent splits, which resulted in the formation of new family units (AC, DC, LC, LD, PC, QC) (see table 13.1). Eight more families are currently in the process of fission.

Pubertal male family members tended to join and leave their natal family repeatedly during the year or so before they became independent, and their presence or absence changed the family composition from sighting to sighting (see Lee et al., chapter 17). Males in the process of dispersal were excluded from analyses of family cohesiveness.

Average cohesiveness for families ranged from 15 to 95 percent, with a mean of 54.7 percent ($N = 30$). Year by year, cohesiveness could vary from 10 to 100 percent for an individual family. For some well-known families such as the EBs, GBs, and TDs, average cohesiveness was well above 50 percent (figure 13.2) and varied little over time (figure 13.3), while for two families, the DBs and FBs, cohesiveness dropped in the second decade and was then restored in the third decade.

Small families with only one adult female were very cohesive, with most of these families sighted with all members present 100 percent of the time. By comparison, larger families were significantly less cohesive ($r = -0.822$, $N = 30$, $p < 0.001$). Cohesiveness declined for 21 out of 30 families over the study period, which was probably associated with the general increase in family unit size over time (e.g., figure 13.1).

A number of factors may act independently or in combination to determine the degree of cohesiveness of family units. Some of these factors are environmental (see below), while others are demographic and social. Stability within the family is disrupted by the death or illness of a matriarch or other family members. The overall increase in family unit size may make it harder to coordinate activities and associations, while the birth of calves can either increase cohesion when many females share the same needs for food and water or promote the breakup of associations when the needs of females differ or conflict. Reproductive state of the female, including estrus, is thus an important short-term influence on grouping dynamics. When females are in estrus, they either attract numbers of bulls to their group, or very occasionally, they will leave their family and associate with bulls (see Poole et al., chapter 18). Estrus is a relatively rare event (experienced for 3 to 10 days once every four

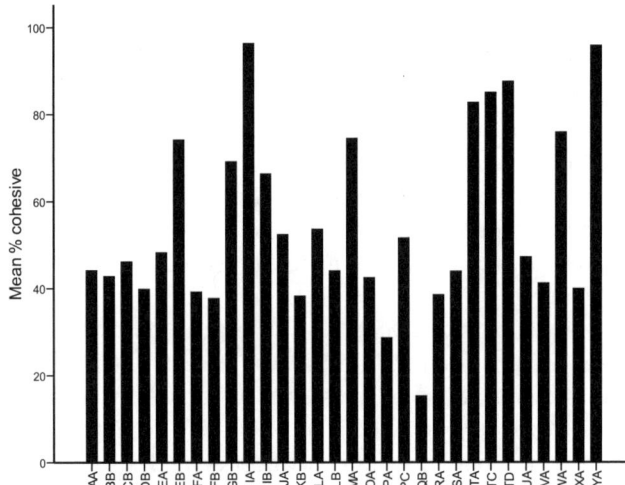

Figure 13.2 Mean family unit cohesiveness (percent sightings with all females present in the same group) for 30 families with more than 700 sightings.

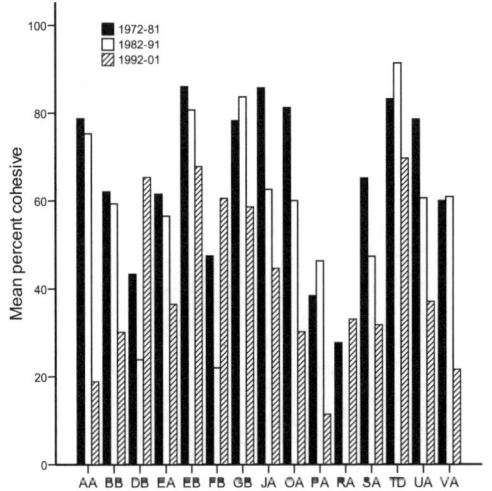

Figure 13.3 Mean family cohesiveness by decade for 15 families that were well known from the start of the study.

years), and thus its effect on family cohesion or gregariousness is likely to be very short term. Finally, the presence of females not born in the family but bonded to family females may lead to sub-grouping and again to loss of cohesion.

The varying levels of cohesiveness over the decades suggest that environmental conditions alone were not the overall determining factor since all the families experienced the same ecological conditions. Demographic events, combined with subsequent increases in family size, appear to have played a more important role for those families whose cohesiveness declined over time. For the two families whose levels dropped in the second decade and then rose again in the third, other social and demographic factors may have come into play (box 13.1). In the case of the DB family, one adult female (Delia) left after a transitional period. In the FB family, two adult females and several calves were killed by spearing in the third decade, which reduced the size of the family. The spearing and losses appeared to consolidate the remaining members.

Loss of Matriarchs. Most family units have lost their matriarchs at some point during the course of the 30-year study, and all but two of the families have lost at least one female over the age of 9 (potential age of first breeding). Only 14 groups retained the same matriarch as was present in 1972. Four families have lost 3 matriarchs over the period; 23 have lost 2 matriarchs; 18 have lost 1 (table 13.1). Although there was no direct relationship between the matriarch's loss and the cohesiveness of the remaining members over the next decade, there was an interaction between cohesiveness, family unit size, and loss of matriarch. If the matriarch died and the family was left with one adult female and her offspring, that unit remained very stable. If, on the other hand, a matriarch died leaving more than one adult female in the unit, the family tended to break up into mother-offspring units and thus lose cohesion (see table 13.1). Often this type of fission occurred for several months and even a year or more after the death, but then the family regrouped.

Permanent Fission of Family Units

Generally, long-lasting permanent splits occurred when a sub-unit within the family gradually spent more and more time away from the designated matriarch's section of the family until that sub-unit was found with its former family members at the same "background" level, as with any random family in the population. We have previously speculated that this type of fission—when two family units are formed from one—is the mechanism by which bond groups are created (Moss and Poole 1983; Moss 1988). Up until the early 1980s, Amboseli family units remained intact, with the exception of the unusual event of Delia and Quilla leaving their families and joining to create a new family of their own (see box 13.1). The first record of fission was the largest family in Amboseli, the PAs, who at 28 members broke more or less in two. This split appeared to make social and ecological sense: the family became too big to maintain coordinated travel, feeding, resting, and infant care, and as a result the social binding forces weakened to the breaking point. However, this split of a large family in almost neat halves is the only time that it has occurred. Elephant social life turns out to be even more intricate than we had previously described, and fission in families appears to come about under a variety of circumstances, is often not predictable, nor does it always make sense to observers. Much of the biologically idiosyncratic behavior that we observe could be related to individual characteristics or elements of personality (see box 13.2).

BOX 13.2 PERSONALITY IN ELEPHANTS
Phyllis C. Lee

Personality in animals has received recent attention as a possible mechanism for understanding animal welfare and well-being in captivity (Weiss, King, and Hopkins 2007) and as a descriptor of individual traits underlying adaptive responses to those social conditions that require co-operation or operate via games strategies (Dall, Houston, and McNamara 2004). Adjectives that describe behavior or aspects of temperament and that have been applied to the study of individual characteristics or personality in a variety of non-human primates (Stevenson-Hinde and Zunz 1978; Murray 1995, 1998) were used here to explore whether elephant females could be statistically discriminated on the basis of consistent individual traits or "personalities." Four raters were asked to score individual female elephants on each of 27 adjectives that either describes traits related to sets of behavior (e.g., "playful," "protective") or characteristics such as "irritable." Each rater scored the females independently, using a 7-point Likert scale (from least applicable, seldom seen, to most applicable, most common).

The four observers rated 11 adult female individuals (aged over 10 years) from the well-known EB (Echo) family unit. These observers all had considerable familiarity with the family over a period of years. In order to explore whether raters were consistent in how they internally defined and then applied these descriptive adjectives to each elephant, scores were

Table A

	C1		C2		C3
(% Variance)	46.9%		28.4%		12.9%
	Sensitive to insecure		Slow to playful		Irritable to gentle
Sensitive	.948	Slow	.858	Irritable	.647
Popular	.866	Strong	.857	Excitable	.554
Intelligent	.864	Effective	.821	Aggressive	.424
Protective	.847	Maternal	.776	Active	.414
Confident	.823	Permissive	.713	Confident	.410
Social	.822	Aggressive	.647	Opportunistic	.381
Gentle	.813	Equable	.486	Curious	.353
Opportunistic	.801	Intelligent	.365	Intelligent	.222
Equable	.759	Protective	.338	Strong	.209
Curious	.729	Irritable	.317	Social	.113
Predictable	.630	Popular	.272	Permissive	.106
Playful	.606	Confident	.247	Playful	.104
Permissive	.583	Tense	.225	Popular	.075
Active	.563	Predictable	.194	Tense	.049
Effective	.532	Apprehensive	.146	Sensitive	.036
Maternal	.254	Sensitive	.124	Predictable	.008
Excitable	.241	Opportunistic	.114	Protective	.006
Strong	.070	Gentle	−.011	Effective	−.001
Slow	−.080	Insecure	−.256	Slow	−.172
Fearful	−.520	Fearful	−.272	Maternal	−.343
Aggressive	−.573	Social	−.384	Equable	−.405
Tense	−.661	Curious	−.569	Insecure	−.433
Apprehensive	−.680	Excitable	−.603	Fearful	−.478
Irritable	−.684	Active	−.662	Apprehensive	−.563
Insecure	−.772	Playful	−.767	Gentle	−.573

compared, adjective by adjective (listed in box 13.2, table A), across all the raters. Most raters gave similar scores to adjectives for each of the eleven individuals, and thus they tended to be highly concordant in their scores. The exceptions were Echo, the matriarch, who appeared to be more "difficult" to rate consistently, and Edwina, who was somewhat idiosyncratic in her behavior and thus there was less uniformity in her ratings (box 13.2, table B). In general, in this very small sample of elephants and raters, there appears to be sufficient consistency in the scores of adjectives to make it worth exploring further whether these adjectives usefully typify the personality of individual elephants.

Based on the high degree of similarity among raters, a mean of the rater's scores was calculated for each adjective for every female. These mean adjective scores were then entered into a Principle Components Analysis (PCA) as a mechanism to discriminate among adjectives and assess any consistencies in traits for each female. In an initial correlation matrix of all adjectives, three were found to be unrelated to any other ("solitary," "deferential," and "eccentric"). These were subsequently removed from analysis as being unlikely to contribute to any associations among the adjectives or to general descriptions of personality characteristics. The PCA produced three dimensions of associated and inter-correlated adjectives that together explained 88 percent of the variance in scores among individuals (box 13.2, table A), although studies of personality in chimpanzees and humans tend to produce a typical "5-factor" personality description consisting of Extraversion, Conscientiousness, Agreeableness, Neuroticism, and Openness (Weiss et al. 2007). In this sample of elephant females, there was considerable overlap between the adjectives that statistically describe the three components. The PCA assumes that there will be no interactions among variables; thus at this stage, the results of the PCA need to be interpreted with caution (the PCA correlation matrix was not positive). Furthermore, only 11 individuals were rated, which is a very small sample in any statistical attempt to analyze individual characteristics (PCA would conventionally require three to five times as many subjects as adjectives). The basic assumption, however, is that correlations between variables are the result of sharing common factors or components in the analysis, and thus we regard the PCA as highly suggestive of those personality factors contributing to underlying variance among individuals in such traits.

Component 1 explains 46.9 percent of variance and was defined using the loadings of adjectives on the component, from most or strongest positive to least or strongest negative. This component is labeled as "Sensitive to Insecure." Component 2 explains 28.4 percent of variance and is labeled "Slow to Playful." Component 3 explains 12.9 percent of variance and is "Irritable to Gentle."

Each female was given an individual z-score for each of the three components (her rating on each adjective contributing

(continued)

Table B

Name	Status	Age	Kendall's concordance (df = 3)
Echo	Matriarch	59	$X^2 = 2.9$, $p = 0.41$
Erin	Eldest daughter of Echo	35	$X^2 = 9.8$, $p = 0.021$
Enid	Daughter of Echo	22	$X^2 = 12.6$, $p = 0.006$
Eliot	Daughter of Echo	19	$X^2 = 6.3$, $p = 0.099$
Ebony	Daughter of Echo	10	N/A
Edwina	Daughter of Erin	12	$X^2 = 4.7$, $p = 0.19$
Eleanor	Daughter of Erin	19	$X^2 = 11.8$, $p = 0.008$
Ella	Sister to Echo	39	$X^2 = 18.7$, $p < 0.001$
Emma	Daughter of Ella	17	$X^2 = 25.2$, $p < 0.01$
Eudora	Daughter of Emily	32	$X^2 = 25.0$, $p < 0.001$
Elspeth	Daughter of Eudora	16	$X^2 = 15$, $p = 0.002$

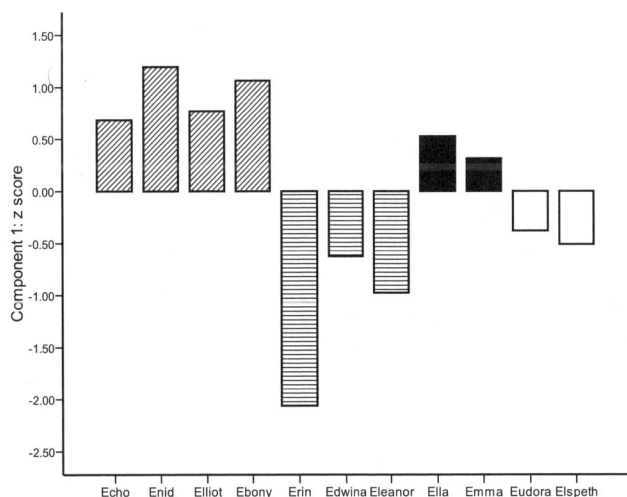

Box 13.2, figure A Individual z-scores for Component 1. Females are grouped by matriline, with the matriarch and her younger daughters first, followed by Echo's eldest known daughter, Erin, her daughters, and then the other females in the family.

BOX 13.2 (continued)

to each component was standardized against the mean rating for that component). These z-scores were uncorrelated across individuals, suggesting that the component may indeed represent different unique individual aspects of "personality" or temperament for female elephants. Individuals scoring highly on Component 1 can be suggested as representing leadership, social engagement, and integration within the family unit as a whole. The Matriarch, Echo, and her daughters are all high on Component 1, as are females from the next most dominant family, Ella. Erin stands out as a clear negative outlier, along with her daughters (box 13.2, figure A), with high insecurity and low sensitivity to others. Component 2 may represent traits associated with the age of the individual (box 13.2, figure B) and thus could be maturational rather than associated with stable individual characteristics. Since dominance in particular is age-graded in female elephants (Archie et al. 2006), it would be expected that at least some individual traits are well described by age. Component 3 appears to reflect some level of social irritability, and again Erin is an outlier on this component (z-score of 2). The youngest daughter of Echo is also high on C3, suggesting that these are apprehensive and excitable females (box 13.2, figure C).

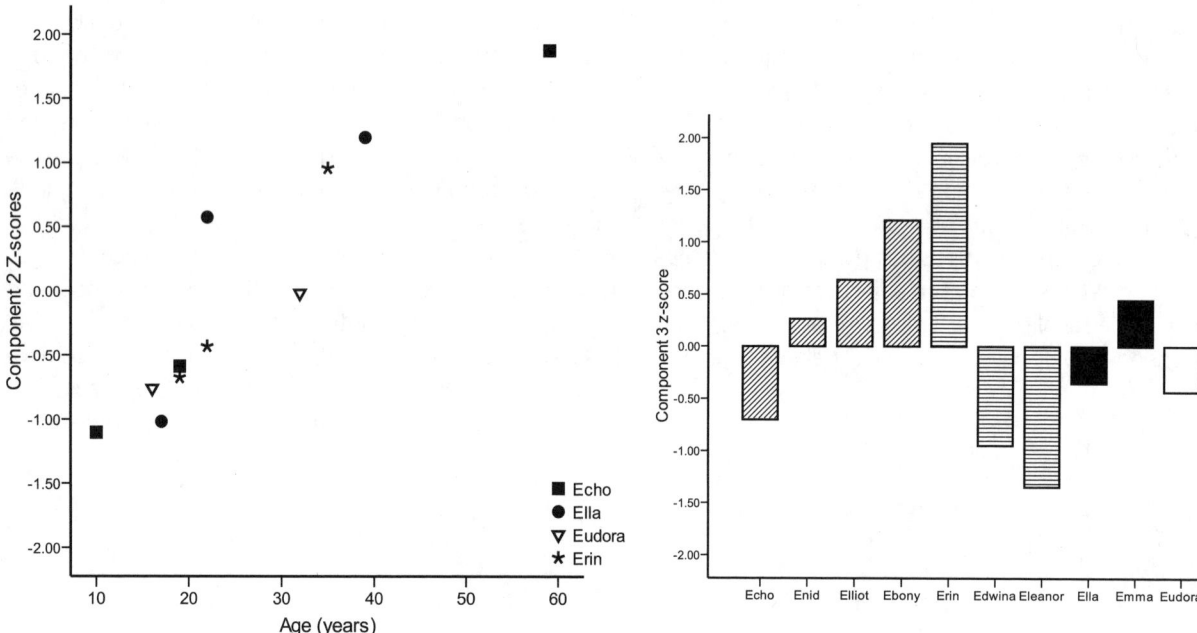

Box 13.2, figure B The relationship between individual z-scores from PCA analysis and age for Component 2, "Slow to Playful."

Box 13.2, figure C Individual scores for Component 3, "Irritable to Gentle," with females again grouped by mother with oldest first.

Flexibility within Families

Short-Term Fission. Each female appears to have a choice as to whether she associates with the other members of her family or not, irrespective of her status as a family member. She can forage with just her immediate offspring or other close relatives. In this way, she may be weighing the costs and benefits of grouping. However, a lone female was a very rare occurrence in Amboseli. The only recorded cases were of either estrous females or sick or very old individuals who moved on their own for a few weeks or months before dying. Temporary breakups of families occurred most often during the dry season and particularly during times of severe drought. An adult female with her dependent offspring may be able to avoid competition or the energy costs of foraging by feeding and moving in a much smaller group (see Mutinda, Poole, and Moss, chapter 16). She may temporarily break off from the family when she has a very young calf who cannot go into certain habitats, such as the deep swamps, or cannot move long distances in the course of a day.

However, a female with a young calf who splits from her family increases the risk that her calf will be taken by predators such as hyenas. In Amboseli, the hyena population grew from fewer than 40 to over 250 in less than 10 years. Since the increase, far fewer females have been seen moving with just their young calves. While we are unable to quantify the losses to hyenas (as most predation is unobserved),

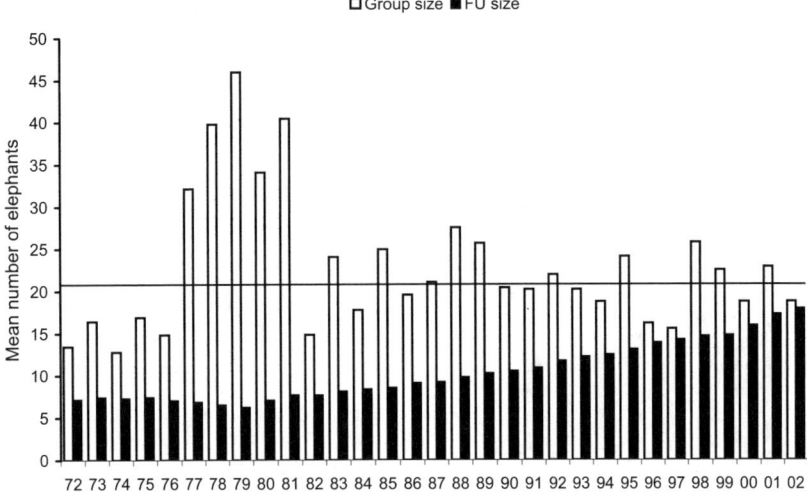

Figure 13.4 Mean group size by year plotted with average family unit size over 30 years. The overall average group size is indicated for comparison.

around 2 percent of calves ($N = 40$) die as neonates, and some of these deaths may be predator-related. Both lions and hyenas have been observed making attempts on young calves. Younger females who split to reduce competition also lose the benefit of age and experience of the matriarch who acts as the repository of knowledge for the family (McComb et al. 2001; and see Mutinda, Poole, and Moss, chapter 16). These older females know migration routes, where to find the last food and water, and how to avoid predators, particularly in the case of Amboseli, Maasai warriors. A larger group with more allomothers to care for the young calves also leads to greater calf survival (see Lee and Moss, chapter 14).

Environmental and Demographic Influences on Sociality

Temporal Changes in Grouping Dynamics

Over the 30 years of the study, annual mean female group sizes have fluctuated considerably, starting with relatively small groups during the dry years of the early 1970s (mean = 14.8), peaking at 45 in 1979, when elephants were generally spending much of their time within the protected area (see Lee, Lindsay, and Moss, chapter 6), and then leveling off to about 20 from 1991 onward (figure 13.4). At the same time, as noted above, the average family unit size grew from 7 to over 18. That mean group size did not increase in proportion to mean family size suggests that there was an optimal group size for the Amboseli elephants, which averaged around 20 individuals. By 2002, the average family size was close to that of the average group. As expected, as family size increased over the years, the average family unit spent significantly more time on its own rather than in a group with other families ($r = -0.758$, $N = 30$, $p < 0.001$).

Environmental Influences on Sociality

Grouping patterns of the Amboseli elephants are influenced by environmental conditions. During periods of low rainfall, extended dry seasons, and reduced food biomass, average group size tended to be smaller, while the reverse was true for periods of high rainfall. However, the correlation was neither strictly linear nor particularly strong ($r^2 = 0.144$, N years = 31, $p < 0.01$), suggesting that factors other than simple foraging constraints operated on group size. During and after the wet season, family units tended to form large aggregations in grassland or savannah habitats where vegetation was abundant and relatively evenly distributed (Western and Lindsay 1984; Poole and Moss 1989; see also Croze and Moss, chapter 7, and Lindsay, chapter 5). Joining such aggregations were the adult independent males. Adult males made up 12 percent of the total size of mixed groups (see also Poole 1982; Lee et al., chapter 17). The size of groups that contained only family units was smaller on average (median = 11, range 1–420, $N = 11,374$) than were those groups that contained both families and associated independent males (median = 20, range = 2–550, $N = 10,312$). Over 6 percent of mixed-sex groups contained more than 100 elephants, by comparison with 0.2 percent for cow-calf–only groups. Within a year, mixed group sizes responded more strongly to rainfall, with the largest groups observed in the wettest months from December to May (figure 13.5). There was less effect of food availability indicated by rainfall on the smaller sized cow-calf-only groups,

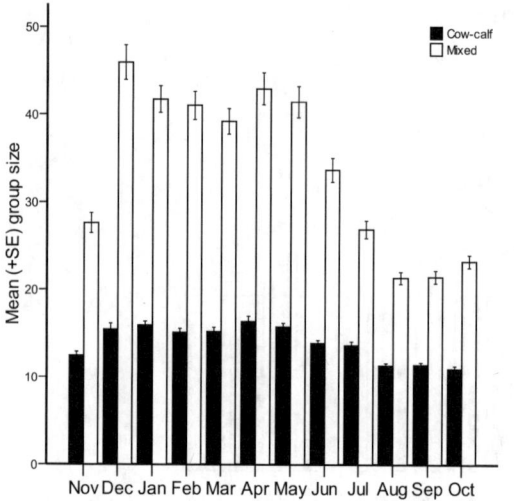

Figure 13.5 Thirty-year mean monthly size of cow-calf and mixed groups.

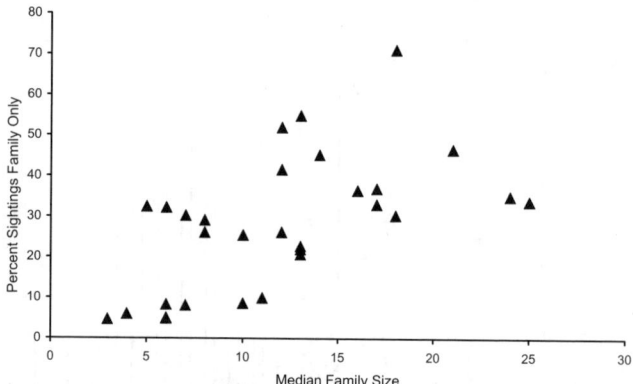

Figure 13.6 The overall percentage of sightings that a family was observed in groups with only that family present in a group, plotted against median family unit size over the study.

although both cow-calf and mixed groups showed the wet season highs in group size. We suggest that female elephants will group whenever possible and will do so in the largest groups that can be sustained.

Cohesiveness within a family tended to decrease only slightly when conditions were dry (dry season intensity index [DSI] and mean cohesiveness; $r = -0.322$, $N = 22$, $p = 0.12$). Since families in cow-calf–only groups find themselves in groups that are typically only just larger than their family alone, there may be relatively low pressure on families in such groups to fragment in response to low food availability. Rather, the pressure will be on families to spend less time with other families, affecting tendencies for gregariousness more than cohesion (see below). However, at certain times of the year (for example, during the driest times), a female elephant might associate with only her immediate family members or even just her own offspring, while at other times, such as just after the rainy season, she might associate with many of the other females and calves in the population. The largest group ever sighted in Amboseli was 550 in May 1981, which at that time represented over 90 percent of the entire population.

Determinants of Gregariousness

Families spent an average of 72 percent of time ($N = 30$) in groups with individuals from other families, although gregariousness varied considerably from family to family and from year to year during the study. There was a marked decrease overall as family size increased ($r = -0.768$, $N = 30$, $p < 0.001$; figure 13.6). On average across all years, 60 percent of all observed female groups contained individuals from more than a single family. Thus, most families were actively choosing to associate or group with other families.

The most gregarious families were the MAs (4.9 percent sightings alone) and YAs (4.7 percent alone). Both of these families were small, and both were tightly bonded with other larger families. The least gregarious was the AA family, with 71 percent of sightings in a group with only its own family members. Why might some have been less gregarious while others were highly social?

One major determinant of sociability was family size. Large families tended to be less gregarious (see also Mutinda, Poole, and Moss, chapter 16) as well as less cohesive, possibly because there is an optimal size at which a family can maintain spatial coordination and simultaneous activities. If this size is close to that of the largest families, then those families will tend not to be found in groups with other families, except during those times of year (wet seasons) or in years (non-drought years) when the average group size that can be sustained is high. Across all families, gregariousness was a function of family size, but once the effect of size was removed there was only a slight trend for cohesiveness to be related to gregariousness (stepwise multiple regression, gregariousness, and size: $r^2 = 0.341$, $t = 3.8$, df = 29, $p = 0.001$; gregariousness and cohesiveness controlling for size: $t = 1.36$, df = 29, $p = 0.18$). Sociable families were not always those who were more cohesive; some families were simply gregarious.

Because families vary in their cohesiveness, gregariousness, and their social and demographic histories, all of which underlie the patterns of fission and fusion observed on a daily, seasonal or annual basis, a family-by-family exploration of these factors was made, including the annual ecological estimator of food abundance—Dry Season Intensity (DSI) (table 13.4). For 8 families, their cohesiveness was unexplained by family size, food abundance estimated by DSI, or their gregariousness. For 16 families, cohesiveness was a function of family size, with greater cohesion at

Table 13.4 The statistical relationships of effects of ecological conditions, family unit size, and gregariousness on family cohesion for 30 family units with more than 700 sightings

Family	Significant factors in the model	Relationship	r^2	df	p
AA	Family size, DSI	Negative	0.849	2,20	—
BB	Family size, Solitary	Negative	0.711	2,21	—
CB	Family size	Negative	0.325	1,17	0.013
DB	Family size	Positive	0.593	1,20	—
EA	Family size, DSI	Negative	0.571	2,20	—
EB	Family size, DSI	Negative	0.702	2,21	—
FA	None				
FB	Family size	Negative	0.461	1,21	0.001
GB	Solitary	Negative	0.550	1,19	—
IA	None				
IB	DSI, Solitary	Negative	0.745	2,17	—
JA	Solitary	Negative	0.476	1,19	0.001
KB	Family size	Negative	0.574	1,17	—
LA	None (positive correlation with family size)				
LB	Family size	Negative	0.840	1,17	—
MA	Family size, solitary	Negative	0.720	2,15	—
OA	Family size	Negative	0.582	1,18	—
PA	Solitary	Negative	0.760	1,19	—
PC	Solitary, (DSI)	Negative	0.481	1,17	0.001
QB	Family size	Negative	0.309	1,17	0.017
RA	None (negative correlations family size, solitary)				
SA	None (negative correlation family size)				
TA	None				
TC	Family size	Negative	0.323	1,14	0.027
TD	Family size	Negative	0.314	1,17	0.016
UA	Solitary	Negative	0.740	1,19	—
VA	Family size, DSI	Negative	0.705	2,19	—
WA	None (negative correlations family size, solitary)				
XA	Family size, solitary	Negative	0.509	2,17	0.005
YA	None				

Note: Stepwise multiple regression analysis on percent sightings all together (cohesive) as dependent, with percent sightings family only (Solitary, i.e., low gregariousness), Dry Season Intensity (DSI), and Family size as independent factors in the regression model.

small sizes. For 6 families, cohesion was greater when DSI was low, suggesting a tendency to split up during bad years. These latter 22 families were either those that also split more as a result of having a large family or that initially tended to be relatively less gregarious. Finally, for 9 families, lower cohesion was also associated with more time spent alone in groups with no other families. In summary, no single factor explained the ways in which females associated either with other members of their own family or with other family units. Diverse factors were important for different families, and in effect a complex, non-linear and individualistic pattern of social dynamics existed for each family.

Associations among Families: Time and Space

The group in which a family member finds herself changes in both size and composition from day to day and even from hour to hour (see Mutinda, Poole, and Moss, chapter 16). As previously noted, females can find themselves in groups with no other families or potentially with all other families in the population. While the average female will experience a group size of 20 individuals (including independent males) with a range of 1 to 550, the actual size of groups experienced was variable (figure 13.7). During the first decade of the study, average group size experienced was significantly larger than later in the study, as we discussed above (see also figure 13.1). As shown for gregariousness, no single factor influenced the group size experienced by families in the first decade. However, as expected, as average group size declined in subsequent decades, the size and sociability of the family began to play a larger role in the group size the family experienced (1982–91: ANOVA overall stepwise model of family size and family gregariousness on group size experienced $F = 3.5$, df =3,30, $p = 0.04$; independent effect of family size, $t = 2.65$, $p = 0.013$. 1991–2002: ANOVA overall stepwise model $F = 8.5$, df = 2,30, $p = 0.001$; independent effect of gregariousness, $t = 4.12$, $p < 0.001$; family size, $t = 2.28$, $p = 0.030$). With an increase in the size of a family, its propensity to associate with other individuals or families appeared to decline.

Whether a family was gregarious or not and whether what drives the level of gregariousness was social, demographic, or environmental, there was consistency in who associated with whom among families. The groups formed were not made up of random aggregations of family but had components of both spatial consistency and social partnerships. Using the same index of association, families with over 200 sightings were hierarchically clustered into a dendogram representing the most associated family units over the entire 30-year period. Statistically, there appeared to be five distinct clusters (figure 13.8), which were both socially and, to some extent, geographically linked (see Croze and Moss, chapter 7; figure 7.14).

1. TA, TB, TC, TD, the T bond group (see Moss 1988), which extensively used the eastern areas of the basin.
2. DA, DB, EA, EB, QB, RA, SA, XA, used the central area of the Park together with AA, FA, FB, HA, and KA, who centered around the western edge of the main Enkong'o Narok swamp.

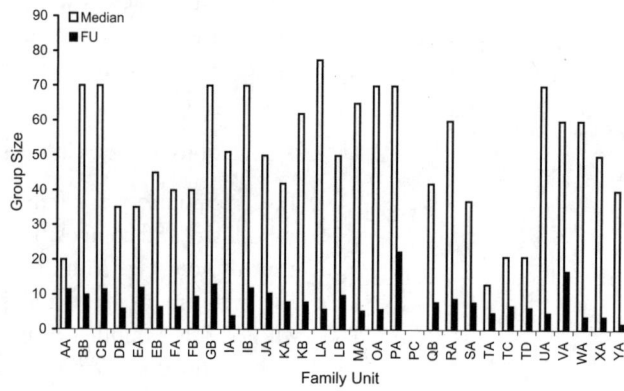
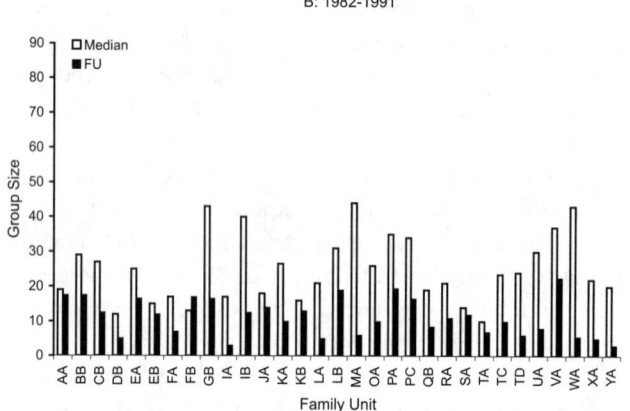
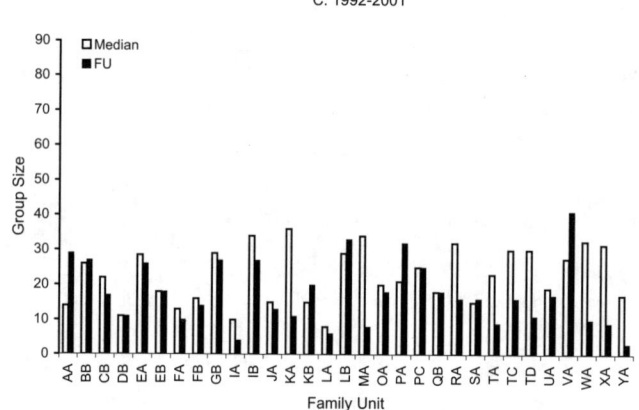

Figure 13.7 Family unit sizes in relation to the median group size experienced by the family for each decade: (a) 1972–81, (b) 1982–91, (c) 1992–2002.

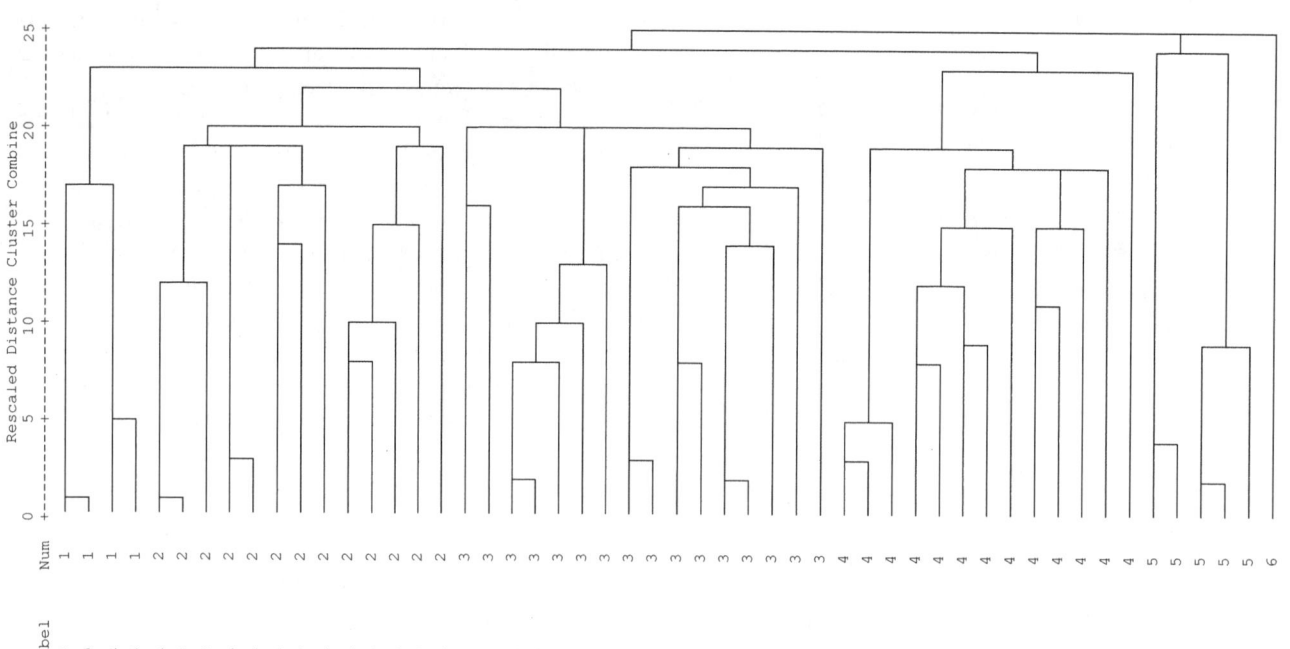

Figure 13.8 Hierarchical cluster diagram using average linkage between groups.

3. KB, OB, a bond group that used the southern and central areas of the Amboseli basin together with BB, CA, CB, GB, IB, LA, LB, MA, OA, PA, PC, UA, VA, WA, who used the eastern Longinya swamp and ranged into basin areas to the north and south.
4. AB, BC, GC, HB, JB, NB, MB, PB, WB, ZA, ZB, ZC, which ranged to the west and into Tanzania along with DC, a small family, usually seen alone in the west.
5. IA, IC, small families sighted only with each other, often near the main Park's causeway or to the south, associated with JA, YA, SB, a tightly meshed western bond group.
6. LC, an idiosyncratic family that had split from its other members (LA of cluster 3) and used areas south and east of the Park.

As we have described, over time families constantly changed in composition, size, and underlying sociability. Did these demographic processes affect who were associates over time? The top associate of each family, representing the most frequent other family associated in groups, is presented in table 13.5. Some families were consistent in their top associate over the entire period, while others changed their associates as their range within the basin shifted or as demographic processes, such as family splits or extinctions, acted on previous associates.

Conclusions

The long-term data on association patterns among adult female elephants in Amboseli reveal persistent, strong social bonds between individuals. Females originally assigned to membership in a family unit at the beginning of the study maintained, with very few exceptions, consistent and long-lasting relationships with their fellow family members. The DNA studies have shown a high level of relatedness among the females within a family (see Archie, Moss, and Alberts, chapter 15), and thus they form a kin unit, which contributes to the survival and reproductive success of its members. Nevertheless, while social and kinship bonds are the glue of elephant society, there are strong forces at work that allow the society to be remarkably flexible (table 13.6).

Kinship, friendship, matriarchy, personality (box 13.2), history, sex, age, demographic events, mating strategies, and environmental conditions are all determinants of and influences on fission and fusion among elephants. A female born into an Amboseli family will more than likely remain in that family for the rest of her life, but she can make choices about how and when she associates with the other members. Some of these choices will be based on her history within the family. If her mother has died and she had no sisters she may become peripheral and move on her own more frequently with just her offspring. On the other hand, the daughter of a living matriarch is very unlikely to split from the family, even for short periods.

The option to split may be taken under specific conditions when there are conflicts of interest between females.

Table 13.5 Bond group families, overall top associate, and top associates decade by decade for 30 family units

FU	Bond group	Top associate overall	1972–81 Top associate	1982–91 Top associate	1992–2001 Top associate
AA		FA	FB	FA	FB
BB	UA	UA	UA	UA	UA
CB	OA	OA	OA	OA	OA
DB	DA, QB	DA	DA	DA	VA
EA	EB	EB	EB	IB	TA
EB	EA	EA	EA	EA	EA
FA	KA, HA	KA	HA	KA	KA
FB		KA	KA	IA	AA
GB	IB	IB	IB	IB	IB
IA	IC	IC	RA	IC	IC
IB	GB	GB	GB	GB	GB
JA	YA, SB	YA	SB	YA	YA
KA	HA, FA	HA	FA	HA	HA
KB	OB	OB	GB	OB	OB
LA	MA, WA, CA, VA	VA	VA	MA	LC
LB		BB	PA	PA	TA
MA	WA, VA, CA, LA	WA	WA	WA	WA
OA	CB	CB	CB	CB	CN
PA	PC	BB	BB	PC	IB
PC	PA	PA		PA	IB
QB	DA, DB	DA	DB	DB	DC
RA	XA	XA	XA	XA	XA
SA		RA	RA	EA	MB
TA	TB, TC, TD	TB	TB	TB	LB
TC	TD, TA, TB	TD	TD	TD	
TD	TC, TA, TB	TC	TC	TC	
UA	BB	BB	BB	BB	BB
VA	MA, WA, CA, LA	CA	LA	MA	MA
WA	MA, CA, LA, VA	MA	MA	MA	MA
XA	RA	RA	RA	RA	RA
YA	JA, SB	JA	SB	JA	JA

Table 13.6 Summary of factors leading to fission and fusion of family units

	Fission	Fusion
Environmental	Low rainfall Low food availability Dispersed patchy foods Competition	High rainfall Abundant and widespread resources Low competition
Demographic	Death of matriarch Death of other adult females Birth of calf	Death of individuals with resulting reduction in number in family Death of all adult females
Social	No close relatives No strong affiliative bonds Low ranking	Leadership of older females Anti-predator benefits Resource defense

For example, when rates of food intake drop too low to provide sufficient energy, then it would be expected that families or even individual mother-calf units from families would forage on their own (see Mutinda, Poole, and Moss, chapter 16). But other constraints may operate on individual decisions about grouping. If there are females who need access to water for milk production (e.g., during the first 6 to 12 months of a calf's life, when calves are suckling at the highest frequency and their independent feeding and drinking is limited), then these females may need to travel to certain locations or forage in proximity to water more than do females without young calves. And finally, very young calves slow the rate of movement of a foraging family; more mobile females might split off so as to forage more efficiently.

Elephant sociality does not end with the family, irrespective of its importance. The society is complex and multi-tiered, extending from the family to bond groups to clans to sub-populations and to the whole population. Elephant families associate at all these levels. In Amboseli, some families maintained strong associative bonds over the entire period, despite demographic and ranging changes. These bond group affiliates contribute to the social and spatial clusters observed. It is thus clear that some families affiliate by choice rather than spatial propinquity.

Bond group affiliation appears to be based partly on kinship (see Archie, Moss, and Alberts, chapter 15) and partly on history, experience, and friendship between matriarchs. Whether or not families are part of a bond group influences their overall levels of sociability. When families have consistently high levels of association with particular other families, those families will be more gregarious. Some social and demographic factors promote gregariousness. If social companions (e.g., allomothers, allies) are not available within a family, then small families may fuse either temporarily or more permanently to ensure access to companions (see also Mutinda, Poole, and Moss, chapter 16). The age and status of the matriarch may also influence the tendency to temporary or even permanent fusion. Older matriarchs attract families with younger leaders (see Mutinda, Poole, and Moss, chapter 16). Their experience and accumulated wisdom act as a repository of knowledge for families, bond groups, and even the population (McComb et al. 2001). In addition, we suggest that the particular personality of a matriarch can contribute to the large inter-family variation observed in patterns of sociability.

Ecological influences appear to be less important as determinants of group size, structure, or affiliative patterns than might be expected from other elephant populations (e.g., Dudley et al. 2001; Foley 2002; Wittemyer et al. 2005). Female elephants in Amboseli shift their food types and habitat use in response to seasons less as energy maximizers and more as social maximizers (Lindsay 1994). Thus we see a pattern, in times of higher food availability, of very large mixed-sex aggregations, which are sustained for as long as is socially possible. Rather than rapidly shifting to smaller groups or habitats with higher-quality forage in direct response to declining food availability and quality, the largest possible group size was maintained even with some cost to energy intake (Lindsay 1994). Once these costs pass some threshold, however, as at the end of the dry season or in exceptionally dry years, groups and families fragment into very small units. In Amboseli, it can be suggested that food availability acts more as a social enabler than as a constraint for much of the time. The year-round availability of swamp forage may be a significant factor in minimizing ecological constraints (see also Lee, Lindsay, and Moss, chapter 6).

A number of non-food-related benefits can be suggested for elephants grouping with individuals other than family members, such as contact with mates and opportunities for mate choice (Poole 1982; Moss 1983; Poole and Moss 1989; Rasmussen 2005). Estrous females are easier to locate when in a large aggregation than when dispersed, although controlling access may be more difficult. Younger sexually active males may thus be able to obtain clandestine copulations. Females also have greater choice when several musth bulls are attracted to an aggregation, and there may be some synchrony of estrus resulting from females observing and picking up olfactory cues from estrous females in the same aggregation.

Large groups enhance socialization opportunities and opportunities to gain knowledge through meetings with familiar and unfamiliar individuals (Lee 1986; McComb et al. 2003), and provide opportunities for learning and the transmission of information (Lee and Moss 1999). Other benefits may derive from a general reduction of the risks of mortality for calves and adults when in relatively larger groups. Thus, when all these potential benefits are operating, they may partially outweigh energy restrictions due to foraging in larger groups.

The question remains: why should females be social given that there are foraging costs, conflicts of interest over the kinds of resources required, over energy costs to ranging, and over the kinds of risks posed by different habitats? We suggest that elephants' complex communication about other individuals and information exchange within and between families improves both survival and reproductive success (see Poole and Granli, chapter 8; Poole, chapter 9). Sociality is thus strongly promoted in a context where information about others and environments is essential. In addition, within-family shared infant care, protection, and active defense enhances calf survival (see Moss and Lee, chapter 12; Lee and Moss, chapter 14; Archie, Moss, and Alberts, chapter 15). There may be advantages to sharing the care of calves even within larger groups.

The benefits of living in a family consisting of more than just a single female and her offspring appear to be a driving force in savannah elephant social structure, and we suggest that predation on young animals is a significant factor. Among forest elephants the average group size is two to three (Theurerkauf, Ellenber, and Guiro 2000; Morgan and Lee 2007; Fishlock, Lee, and Breuer 2008), although family size is not well known. Males leave their families at 4 to 5 years old and females leave at about 10 to 12 years. A typical forest elephant family consists of a female and two offspring. Andrea Turkalo, who has been studying forest elephants for over 20 years, attributes small family and group size to the lack of significant predators in the forest (A. Turkalo, personal communication), although competition for dispersed, high-quality foods may also play a role. By comparison, Asian elephants, also forest dwellers, live in larger families typically consisting of two or more adult females and their offspring (Sukumar 2003). In almost all areas where Asian elephants live, there are or were tigers.

In summary, female African elephants live in a remarkably complex social world in which there are long-lasting bonds between individuals. At the same time, each individual has the ability to remain flexible in her choice of associates and group size. Her choices are influenced by a combination of environmental, demographic, and social conditions as well as by individual characteristics such as leadership, personality, and experience.

Chapter 14 Calf Development and Maternal Rearing Strategies

Phyllis C. Lee and Cynthia J. Moss

LONG-LIVED SPECIES of mammals experience and interact with an environment that changes around them even as they themselves grow, mature, and learn. How they cope with change and how patterns of behavior emerge and develop are questions of special interest when the period of immaturity is prolonged. This mosaic of patterns and processes in physical and social development is one of the more intriguing features of very long-lived species and sets them apart from other animals. Species with brief, sensitive periods for learning, with rapid reproductive maturation, or with abrupt and well-defined transitions between developmental phases such as weaning in small mammals or fledging in birds represent one end of a developmental spectrum. By contrast, for species like elephants with a period of lactational dependence of 4 or more years who reach reproductive maturity at 10 to 15 years for females and 20 to 30 years for males, it is difficult to be precise about the length of the developmental period.

As we have discussed elsewhere (Lee and Moss 1999), the initial social experience of male and female elephant calves is similar and occurs within a family, but as adults, reproductive and social strategies are strikingly different for the two sexes. Do calves of both sexes "construct" their own early experiences, and how do these early experiences produce sex-specific outcomes among juveniles and adults? Sexually differentiated hormones, metabolic processes, and growth rates drive innate tendencies that impact on the experiences of males and females in ways that cause each sex to seek different social contacts, engage in different activities at different rates and times, and generally be motivated to pursue sexually distinct developmental pathways from an early age. At the same time, mothers, allomothers, and other family members treat male and female calves differently. Interaction between the social context and individual characteristics facilitates sexual differentiation during development, and this sexual differentiation has major consequences for subsequent social and reproductive success.

In this chapter, we are concerned with describing normative elephant development and by extension, the capacity for the modification of behavior or developmental plasticity as a function of individual experiences over a prolonged period. Specifically, we address the following:

- We illustrate sex-specific developmental patterns over the first five years of life and discuss the social, developmental and hormonal influences on sex differences. We ask how and when the major sex differences in social and reproductive behavior arise during the lengthy period of development.
- We explore separately behavior with functions specific to the immature phase—juvenile specializations such as play—and behaviors that emerge and are built upon to form adult behavior, such as foraging skills (scaffolding *sensu* Bateson [1981]).
- We speculate on the long-term consequences of behavioral development in relation to general elephant patterns of social and reproductive maturity. If experiences during development vary between individuals, does this affect their survival or reproductive strategies? If so, how?

Table 14.1 Number of focal hours and focal scans for activities and partner preferences at each calf age

	Males		Females	
Calf age	Scans	Focal Hours	Scans	Focal Hours
0–6 months	361	26	417	25
7–12 months	146	22	138	22
13–24 months	153	27	262	28
25–36 months	164	24	250	31
37–48 months	385	20	447	29
49–60 months	242	17	158	10
60-plus months	136	—	85	—

The analyses we present here are derived from a variety of types of observations (see appendix 1). Scan samples of activity and proximity ($N = 3,338$) were made on 237 calves aged between birth and 9 years old. A further 272 hours of focal samples on 124 calves between birth and 5 years old recorded calf interactions with the mother and others, suckling, play, and approaches and leaves by the calf and others. Growth and size are assessed from hind-foot length or shoulder height measures on 984 elephants of all ages, while data on inter-birth intervals and reproductive parameters are derived from the general demography data. Sample sizes at each age are provided in table 14.1.

Patterns and Sequences of Calf Development

While the concept of developmental milestones is a familiar one in human psychology, biologists tend to think in terms of major transitions between arbitrary phases (weaning, fledging, onset of reproductive activity). Behavior can either be part of a continuous process, or it can represent a discontinuity—a jump or break—in behavioral organization (Bateson 1981), which may illuminate the underlying controls on development. In table 14.2, we outline the major biological and behavioral transitions during elephant development leading to adult competence.

There are a series of milestones at around 24 months, the age associated with survival if a calf is orphaned. Full weaning, defined by the presence of a new sibling and the cessation of suckling, occurs over two years after this metabolically competent age. Social processes such as independence from the mother, integration within the family, and exploring the wider social context of the population change gradually as calves age. Extended social contact can facilitate the learning of key environmental coping behaviors such as how to survive dry seasons with shifts and reductions in food types requiring movement between habitats and areas. Social learning contexts thus can prolong the period of mother-calf interaction beyond the minimum threshold for sustaining a dependent calf's survival (van Noordwijk and van Schaik 2005). Body mass almost doubles between the first stage of metabolic efficiency at 24 months and suckling cessation at 55 months, and shoulder height increases by up to 50 cm (table 14.2 and figure 14.13). The period between 24 and 48 months is thus a period of rapid growth that requires major energetic input; growth faltering or energy constraints pose extra risks to calves in poor seasons or poor years if they are forced to forage independently.

From the mother's perspective, she needs to ensure that her calf survives, grows, and becomes socially integrated. Calf survival makes a greater contribution to lifetime reproductive success than does a female's rate of reproduction (see Moss and Lee, chapter 12) since rapid rates are associated with calf death and lost time and effort in a "wasted" reproductive event. Continued investment in calves, both in terms of lactation and social interaction, benefits mothers more than a short inter-birth interval does. On average, the first calf will be between 25 and 41 months (median 32 months) when the next calf is conceived. This timing coincides with the transition to completely independent foraging by the calf as well as with the period of suckling conflict between mother and calf. The variation in investment periods may be due to extended care allocation in order to ensure survival and growth. If social and energy requirements differ between the sexes of calves, then the mother's allocation of care will also differ between sons and daughters.

Calf Sociability and Proximity

Proximity to Mothers

Elephant calves are dependent on their mothers and other family members for social support, survival, and learning. If a mother dies when a calf is under 24 months, calves almost always die. Being orphaned between 2 and 9 years still reduces the chances of survival (Cox regression: mean survival = 13 years versus 24 years when mothers survived; log rank comparison = 96.55, $p < 0.001$). Thus, for a calf, the mother remains a crucial and central individual for the first 5 to 10 years of life, and calves make frequent vocal and physical contact with their mothers.

Elephant mothers are some of the most tolerant and patient mothers among mammals, allowing and indeed facilitating close proximity. In the early post-natal phase, calves

Table 14.2 Developmental "milestones" for elephant calves from birth to onset of reproduction

Trait and appearance	Use of trunk for foraging and drinking	Tusks appear beyond gums	Metabolically independent foraging (med age at conception of next calf)	Equal time close and far from mothers	Play peak (weaning trough age)	Weaning Age (end suckling) in months (median IBI)	Weaning conflict: peak period of rejections	Independence from family unit	Puberty (sperm production; first estrous)	First calf conceived
Males										
Age Range	3–4 mo foraging; 8–9 mo drinking	20–24 mo	24–36 mo (33 mo; 26–43 mo)	31–36 mo	0–6 mo (37–48 mo)	32–114 mo (med = 55)	31–36 mo	10–17 yrs (mean = 14 yr)	10–12 yrs	21–24 yrs
Height (Birth: 95 cm)	100–115 cm	120–130 cm	130–150 cm	130–150 cm	90–110 cm	150–160 cm	130–150 cm	225–250 cm	190–230 cm	226–270 cm
Mass (Birth: 120 kg)	~150 kg	~450 kg	~500 kg	~650 kg	~140 kg	~850 kg	~650 kg	~2,000 kg	~1,600 kg	~3,000 kg
Molar stage (Birth: M2 visible in jaw)	M3 forming	M3 in wear	M3 in wear	M3 worn M4 forming	M2 in wear	M4 formed but unfused	M3 worn M4 forming	M5 coming into wear	M4 in wear	M5 anterior eroded, M6 forming
Females										
Age Range	8–9 mo	22–34 mo	24–36 mo (30 mo; 24–39)	37–42 mo	0–6 mo (25–36 mo)	26–113 mo (med = 52)	10–12 mo and 25–30 mo	N/A	8.5–16 yrs (mean = 14 yrs)	8.5–14 yrs
Height (Birth: 80 cm)	100–116 cm	120–130 cm	110–130 cm	130–140 cm	80–105 cm	140–170 cm	130–150 cm		195–230 cm	195–230 cm
Mass (Birth: 100 kg)	~200 kg	~350 kg	~350 kg	~500 kg	~150 kg	~700 kg	~400 kg		~1,200 kg	~1,200 kg
Molar stage (Birth: M2 visible in jaw)	M3 forming	M3 worn	M3 worn	M3 worn M4 forming	M2 in wear	M4 visible but unfused	M3 worn M4 just forming		M4 worn M5 fused	M4 worn M5 fused

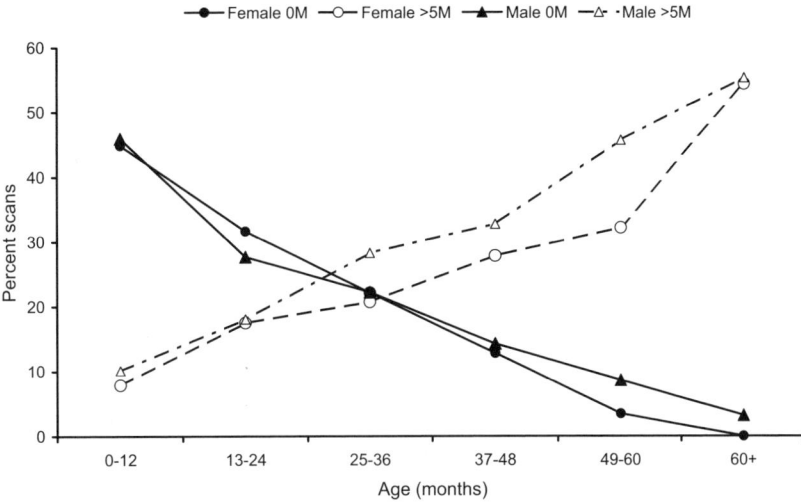

Figure 14.1 The percentage of scans with male and female calves close to the mother (within contact distance) or far from the mother (>5 m) against calf age.

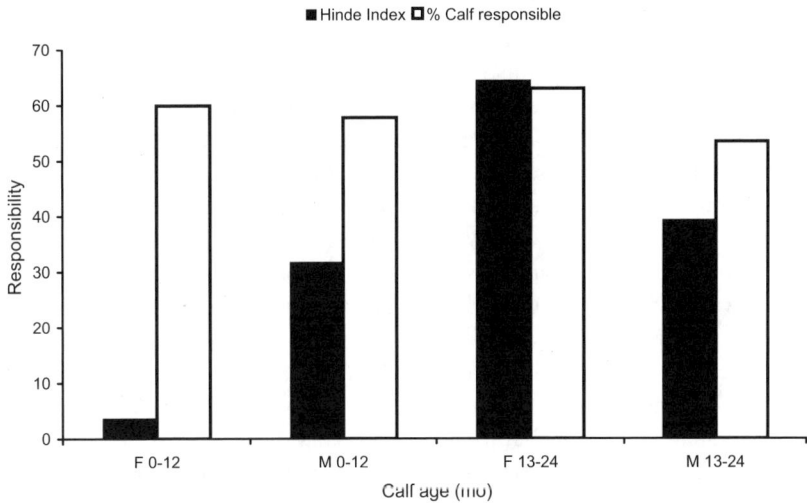

Figure 14.2 Responsibility for the maintenance of proximity within 5 m for male and female calves in the first two years of life. For Hinde's Index, positive values represent an infant who approaches its mother more than it leaves her; a negative index represents a calf who moves away more than it approaches. The "relative" values are Brown's (2001) index illustrating the relative responsibility for changes in proximity between mother and calf. Positive values indicate that the calf is relatively more responsible for all changes in proximity; negative values indicate that the mother is relatively more responsible.

of both sexes remain very close to their mothers (effectively within a trunk length; figure 14.1). Between 0 and 6 months, they spend 56 percent of the time within contact distance, and as calves age, they spend less time at very close distances to the mother and increasing amounts of time with the mother when she is more than 5 m away. It is not until a calf is well over 24 months of age that it is equally likely to be more than 5 m away from the mother as to be close to her and sons are more likely to be found at a distance than are daughters ($X^2 = 15.4$, df = 1, $p < 0.01$). The question is which partner in the mother-calf dyad moves away and which moves back into proximity, as this defines the relative responsibility that each takes in maintaining close proximity.

Throughout the first two years of a calf's life, changes in proximity within a 5-meter radius are relatively equally shared between calves and mothers, with few sex differences (figure 14.2). Male calves take slightly greater responsibility than do mothers for changes in proximity (responsible for 55.6 percent of changes) and thus appear to be more adventurous in seeking contact and proximity with other individuals as well as more responsible for returning to their mother, especially in order to suckle. Young female calves control proximity to their mothers in relation to their interactions

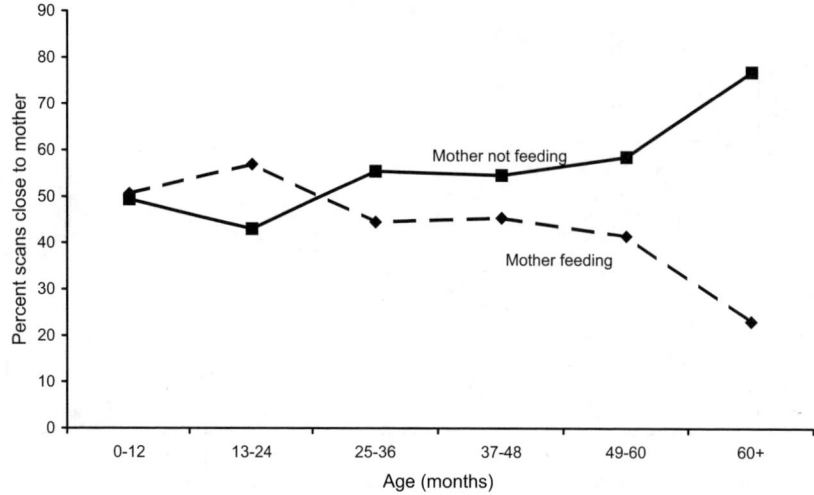

Figure 14.3 Distribution of scans when the mother was feeding (diamonds) or engaged in activities other than feeding (squares); males and females combined for calves close (contact distance) to mother.

with others. Female calves confidently leave their mothers in order to spend time near other family members, especially older juvenile females. Age changes in responsibility are probably context specific; while suckling, resting, or traveling, both sexes tend to remain with mothers, but when playing or interacting, males tend to seek novelty.

Calf proximity is influenced by maternal activity. When mothers are feeding, calves over the age of 24 months are likely to be more than 5 m away, while when not feeding, calves are closer (figure 14.3). Older calves may maintain close proximity in order to suckle at those times when the mother is relaxed rather than foraging. In order to feed efficiently while moving, adult females may travel at a pace that is too quick for a feeding young elephant, leaving older calves behind. For younger calves who are not sustaining themselves through independent foraging, keeping up with a mother when she is feeding may be less problematic. In addition, these young calves may rely on their close proximity to their mother in order to observe and sample maternal foods—socially mediated food learning—prior to becoming dependent on their own foraging (see also Lee and Moss 1999).

Proximity to Others

While proximity to the mother is a feature of the dynamics of the mother-calf relationship, proximity to other individuals reflects the highly social and cohesive nature of elephant families (see Moss and Lee, chapter 13). Within families, individuals generally maintain very close proximity. Calves tend to be found close (within physical contact distance) to a non-mother elephant for around 20 percent of time, and this tendency changes only slightly with age for males (figure 14.4). After two years of age, females tend to increase the time spent in close proximity to a neighbor, which probably reflects their increased attraction to and interaction with younger calves. Overall, only about 10 percent of time is spent with another elephant more than 5 m distant (mean females = 8 percent, males = 10.3 percent). Young calves, older juveniles, and even adolescent individuals are firmly embedded in the family social context.

The neighbors of male calves differ from those of females after the first year of life. The nearest non-mother neighbor of a male calf is more likely to be a novel, non-family male (7 percent, $N = 1{,}588$) than is the case for females with novel peer females (2.7 percent, $N = 1{,}758$, $X^2 = 31.8$, df = 1, $p < 0.001$). Males thus may be at greater risk of becoming separated from the family or encountering other environmental hazards and predators. There is also considerable variation among individual calves in their proximity to mothers, family members, and novel elephants, and this variation is likely to be associated with factors such as family unit size and composition, the presence and number of allomothers, family dominance relative to other families, the habitat or environmental context, and the activity state of the mother. In addition, whether the calf has same- or opposite-sex peers within the family will strongly influence his or her patterns of social interaction: males lacking peers may seek them out from other families, generating the novel neighbor pattern observed. Females may simply prefer other family members and have fewer preferences for their partners' sex.

Calf Exploration, Interaction, and Learning

Both sexes of calves explore their social and physical environment as a mode for learning about their world as well as generally improving skills such as the manipulation of objects

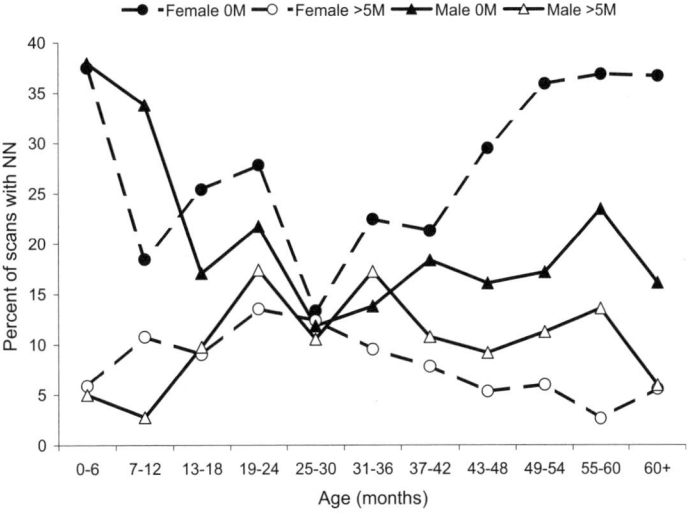

Figure 14.4 Percentage of scans with a nearest neighbor close (0 m or contact distance) or far (>5 m) for male and female calves at each age.

with their trunks. Food is gradually incorporated as a source of energy, with independent feeding (as opposed to simply "tasting", handling, or exploring foods) becoming an energetically significant contribution after about 9 months of age (Lee and Moss 1999). By the age of 20 to 24 months, calves are probably able to sustain much of their metabolic requirements through independent feeding, and the time spent feeding (while seasonally variable) tends to stabilize at around 60 percent after 24 months. As we note in chapter 12, calves orphaned at less than 24 months are unlikely to survive. The youngest female orphan to survive was 18 months, and the youngest male was 19 months. Both had supportive elder sisters and were orphaned in years of high rainfall and food availability. Food types need to be learned, as do sources of water, safe areas, and risks posed by the environment and potential predators. The context for food learning is primarily social and occurs within the family as a function of exploratory contacts with family members (Lee and Moss 1999). The first three sets of teeth are small and progress through the jaw rapidly. Calves are also short relative to the size of bush or tree foods, and because the strength and length of the trunk is limited, calves are therefore unlikely to be able to be forage independently and effectively on resources other than low browse, shrubs, grasses, or herbs until they are over five years. Elsewhere we have suggested that physical constraints are significant factors in food choice and nutrient intake for calves, and these constraints influence the pace of learning as well as opportunities for exploration (Lee and Moss 1999).

Calf Interactions: Aggression and Affiliation

Elephant calves have a wide variety of social contacts with non-mother adults and other calves (figure 14.5). Relaxed, friendly, and supportive contacts between calves and non-mother elephants are frequent, while the most common form of calf-calf interaction is play. Aggression or dominance-subordinance interactions involving calves are rare. Aggression initiated by calves is often directed at novel objects in the environment, such as birds or other mammals. Calves aggressively threaten with vocalizations, headshakes, and charges, and they chase small mammals such as warthogs or monkeys. They tend to avoid situations where others might attack them, and if harassed, chased, or poked, a distress trumpet or bellow will usually bring other elephants to their aid. Since most interactions involving a calf are friendly or supportive, young calves do not appear to have a specific strategy for the acquisition of dominance or status, unlike mammals such as primates (Pereira and Fairbanks 1993; Bercovitch 2000). Elephant dominance is primarily size or age related (Lee 1987; Moss 1988; Foley 2002; Archie et al. 2006; see box 12.2), and thus the youngest individuals would theoretically be the most subordinate. However, close mother-calf and calf-other associations protect calves from dominant elephants and probably facilitate access to food or water that calves could not obtain by virtue of their own status. Status differentiation may be related to behavior exhibited during adolescence, especially for dispersing males (see Lee et al., chapter 17).

Exploration of others, and others' explorations of calves, can be distinguished as greetings (often with a vocal component; see Poole, chapter 9), friendly body contact (rubbing or comforting), food explorations, and invitations to play (Lee 1987; Lee and Moss 1999). Greeting others is the most common calf interaction, occurring at the highest rate in the first 24 months of life and at twice the rate among female calves (Lee 1987). Other friendly contacts between calves

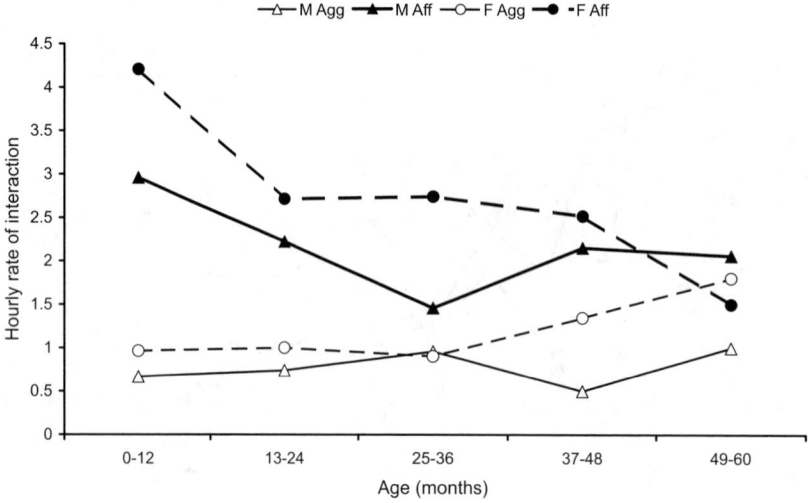

Figure 14.5 The hourly rate of friendly (closed symbols: greet, rub, investigate food, invite, play) and aggressive (open symbols: push, poke, threat, shove, chase) social interactions between calves and non-mother individuals (circles = females; triangles = males).

and others appear to be forms of testing—testing of a novel individual, an unusual state (distress, arousal, sexual), or another individual's foods. Rubbing against others, investigations of food, and invitations to play do not differ significantly between the sexes nor is there a consistent change with increasing age. While adult and adolescent males show relatively little interest in young calves (<24 months), when males do contact young calves, these contacts tend to be brief greetings (see Lee 1987). Juvenile and adolescent females interact most frequently with young calves and direct their greetings and comfort interactions to these calves. Interactions between the youngest calves tend to be reciprocal, and play invitations occur preferentially with peers. Overall, female calves are involved both in more friendly and more aggressive interactions than are males ($X^2 = 84$, df = 4, $p < 0.001$).

Calf Play

Elephant play is complex and varied. Occasionally, adults play with objects or engage in lone locomotor play (mock charges, running, spinning) accompanied by loud "play trumpeting" (see Poole and Granli, chapter 8). Adults of both sexes engage in gentle sparring (head-to-head or tusk-to-tusk contact with trunk pushing). Young and fully adult males will spar more forcefully and include whole-body pushing, but this behavior seldom escalates into aggression and is usually accompanied by nonaggressive ear and head postures (Poole and Granli 2004). Sparring can produce a state of high arousal but seldom results in wounding or retreats. Thus, it might be considered a form of adult play or status (weight, size, dominance) assessment.

Among calves, play can be categorized as object play, lone locomotor play, gentle contact play, and escalated contact play (defined in table 14.3). The most common forms of play change with age, with subtle differences between the sexes. Young calves of both sexes engage in escalated contact play as well as vigorous object and lone play. Lone play or object play was rarely seen after the first year of life. Rates of play are highest in the first six months of life (figure 14.6) and are relatively constant after this age. Older males (60-plus months) play more than do older females, and males in general spend a higher proportion of their daily activity budget in play ($X^2 = 7.6$, df = 1, $p < 0.02$) and have higher hourly play rates than do females (figure 14.6). Males and females both engage in gentle and escalated play for a similar proportion of bouts, but as calves age, the play types become more sex-specific.

There are obvious differences between the sexes in their play partners (figure 14.7). Males over 24 months play more with partners matched for age and sex (10.2 percent of partners for males, 5.2 percent females, $X^2 = 4.7$, df = 1, $p < 0.05$). Males also preferentially seek novel partners who are not members of their family (42.5 percent males, 22.2 percent females, $X^2 = 15.7$, df = 1, $p < 0.001$). To some extent, a preference for a matched-sex peer might necessitate seeking partners outside the family if few such peers are available within a family. However, this is not necessarily the case since synchrony of births within families can produce a disproportionate availability of age mates, at least 50 percent of whom will also be matched in sex. Thus, the trend for males to seek non-family play partners appears to be the result of active choice.

For immature elephants, as for many other mammals

Calf Development and Maternal Rearing Strategies 231

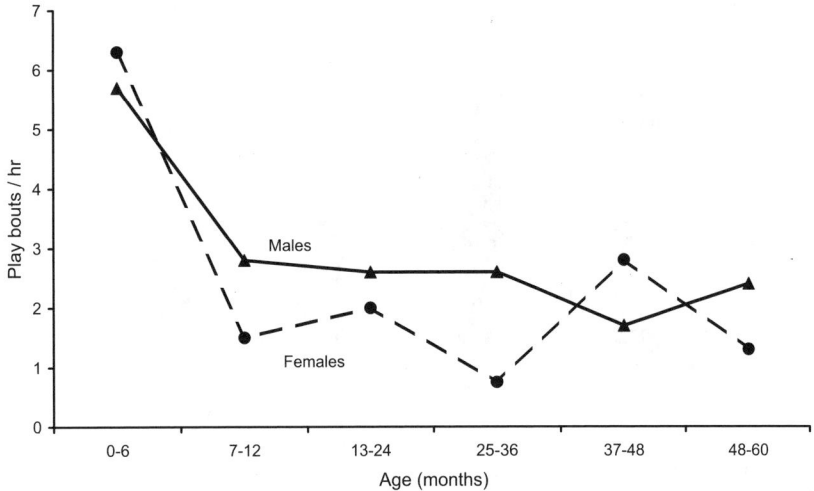

Figure 14.6 Rates of play (bouts/hr) for male (triangles) and female (circles) calves at each age.

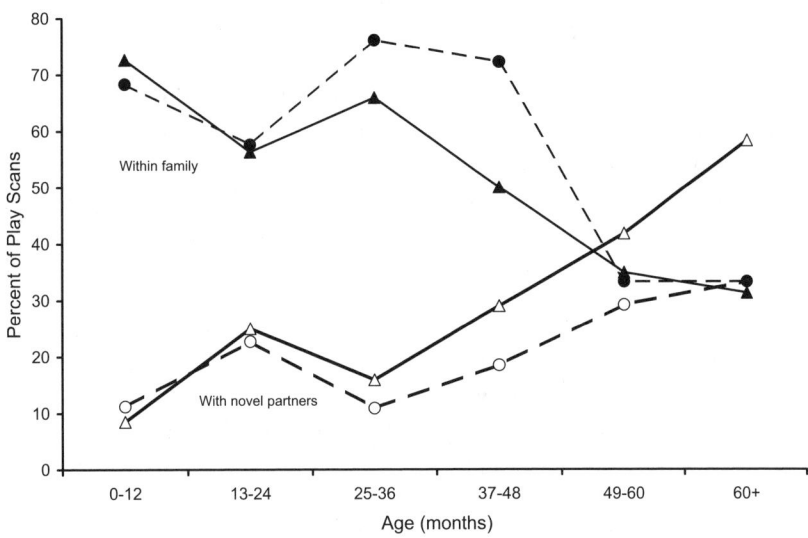

Figure 14.7 Familiar (family—filled symbols) versus novel play partners (open) for males (triangles) and females (circles) at each age (*n* scans male play = 400; *n* scans female play = 363).

(Fagen 1981; Bekoff and Byers 1998), play appears to serve a variety of functions. Bear cubs who play more have a higher probability of survival associated with rates of play alone (Fagen and Fagen 2009). These functions shift over the course of development and differ between males and females. Motor skill development is particularly apparent in locomotor and object play by the youngest age groups. Gentle contact play may have a twofold function. During gentle play among younger calves, social contact occurs in a physically undemanding context, allowing partners to gain knowledge of each other as individuals and initiate assessments of relative size and strength. Older males use vigorous play contacts in order to size up others, and assessment priorities shift to novel individuals in a relatively secure nonaggressive context. For older juvenile females, play becomes a mechanism for contacting young calves, serves to maintain close proximity with calves, and enhances experience in interaction with young calves. Play in this context might be a form of allomothering, stimulating younger calves while ensuring that an older individual is in close caretaking proximity to the most vulnerable age group. These allomothering play interactions in which older individuals play with calves under 24 months of age are clearly sexually differentiated: older females are far more likely to engage in play with young calves than are same-aged males (figure 14.8).

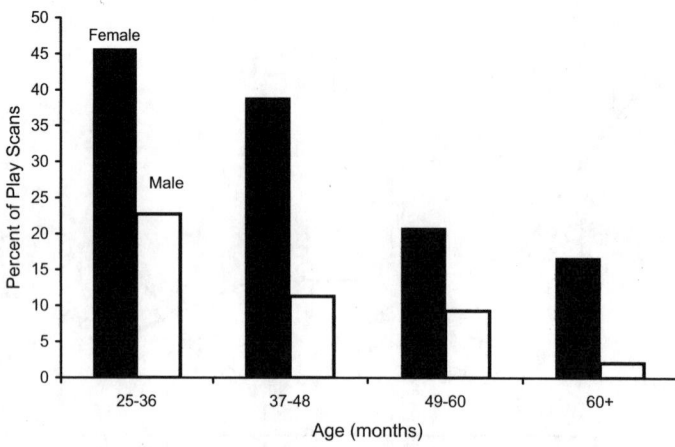

Figure 14.8 Allomothering play—gentle play between calves of either sex aged 0–24 months and older male or female individuals.

Is elephant play costly? Play occupies 1 to 12 percent (median = 2.7 percent, IQI = 2.2–3.3 percent) of an elephant calf's total activity budget. However, it represents 3.8 percent (range 2.2–15.8 percent) of time not spent resting and thus may add to total energy expenditure. If play were very costly, it would be expected to decline or drop out of the repertoire during periods of energy limitation (Lee 1984; Miller and Byers 1991). For calves under 24 months, no significant difference could be detected in rates of play between seasons or between normal and drought years, although the age-matched sample sizes are very small. In part, mothers might be sustaining the additional energy costs of play through lactation and effectively buffering calves under the age of 24 months from seasonal constraints on play. For older calves, play does appear to decline during times of stress. During the period when mothers limit suckling or weaning conflict, when rates of successful suckling are declining most rapidly, the rates of play drop off markedly, although play rates recover after this "weaning trough." The decline happens between 25 and 36 months for females but not until 37 to 48 months for males (see figure 14.6), ages that match both a drop in suckling rates and an increase in the rate of maternal rejection from suckling (Lee and Moss 1986; table 14.2). Although play consumes energy, its costs are short term while potential social benefits may be long term. Thus, diverting energy away from growth to play is unlikely to have post-weaning survival costs, as has been suggested for antelopes (Byers 1997; Miller and Byers 1991). Furthermore, much play when vulnerability to mortality is highest (0–24 months) occurs between young calves and allomothers or in proximity to allomothers, so play is unlikely to have survival costs due to increased exposure or vulnerability to predators or other environmental risks.

Maternal Rearing Strategies and Maternal Investment

Reproductive success for elephant mothers, as for most mammals (Clutton-Brock 1988), is a function of four components: age at first reproduction, survival during the reproductive period (longevity), reproductive rate, and infant survival (see Moss and Lee, chapter 12). Of these four elements, the last two are most strongly influenced by a mother's allocation of resources, energy, and time toward her existing calf. The trade-off between current and future infants is the classical dilemma of maternal investment (Trivers 1972), and when a trade-off exists, it is predicted to be biased toward the sex of the calf that provides the greatest fitness benefits to the mother.

Elephant rearing strategies are the result of a complex interaction between maternal age and size, which equate with a mother's general physical condition (see Moss and Lee, chapter 12); the sex of the calf, which determines its needs for energy intake in order to sustain growth; and the mother's past experience of rearing calves, which we suggest is a learned response to calf demands. This last element appears to be vital: mothers have to learn to maintain close proximity to calves, keep a watchful eye for hazards and predators, and respond to the suckling demands of their calves over a series of reproductive events (Lee and Moss 1986; and see below).

Assessing investment is problematic in that the majority of behavior directed toward infants is caregiving. While maternal patterns of care allocation have been found to contribute to variation in reproductive rates among females, we do not yet have a means of measuring the energy costs of caregiving behavior. Direct observations of how the allocation of maternal care influences infant growth and survival are rare although links between care allocation and

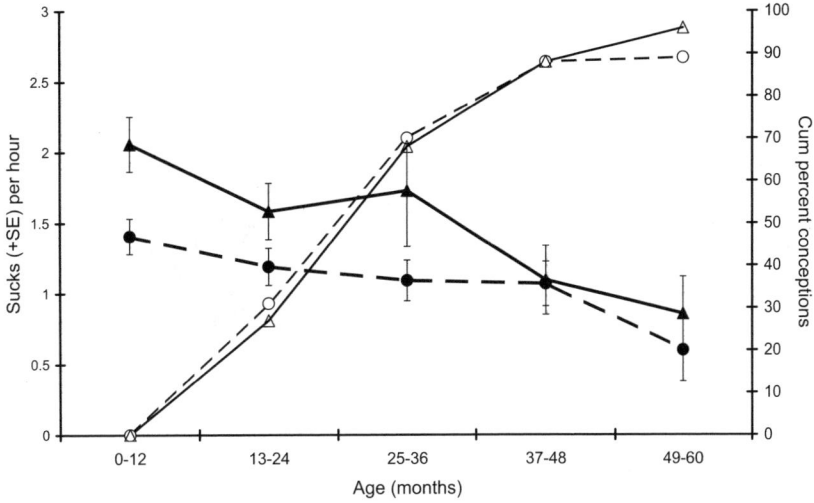

Figure 14.9 Mean hourly rate of suckling by age and sex (males = 135; females = 145; filled symbols) is plotted against the age of the calf when the mother reconceived (males = 58; female = 73; open symbols). Males = triangles; females = circles.

growth have been demonstrated for some sexually dimorphic mammals (elephants: Lee and Moss 1986; deer: Blaxter and Hamilton 1980; Clutton-Brock, Albon, and Guinness 1982; Kojola 1993; sheep: Hogg, Hass, and Jenni 1992; Festa-Bianchet 1988; Bérubé, Festa-Bianchet, and Jorgenson 1996; ungulates: Pélabon et al. 1995; pinnipeds: Trillmich 1996) but not for others (reindeer and caribou: McEwen and Whitehead 1972; bison: Green and Rothstein 1991; antelope: Byers and Moodie 1990; seals: Lunn and Arnold 1997; Ono and Boness 1996). Maternal patterns of care allocation contribute to variation in reproductive rates among females.

Many explorations of differential investment focus on sex ratio at birth (allocation of effort to the production of one sex over another). Amboseli elephants, irrespective of parity, maintain an equal birth sex ratio with no differential investment through the production of sons or daughters (see Moss and Lee, chapter 12). There is, however, differential investment in the timing of care allocation between sons and daughters, depending on maternal experience and age.

Lactation is probably the most energetically costly interaction between mothers and calves and is estimated to represent 75 percent of the costs of a reproductive event (Byers 1997). During poor seasons or years, mothers of calves under 24 months are at greater risk of mortality as a consequence of lactation costs (Lindsay 1994; see also Lee et al., chapter 17). Furthermore, suckling at high frequencies inhibits the ability of a mother to return to estrus for at least 9 to 12 months due to either hormonal insufficiencies or a complete block on reproductive hormone production. The probability of conceiving the next calf is related to a reduction in suckling frequencies to less than 1.5 to 2 per hour (Lee 1986; figure 14.9).

Male and female calves have different metabolic energy requirements as a function of their birth mass, growth rates, and energy expenditure while traveling in association with the mother. The mother must meet these needs or risk stunting her calf or causing its death—care-dependent infant mortality. Although suckling frequencies may not accurately reflect milk intake (Cameron 1998), the correlation between infant growth and suckling frequencies in several ungulates make this a useful proximate measure (Birgersson and Ekvall 1997; Byers 1997). Furthermore, the association between suckling and a delay to estrus suggests that suckling remains a reasonable proximate measure of care allocation for elephants.

The highest growth velocities are found early in life, and male calves have higher growth rates than do females throughout the lactation period (see below). Males demand more, are more successful at gaining access to the nipple, and have a higher total time spent suckling than do female calves (Lee and Moss 1986, 1999). Seeking access to the nipple is a function of male requirements for growth and thus is driven by the internal physiological needs of sons. Mothers appear to recognize and respond to these needs, but their ability to do so is a function of prior experience with calves. Inexperienced mothers respond to the demands of sons and daughters equally, while experienced mothers bias their allocation of milk, or at least time in nipple contact, toward sons (ANOVA on duration controlling for age; birth order $F_{2,280} = 2.44$, $p = 0.08$; ANOVA on frequency controlling for age; sex $F_{1,279} = 4.84$, $p = 0.02$; figure 14.10). Behaviorally, while mothers are generally tolerant,

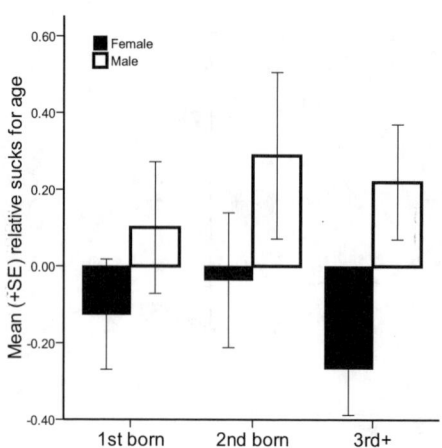

Figure 14.10 Age-specific frequency of suckling by sons and daughters for mothers of different experience (n 1st 37 m, 39 f; n 2nd 17 m, 30 f; n 3rd 56 m, 56 f).

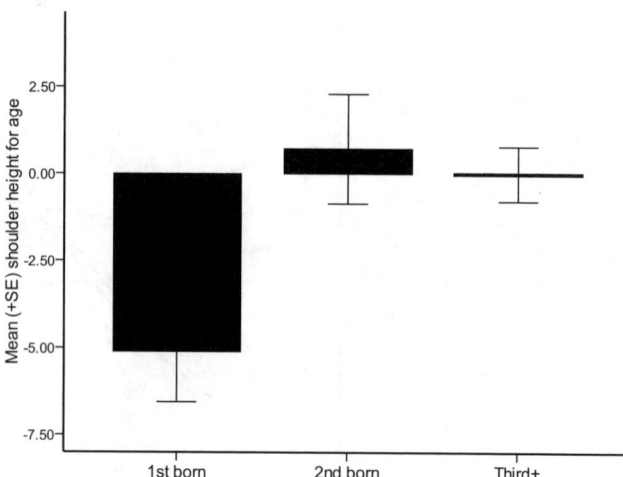

Figure 14.11 Relative shoulder height for age (computed separately for each sex) first-born calves ($n = 125$), second-born calves ($n = 73$), and third or greater ($n = 420$).

the meshing of mothers with calf demands is not an easy or rapid process. Mothers with first calves need to learn to respond to the suckling requests of calves as well as schedule their own activities in order to maximize the efficiency of allocation of lactation.

Mothers must be able physiologically to sustain calf needs through increased energy intake or depletion of maternal body reserves or by limiting their own growth while lactating. Thus, experienced mothers who are larger and older or mothers who are in good physical condition should be better able to sustain calf growth and respond to calf demands than would smaller, still-growing, or low-condition females. Further, these differences between mothers, theoretically, should be reflected in the growth and survival of male and female calves. Primiparous mothers have the lowest calf survival overall (see Moss and Lee, chapter 12). Mortality is particularly high for daughters in dry times, perhaps because the sons of inexperienced mothers gain access to the nipple for slightly longer periods than do daughters (post hoc comparisons, $p < 0.05$).

The greater responsiveness to male demands and increased time spent suckling sons suggests that they do "cost" more, and longer inter-birth intervals after sons reflect these costs. In sum, elephant mothers do not appear to invest differentially by sex in the production of calves but do invest differentially in the allocation of care between sons and daughters, with time costs in the form of a delay to subsequent reproduction. The cost of each sex differs between first-time and experienced mothers, as do the potential benefits. Thus, primiparous (small, growing, and inexperienced) mothers invest more time in daughters who may be more likely to survive even in bad times than are more costly sons, and these daughters can subsequently contribute to a mother's next reproductive event by acting as an allomother. Experienced mothers, by contrast, are more responsive to sons and appear to invest in these sons for a longer period compared with their daughters. For inexperienced mothers, the consequences of their poor allocation of care are reduced calf growth (ANOVA, $F_{2,212} = 4.05$, $p = 0.02$; figure 14.11) and high mortality especially of sons.

Maternal Investment in a Social Context: Families and Allomothers

The context for early development in elephants has both environmental and social components, which interact with rates of change, risks of death, and processes of skill acquisition. We explore the influence of the family on mother-calf relationships. Do family composition and size influence the nature of mother-calf interactions? Given the relatively few samples where calves were intensively followed over the long term, this question is difficult to answer. Larger families, especially those with more allomothers, are better at ensuring the survival of their calves (Moss and Lee, chapter 12). Mothers in larger families might be more "relaxed" with their calves, since there are many others with whom to share the burden of attentiveness. Such mothers may be able to feed more effectively and thus suckle their calves with greater energy efficiency. While such factors obviously influence the rearing strategy and success of individual mothers, they are difficult to detect in an analysis. However, the personality profiles of females (see box 13.2) suggest that some females are indeed relaxed, while others are more tense and

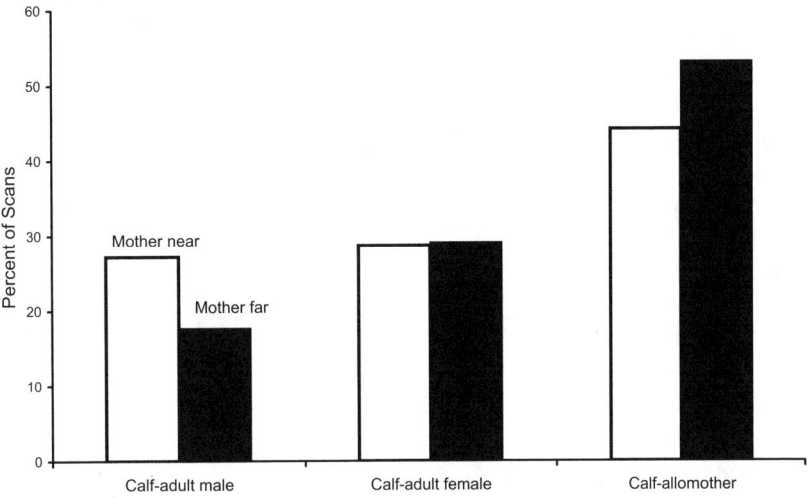

Figure 14.12 Time spent in close proximity to another elephant over two years old by calves aged 0–24 months when the mother is >5 m distant as opposed to close.

apprehensive, with potential consequences for their interactions with their calves.

One of the more interesting features of elephant society is the extensive network of non-mother caretakers—allomothers—the majority of which are female. Of non-play interactions between calves (0–24 months) and elephants over 5 years old, 76 percent are with females ($N = 135$). Furthermore, juvenile, adolescent, and nulliparous adult females are responsible for the majority (85 percent) of all responses to a calf signal of distress, either vocal (trumpeting, bellowing, protest rumbling) or physical (running, ear-flapping, or head-shaking). These attending females are called allomothers (see also Lee 1987). Allomothers range in age from 2-year-old juveniles to grandmothers but are most frequently juveniles over the age of 5 years and adolescent or nulliparous adult females under 15 years rather than lactating mothers (Lee 1987). Their primary role is that of guardian rather than provider of milk since suckling by allomothers is very rare (<4 percent of all suckling events; see Lee 1987). When suckling from other females does occur, it tends to be restricted to grandmothers or females who have just lost their own calves. Allomothers contact, greet, and reassure calves when they are separated or distressed or if they vocalize. These females pay attention to young calves, herding them away from predators, trouble, or dangerous obstacles and generally monitoring the calf's social and physical environment. Allomothers maintain close proximity to calves, an especially marked tendency when mothers are at a distance (figure 14.12). A calf may gain confidence as a result of being near an allomother and thus be more willing to move away from the mother, and the mother is more willing to move off and pursue her own activities when an allomother is present. Very young calves tend to lie down to sleep at every opportunity and thus risk being left behind by a moving family. Keeping an infant traveling with the family appears to be one important function of allomothers, which minimizes the chances of a calf being separated from the family. Allomothers also defend calves on the rare occasions when calves are poked, pushed, or aggressively contacted by other elephants.

Despite a lack of nutritional input from allomothers, calves born into families with several allomothers have higher survival rates (Lee 1987, 1989; see also Moss and Lee, chapter 12). Adoption of weaned orphans by another member of their immediate family has been observed on several occasions. As we note, mothers who give birth to a daughter in their first natality could conceivably pave the way for higher infant survival among their subsequent infants. Although the sex ratio at birth is not skewed toward females for primiparous females, the lower survivorship of sons produces a slight skew in sex ratios of the calves of primiparous females that survive to age 5, and therefore these female calves will be potentially available as allomothers ($X^2 = 2.1$, df = 1, $p = 0.07$ one-tailed). It is also possible that allomothering establishes a reciprocal relationship between the calf that receives care and the nulliparous female who provides care. With an age gap on average of 5 to 10 years between calves and allomothers, a female calf under 2 years will be available to act as an allomother for the first calf of her "partner" female at first reproduction between the ages of 10 and 15 years. Thus, a calf may subsequently aid the female who provided her with care when that female reproduces.

One primary function of the family unit is the defense of

infants in a context of high environmental risks and predators. The presence of allomothers does not appear to enhance growth rates of calves, as might be expected if they acted specifically to minimize foraging costs to mothers and thus released additional energy for calf growth (Whitehead 1996). If their primary role is protective, then in environments where hazards are fewer, a tightly bonded family, which facilitates allomothering behavior both across generations and between cohorts of siblings, might be less necessary. This might produce family structures more similar to the small fragmented units seen in forests and among Asian elephants (Fernando and Lande 2000; Sukumar 2003).

From the allomother's perspective, when immature, she will learn how to ensure that calves move with the family, what the risks and environmental hazards are to calves, and what contact and support calves need at young ages. Despite all this experience, primiparous females still are less competent mothers than are the older, experienced females (see Moss and Lee, chapter 12). It can be suggested that in the absence of allomothering experience, first-time mothers would be disastrously unprepared, as seen in zoo elephants (Clubb and Mason 1998; Clubb et al. 2008).

Long-Term Consequences of Early Social and Physical Development

We have described processes of social development among elephants aged between birth and 5 to 10 years old in relation to age and sex and in the context of interactions with mothers and others. We address the final question that we raised at the outset: can early development be linked to subsequent social and reproductive success?

Physical Development

One of the more significant influences on the health, survival, and success of elephants is that of early growth. Maximum growth rates are seen in the first 10 years of life; males grow faster than females during this period (figure 14.13), carry on growing for twice as long as females, and end up being 1.5 times taller than females (Lee and Moss 1995). Rapid male growth may be unassociated with hormonal activation, since the onset of spermatogenesis occurs at around age 10 with no increase in growth velocity. However, young males (15–25 years) have relatively high testosterone levels and periods of sexual activity (Rasmussen 2005), which may be associated even early in life with the development of sex-specific behavior and high growth rates.

For both males and females, low growth rates in early life are associated with an increased probability of mortality

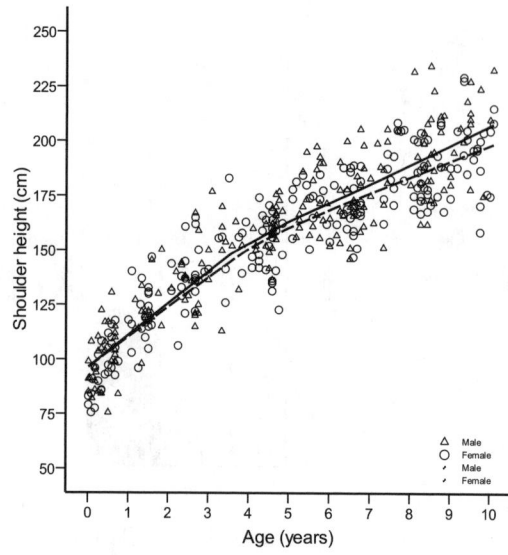

Figure 14.13 Measures of shoulder height for male ($n = 152$) and female ($n = 154$) elephants under 10 years of age (lowess curves fitted as trend line).

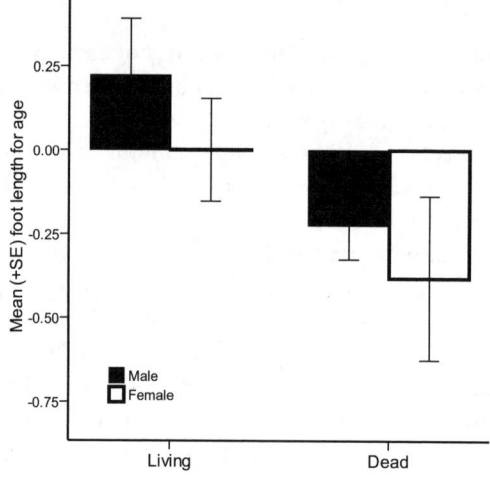

Figure 14.14 Relative size for age (from foot length) in the first 10 years of life for male and female elephants who either survived or died during this period.

(figure 14.14). Growth in the first 5 to 10 years appears to have a major impact on later success in reproduction for both sexes. It sets the pace for subsequent growth and possibly determines the onset of different developmental and reproductive stages. Calves born in "bad" years tend to have reduced growth rates and higher age-specific mortality throughout their lives. Females born in drought periods have a later age at first reproduction. Males who are large for their age (often those born in good years) have an earlier age at first musth. The effects of early growth, both positive and negative, are features of elephant biology and behavior, with consequences that potentially persist throughout an elephant's lifespan (Lee, Smith, and Moss 2009).

Social Development

As we discuss above, the first 2 to 5 years of social and environmental experience in an elephant's life has the potential to affect much of its later behavior. The learning of skills related to foods, habitats, and "friends" and "enemies" must play some role in the success of an adult. But, as is true of most studies of early development, it is difficult to demonstrate specific outcomes (Altmann 1980; Caro and Bateson 1986; Pereira and Leigh 2003).

All calves play and interact within their families; thus, there is no "deprived" group for comparison. Having a large or small family could affect some processes of development, since family size is a major factor in social dynamics. However, elephants are by no means limited to interactions within their families; they gather consistently in bond groups and seasonally in large herds (see Moss and Lee, chapter 13). The breadth of social opportunities is particularly clear in the case of older juvenile male-male play. Males actively seek out novel partners, crossing family unit and bond group social boundaries in order to engage with these individuals. If this level of flexibility is present for the majority of social interactions, what results is variation in social context and survival and reproductive tactics as a function of individual decisions or strategies rather than as a consequence of external constraints. Individual decisions determine the timing and process of male dispersal from the natal family (Lee and Moss 1999; see also Lee et al., chapter 17). The huge variation observed in the age of departure and how long this process takes is probably a function of individually distinctive patterns of physical and hormonal development placed into a social context of contact with familiar and novel peers, maternal status, and other family characteristics.

Knowledge plays a key role in elephant success; older females with greater experience act as repositories of social and environmental information (McComb et al. 2001; see also McComb et al., chapter 10). Experienced mothers have rearing strategies that differ from those of less experienced mothers, with consequences for the survival and growth of their calves. For elephants, more practice clearly enhances some social skills, especially those associated with the complex phenomenon of behavioral meshing between a mother and infant. Despite females having early experience as allomothers, mothers must learn to assess the needs of their calves, judge the "honesty" of these signals of need (e.g., Hauser 1993), pace their responses so as not to place themselves at risk of physical depletion, and strike a balance between staying close to calves for their protection while maximizing their own rate of food intake, which may require more rapid movement than is optimal for the calf. Size, condition, access to resources, and knowledge all increase with maternal age. As a result, rearing strategies shift, become refined, and have better outcomes for both mothers and calves.

Conclusions

We have described the developmental patterns of males and females over the first five years of life, noting those times and contexts where the sexes differ in behavior. Given that elephants experience a relatively long period (3–5 years) of dependence on the mother and an even longer period (8–10 years) of social dependence, we have described behavior relevant specifically to this immature phase (juvenile specializations) and the behavior upon which later adult social competence is built (scaffolding). Finally, we can ask: how does behavioral plasticity arise, and what are its consequences over the 65-plus years of an elephant's lifespan? A period of development extended over 10 years produces different kinds of problems and both forces and enables different solutions to these problems among elephants and other long-lived species such as whales (Whitehead and Mann 2000) or humans. These problems and their solutions contrast with strategies of rapid development among smaller mammals, where fewer options for developmental processes and responses are available and most mortality occurs within a brief window prior to age of the onset of reproduction. It is this enormous complexity in the context for development that gives the study of elephant immaturity, maternal rearing strategies, and social contexts for development the potential to illuminate patterns and processes that are of evolutionary significance.

Chapter 15 Friends and Relations: Kinship and the Nature of Female Elephant Social Relationships

Elizabeth A. Archie, Cynthia J. Moss, and Susan C. Alberts

The Evolution of Social Relationships

Across all animals, only a few species live in societies where individuals live together in permanent social groups. These stable and predictable associations promote the evolution of close and enduring social relationships among group members (Alexander 1974; Trivers 1971; West-Eberhard 1975). Such relationships are found in several species of non-human primates as well as some cetaceans, social carnivores, cooperatively breeding birds, and Asian and African elephants (Emlen 1982, 1994; Jennions and Macdonald 1994; Moss 1988; Silk 1987; Sukumar 2003; Whitehead 2003; Wrangham 1980). In all of these species, social relationships are thought to help individuals solve the problems of survival and reproduction; social partners work together to obtain resources, avoid predation, reduce parasite loads, gain mates, and raise offspring (Emlen 1982; Stacey and Ligon 1991; Vehrencamp 1983; Wrangham 1980, 1987).

Three major questions about the evolution of social relationships have been addressed in the literature. First, what kinds of benefits—in terms of Darwinian fitness—do animals receive from close social relationships? Second, do all partners benefit equally, or are some individuals being altruistic (experiencing direct fitness costs while they provide benefits) to their partners? Third, if some relationships are altruistic, how does that altruism evolve, given that an altruistic animal would have lower lifetime reproductive success than a non-altruistic animal, all else being equal?

A variety of answers have been proposed for these questions. If social relationships directly help animals survive and reproduce, the benefits of social relationships will accrue directly via an individual's own reproductive success. In addition, because most social groups are also kin groups, the benefits of close and enduring social relationships will also often accrue via indirect fitness benefits (kin selection). That is, to the extent that social relationships assist relatives, individuals will pass on their genes through the reproductive success of their genetic relatives (Hamilton 1964a, 1964b; West-Eberhard 1975).

These benefits of sociality, whether direct or indirect, may accrue to all social partners, for instance, when cooperative hunters work together in kin groups to kill large prey (Creel and Creel 1995; Girman et al. 1997). Alternatively, social interactions may be truly altruistic, i.e., initially costly to the direct fitness of one social partner but beneficial to the other partner. Altruism can evolve if it is traded reciprocally (Axelrod and Hamilton 1981; Trivers 1971) or for other services, as in a biological market (Barrett et al. 1999; Noë and Hammerstein 1994), or if altruistic animals reap the rewards of collective action (Nunn and Lewis 2001). Additionally, altruistic behaviors may be maintained partially or entirely by kin selection; that is, they may evolve because the altruist is genetically related to her partner, and thus she benefits indirectly by helping her kin reproduce (Hamilton 1964a, 1964b; Maynard-Smith 1964; West-Eberhard 1975).

What remains unknown is the extent to which different answers are true for different species. In particular, the generality of kin selection as an explanation for altruism is still a major question (Chapais 2001; Chapais and Belisle 2004; Clutton-Brock 2002). Hence, in order to understand which evolutionary mechanisms have been important in the origin and maintenance of social relationships, we need to know whether close social partners are also close relatives. Measuring the correlation between social structure and population genetic structure has only become feasible in the last 15 years with the advent of molecular techniques, including polymerase chain reaction (PCR), that allow relatively rapid and accurate assessment of individual genotypes. The results are revolutionizing the study of social behavior (see box 15.1).

Sociality in African Elephants

Elephants—both Asian and African—live in flexible, multi-tiered, fission-fusion societies (Douglas-Hamilton 1972; Moss and Poole 1983; Sukumar 2003; Wittemyer et al. 2005; Poole and Moss 2008; see Moss and Lee, chapter 13;). That is, social groups (also called families, family units, or core groups) are composed of discrete, predictable sets of individuals, but over the course of hours or days, these groups may temporarily divide and reunite, or they may fuse with other social groups to form much larger social units. The fluidity of this social system is an unusual context for the close and enduring social relationships that form among female elephants. In fact, this combination of social traits—close and enduring social relationships and fission-fusion sociality—is found only in elephants, in a few cooperatively hunting carnivores like hyenas and lions, possibly in some cetaceans, and in a few primates, including chimpanzees and humans (Dunbar and Dunbar 1975; Goodall 1986; Kruuk 1972; Kummer 1968; Schaller 1972; Whitehead 2003). In these taxa, the fission and fusion of groups may mean that social relationships are unusually complex because individuals interact with many animals from different social groups across the population, and social partners may not always be together in the same group (Connor et al. 1998; McComb et al. 2000, 2001; Poole and Moss 2008; see Moss and Lee, chapter 13). This social complexity

---

BOX 15.1 NON-INVASIVE GENETIC SAMPLING

Researchers who want to measure genetic relationships among individuals often require samples of DNA, and these can be extremely difficult to obtain from large wild mammals, like elephants, without disturbing them. A major advance in field studies of social behavior in the past 15 years has been the development of methods that allow us to conduct genetic analyses on DNA extracted from dung. In particular, small, polymorphic genetic markers, such as microsatellites (short, repetitive sequences of DNA that vary in length across individuals), make it possible to amplify DNA from dung samples where the template DNA is often degraded into short fragments. Furthermore, species-specific primers allow us to amplify these markers from the DNA of only the target species—exclusive of the DNA from other organisms, like gut bacteria or the plants that elephants eat, which are also present in dung. However, non-invasive genotyping presents a number of problems, including extraordinarily low quantities of template DNA or the presence of compounds that inhibit PCR reactions, but a number of researchers have developed robust methods for successfully combating these problems (Broquet and Petit 2004; Morin et al. 2001; Navidi, Arnheim, and Waterman 1992; Taberlet et al. 1996).

Specific descriptions of the methods we used to construct accurate genotypes of the Amboseli elephants can be found in Buchan et al. (2005) and Archie, Moss, and Alberts (2006). Briefly, we collect fecal samples from known individuals, almost always within a few minutes of defecation.

These samples are stored in ethanol at ambient temperature in the field, sometimes for over a year, until we extract the DNA. From this DNA, we amplify 11 microsatellite markers, i.e., highly variable and highly repetitive sequences of nuclear DNA (Archie, Moss, and Alberts 2003, Archie et al. 2006; Nyakaana and Arctander 1998). We also sequence a highly variable region of the mitochondrial genome—which is only inherited through the mother—called the control region, or "d-loop" (Greenwood and Paabo 1999; Nyakaana and Arctander 1999). Whenever possible, each genotype is replicated from an independently collected fecal sample, and in all cases, genotypes are replicated several times from the same fecal sample to ensure that we construct complete and error-free genotypes (Archie, Moss, and Alberts 2006; Buchan et al. 2005).

The mitochondrial d-loop data allow us to group animals into discrete, distinct maternal lineages. If two animals share a d-loop sequence, they are not necessarily closely related, but they did necessarily share a common female ancestor sometime in the evolutionary past. The microsatellite data allow us to assign estimates of relatedness to each pair of females in the population (Queller and Goodnight 1989). That is, for each pair of females, the microsatellite data produce an estimate of r, the coefficient of relatedness, which is a measure of the proportion of genes shared between the pair. For reference, r between parents and their offspring is 0.5 because offspring get one half of their genes from each parent, while r between full siblings is also 0.5 and r between half-siblings is 0.25.

may be one reason why elephants have unusually large brains and apparently high intelligence (Byrne and Whitten 1988; Connor et al. 1998; Jolly 1966; Rensch 1957; Plotnik, de Waal, and Reiss 2006; Poole and Moss 2008; and see Byrne and Bates, chapter 11). The social complexity hypothesis posits that high intelligence has evolved to help individuals cope with the labile and often difficult-to-predict behavior of group members (Jolly 1966).

Many features of close and enduring female social relationships are remarkably similar across a wide variety of social animals. In particular, the bonds that form among female elephants closely resemble those seen in several species of non-human primates and carnivores (Douglas-Hamilton 1972; Moss 1988; Silk 1987). Like many of these species, female elephant social partners preferentially rest, feed and travel together, and engage in a wide variety of affiliative behaviors like social rubbing, gentle touching with their trunks, and contact-calling in order to maintain proximity (Charif et al. 2005; Dublin 1983; Lee 1987; Moss 1988; Poole et al. 1988; Poole, chapter 8; Poole and Granli, chapter 9). Because elephant social partners are sometimes separated by large distances, their contact calls have an infrasonic component that can be individually recognized from as far away as 2.5 km (McComb et al. 2003). Reproductive and other socially relevant information may be transmitted at 4 to 10 km (Langbauer et al. 1991; Larom et al. 1997). The oldest females in the population recognize an unusually large number of these calls and may be able to interpret the calls of around 100 other females from an average of 14 other families in the population (McComb et al. 2000). After social partners who have been separated find each other again, they engage in greeting behaviors; individuals trumpet loudly and rumble, spin around, defecate and urinate, stream from their temporal glands, and reach their trunks into each other's mouths (Douglas-Hamilton 1972; Moss 1988; Poole, chapter 8; Poole and Granli, chapter 9).

Female elephants may benefit directly from the affiliative interactions described above, but these interactions might also reflect the strength of cooperative social relationships (i.e., those that are presumed to be mutually beneficial—either via kin selection or other mechanisms described above—and might have substantial impacts on individual fitness). Such cooperative relationships include the sharing of ecological and social knowledge, cooperative defense of calves against predation, allomothering, and coalitions in order to obtain resources (Archie, Moss, and Alberts 2006; Dublin 1983; Foley, Papageorge, and Wasser 2001; Lee 1987; Lee and Moss 1999; McComb et al. 2001; Sukumar 2003; Poole and Moss 2008; Mutinda, Poole, and Moss, chapter 16; Moss and Lee, chapter 12). We do not yet know which evolutionary mechanisms are most important in the origin and maintenance of these cooperative social relationships, but it is possible that female elephants benefit from these relationships by both increasing their own reproductive success and helping their relatives to do so.

Hence, in this chapter, we review the role that kinship plays in the spatial and social relationships that occur among adult female African elephants in Amboseli.

- First, we describe the range of genetic relationships that occur among adult females that live in the same family group. Long-term observations indicate that nearly all female elephants are matrilocal (Douglas-Hamilton 1972; Moss 1988). We use genetics to confirm this observation.
- We then describe the correlation between kinship and the patterns of fission and fusion within families. The degree to which kin are predictably together in the same group determines the opportunities that individuals have to influence their indirect fitness.
- Next, we move beyond family-level association patterns and describe the degree to which kinship predicts fission and fusion between family groups across the population. Some families that form close social ties with each other (i.e., families that constitute bond groups) are hypothesized to have once been part of the same original family that underwent a more permanent fission at some point in the relatively recent past (Douglas-Hamilton 1972; Moss 1988).
- In the last section, we investigate whether kinship influences affiliative, cooperative, and competitive social relationships by first discussing whether female elephants seem to discriminate among affiliative and cooperative social partners on the basis of kinship and then investigating whether kinship predicts dominance rank relationships.

Genetic Relatedness and the Fission and Fusion of Female Elephant Groups

Genetic Results: Fission and Fusion within Families

Our first question was whether genetic analysis would support the hypothesis of female matrilocality, i.e., the inference drawn from observations that females remain with their maternal kin (Douglas-Hamilton 1972; Moss 1988; Moss and Poole 1983). While males disperse from their natal group at maturity (Lee et al., chapter 17), females are thought to be matrilocal. In order to address this question, we sequenced the mitochondrial d-loop and found four distinct d-loop sequences among the Amboseli elephants. These essentially represent four distinct and relatively ancient "matrilines" that have grown and spread (Georgiadis et al. 1994). In Amboseli elephants, almost all females in any given family share the same mitochondrial haplotype (Archie, Moss, and Alberts 2006). This result confirms that, indeed, females tend to remain with their maternal relatives. Female Asian elephants also seem to be matrilocal because family groups of Asian elephants also have uniformity of mitochondrial d-loop sequences (Fernando and Lande 2000).

While female elephants are matrilocal, movement of females between family groups does occur in Amboseli elephants on rare occasions. We estimate that no more than 10 percent of families in Amboseli contain a female immigrant (Archie, Moss, and Alberts 2006). In many cases, we suspect that these females immigrated because they were the sole surviving adult member of their natal group, which was destroyed by illegal hunting by humans. Instead of living alone, these females began to associate on the periphery of another family and eventually became integrated into that family (see box 13.1), even in the absence of close relatives (Archie, Moss, and Alberts 2006). However, females that lost their natal group are not the only immigrants. In at least one rare case in Amboseli, two females left their respective natal groups, which were still intact, and formed a new family group together (Moss 1988). We were unable to collect genetic samples from these females, so their genetic relationship remains unknown.

Next, we characterized the patterns of genetic relatedness between females that live in the same family group by measuring the extent to which they share alleles at microsatellite loci. We found that average pair-wise genetic relatedness within elephant families in Amboseli was 0.14—on the order of aunt-niece relationships (figure 15.1; Archie, Moss, and Alberts 2006). This reflects a high level of relatedness within families and again supports the observation that females are matrilocal.

We also found that mother-offspring pairs were the closest genetic relationships within families; full sibling relationships were rare (Archie, Moss, and Alberts 2006; Hollister-Smith 2005; Hollister-Smith et al. 2007; Archie et al. 2008). The next closest genetic relationships occur among maternal, and occasionally paternal (Archie, Moss, and Alberts 2006; Hollister-Smith 2005; Hollister-Smith et al. 2007), half siblings. A large range of other kin relationships also occurred within families in Amboseli (figure 15.1), commonly including aunt-niece, first cousin, and grandmother-granddaughter relationships. Interestingly, some females had no close relatives in their family. A few of these females were immigrants, as described above, but females more often had few or no close kin (i.e., relationships closer than cousins) because of natural demographic events, like deaths and the chance births of many male offspring.

Despite the range of kin relationships that occurred

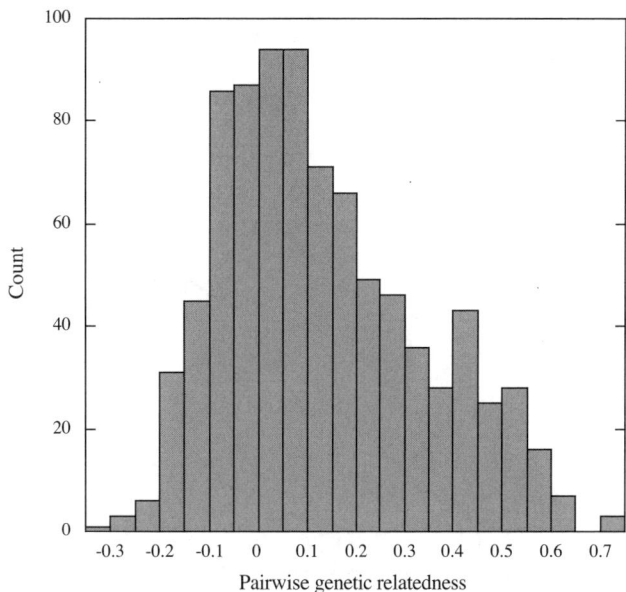

Figure 15.1 Histogram of pair-wise genetic relatedness among 866 pairs of female elephants, where each member of the pair belongs to the same family group (N = 44 families). Pair-wise genetic relatedness values are estimated from individual genotypes at 11 microsatellite loci.

within elephant family groups, when family groups fissioned temporarily (i.e., over the course of hours, days, or weeks), close relatives tended to remain together. Average genetic relatedness among groups of adult females that spent more than 90 percent of their time together in the same group was 0.39 (Archie, Moss, and Alberts 2006). This indicates that the most closely associating females were very often first-order maternal relatives (i.e., mothers and offspring and maternal siblings). As individuals expanded their social relationships to include other family members, average relatedness within the group declined. For instance, average relatedness within groups of females that spent 80 percent of their time together was 0.32; among females that spent 60 percent of their time together, it was 0.28 (figure 15.2; Archie, Moss, and Alberts 2006). These results indicate that despite living in a fission-fusion society, female elephants remain with their closest kin. Thus, the close and enduring social relationships among female elephants are susceptible to kin selection (Archie et al. 2006).

Genetic Results: Fission and Fusion between Families

We found strong evidence that bond groups in Amboseli usually form as a consequence of the permanent fission of a family. Families in the same bond group almost always had the same mitochondrial haplotype. Furthermore, the matriarchs of these closely associating families tended to be close genetic relatives—on the order of first cousins (figure 15.3a; Archie, Moss, and Alberts 2006). However, the

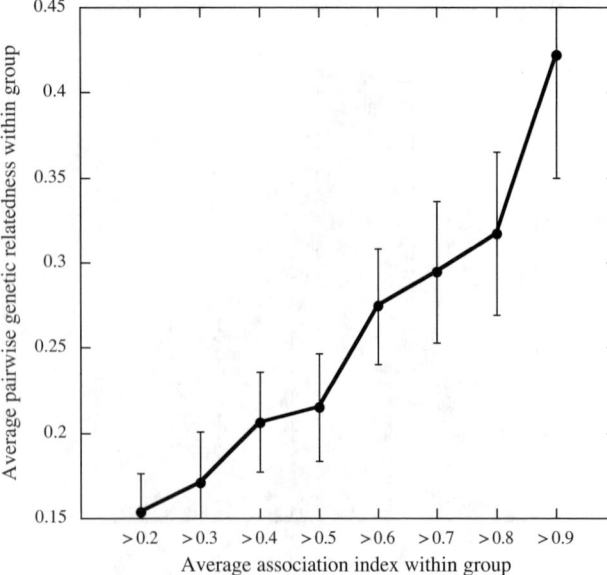

Figure 15.2 Plot of the average pair-wise genetic relatedness within a group of female elephants from the same family group as a function of the average association index within that group. Standard errors are calculated by jackknifing across loci.

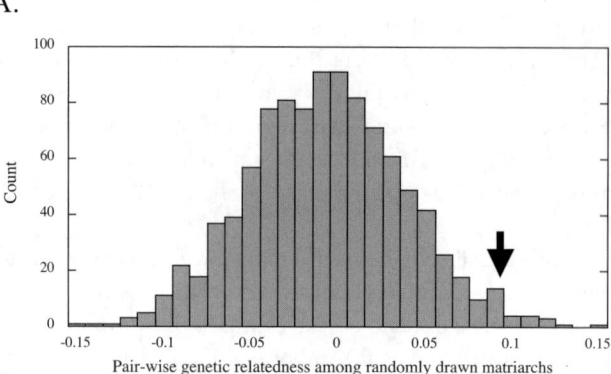

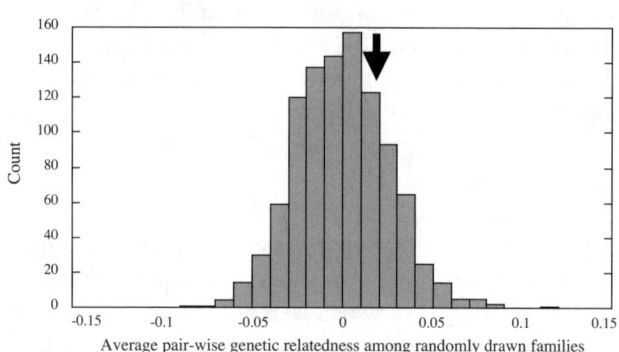

Figure 15.3 (A) Bars represent the distribution of average pair-wise genetic relatedness between matriarchs for 1,000 pairs of families randomly drawn from the Amboseli population. The arrow indicates the average pair-wise genetic relatedness among matriarchs that belong to the same bond group but not the same family. (B) Bars represent the distribution of average pair-wise genetic relatedness between adult females from 1,000 pairs of families randomly drawn from the Amboseli population. The arrow indicates the average pair-wise genetic relatedness among females that belong to the same bond group but not the same family.

average pair-wise genetic relatedness between adult females that were bond group members was not significantly higher than expected compared with average pair-wise genetic relatedness between pairs of families drawn randomly from the population (figure 15.3b, Archie, Moss, and Alberts 2006). The fact that most bond group members are not close genetic relatives probably reflects the fact that female deaths and male gene flow quickly erase the signal of genetic relatedness between families. Together, these pieces of evidence indicate that when family groups of elephants fission permanently (see box 13.1), the members of those newly formed families continue to associate with each other even after many of the closest maternal kin have died (Douglas-Hamilton 1972; Moss 1988).

Consistent with the observation that matrilineal fission characterizes both temporary and permanent family fissions, maternal kin relationships predict the fission and fusion of groups across the entire population. Family groups that shared the same mitochondrial haplotype were slightly, but significantly, more likely to fuse into a temporary group (Archie, Moss, and Alberts 2006). This result held true even when bond groups were excluded from the analysis. Further, these matrilineal affinities occurred even though none of the members of these families—including the matriarchs—were close genetic relatives as measured by microsatellite markers (Archie, Moss, and Alberts 2006).

Families that share mtDNA may fuse into temporary groups for two reasons. First, although these families underwent a more permanent fission at some point in the past, they may continue to share similar home ranges and therefore may have a greater chance of meeting and forming a temporary group. Second, these families may continue to associate through associative learning; when calves are young, they may learn which females and families in the population are familiar social partners. In this way, the association patterns persist long after the females who were related and part of the original family group have died. Among Amboseli elephants, only 7 of 50 family groups had fissioned permanently in the first 30 years of the study (see Table 13.1). Assuming a constant rate of family fission, this finding indicates that these patterns of spatial association among families that share mtDNA may have persisted for hundreds of years. There is strong evidence that associative learning, mediated by olfactory, auditory, or visual cues, is an important feature of elephant social organization (Lee and Moss 1999; McComb et al. 2001; Moss 1988). In particular, older females are thought to have accumulated the most knowledge about which social partners are familiar associates and where the best resources can be found when food and water are limited (Foley 2002; McComb et al. 2001; Moss 1988; Mutinda, Poole, and Moss, chapter 16). Families that contain one of these experienced, older matriarchs have higher calf survival rates than do families with younger, less experienced matriarchs (McComb et al. 2001; Moss and Lee, chapter 12).

We have shown that, in Amboseli elephants, maternal relationships are an important predictor of the patterns of fission and fusion among family groups (Archie, Moss, and Alberts 2006); however, Charif et al. (2005) found a different pattern in Sengwa Wildlife Research Area, Zimbabwe. Sengwa elephants are the only other African elephant population that researchers have investigated with regard to the relationship between kinship and the patterns of fission and fusion among family groups (Charif et al. 2005). Although bond groups were not identified explicitly by Charif et al. (2005), the authors did identify eight pairs of females that exhibited coordinated movements across the habitat, despite being from different families. These associations may be similar to the bond groups we observe in Amboseli elephants, but Sengwa bond groups differed from Amboseli bond groups in that they were not more likely to have the same mitochondrial DNA haplotype as expected by chance; hence, they were not maternal kin.

One possible reason for this discrepancy is that culling has had an enormous impact on the Sengwa population (Charif et al. 2005). Between 1978 and 1986, approximately 1,400 elephants were culled from Sengwa, which appears to have reduced the total population by as much as 75 percent (Osborn 1998). Culls attempted to remove entire family units, and as a result, many families may have lost their original bond group partners. These families appear to have formed associations that resemble bond groups but with families that are not maternal kin. The fact that matrilineal kinship does not predict association in Sengwa elephants suggests that bond group partnerships are somewhat flexible. Such non-kin-based associations between families also sometimes occur in Amboseli elephants, but these associations are not as close as those observed among families that share mitochondrial DNA (Archie, Moss, and Alberts 2006). The discrepancy between Amboseli and Sengwa elephants is strong evidence that poaching, culling, and other human interventions can destroy the degree to which maternal relationships predict social relationships in populations. If female elephants discriminate among social partners on the basis of kinship and if social partners strongly influence each other's reproductive success, there may be serious implications for culling practices that remove whole families or parts of families and for captive management procedures that involve moving elephants from one captive facility to another.

Kinship and Female Elephant Social Relationships

It is clear from the analysis of fission-fusion patterns that females preferentially remain with their close kin during both permanent and temporary fissions. This pattern might lead us to expect that the relationships that females have with their close kin are very distinct from their relationships with more distant kin or with non-relatives. Indeed, in Amboseli elephants, kinship appears to determine, at least partially, the nature of female social relationships. For instance, female elephants directed most of their affiliative social interactions (i.e., rubbing and greeting) toward their closest maternal relatives rather than toward more distantly related group members (figure 15.4; Archie 2005; Lee 1987; Poole, chapter 8). Furthermore, pairs of adult females that were more closely related were more likely to form cooperative coalitions in order to obtain resources than were pairs of more distant kin (Archie 2005). Kinship may also influence other types of cooperative relationships. Poole (chapter 8) describes how relatedness determines the pattern of "little-greetings" within a family. Lee (1987) reported that siblings commonly act as allomothers, and it may be that the contributions of allomothers vary with respect to their relatedness to the calf. It is also possible that the average relatedness of the group predicts the likelihood that a family will bunch in response to a threat or that knowledge about resources is shared among social partners, but these hypotheses are as yet untested.

However, one important caveat to the finding that female elephants tend to form their closest social relationships with their maternal kin is that females do not exclude non-kin from cooperation and affiliation. In fact, females without any close kin in their family were just as likely to have close neighbors, receive affiliation, or engage in a cooperative coalition (two or more females cooperating to aggress against another individual or individuals) with another group member as were females that had many close maternal kin (Archie 2005). One reason for this pattern is that kinship does not appear to be the only important component of female elephant social relationships. While kinship explained some of the variability in affiliative relationships between females, the amount of time two animals spent within 5 m of each other was actually a better predictor of how often they engaged in social rubbing or greeted each other after being separated (Archie 2005). More importantly, whether a given pair was very affiliative and spent a lot of time together was a much better predictor than kinship of whether the elephants formed a cooperative coalition (Archie 2005).

There are many reasons why kinship may not be the only or the most important component of affiliative and cooperative social relationships among female elephants. For instance, elephants may not be very good at true kin

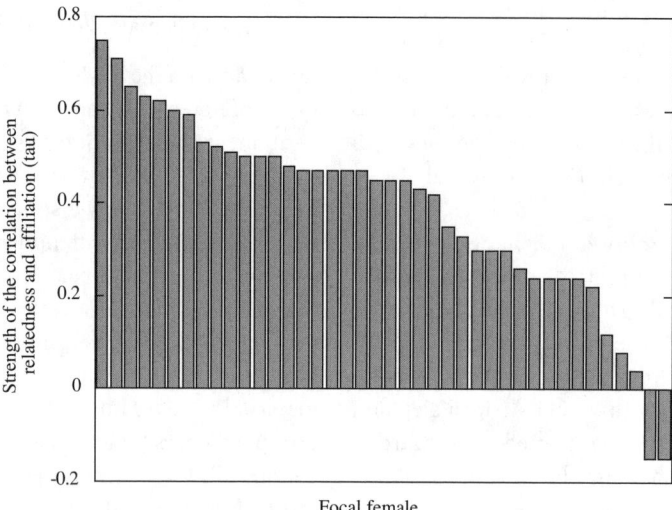

Figure 15.4 Bars represent the strength of the correlation (tau) between genetic relatedness and the level of affiliation with a given focal female and her adult female family members. The correlation between relatedness and affiliation was positive for 38 of 40 focal females. Correlations were calculated using a nonparametric row-wise correlation procedure (see Archie 2005 for details).

recognition, though we think this is very unlikely given their good memories for social relationships and the fact that elephants could use vocal or olfactory cues to recognize kin (McComb et al. 2000, 2001, 2003; Moss 1988; Poole et al. 2005; Rasmussen and Krishnamurthy 2000). Another possible explanation is that female elephants actually benefit from expanding their social relationships beyond their closest maternal relatives. In particular, several of the cooperative activities that female elephants engage in may be more effective or more beneficial when larger numbers of individuals participate (Mutinda, Poole, and Moss, chapter 16). For instance, larger groups may form more effective defensive bunches against predators. In addition, larger groups may be better able to maintain access to resources against other groups. Because female elephants have slow reproductive rates, their matrilines are relatively small. Hence, female elephants may benefit greatly from expanding their social relationships beyond their closest maternal relatives to include non-kin.

Dominance Relationships within Family Groups

While affiliative relationships among highly social females often appear similar across a wide variety of species, one aspect of social relationships that is quite variable is the pattern of dominance rank relationships. In some species, dominance relationships among females are egalitarian; dominance interactions may be rare, and when they occur, the outcome is unpredictable. In other societies, aggression is more common, and the outcome of dominance rank relationships is relatively predictable such that females that are older, larger, or more aggressive are more likely to win a given conflict. Some of the most interesting dominance rank relationships occur in "female-bonded" primates, like baboons and macaques, and in spotted hyenas. In these societies, kin form close social relationships and help each other in conflicts, which leads to the development of "nepotistic" dominance hierarchies in which daughters inherit a rank close to that of their mother and whole matrilines outrank other matrilines within the same social group (Engh et al. 2000; Gouzoules and Gouzoules 1987; Holekamp and Smale 1990).

Because elephants also form close and enduring relationships with kin, their social relationships resemble, at least superficially, the relationships that occur in "female-bonded" societies. As a result, one might expect that they also have nepotistic dominance hierarchies. However, while female elephants do have dominance hierarchies and clear patterns of rank relationships (see also box 12.2), kinship has no influence on dominance rank relationships among female elephants (Archie et al. 2006). Instead, dominance rank relationships are almost entirely ordered by age and therefore size, with very few reversals (Archie et al. 2006). Hence, in contrast to nepotistic hierarchies in which younger, smaller animals necessarily outrank older, larger animals from lower-ranking matrilines, female elephants rarely win dominance interactions against the age/size hierarchy of the group. These patterns of dominance rank relationships also support the hypothesis that kinship is not the only important component of social relationships among female elephants. Instead, the presence of a clear age-ordered hierarchy may simply reflect the fact that conflict may be dangerous for elephants because of their large body size and formidable weaponry; a well-defined set of rank relationships may reduce the risk of injury by reducing the probability of escalated fights.

Conclusions

Female elephants are similar to social carnivores, some cetaceans, and many non-human primates in that they form close and enduring social relationships with other female social group members. In many of these societies, the most affiliative and cooperative relationships form preferentially among kin, and elephants are no exception. However, elephants differ from many of these social species in several important ways. First, their social relationships are unusually fluid and rival those of chimpanzees and humans in their complexity. Second, elephants have the opportunity to interact with a very wide range of social partners, more so than social canids and chimpanzees and probably similar to some cetaceans. Third, in elephant family groups, kinship has no bearing on dominance rank relationships. Instead, rank is predicted by age or size. Together, these features of elephant relationships suggest that female elephants balance a complicated array of selective forces and that social relationships with kin are just one way that female elephants solve the problems of survival and reproduction.

Chapter 16 Decision Making and Leadership in Using the Ecosystem

Hamisi Mutinda, Joyce H. Poole, and Cynthia J. Moss

Free-living elephants must find food, water, and mates in a large, ecologically unpredictable, and socially dynamic landscape. As elephants compete for these resources, dominant families, parties, and individuals will access higher-quality food resources and in the case of males, a greater number of mates than those who are lower in rank (Poole 1989b; Archie et al. 2006; Hollister-Smith et al. 2007; Wittemyer et al. 2007). As elephants attempt to maximize their intake rate, they must make frequent decisions regarding when and where to go, with whom, and how long to stay in particular group configurations and specific locations. As predicted by optimal foraging theory (e.g., Belovsky 1978; Sorensen 1984), elephants must decide which groups and locations to gravitate toward or avoid so as to minimize within- and between-group competition and reduce encounters with (mainly human) predators. When water and forage are widely available, such as during the wet season and in open grassland, scramble competition ensures that individual gains are limited only by speed of intake. During the dry season, however, the uneven distribution of resources may result in intense contest with dominant families, parties, and individuals, benefiting more than others.

While foraging and mate-searching decisions by adult males are tailored to maximize individual fitness (Poole 1989b; Poole and Moss 1989; Hollister-Smith et al. 2007), those taken by females are likely also to be associated with increasing inclusive fitness, as is the case with many species living in kin-based social groups (Armitage and Schwartz 2000; Lyon and Eadie 2000). Thus, among family groups, there is a constant conflict between affiliative behaviors, which increase inclusive fitness (the "pull" factors), and those that reduce inter-group and intra-family competition (or "push" factors). Decisions by adult females are, consequently, aimed at balancing the negative effects of group living against the improved chances of survival and reproduction offered by sociality (see Moss and Lee, chapter 13).

Matriarchs, who are the oldest and most experienced members of the family, have a strong influence over the decisions taken by family units (Moss 1988), often appearing to determine group movement. Although other members of the family might initiate a movement or propose the direction of routine travel (Mutinda 2003; see Poole, chapter 9), the matriarch's own activity and behavior typically determines the group's departure time and direction from particular locations (Poole and Moss 1989).

Amboseli matriarchs differ in age and experience, and they lead family units of varying sizes (see Moss and Lee, chapter 13) and social rank. Each family unit's home range overlaps with the ranges of many other families (see Croze and Moss, chapter 7), and each must compete for the limited resources found within this habitat. While larger families may experience higher levels of stress caused by within-family competition, these generally dominant families may suffer less from the negative consequences of "despotic foraging" encounters with other families (e.g., Dublin 1983). We suggest that family units adopt strategies individually tailored to allow them to cope effectively with the level of competition experienced. Consequently, decisions taken by high-ranking families are likely to differ from those taken by low-ranking families. In addition, the members of a family are not always together; hence, individuals may find themselves making decisions with regard to their position within fragments of families, or "parties" (*sensu* Chapman, Wrangham, and Chapman 1995; see box 1.1).

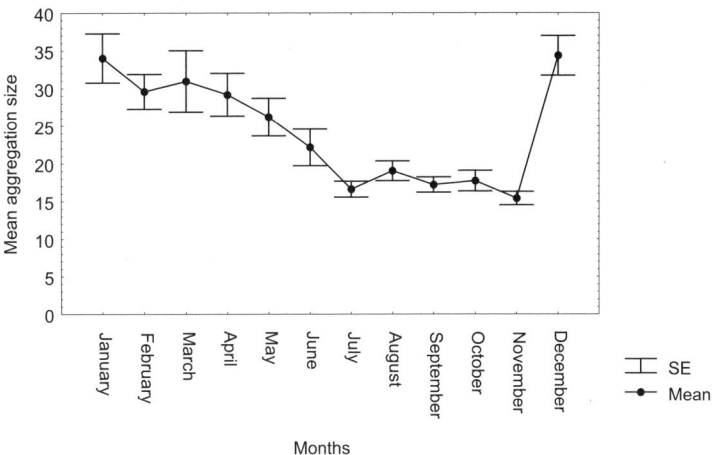

Figure 16.1 Monthly variation in mean (±SE) aggregation size (more than three families present). Data are from 15-minute scan samples taken during the morning (0700–1330 hrs) and afternoon (1540–1715 hrs) sessions.

If leadership is critical in determining elephant movements, and if experienced matriarchs play a decisive role in such decision making, we would expect to find stable, more cohesive families led by older matriarchs. In addition, we anticipate finding these families in aggregations (groups consisting of two or more families), since other families seeking such leadership qualities may wish to associate with them. Furthermore, since elephants appear to be capable of judging their own and others' relative social experience (McComb et al. 2001; see McComb, Reby, and Moss, chapter 10) and are more likely to take cues from experienced individuals, we would expect movements initiated by older matriarchs to have a larger following than those initiated by younger matriarchs. Moreover, since families led by older matriarchs are generally dominant to those with younger matriarchs (Moss 1988; Wittemyer et al. 2007), we expect dominance rank to play a pivotal role in the configuration of families within processions.

In this chapter, we explore the following:

- How matriarch leadership and rank influence group movements and the dynamics of short-term group fission and fusion as families move between habitats.
- The pattern of trail use among families.
- How trail use relates to leadership, resource acquisition, and the mitigation of resource competition.

Aggregation Formation

Joining and Leaving Aggregations

Amboseli elephants exhibit a definite pattern of grouping and movement. Approximately one month after the onset of the November "short rains," elephants began to aggregate, forming large groups (see figure 16.1). Our intensive study of decision making and leadership between August 1998 and September 2000 revealed that large groups tended to form prior to 1030 hours in open habitats as elephants moved from the *Acacia* woodlands toward the swamp-edge grazing arenas. It appeared that family groups waited (see waiting behavior in Poole and Granli, chapter 8) for each other and by doing so, formed large aggregations before they made a final move to the swamp edge. As elephants approached the swamp-edge grasslands, the numerous multi-party and multi-family groups broke down into smaller groups composed of single-family units or parties, which then entered the swamp at different locations. Groups remained relatively small between 1030 and 1530 hours while foraging in the deep swamp. As they came out, large aggregations re-formed along the swamp edge as families and parties grazed, socialized, and waited for one another before departing for the woodlands as large multi-family aggregations once again (figure 16.2).

Average aggregation size differed significantly among the sampling periods throughout the day (ANOVA: $F_{11,336} = 3.95, p < 0.05$). The mean size observed after 1530 hours was significantly larger than that observed during the morning session (*t*-test for independent samples: mean size, am = 138±101, N = 278; $t = 5.12$, df = 348, $p < 0.001$; mean size, pm = 210±120, N = 72; see figure 16.2). The aggregations that formed in the late afternoon remained consistently larger, only beginning to break up toward sunset.

Effects of Family Size and Age of Matriarch on Aggregation Formation

Large families, especially those with most or all members present (see also Moss and Lee, chapter 13), tended to

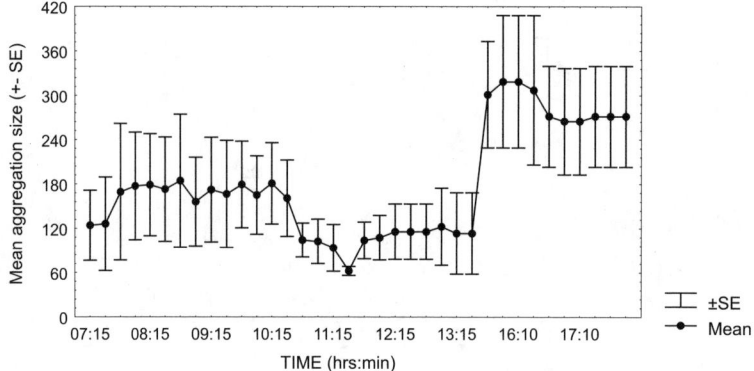

Figure 16.2 Mean (±SE) aggregation size by time of day.

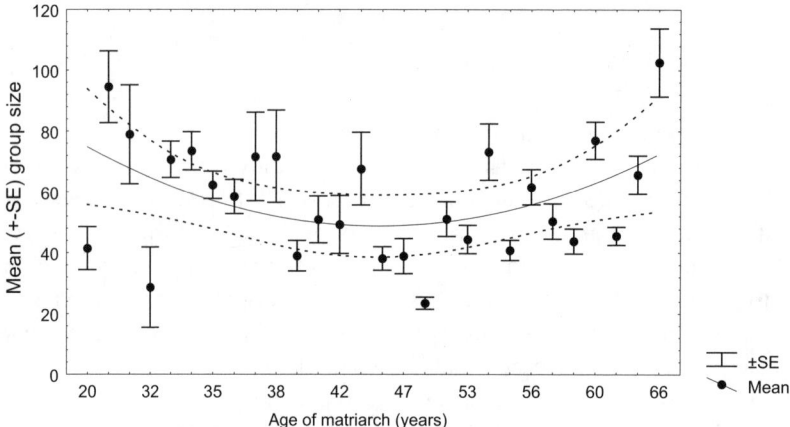

Figure 16.3 Mean (±SE) group size in relation to age of the matriarch among Amboseli elephants. The polynomial regression line is shown, with confidence intervals.

remain as single-family groups, while small families or parties formed larger aggregated groups by associating with other families or parties. The number of family units associated with a particular family declined as the number of family members present increased ($r = -0.60$, df = 3,381, $p < 0.001$).

On average, aggregation size was unrelated to the average size of those family units that were present ($r = -0.043$, df = 3,350, $p > 0.05$). Within aggregations, individual families are typically discernable as distinct socio-spatial units within the larger group. This pattern is particularly true of families with older matriarchs who tend to remain as discrete units or sub-groups within an aggregation. The relationship between matriarch age and the size of aggregations with which their families associated was complex. A second-order polynomial regression ($\beta_0 = -0.856$, $t_0 = -5.66$, $\beta_1 = 0.793$, $t_1 = 5.25$, $p < 0.001$) described a decline in mean group sizes with matriarch age up to 45 years, and thereafter, mean group size increased with the advancing age of matriarch (see figure 16.3).

Family Size and the Matriarch's Role in the Initiation of Movement, Group Formation, and Fragmentation

Movement initiation. Within aggregations, the vast majority of movement initiations (91.9 percent, $N = 503$) did not lead to movement of the entire group but rather affected only a segment of the group. Non-matriarch adult females initiated 18.9 percent ($N = 95$) and matriarchs initiated 81.1 percent ($N = 408$) of these movements. Of initiations made by non-matriarchs, 74.7 percent ended in only a section of her own family moving, while 97.6 percent of initiations made by matriarchs resulted in movements of her entire family unit as well as adjacent groups. In other words, decisions taken by matriarchs regarding movement were more likely to be followed than were decisions made by non-matriarchs.

In all of the 41 cases in which entire multi-family aggregations moved in response to an initiation, the oldest matriarch present was responsible. Entire group responses were observed mainly in aggregations containing fewer than

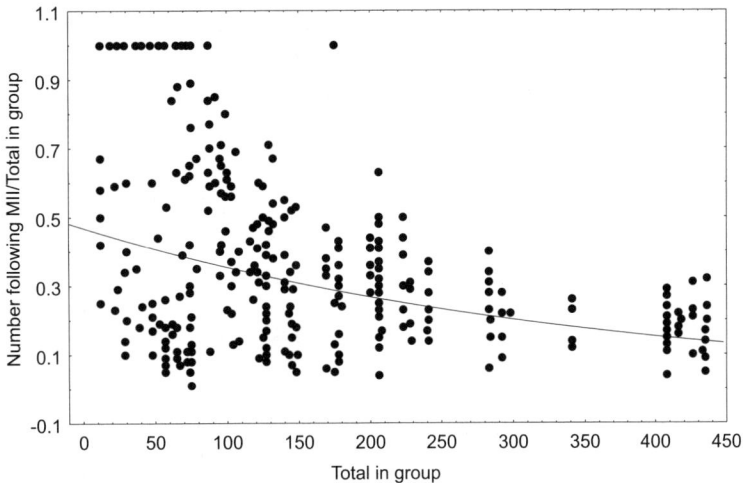

Figure 16.4 The relationship between the proportion of individuals moving following a movement initiation as a function of the total size of the aggregation. (MII = movement initiating individuals). The exponential regression line is shown ($y = 0.459 \cdot \exp[-0.003 \cdot x]$).

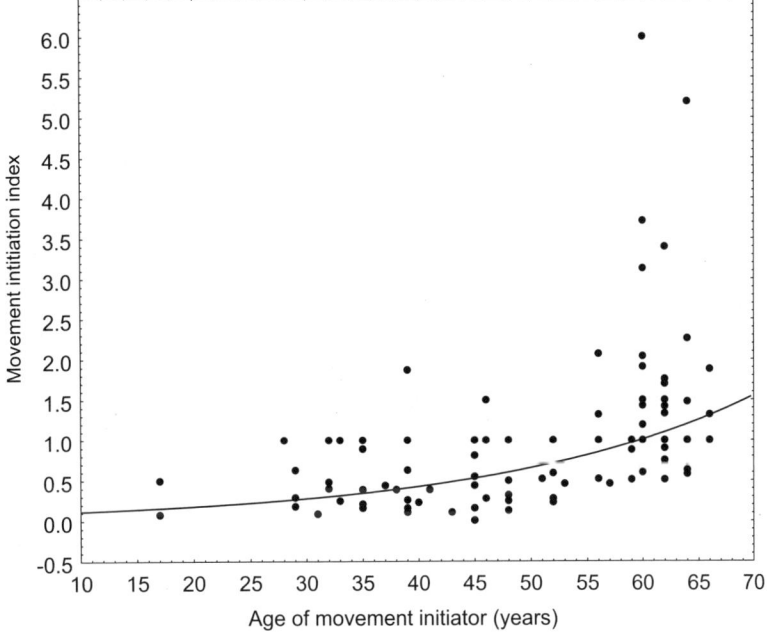

Figure 16.5 The relationship between movement initiator's age and the family size movement initiation index. The exponential trend line ($y = 0.079 \cdot \exp[-0.042 \cdot x]$) is shown.

175 elephants (mean±SE = 48.7±4.1). The proportion of elephants following a movement initiation declined significantly with increasing size of aggregations, and entire group movement rarely occurred when aggregations were larger than 200 individuals ($r^2 = 0.49$, df = 502, $p < 0.001$; see figure 16.4).

The age of the initiator, whether a matriarch or not, significantly influenced the proportion of family members that followed her movement ($r^2 = 0.55$, $t = 14.71$, df = 502, $p < 0.001$). Elderly individuals, 60-plus years, tended to have greater success initiating large-scale movements than did younger individuals (figure 16.5).

Short-term social fission and fusion. As a group moved through the course of the day, its size and composition changed as individuals, parties, and families joined and departed. Small families and parties were more likely to approach large families than the reverse. Large families either were indifferent or avoided other groups (table 16.1). Moreover, families with young matriarchs were more likely

Table 16.1 Frequency with which family units in an aggregation approached each other in relation to family size

Observation category	Count	%
Size of approaching family < size of approached family	584	82.4
Size of approaching family > size of approached family	103	14.5
Size of approaching family = size of approached family	22	3.1
Total	709	100

Note: These distributions were significantly different from those expected from an equal distribution ($X^2 = 259.9$, df = 2, $p < 0.001$).

Table 16.2 Frequencies with which family units within an aggregation approached each other in relation to the age of matriarch

Category	Count	%
Age of matriarch of approaching family unit < age of matriarch of approached family unit	566	79.8
Age of matriarch of approaching family unit > age of matriarch of approached family unit	136	19.2
Age of matriarch of approaching family unit = age of matriarch of approached family unit	7	1.0
Total	709	100

Note: These counts were significantly different from those expected from an equal distribution ($X^2 = 244.6$, df = 2, $p < 0.001$).

to approach and follow adjacent units led by older matriarchs than the reverse (table 16.2). Family units led by matriarchs who were younger than 44 years old were responsible for 50 percent of the approach events ($n = 355$), while 93 percent of cases ($N = 659$) in which family units were approached were led by matriarchs 44 years old or older.

Large families and those with older matriarchs tended not to join other families, while smaller families or those with young matriarchs were more gregarious and sought out the company of family units with older matriarchs (Kruskal-Wallis H (9, $N = 2079$) = 272.18, $p < 0.05$; see also Moss and Lee, chapter 13). Furthermore, parties without their matriarch present were frequently found in multi-family groups, whereas parties traveling with their old (age 59 years and above) matriarch were more likely to remain alone as a sub-group or to be joined by family units that had young matriarchs (Kruskal-Wallis H (9, $N = 3,300$) = 37.44; $p < 0.05$). These results suggest that both the presence and age of a matriarch have strong influences on the pattern of daily fission-fusion events.

Dominance Relations, Leadership, and the Use of Trails

Dominance in Relation to Matriarch Age and Family Size

Ideal interference models predict asymmetry between competing foraging individuals and groups in resource acquisition (Sutherland 1983; Sutherland and Parker 1992; Bautista, Alonso, and Alonso 1995; Moody and Houston 1995; Holgrem 1995; Houston, McNamara, and Milinski 1995; Lessells 1995; Tregenza, Parker, and Thompson 1996). Subordinates adopt foraging tactics that enhance intake rate while ensuring minimum conflict with dominant individuals (Csermely and Wood-Gush 1986; Schmidt and Hoi 1999). In female elephant groups in Amboseli, agonistic interactions increase during the dry season (Andelman 1985). Stress levels, as measured by fecal cortisol hormone, change inversely with rainfall, increasing with group size and decreasing with an increase in dominance rank (Foley, Papageorge, and Wasser 2001), suggesting that elephants in small families and those with older matriarchs suffer less from the stress of within-group competition.

Matriarchs are the highest-ranking individuals in a family unit (Moss 1988). Between family units, older matriarchs tend to rank above younger matriarchs (Moss 1988; Wittemyer and Getz 2007). While aggression within and between family units does increase during dry periods (Andelman 1985), aggressive encounters are relatively infrequent in Amboseli (Archie et al. 2006). Matriarchs assert their authority in numerous subtle ways (Moss 1988; Poole and Moss 1989), including use of gestures and acoustic signals (McComb et al. 2001; see Poole and Granli, chapter 8; Poole, chapter 9). Animal societies in which conflict is resolved in a subtle, non-aggressive manner are referred to as egalitarian (Hand 1986). In addition to the assertion of rank, the elephants of Amboseli appear to resolve many differences in a more egalitarian manner, often involving many individuals in decision making (Poole, chapter 9).

Relatively few overtly aggressive encounters were observed within or between groups in large aggregations ($N = 137$). While non-matriarchs initiated 70 percent of the observed agonistic interactions between family units, matriarchs intervened on behalf of the recipient in 18.3 percent of these events. Family size also affects inter-family dominance rank, with larger families ranking above smaller families (Foley, Papageorge, and Wasser 2001). In Amboseli, too, individuals who won came from larger families than those who lost (paired t-test: $t = -4.99$, df = 137, $p < 0.001$; see tables 16.3 and 16.4). Furthermore, in 76.7 percent of cases ($N = 137$), winning individuals came from families led by matriarchs who were older than the matriarch of the loser.

Table 16.3 Frequencies and percentages of winners and losers of contests when age of contestants, age of matriarch, and family size were considered

Contest categories	Factor 1		Factor 2		Factor 3	
	Frequency	Percent	Frequency	Percent	Frequency	Percent
Winners = Losers	7	5.1	8	5.8	4	2.9
Winners > Losers	95	69.3	105	76.7	98	71.5
Winners < Losers	35	25.6	24	17.5	35	25.6
Total	137	100.0	137	100.0	137	100.0

Note: Factor 1 = relative age of contestants; Factor 2 = relative age of matriarch of family from which contestants came; Factor 3 = relative size of family from which contestants came.

Table 16.4 Comparisons of relative age of winners and losers in contests by individual, by contestants' matriarch age, and by family size

Analysis categories	Mean	SE	t	df	
Age of losers	41.1	1.05	−5.46	136	$p < 0.001$
Age of winners	47.4	1.09			
Age of matriarch of family unit of losers	45.9	1.03	−6.78	136	$p < 0.001$
Age of matriarch of family unit of winners	53.6	0.94			
Size of family unit of losers	16.2	0.74	−4.99	136	$p < 0.001$
Size of family unit of winners	20.9	0.81			

Note: Results from paired *t*-test on ages and family size of winners and losers in 137 dyadic contests.

Partitioning of Resources through the Use of Bull Areas and Clan Areas

The cost of traveling to feeding grounds limits optimal foraging and ideal free distribution (pigeons: Baum and Kraft 1998; fish [*Aequidens portalegrensis*]: Tregenza and Thompson 1998). Behavioral observations in Amboseli suggest that both travel costs and the risk of predation affect elephant movements and aggregation (Kangwana 1993; Mutinda 2003). When high-quality foraging patches are widely dispersed and the energetic cost of travel is high, foragers may be "forced" to use nearby but lower-quality patches (Korona 1989; Astrom 1994; Baum and Kraft 1998). Although the elephant's energetic cost of locomotion has been estimated to be one of the lowest for terrestrial species (Langman et al. 1995), energetics clearly influence the seasonal ranging and aggregation patterns observed among elephants in Amboseli (Western and Lindsay 1984; Moss 1988; Lindsay 1994; Wall, Douglas-Hamilton, and Vollrath 2006).

Both male and female groups have particular areas in which they spend the majority of their time (Poole 1982; Moss and Poole 1983; Moss 1988; Croze and Moss, chapter 7). We refer to those areas used by particular males as *bull areas* and those used by family groups as *clan areas*. Amboseli's elephants have effectively partitioned the ecosystem, with different areas being primarily used by particular individuals or families. Nevertheless, this system of tenure is very flexible, allowing spatial overlap and temporal shift of range.

In addition, detailed observations at Empaash-Oldepe (figure 16.6), an area outside the National Park's southern boundary used by members of several clans, showed that although several elephant families might be moving through the area simultaneously, they might well have different final destinations. As they moved from one foraging area to another, they aggregated, forming a relatively cohesive unit as the group moved along, but fragmented as each party or family arrived at its respective destinations. The mean group size at Empaash-Oldepe (39.79±1.52 SE) was significantly higher than the mean group size at final destinations (29.42±0.824 SE; $t = 6.49$, df = 428; $p < 0.001$). Moreover, group size at both the place of origin and destination showed significant monthly variation, reflecting seasonal ecological changes as well as the shifting family groups using the area (ANOVA, group size at origin, $F_{7,543} = 12.04$; group size at destination; $F_{7,404} = 10.25$, $p < 0.001$; see table 16.5).

Seasonal Use of Empaash-Oldepe

Eighty-one percent of elephant groups ($N = 554$) coming from the Empaash-Oldepe area went to Longinye swamp, while the remainder either went to the Ol Tukai Orok/Enkongo Narok swamps or remained in the Oldepe area and moved further away from the Park. Most family groups preferred moving into Longinye swamp in all months except November and December, when they tended to shift to the Ol Tukai Orok/Enkongo Narok swamps. During the

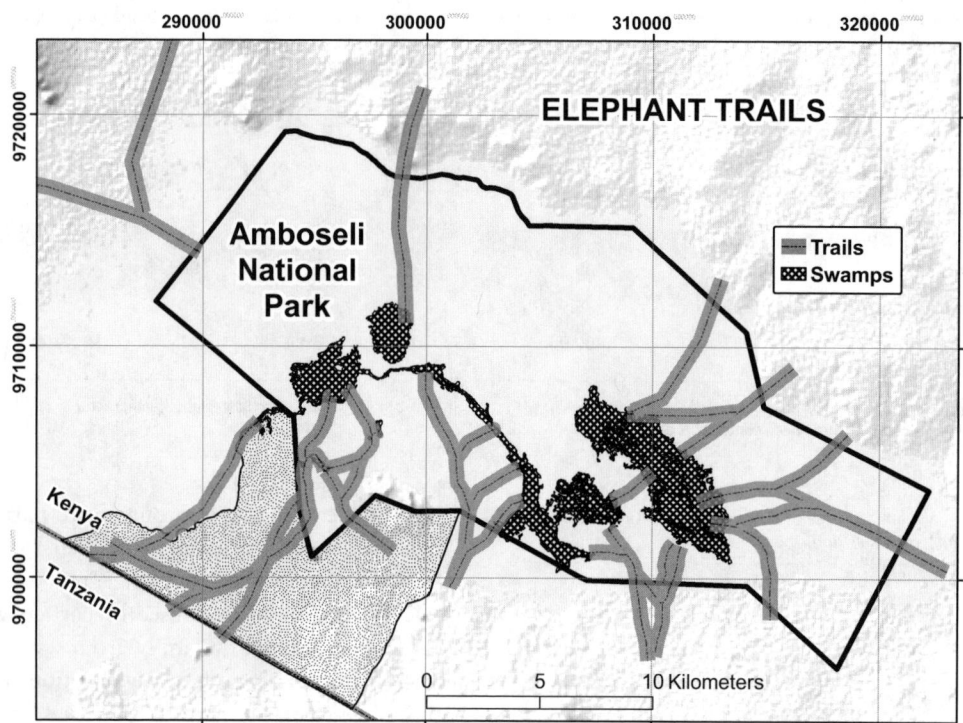

Figure 16.6 Map of the trail system used by elephants in Amboseli.

Table 16.5 Mean monthly group size (±SE) at first sighting and at swamp destination

Month	Mean at first sighting	N	SE	Mean at swamp destination	N	SE
November	28.9	105	3.2	24.5	83	2.1
December	68.4	60	89.5	45.4	54	21.6
January	49.5	60	30.4	33.8	48	8.6
February	38.0	48	12.3	30.0	43	5.1
March	64.8	8	34.7	55.6	5	6.7
August	32.9	46	15.5	29.12	43	7.6
September	30.7	125	3.9	23.7	78	2.4
October	25.6	99	3.2	22.7	58	2.6

drier months (August–November), approximately 30 percent of groups sighted remained in the *Acacia* bushland in Empaash-Oldepe to browse and moved further away from the Park for a day or two before returning to the swamps in the Park to drink and graze (figure 16.7).

Trails as Sources of Public Information

In order to use the available habitat patches efficiently, elephants use trails. Elephant trails have been observed wherever elephants are found throughout Africa (Dublin 1983; Fay and Agnagna 1991; Ruggiero 1992; Vanleeuwé and Gautier-Hion 1998). Trails consist of an extensive network of larger and smaller paths that are regularly used as elephants move through their range. Vanleeuwé and Gautier-Hion (1998) classified larger elephant trails in the forest as "boulevards" and smaller trails as "foraging paths."

Trails, along which the scents of other individuals have been deposited, can be seen as a source of public information (Mutinda 2003). Information may include identity, number, reproductive state, condition, and direction taken of other individuals who have passed along the trail earlier. Optimal foraging decisions require using information regarding the location of quality forage and how to minimize risks and competition. The assimilation and use of information gathered from the trails lead to increased foraging efficiency and contribute to the dynamics of social foraging.

Effect of Group Size and Residence on Trail Use

Elephants in Amboseli use trails to access various habitat types (e.g., figure 16.6). For example, groups of elephants coming from Empaash-Oldepe accessed Longinye and Ol Tukai Orok swamps using distinct trails. The trail chosen by a particular group appeared to depend upon that group's size as well as whether or not the group's members were from clans typically resident in that area. Elephants in large

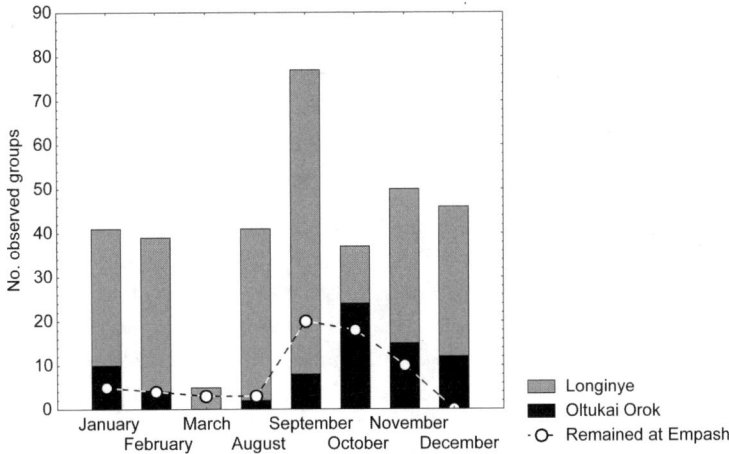

Figure 16.7 The monthly changes in number of elephant groups that went into the Ol Tukai Orok/Enkongo-Narok and Longinye swamps and those that remained or moved further away from the Park.

aggregations and elephants from other clans consistently preferred the large, heavily used "trunk-trails" or boulevards. Resident families, coming to the swamps singly or as a bond group, preferred using minor, less heavily used "family trails" (Spearman $r_s = 0.182$, df = 672, $p < 0.001$). Consequently, there was a significant correlation between aggregation size and the frequency with which marked trails (see methods in box 16.1) were used (Kendall's tau, $b = 0.119$, $t = 5.039$, $p < 0.001$). Table 16.6 shows the mean group size and the number of families using each of the marked trails. Furthermore, figure 16.8 shows that L3, and L4 were trails of choice for multi-family groups entering Longinye compared with other L-labeled trails (Longinye swamp/Empaash woodland habitat), which tended to be used by single families.

Seasonal Variability in Trail Use

The frequency of trail use from Empaash-Oldepe in the vicinity of Njiri, an area referred to as the "V" (see figure 16.6), showed significant seasonal variability (Kruskal-Wallis H (9, $N = 372$) = 20.00, $p < 0.05$). Family units used certain trails during the dry season and then shifted to others located 200 m to the northwest once new growth was established approximately one month after the rains began; we refer to these as the V-trails (table 16.7). The move from dry season to wet season trails was dramatic; only 5.3 percent of groups observed in November ($N = 57$; rain began mid-November) used the wet season V-trails, while 47.4 percent of groups observed in December ($N = 95$; wet season growth established) used these V-trails. In January, use of these same trails remained high at 40.4 percent ($N = 52$) but by February, it had declined to 31.4 percent ($N = 51$).

Table 16.6 Differences in mean group size and number of family units using marked trails

Marked trails	N (observations)	Group size on trail		N (families using trail)	
		Mean	SE	Mean	SE
L2	34	28.0	2.8	1.66	0.14
L3	58	28.9	1.9	2.03	0.14
R1	18	28.7	4.2	2.39	0.38
L4	79	35.0	3.1	2.03	0.18
R2	35	32.9	4.4	2.14	0.23
L1	14	22.2	3.8	1.36	0.17
L5	20	27.1	3.2	1.76	0.18
L17	7	24.4	4.7	1.33	0.21
L9	4	22.3	4.4	1.50	0.29
L15	15	29.6	4.1	1.43	0.14
L16	14	27.9	2.6	1.14	0.10
R3	29	23.3	2.6	1.48	0.12
L6	11	22.7	4.1	1.82	0.23
OL	58	33.5	2.3	2.00	0.14
R4	10	27.4	4.9	1.78	0.55

Note: Variation among trails in group size was not significant, but differences in the mean number of family units using each trail were statistically significant ($p < 0.05$).

Use of the V-trails by the few family units that used this area during the dry season remained consistently below 10 percent during the drier months. Families that abandoned their wet season V-trails as the dry season progressed returned to use the established trails they had used the previous dry

BOX 16.1 TERMINOLOGY AND METHODOLOGY

Specific Terms Used in This Chapter

Age power index: The number of individuals that followed a movement initiator divided by the family size of the individual initiating the movement (so called because of the influence that age of matriarch has on the movement of a family group).

Approach: A family or section of a family unit moves toward other families, and their members mix to form a homogeneous group.

Avoid: When a family or party moves away from another group or when the direction of its movement is deflected away from another family within the same aggregation.

Movement initiation index: Number of individuals that follow a movement initiator divided by the total number in the group.

Movement initiator: An individual whose movement is followed by other group members within one minute.

Methods Used in Data Collection

Movement and Association Dynamics Within Large Aggregations

The aim of the observations was to assess how age of matriarch and family size determined the formation of aggregations, family associations within large aggregations, and the initiation of movements. Dominance interactions between families were also observed.

Large groups composed of at least three families were chosen in order to maximize opportunities for observing inter-family and inter-sub-group dynamics, increased movement initiation as family groups interacted with each other, and increased dominance contests as competition increased with group size.

Opportunistic searching for large aggregations began at 0700 hrs. Once a suitable group was found, it was followed for as long as possible. Observations were terminated when all members of groups went into the swamp or when they entered thick swamp-edge vegetation and were out of sight.

Continuous and scan sampling procedures were used in data collection (see appendix 1). Once a suitable group was located, the first 10 minutes were spent on intensive scrutiny of the group to accurately establish the numbers of individuals in the whole group, the identities of associating families, and whether any family was split. This scrutiny was important because it made estimates more accurate since the

---

season. Seventeen out of 20 (85 percent) marked trails that had been abandoned during the wet season were back in use the following dry season. Elephant families were very precise in the reuse of old trails; in 70 percent of observations, they deviated by less than 2 m from the paths that they had used in the previous season.

Social Organization in Trail Processions

As noted above, trails are a source of public information, allowing for the exchange of olfactory signals between individuals and groups. Information found on trails aids the receiver in assessing the rank of the signaler (Kappeler 1993; Gese and Ruff 1997; Poole 1989a, 1989b; Poole and Moss 1989), the quality of habitat from which the signaler originated (Galef and Buckely 1996; Galef and Giraldeau 2000), and the signaler's sexual state (Poole 1987; Poole and Moss 1989). Scent signals also help maintain contact between group members (MacDonald 1985).

Despite the potential for information exchange, there are still constraints on decision making as a function of characteristics such as food patch renewal rates, risks, and travel costs (Abrahams 1986; Gray and Kennedy 1994a, 1994b; Kennedy and Gray 1993; Kennedy, Shave, and Spencer 1994; Spencer, Kennedy, and Gray 1995). Highly social elephants may circumvent some of these problems by exploiting the vast knowledge and experience acquired by older individuals, especially matriarchs (McComb et al. 2001), allowing them to make decisions related to movement, aggregation, and group defense (Moss 1988; Mutinda 2003).

The matriarch's role and position. Elephant society is organized in a complex hierarchy (Moss and Poole 1983; Moss 1988; Archie et al. 2006; Wittemyer, Douglas-Hamilton, and Getz 2005; Wittemyer et al. 2007). Matriarchs determine how groups associate, where and when to depart from visited habitats, and which groups to avoid and join (Moss 1988; McComb et al. 2000). They exert their influence through a complex repertoire of acoustic signals (Poole et al. 1988; McComb et al. 2000; see Poole, chapter 9),

parameters were unlikely to change dramatically during the course of sampling. Scan samples lasting 2 minutes and taken at 15-minute intervals were started 5 minutes after the initial scrutiny in order to allow the group to settle.

Continuous sampling methods, each bout lasting 30 minutes and separated by 10-minute intervals, were used to study movements of family units within the groups. Specifically, families that made approaches and those that were approached were recorded. Additionally, family units that avoided others and those that were avoided and individuals and family units that initiated movements and those that followed were recorded. A movement initiation was determined by observing and recording the number of individuals that started moving within one minute after movement was initiated by the focal female.

Using the dominance-subordinance behavior described by Moss (1983, 1988) and Poole (1987a, 1989a, 1989b), dyadic contests between females were observed and recorded whenever they occurred. Winners and losers in any contesting pairs were recorded in relation to the characteristics of the family from which they came.

Habitat Selection, Use of Trails, and Social Dynamics Within Trail Processions

Between 0700 hrs and 1400 hrs, elephant groups using the *Acacia mellifera* and *Acacia nubica* woodland in the Empaash-Oldepe area south of the park were observed. This particular area was selected because from here elephant groups have the option of going to either Longinye or Ol Tukai Orok swamps in the Park or staying out and moving even further into community-owned land.

Once groups were located, the number of individuals and their identities and those of their families were established and recorded against the time of first sighting. Scan sampling was conducted every 15 minutes as the group moved; all the leading individuals and those that came last in the procession were recorded. The groups were monitored until 1400 hr in order to establish their destination.

At the swamp edge, trails were marked using aluminium metal pegs in order to minimize losses due to decay and were covered by rocks in order to avoid detection and possible removal by elephants. Markers were placed at least 300 m from water holes in order to minimize disturbance to elephants in their natural drinking and wallowing sites.

As elephants came to the drinking places, the number of individuals and the identities of families constituting groups using labeled trails were recorded. Further, the position of each family in the procession was recorded as 1 for leaders, 2 for second position from the front, and so on, up to fourth position, and the names of the leading and last individual within the family were recorded. The time at which each group crossed trail markers and moved to drinking places was also recorded.

All statistical analyses follow procedures described in appendix 1 for parametric and non-parametric hypothesis testing.

which are further enhanced by olfactory signals deposited on the trails they regularly use (Poole and Moss 1989).

Such leadership is critical in the effective use of large ranges. The home range of Amboseli families is about 570 km^2 (range 95–1,690; see Croze and Moss, chapter 7), comparable to the ranges of the adjacent Tsavo elephants (350–3,744 km^2; Leuthold 1977) and those of the Laikipia-Samburu ecosystem (102–5,527 km^2; Thouless 1996). Leadership is required in making reliable projections regarding foraging opportunities, potential intake rates, possible competition levels, and the risks involved in accessing new areas or remaining in suitable habitats. Since elephants are long lived, such socially learned skills are likely to be passed on from generation to generation and from family to family.

Yet relative rank plays an important role, too (Wittemyer and Getz 2007). While subordinate individuals and groups who choose to associate with other family units may benefit from an experienced matriarch, they may also be forced to adopt behavioral strategies to minimize conflict with more dominant individuals and groups. The low frequency of agonistic encounters observed during this study suggests that subordinate individuals and families minimize conflict by avoiding contesting over resources when they associate with dominant family units or individuals. In addition, choosing to follow the initiation of movement by older leaders may act as a further mechanism for conflict reduction.

Status played a clear role in the position of individuals in processions. At the start of daily journeys to the swamps, matriarchs led groups in 59 percent of cases ($n = 203$). Non-matriarchs led in 41 percent of cases, of which 23 percent ($N = 84$) were matriarchs' daughters. While matriarchs took the lead in initiating a movement and starting the procession, they later tended to revert to the rear of the procession, and the lead position was replaced by lower-ranking individuals or families (see below).

As single-family and multi-family groups moved in a procession along a trail and approached water or wallowing sites, the adult females at the head of these processions lifted their trunks and sniffed in the direction of the

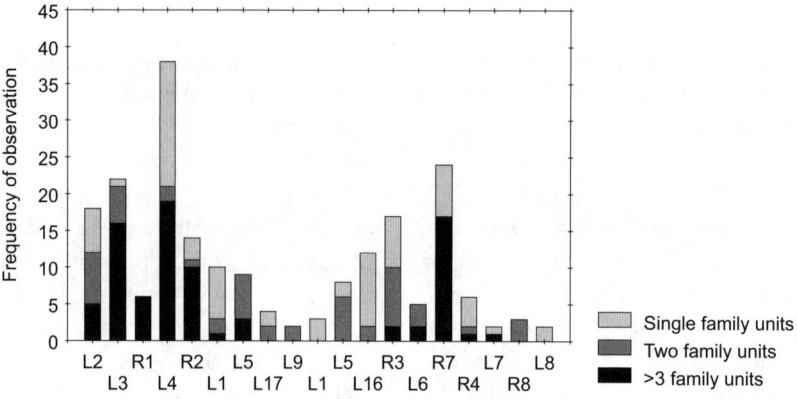

Figure 16.8 Frequencies of use of marked trails by elephants, either as single families or as aggregations.

Table 16.7 Frequency of use of marked trails in different months

Month	L2	L3	R1	L4	R2	L1	L5	V	L17	L9	LI	L15	L16	R3	L6	OL	R4	L7	R5	L8	Total
November	4	7	1	13	9	1	2	3	1	0	0	3	4	3	3	13	2	1	0	0	70
December	1	3	8	8	9	0	3	49	0	0	0	2	0	7	1	1	2	0	2	0	96
January	3	5	1	7	2	1	0	21	1	1	0	0	1	4	0	24	1	0	3	1	76
February	4	5	1	8	3	2	1	16	1	0	0	2	1	0	3	9	2	0	1	1	60
March	0	3	1	2	1	0	1	4	0	0	0	0	0	0	0	0	0	0	0	0	12
August	10	17	1	13	2	4	6	1	1	1	3	2	1	1		0	0	0	0	0	63
September	9	17	1	19	0	4	6	2	0	0	0	3	3	3	3	9	2	1	1	0	83
October	4	3	3	9	8	2	2	3	3	2	0	3	4	11	1	2	1	1	0	0	62
Total	35	60	18	79	35	14	21	99	7	4	3	15	14	29	11	58	10	3	7	2	524

Note: November–December = wet months; January–February = inter-rains with high grass production; March = start of the long rainy season; August–October = dry months.

swamps. The leading families in aggregations tended to be smaller in size than those coming last in the processions. These leading families were also smaller than those moving in single-family groups. Furthermore, families in the lead of long processions tended to be led by matriarchs who were younger than those whose family units took up the rear of processions. They were also younger than those in families that went to the swamp alone. These differences in mean family size and age of matriarch in different positions in the processions were significant (family size: ANOVA, $F_{4,1018}$ = 50.22, $p < 0.001$; age of matriarch ANOVA, $F_{4,1018}$ = 97.47, $p < 0.001$; see table 16.8).

Families that joined processions as a party were also found in the lead position at significantly higher frequencies than were entire families (median test, χ^2 = 16.99, df = 9, $p < 0.05$). However, the rank of fragments of families that included the family matriarch did not appear to alter.

Procession order within single-family groups. Among single-family group processions, females under age 40 led the group in 80.2 percent ($N = 999$) of incidences, with 72 percent being 15 to 20 year olds. While females over 40 years old led in only 19.8 percent of the single-family processions, they took up the rear position in 58.6 percent of cases. Age was a significant factor determining leader and rear-guard positions (see also rear-guard behavior in Poole and Granli, chapter 8) in processions (Kruskal-Wallis H (6, $N = 999$) = 45.16, $p < 0.05$).

Social status in processions. The leaders of both single-family and multi-family processions were principally non-matriarchs (67.3 percent; $N = 1,040$). Matriarchs and their eldest daughters were observed leading only 18.8 percent and 13.9 percent of the occasions, respectively. Across all procession types, matriarchs were much more likely to be observed in

Table 16.8 Characteristics and positions of single-family and multi-family groups

Group type	Position of family within aggregation	N	Family size		Age of matriarch	
			Mean	SE	Mean	SE
Single-family group	Sole	250	20	0.5	51	0.6
Multi-family groups	4th from front	44	21	0.5	55	0.5
	3rd from front	127	15	0.4	45	1.7
	2nd from front	294	15	0.6	42	0.6
	Leading	308	13	1.2	38	1.0

Note: Mean (±SE) family size and mean (±SE) age of matriarchs in relation to family position in processions as groups moved along trails 100–200 m away from the swamp.

last position (61.1 percent; $N = 1,040$), a spot very rarely taken by their adult daughters (1.1 percent) but frequently taken up by other adult non-matriarch females (36.8 percent). The difference in the frequencies with which matriarchs, their daughters, and other adult non-matriarch females either led or came last in a group moving toward the swamp was significant (Kruskal-Wallis H (4, $N = 1,040$) = 231.08, $p < 0.05$).

Discussion

From a foraging perspective, an elephant matriarch should lead her close relatives to join other groups when and where food is abundant (Charnov 1976) and avoid associating with other groups when despotic interference might result in reduced feeding rates (de Waal 1989; van Schaik 1989). Yet, elephants must balance their foraging needs against the many other social and survival benefits of group living (e.g., Poole and Moss 1989; see Moss and Lee, chapter 13; Archie, Moss, and Alberts, chapter 15; Poole, chapter 9). In Amboseli, although decision making is often a group effort (see Poole, chapter 9), the matriarch is primarily responsible for these intricate decisions, and her leadership role and judgment are affected by factors such as her age and experience, her family's size, her own and her family members' reproductive status, the presence of young calves, the season, and food and water availability.

Amboseli's elephants benefit in numerous ways from being together, and they tend to aggregate whenever food is plentiful (see also Moss and Lee, chapter 13). Nevertheless, they must weigh survival, social, and reproductive pull factors against the push factor of foraging competition, both contest and scramble. A matriarch's decision to break away from an aggregation is likely to be designed to minimize feeding interference and maximize intake rates. The organization of aggregations into distinct sub-groups, each composed of families or fragments thereof, facilitates the reduction of between-family foraging conflicts while retaining the survival, social, and reproductive benefits of being in a larger group, one often containing a more experienced matriarch.

Scramble competition ensures that individuals benefit equally from resources in productive patches, while in contest competition, high-ranking individuals, or "despots," gain more than do other members of the group. Our observations in Amboseli as well as conclusions from studies of other habitats and populations suggest that elephants form the largest aggregations in habitats that provide a uniformly distributed source of food, especially in grassland habitats (see also Lindsay, chapter 5). In these productive habitats, each individual in a large group has access to an evenly distributed source of food, unlike in forest or dense bushland habitats in which elephants must displace each other in order to compete over individual patches.

Another factor may come into play here, too. In Amboseli, Oldepe-Empaash straddles the National Park boundary and is an area both inside and outside the Park that is frequented by the Maasai and their livestock. The Maasai are the greatest threat to adult elephants in this population (Bates et al. 2007; see also Sayialel and Moss, box 19.1; Kangwana, chapter 20). In many wild and domestic herbivore communities, grouping and foraging in open areas is associated with ease of predator detection (Burger, Safina, and Gochfield 2000).

The Oldepe-Empaash aggregations invariably split up as the elephants approached the more productive swamp and swamp-edge habitats (Deshmukh 1984; Western and Lindsay 1984; Lindsay 1994). We suggest two reasons for this behavior. First, elephants are likely to feel safe from predators, including human predators, once they are in the deep swamps. Second, by presenting a physical obstacle to socializing, the deep swamps become a foraging-only habitat. Thus, while in the swamps, elephants obtain very little social benefit from being in a large aggregation; instead, they maximize their intake by foraging in small groups and later aggregate to socialize as they come out of the swamp.

Aggregations offer many additional benefits such as effective mate searching (Poole and Moss 1989), the maintenance of kinship bonds and other social relationships (Moss 1988; see Moss and Lee, chapter 13; Archie, Moss, and Alberts, chapter 15), social learning (Lee and Moss 1999), and potential benefits accrued from others' longer experience and greater knowledge (McComb et al. 2001). Poole and Moss (1989) showed that estrous females were more likely to be found in large aggregations and argued that musth males who associated with these assemblages expended less time searching. Furthermore, large groups permit social interactions between calves and juveniles of different families that are vital for social development and calf survival (Douglas-Hamilton 1972; Lee 1986; Moss 1988; Poole and Thompsen 1989; Lee and Moss, chapter 14). It is therefore likely that matriarchs lead their family units to large aggregations in order to maximize those social and sexual interactions that are so vital to the very fabric of elephant society. By ensuring that they aggregate during times of high food abundance, they mitigate the per capita costs to foraging.

Family size and matriarch age appeared to affect elephant assemblages in different ways. Small families were more likely to join aggregations and remain in them for longer than were larger families. Smaller families receive greater relative benefits (increased social, anti-predator, and breeding opportunities) from joining other groups than do families who are already large themselves. Yet the effects of matriarch age on aggregation size, structure, movement initiation, and approach-avoid interactions are not straightforward and may reflect elements of personality or individuality among the matriarchs and the other members of the family.

Intelligent, large, and long-lived, advanced age among elephants translates into experience (McComb et al. 2001), dominance (Poole 1989a; Archie et al. 2006), and reproductive success (Poole 1989b; Moss 2001; see Moss and Lee, chapter 12; Poole et al., chapter 18.) The older the matriarch is, the greater her influence on overall aggregation dynamics. Her long-gained experience means that she is better able to distinguish between friend and foe; more capable of making successful social, survival, and reproductive choices (McComb et al. 2001; Moss and Lee, chapter 12); and more likely to lead the rest of her family to productive, low-risk foraging sites (Douglas-Hamilton and Douglas-Hamilton 1975; Dublin 1983; Moss 1988). Through social learning (Lee and Moss 1999; Poole and Moss 2008), the knowledge and experience of an older matriarch is passed on to members of her own family as well as to members of families who are close associates.

An older matriarch can be viewed as an important resource, especially for those families with young, inexperienced matriarchs. This view is supported by the observation that family units led by very young and very old matriarchs were found in larger than average group sizes and that family units led by the youngest matriarchs sought out the company of family units with very old matriarchs.

One question arising from these observations is, what happens when a matriarch dies or is absent? Do families without matriarchs behave differently? Our observations suggest that they do (see also Moss and Lee, chapter 13; Croze and Moss, chapter 7). Parties without matriarchs were more likely to be found in aggregations than were parties with matriarchs. This propensity helps to explain why elephant populations that have undergone large-scale poaching tend to form aggregations (e.g., Poole 1989c; Barnes and Kapela 1991; Abe 1995). As poaching focuses on older individuals with larger tusks, matriarchs and other experienced individuals are targeted, and the remnants of these families cluster around the few remaining experienced females.

In Amboseli, elephant aggregations were composed of a number of clusters, each made up of a family, a party, or a fragment of a family. The oldest matriarchs in these aggregations were often able to influence not only their own family but also the entire group, which might number as many as 175 individuals. In these cases, the oldest, most influential matriarch functioned as a "matriarch of matriarchs."

Matriarchs were primarily responsible for the initiation of movement within aggregations, and while other adult females also contributed, their influence appeared to be limited to only a few individuals within the party. The existence of numerous movement initiators within an elephant assemblage creates a problem of stability. It is probably for this reason that such large assemblages of elephant groups are relatively short lived.

One reason that has been proposed for the instability of social groups is the increased occurrence of aggressive interactions as food supplies diminish (Alonso et al. 1995; Koenig 2000) or group size increases (Blumstein, Evans, and Daniel 1999). Among female elephant groups in Amboseli, aggressive encounters between individuals are more common in large aggregations (Moss 1988). Matriarch age and group size are two factors influencing the social rank of family units (Moss 1988). A question arising from these observations is, what role does numerical strength play?

Members of family units exhibit highly affiliative behavior (Moss and Poole 1983; Moss 1988), including the formation of coalitions against other groups (Moss 1988; personal observations) and against potential threats (Sikes 1971; Douglas-Hamilton and Douglas-Hamilton 1975; Kangwana 1993). In evolutionary terms, in order for such groups to exist, the fitness benefits accrued by sociality should outweigh the costs of reduced feeding rate. Social groups can secure these benefits in three ways. First, groups might choose to move into rich habitats as predicted in the Ideal Free Distribution theory (Fretwell and Lucas 1970) in order to reduce the energetic cost of searching (e.g., free-ranging baboons: Altmann and Muruthi 1988; Muruthi, Altmann, and Altmann 1991). Second, social groups might fragment as resource quantity and quality shrink (Western and Lindsay 1984). Finally, social units might avoid amalgamating to form larger groups, especially with individuals that are higher in rank (e.g., female elephant groups in

Kruger National Park, South Africa: de Villiers and Kok 1997). It appears that Amboseli's matriarchs combine all three strategies in their decision making.

During the dry season, some family units remained in an aggregation only while the group was in particularly rich habitats, separating off once they were in the drier woodlands. In the wet season, however, when forage was abundant, family groups remained aggregated in both the open grassland and woodland habitats, only splitting into smaller groups at the swamp edge.

The assembly of elephant groups as they began their trek from the bushland toward the swamps was a very common observation. Certain family units appeared to wait for others, gradually forming one or several aggregations before starting to move across more open habitat toward the swamp edge. While the value of grouping before moving into different habitat types is not known, it can be speculated that greater numerical strength provides security in potentially dangerous areas, particularly outside the Park (Kangwana 1993).

Elephants moving from one location to another generally used an extensive and intricate network of trails, including many minor foraging trails that lead to the major between-habitat link-trails. Yet, why each family used the maze of trails as it did is still not understood. Our observations suggest that elephant groups understood when to use the different trail types (personal foraging versus more public boulevards; Vanleeuwé and Gautier-Hion [1998]) as well as when to shift to trails in different areas depending on the season. A set of conventions appeared to govern trail use. Certain trails were available for use by all, while only particular family groups used others. Elephants use their trunks to sniff as they walk along trails, and we speculate that they are able to make use of odor secreted from glands in their feet as well as scents in urine and dung (Rasmussen and Krishnamurthy 2000). In this way, they are able to identify the "ownership" of different trails as well as the identity, age, sex, and reproductive state of individuals who have gone before them.

The observation that families reoccupied previously used trails after shifting away from them seasonally suggests that they were able to recognize the exact location of their old trails even after a period of disuse. Since chemical markers are ephemeral and therefore not likely to be of particular use in relocating old trails, we speculate that their excellent memories (Rensch 1957; Shoshani, Kupsky, and Marchant 2006) were put to good use in the relocation.

Dominance relationships appear to influence the position of individuals in processions (lower-ranking families toward the lead and higher-ranking families toward the rear) along the trails as well as determine the accessibility of resources at the point of destination. It may be for this reason that aggregations often disassembled as they approached the swamp-edge habitats. After a long march, elephants arrive at the swamp edge hot and thirsty, and access to positions for drinking and mud-splashing will be of prime importance.

In order for trails to be useful to elephants, they must provide information as well as an efficient route for travel between foraging patches. Scents deposited (actively or passively) by previous trail users (e.g., in urine, dung, mucus excreted from the nostrils, and secretions from the temporal gland and interdigital glands) will act as cues as to identities, age, sex, rank, and reproductive states as well as provide information about the quality of habitat these trails users have recently visited. We suggest that matriarchs incorporate this information as they decide where to go and which trails to use. Furthermore, it appears that once the destination habitat is clear to every member, the matriarch or another high-ranking, experienced individual moves to the rear-guard position (Douglas-Hamilton 1972; Poole and Granli 2003, 2009; see rear-guarding in Poole and Granli, chapter 8) while another relatively high-ranking individual occupies the lead, thus protecting the family as it moves in procession from one habitat to another.

Dominance rank alone may not explain the positions of members in family processions. The observation that daughters of matriarchs led processions on numerous occasions may further show the influence of matriarchal leadership in the exploration and use of diverse habitats within Amboseli.

Elephant foraging and social relationships involve complex decisions regarding how to balance energy requirements, social and reproductive needs, and anti-predator behavior. For an intelligent and long-lived animal, the role of experience and knowledge is foremost in the making of these crucial decisions for survival, and it thus seems fitting that a virtual encyclopedia of knowledge is written across the elephants' range in the form of a network of information-laden trails.

Chapter 17 Male Social Dynamics: Independence and Beyond

Phyllis C. Lee, Joyce H. Poole, Norah Njiraini, Catherine N. Sayialel, and Cynthia J. Moss

THE SOCIAL life of a male elephant can be divided into distinct periods, and males face specific social challenges in each (figure 17.1). The first period, that of dependence on the mother and allomothers, is an experience shared with female peers. Even in the first two years of life, however, males are treated differently by their mothers and others and interact with peers in a way that sets them apart from female calves (see Lee and Moss, chapter 14). The juvenile stage sees the beginning of socio-sexual differentiation, and male interactions focus on male agemates. From the ages of 5 to 15 years, the types of exploration, play, associations, and physical maturation differentiate young males from female peers. Males depart from the family at an average age of 14 years (figure 17.2), and a male's behavior sets the scene for social and reproductive outcomes over the next 45 years of his lifespan. After independence, the young male explores a variety of feeding areas, assesses his power against peers in both relatively friendly and more aggressively competitive interactions, and sometimes associates with females for companionship, knowledge, and potential sexual interactions (Poole 1982; Ganswindt et al. 2005). After 20 years of age, he enters an annual musth phase (Poole 1987a, 1989a, 1989b; Poole et al., chapter 18) of highly competitive interactions with other males and sexual interest in females. Between musth and non-musth phases, his social dynamics shift from searching for females and competing with other males to occupying a specific feeding and social bull area (Poole 1982; Moss and Poole 1983).

In this chapter, we explore the changing social dynamics of male elephants with age and address how they potentially impact on reproductive success. Age-graded reproductive strategies are well documented for male mammals (see Poole et al., chapter 18). There are, however, relatively few studies that focus on the social processes or transitions between being immature and becoming a reproductively competent adult. The period of adolescence is critical to understanding male success (Bercovitch 2000; Setchell and Lee 2004), since it influences both physical and social development. How does a young male elephant growing up in a female-centered world acquire sex roles and sex-specific behavior appropriate to a male? Males must solve such behavioral problems in order to attain a position within the mature male social hierarchy. What underlies the process of establishing a male social strategy—that of becoming an independent and relatively solitary individual—within the context of fission-fusion sociality for family units? We explore the process of becoming "male" in relation to social dynamics during three life stages (see figure 17.1): (1) adolescence and dispersal, (2) the period when sexually active males establish their ranges and associations, and (3) the stage at which males associate and engage with other mature males and family groups and exhibit annual musth cycles.

Males interact frequently with other elephants. Young males face feeding competition from females and other males, and all males compete over access to females. In bull-only groups, older, larger males supplant younger males in competition for food, and in mixed groups containing an estrous female, older males threaten young males and rival large males (Poole 1982). Rates of aggression among mature adult males tend to be high (7.5 events/hour when in

Male Social Dynamics: Independence and Beyond 261

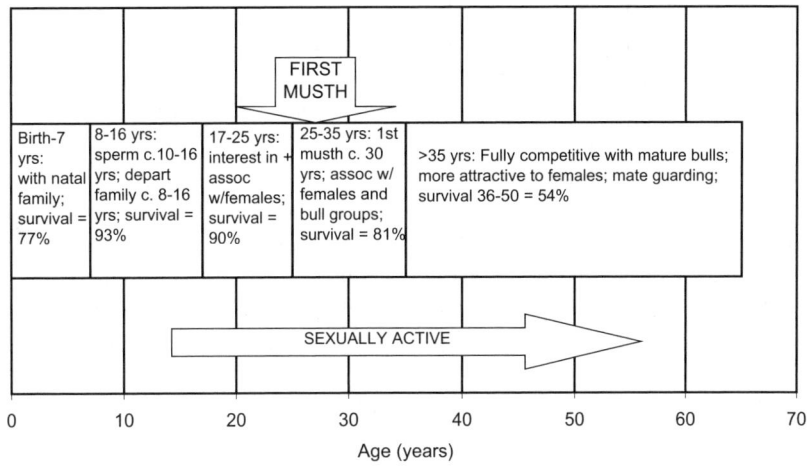

Figure 17.1 The life stages of an Amboseli male elephant.

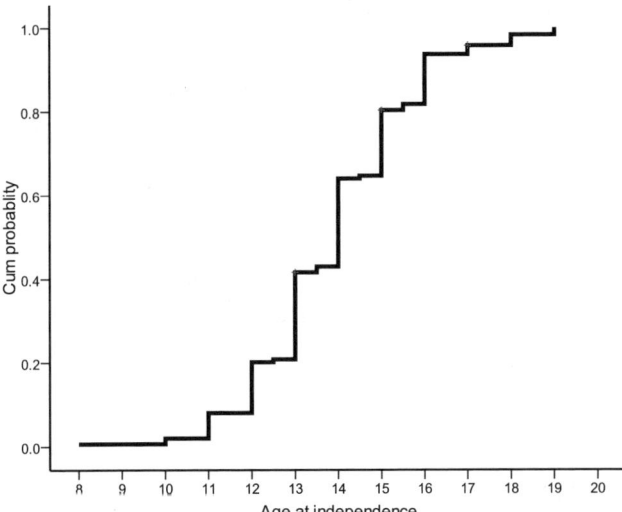

Figure 17.2 Probability of dispersal from the family, by age (N = 155, mean = 14, 95% CI = 13.82–14.18 Kaplan-Meier survival analysis).

musth and 3.5 events/hour when not in musth; Poole 1982). Competition obviously affects male grouping and associations, which we explore further in this chapter.

We first address how young, sexually active, non-musth males associate with other elephants and then discuss how being an older musth male affects sociality. We have separated these early (approximately 20–30 years) and later phases (generally >30 years) for several reasons. Early in life, males have high energy costs due to growth and a high rate of mortality; with age, the growth costs diminish, but competitive costs increase. Here, we explore social dynamics from a developmental perspective—as a process of long duration—and with respect to the attainment of full socio-sexual functioning.

Our aims in this chapter are to

- describe the process of departure from the natal family and the establishment of a male social milieu independent and distinct from that of families;
- portray the social dynamics of males as they enter into sexual activity and relate male sociality to reproductive state as males mature; and
- place the age-graded social dynamics into a context for understanding processes and patterns in male reproduction.

Determining Male Associations

Data on 534 males aged 8 years or over contribute to our analyses. We observed the process of dispersal in 155 males from 44 families born between 1967 and 1985 and aged 8 to 20 years in the course of the study. There were 205 older post-independence males, of which 43 survived until 2002 and 134 had been observed in musth. Male grouping and associations derive from sightings data (see appendix 1). Over the 1975–2002 period, we sighted the individual males used in these analyses an average of 19 to 20 times per year (range 1–60). As males aged over the course of the study, they changed age-class, and thus the same individual may be represented longitudinally. Males at the start of the process of leaving their natal units, from around 8 to 16 years, were termed transitional males and are the focus of the first analyses. The presence or absence of transitional males and juveniles with their natal families was recorded during monthly family censuses. Absent natal males were noted as "away" from the family. If the male was subsequently sighted and confirmed as away, his group size and group composition were recorded. The percentage of sightings away from the family was calculated for each individual out of the total number of sightings with or away from the family plus the number of separate, independent sightings of the family without the male present. During the intensive period of data collection on transitional males (1983–85), a mean of four censuses of each family (range 3–8) were made each month. The average number of censuses plus sightings for the 155 transitional males from 1975–97 was 14 in each year (range 5–26; 0–3 per month).

For the older males with known family histories, groups containing family females were analyzed separately. A subsample of the sightings of 13 males with known dispersal histories was used to examine changes in association with natal families over a 20-year period. In all sightings, mature males were recorded as solitary, in bull-only groups, or in mixed-sex groups; additionally, information regarding whether an estrous female was present and whether the male was in musth was recorded. All identified families and other males were noted. Indices of association (Sorenson's half weight; see appendix 1) were calculated for 31 well-known mature males in association with other males (mean sightings with other males = 142; SD = 40.5) and for 21 musth males in association with family units when not in musth (median sightings with each of 49 families = 18). The simple index of association (excluding time solitary) was used for associations of males with family units when a male had been seen in mixed groups for more than 50 events and a family had been seen at least 100 times over the course of the study.

The Social and Ecological Context of Dispersal among Young Male Elephants

Natal dispersal occurs in most birds and mammals (Greenwood 1980; Chepko-Sade and Halpin 1987; Pusey and Packer 1987). There are only a few exceptions, such as some cetaceans (killer whales: Baird 2000; bottle-nose dolphins: Connor et al. 2000), to the "rule" that prior to breeding, one or both sexes disperse from the area or group where their parents reside. For many years (e.g., Sikes 1971; Douglas-Hamilton 1972), it had been assumed that male elephants were forced out of their families as they matured sexually. Independence, we suggest, is not an event imposed upon a young male but rather a process shaped by a male's decisions, experiences, physiological state, and the nature of his relationships with other family males and females. The timing and the process of how young males leave their families are the results of interactions among a number of factors—their mother's dominance, whether they have peer males for social interaction, and their physical maturation (size and strength) for age (see Lee and Moss 1999). Success or failure during the dispersal phase could be key to understanding resulting patterns of male social behavior. The process of dispersal is briefly outlined here.

Inbreeding and Dispersal

One of the main suppositions underlying dispersal is that it functions to minimize inbreeding between closely related individuals. The genetic consequences of dispersal have been assessed for some primates (baboons: Alberts and Altmann 1995; chimpanzees: Morin et al. 1994; Gagneux et al. 2001), social carnivores (lions: Spong et al. 2002; hyenas: van Horn et al. 2004; coatis: Gompper, Gittleman, and Wayne 1998), seals (Amos et al. 1995), cetaceans (Amos, Schlotterer, and Tautz 1993), and birds (e.g., Petrie, Krupa, and Burke 1999; Coltman 2005; Double et al. 2005). While genetic heterogeneity is introduced by the movement of individuals between groups or geographic areas, these studies suggest that nearby individuals tend to be related and thus that kin clustering is still an important feature shaping behavior and territory structure, with mates chosen from further afield. Knowledge of how non-random genetic patterns or clusters of kin arise through dispersal behavior is leading to new models of social evolution. How, then, do the advantages of shifting groups or settling on territories relate to elephants whose ranging patterns cover huge areas and whose contact with other individuals is based on social knowledge (McComb et al. 2001; see McComb, Reby, and Moss, chapter 10) as well as on regular and opportunistic encounters with more than 1,000 other individuals?

Among savannah elephants, female dispersal is the exception, and large families of female kin develop over time (see Archie, Moss, and Alberts, chapter 15), while among African forest and Asian elephants, both sexes can disperse, resulting in smaller units (Turkalo and Fay 1995; Sukumar 2003). Whether males, females, or both sexes depart from their family, they may remain in some degree of social contact. Forest elephant females, for example, aggregate at *bais* (salt pans: Turkalo and Fay 1995; Fishlock, Lee, and Breuer 2008) or along coastal strips (Morgan 2001), where social relationships can be maintained. In Amboseli, males exploit bull areas that are distant from the majority of the female population, but when these males are sexually active, they return to core female areas, including their own natal area, to find mates (Poole 1982). Are the patterns we see in Amboseli elephants those of genetic dispersal or simply familial dispersal or "departure" from close social proximity to relatives?

A large musth male in Amboseli can easily cover up to 30 km in a day (Douglas-Hamilton, Krink, and Vollrath 2005; Douglas-Hamilton, box 7.1), encountering perhaps

50 different families. Thus, if he remains within his natal population, there is a probability that he will encounter a female family member at some point when both he and she are reproductively active. The chance of breeding with family females (matrilineal kin) is still small, however (Archie et al. 2007). A female may conceive her first calf at 12 years, with a 50 percent chance that it will be male (Moss 2001). The average longevity of a female is over 40 years; thus, by the time her first-born son reaches the age when he is regularly in musth (~35 years), there is a greater than 60 percent probability that his mother will be dead or at the end of her reproductive career. Fewer than 40 percent of males survive to the age of first musth, reducing probabilities of inbreeding in the population as a whole even further (Moss 2001). A male is probably more likely to mate with his maternal half sisters or female kin derived through the paternal line than with his mother, but the chances of a male breeding with his sisters should be equal to the chances of him breeding with any individual female of her age group. This probability will be a function of the male's geographical overlap with the range of his family. Recent genetic analyses reveal that there is inbreeding avoidance due to behavioral mechanisms for kin recognition (Archie et al. 2007). The roaming behavior of musth males (Poole 1982) could act as a further mechanism to reduce inbreeding.

The Transition to Independence

Independence from the natal family does not occur as a single event but rather is a gradual process with a high degree of inter-individual variability. There is no consistent association with a specific age, such as the onset of spermatogenesis at age 10 to 12 years. We defined the onset of the independence process as the first sighting of the male on his own, in a male group, or with unfamiliar females when his family unit was known to be completely separate (and usually a considerable distance away) from the group containing the male. This subjective "first independence event" appears to reflect a starting point in the male's process of independence (Lee and Moss 1999). After this event, only 53.2 percent of sightings were with the family (N observed first events = 92). We defined independence from the family as the age at which a male was sighted for over 80 percent of the time with elephants other than those of his natal family unit, with no return to higher levels of association in the following three-month period.

The mean age of independence was 14 years, with a range of 8 to 19 years (figure 17.2). Our sample could represent a portrait of young males dispersing during a period of population increase. We have thus compared the final dispersal age of a small number of males born early in the study with those of later-born males. Males born before

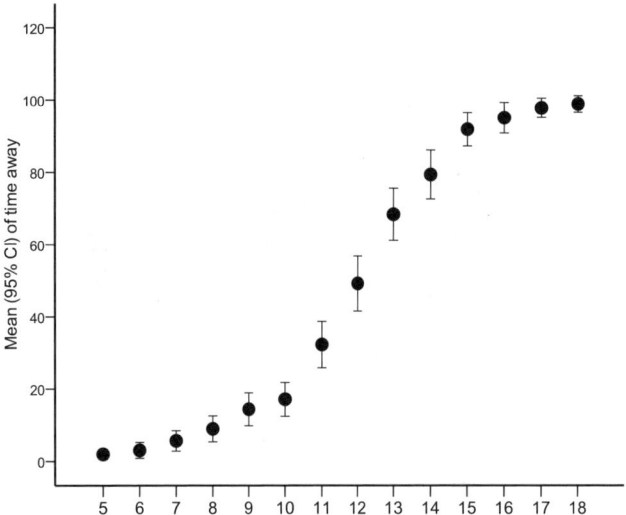

Figure 17.3 Individual variation in time away from the family, by male age in years (mean and 95% CI; N varies from 44 to 94 at each age).

1970 ($N = 18$) dispersed at an average age of 15.5 years, while those born after 1980 ($N = 80$) were independent at 14, while males born after 1990 with larger family and population sizes appear to be dispersing between 8 and 10 years of age. Whether this latest cohort of males left at a younger age by choice or as a result of increased feeding competition in larger families is a question that needs further exploration.

Individual variation in time away from the family was marked. Juveniles as young as 5 years old spent time away from their family in up to 25 percent of sightings, while some 18-year-old males were still with the family in 50 percent of sightings (figure 17.3). The shift to independence could occur as quickly as in five months' time or take as long as 8 years. We used the time that an individual spent away from his family relative to the mean for his age to describe the overall "speed" of the independence process for 132 males, with reference to a calculated logistic curve for age (see also Lee and Moss 1999). We determined residual values for each male and classified positive residuals as "quick"; those that were within one standard deviation of zero were "average," and negative residuals were classified as "slow." Quick males tended to start the transition at a later-than-average age and then simply left the family (figure 17.4). Males who took a long time to compete their dispersal started younger, appearing to come and go from the family over a period of years, but on average they left at a younger age (ANOVA $F_{2,129} = 16.5$, $p < 0.001$).

Males were individual and idiosyncratic in terms of the age at which they finally left their families. We examined family composition—its size, cohesion, and gregariousness (sociability with other families; Moss and Lee, chapter

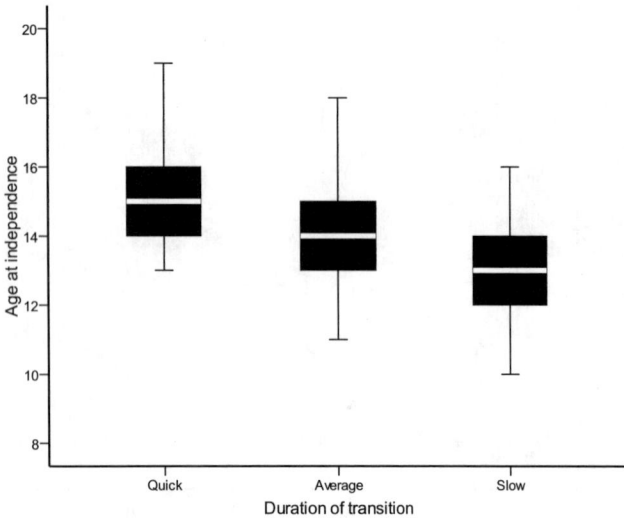

Figure 17.4 Age at independence (median, IQR, and 95% CI) in relation to the speed of the transition (N = 34 quick, 54 average, and 44 slow).

Table 17.1 Family and individual characteristics associated with the process of independence for transitional males

Model	F	P
Main effects		
Speed (number of months from first event to full independence: slow, average, or fast)	5.32	0.007
Number of peer males	2.44	0.095
Matriarch (mother was matriarch)	2.86	0.040
Interactions		
Number of peers · speed	3.52	0.011
Number of peers · matriarch	2.69	0.028
Mother survived · speed	4.68	<0.001
Mother survived · number of peers	2.96	0.009
Mother survived · matriarch	4.71	0.000

Note: All variables are coded as categorical variables (ANOVA results, N = 132).

13)—as well as male characteristics such as relative body size as potential determinants of male age at independence and the duration of transition.

We started with an assumption that at around the age of 10 (see figure 17.3), males are likely to initiate the process. Therefore, a male's individual and family characteristics were determined for age 10. If males leave their families to gain experience with age-mates, then the number of peer males available in a family could influence the timing of departure; peers were defined as family males aged within two years of the subject males. Since a male's size might predict his willingness to incur social or physical risks, we used a measure of relative size. Using residuals of foot length for age or shoulder height for age (see appendix 1) closest to age 10, we classed males as small, medium, or large for age.

Large families tend to be less cohesive and to fragment more (Moss and Lee, chapter 13), which may also promote independence. Family size (see figures 13.1 and 13.4 for Amboseli family sizes) at male age 10 was categorized as small (under 10), medium (10–17), and large (greater than 17). The gregariousness of a family also potentially influences opportunities for interaction with novel peers. We used the family sociability score (percentage of sightings with other families) at age 10. For males where family gregariousness was unknown at age 10, we used mean family gregariousness over the study period. Maternal status may predict the risk of aggressive behavior toward a young male, which could hasten his departure. Status of the mother was defined as a matriarch considered capable of defending her son, a high-ranking female who was also dominant, or a subordinate female whose son might be at risk of being the target of aggression from his family. Whether the mother had died was also noted. Finally, rainfall and its distribution were used to assess food availability in relation to dispersal events.

We examined these factors for effects on age at independence. Due to the study's small sample size, we cannot yet make definitive statements, but several trends can be suggested (table 17.1). The presence of a mother may be a protective influence, enabling sons to depart when they are ready. If the male's mother was a matriarch, he was more likely to begin the process at a younger age but took longer to complete it (figure 17.5). If in addition to having a matriarch for a mother, a male was from a small family with no peers, he was likely to leave at a young age and complete the process rapidly. Males whose mothers were alive also completed the transition quickly but left at older ages. It could be argued that males who leave early and quickly are socially confident. While family size and sociability provide opportunities for males to interact with others, and a male's relative size and dominance could influence peer interactions, no significant effects of body size, family size, or gregariousness were found. Further, there was no obvious ecological influence; that is, males were as likely to leave early or late when food was limited or abundant. As we have noted elsewhere (Lee and Moss 1999), no single factor or interaction of factors explains why and when males depart; they appear to be making decisions rather than responding to predictable social, demographic, or ecological contexts. Rare events, such as a bout of targeted aggression from older family females, might precipitate the process for some males, while a propensity to associate with specific peers who are dispersing might be a major factor for other males.

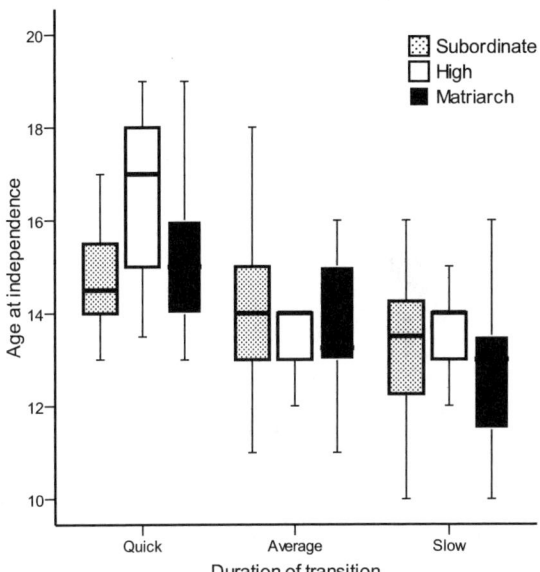

Figure 17.5 Age at independence (median, IQR, and 95% CI) as a function of maternal status (mother was matriarch, $N = 49$; mother was high ranking, $N = 11$; mother was subordinate, $N = 58$) and the speed (quick, average, or slow) at which the transition took place.

Association Patterns of Transitional Males

To some degree, young males are constrained in their potential associations and interactive partners by the association preferences of their natal family. Moving away from the family facilitates greater flexibility and choice of associates. If peers are important social partners for assessing individual strength and establishing competitive hierarchies, then we might expect transitional males to seek out groups containing peers.

Prior to independence, males tend to have family and unfamiliar male peers as nearest neighbors more than might be expected on the basis of availability (Lee 1986; Lee and

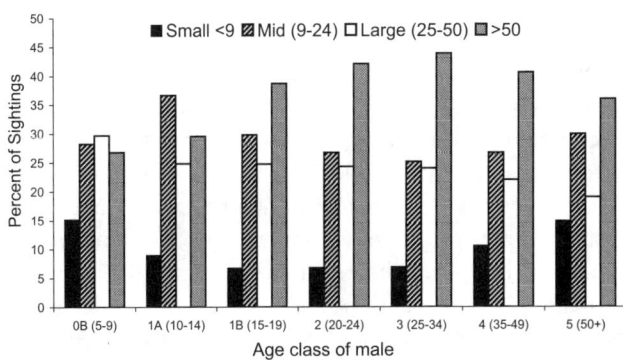

Figure 17.6 Mean percentage of sightings where males of each age-class were found in small (2–8), medium ("mid," 9–24), large (25–50), or very large (>50) mixed-sex groups (N sightings: 0B = 138; 1A = 2,687; 1B = 5,723; 2 = 4,667; 3 = 4,644; 4 = 1,830; 5 = 147).

Moss 1999). After independence, young males can group so as to maximize their contact with similarly aged males. Although some 25 percent of the population consists of independent males, they compose on average only 12 percent of mixed-sex groups. From the females' perspective, a small group is likely to have few independent males present, while large groups (>50 elephants) almost always contain one or more independent males. However, from the male's perspective, he can join female groups at a frequency commensurate with a group's size or occurrence or he can selectively join groups of different sizes. Thus, if young males preferentially seek peers, they should be found in groups in which the availability of other males is consistently higher. Relative to older males, those under 20 years of age are found more frequently in smaller-sized groups (figure 17.6) in which there will be 1 or 2 peers available (range of number of males in small groups is 0 to 9). Even in large groups, only 4 to 5 other males (range 0–24) are likely to be present, although the largest groups (those over 50) can contain many males (median = 15, range 1–70). Younger males may minimize the competition experienced from both males and females by associating with small and mid-sized groups.

Despite the above suggestion that young males avoid competitors, males aged 10 to 19 years do associate with groups containing more males than would be expected from the statistical relationships between group size and male numbers ($N = 5,702$, $r^2 = 0.556$, $p < 0.001$; post hoc, residuals by age category $p < 0.001$). In small to medium-sized groups, males represent a higher proportion (25 to 45 percent) of the group. Although more numerous in large groups, they represent a smaller proportion (12 percent). Peer-peer contact and establishing familiarity through association appear to be priorities of transitional males, and they seek out groups that allow these contacts and close association rather than extremes of competition.

Even after dispersal, males associate with their family, particularly in large aggregations. We tracked the associations of 13 well-known males with their natal families over a period of 20 years after independence. Irrespective of the actual age at independence, in the first five years after independence, males were still found in groups with their families in almost 14 percent of their sightings. After 10 years, their association in groups with their natal family occurred in only 4 to 7 percent of sightings (figure 17.7). The observed age-related frequencies of association with families may represent a "background" probability of encountering female kin for males in this population.

Aggression toward Family Males

Within families, elephants engage in aggression relatively infrequently (median = 0.17 events/hour, $N = 310$ focal

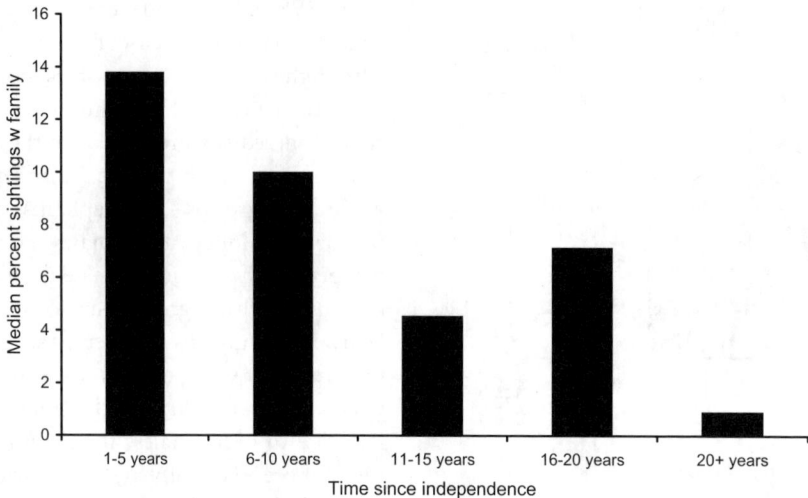

Figure 17.7 The median percentage of sightings that known family males were seen in groups with their families over a 20-plus-year period after independence.

hours). Aggression by females directed toward family males does not increase markedly with age (Lee 1986; see Lee and Moss, chapter 14). High-level aggression is, however, targeted toward non-family transitional males (85.7 percent of 14 events), at least until they are about 20 years old and noticeably larger than the adult females. As males grow to be the same size or just bigger than adult females (16–20 years), they may use their newly gained physical status to harass females. Adult females have been seen to form coalitions of two or more individuals to attack or drive away irksome non-natal males. Such coalitions typically involve assistance from the matriarch. Natal males, on the other hand, appear to be tolerated and even greeted upon rejoining the family after being away. For the 13 males tracked over the long term in figure 17.7, we have no records of family aggression upon reuniting.

Male Sociality beyond Independence: The Social and Ecological Context

Once males leave their families, their choice of association partners and areas for foraging is unconstrained by family decisions. They can use social and spatial strategies to maximize their access to food and reproductively active females and minimize the risks of competition and mortality. Independent males (15–60 years old) are frequently seen in groups containing females; of all sightings of bulls in the central areas of the Amboseli basin, 57 percent were in mixed groups, 30.5 percent were in all-male groups, and 12.5 percent were alone. A strong age-dependent pattern emerges with the time spent in the company of females, declining with increasing age and the time spent alone or in association with other males (Poole 1982; figure 17.8).

That older, larger males spend less time with females than do younger males has been noted anecdotally and has given rise to the belief that older males are reproductively senile. They are therefore considered by some to be suitable targets for trophy hunters, and that their loss would not be detrimental to the population. This contention could hardly be further from reality. Older males are, in fact, more successful at guarding and mating with females than are younger ones (Poole et al., chapter 18). They focus on female groups when they are in musth and in peak condition and are most likely to locate receptive females (Poole 1989a, 1989b). Outside sexually active periods, older males withdraw to habitats known as bull areas (Moss and Poole 1983) in order to regain condition and avoid potentially aggressive interactions with sexually active musth males (Poole 1999a).

Most large males spend over 50 percent of their nonmusth time in a geographically consistent bull area (mean = 69.8, range 50–92 percent of sightings, N bulls = 39, N sightings = 4,350). Some males (15.4 percent) have less fidelity to one bull area, but even they are not sighted evenly across all areas. After independence, males gradually develop a preference for a particular bull area. Bull areas tend to be distinct from habitats favored by cows and mixed groups, although one bull area used by a small proportion of males (12.8 percent) is relatively central. Males using this central area tend to be from slightly older age cohorts. Males over 40 years return to female core areas primarily when they are in musth, which represents only about 10 to 25 percent (mean = 17 percent; Poole et al., chapter

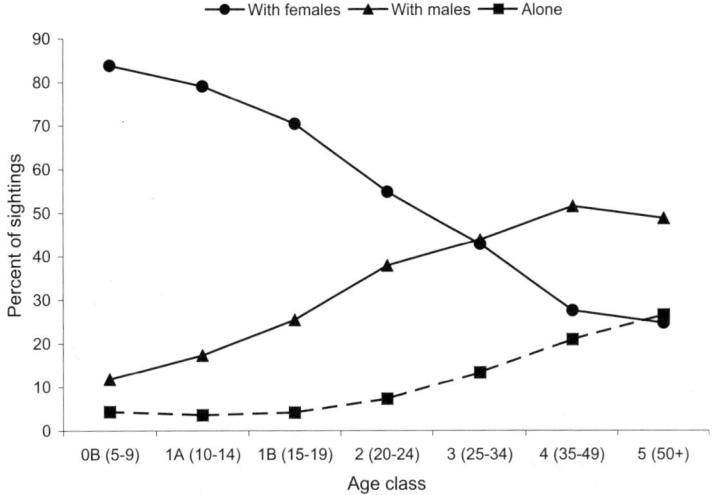

Figure 17.8 The percentage of individual bull sightings alone, in groups with other males, or in groups with females, by age category (N as figure 17.6).

18) of time. Young, sexually active males frequently roam into female areas searching for mating opportunities or for additional sources of food and water. The three bull areas within the Park show up strikingly in the musth and non-musth distributions of three selected males (figure 17.9): Western male (M13), Eastern male (M7), and Central male (M22). Over time, bull areas have shifted to outside the Park boundaries, as the males have reduced the available browse within the Park, moving south into Tanzania, further east to Kimana, and into the southern bushland on the slopes of Kilimanjaro. At this stage, the ecological profiles of the different bull areas remains to be characterized, but they are typically high-browse habitats.

Older Males in Mixed Groups

Do older males make decisions about the size of female groups with which they associate? Younger males tend to associate with smaller female groups, while mid-age sexually active males are found in large female groups (ANOVA: $F_{6,20680} = 45.8$, $p < 0.001$). The oldest males (50-plus years), by contrast, associate with relatively smaller female groups (see figure 17.6) but also with very large groups. Large males, particularly when in musth, associate with small mixed groups when they contain estrous females.

Very small mixed groups (fewer than 10 elephants) have a high proportion of males and are likely to contain a musth bull, specifically when an estrous female is present. In small groups, however, males under 25 years are also present in over 20 percent of sightings with estrous females. Musth males additionally associate with very large groups (>50) in order to reduce the time spent searching for estrous females

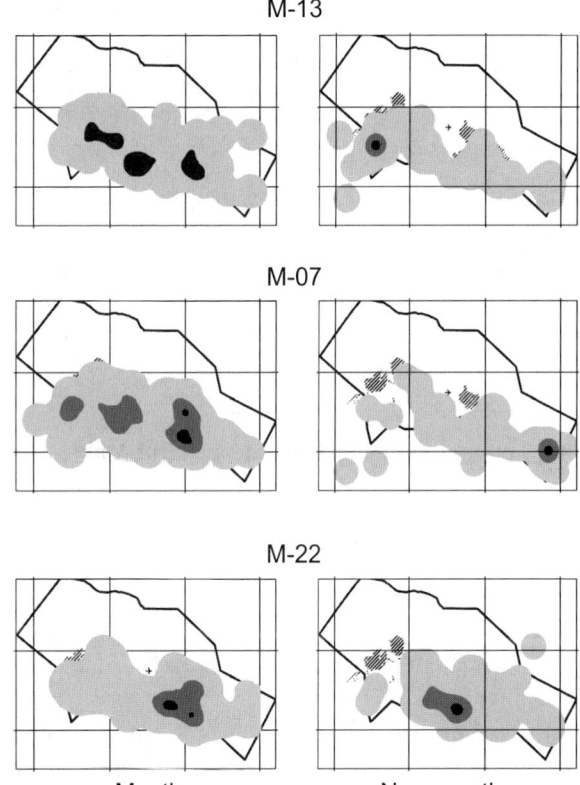

Figure 17.9 Musth and non-musth distributions of three older males, M13, M7, and M22. Western Male M13: Musth and non-musth. Kelunyiet Male M7: Musth and non-musth. Ol Tukai Orok–Southern male M22: Musth and non-musth.

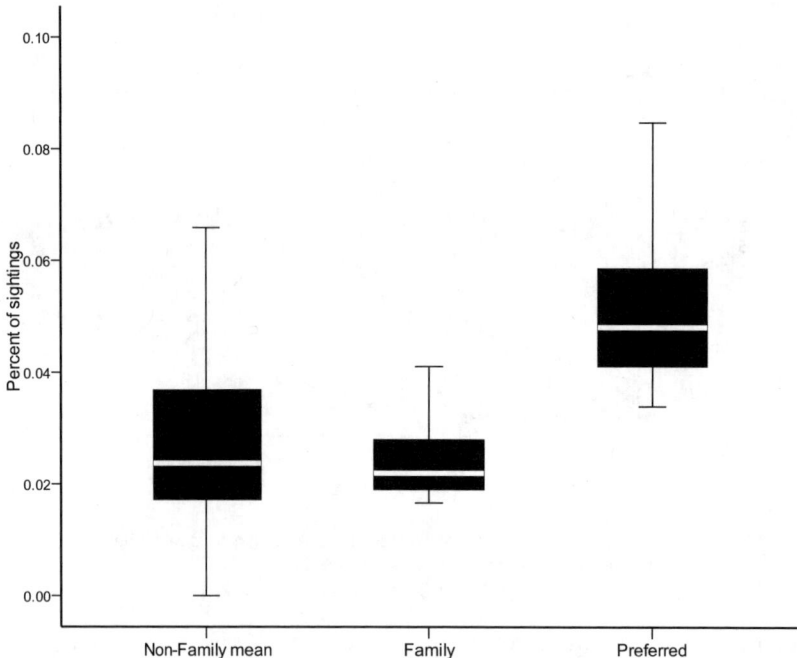

Figure 17.10 Median (IQR and 95% CI) percentage of sightings where independent males from known families were in groups with their natal family, with an "average" family in the population, and with their most frequent family associate ("preferred family").

(Poole and Moss 1989). Over 50 percent of sightings of musth males are in groups larger than 30 while the median size of mixed groups lacking a musth male is 18. The larger the group, the more attractive it is to high-ranking, sexually active males: half of the groups with more than 50 individuals contained one or more musth males.

Family Preferences among Fully Independent Males

Given that fully independent males roam in search of females and food, it has been assumed that they form few long-lasting bonds with other elephants, whether male or female. Here we examine male associations with family groups, looking first at mature males whose natal families were known and then at the older males whose natal families are unknown.

Most pre-musth age, sexually active males associated with each of the families in the population at least once (mean = 53.9 ± 2.9 out of 56 possible families). These males show some preferences for families that they encounter in their normal ranging area. The home range of a juvenile male naturally falls within its family's home range until independence, at which point a male's associations and therefore his distribution begins to shift (Croze and Moss, chapter 7). Many young males, however, show a strong association with one or more families from their mother's original clan area before settling into a bull area. This phenomenon is illustrated by the percentage of time a young male spends with his top family associates (his "preferred family") compared with the time spent in association with his natal family (figure 17.10).

Older males associated with most of the available family groups. No consistent association preferences were found among the 49 families and 21 musth-age males that were observed over the entire study. The top family associate for 14 of the 19 older males tended to range in areas close to the male's bull area when not in musth. During their sexually active period, males appear to focus on as many families as they can contact across wide areas of the habitat. During their non-sexual phase, they encounter females from or near their bull areas on the relatively rare occasions when they join groups with females.

Males in Male Groups

Very few studies have focused on social relationships between individual males (but see Croze 1974; Evans and Harris 2008). Although it has been argued that male elephants do not form long-term friendships, they are generally gregarious by nature, and sexually inactive males in Amboseli form groups with other males that range in size from 2 to 40 individuals (figure 17.11). The question

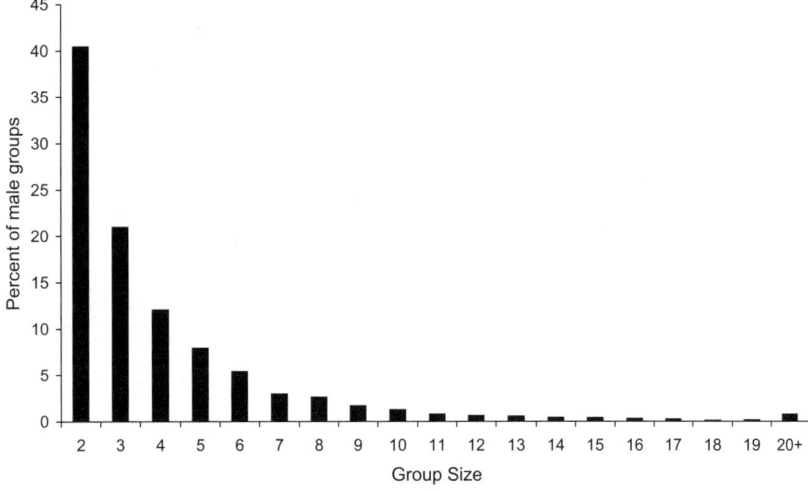

Figure 17.11 Frequency distribution of male-only group sizes for males when not in musth ($N = 3,462$).

is whether males form these groups with the same companions over time or if their associates are chosen at random.

Associations among Younger Mature Males

Young, recently independent males spend more of their time in the company of females (see figure 17.7) than in bull groups. For this young age set, we examined male-male associations in all group types when there were more than 50 sightings of an individual male. Most males associate with one close partner within a cluster of 3 to 6 other males (figure 17.12). Since all the males in this analysis are relatively close in age, age-matched clusters were not expected, nor were they observed. Males from the same families are found within the same clusters (e.g., GB males), while males from bond groups (GB/IB, EA/EB) are not closely clustered. Males whose mothers were from similar geographical areas (see Croze and Moss, chapter 7) do form large, loose clusters (JA, PB, AA, QB, ZA, ZB males). The fact that these clusters are based primarily on sightings of males when they were with females suggests that there is either a social selection ("friendships") or spatial overlap due to males residing in the same areas after dispersal and simultaneously forming groups with females and with each other.

We found one top associate who was distinct from other, looser male associates in the clusters ($t = 16.13$, df = 30, $p < 0.001$; figure 17.13). For four males, this top associate was a male from the same natal family, but there were no other top associates from even the same bond groups. Again, since this sample represents males primarily associating with each other in female groups, these associations may not reflect male friendships or choices. Male associations in the context of females could be structured by competition rather than friendship.

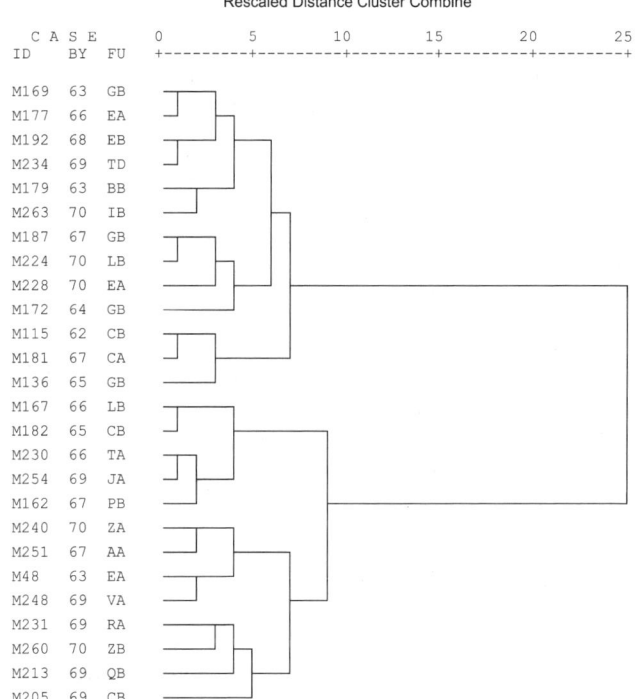

Figure 17.12 Cluster of male-male associations for males born after 1962 and from known families (Ward's method for linkage, squared Euclidean Distance). Male birth year (BY) and natal family (FU) are given.

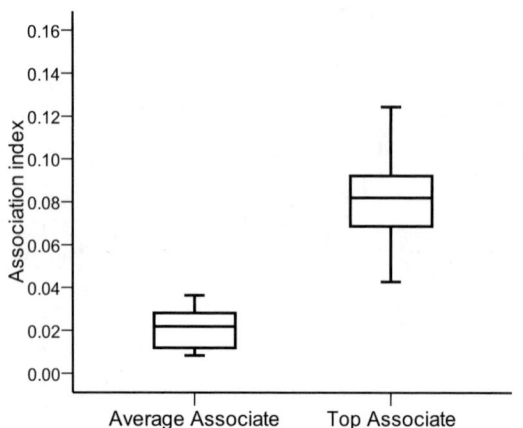

Figure 17.13 Average and top association indices (median, IQR, and 95% CI) for 31 young mature males with well-known histories when in groups with other males.

Associations of Older Mature Males

Groups of non-musth males could simply be based on loose associations between males who happen to be in the same place at the same time, contingent on sharing similar bull areas. Alternatively, associations between these males could be non-random. With more than 30 years of data on association patterns among adult males, we explored whether older males form lasting friendships with other males and if so, what these bonds are based upon. It is worth noting that sharing a bull area is a decision in itself, since bull areas differ from the family areas where males were raised (see Croze and Moss, chapter 7). Males may move into particular areas because they have associates who live there; alternatively, they may form associations with males who share their areas after choosing specific habitats or ecologically valuable areas.

Some older males are "gregarious"; in other words, they spend most of their sexually inactive, non-musth time in the company of other males, while others are less gregarious, showing a more solitary nature when out of musth. For example, M126 spent 2.3 percent of his sexually inactive time alone, while M80 spent 75.6 percent of his sexually inactive time on his own. The median time spent alone by sexually inactive large males was only 9.8 percent.

We selected males who were older than 20 years in 1980, who survived for at least 10 years, and who were seen more than 40 times in bull-only groups when not in musth between 1972 and 2002 ($N = 39$). We determined association indices for all 39 males and used these indices to cluster males. There are few close symmetrical partners, although males do appear to have stable associates, with clustering among males who use the same bull areas and who are similar in age (figure 17.14). The clusters of mature

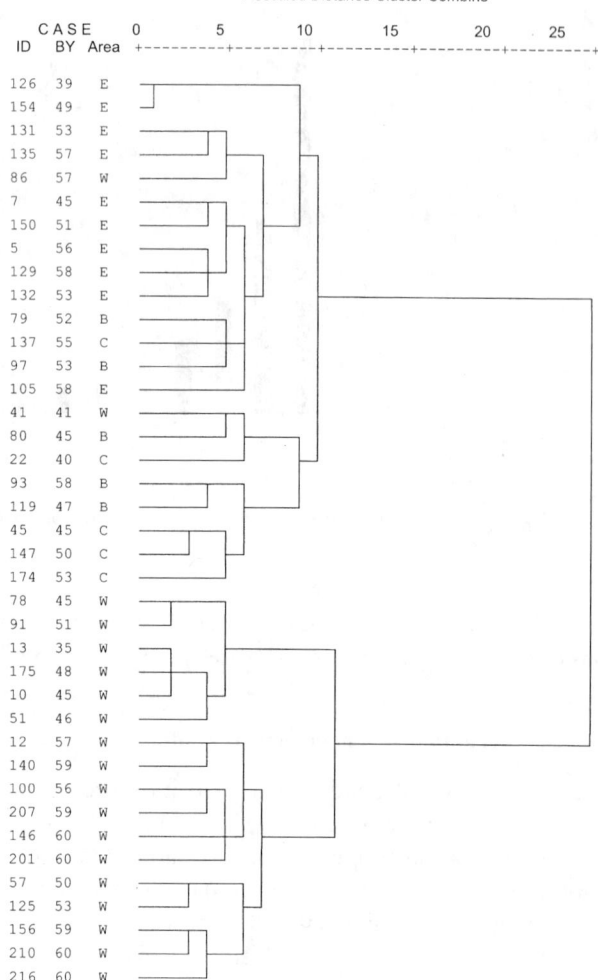

Figure 17.14 Clusters of older males outside musth when in bull groups (Ward method for linkage). Birth year (BY) and general bull area are indicated (E = East, W = West, C = Central, B = broad ranging).

males are linked less closely than are those of younger males (see figure 17.12).

Gregarious males can be either "promiscuous" (i.e., spend a lot of time with many individuals) or they can be more "discriminating," appearing to choose particular males to associate with. Gregarious males can also form close relationships with different males at varying stages of life. Age alone seems to explain relatively few of the patterns of association. While older males tend to have fewer total associates ($r = -0.33$, $N = 39$, $p = 0.04$), there is less of a focus on specific partners as males age ($r = 0.43$, $p = 0.006$). Large, older males tend to associate fairly evenly with other males, although each has a top associate. Since the top associate is unique to each male (and may not be the closest male emerging from the clusters of all associates), we examined top associates for potential friendships

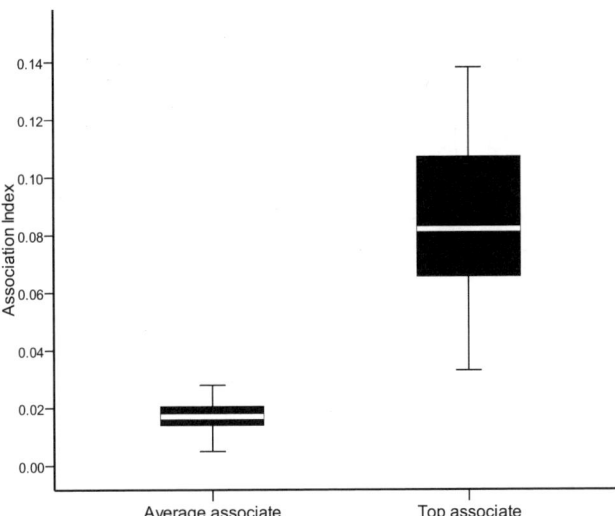

Figure 17.15 Average and top association index (median, IQR, and 95% CI) for 39 large males with other males.

(figure 17.15). The top associate is indeed distinctive, with far higher than expected association indices compared with the mean ($t = 11.77$, df = 38, $p < 0.0001$). Large males have at least one friend, who is typically close in age. The median age difference between top associates is ±2 years, with a range of 0–15 years, while the range in age differences among males in the entire sample is 0–23 years.

Males of similar ages might be expected to compete directly with each other in order to come into musth in those months when more females are in estrus. In addition, if males are simultaneously in musth, age-matched associates potentially could be significant rivals. Males, however, were not often seen in musth in the same months as their top associate (median percentage of months simultaneously observed in musth = 2.5, range 0–10.6 percent, mean = 2.8±2.4). This overlap is close to that expected by chance (2.1 percent) from a mean of 17 percent of time in a year spent in musth (see Poole et al., chapter 18). Thus, while top associates may not be actively avoiding each other's musth months, well-matched males who minimize their competition for best musth months may then be able to preferentially associate outside their musth periods.

Conclusions

The social life of male Amboseli elephants is structured by age, preferences for residence areas or habitats, individual friendships, and sexual states. It is thus complex and diverse, and we are only now starting to understand the timing of events, social preferences, and effects of sexual activity that go into making up a male's social experiences. Males in populations elsewhere may face the same age and sexual state decisions but have different opportunities to express individual friendships with either females or other males. They may either range over a much greater area (Douglas-Hamilton, Krink, and Vollrath 2005; Rasmussen 2005; Evans 2006) or be far more constrained in their movements by the distribution and extent of preferred habitats (Tchamba et al. 1994).

The age-sequence of changing social dynamics is clear; males between the ages of 8 to 18 years establish themselves in a male social milieu. They can form temporary affiliations with favorite family units or reside in a specific area shared with females and other younger males. With age, they shift toward a better-established bull area but still associate more with female groups as part of their phase of sexual activity. These pre-musth, sexually active males tend to seek out large groups of females rather than specific family units and often find themselves in association with many other male competitors. Clustering among sexually active males appears to be more the consequence of co-association within these large mixed groups and thus could reflect competitive relationships rather than affiliation. In addition, the males in mixed groups could share an association because they seek groups in specific female core areas with which they are familiar.

Once males enter into their annual musth sexual cycle, they show greater fidelity to a specific bull area. We do not yet know why males reside in different bull areas, nor do we know the habitat or foraging benefits of the bull areas. Bull areas do appear to be vital to regaining any condition lost during the annual musth cycle (Poole 1989b), but whether this is due to minimized competition costs from females or specific nutritional advantages of the habitats remains to be determined.

Within bull areas, some large musth males have strong affiliations (low or non-existent aggressive interactions) and associations with other males, which appear to be manifest as friendships. Some males are generally gregarious, while others are highly selective. Some are sociable, others less so. As we note elsewhere throughout this volume, the individual characteristics of elephants are marked; these are animals with distinctive traits and personalities, and their individuality makes it difficult to generalize about their behavior. As males age, their social, ecological, and reproductive strategies change in predictable ways, but their individuality remains paramount.

Chapter 18 Longevity, Competition, and Musth: A Long-Term Perspective on Male Reproductive Strategies

Joyce H. Poole, Phyllis C. Lee, Norah Njiraini, and Cynthia J. Moss

REPRODUCTIVE SUCCESS has been strongly linked to longevity in a number of species (e.g., fulmars, 60 percent of total variance: Ollason and Dunnet 1988; cetaceans: Connor, Read, and Wrangham 2000; lizards: Lopez, Aragon, and Martin 2003; orangutans: Atmoko and van Hooff 2004). For both male (Poole 1989b; Rassmussen 2005; Hollister-Smith et al. 2007) and female elephants (Moss and Lee, chapter 12), longevity is a key factor in reproductive success. As is typical for polygynous and sexually dimorphic mammals (e.g., mandrills: Setchell, Charpentier, and Wickings 2005; red deer: Clutton-Brock, Albon, and Guinness 1982; elephant seals: Le Boeuf and Reiter 1988; lions: Packer et al. 1988), variance in male reproductive success is greater that that of females. Thus, an understanding of what underlies variation in success throughout the reproductive lifespan of a male elephant should provide insights into the factors contributing to reproductive success and how these differ between individual males.

During his lifespan, a male elephant may initially become sexually mature when he is too small to compete effectively. Later, his reproductively active annual cycle, musth, may or may not coincide with the conception cycles of females. He may encounter times when there are many receptive females and times of scarcity or he may engage with many or with few male competitors. All these factors influence the success of an individual male in terms of access to receptive females and even determine whether he is likely to die in contests with other males. Thus, decisions or behavior in any annual cycle can drastically influence overall reproductive success, while there are likely to be other contexts in which a decision taken one year will have little effect on reproduction in subsequent years. All in all, the impact on reproductive performance of either variation or consistency in the behavior of individual males is poorly known for mammals (but see Coltman et al. 1999a), especially in the context of a potential for 30-plus years of sexual activity. Here, patterns and parameters of musth and the reproductive strategies of individual male elephants are outlined from a long-term perspective. In the course of this long-term study, we are just beginning to explore individual variation in reproductive outcomes.

The Context of Musth

Considerable background information exists on the mechanisms and controlling factors for male elephant reproduction, from detailed descriptions of musth as a period of heightened sexual and aggressive activity among male Asian (Jainudeen, Eisenberg, and Tilakeratne 1971) and African elephants (Poole and Moss 1981; Poole 1982; Hall-Martin and van der Walt 1984; Poole et al. 1984; Hall Martin 1987; Poole 1987a; Poole 1989a, 1989b; Poole and Moss 1989; Poole 1999a; Kahl and Armstrong 2000) to hormone profiles of musth and non-musth males during development and as adults (Poole et al. 1984; Hall-Martin and van der Walt 1984; Ganswindt et al. 2005). Contests and mating behavior are also well documented; musth males interact aggressively with other large adult males, particularly those in musth, and spend much of their time searching for, attempting to gain access to, or guarding estrous females

(Moss 1983; Poole 1987a; Poole 1989a, 1989b; Poole and Moss 1989). While non-musth males can and do mate successfully (Ganswindt et al. 2005; Hollister-Smith et al. 2007), male-male competition and female choice combine to enhance mating success for those in musth (Poole 1989a, 1989b). Thus, while musth does not preclude mating activity and a successful reproductive strategy on the part of non-musth males, a male who lives to be over the age of 30 relies on musth as his primary reproductive strategy. Musth is a strategy associated with longevity. Our perspective here is to explore in detail and over the long term the relations among males when they exhibit musth, how long they stay in musth, and how their musth timing relates to access to reproductive females as well as to the potential for contests between males.

Age-graded reproductive strategies are common in male mammals (Clutton-Brock 1988). Reproductive potential, considered to be the capacity for spermatogenesis, occurs at around 10 years of age in elephants, but males do not begin to produce sperm in quantity until about 17 years old (Laws 1969). Males between the ages of 15 and 20 years show interest in estrous females but weigh only about half that of full-grown males (Laws 1966). Since males grow in stature and mass throughout their lives (Lee and Moss 1995; and see below), older males are larger and generally dominant to younger males (Poole 1989a). Owing to intense competition from older males (Poole 1989a) and strong female preferences for older individuals (Moss 1983; Poole 1989b), young males have limited mating opportunities. Most successful matings by younger and non-musth individuals occur during early and late estrus (Poole 1989b) and are less frequent during mid-estrus, when conception is most likely to occur. The fact that young and non-musth males are actively competitive around estrous females does, however, indicate that on occasion their efforts pay off (see box 18.1).

Males begin exhibiting specific sexually active periods around 25 years of age. By sexual "activity" and "inactivity," we refer to males who exhibit a clear, consistent, and alternating pattern of associating for several months with female groups (or alone searching for them) and then a longer period of the year in the company of other males or alone (Poole 1987a; Ganswindt et al. 2005; see Lee et al., chapter 17). Prior to this age, males engage in sexual interactions with females and, given the opportunity, will mate, but these are transitory, rare, and opportunistic events. By the age of 30, the majority of males have experienced their first musth period, which occurs at some point within their sexually active period.

Musth is characterized by distinct displays, swollen and secreting temporal glands, the dribbling of strong-smelling urine (Poole and Moss 1981; Poole 1987a; Kahl and Armstrong 2000), and in the African genus, a very low-frequency vocalization, the "musth rumble" (Poole 1987a; Poole et al. 1988; Poole 1999a). During musth, male elephants experience dramatic surges of circulating testosterone (Hall-Martin and van der Walt 1984; Poole et al. 1984; Ganswindt et al. 2005). Among young males, musth may last days or weeks, while among older males, musth lasts several months (Poole 1987a). In general, males with longer musth periods lose more condition than do those males with shorter periods (Poole 1989a). As will be discussed in detail here, musth periods are asynchronous, and each male exhibits his own particular cycle (Poole 1987a).

Musth has a striking effect on the relative dominance ranks of males (Poole 1989a). Typically, male dominance is based on body size. However, with few exceptions, a musth male, whether large or small, ranks above all non-musth males. Thus, small, sexually active musth males may meet, interact with, and dominate larger, sexually inactive non-musth males. Musth males tend to interact aggressively whenever they meet. Between musth males, dominance is also size related, and thus escalated contests occur between musth males who are closely matched in both size and condition. Serious fights also take place when the normal rules that predict status are perturbed due to loss of condition. Since the musth periods of males are asynchronous, a large male in poor condition toward the end of his musth period may meet a smaller male in peak condition at the beginning of his period, and the smaller male may then challenge the larger male. The presence of an estrous female is not a prerequisite for a fight (Poole 1989a), nor are outcomes predictable.

In this chapter, we aim to relate individual patterns of musth to opportunities for mating and the factors that constrain those opportunities. Specifically, we intend to

- assess the effects of longevity on male reproductive potential,
- examine musth timing and how it relates to female availability and male age, and
- explore how competition with other males influences probabilities of access to estrous females for males at different ages.

Quantifying Reproductive Parameters

The results we discuss here are derived from observations of male elephants between 1976, when musth was first described (Poole and Moss 1981), and 2002. Musth state was recorded whenever males were sighted (see appendix 1). Our goal in this section is to outline some of the parameters associated with male reproductive performance, which necessarily incorporates the female perspective (Moss and Lee,

BOX 18.1 GENETIC PATERNITY ANALYSIS OF THE AMBOSELI ELEPHANT POPULATION

Julie A. Hollister-Smith, Joyce H. Poole, Cynthia J. Moss, and Susan C. Alberts

As noted by the authors of this chapter, musth has a dramatic effect on many aspects of male reproduction. Musth temporarily elevates males' dominance rank (Poole 1989), musth males achieve higher mate-guarding success than non-musth males (Poole 1989a; Poole and Moss 1989), and estrous females prefer large musth males as associates (Moss 1983; Poole 1989a). Studies in some species have shown, however, that dominance rank, mate guarding, and even mating itself are not always perfect predictors of reproductive success (Coltman et al. 1999a; Eady and Hardy 2001; Hughes 1998; Pemberton et al. 1992; Preston et al. 2001). Therefore, genetic determination of paternity is important, particularly in species such as elephants, in which few mating episodes are observed.

Establishing Genotypes

We focused our collection efforts on the ca. 1,200 elephants that were recognized members of the Amboseli study population. For nearly 85 percent of sampled individuals, we collected multiple independent fecal samples and for the remaining 15 percent, a single fecal sample. In addition, we collected tissue samples for roughly 10 percent of all sampled individuals.

We established genotypes for individuals at eight tetranucleotide microsatellite loci as described in Hollister-Smith (2005). Our genetic data set included (1) 114 males estimated to be aged 20 years or older at the time of sample collection (approximately 66 percent of potential fathers), (2) as many calves as possible from each of the 52 families in the population (N = 403 calves), and (3) their mothers whenever possible (we sampled mothers for 329 calves). The data set included several mothers with multiple offspring. Because multiple generations were genotyped within a family, a female may be counted as both an offspring and a mother (Hollister-Smith 2005). The Amboseli elephant population showed high genetic diversity (Hollister-Smith 2005), which increased the power of paternity assignments despite incomplete sampling of the population. No individual born before 1967 was considered for parentage assignment because low-sampling intensity of offspring mothers and candidate males from this period made it unlikely that paternities could be assigned with high confidence.

Assigning Paternity

For all sampled individuals born after 1967, we assigned paternity using a three-step process (Hollister-Smith 2005). This process relied heavily on the software program CERVUS 2.0 (Marshall et al. 1998), which uses a maximum likelihood approach. We supplemented the CERVUS analysis with a detailed review of all paternities assigned with high confidence (Hollister-Smith 2005). In order to be conservative, we accepted only those paternities assigned with 95 percent confidence that were either perfect matches; that is, the identified male matched at all eight loci and was the only male that did so, or in five cases the male had one to two mismatches with the calf but at homozygous loci. This process resulted in paternity assignments for 131 of the original 403 calves sampled (Hollister-Smith 2005).

Measuring the Relationship Between Age and Paternity and Between Musth and Paternity

Ages for all males were extracted from the Amboseli Elephant Research Project database and were used in assessing the relationship between age and reproductive success. In order to examine the effects of musth on reproductive success, we identified the month in which each calf was conceived and then examined musth records during the month the offspring was conceived plus the month previous to and the month following conception. This three-month window helped identify the musth state of fathers who had incomplete musth records and allowed for variation in gestation length. For each offspring conception, fathers were then grouped into one of five different categories depending on whether the male was seen during the month and whether he was observed in musth or not in musth (Hollister-Smith 2005). Analysis of the relationship between musth and paternity was limited to 119 offspring because detailed musth observations were unavailable beyond 1998.

Paternity Success

Age and Reproductive Success

Older males had higher reproductive success than did younger males (Hollister-Smith 2005; box 18.1, figure A). Males showed a steady increase in the number of offspring conceived, from first reproduction until their early 40s when they reached a plateau and remained at a high reproductive output for roughly 15 years. Reproductive output of males showed a modest decline by their mid-50s. We found no evidence of male reproductive cessation or even marked decline in reproduction with age (box 18.1, figure A). However, not all males over age 50 are equally successful, and relatively few males in the population live to this advanced age.

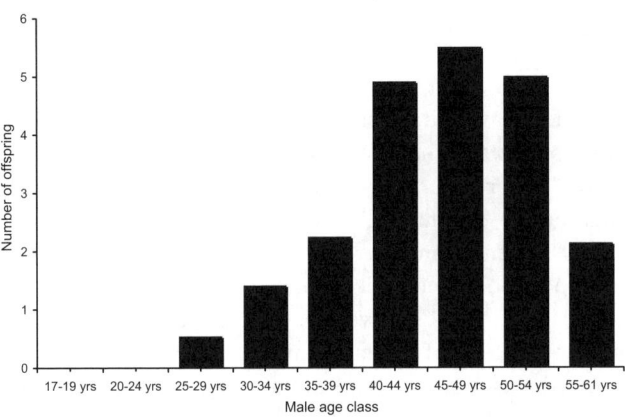

Box 18.1, figure A Mean number of offspring produced by a male in an age-class. This number was calculated by placing the 131 offspring into the appropriate age-class of the father at the time of conception and then dividing the sum for each age-class by the average number of genotyped males in that age-class. The average number of adult males in the age class was calculated for the 10-year period with the best observational and demographic data on males in the population (1993–2002).

Further, many different males of varying ages sired offspring. One-third of the males we genotyped (38 of 114) sired at least one calf (Hollister-Smith 2005). Reproduction was therefore not monopolized by a small subset of males. It was, however, quite skewed, with just three males responsible for approximately 30 percent of the assigned calves. Males as young as their mid-20s sired some offspring, although this was not a common event; eight males in their 20s each sired a single calf (Hollister-Smith 2005).

The pattern of continued high male reproductive output into old age is unlike that of most mammalian species. Most male mammals show a sudden rise in reproductive output with maturity, achieve high reproductive success for a very limited period during their physical prime when at their highest dominance rank, and then show a steady decline with increasing age (Clutton-Brock, Albon, and Guinness 1988; Le Boeuf and Reiter 1988; Packer et al. 1988; Owen-Smith 1993; McElligott, Altwegg, and Hayden 2002; Bercovitch et al. 2003; but see Roed et al. 2002).

Musth and Reproductive Success

We found that males of all ages were significantly more likely to be in musth at the time of offspring conception (box 18.1, figure B). This finding provides strong support for the long-held hypothesis that musth confers a male mating advantage. Moreover, this mating advantage results in true reproductive gain. A musth male's advantage may be due to male-male competition for access to females or to female preference for musth males, or both. Non-musth males are not completely precluded from mating, as some non-musth males do achieve reproductive success.

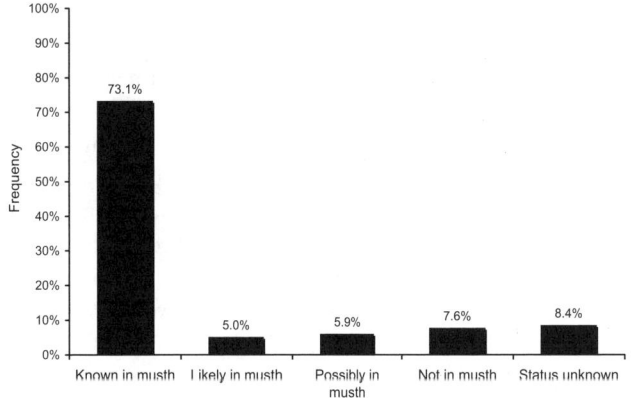

Box 18.1, figure B Proportion of fathers in musth at the time of calf conception (n = 119 calves and 36 fathers). The 36 fathers ranged in age at the time of conception from 26 to 59 years old. Musth records were utilized to place fathers into one of five categories at offspring conception: (1) Known in musth—fathers observed in musth during the month of offspring conception, (2) Likely in musth—fathers not seen in conception month but observed in musth either the month before or after, or both, (3) Possibly in musth—fathers seen not in musth during conception month but observed in musth either the month before or after, (4) Not in musth—fathers were observed in conception month not in musth and were either seen not in musth or were not seen the month before or after, and (5) Status unknown—fathers not seen during the conception month and not seen either the month before or after.

We conclude that in typical wild elephant populations, the older males sire the largest proportion of offspring, all adult males in the older age classes (over 25 years) are reproductively active, and most males are in musth when they sire calves.

chapter 12). Since only approximately 50 percent of estrous observations result in a birth (Poole 1989b), and since observers miss a proportion of estrous periods, we have used conceptions as a reproductive measure. We calculated the number of conceptions occurring by week in each year. Next, we determined the duration and timing of musth for each male each year. Then, we combined these perspectives and added the dimensions of competition and choice.

In order to calculate which males got the timing of their musth period "right" in relation to an ability to inseminate females, we needed an accurate figure for the duration of gestation. Gestation has been estimated as 656 days (Moss 1983) or 660 days (Laws 1969). Since there was still some uncertainty as to its duration, gestation length was recalculated by examining only those births for which both the mother's estrous behavior was observed and the date of birth was known to within 24 hours ($N = 49$).

Gestation lengths calculated as specified above ranged between 515 and 679 days. The two calves with the shortest gestation (515 and 625 days) both died immediately after birth; we assume they were premature and thus removed them from subsequent analyses. We have excluded all gestation intervals longer than 679 days ($N = 12$), since these calves must have come from females who did not conceive during their observed estrus but rather during a subsequent estrus (which occurs at three-month intervals; Hodges 1998). The median duration of gestation among the remaining 35 females was 661 days (inter-quartile interval [IQI] 656–669, range 642–679). One complication is that females can exhibit an anovulatory period of estrous-like behavior approximately three weeks prior to true estrus (Poole and Moss 1989), which is associated with LH surges (Kapustin et al. 1996). Since some observations of estrous behavior were likely to be associated with pre-estrous periods, we have eliminated those intervals longer than 675 days ($N = 5$). The median gestation length of the remaining 30 observations is 660 days (IQI = 655–665, range 642–675), and this figure is used to calculate conception dates for 642 births (figure 18.1).

Reproductive Timing

Temporal Patterns of Conception

A total of 247 females were observed in estrus and subsequently gave birth. In an additional 395 cases, the dates of estrus and conception were calculated from birth dates using the gestation length of 660 days. These dates of conception were then assigned to weeks within years and compared with male sexual activity.

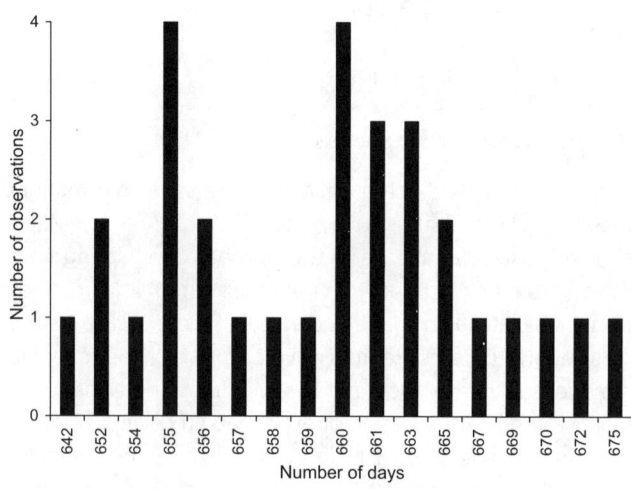

Figure 18.1 Gestation length as determined from observations of estrus and births in Amboseli ($N = 30$).

In some years associated with food abundance and variation in the number of females available to conceive (Lee, Lindsay, and Moss, chapter 6), there were far more conceptions than in other years. Within years, some weeks were "better" (i.e., they contained more estrous females) than others, and finally, over all years some weeks tended to be "better" than other weeks (figure 18.2). The number of conceptions per week ranged from 0 to 10 (med = 1; IQI = 1–10; $N = 1,219$ weeks). Over 23 years, week 4 (first peak at the end of January) through week 30 (first major decline) was more likely to contain an estrous female than was week 31 through week 3 (Mann Whitney $z = -6.12$; $p < 0.001$).

Week 44 (end of October/start of November) typically represents the onset of the rains leading to grass growth and estrous activity about 12 weeks later. Weeks 32–3 (August to mid-January) contained a median of 8.5 estrous females (IQI 5–15, range = 3–21, $N = 26$) while weeks 4–31 (late January through July) contained a median of 33 estrous females (IQI = 29–38, range = 19–46, $N = 27$). Thus, males whose musth periods fell in late January through July were likely to overlap with more estrous females than were those males whose musth periods fell between August and mid-January.

Temporal and Age-Specific Patterns of Musth

An individual male's musth timing is conspicuous in its patterning and consistency across years (figure 18.3). For males with incomplete records, we used their long-term pattern to estimate the week of onset and the termination of musth. In these cases we established the musth period around the central dates that a male was actually observed in musth,

Longevity, Competition, and Musth: A Long-Term Perspective on Male Reproductive Strategies

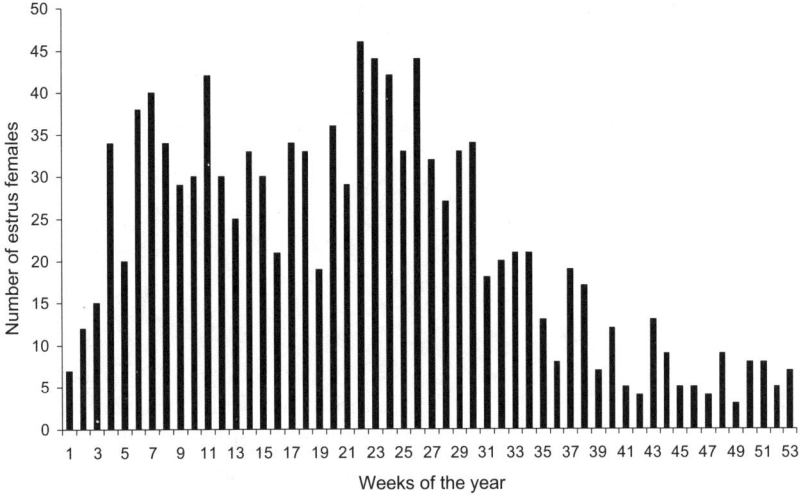

Figure 18.2 Number of estrous females who conceived by week of year (1–53) over 23 years.

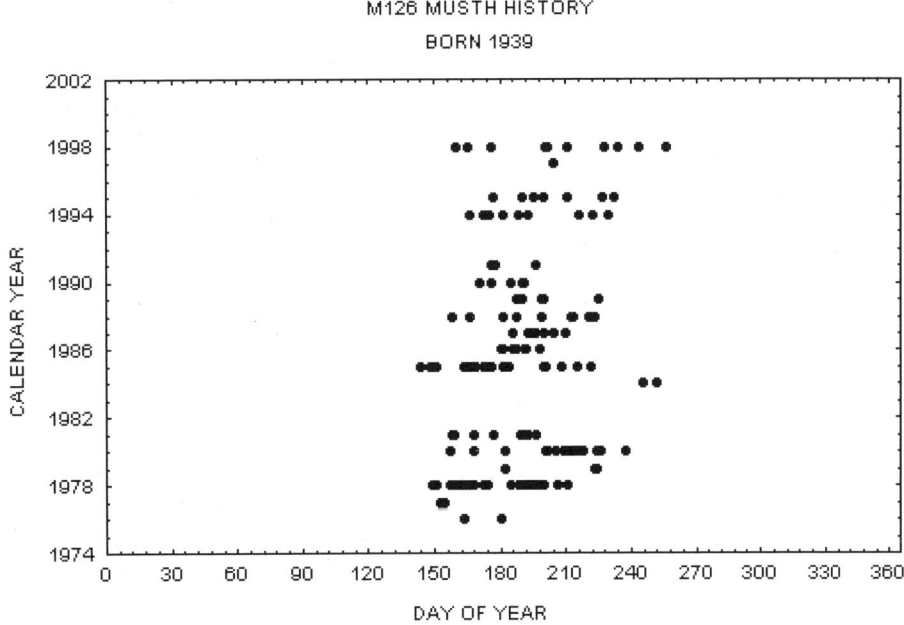

Figure 18.3 Observations of musth for M126 between 1976 and 1998, showing clear consistency in both the timing and duration of musth.

taking into account the timing of any missing data. The very long-term records and the remarkable consistency of musth onset between years for individual males made the determination of musth timing obvious for most males.

Musth duration was calculated as the number of weeks between the first and last day a male was seen in musth (Poole 1987a). In some years, males were not observed throughout their entire musth period, although they were seen at intervals before, during, and after musth. Durations for individuals with incomplete records (i.e., sightings of a male in musth separated by no sightings of the male at all) were predicted from a survival analysis of the median duration of musth for age taken from those years in which more complete weekly observations were possible (1980, 1981, 1983, 1985, 1986, 1987, 1989; see table 18.1). A musth period interrupted by any observation of the male out of musth was considered to be two separate musth periods. Duration in these cases was considered to be the longest of any of the musth periods observed in a year.

Patterns of musth are age-dependent. In 1982, Poole

Table 18.1 Duration of musth in weeks for males in different age groups

Age of male in years	16–25	26–35	36–40	41–45	46–50	51–60
Observations of musth/year	14	79	54	48	10	12
Median duration in days	2	13	52	69	81	54
Interquartile intervals in days	1–6	2–32	39–74	55–80	41–98	32–69
Range in days	1–19	1–80	10–131	21–109	22–126	22–113

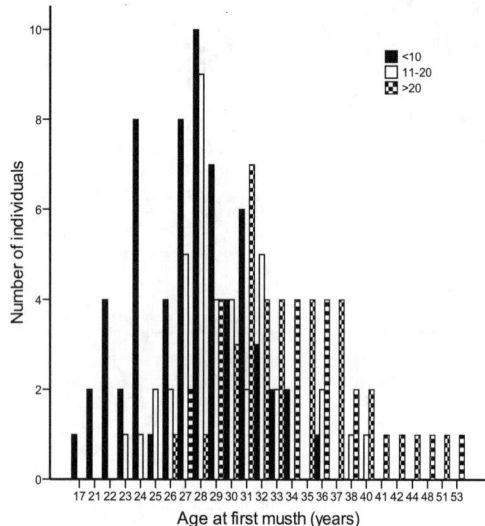

Figure 18.4 Frequency distribution of known age at first musth for males who were 10 years old or younger in 1972 (N = 65), males between 11 and 20 years of age in 1972 (N = 41), and older males (N = 27).

reported the age of first musth to be 29 years based on a sample of 30 males. Now, with over 30 years of data on 135 individuals and using proportional hazards analysis, the median age at first onset has been determined to be 30.4 years. The youngest male seen in musth was 17 years of age, while the oldest male seen in musth for the first time was 51 years (figure 18.4).

Males born later in the study (after 1961) have come into musth at significantly younger ages than males born earlier (earlier than 1962; Gehan statistic = 25, $p < 0.001$, $N = 135$). During the early years of the study, our observations of musth males aged 35 and over were unlikely to have represented the male's first musth period (for example, Male 103 who was first recognized as being in musth at age 51 in 1978). Excluding records of males who were older than 20 years at the start of the study, however, does not eliminate the significant effect in the survival analysis. Males who were under 10 years old at the start of the study came into first musth at a median age of 29 years, while males aged between 11 and 20 years at the start of the study first came into musth at a median age of 30.1 years (Gehan statistic = 8.44, df = 1, $p = 0.0037$), suggesting that the trend is not simply one of observational accuracy. As noted previously for female reproduction (see Lee, Lindsay, and Moss, chapter 6), ecological conditions associated with range expansion appear to have enhanced reproductive potential over the duration of the study; higher rates of reproduction, higher calf survival, and increasing population size were all associated with earlier ages of reproduction for males and females.

Males 17 to 63 years of age have been observed in musth. The duration of musth increases with age until approximately 50 years, when it declines slightly (figure 18.5). Among males 17 to 25 years old, musth typically lasts a few days to a week and may occur several times during their sexually active period, which can last for 3 to 4 months (Poole 1982). Males aged 26 to 35 years tend to come into musth for approximately 2 weeks and may come into musth several times in a year, again in an extended sexually active period. Over age 35, males are in musth typically only once a year: for 4–8 weeks for males aged 36–40 years old, 8 weeks for males 41–45, 10 weeks for males 46–50, and less than 8 weeks for males over 50. Musth duration is longest between the ages of 45 and 50. The relationship between a male's age and duration of musth is highly significant (for known duration only: $r_s = 0.73$, $N = 187$, $p < 0.001$; figure 18.5).

Growth is a key factor influencing male reproductive performance. Male elephants gain 4 to 7 cm per year in height until they are in their 30s, after which growth begins to slow. Based on both footprint and shoulder height measurements, growth continues until after 45 years of age (figure 18.6). The younger ages of first musth (ages 17–25) coincide with a major decrease in the growth velocity as males reach 20 years old. At this age, they are also larger than almost all adult females and potentially able to mount an older estrous female. We suggest that there is a size and strength constraint on the capacity for males to mate, not to mention the need for experience in technique. The allometry of growth suggests that penis size is a further constraint for small or younger males on a mating that results in a conception (Poole 1982).

After the age of 40, males are almost twice the mass of males aged 20 and over 30 percent taller. They no longer

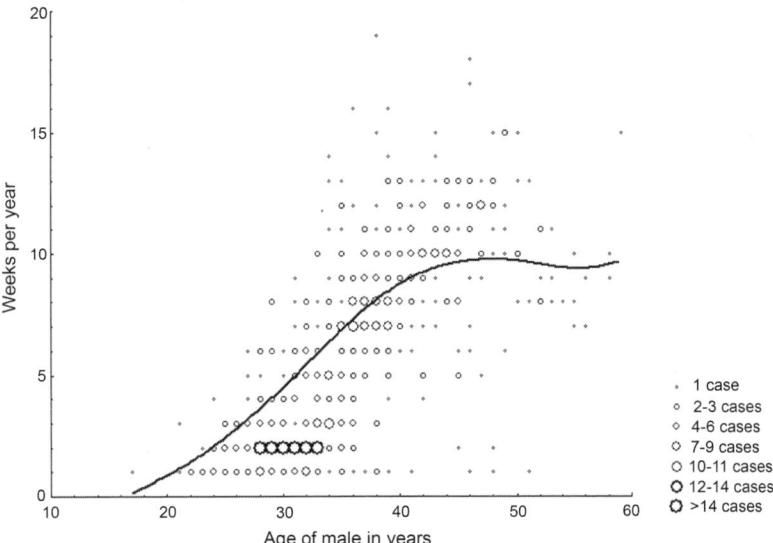

Figure 18.5 Duration of musth (in weeks) by male age (lowess curve fit); all known and estimated musth durations included.

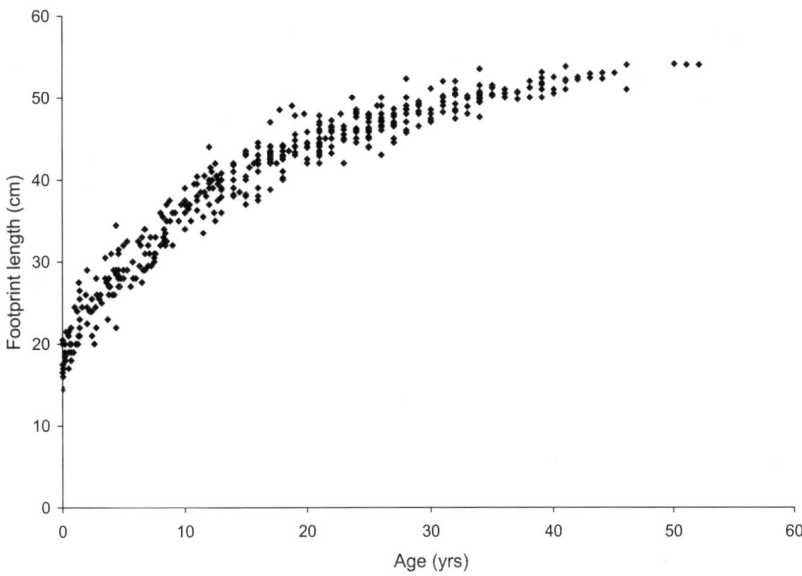

Figure 18.6 Growth in male hind-foot length with age ($N = 252$).

allocate as much of their energy to growth and may have the capacity to invest in gaining condition so as to sustain an extended musth period. We also suggest that the trend for a younger age at first musth may be related to an increased early growth rate and improved access to food early in life. Being relatively large for age (as determined from residuals of growth in foot length against age) tends to be associated with an earlier age of musth, especially for males born after 1962 (ANOVA: $F = 2.91$, df = 2,227, $p = 0.057$).

Male Success: Longevity, Competition, and Priority of Access

Over the course of the study, the number of males observed to be in musth each year has steadily risen (figure 18.7). The first observation of a male in musth (M28) took place in 1974. By 1979, 10 males were known to have come into musth; by 1981, 30 males; and by 1998, more than 50 males. The increase is due in part to gradual overall

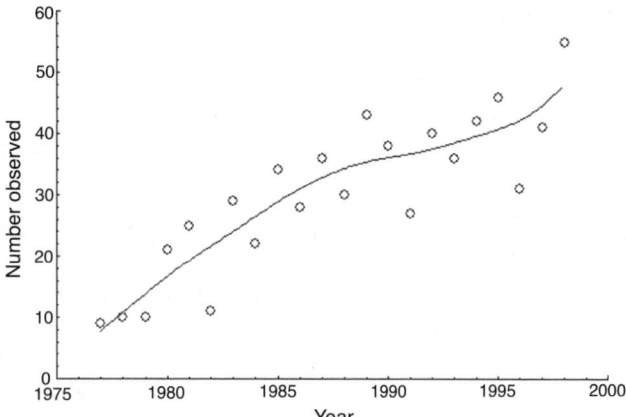

Figure 18.7 Numbers of males coming into musth, by year over the study period (non-linear curve fit).

age-classes, overall median life expectancy for males was 27.17±1.06 (SE). For human-caused deaths only, life expectancy was 34.92±1.61, while for natural mortality, it was 42.50±3.71 (see box 6.1). Losses due to human causes probably reflect male strategies of risk-taking throughout their lives, which both places them in closer proximity to people and in competitive contexts with other males. The key observation relevant to reproductive strategies is that, as noted by Moss (2001), only half of males born will survive to peak competitive ages of musth, whatever their cause of death.

Timing of Musth and Longevity

The median number of males seen in musth per week over the study was three (IQI = 1–5, range = 0–19, N = 1,219 weeks). Individual males were seen in musth throughout the year, with more in musth from late January to July (weeks 3–30) than from August through mid-January (weeks 31–2). This seasonal pattern was similar across the years of the study (figure 18.8), although inter-annual variation was high due to variation in female availability and the condition of males. During wetter periods, there were significantly more males in musth (z = 4.64, $p < 0.001$; D–J: 97, IQI = 77.5–101.5, range = 27–136, N = 32 weeks vs. J–N: 49, IQI = 30–66, range 24–95).

The number of males observed to be in musth by week was positively correlated with the number of conceptions by week (r_s = 0.45, $p < 0.001$, N = 1,217 weeks). High-conception weeks are mirrored by peaks in the number of males observed to be in musth in the same weeks (figure 18.9). There were, however, individual weeks when males "got the timing wrong"; for example, there was one week in 1997 when there were 19 males in musth and no females seen in estrus.

From the perspective of an individual male, success relates to the number of conceptions that occur during his

population growth but more importantly, to the dramatic increase in the number of males of reproductive and musth age. In the mid-1970s, there were very few males of musth age due to poaching and sport hunting in the surrounding ecosystem in the late 1960s and early 1970s. In 1976, there were only 19 males 30 years or older in a population of 496 elephants; in 2002, there were 70 males in the same category in a population of 1,170 elephants. While not a significant proportional change (3.8 vs. 6.0 percent, X^2 = 2.78, p = 0.096), it is socially important in terms of absolute numbers of mature musth males encountering each other.

Male Lifespan

Moss (2001) found that the average male's life expectancy at birth was 24 years, with deaths attributable to illness, accident, predation, drought and human-caused mortality. Based on the current sample, with the addition of six years of data specifically capturing deaths among the older

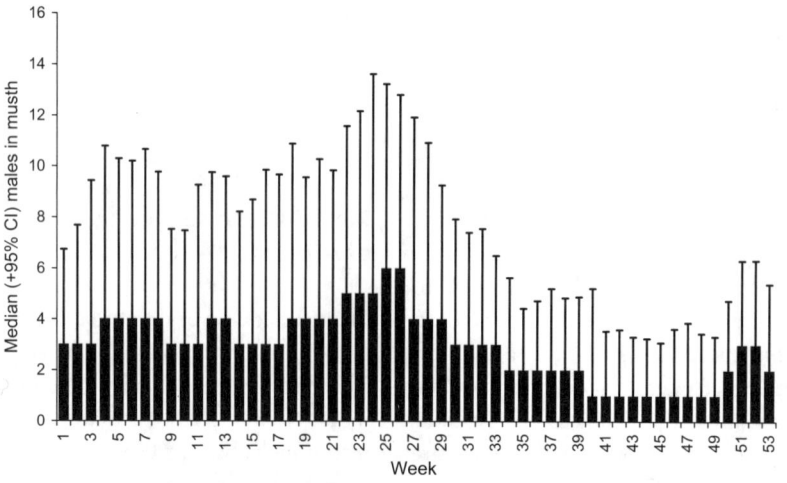

Figure 18.8 Median (and range) of number of males in musth in each week (N = 1,219 weeks).

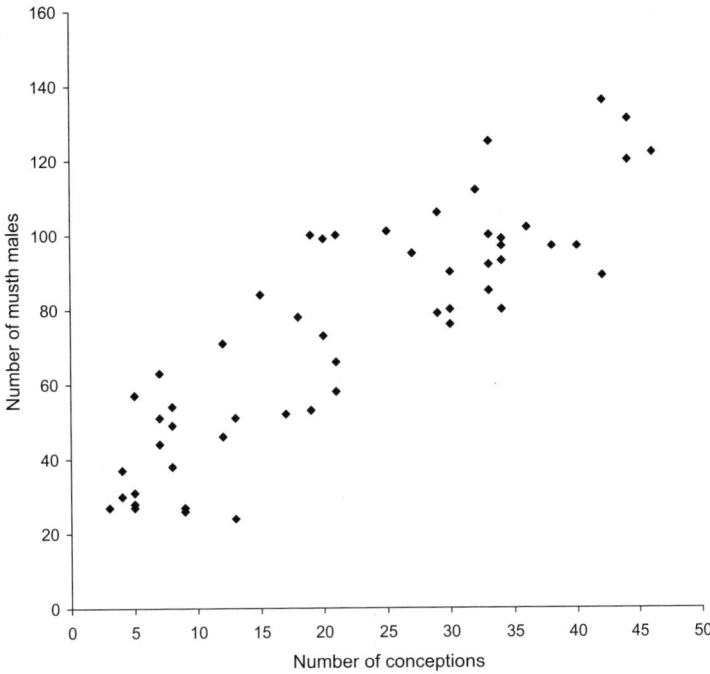

Figure 18.9 Total conceptions against total number of males in musth, by week across 23 years.

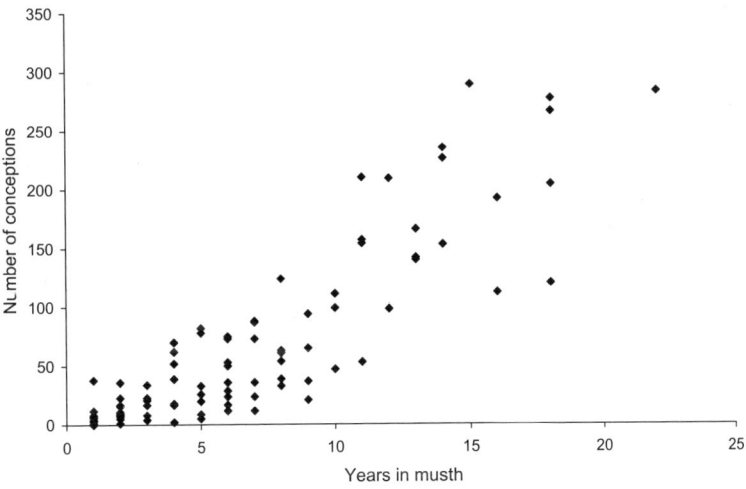

Figure 18.10 Number of years that a male was observed in musth and the total number of conceptions with which a male's musth periods coincided.

total musth period. In an "average" bull's lifetime, 23 conceptions coincided with the total number of weeks that he was in musth, with considerable variance (med = 23.5, IQI 6–75, range 0–289, N = 114). The range of variation is due to factors such as the number of females available in any year; how many weeks an individual has spent in musth over his lifetime, which is related to age or mortality (in other words, his presence in the sample); and the timing of an individual's musth periods, which may coincide with good or bad ecological periods. For example, the 1997–98 El Niño year was extraordinary in terms of the number of females conceiving (N = 110), the number of young males coming into musth for the first time (N = 7), and an overall total of 105 males observed in musth.

The longer a male survives and the older he becomes, the more "successful" he has the opportunity to be. Thus, a strong relationship exists between the number of years that a male has been seen to be in musth and the total number of conceptions that occurred during his musth periods (r_s = 0.89, $p < 0.001$, N = 112; figure 18.10). Furthermore, the

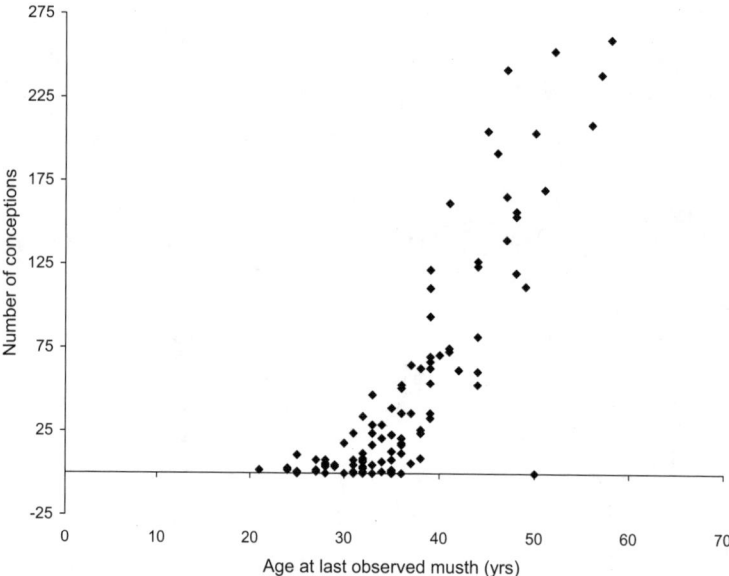

Figure 18.11 Age at last observed musth (reproductive longevity) and the total number of conceptions with which a male's musth periods coincided over his reproductive lifespan.

age of a male at his last observed musth correlates with the cumulative number of conceptions that occurred during his musth periods ($r_s = 0.83$, $p < 0.001$, $N = 105$). Longevity (or relative old age) accelerates a male's potential for reproductive success rather more than male quality.

Based on the duration of their musth period and success in guarding estrous females (Poole 1989b), males reach their peak reproductive period after the age of 40. As we note earlier, the median life expectancy of an Amboseli male is 27 years from all causes of death, and 76 percent of males die before reaching age 40. If a male reaches 27, he still has only a 47 percent chance of surviving to the age of peak reproduction; males reaching 40 will have on average 7.6 years of additional reproductive life. The years after 40 are likely to be the most reproductively important years. Males who delay coming into musth until a later age (e.g., older than 30 years) may not experience much loss of reproductive success as long as they survive. However, they are still limited in their experience with females and contests with males. Even males whose musth periods coincide with few estrous females are likely to have some success if they survive over a long period. Males who die under the age of 40 have the least opportunity for reproduction; they are in musth for short periods and coincide with few estrous females (figure 18.11), and due to competition with older, larger males, they may be even more limited in their reproductive success.

Competition

While longevity accounts for a major portion of a male's lifetime reproductive potential (with variance from 0 to >250 possible conceptions), the timing of a male's musth period also plays an important role. Variance in reproductive performance is high at any age, and at least some of these extremes may result from a male's timing. How well are males able to predict when they "should" be in musth in order to maximize their access to fertile females? Based on earlier work (Poole 1982, 1989a) the onset of musth in younger males appears to be "switched on" by the appearance of an estrous female and "switched off" by the presence of or aggressive competition from higher-ranking males (Poole 1982, 1989a; Slotow et al. 2000). Older males by contrast, have established musth times, and they respond to ecological constraints by skipping a whole year, or they may drop out of musth early during a bad year or after losing a battle with another male. These short-term facultative decisions may be less influential, however, than timing access to females "correctly" over the long term.

If a male is likely to encounter more estrous females from late January to July than from August to December, why do some males come into musth in the latter half of the year at all? One explanation may be high levels of competition between males for access to estrous females in the first half of the year. Does the presence of a larger number of males predict greater competition, or is competition the result of the number of males relative to the number of estrous females? Each male's potential access to receptive females was explored in relation to the number of males simultaneously present in a group with an estrous female. We assumed that each male present had equal access to the estrous female. Based on sightings data from 1983 to 2002, the number of males associating together in a group containing an estrous female ranged from 1 to 40, with a median of 5 (figure 18.12).

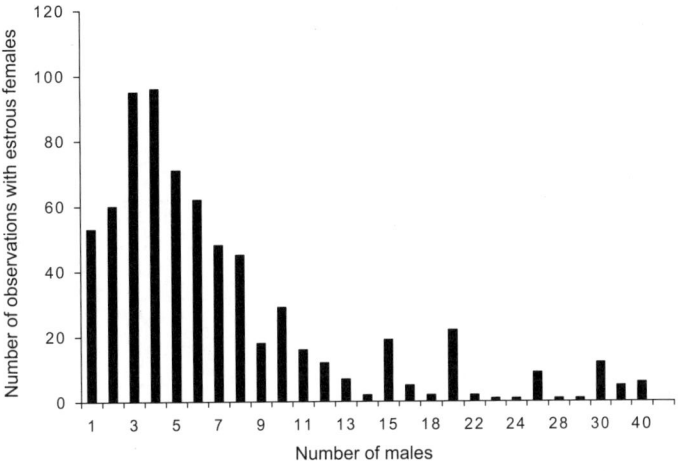

Figure 18.12 Number of males associating in a group containing an estrous female in which at least one male in the group was in musth ($N = 693$).

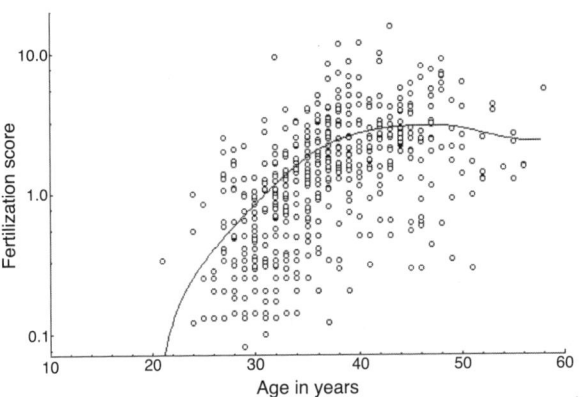

Figure 18.13 Annual fertilization probability score for individual bulls as a function of age (non-linear curve fit).

Males in musth from December to June experienced significantly higher levels of competition due to the simultaneous presence of other males than did males in musth from July to November (Mann-Whitney $z = 5.84$, $p < 0.001$, December–June: med = 6.0 males, IQI 4–10, range 1–40, $N = 532$, July–November: med = 4.0 males, IQI = 3–6, range 1–30, $N = 161$). As we suggest above, males who came into musth in the latter half of the year appeared to have a strategy of minimizing competition.

In the context of competition, we explored the proportion of potential fertilizations accounted for by each male. We calculated an annual fertilization probability as the number of conceptions that occurred during each male's annual musth period divided by the number of males simultaneously in musth, which can be taken as an estimate of the number of other potential "fathers." Our estimates suggest that by age 45, most males had a probability of fathering two offspring each year, though some males were notably more successful (figure 18.13). Older males who have been observed in musth for many years have a higher fertilization probability ($r_s = 0.91$, $p < 0.001$, $N = 112$). Total scores range from 0 for males observed in musth for only one year to 62 for a male (M126; see table 18.2) observed in musth for 23 years.

Priority of Access

Age-related male dominance (Poole 1989a) and female choice (Moss 1983; Poole 1989b) are key determinants of the mating and guarding success of males. Determining the effect of priority-of-access models, where a dominant individual is predicted to obtain disproportionate access to mates (Ellis 1995), depends on being able to monitor and assess the outcome of dyadic interactions (Alberts, Watts, and Altmann 2003). In typical dyadic contests, musth males win in contests with non-musth bulls (Poole 1989a). Furthermore, as we note above, older musth males win in contests with younger musth bulls. Male condition (Poole 1982, 1989a) is an additional predictor of contest outcome. However, accurate assessment of priority-of-access model outcomes with regard to male mating success depends on genetic paternity (Alberts, Watts, and Altmann 2003). In other populations of elephants, musth males have been found to contribute disproportionately to paternity (Whitehouse and Harley 2001; C. Foley, personal communication). Genetic paternity data from Amboseli show a similar pattern (Hollister-Smith et al. 2007; see box 18.1). Of 70 resolved paternity exclusions, 52 calves were fathered by males in musth and 7 were fathered by males known to be non-musth when the conception occurred. Of this latter group, 5 had never been seen in musth. The musth state of males for the remaining 11 paternity exclusions could not be determined due to incomplete sightings records; however, all but 2 of these males had been observed in musth previously (box 18.1; figure 2).

Table 18.2 Individual variation in reproductive success among the oldest males

Male	Birth year	Death year	Age at last musth (2002)	Total number of weeks in musth (to end 1998)	Total number of conceptions that musth periods coincided with (to end 1998)	Total conception probabilities (to end 1997)	Total sightings of guarding (1980–2002)
126	1939		63	210	283	61.99	26
22	1940		62	182	266	51.19	50
7	1945		57	192	277	42.93	33
13	1935	1991	55	150	209	59.22	50
51	1946	2001	53	100	210	23.59	12
154	1949		52	168	120	49.06	16
147	1950		51	138	289	35.11	20
10	1945	1995	50	194	204	35.31	28
41	1941	1990	49	84	111	26.75	7
78	1945	1994	49	147	112	24.51	20
132	1953		48	91	153	20.95	10
45	1945	1993	48	86	154	28.69	11
80	1945	1993	47	91	157	29.84	14
119	1947	1994	47	123	166	28.38	12

We now ask a simple question about priority of access and female choice: Is an estrous female more likely to associate with a musth male than when she is not in estrus? Of the 533 groups with an estrous female present, 86.4 percent contained a musth male. By contrast, musth males were present in only 17.4 percent of cow-calf groups lacking an estrous female ($N = 4{,}675$). Musth males and estrous females have a preferential association ($X^2 = 808$, df = 2, $p < 0.001$). In the context of priority of access, it is significant to note that groups containing estrous females were more likely to have more than one musth male in association (34.3 percent, $N = 478$) than were groups without estrous females (14 percent, $N = 812$, $X^2 = 57.8$, df = 1, $p < 0.001$).

When a musth male associates with a group containing an estrous female, he also attempts to guard that female. A guarding male maintains a close proximity to the female; he may move alongside her or attempt to influence her choice of direction. He also monitors the other males present and will attempt to ensure that they do not approach the female. He is responsible for the maintenance of proximity to the female and does not avoid any other male present. Guarding behavior by males is age specific. Prior to the age of 35, males were seen guarding on very few occasions (med = 0, IQI = 0–1, range = 0–8, $N = 28$). Between the ages of 35 and 44, males began to achieve some guarding success (med = 2, IQI = 1–5, $N = 58$), and although there was considerable variance in the degree of success (range = 0–14), this age group was significantly more successful than the younger group ($z = -4.28$, $p < 0.001$). Males 45 and over were, however, much more successful than the middle-aged bulls (med = 14, IQI = 10–20, range 5–50, $N = 21$, $z = -6.3$, $p < 0.001$). Over a male's lifespan, there is a strong positive relationship between age at last observed musth (a combination of age and longevity) and the number of estrous females that an individual male was seen guarding ($r_s = 0.78$, $p < 0.001$, $N = 107$; figure 18.14).

Among the older males, however, the variance between the extremely successful and the moderately successful was even more pronounced. What accounts for some males' extraordinary success and others equally surprising failure? In table 18.2, we highlight the individual variation among the oldest males in terms of the reproductive factors discussed in this chapter. Males appear to have different strategies: they can attempt to achieve priority of access and enhance female choice via guarding (e.g., M22 and M13); they can opt for lengthy musth periods but not at the maximal time of year (e.g., M154); they can have musth at times of peak female availability but then experience more competition (M147); or they can live for a long time (the difference between M41 and M22). Figure 18.15 illustrates the opportunistic shifting of musth timing with age and dominance so as to maximize mating opportunities (e.g., M7).

Longevity, Competition, and Musth: A Long-Term Perspective on Male Reproductive Strategies 285

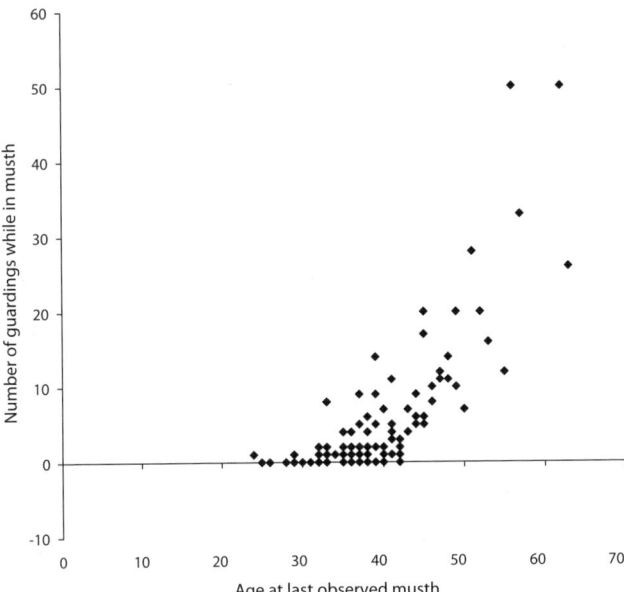

Figure 18.14 Total sightings of individual musth males guarding estrous females against the age of the male at his last observed musth.

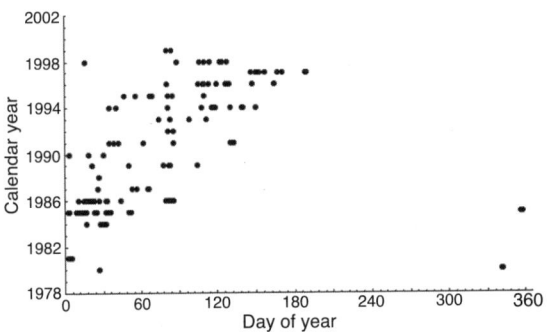

Figure 18.15 The musth history of M7, a successful male born in 1945 who died in 2005.

This change in the timing of an individual's musth period is one of the more interesting features of his "strategy" and is as yet relatively poorly understood. While some males maintained the same musth period throughout the study (e.g., M126; figure 18.3), others shifted over time. What this change suggests is twofold: first, that the hormonal mechanisms controlling the onset of musth are facultatively responsive to the physical and social environment of individuals and second, that males may have complex assessment strategies influencing these mechanisms.

One additional influence on the musth patterns of individual males pertains to demographic context beyond the simple number of competitors. Over time, competitors mature, gain experience, lose weapons such as ends of tusks, shift their musth period, and die. Each male's reproductive decisions are complexly embedded in this shifting web of competitors, which is layered onto female reproductive patterns. Thus, as is so typical of elephants, no single factor or element can describe lifetime reproductive success of individual males.

Conclusions

Understanding the reproductive strategies of male elephants requires an exceptionally long-term perspective. Elephants are a species whose males begin to reproduce with consistent success only by the age of 40—an age at which three-quarters of individuals will have already died. Thus, as we note throughout, knowing the factors that constrain or promote longevity are key to understanding an individual male's reproductive success.

The importance of reproductive lifespan as a determinant of male reproductive success has been emphasized for some species (see Clutton-Brock 1988), but longevity has an importance over and above reproductive lifespan, specifically for musth strategies among elephants. As discussed in detail elsewhere, musth is an honest physiological signal that has either condition or mortality costs, or both (Poole 1989a). Over the years of the study, at least four males were known to have died as a result of fights during musth. There are further mortality costs possibly associated with condition loss as suggested by a higher probability of death during musth periods compared with non-musth periods (Hall-Martin 1987). This costly strategy can be sustained best by larger, older males who are in good condition. Studies of corticosteroid stress hormones in free-ranging musth and non-musth males (Ganswindt et al. 2005) found no elevation in cortisol associated with musth periods. It can be suggested that the only males able to sustain musth are those in peak condition who have sufficient experience to cope with any physical and behavioral stressors. Increased size, mass, condition, and experience are associated with increasing age; therefore, longevity underlies both the maintenance of the musth strategy and the overall success of males.

Other features contribute to variation between individual males who survive to musth age. The point at which males come into musth during the year, how long they can remain in musth, and the serendipitous (or perhaps just less predictable) numbers of females in estrus while they are in musth all contribute to their annual success or failure. Some of the variation is size and age-dependent, but there are sufficient differences between males in their size for age and their condition in any year to make age unreliable as the only predictor of variance.

Finally, that some males may be more attractive to females than are others is a topic that remains unexplored here but deserves consideration. In species in which female choice and cooperation in mating attempts underlie the success or failure of a male (see Hoelzel et al. 1999; Engh et al. 2002), chosen males, whatever their attributes, may simply do better over the long term than males who rely solely on contest strategies. Further genetic assessments of paternity should contribute to understanding the consequences of mate choice. It is, however, over the very long term that reproductive payoffs accrue to male elephants. Age-graded models of reproductive strategies for such a long-lived species need to include an unusually extensive time dimension in order to appreciate the inherent stochasticity in social, ecological, and demographic factors acting in concert to influence male elephant behavior.

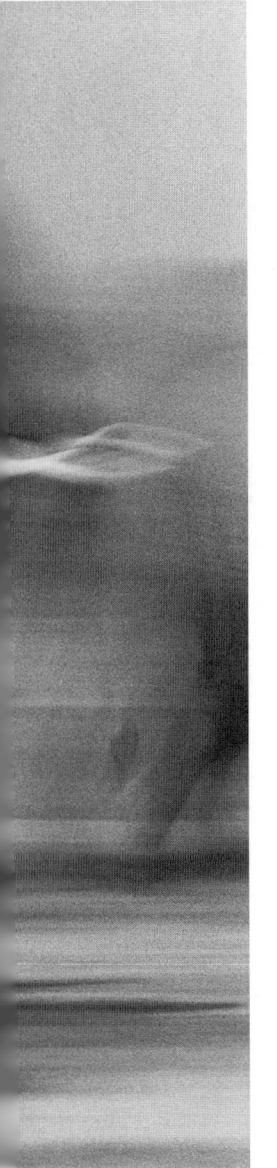

Part 5
Elephants in the Human World

Section editors: Cynthia J. Moss and Kadzo Kangwana

Chapter 19 The Maasai-Elephant Relationship: The Evolution and Influence of Culture, Land Use, and Attitudes

Christine Browne-Nuñez

The African elephant evokes varied and deep emotion in people around the world (Adams and McShane 1992; Dublin 1994; Naughton-Treves, Rose, and Treves 1999). It is "characterized by the most extreme attitudes" (Dublin 1994, 5). To some, it is the symbol of wisdom, strength, and good luck. For others who see the elephant only in zoos, in books, and on television, it is a gentle giant, intelligent and nurturing, maintaining close family bonds. For people who live alongside elephants and those who work to manage and conserve them, the story is more complex.

One of the greatest challenges in conservation today is how to balance local concerns of security and development with international interests in conservation of threatened species (Treves et al. 2006). After the ban on ivory was implemented in 1990 and elephant populations across Africa began to recover, the incidence of human-elephant conflict began to rise. With this change in circumstances, wildlife managers and conservationists started looking for new ways to manage and conserve elephant populations. Several approaches have emerged to address this and other modern conservation challenges. These methods include community-based conservation, integrated conservation and development programs, and collaborative management (Mburu, Birner, and Zeller 2003; Wells and Brandon 1992; Western and Wright 1994). Such programs are part of an effort to move away from the "fences-and-fines" approach, or "fortress conservation," by involving local people in conservation (Adams and Hulme 2001; Bauer 2003; Gibson and Marks 1995; Holmes 2003; Songorwa 1999), with the goal of affecting the conservation attitudes and behaviors of local people (Abbot et al. 2001; Infield and Namara 2001). Amboseli is the setting for some of the earliest community conservation programs in Africa.

As the Amboseli landscape has evolved over time, so has the relationship between elephants and the Maasai. Understanding the dynamics of their interactions is critical to conserving the Amboseli elephants and ensuring the well-being of the human population with whom they coexist. This chapter considers the Maasai-elephant relationship by

- examining the attitudes and behaviors of the Maasai toward elephants over time by reviewing historical accounts of early European travelers through Maasailand,
- discussing the results of recent attitudinal research in the Amboseli ecosystem, and
- evaluating the notion of the Maasai as conservationists and considering the importance of culture, livelihood activities, conservation interventions, and land-use change in determining the future of the Maasai-elephant relationship in Amboseli.

Attitudes, Behavior, and Wildlife Conservation

There have been a number of attitudinal studies done in Africa regarding wildlife conservation (Akama, Lant, and Burnett 1995; Ali 2006; Browne-Nuñez 2010; DeBoer and Baquete 1998; Gillingham and Lee 1999; Hill 1998; Infield 1988; Kangwana 1993; Kideghesho, Roskaft, and

Kaltenborn 2007; Kioko 2004; Lindsey, du Toit, and Mills 2005; Mordi 1987; Newmark et al. 1993; Omondi 1994; Parry and Campbell 1992; Pennington 1981; Sitati 2003). Four studies (Browne-Nuñez 2010; Kangwana 1993; Kioko 2004; Sitati 2003) specifically investigated attitudes toward elephants in Maasailand. Two of these studies were conducted around the Amboseli National Park (results discussed later). Many of these studies found that people hold positive attitudes toward wildlife and the concept of conservation but are more negative toward wildlife authorities and conservation policy (Gillingham and Lee 1999; Infield 1988; Kangwana 1993; Newmark et al. 1993; Parry and Campbell 1992). Where attitudes are found to be negative, it is often due to the costs local people incur as a result of conservation policy or wildlife damage such as loss of land or other constraints on livelihood activity (Anthony 2007; De Boer and Baquete 1998; Gadd 2005; Gillingham and Lee 1999; Kideghesho, Roskaft, and Kaltenborn 2007; Newmark et al. 1993; Parry and Campbell 1992) and loss of crops and livestock (De Boer and Baquete 1998; Infield 1988; Marquardt, Infield, and Namara 1994; Mugisha 2002; Naughton-Treves 1998; Newmark et al. 1993).

Methods suggested for improving attitudes toward wildlife include education and outreach, increasing benefits to local people, and involving local communities in conservation (Hulme and Murphee 2001; Parry and Campbell 1992). Positive attitudes have been found in communities that recognize benefits related to wildlife such as having a household member employed in a wildlife-related job (Anthony 2007; Infield 1988; Lewis, Kwaweche, and Mwenya 1990), having access to game meat (Infield 1988; Kideghesho, Roskaft, and Kaltenborn 2007), and generation of foreign exchange from tourism (Gadd 2005; Newmark et al. 1993; Pennington 1983; Weber 1987), among others (Abbot et al. 2001; Bauer 2003). Programs that are successful in providing wildlife-related benefits to local communities not only highlight the importance of providing benefits that are valued by the communities but also demonstrate a clear link between the benefits and the wildlife resources (Gadd 2005).

Although there has been some success, a consistent link has not been demonstrated between improved or positive attitudes and conservation interventions (Kideghesho, Roskaft, and Kaltenborn 2007; Parry and Campbell 1992; Wells and Brandon 1992). Some reasons for the lack of success include the following: costs exceed benefits (see Emerton 1998; Hulme and Murphree 2001), broken promises (Kangwana and Browne-Nuñez, chapter 3), lack of awareness of source of benefits (Archabald and Naughton-Treves 2001; Kangwana 1993), unrealized expectations (Gadd 2005; Songorwa 1999), limited community involvement (Parry and Campbell 1992; Songorwa 1999), lack of socioeconomic information (Wells and Brandon 1992), and lack of understanding of the link between conservation and development (Barrett and Arcese 1995; Newmark and Hough 2000; Songorwa et al. 2000; Wells and Brandon 1992).

The limited success of conservation may be attributed in part to a lack of understanding of the local context of conservation (e.g., the way local people value natural resources). Social science offers the tools for understanding these issues. Prior to developing new programs or modifying existing programs, there needs to be an "understanding of the community's history and current dynamics, particularly in relation to the authority structures which influence peoples' behavior and the patterns of resource use which form the basis for both conflicts and opportunities in wildlife management" (Kiss 1990, 25). The knowledge and experience derived from research and conservation programs in Amboseli can aid in the development of new conservation and development initiatives as we adapt our methods in this ever-changing environment.

The Elephant in Traditional Maasai Culture

As mentioned in chapter 3, the elephant appears in many Maasai stories and proverbs. Some stories demonstrate the cultural value of the elephant, such as the tale of the relationship among humans and elephants and the following legend that describes the elephant's role in aiding the Maasai in acquiring their highly valued cattle:

At the beginning of the Earth, the elephant, along with a member of the Dorobo tribe and a snake, was one of three things *Enkai* (God) found upon the land when he came to prepare the world. In this story, the Dorobo kills the snake for breathing on him and the elephant for muddying a waterhole that was to be used by his cow. The elephant's baby leaves the Dorobo for another land where he finds a Maasai. The young elephant takes the Maasai to the home of the Dorobo. The Maasai overhears a conversation between Enkai and the Dorobo. The next day the Maasai goes to where the Dorobo had been instructed to go and, acting as the Dorobo, is given a herd of cattle (Hollis 2003).

Other stories are not as exalting and often involve the elephant being deceived by the clever hare and other small, less powerful animals (Greaves 1996b; Hollis 1905; Kipury 1983). Such stories, while portraying the elephant as an easy mark, also portray it as a creature who "breeds no malice, but his mere size and little brain is a cause for constant ridicule" (Kipury 1983, 24).

The Maasai and the related Samburu people have a number of traditional beliefs and practices associated with elephants. Members of both groups have reported customs such as placing green branches or grass atop the remains of or in the orifices of the skull of a dead elephant (Browne-Nuñez 2010; Kangwana 1993; Kioko 2004; Kuriyan 2002). The Maasai have explained this practice as a way of appeasing the spirit of the elephant—the only other creature with a soul like a human (C. Moss, personal communication). Elephants are seen as having physical characteristics similar to those of humans, such as two breasts (Browne-Nuñez 2010; Kioko 2004; Kuriyan 2002), a trunk that operates like an arm, and comparable skin (Kuriyan 2002). Maasai believe it is good luck to find an elephant placenta (Browne-Nuñez 2010; Chadwick 1992; Kangwana 1993; Kioko 2004; Sitati 2003). It is thought that if one finds an elephant's afterbirth and then constructs a temporary *boma* and sleeps there overnight with his animals, he will become rich in cattle (Chadwick 1992; George Lupempe, personal communication; Sitati 2003).

Early Maasai used products made of ivory such as tobacco and snuff containers, fly-whisk handles, upper-arm bands, *rungus* (clubs), earlobe stretchers, and clappers hung around the necks of domestic animals (Bernsten 1976; Blackburn 1982; Hollis 1905; Kasfir 1992; Merker 1910; Mol 1981; Thomson [1885] 2006). These items no longer seem to be in frequent use today (Browne-Nuñez 2010; Kasfir 1992). Plastic containers have replaced snuff containers and earlobe stretchers, although one can occasionally observe a Maasai man wearing an ivory ring or see a pendant hanging from the neck a domestic animal (personal observation; George Lupempe, personal communication). Elephant products have reportedly been used by the Maasai for medicinal purposes (P. E. Glover in Mol 1981).

When discussing the use of elephant products, two questions arise: (1) how did the Maasai acquire these items and (2) what does their use indicate, if anything, about Maasai attitudes toward elephants? With pastoralism being the primary mode of subsistence, the Maasai have rarely hunted wild animals, viewing them as their second cattle upon which they could rely in times of severe famine (Baumann 1894; Berger 1996; Deihl 1985; Homewood and Rodgers 1991; Smith 1907; Waller 1976; Western 1997). It is even claimed that they have contempt for those, such as the Dorobo, who regularly hunt and consume wildlife (Kipury 1983; Saitoti and Beckwith 1980; Thomson [1885] 2006). In Laikipia, Kenya, some Maasai report having cultural taboos against consuming the meat of elephants because of their similarities to humans (Gadd 2005), but there are historical accounts of Maasai hunting wildlife, including elephants (Berger 1993; Hollis 1905; Huntingford 1953). For instance, in the early 20th century, Maasai boys reportedly killed elephants and would take only the tusks to exchange for cattle (Hollis 1905).

An exception to the rarity of hunting is the *ilmoran*, or Maasai warrior, tradition of hunting lions (*olamayio*) to demonstrate bravery and courage (Mol 1981; Saitoti and Beckwith 1980). The warriors also hunt elephants to retaliate against them when they injure or kill livestock or people (see box 19.1). The purposes of these hunts have been the same as for the lion hunt. Additionally, in more recent decades, spearing has been political in nature, with wildlife being targeted in order to demonstrate unhappiness with government policy (Lindsay 1987; Peluso 1993; Western 1982a).

Other Maasai methods for obtaining ivory included finding it in the bush (Mol 1981) and acquiring it from their hunting neighbors, including the Waboni of the coast (Beachey 1967) and the various groups that fall under the name Dorobo (Blackburn 1982; Huntingford 1953; Mol 1981). As outlined in chapter 3, ivory acquired from the Dorobo and Waboni enabled the Maasai to participate in the lucrative ivory trade during the 19th century (Beachey 1967; Bernsten 1976; Waller 1976). Waller (1976, 533) reports that after a period (1884–94) referred to as The Disaster when two disease epidemics severely depleted livestock herds, the Maasai of Loitokitok were "kept alive by hunting, cultivating, and by selling ivory and rhino horn to traders."

More recent discoveries of Maasai artifacts made of ivory and other materials are the source of debate. In his book, *The Art of the Maasai*, Turle (1992) claims that numerous ivory pieces are Maasai cultural artifacts. These include ivory pipes used by *laibons* (diviners or medicine men) to administer medicine, elephant vertebrae used for grinding bowls, elephant pelvic bones used for stools, and ivory *rungus*. Critics of Turle's collection point to the numerous inaccuracies written about Maasai culture; they point out that anthropologists, who have spent years studying Maasai culture, have never come across such artifacts, and they ask why Turle never describes seeing the artifacts in use in his encounters with Maasai (Pido 1994). Turle (1992, 131) himself states that these artifacts "seem to have sprung into view from a past without clues." While experts debate these more recent "discoveries" (Blackburn 1996; Kasfir 1995; Kurtis 1994; Kratz 1996; Pido 1994, 1995; Schildkrout 1996), there is no doubt that elephant products were used by the Maasai in the past and to a lesser degree today. For whatever reason, ivory has been replaced by other materials.

Sitati (2003) reports other traditional values and uses of

BOX 19.1 CONSOLATION FOR LIVESTOCK LOSS: A CASE STUDY IN MITIGATION BETWEEN ELEPHANTS AND PEOPLE
Soila Sayialel and Cynthia J. Moss

Negative human-elephant interactions, or as it is more often called, human-elephant conflict, is a widespread and increasing problem in almost all areas of Africa and Asia where elephants range. The people and elephants of the Amboseli ecosystem have not been spared. In this chapter, we would like to present one case study of the history and attempted mitigation of negative human-elephant interactions.

Background

In November 1996, a series of events began that accelerated and exacerbated elephant-human conflict in the Amboseli area. The first and most serious was the death of an elder by an elephant. It resulted in 47 *ilmoran* (Maasai warriors) massing and entering the Amboseli National Park to spear elephants and buffaloes.

The Kenya Wildlife Service (KWS), the body responsible for all wildlife inside and outside protected areas in Kenya, proposed the setting up of a Conflict Resolution Committee. The members were elected, and their first request was to have an elephant shot every time a cow was killed. (During harsh dry seasons and years of drought, livestock and wildlife are often crowded into the remaining areas of the ecosystem with food and water [see also Kangwana, chapter 20]. Under these circumstances, when elephants inadvertently get mixed in with herds of domestic animals, they may kill a cow, sheep, or goat. It appears to occur mostly out of fear and confusion. Traditionally, the Maasai retaliated for these deaths by spearing elephants. Such action only aggravated the situation, making the elephants more frightened and aggressive.) KWS agreed to the Maasai request.

On January 15, 1997, just outside the Park in an area known as *Ilmarn*, an elephant killed a cow. The owner of the cow reported the incident at the Park headquarters, and in keeping with the new KWS directive, a warden and rangers were sent to the spot to kill the offending elephant. Nearby they found four bulls and a family of elephants known as the TAs. Eyewitness reports differ over which elephant killed the cow and over what happened during the next two hours, but it is known that the family was chased for two hours before the tuskless matriarch was shot and killed. Despite the fact that there were tusk marks next to the dead cow, the warden and rangers were satisfied that they had killed the correct individual.

The dead matriarch, known to us as Tuskless, was without doubt the most viewed and the most photographed elephant in all of Amboseli. Her family was remarkably habituated to tourists and vehicles and spent most of its time in the central part of the Park close to the tourist lodges. The family, and Tuskless in particular, had been photographed literally thousands of times and had appeared in well over 100 documentary films made by British, American, French, German, and Japanese filmmakers. In addition, because they spent time near the lodges, the elephants were viewed on a daily basis to the great delight of the tourists. We had no doubt that Tuskless created more wonder and joy for visitors than any single wild elephant anywhere in the world.

At the same time, the Maasai owner of the cow lost a valuable animal. But it seemed to us that the shooting of elephants in retaliation for the killing of cattle was a dangerous and entirely negative approach. The Maasai owner had still lost his cow; Kenya had lost a valuable elephant; the elephants became agitated and aggressive, leading to more deaths of cows and most likely people as well. No one benefited in the end.

Unfortunately, the situation in Amboseli soon became even more volatile because a drought developed. There were thousands of cattle coming into the Park each day and thousands more on the periphery. As usual, most of the boreholes and pumps just outside the Park were out of action, and thus the main source of water was in the Park. Under these circumstances, KWS allows cattle to come into the Park to drink. Therefore, elephants, people, and livestock were meeting frequently and the situation began spiraling out of control.

Trying to find a solution

The relationship between the management of the Park and the local Maasai community was at an all-time low. Even though the Amboseli Elephant Research Project (AERP) is primarily a scientific research project, we felt that we had an obligation to both our study animals and the community with whom we lived. We began working almost exclusively on finding a way out of the downward spiral. The Maasai and elephants had been coexisting for hundreds, probably thousands, of years, and we believed that they could continue to do so without this sort of tit-for-tat killing. Continuing the retaliation policy would negatively impact both elephants and Maasai. After months of discussion with Maasai group ranch leaders and KWS regional, Park, and community wardens, we concluded that paying for livestock killed by elephants was the only way to relieve the conflict. It soon became clear that the AERP, as an independent non-governmental organization, was the best body to take on the task.

Getting the program accepted by the entire Maasai community and KWS was something else altogether. Somewhat innocently, we assumed that everyone would want the program. Geography and politics presented some difficulties. In 1996, a new organization, the Amboseli Tsavo Group Ranch Conservation Association, had been formed to try to coordinate development activities, particularly those relating to natural resources and tourism, on the seven group ranches in the Amboseli ecosystem. Since this new group was representing the entire community, we began our discussions with its officials. At this time, AERP functioned under the umbrella of the African Wildlife Foundation (AWF). Therefore, negotiations with the communities involved AERP and AWF as well as KWS.

When it came to discussing the scope of the program, we discovered a substantial geographical hurdle. The seven group ranches covered an area of 5,200 km^2, stretching 110 kilometers from the western border of the Olgulului Trust Group Ranch to the Chyulu Hills and 75 kilometers from the Tanzanian border to the Eselenkei River in the north (see the map in chapter 2, figure 2.1). It would be logistically impossible for us to cover this entire area. If a cow was killed in Rombo or Mbirikani, it could take the Maasai owner several hours to notify KWS, and it would take our staff with, at that time, one functioning vehicle at least two hours to get to the site and another two hours to get back. It was simply not workable. Therefore, we proposed that the program apply only to the Olgulului-Ololarashi, Eselenkei, and Kimana Group Ranches—the three ranches that surrounded Amboseli National Park. This proposal was vehemently rejected. Having recently created the Group Ranch Conservation Association, the leaders felt that they had to stick together and all should be included.

A second hurdle was more complicated. Certain leaders wanted to be able to gain political clout from bringing the program to the people. They wanted us to raise the money and give it to them to administer. For many reasons, we did not feel that this system would work, and we refused.

In the meantime, we still did not have the approval of KWS, which was worried about setting a precedent and raising expectations in all areas where there was human-elephant conflict. Here was a good idea that the people very much wanted, but much to our naive surprise, we could not reach an agreement. One of the authors is from the local Maasai community, and she began quietly talking to influential people and representing AERP at various meetings, including the Conflict Resolution Committee. AWF staff on the ground also worked to find a solution. We continued to have discussions with KWS.

On April 14, 1997, we had what seemed like a final meeting with the Amboseli-Tsavo Group Ranch Association. Members concluded that they wanted to set up their own scheme for all seven group ranches. We agreed to try to help them, and they were to see a lawyer and come to us with a proposal.

Five months went by during which the situation with regard to the Maasai, wildlife, and KWS deteriorated. The spearing of elephants and Maasai entering the Park to kill animals continued. In August, tragically, a man was killed by an elephant south of the Park. That elephant was tracked and killed by KWS, and all agreed, including us, that it was the right elephant and the right thing to do. Nevertheless, during 1997 alone, 14 elephants were speared, of which 8 died as a result of their wounds.

The Amboseli-Tsavo Group Ranch Conservation Association finally presented a proposal to AWF in October 1997. The scheme was far too expensive and far too grand in attempting to cover all sorts of additional contingencies. Subsequent efforts to meet and discuss the proposal failed.

The situation went from bad to worse. In December, after lions killed a cow, the warriors entered the Park and killed a buffalo and a warthog. When KWS tried to disperse the warriors, a confrontation developed, with KWS rangers eventually shooting over the heads of the *moran*. Maasai elders tried to stop the *moran*, and bitter words were exchanged, including a curse of the elders by the warriors. A curse of this kind in Maasai culture is very serious and led to problems between the age groups that did not get resolved for some time. (The curse was eventually lifted by the spiritual leader of the entire *Ilkisongo* section to which the Amboseli Maasai belong.)

In February 1998, two goats were killed by a lion. The next day, 44 armed *moran* went to Park headquarters and threatened to enter the Park and kill all the lions unless KWS paid for the goats. The KWS community warden held a meeting with the *moran* and some members of the Conflict Resolution Committee and calmed them down. The Maasai demanded a meeting with KWS and AWF and wanted to know where the promised money was for elephant-caused livestock deaths.

Finally, on March 2, 1998, the Amboseli Tsavo Group Ranch Conservation Association leaders, KWS, and the authors met at Park headquarters. After prolonged discussion, the association agreed that the consolation scheme could go ahead with just the three ranches. We promised to start the scheme as of the next day, March 3, 1998, and back date the payments to February 1997, when we first proposed a consolation program.

(continued)

BOX 19.1 (*continued*)

The Program Design

After much consultation, we proposed the following program:

- When an elephant kills a cow, sheep, or goat, the owner or his representative would not move the animal and would report to KWS Headquarters as soon as possible.
- A verification team would then be assembled consisting of a representative from the Maasai Group Ranch Conflict Resolution Committee, a warden or ranger from KWS, and a member of AERP. This team would go to the place where the animal was killed and determine (1) if it was outside the Park and (2) if it had been killed by an elephant. We would not be able to pay for livestock within the Park because, theoretically, livestock is not allowed inside the Park, and thus KWS would be put in a difficult position.
- If it was determined that the animal was killed by an elephant, the owner would be paid as follows: 15,000 Kenya shillings for a cow; 5,000 for a sheep or goat. (These figures are higher-than-average livestock prices offered locally or at markets.) The payment would be made within a few days and, most important, it would be made directly to the owner of the animal.
- The payment and in fact the entire program would not be called "compensation" but rather "consolation." It was our way of saying, "We're sorry for your loss." This distinction, which was well accepted, solved the potential problem for KWS.
- Separating cattle deaths from human deaths and crop losses was justified on the basis that human deaths are compensated for by the government, and crop damage is something that the government was considering including in a new wildlife bill.

Outcome

Livestock killed and elephants speared

As of the end of 2007, the consolation program had been running for 11 years, including the backdated payments made in 1997, and a total of 3,297,000 Kenya shillings ($43,960) had been handed out to individual owners of cows, sheep, and goats. The following table illustrates the payment history.

Year	Number of Animals Killed	Total Consolation Payments for Year	
		Kenya shillings	U.S. dollars
1997	7	75,000	1,000
1998	1	5,000	67
1999	4	30,000	400
2000	33	475,000	6,333
2001	3	35,000	467
2002	33	322,500	4,293
2003	18	260,000	3,467
2004	44	570,000	7,600
2005	51	695,000	9,267
2006	42	540,000	7,200
2007	26	290,000	3,867
Total payments		3,297,000	43,960

Spearing has not stopped by any means, but retaliatory spearing as a result of livestock being killed has halted completely in the group ranches covered by the program. Unfortunately, there are many reasons why Maasai spear elephants, including proving bravery as part of the culture of the warriors, political protest, or anger at another situation (see chapters 19 and 20 for more details). Therefore, spearings did continue after we started paying for livestock, but for different reasons. In 1998 and 1999, incidents of both livestock loss and elephants

elephants and elephant products in the Trans Mara District of Kenya, which include the following:

- Elephants locate salt licks and water for livestock use.
- Elephants thin forests.
- Elephant bones are used to treat trypanosomiasis.
- Elephant fat is used to treat skin disease and is mixed with herbs to increase growth in babies.
- Elephant dung is burned in order to smoke out bees for honey harvesting and to treat measles.
- Elephant tails have been used to make bangles.
- Ivory is a source of money.
- Elephants deter cattle rustlers.
- Elephant trails are used for ceremonies, as they are believed to be without obstacles.

speared were relatively low. The main factor here was that there was exceptionally high rainfall, and elephants and people were well spread out and able to avoid one another.

Then, in 2000, there was another drought, and 31 cows and 2 goats were killed and 10 elephants were speared and killed. One elephant family alone lost 4 members in a single attack. These spearings were not related to livestock killed but to young boys trying to prove their bravery. The Maasai elders and group ranch leaders were very angry with the boys and punished them. The leaders made it clear to us that they did not want the consolation program jeopardized. To them, the program was successful and valuable.

Since that time, the numbers of livestock killed each year has gone up, especially in the last three low-rainfall years, resulting in harsh competition between wildlife and people. What is significant is that while the number of animals killed has increased, the number of elephants speared has not. It has remained around the same, starting in 2000 with an average of 8.3 elephants speared each year, of which an average of 5.2 have died. The year 2005 was the worst yet in terms of livestock killed: 45 cows, 2 sheep, and 3 goats were killed by elephants, and a total of 695,000 Kenya shillings ($9,267) was paid out. However, as a result of the consolation payments and the conflict resolution work of AERP's project manager, there was not a huge increase in elephants speared that year. Ten elephants were speared, of which 5 died. See figure A.

Conclusions

Although the consolation program has not stopped spearing altogether, it has radically changed our relationship with the local community. They are appreciative of the payments and of our understanding and concern for the losses they incur. In addition to the consolation program, we have a scholarship program, which sends young men and women from the community to university and girls to secondary school. Such initiatives have resulted in a relationship that is open to dialogue and compromise. The relationship between the Maasai and the elephants is by no means perfect, but the consolation program provides one way to promote greater tolerance of elephants by mitigating some of the losses sustained.

As far as the elephants are concerned, the reduction in retaliatory spearings has allowed an expansion of elephant range (see Croze and Moss, chapter 7). In every workshop we have had on biodiversity in Amboseli, the conclusion reached is that the elephants must be allowed to maintain their large-scale movements if the Park and its wildlife are going to survive and prosper. We have seen a significant movement out of the Park, but if elephants are speared and poached outside the Park, they will resume their earlier concentration within its relatively secure boundaries. Clearly, a positive relationship with the local communities must continue in order to secure the long-term future of the elephants and the ecosystem.

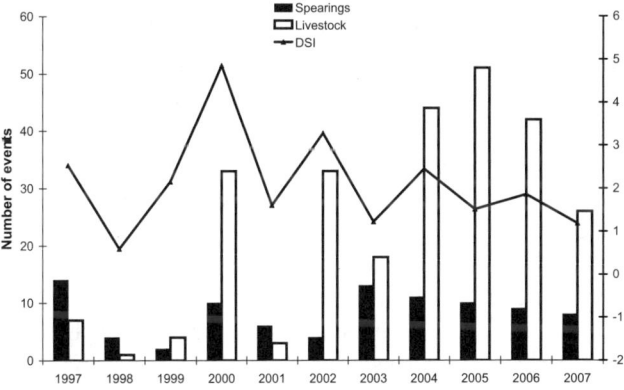

Box 19.1, figure A Spearing, livestock killed, and dry season intensity index, 1997–2007.

It is evident that wildlife, including the elephant, is a valued element of traditional Maasai culture. Values in nature have been categorized to include cultural, utilitarian, economic, aesthetic, and religious values among others (Kellert 1976, 1996; Rolston 1988). For the Maasai, these values vary by species (Browne-Nuñez 2010). The influence of the ever-present religiosity of the Maasai (Berger 1993; Browne-Nuñez 2010; Goldman 2003; Mol 1981; Saitoti and Beckwith 1980) is illustrated by the terms used to describe some wildlife. For example, small plains animals are referred to as *inkineji e Nkai*, the goats of God, or *inkishu e Nkai*, the cattle of God (Mol 1981). Furthermore, there is a degree of fatalism present in Maasai culture. It is not uncommon for Maasai to state that because God put them on earth together with wildlife, they must live together (Browne-Nuñez 2010). The oral literature and extensive use of elephant products provide examples of cultural and utilitarian values. Additionally, some cases of the spearing of elephants by the *ilmoran* demonstrate a value similar to what Rolston (1988) termed a character-building value.

The Maasai, Pastoralism, and Wildlife Conservation

"Being Maasai." Over time, there have been various characterizations of the Maasai, their culture, and their impact on wildlife. From guardians of wildlife to irresponsible, overgrazing herders, these descriptions contain various degrees of accuracy and have resulted in several stereotypes. Some of these false perceptions are the bases of the modern and often romanticized image of the Maasai (Ole Ndaskoi 2006). While a complete discussion of the Maasai identity is not within the scope of this chapter (see Spear and Waller 1993 for a more exhaustive treatment), a brief overview provides a foundation for understanding the relationship between Maasai and elephants.

Defining what it is to be Maasai is complex, as "there are a series of political, cultural, and ecological divisions and subdivisions of the Maasai" (Talbot 1972, 702). Maasai interactions with their environment have been dynamic over space and time, depending on environmental and social conditions (see also Kangwana and Browne-Nuñez, chapter 3). For instance, the Maasai of Ngorongoro have dealt with different environmental and political forces from the Maasai of Amboseli. Also, in times of disease and famine, Maasai have turned to other ethnic groups and adopted other means of subsistence; when they have become impoverished, they have sought employment in order to purchase livestock and return to pastoralism; when political pressure has required it, they have made superficial or temporary changes with their culture intact (Knowles and Collett 1989).

Europeans first became aware of the Maasai through the writings of Krapf (1854, 1860), who obtained his information from caravan leaders and a Wakwavi slave in Mombasa (Knowles and Collette 1989). He described the Maasai as hostile savages who were feared by their neighbors; subsisted strictly on milk, butter, honey, and meat (including hunted wildlife); and detested agricultural foods (Collett 1987). Others described the Maasai as "warlike" (Lugard 1893) and as pure pastoralists (Hinde and Hinde 1901; Hollis 1905). Thomson ([1885] 2006), who later traveled though Maasailand, provided a different, somewhat more accurate portrait. He noted that the Maasai were not hunters, and wildlife displayed little fear toward humans. Additionally, he reported that women, children, and married men consumed agricultural products that were acquired through trade from neighboring agriculturalists.

Thomson's portrayal was not the one adopted by the British colonial administration, who concurred with Lugard and Krapf in viewing the Maasai as hostile and purely pastoral (Collett 1987; Knowles and Collett 1989). Such inaccurate accounts along with the belief that the Maasai failed to use their land properly (Eliot 1905; Lugard 1893) led to a history of policies resulting in land alienation and the breakdown of traditional Maasai culture (see Kangwana and Browne-Nuñez, chapter 3; Croze, Lindsay, and Moss, chapter 22).

Defining a Maasai identity is also dependent on how one views ethnicity. Galaty (1993b, 174) states that the term Maasai marks "two different cultural realities, the first being the gamut of Maa-speaking people and groups in East Africa, the second [being] the set of central and primarily pastoral Rift Valley Maa-speaking sections for which the term marks their distinctiveness." As stated in chapter 3, this second "reality"—the Maasai as "people of cattle"—is the most widely adopted, but as Galaty (1993b, 179) states, "to speak Maa is increasingly equated with being Maasai, pastoral or not." Therefore, as one considers the values, attitudes, and behaviors of the Maasai, it is important to keep the complexity and fluid boundaries of the Maasai ethnic identity in mind.

Maasai Pastoralism. Pastoralists have received much of the blame for "desertification," with critics pointing to the theory of the "tragedy of the commons" (Hardin 1968), citing overstocking and overgrazing as the primary—if not the only—causes of desertification (Lamprey 1983). Today, most who have studied and worked with the Maasai see this as a faulty contention, given that the Maasai have always had a system in place to control access to resources communally. Moreover, research shows that the ecological dynamics of savannah ecosystems have been misunderstood, and concepts such as equilibrium and carrying capacity used to describe temperate zones are not able to explain the functioning of savannah ecosystems (Behnke 1994; Homewood and Rodgers 1991; Little 1996; Croze and Lindsay, chapter 2, box 2.1). The complex relationships between humans and African savannahs are better understood using theoretical concepts such as disequilibrium and disturbance, given that these ecosystems are inherently unstable (Little 1996).

Although some studies have concluded that there is competition between wild ungulates, such as wildebeests and zebras, and domestic animals (Fritz, De Garine-Wichatitsky, and Letessier 1996; Lamprey and Reid 2004; Prins 1992; Voeten and Prins 1999; Young, Palmer, and Gadd 2005), the idea of the compatibility of pastoralism and wildlife is increasingly supported (Deihl 1895; Goldman 2003; Little 1996; Nelson 2000; Reid et al. 2003). There is substantial evidence that light-to-moderate livestock grazing increases rangeland productivity (Mearns 1997) and that "the customary pastoral approach to resource management constitutes an efficacious adaptation to ecological stress" (Behnke 1994, 21). Other evidence of compatibility is offered by Homewood and Rodgers (1991), who found that pastoralist land use in the Ngorongoro Conservation Area was not a threat to wildlife populations or the environment. In fact, research has demonstrated that removal or decline of pastoralist populations can have negative effects on vegetation (Conant 1982; Lamprey and Waller 1990; Mearns 1997).

Speaking at the 2003 World Parks Congress, wildlife veterinarian Richard Kock stated, "Traditional pastoralism has very positive benefits to the environment and wildlife. You only have to look at distributions of wildlife to see that they are often associated with pastoral systems. So, should we be doing wildlife? Should we be doing livestock? Should we be doing both? I think it's clear [researchers] have done good work on this over the years and have shown how the benefits—economically, environmentally—of mixed systems are there." Mearns (1997) qualifies this position stating, "Under the right conditions, production systems relying on mobile livestock represent the most sustainable way to utilize arid rangelands and ought to be supported and enhanced through policy intervention designed to give greater decision-making power to local producer groups."

Today, the nature of pastoralism is changing. Pastoralists of East Africa are diversifying their livelihood activities, depending less on their livestock than in the past (McCabe 2003; McCabe, Perkin, and Schofield 1992). This change is a result of several factors, including loss of grazing land resulting from conservation policy, subdivision of the group ranches, and immigration of people from farming groups (Campbell et al. 2000; Kangwana and Browne-Nuñez, chapter 3). The "ideal model of pastoralism" is disappearing, if it ever existed at all (Hogg 1987, 293). As the Maasai of Amboseli diversify their livelihood practices—grow crops, rent land to cultivators who are members of other ethnic groups, develop small enterprises, and so on—the challenge for conservationists is to work with the Maasai to develop conservation strategies that will enable wildlife to continue to utilize the land outside the park. An important question is whether there exists a conservation ethic or attitude among the Maasai that can serve as the foundation of local conservation efforts.

The Maasai as Conservationists. Many have described the relationship between the Maasai, wildlife, and the environment they share as one of peaceful or harmonious coexistence (Amin, Willetts, and Eames 1987; Berger 1993, 1996; Homewood and Rodgers 1991; IUCN 1987; Kinyua, van Kooten, and Bulte 2000; Kenya Wildlife Service 1990; Lovatt Smith 1986; Ole Dapash 2002; Orindi and Huggins 2005; Parkipuny 1997; Saitoti and Beckwith 1980). Further, the Maasai have been described as the "custodians of Kenya's wildlife" (Asiema and Situma 1994), the "greatest preservationists in Africa" (Simon 1963), "historic vanguards of conservation" (Parkipuny 1997), "par excellence conservationists" (Richard Leakey cited in Horgan 1989), and "more valuable to the cause of conservation than a whole army of paid game scouts" (Fosbrooke 1972).

The notion of harmonious coexistence is not accepted by all. Prins (1992) questions the harmony theory, pointing out that many pastoralists across Africa, including the Maasai, have at some time hunted and that livestock competes with wildlife. Some simply view coexistence as a romantic notion that was made possible by the fact that the pastoralists lacked the resources for change and had low population levels and low production demand that allowed them simply to ignore wildlife (Adams and McShane 1992; Norton-Griffiths 1998). Richard Leakey states, "I don't know that people anywhere have ever lived in harmony with wild animals, despite our wishful belief that once this was so, but the Maasai came as close as anyone ever has" (Leakey and Morell 2001, 146). Early accounts of harmonious coexistence might be explained, in part, by a series of disasters that occurred in East Africa in the late 19th century, when disease wiped out most of the region's cattle population and, together with colonial warfare and famine, drastically reduced pastoral populations (Enghoff 1990; Talbot 1972). This snapshot in time showed the land filled with wildlife and a sparse human population.

If we reject the notion of harmonious coexistence, we are still left to consider the Maasai as conservationists. Likely the most frequently offered evidence of Maasai conservation is the existence of large numbers of wildlife on their land (Asiema and Situma 1994; Bulte et al. 2006; Mol 1981; Organ and Fosbrooke 1963; Parkipuny 1989; Simon 1963). More than half of Kenya and Tanzania's wildlife is found in Maasailand (Ole Parkipuny and Berger 1993). Explanations of this phenomenon include the compatibility of pastoralism and wildlife conservation, Maasai attitudes toward wildlife that permitted coexistence, low population densities that allowed enough space for coexistence (Mol 1981), and discouragement by the Maasai of the intrusion of others (Fosbrooke 1972), such as poachers (Beachey 1967; Homewood and Rodgers 1991; Western 1997) and agriculturalists (Organ and Fosbrooke 1963).

A useful approach for considering conservation among the Maasai and other indigenous societies is to ask whether there is intention to conserve. Hunn (1982) distinguishes between deliberate or true conservation and unintentional or epiphenomenal conservation. Deliberate conservation occurs when people pay an "enduring cost in the present so that some benefit will be realized in the future" (Alvard

1993, 358). Conversely, epiphenomenal conservation can be an artifact of low human population densities, high mobility, limited technology, abundance of or low demand for a resource, or security issues (Ruttan and Borgerhoff Mulder 1999; Smith and Wishnie 2000).

Although it is clear that Maasailand is home to abundant and diverse wildlife, many would ague that this is indeed an artifact of traditional Maasai pastoralism and low population densities. While the livelihood strategies and cultural institutions of the Maasai aim to manage grasslands, when possible, there is no evidence that conservation of wildlife is a principal goal of the Maasai. Mol (1981, 27) supports the idea of the Maasai as "preservationist . . . rather by accident than by purpose." He states, "From all this we can conclude that the Maasai did not interfere with the position of the game on their land. On the other hand it is also clear that the Maasai did not take any positive steps to maintain and preserve the game. The game happened to exist peacefully alongside the Maasai . . . before [human] population pressure began to interfere with the status of the game." (Mol 1981, 26)

From Early Attitudes toward Elephants to Modern Attitudes and Interactions

Early Attitudes

The abundant wildlife found in Maasailand has also been offered as evidence of positive or at least tolerant attitudes toward wildlife (Mol 1981; Ole Dapash 2002; Parkipuny 1997; Simon 1963). Simon (1963, 93) states that "it is entirely due to [the Maasai] that substantial herds of plains game still exist in southern Kenya and northern Tanganyika. We owe an immense debt of gratitude to the Masai for being the only East African tribe to adopt an indulgent attitude towards wild animals." Many authors have anecdotally described a traditionally tolerant attitude (Campbell et al. 2000; Capone 1972; Darling 1960; Fosbrooke 1972; Kipury 1983; Ndaskoi 2006; Western 1997). One proposed explanation for Maasai tolerance of wildlife is their reliance on wildlife as their second cattle (Western 1994). Myers (1973) described the relationship between Maasai and wildlife as one of indifference. This descriptor corresponds with the view of the Maasai as "accidental" conservationists in that they did nothing to threaten wildlife, nor did they do anything to conserve it. An example of this laissez-faire attitude is that the Maasai took no action when soldiers in World War I killed large numbers of wildlife along the Kenya-Tanganyika border (Lovatt Smith 1997). It could be that tolerance or apparent indifference is based on the fatalistic orientation of many Maasai.

Not everyone agrees that the Maasai only tolerated or were simply uninterested in wildlife. Some who have studied and worked with the Maasai have argued that they have demonstrated positive feelings toward wildlife and recognize its intrinsic value (African Wildlife Foundation 1999; Homewood and Rodgers 1991; Kipury 1983). When discussing the possible exclusion of the Maasai from the Ngorongoro Conservation Area, Homewood and Rodgers (1991, 248) state, "The Maasai respect for wildlife and the strong aesthetic as well as practical sense of their environment are such a natural basis for local conservation support that it is counterproductive as well as hypocritical and unethical to exclude them." Further, some argue that the Maasai love of wildlife is evidenced in their children's songs and chants and that because of this positive feeling, they dislike other people who kill wildlife and "destroy what should be left to exist for its own aesthetic value" (Kipury 1983, 4). Again, these claims could be supported, in part, by the earlier discussion of values of wildlife.

A concern of conservationists working in Maasailand today is how much of the positive attitude toward, or at least tolerance of, wildlife remains. In the latter part of the 20th century, several authors noted the changing attitudes of the Maasai (Darling 1960; Mol 1981; Myers 1973; Simon 1963; Western 1994). Simon (1963) asserts that two beliefs held by the Maasai at the time were the basis for a change from tolerance to "veiled hostility": (1) that wild animals consume resources that should be for cattle and (2) that Europeans were casting "envious eyes" in the direction of lands that were rich with wildlife. Others have attributed a change in attitudes to increased competition between the Maasai and wildlife brought about by rising human and livestock populations (Darling 1960; Mol 1981; Myers 1973).

Western (1994) describes three phases of the development plans for Amboseli that influenced attitudes from the late 1970s through the early 1990s. Phase 1, the implementation phase, which included a water pipeline and imposition of a wildlife utilization fee, had early success, including the generation of funding for the construction of the first school in the area and a decline in spearing and poaching of wildlife. After 1981, the plan began to fail. The water pipeline stopped operating, the wildlife utilization fee was no longer being paid, relations between the Maasai and wildlife officials deteriorated, and spearing of wildlife, especially rhino, increased. The next phase saw the Maasai of Olgulului Group Ranch take an active role in a number of wildlife-related initiatives, such as assuming responsibility for the public campsite, constructing a fence at Namelok, and negotiating a tourist concession (Western 1994). There was a renewed tolerance for wildlife, and wildlife numbers, especially elephants, increased in Amboseli when poaching was rampant in other parts of the country.

The third phase came with the creation of the Kenya Wildlife Service (KWS) in 1990. Western (1994) cites some early successes of this phase that include the building of new schools and hiring of Maasai game scouts (paid for with monies from the revenue-sharing program), Olgulului's resistance to subdivision in order to protect wildlife corridors, plans for electric fences to protect crops, improved relations between KWS and the Maasai, low poaching levels, and "fairly positive" attitudes toward wildlife (Western 1994). This last outcome is substantiated by Kangwana's 1991 attitude survey (discussed below).

Measures of Attitudes

As part of her study of elephant-Maasai interactions around Amboseli National Park, Kangwana (1993) conducted an exploratory survey of Maasai attitudes toward wildlife and the park. Most respondents were not in favor of removing wildlife from the area, indicating "that the overall attitude towards wildlife is, if not positive, then at least tolerant" (95). Surprisingly, agriculturalists were found to be more positive toward wildlife and the park than pastoralists. One possible explanation offered by Kangwana (1993) is that agriculturalists were less marginalized, having diversified livelihood activities (i.e., crops and livestock) and more access to water, as most lived in the better-watered Kimana Group Ranch. These individuals also tended to be more educated and aware of the benefits of conservation. Elders were found to be more positive than *ilmoran* and women, with women being the most negative. Elders and agriculturalists were also more likely to be aware of benefits brought by wildlife, although respondents in general were more likely to believe that wildlife benefits the country of Kenya rather than the Maasai as a group or themselves as individuals. Additionally, some respondents who were receiving wildlife-related benefits did not recognize them as such, highlighting the importance of increasing awareness of the link between wildlife and benefits through education.

Overall attitudes toward elephants were found to be positive, with 58 percent of the people categorized as pastoralists and 47 percent categorized as agriculturalists saying that elephants should not be removed from the area. Those with a positive attitude believed that elephants were beneficial to the area, were gentle, and had always been there. Respondents with a negative attitude believed they should be removed because they were dangerous and destroyed crops. Elders had more positive attitudes than others who were surveyed. Kangwana posited that the elders were more aware of the benefits (e.g., opening up grazing land for cattle) of elephants than were younger individuals. This view is supported by the Maasai saying, "Cows grow trees, elephants grow grasslands" (Western 1997).

Change, Conflict, and Intervention

Many changes have occurred in the greater Amboseli ecosystem since Kangwana's study. The most critical is that of land-use change. In addition to the influence of colonialism, conservation, and agriculture, the human population is continuing to grow. With rapid population growth resulting from immigration (table 19.1), the rain-fed areas have become settled, which has reduced the area available for dry-season grazing and access to water for livestock and wildlife (Campbell et al. 2000). Intensifying the land-use problem is the more recent subdivision of the group ranches. As of 2006, of the 52 group ranches in the Kajiado District, 32 had completed subdivision and 15 were in the process and subdivision, leaving only 5 that had not started the process (BurnSilver and Mwangi 2007). The two group ranches that immediately border Amboseli, Olgulului/Lolarashi, and Kimana, which were surveyed by Kangwana in 1991 and Browne-Nuñez in 2005, were among the last to begin subdivision. Kimana completed the process in 2005, and in 2007, the rain-fed and irrigated agricultural areas of Olgulului were being subdivided.

Changes in land use, along with the growing human and elephant populations, have significant implications for Amboseli's elephants. Conversion from pastoralism to agriculture is a critical threat to elephants as it contributes to habitat loss and decreased tolerance (Gadd 2005). Human-elephant conflict is a growing problem—perceptual, political, and actual—in Kenya and across Africa (Ngure 1992; Waithaka 1993; Hoare 1995; Tchamba 1995; Thouless and Sakwa 1995; Barnes 1996; Lee and Graham 2006) and threatens the conservation of elephants. Most incidents of conflict involve crop raiding or competition for water and grazing resources. Deaths and injury to both humans and elephants occur as a result of the negative interaction. Mortality among Amboseli's elephants caused by local

Table 19.1 Population of Kajiado District, 1969–99 (adapted from Campbell et al. 2000)

	Kajiado District			Kenya
Year	Population	Inter-census growth (%)	Average annual growth (%)	Average annual growth (%)
1969	85,093[a]			
1979	149,005[a]	75.1	5.76	3.8
1989	258,659[a]	73.6	5.67	3.4
1999	405,000[b]	56.6	4.58	2.9[a]

[a]Kenya Central Business of Statistics, Population Census, 1969, 1979, 1989.
[b]Provisional Results of the 1999 Population and Housing Census, Ministry of Finance and Planning, February 2000.

people may be the result of retaliation for losses incurred directly from conflict but can also be a demonstration of unhappiness with a social or political situation. Historically, levels of spearing were associated with the circumstances of the *moran* age set. For instance, in 1984, when there was an initiation of new *ilmoran*, there was an increase in the level of spearing (Lindsay 1987; Kangwana 1993). The opposite occurred in the six-year period prior to the 1984 initiation. In 1978, *ilmoran* were initiated into elderhood, and there was not another *moran* initiation until 1984. Western (1982a) attributes this low spearing level to positive attitudes resulting from the receipt of benefits allocated to the Maasai by the development plans for Amboseli.

Conflict with wildlife can cause negative attitudes toward wildlife and reduce support for conservation (Newmark et al. 1993; De Boer and Baquete 1998; Naughton-Treves 2001) and can affect the long-term success of conservation programs (Webber, Hill, and Reynolds 2007). Although the elephant is the species most often reported as being a problem by local people (Gadd 2005; Parry and Campbell 1992), it is not always the proven cause of most damage. Naughton-Treves, Rose, and Treves (1999, 7) offer two case studies that demonstrate "that elephants create distinctive, highly localized crop damage patterns that are cataclysmic for the affected individual farmers, but insignificant to the regional farming economy." Although smaller animals tend to produce greater economic loss, they are better tolerated than elephants (Gillingham and Lee 1999 2003; Hoare 2000; Naughton-Treves, Rose, and Treves 1999; Naughton-Treves and Treves 2005). This attitude can be explained, at least in part, by the fact that elephants are a high-profile species, and they can be more dangerous to people than can other "pest" species (De Boer and Baquete 1998; Naughton-Treves, Rose, and Treves 1999; Sitati et al. 2003).

In interviews conducted with Maasai around Amboseli in 2005, when asked which wildlife species cause problems, 44 percent of individuals surveyed reported they had personally experienced problems with elephants and 21 percent said someone else in their *boma* had had a problem (Browne-Nuñez 2010). Later in the same interview, as survey questions became more focused on elephants, 65 percent of respondents stated that elephants had caused them problems. Part of the explanation for this increase is that some respondents consider other household members' elephant problems to be their own problems. Another possible explanation is an awareness of the value of the elephant to others (e.g., the Kenyan government, tourists, researchers, and conservation organizations). The elephant is the symbol of conservation found on uniforms and vehicles of government and nongovernmental organizations. Elephants receive a great deal of attention from these groups, and when one is killed, there is an instant reaction (killing other pest species does not elicit the same kind of reaction). Individuals suffering from wildlife damage, particularly from large species such as the elephant, are often hoping for compensation, which may lead to an increase in damage reports (De Boer and Baquete 1998; Gesicho 1991; Mascarenhas 1971 [cited in Naughton-Treves, Rose, and Treves 1999]).

While costs to agriculturalists are the chief complaint in many parts of the elephant's range, the leading costs reported by people living around Amboseli National Park in 2005 were human and livestock injuries and death (Browne-Nuñez 2010). Additionally, 42 percent of respondents who believed that local people incur costs related to problems with elephants complained about damage to trees. Only 8 percent cited crop damage as a cost. Other costs included competition for water and grazing and damage to property such as water pipelines, fences, and houses. In addition to the direct costs of living alongside elephants, there are opportunity costs. "These costs are difficult to quantify but may outweigh the direct costs of agricultural damage and be a major component of conflict as it is perceived by local people" (Hoare 2000, 35, citing Dublin, McShane, and Newby 1997). Survey respondents reported such costs, including the inability to walk freely in the group ranches, interference with children attending school, and sleepless nights (guarding crops or property) (Browne-Nuñez 2010).

There have been numerous interventions initiated around Amboseli to improve attitudes toward elephants and mitigate conflict. These have been implemented by several organizations, including KWS, the Amboseli Elephant Research Project (AERP), and the African Wildlife Foundation (AWF). While these interventions vary in their approaches and objectives, all have the long-term goal of ensuring continued tolerance of elephants outside of the relatively small protected area of the park.

In 1998, after a period of intensified conflict and long negotiations among community representatives and members of KWS and AERP, the elephant research project implemented a consolation scheme that pays for livestock killed by elephants outside the park (see box 19.1). Twenty-four percent of local people, mostly men, who were surveyed in 2005 were aware that AERP has a consolation program for livestock losses, with 73 percent of these individuals evaluating AERP as being slightly good to very good (Browne-Nuñez 2010).

In order to address human-elephant conflict in the nearby agricultural areas, an electric fencing project was started by KWS in two areas in 1996. The two solar-powered fences, at Kimana and Namelok, were financed by the European Union and completed in 2000. The fences have had limited efficacy because they are often in disrepair due to vandalism and elephant damage. They have not solved the problem of human-wildlife conflict, as conflict has shifted to the Loitokitok-Olchorro area and made wildlife more vulnerable to poaching in nearby Tanzania (Ntiati 2002). Additionally, there is confusion among community members as to who owns the fence and who is responsible for its maintenance (Browne-Nuñez 2010; Kioko 2004). Despite these issues, most of the 2005 survey respondents (99 percent) living within the fences believed the fences do a good job (Browne-Nuñez 2010).

Another significant change around Amboseli is the growing number of cultural *bomas* established in the Olgulului Group Ranch along the southern park boundary. These business entities have been established by Maasai who are interested in entering the cash economy. Here, tourists can tour a "traditional" Maasai homestead, buy beadwork from the women (and other handicrafts brought in by outside sellers), and observe demonstrations, including dancing, singing, the building of fires, and home maintenance. While there are perceived and real benefits for the Maasai who participate, such as the provision of an alternative economic activity, these *bomas* are fraught with problems. A major issue that hinders their economic success is the problem of the tour drivers who pocket most of the money charged to the tourists for entrance into the *bomas* (Browne-Nuñez 2010; Onetu 1998; Ritsma and Ongaro 2002). Cultural *boma* members believe there is nothing that can be done about this situation, concluding that if they protest, drivers will take their groups to other *bomas*, so a little money is better than none at all.

A further criticism of the cultural *bomas* is their contribution to the erosion of Maasai culture (Onetu 1998; Ritsma and Ongaro 2002). While this claim is also directed at tourism in general, the cultural *bomas* are specifically blamed for splitting families, as women leave their homes to work in the *bomas*. Related to this problem is the concern among community members that cultural *bomas* provide a setting for prostitution—a new occurrence in Maasai culture—and a rise in the incidence of sexually transmitted diseases (Onetu 1998). Additionally, cultural *bomas* "act as collection points for idle sitters who in turn participate in petty crimes and other unapproved social behaviors" (Onetu 1998, 47). From a health and safety perspective, the cultural *bomas* have led to an increase in disease, as large numbers of people concentrate in small areas (Akama 2002; Onetu 1998). Finally, because they obstruct elephant corridors, they actually increase, rather than decrease, conflict (Douglis 2001; Ritsma and Ongaro 2002).

In 1999, AWF proposed several activities aimed at "ensuring that the Maasai areas surrounding Amboseli National Park are 'friendly' to elephants (and other wildlife) by reducing conflicts between humans and elephants" (AWF 1999). Among these activities was an outreach program that would induce "change favorable to wildlife conservation," including improving attitudes toward elephants through meetings, workshops, and educational tours. At the time of the 2005 survey, only 7 of 567 respondents identified AWF as an organization that helps people who have problems with elephants, while 27 stated that AWF helps elephants. When asked directly about AWF's activities in the area, many Maasai complained that they were unsure exactly what members of the organization do, only that they drive their cars around and hold meetings (Browne-Nuñez 2010).

Other activities and projects are in progress in the ecosystem that are designed to ensure that elephants and other wildlife are able to continue to share group ranch lands with the Maasai. Many of these involve providing alternative forms of income for local Maasai through employment in the conservation and tourism industries. For example, the Selenkei Group Ranch has set aside an area for a private wildlife sanctuary and rents land to Porini Ecotourism for a luxury tented camp. In addition to rent, the community receives an entry fee for each visitor. The Kimana Group Ranch also has a wildlife sanctuary and lodging, although local Maasai complain that not enough Maasai are employed in these establishments and that the jobs that are available are usually low-level, degrading positions (Browne-Nuñez 2010; Ole Dapash 2002; Onetu 1998). This complaint is a criticism of the positions held by most Kenyans working in the tourism industry (Sindinga 1994). Maasai are believed to be the least represented in the industry, even in the lowest-level jobs (Akama 1999). Another serious problem is the equitable distribution of benefits from tourism-generated revenues. Local authorities such as the county council often are the recipients of tourism money, with little going to the individual group ranch members. Resolutions to these issues are needed in order for tourism to be the viable alternative economic activity that government and conservation organizations promote.

Current Attitudes

In the 2005 survey, several questions were asked concerning attitudes. First, respondents were asked to arrange cards with drawings of wild animals into piles according to whether they liked or disliked the animals. Most respondents (57 percent) placed the elephant card in the "dislike" pile, while 42 percent of respondents placed the elephant card in the "like" pile (the remaining 1 percent were either neutral or said they liked all animals). When asked specifically if they liked seeing elephants in their group ranch, there was a 50/50 split. The most common reason for liking to see elephants in the group ranch is the belief that elephants attract tourists and therefore generate revenue. This most likely explains why some who report not liking elephants say that they like to see them in their group ranch. The top reason for not wanting elephants in the group ranch is the danger they pose to people.

As in the survey conducted in 1991, where respondents were asked if elephants should be removed (Kangwana 1993), respondents in 2005 were asked about a hypothetical vote on whether to allow elephants to continue to enter their group ranch. A majority of people (53 percent) said they would vote to allow elephants to continue to be in the group ranch, the same percentage who were against removing elephants in 1991, while 46 percent said they would vote not to allow them access. In 1991, 45 percent of respondents stated that elephants should be removed or confined to the park (Kangwana 1993). Again, the top reason for a positive response was tourism revenue, and the top reason for a negative response was the problems caused by elephants. Individuals with an awareness of organizations that help people when they have a problem with elephants were more likely to say that they would vote to allow elephants in the group ranch. There was not a significant difference in voting intention between those who are primarily agriculturalists or pastoralists, between age groups, or between group ranches, but there was a significant difference between men and women, with more men being favorable to allowing elephants in the group ranches.

As mentioned earlier, it is essential to understand the complexity of the attitude-behavior relationship. Knowledge of attitudes alone is not sufficient for making policy decisions, as attitudes are not always straightforward. For example, 25 percent of respondents who stated they do not like to see elephants in their group ranch stated they *would* vote to allow them in the group ranch. Of these individuals, 85 percent cited tourism revenue as their reason for toleration. Conversely, 18 percent of respondents who stated they do like elephants in their group ranch stated they *would not* vote to allow them in the group ranch. Seventy-three percent of these respondents would vote against elephants because of the danger they pose. Here it is evident that attitude alone does not predict behavior or, in this case, behavioral intention. Other variables, such as economic benefits and fear of human-elephant conflict, contribute to the level of tolerance of elephants.

As the context of Maasai-elephant interactions continues to evolve in the greater Amboseli ecosystem, it is important to understand to what degree Maasai behaviors toward elephants are a consequence of theoretically predicted attitudinal variables and the extent of influence of more dynamic situational variables such as socioeconomic status, prior experience, and land use. This understanding can contribute to the development of strategies to maintain the coexistence of the Maasai and elephants.

The Future of the Maasai and Elephants of Amboseli

Writing about the fate of Amboseli's elephants, Chadwick (1992, 85) stated, "The future of Amboseli's giants clearly hinges upon the attitudes and land use practices of the people surrounding the reserve." The future of the people of Amboseli is also dependent upon the fate of the elephants.

Maasai and wildlife have always competed to some degree but have also been very interdependent (Deihl 1985). Perhaps the best summary of the symbiotic relationship between the Maasai and wildlife was given over 30 years ago by a Maasai elder addressing a group of representatives from the World Bank (Western 1997):

Yes, the Maasai do intend to accommodate wildlife . . . as we always have. We protected wild animals from hunters, and wildlife protected us from drought, so we see it as fatal that we should not be allowed to move back into Amboseli. We will leave the park to show our good intentions. We will allow wild animals onto our land, stop our young men from harassing them, and discourage poachers. But we expect that when it is seen that we cannot survive without Amboseli, we too shall be extended the same treatment as the animals and that we will derive the benefits from the park. If such reciprocity is not shown, we cannot make assurances for the future of wildlife outside Amboseli National Park. And if wild animals are restricted to the park, their numbers will fall. If we are excluded, our livestock will die. Coexistence is the essence of survival for us both.

This tone of cooperation and goodwill is still evident among the Maasai of Amboseli today. Although a large amount of modern tolerance of elephants appears to be explained by the perceived role of elephants in attracting tourists to the area (economic value), there is an element of cultural value present today. Traditional tales of the elephant are still told by the Amboseli Maasai. During survey follow-up interviews conducted in May 2005, senior elders told many of the stories and proverbs of the elephant described earlier.

The Maasai of Amboseli have indeed been "people of cattle." Some have argued that Maasai throughout history have always made changes when necessary with the intention of returning to pastoralism (Knowles and Collett 1989). This notion is supported by research in northern Tanzania, where Maasai stated they began farming in order to save their livestock, that is, so that fewer animals would have to be sold in order to obtain food (McCabe 2003; O'Malley 2000), but this view is not universal among all Maasai. As one elder living near Tarangire National Park, Tanzania, stated, "You cannot expect us to remain in history. Many projects come and recommend we remain pastoralists, but we have discovered the new foods. Now we want to grow crops and keep our cows" (in Kangwana and Ole Mako 2001, 159).

If modern Maasai do indeed diversify their livelihood activities with the intention of returning to transhumant pastoralism, would such a transition even be possible in the greater Amboseli ecosystem, given the many changes, namely subdivision, in group ranches? The answer is obviously no. While it is impossible to turn back the clock and return to the days when the Maasai, their livestock, and wildlife coexisted, unrestricted by boundaries created by outside policy, it may be feasible to nurture what remains in terms of traditional values and beliefs, develop sustainable opportunities for livelihood diversification, and encourage land-use planning that will minimize further loss of land available for wildlife and livestock.

Understanding, respecting, and including local people in decision making are requirements for successful conservation. In Amboseli, where culture and land use have played significant roles in the conservation of diverse and abundant wildlife populations, this holds especially true. "Aspects of Maasai communal social organization, their productions system, and their culture are valuable human resources that can be a foundation for a modern livelihood which integrates livestock keeping with wildlife management and tourism" (Berger 1996, 175). Although changes in land tenure—communal property to private property—have weakened these "human resources," many conservationists recognize these resources and are working to include the local people as partners in conservation.

Education can be an important part of conservation programs, although there are caveats as to how it should be done. The goal of environmental education programs is to influence behavior. As discussed in this chapter, many variables influence behavior; therefore, addressing these other variables must be a component of conservation programs. There is a large body of literature on the efficacy of environmental education that cannot be adequately reviewed here, but the following excerpt from Byers (1996, 86) provides insight into the best approach to environmental education:

Modern environmental education recognizes that environmental behaviors are influenced not only by knowledge, but also by values, options, skills, and many other motivating factors (Hungerford and Volk 1990; Wood and Wood 1990). [It] attempts, therefore, to communicate more than just knowledge. It is a "process that enables people to acquire knowledge, skills, and positive environmental experiences in order to analyze issues, assess benefits and risks, make informed decisions, and take responsible actions to achieve and sustain environmental quality" (North American Association for Environmental Education 1993). . . . [It] is concerned with communicating environmental values and ethics, not just knowledge and information. (Caduto 1985)

The results of the 2005 attitude survey show that while the cultural value of elephants does not appear to be highly prevalent among today's Amboseli Maasai, there is some degree of appreciation (e.g., beauty, role in ecosystem modification, etc.) and theistic value as demonstrated by the statement of an old woman in Kimana group ranch: "God created elephants and they belong to this earth just like us" (Browne-Nuñez 2010). An environmental education program in Amboseli-area schools and the greater community based on traditional knowledge, beliefs, and values as well as on science would not only increase knowledge of and cultivate positive attitudes toward elephants but also perhaps aid in the preservation of the culture that has been credited with conserving the diverse wildlife of the greater Amboseli ecosystem.

Economic Opportunities and Other Benefits

Improving knowledge and attitudes alone is not enough. As pastoralist families move closer to poverty, the need to diversify their livelihood activities increases. Many have turned their hopes to tourism as a source for alternative income. In the last few decades, there has been great emphasis on creating economic incentives for conservation in rural areas. This approach is based on a growing, but debated, effort to link conservation and development, but "most of Africa's protected areas do not and almost certainly will not contribute significantly to reducing poverty" (Infield 2001, 800). While the Maasai recognize the benefits or potential benefits of tourism, many do not see it as benefiting them personally or their household (Browne-Nuñez 2010). Although a relatively small percentage (16 percent) mentioned selling handicrafts as the top personal benefit, many saw it as a benefit to the Maasai in general. About 10 percent mentioned school bursaries as a benefit to their household.

These results have implications for elephants, given that the most-cited reason for tolerating elephants is generation of tourism revenue. Amboseli is one of Kenya's top earners among parks in terms of annual gate fees (Bulte et al. 2006; Ole Dapash 2002), but benefits have been inconsistent, insufficient, and inequitably distributed (Mburu, Birner, and Zeller 2003). These problems and others previously mentioned that are related to tourism in Amboseli need to be rectified if tourism is to be promoted as a viable alternative livelihood activity. There are, however, examples of ecotourism enterprises in the greater Amboseli ecosystem that not only limit the impact on the natural environment but respect local culture and provide more equitable sharing of benefits. These include the Oldonyo Wuas Camp in the Mbirikani Group Ranch and the Porini Camp in the Eselenkei Group Ranch. Additionally, some organizations working in the ecosystem, such as the Amboseli Trust for Elephants and the Maasai Wilderness Conservation Trust, provide benefits to the local community such as scholarships for school-age children and university students. There are also consolation and compensation schemes such as the one implemented by the trust and a scheme for lion damage operated by the Lion Guardians Project. Where benefits do exist, they should be promoted to increase awareness in the community.

Land-Use Planning

Recent research shows that pastoralists are unsure of the economic viability of individual land parcels after group ranch subdivision and are working to "re-aggregate their access to resources through pasture sharing and swapping mechanisms" (BurnSilver and Mwangi 2007, 4). These arrangements are based on reciprocity or, less commonly, monetary exchange. These findings refute the inevitability suggested by earlier research that changes in land tenure and livelihood activities would weaken social and cultural relationships among pastoralists (Kituyi 1990; Ensminger and Rutten 1991 [cited in BurnSilver and Mwangi 2007]). It is still early in the process of transition from communal to individual ownership. Challenges to these new arrangements may become more evident. In the 2005 survey, 70 percent of respondents named pastoralism and 25 percent named cultivation as their primary economic activity. When asked what they thought their primary livelihood activity would be in five years' time, only 57 percent said pastoralism, while 35 percent predicted they would primarily be farmers (Browne-Nuñez 2010).

Conclusions

The welfare of the Amboseli elephants has long been intertwined with the local human population. As noted here and by other authors (Campbell et al. 2003), the social context of Amboseli is changing. The human population is no longer a homogeneous society but rather comprises a diversified Maasai population with an ever-increasing population of cultivators from other ethnic groups. As the human dimensions of conservation in the greater Amboseli ecosystem continue to evolve and become more complex, local, national, and international stakeholders need to collaborate in order to continually monitor and adapt conservation and development strategies to ensure the well-being of both species.

Chapter 20 The Behavioral Responses of Elephants to the Maasai in Amboseli

Kadzo Kangwana

The Amboseli elephants have a long history of coexistence with the Maasai. The current range of the elephants extends well beyond the boundaries of Amboseli National Park (see Croze and Moss, chapter 7) into land that is soon to be subdivided into individual plots and was previously owned communally by the Maasai as group ranch land. The land outside the protected area is dotted with Maasai settlements and is used by the Maasai for grazing livestock and, where water is available, to grow a variety of crops. Some of the land is also leased to tourist operators.

To maintain the Amboseli elephant population at any level, the elephants need access to land outside the protected area. Individual landowners and conservationists need, therefore, to work together to set aside land for wildlife access and to delineate corridors for the movement of elephants. The sharing of a range and resources by the Maasai and elephants in the Amboseli ecosystem makes it necessary to understand the dynamics of the human-elephant interface, examining both how people are impacted by elephants (see Browne-Nuñez, chapter 19) and how elephants are impacted by people. We must also take our understanding of the impact of people on elephants beyond quantifying elephant population loss and determine how elephants respond to people. How do elephants share their range with people? What strategies do these highly intelligent animals adopt when forced to share resources with people? What is the impact of the presence of people on the complex social networks of elephant families and bond groups?

This chapter addresses these questions. The aims of this chapter are to describe how the Amboseli elephants

- use areas of Maasai settlement,
- share sources of water with the Maasai, and
- respond to meeting Maasai in their range.

The Maasai have a tradition of spearing wildlife in order to prove bravery, and the Amboseli elephants have been victims of both this ritual spearing and spearing carried out as political protest against Park authorities (see Kangwana and Browne-Nuñez, chapter 3). Examining elephant responses to the Maasai in the context of Maasai spearing of elephants is therefore an important component of this study.

Increasing our knowledge about what takes place when elephants and humans meet in Amboseli will assist us in making the adjustments necessary to ensure coexistence into the future and will contribute to the broader debate on elephant-human interactions throughout Africa. This chapter therefore ends with a discussion of how the observed responses of elephants to the Maasai might influence management decisions as well as a discussion of the issues surrounding the traditional Maasai practice of spearing elephants. Understanding elephant responses to the Maasai was of interest in the early 1990s when the fieldwork for this study was carried out, and it is of even greater importance today as the human landscape of the Amboseli elephants changes and the challenges of conserving the elephant population increase in complexity.

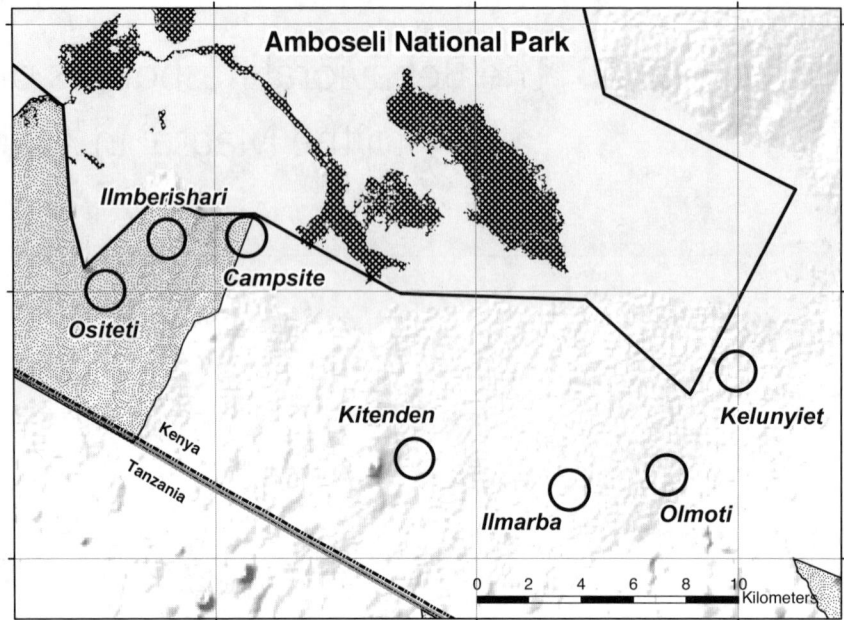

Figure 20.1 Map of areas where dung counts were carried out during the study.

Elephant Distribution and Maasai Settlement

The *Acacia* woodland just south of Amboseli National Park is occupied by the Maasai and used regularly by the Amboseli elephants. Some of the homesteads in this habitat are used throughout the year by the Maasai, making it possible to examine how elephants use permanently occupied areas. In this habitat, there are also temporary homesteads that are occupied by the Maasai only in the dry season, which made it possible to investigate whether occupancy of a homestead influences the use of that area by elephants.

The role of Maasai settlement in determining patterns of elephant distribution was examined using a series of dung counts. The technique of counting dung along transects has been developed and used extensively in the tropical rain forests of Africa (Barnes and Jensen 1987; Barnes, Alers, and Blom 1989; Barnes et al. 1991) and elsewhere (Kibale: Wing and Buss 1970; Kasungu: Jachmann and Bell 1984; Marang: Reuling 1991) to determine elephant densities. This method was adapted and used to investigate the effect of Maasai settlement on elephant distributions in Amboseli. While the workers who developed dung count methods used them as indexes of elephant density, in this study dung density was used as indices of occupancy or relative use of an area: high dung density reflects high elephant usage and low dung density reflects low elephant usage.

In conjunction with dung counts, experiments on dung decay were carried out to determine how long dung lasted after it had been deposited and hence to facilitate the interpretation of the abundance data from the dung counts. In the wet season, all elephant dung was washed out within six weeks of the beginning of the experiment, while in the dry season 64 percent of the observed dung piles persisted for more than 20 weeks. These results were consistent with data collected in different savannah environments (Reuling 1991) showing that dung disappears most rapidly during the wet season and that decay rates were almost negligible during the dry season.

Permanent Homesteads

Dung counts were carried out relative to five groups of permanent homesteads in the *Acacia* woodland habitat south of the protected area: Ositeti, Olmoti, Kelunyiet, Kitenden, and Ilmarba (figure 20.1). From each of the five groups of homesteads, dung counts along transects were carried out at the homestead and then at 0.5, 1.0, 1.5, 2.0, and 2.5 km from the homestead along a line in a due north direction and in one other randomly chosen direction. Each transect involved walking a 250 m long and 30 m wide strip (perpendicular to the line from the homestead), and counting all the dung piles present and their distance along the transect. Dung counts using the same methods were also carried out during both the dry and wet seasons at four locations in the center of the Park where Maasai do not graze their cattle. These were used as controls for the dung counts around Maasai homesteads.

To assess the effect of vegetation on the density of dung, all trees, bushes, and shrubs present in the transect were noted, and the mean percentage ground cover of the transect was estimated using a pin frame. Dung counts were converted to dung density as dung piles per hectare, and data from 89 transects were analyzed.

Variation in dung density was explained by the variables of distance from homestead ($F = 17.894$, $p < 0.001$), location of homestead ($F = 13.89$, $p < 0.001$), and season ($F = 5.96$, $p < 0.05$). Distance from the Park, woody vegetation density, and percentage ground cover did not make significant contributions to explaining the variation in dung density (distance from Park: $F = 3.85$, $p > 0.05$; woods: $F = 0.14$, $p > 0.10$; ground cover $F = 2.35$, $p > 0.10$). There were no significant interactions between any of the variables.

Dung density increased with distance from homesteads up to 2.0 km and then tended to drop at 2.5 km. However, this drop in mean dung density between 2.0 km and 2.5 km was not significant ($t = 1.22$, df = 26, $p > 0.100$). Dung density (adjusted for distance from homesteads and season) varied between homestead locations, being highest in Kitenden and Ilmarba, lower in Ositeti and Olmoti, and lowest near Kelunyiet. Dung density (adjusted for distance from settlements and location of settlements) was higher in the wet season than in the dry season.

The role of water availability in determining elephant use of areas near homesteads was examined by regrouping the five homesteads into those that had a borehole near them (Kitenden and Ilmarba) and those for which the nearest water was in the Park (Ositeti, Olmoti, and Kelunyiet). Mean dung density relative to these two groups of settlements was compared for each season separately using t-tests.

The homesteads with a borehole near them (Ilmarba and Kitenden) had a significantly higher dung density than did the homesteads without a borehole (Ositeti, Olmoti, and Kelunyiet) in the dry season ($t = 3.32$, df = 48, $p < 0.05$). In the wet season, however, the differences in dung densities as a function of proximity to water were not significant ($t = 0.97$, df = 37, $p > 0.10$).

Dung counts relative to permanent homesteads indicated that dung density and hence elephant use of areas increased with distance from a homestead. Elephants did come close to homesteads, and in a few cases, dung was present at the homesteads themselves, but dung density increased sharply between 500 m and 1 km from the homestead. The elephants' avoidance of homesteads could have been due to the lack of feeding habitats close to the homesteads as a result of grazing by livestock. The vegetation measures, tree density, and ground cover used in the model did not, however, have significant effects in explaining the variation in dung density relative to permanent homesteads, suggesting that the elephants' avoidance of homesteads was not driven by variation in vegetation. The increase in dung density with distance from homesteads may also indicate that elephants were avoiding the immediate vicinity of the Maasai and were responding to the Maasai as predators due to the history of Maasai warriors spearing elephants in order to prove bravery and later for political gains.

Dung density also varied as a function of the location of homesteads. The highest dung densities were observed at Kitenden and Ilmarba, where water was available. The dung density at these locations approached that of areas in the center of the Park, despite these two settlements being the furthest away from the Park. As noted above, variation in dung density with location was not mirrored by the woody vegetation density or percentage ground cover, suggesting that vegetation density on its own does not explain the difference in use of these locations. A comparison of dung density for permanent homesteads once they had been regrouped according to water availability showed that in the dry season, elephants may have been using the areas differently, spending more time near the homesteads with boreholes. There was no such relationship in the wet season, when surface water was widespread. This analysis indicated an effect of the availability of water at homestead sites. More work is needed, however, in order to draw conclusions, as dung counts were done at only one of the homesteads with a borehole in the wet season because the second area was inaccessible during the rainy season.

The higher dung density found during the wet season, despite the higher loss of dung due to washing out in the rains, indicates a higher usage of areas outside the Park during the wet season. This observation was consistent with previous work showing that the Amboseli elephants used areas outside the Park during the rainy seasons, when groundwater was readily available and new growth widespread, but they concentrated in the Park during the dry season (Western 1975; Western and Lindsay 1984; Moss 1988). Subsequent to this study, however, the ranging patterns of the Amboseli elephants in the wet and dry seasons have changed greatly (see Croze and Moss, chapter 7).

Temporary Settlements

Two groups of temporary homesteads existed just south of the Park: campsite and Ilmberishari (see figure 20.1). In keeping with the practice of transhumant pastoralism (see Kangwana and Browne-Nuñez, chapter 3), these homesteads were occupied only during the dry season, with the Maasai and their cattle moving to more distant grazing areas during the wet season.

Dung counts were carried out relative to these homesteads in the wet season, when they were not occupied, and during the dry season when they were occupied. Dung counts were done at the homesteads and in two directions at 0.5 and 1.0 km from the settlements, using the same methods described above in the study of elephant use of permanently occupied areas. The distance from the homesteads for which dung counts could be carried out was limited because both homesteads were within 1.5 km of the Park border and, for consistency, counts relative to homesteads were done only outside the Park. The data from these dung counts were converted to dung piles per hectare and analyzed to determine dung density with distance from the homesteads controlling for vegetation in each of the two occupied states of the homestead. It should be noted that the occupied state of the homestead is linked to season. Data from a total of 24 transects were analyzed.

Location of the homesteads ($F = 15.77$, $p < 0.01$), distance from the homestead ($F = 4.07$, $p < 0.05$), and woody vegetation density ($F = 10.13$, $p < 0.01$) all made significant contributions to explaining the variation in dung density around temporary homesteads. Whether or not the homesteads were occupied and percentage ground cover were not significant in explaining variation in dung density (occupied: $F = 0.03$, $p > 0.10$; ground cover: $F = 0.31$, $p > 0.10$). Interaction terms between variables were not significant.

Dung density (adjusted for location of homesteads and woody vegetation density) increased with distance from homesteads. Dung density (adjusted for distance from homesteads and woody vegetation density) was lower at Ilmberishari than at the campsite. Analysis of dung counts around temporary homesteads did not provide evidence that the occupancy of homesteads affected elephant use of the area surrounding it, and it was not possible to separate the effect of season and occupancy. Thus, while we cannot conclude that occupancy of a homestead influenced elephant use of the surrounding area, the data collected relative to temporary homesteads are consistent with those from permanent homesteads indicating that dung density, and hence elephant use of an area, increases with distance from homesteads.

Times of Elephants' Approach to Maasai Homesteads

To further investigate how elephants used areas settled by Maasai, the Maasai were asked, as part of a broader survey of attitudes toward wildlife, whether or not elephants came close to their homesteads, and if elephants did approach the homesteads, when they did so. Of the 92 men and women from 36 pastoralist homesteads who were asked whether elephants came close to their homestead, 42 said that elephants did come close. These respondents were from homesteads to the south at Ilmarba and Olmoti (areas that correspond with homesteads included in the dung counts) and from homesteads to the north of the Park and parts of the Kimana Group Ranch near the Park.

Of the 42 respondents who said elephants did come close, 90.5 percent said that elephants came close to their homestead at night, 4.8 percent said that elephants came close at dawn, and 4.8 percent said that elephants came close at dusk. No respondents said that elephants came close to their homestead during the day. When asked in what season elephants came close to their homestead, 57.1 percent said that they came during the rainy season, 26.2 percent stated they came during the dry season, and 16.7 percent reported that elephants came close to their homestead all year round.

The responses to the questionnaire survey suggested that elephants were coming close to settlements at night when cattle were enclosed within the settlements and the Maasai were asleep. Long-term sightings data of the Amboseli Elephant Research Project confirm a daily pattern of movement of the elephants that substantiates these findings. Elephant ranging patterns have changed over time, but at the time this study was carried out in the early 1990s, elephants moved out of the Park at night and came back into the protected area during the day (see Croze and Moss, chapter 7). The elephants' movement into the Park during the day was associated with a need to drink and forage in the swamps during the middle and hottest part of the day. The elephants may also have been avoiding the Maasai who were grazing their cattle outside the protected area during the day. The elephants leaving the Park at night to feed in the bushlands may indicate that at these times, they were "safe" outside the protected area. From the questionnaire responses, there did not seem to be a consensus as to which season elephants tended to come close to homesteads, and indeed this may have varied from place to place.

Joint Use of a Water Source by Maasai and Elephants

At the center of the conflict between conservation authorities and the Maasai in Amboseli is the need for access to water (Lovatt Smith 1997; see Croze and Lindsay, chapter 2). The protected area surrounds nearly all the permanent sources of open water in over 2,000 km^2. Further, arrangements to supply the Maasai with water outside the Park via facilities such as boreholes and pipelines have been vulnerable to breakdown over the years. Park authorities have allowed the Maasai to enter the Park in designated areas to water their livestock and fetch water for themselves. These selected areas are used by both wildlife and Maasai, especially in the dry season.

One of these areas, Njiri, in the southeast of the Park was chosen to examine how elephants and Maasai share a water source. A dung count taken in this area showed a high density of both elephant and cattle dung. This concentration of

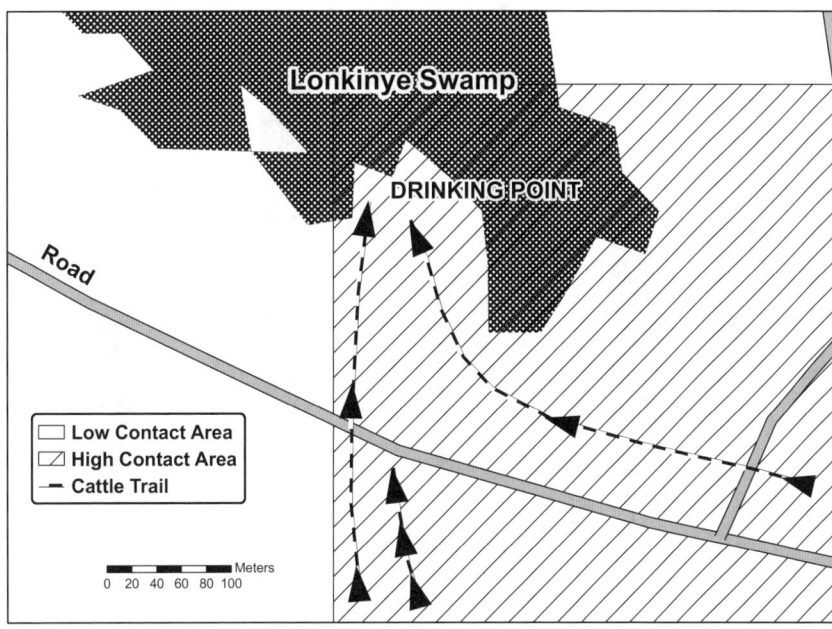

Figure 20.2 Diagrammatic map of the transect (road) used to examine use of a water source by both elephants and Maasai.

elephants and livestock, their very clear paths of entry and exit (see also Mutinda, Poole, and Moss, chapter 16), and the regular sighting of elephants in the daytime indicated that the area as a good one in which to observe dual usage.

A transect through this area was delineated. Because the Maasai did not use the transect after dark, the daytime period from 6.30 a.m. to 6.30 p.m. was divided into four blocks of three hours each. Only two of these time blocks were covered each day in order to reduce dependence between observations. Within the three-hour blocks, an observation was made every hour. Each observation consisted of making an hourly scan drive through the 2 km transect, driving slowly along the road that ran the length of the transect from the starting point and stopping every 400 m to scan the area (see figure 20.2). The scan drive from start to end and the return drive back to the start took on average 20 minutes; each scan drive was counted as one observation. A total of 156 observations were made at Njiri over 31 days in August and September 1991 during the late dry season.

For each observation, the time of day and the presence of elephant groups, their position in the transect, group size, identity, and activity were noted. A group was defined as any coordinated and spatially distinct body of animals (see appendix 1). The presence of herds of cattle, sheep, and goats and their activity and position in the transect were also noted. The livestock herds were always escorted by children or Maasai warriors (*moran*). Movement through the transect by individual elephants, their families, and larger groups as well as by cattle, sheep, and goat herds was also tracked.

The data on livestock sightings were analyzed to determine high-risk times (that is, times when elephants using the transect would be most likely to come into contact with Maasai and their livestock) and high-contact areas in which the elephants would be most likely to meet with Maasai. The classification of the day into high- and low-risk times and the transect into high- and low-contact areas was then used as the basis for analyses of the number of elephant groups coming into the area. Elephant group size was also analyzed as a function of risk time and contact area. Differences in group size between frequent and occasional elephant users of the transect were also examined.

A total of 48 herds of livestock came into the transect over the 31 days of observation. Livestock herds comprised various numbers of cattle, sheep, goats, and donkeys. A total of 128 elephant groups were contacted in the transect of which 95 were family units and 33 were bull groups or individual males. The number of livestock herds and elephant groups coming into the transect by time block is shown in figure 20.3.

Elephants used the transect area throughout the day, but more groups were observed in the last two time blocks of the day ($\chi^2 = 7.12$, df = 1, $p < 0.05$). Livestock came into the transect mostly in the two middle time blocks between 0930 and 1530 hours ($\chi^2 = 40.5$, df = 1, $p < 0.001$). These

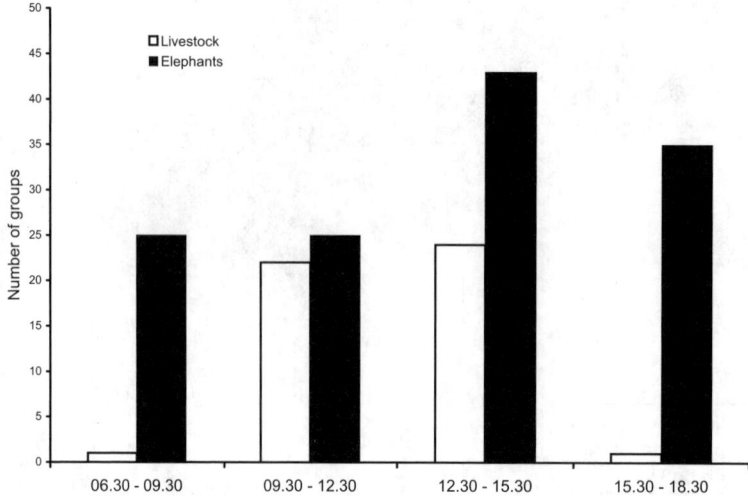

Figure 20.3 Comparison of the number of elephant groups and livestock herds entering the Njiri transect, by time of day.

two middle time blocks were therefore the highest risk times for elephants, during which they had the greatest chance of meeting Maasai.

Elephant Distribution and Group Size in the Njiri Transect

The 128 elephant groups observed within the transect were not distributed equally over the transect area ($\chi^2 = 32.0$, df = 1, $p < 0.001$). Fewer groups ($N = 32$) were sighted in the high-contact area than in the low-contact area of the transect ($N = 96$). Cow-calf groups and bull groups did not differ in their use of the transect; both were more likely to be sighted in the low-contact part than in the high-contact part of the transect ($\chi^2 = 2.2$, df = 1, NS).

The elephants' use of the high- and low-contact areas of the transect varied with time of risk. In the low-contact area, the number of groups seen at low-risk times was less than expected, and the number of groups seen at high-risk times was greater than expected ($\chi^2 = 15.9$, df = 1, $p < 0.001$). For the high-contact area, on the other hand, the number of elephant groups seen at low-risk times was greater than expected, while at high-risk times the number of groups seen was less than expected ($\chi^2 = 21.75$, df = 1, $p < 0.001$).

Median group size for elephant cow-calf groups (11, inter-quartile range (IQR) = 5–25, $n = 95$) was significantly higher (Mann-Whitney: $z = -8.04$, $p < 0.05$) than median group size for elephant bull groups (1, IQR = 1–2, $n = 33$). Bull groups showed little temporal variation in size, and bulls were typically found singly.

Females arrived at the water hole in larger groups during the middle and high-risk times of day (table 20.1) There was a significant difference in cow-calf group size between the first and subsequent time blocks (Mann-Whitney: $z = -4.5$, $p < 0.001$). The largest groups were seen at midday, although median group size in time block 4 was also large, despite this being a low-risk time. This result, however, was due to large groups coming into the transect during the middle of the day and remaining there stationary until the livestock left. Of the 22 groups contacted in the last time block, only 4 entered during this period, and the median size of these groups was 3. It would seem, therefore, that smaller groups were entering the transect in the first and last time blocks, although it is impossible to determine whether this was a function of low risk or simply time of day and thirst.

While aggregating in a large group may reflect the elephants' perception of a higher risk, the median cow-calf group size in high-contact areas of the transect was 4 (IQR = 3–5, $n = 20$) compared with 15 (IQR = 7–27, $n = 75$) in low-contact areas (Mann-Whitney: $z = -4.65$, $p < 0.001$). It might be suggested that larger groups, especially those who wait in areas of lower risk to enter the water hole, provide perceptual security, while avoidance of contact might be achieved by the small groups hurrying in and out to drink at the watering point.

Twenty-two different family units were contacted in the transect over the period of observation. Of these, 11 were observed fewer than five times (5 = median number of first contact observations of a family unit) and were called occasional users, while 11 were observed in the transect more than five times and were called frequent users.

For frequent users, median family unit size was 11 (IQR = 7–22) and median bond group size was 38 (IQR = 24–47). For occasional users, median family unit size was 13 (IQR = 8–18) and median bond group size was 18 (IQR = 9–26). Family unit size did not differ significantly between frequent and occasional users (Mann-Whitney: $z = -0.03$,

Table 20.1 Cow-calf group size in the Njiri transect

Time Block	Median	Inter-quartile range	N
0630–0930	4	3–5	15
0930–1230	13	5–25	23
1230–1530	16	8–27	35
1530–1830	12	5–28	22

NS, but bond group size did (Mann-Whitney: $z = -3.07$, $p < 0.005$). Frequent users of the transect were part of larger social networks.

While frequent users were part of large social networks, their group sizes when they entered the transect did not differ significantly from those of occasional users (Mann-Whitney test: $z = -1.25$, NS). Median group size for occasional users was 9 (IQR = 5–14.5, $n = 32$), while median group size for frequent users was 16 (IQR = 5–22.5, $n = 63$). Frequent and occasional users of the transect used the low- and high-risk areas of the transect similarly ($\chi^2 = 0.59$, df = 1, NS) at similar times of day ($\chi^2 = 3.18$, df = 1, NS). Thus, although the social networks of frequent users were more extensive, all users may have been responding to perceptions of risk by aggregation, whether with friends or with any other family member who happened to be present at the same time in the same area.

Both frequent and occasional users were found in larger groups in the low-contact part of the transect. In the high-contact area, frequent users were found in groups of 3 (IQR = 2.25–5.0, $n = 12$) while in the low-contact area, frequent users' group size was 18 (IQR = 6–29, $n = 51$; $z = 3.41$, $p < 0.001$). Median group size for occasional users in the high-contact area of the transect was 4.5 (IQR = 3.25–5.75, $n = 8$) while in the low-contact area, it was 10 (IQR = 7.25–23.25, $n = 24$; $z = -3.32$, $p < 0.001$). Thus, frequent and occasional users tended to use the area similarly, although frequent users came from family units associated with larger bond groups and tended to be found in slightly larger groups (see also Mutinda, Poole, and Moss, chapter 16).

The need for access to a critical resource can attract elephants to areas also used by the Maasai. However, elephants change their movements within these high-contact areas to minimize their chances of meeting Maasai and their livestock.

Groups observed in the transect used high-contact areas at low-risk times and concentrated in low-contact areas at high-risk times. Cow-calf group size was larger in low-risk areas than in high-risk areas. The smaller number of groups observed in the high-contact area of the transect suggests that one of the social and behavioral strategies used by elephants to avoid Maasai is that of aggregating in low-risk areas in the vicinity of a watering hole or while waiting for access to a point source of water.

Elephants thus show a high level of flexibility in their responses to Maasai, adjusting their use of this relatively small area in order to minimize the chances of direct contact with Maasai. Elephants that used the area frequently were associated with larger bond groups, suggesting that the other social groups encountered by elephants may influence their willingness to risk contact with Maasai. However, the elephants' frequency of use of the transect did not determine how groups were distributed over the transect in relation to contact area and risk time. Interestingly, the elephants often moved into areas adjacent to those with a high-risk of contacting Maasai, forming aggregations of resting families. When the Maasai and livestock left, they would all rush into the watering area. While in the high-risk zone, the elephants were constantly listening and smelling, showing some of the behavioral responses of elephants to Maasai that are explored in more detail below.

Elephant Responses to Sounds of Maasai Cattle

The Amboseli elephants adopt strategies to avoid Maasai both spatially and temporally. It was predicted that the Maasai tradition of *moran* spearing elephants in order to prove their bravery and as retaliation and political protest would have led elephants to fear and avoid direct interactions with Maasai. Anecdotal observations of the elephants over the duration of the long-term Amboseli Elephant Research Project suggest that the sight, sound, and smell of Maasai cause avoidance behavior in elephants. Through controlled experiments in which Maasai-associated sounds were first recorded and then played back to elephants from loud speakers, it was possible to simulate the presence of Maasai in close proximity to groups of elephants. Using this technique, auditory stimuli associated with the presence of Maasai could be presented in a standardized way to different groups of elephants (see McGreggor 1992; Poole, chapter 9; and McComb, Reby, and Moss, chapter 10 for further discussion of the uses of playback). Playbacks were used to address the questions, is avoidance a specific response to Maasai-associated noises, and are there any sex differences in the responses of elephants to the Maasai?

The playback experiments involved recording cattle bells and mooing of cattle from settlements in Amboseli and then playing a 2.5-minute-long stimulus tape of these sounds to groups of elephants. Control experiments were carried out using a 2.5-minute stimulus tape of wildebeest rutting sounds. The sounds were played back from a vehicle placed approximately 100 m from the experimental animals, while observations and a video recording of the elephant responses were carried out from a second vehicle.

The elephants were filmed for 60 seconds before the sound was played to them, for the 2.5 minutes during which the sound was played, and until they had settled down after the sound had stopped or had moved out of sight. Notes were also taken regarding the position of the elephants, the speaker, the video camera, and the elephants' movements in response to the sound. The sound was played to family units only when they occurred as a discrete group without members of other family units present and to bull groups comprising two or more adult bulls.

The control experiments were conducted using the same protocol but presenting the neutral stimulus of wildebeest rutting sounds. This technique was used to ensure that the responses in the main experiment were specific to Maasai-associated sounds and not due to the experimental procedure or the presence of observers. Cattle bells and mooing were played to a total of 13 family units and 10 bull groups, while wildebeest rutting sounds were played to 7 family units and 6 bull groups.

Typical Family Unit Response to Playbacks

Family units typically stopped whatever activity they were engaged in and held still when they heard the sound of cattle bells (table 20.2). They then raised their heads, turned toward the speaker, and turned their heads from side to side as if trying to determine where the sound was coming from, listening and smelling while they did so. Some members of the family units moved toward others, forming a cohesive group. They then turned away from the speaker and retreated, usually in a run (although two of the groups walked away). They ran for up to 300 m before stopping and standing in a clump or defensive formation with the young in the center encircled by older female members of the family. As the group retreated, some members turned to look back in the direction of the speaker, while some paused and listened or smelled with their trunks up.

The family unit members listened to the wildebeest rutting sound and smelled with their trunks down. Members of four of the seven family units paused their original activity while they listened to the wildebeest sound. None of the groups turned to face the source of the sound, nor did they retreat from it.

The likelihood of a family unit stopping activity and holding still on perceiving the sound of cattle bells was not significantly different from their likelihood to be still in response to wildebeest sounds (table 20.2). However, females were more likely to raise their heads in response to cattle bells than in response to wildebeest sounds and were also more likely to turn their heads from side to side in response

Table 20.2 Family unit responses to cattle bells and wildebeest playbacks

	Cattle bells ($N = 13$)	Wildebeest ($N = 7$)	G	p
Pausing activity and holding still	13	4	3.55	—
Raising their heads	12	1	9.33	<0.005
Turning heads from side to side	11	2	4.04	<0.05
Listening	13	5	1.49	—
Smelling				
a) trunks up	7	0	4.27	<0.05
b) trunks down	10	2	2.65	—
Turning heads toward the speaker	11	1	6.29	<0.05
Grouping—moving toward other members of the family or group	10	0	8.98	<0.005
Retreating from the sound				
a) walking	2	0	0.01	—
b) running	11	0	11.19	<0.005
Standing in a clumped group (defensive formation) at the end	12	0	14.02	<0.005

Note: df = 1 in all cases.

to cattle bells than in response to wildebeest. Head turning toward the speaker was also more likely to occur in response to cattle bells than in response to wildebeest sounds. Females were equally likely to show listening behavior in response to cattle bells as in response to wildebeest sounds. Smelling with trunks held up was more likely to be observed in response to cattle bells than wildebeest sounds, while smelling with trunks down was as likely to occur in response to cattle bells as in response to wildebeest. Movement of some of members of the group toward others was more likely to occur in response to cattle bells than in response to wildebeest sounds. All family units retreated from the sound of the cattle bells, while none retreated from the sound of wildebeest.

Two of the family groups retreated from cattle sounds by walking, and 11 retreated by running. Members of 10 of the 13 family units turned to face the sound as they were retreating, and members of 10 of the family units were seen smelling after retreating. Family units were more likely to be standing in defensive formation at the end of a cattle bell experiment than at the end of a wildebeest playback.

Typical Bull Group Response to Playbacks

Bulls in bull groups also appeared wary in response to the sound of cattle bells (table 20.3). They interrupted their

Table 20.3 Bull group responses to cattle bells and wildebeest playbacks

	Cattle bells ($N = 10$)	Wildebeest ($N = 6$)	G	p
Pausing activity and holding still	10	2	5.87	<0.05
Raising their heads	9	0	10.03	<0.005
Turning heads from side to side	4	1	0.18	—
Listening	10	3	3.32	—
Smelling				
a) trunks up	2	0	0.16	—
b) trunks down	7	1	2.48	—
Turning heads toward the speaker	7	0	5.54	<0.05
Grouping—moving toward other members of the family or group	4	0	1.6	—
Retreating from the sound				
a) walking	7	0	5.54	<0.05
b) running	0	0	—	—
Standing in a clumped group (defensive formation) at the end	2	0	0.16	—

Note: df = 1 in all cases.

original activity and stood still, raised their heads, listening and smelling as they did so. Some bull groups turned their heads from side to side, as had been observed for females, and some turned their heads toward the speaker. Some turned away from the sound, and members of four of the bull groups moved toward other members in the group. Not all bull groups retreated from the sound, but those that did retreated in a slow walk. As the bulls retreated, some members of the group turned to look back in the direction of the speaker, while others paused, smelled, and listened during retreat. Males tended to be standing in a more dispersed group at the end of the cattle bell experiment than had been observed for females.

A few of the bull groups listened to the wildebeest rutting sound, pausing their original activity. Smelling with trunks down was observed in one of the bull groups. None of the bull groups turned to face the speaker, nor did any retreat from it.

Males were more likely to pause their activity and hold still in response to cattle bells than in response to wildebeest sounds (table 20.3). They were also more likely to raise their heads in response to cattle bells than in response to wildebeest sounds, although there was no significant difference in head turning from side to side in response to cattle bells and wildebeest.

Head turning toward the speaker was more likely in response to cattle bells than wildebeest sounds. Males were also more likely to listen in response to cattle bells than in response to wildebeest sounds. Smelling with trunks up was as infrequent in response to cattle bells as in response to wildebeest sounds, while smelling with trunks down was as likely to occur in response to cattle bells as in response to wildebeest sounds. Males were more likely to turn away from the speaker in response to cattle bells than in response to wildebeest sounds. Movement of some of the members of a bull group towards other was as likely to occur in response to cattle bells as in response to wildebeest sounds.

Seven out of 10 of the bull groups retreated from the cattle bells, whereas none retreated in response to the wildebeest sounds. All seven of the groups that retreated from the cattle bells walked away. Members of three of the bull groups turned to face the sound as they retreated, while all seven of the groups that retreated stopped and smelled as they retreated. Bull groups were as unlikely to be standing in defensive formation at the end of a cattle bell experiment as at the end of a wildebeest experiment.

Sex Differences in Response to Cattle Bells

All family units and all bull groups stopped their activity and held still in response to cattle bells. There was no significant difference between the proportion of family units and bull groups that raised their heads (G statistic = 0.32, df = 1, NS), turned their heads from side to side (G = 3.23, df = 1, NS), and turned their heads toward the speaker (G = 0.11, df = 1, NS).

All family units and bull groups were observed to be listening. There was no significant difference in the proportion of family units and bull groups that smelled with their trunks up (G = 1.52, df = 1, NS) and with their trunks down (G = 0.01, df = 1, NS). Family units were, however, more likely to turn away from the speaker than bull groups (G = 4.07, df = 1, $p < 0.05$). There was no significant difference between family units and bull groups in terms of likelihood to retreat from the sound of cattle bells (G = 2.33, df = 1, NS). However bull groups that retreated were more likely to do so by walking (G = 5.11, df =1, $p < 0.05$), while family units were more likely than bull groups to retreat at a run (G = 15.14, df = 1, $p < 0.005$).

The latency to retreat time was longer for males than for females (Mann-Whitney: $z = 2.82$, $p < 0.05$). After retreat, family units were much more likely to be seen standing in defensive formation than were bull groups (G = 10.24, df = 1, $p < 0.005$).

Both family units and bull groups showed avoidance behavior in response to Maasai-associated sounds. Family units were more wary of cattle bells than were bull groups, with a higher proportion of them retreating at a run. This response could be due to the vulnerability of females with their young to predation from the Maasai. The pronounced maternal defense and protective behavior of elephants has long been noted (Douglas-Hamilton 1972; Moss 1988; Lee 1987; Oldfield 1988) and might result in the females' immediate avoidance of any contact with Maasai, a trait not exhibited in males. Sex differences also occur in the behavior of Asian elephants: pubertal and adult male elephants incur greater risks of contact with humans than females by crop raiding more frequently than females (Sukumar and Gadgil 1988; Sukumar 1989, 1991). The perceived threats may be greater for females living in stable family groups than for males living singly or in loose associations because more survivors will remember the event, especially the older females with experience of previous attacks. Male spearing experience is more likely to be lost because males are often on their own, and if they are speared and die while alone, the event is "forgotten." This might also explain the greater avoidance of Maasai by females.

The avoidance behavior exhibited by the elephants in response to playbacks suggests that they aim to avoid encounters with the Maasai. The fact that elephants simply move out of the immediate vicinity of the Maasai, running for 300 m at the most and stopping and waiting, indicates, however, that their strategy may be one of avoiding close contact rather than a total spatial avoidance. Such behavior may explain the presence, although decreased, of elephant dung near Maasai *bomas* and the use of these areas at night as well as the avoidance of a watering place by elephants at times when Maasai are likely to use it.

Conclusions

In this study, Amboseli elephants avoided contact with Maasai at three levels. First, the elephants avoided the immediate vicinity of Maasai pastoralist settlements, approaching these areas at night when Maasai and their livestock were enclosed in their homesteads and the threat of meeting people was minimal. Second, the elephants showed significant flexibility when forced to share a water source with the Maasai and their livestock. They used the same watering point at low-risk times and waited nearby, sometimes in large aggregations, when Maasai and their livestock were present at the watering place. Third, the elephants showed a fear response to playbacks of sounds associated with the Maasai and their cattle, with females being more wary than bulls, presumably due to the risk of mortality to their calves.

The elephants' fear response and avoidance behavior have persisted beyond the early 1990s when this fieldwork was carried out. In the late 1990s, elephants were observed forming large aggregations in response to the threat of meeting Maasai (Mutinda, Poole, and Moss, chapter 16), and recent experimental results (Bates et al. 2007) indicate that the Amboseli elephants continue to show a fear response to Maasai. Spearing of elephants appears to have resulted in the elephants generally avoiding contact with the Maasai. A consequence of the elephants' avoidance behavior is that the Maasai and the Amboseli elephants share a range and use resources at different times, reducing the incidence of Maasai-elephant contact and any conflict that may result.

At the time the fieldwork for this study was carried out, Maasai spearing had resulted in only minimal losses to the elephant population (Moss 2001). I made the case for tolerating this traditional practice, which has facilitated the coexistence of people and elephants (Kangwana 1993). I also argued that tolerating the Maasai spearing of elephants may serve conservation interests and be the preferred direction for elephant management in Amboseli (Kangwana 1993). Over the last 15 years, however, the complexity of elephant conservation in Amboseli has increased. Land tenure has changed from communally owned group ranch land to individually held plots (see Kangwana and Browne-Nuñez, chapter 3), making it necessary to renegotiate the availability of land around the Park for elephant conservation. The past 10 years have also seen an increase in the amount of agriculture practiced around Amboseli, making it more critical to explore means and mechanisms by which to solve the problems caused by crop-raiding elephants (see Browne-Nuñez, chapter 19). Although the impact of spearing on the Amboseli elephants seems insignificant compared with these new pressures on the elephant population, the issues surrounding the practice of spearing in its traditional context of *moran* proving their bravery are important to elephant conservation policy.

Human society is not static, and the nature of Maasai society in particular is in great flux. The traditional structures of Maasai society—its age sets and specialized roles (see Kangwana and Browne-Nuñez, chapter 3)—developed at a time when conflict between tribes was prevalent. In this context, it was necessary to dedicate strong, young men to the protection of the tribe, and their feats of courage, such as spearing wildlife, were good preparation for their role.

Present-day Maasai society is becoming more sedentary, entering a commercial economy (see Kangwana and Browne-Nuñez, chapter 3), and is no longer engaged in the tribal warfare of the late 1800s and early 1900s. In this changing Maasai society, it is reasonable to question the necessity of killing wildlife in order to prove bravery. There may well be a need to mark the rite of passage between boyhood and manhood with acts of courage, but while the beauty of humanity's cultural heritage must be preserved, perhaps the form of such rites can be recast. Can bravery be demonstrated without killing wildlife? It may be time to reassess the spearing of wildlife and its consequences.

There are other issues to consider with regard to the spearing of the Amboseli elephants. Specifically, we need to examine the role of culture in defining human experience. Among the Maasai of Amboseli, it is culturally appropriate to kill wildlife in order to prove bravery, but many would argue against this practice. Is culture the ultimate standard by which to judge human behavior, or can other standards prevail? The international community has formulated declarations of universal rights that transcend culture, including the Declaration of the Rights of the Child, the Universal Declaration of Human Rights, and the Declaration on the Elimination of Violence Against Women. We can ask whether human society is far from defining universal standards for the treatment of animals. In the case of violence against women, the international community recognized the detrimental effect of acts of violence on the men that perpetrate these acts (see United Nations General Assembly Resolution 48/104 of December 20, 1993). Are there likewise negative effects of spearing elephants or other wildlife on the young men who spear? This debate is just beginning, and its outcome will affect the Amboseli elephants.

Conservation policy cannot ignore the ethical issues involved in spearing elephants. Many of the ideas that humans hold about the relationship between people and nature were conceived and codified at earlier stages in the development of human society. What was considered appropriate treatment of animals in former times may not be acceptable now, and it is clear that human society is redefining its relationship to animals at many levels. Hunting for sport, while supported by some, is viewed as unethical by others. Society is also establishing new standards for the treatment of animals in zoos, for the use of animals in circuses, and even with regard to practices in the husbandry of animals for food. These changes reflect a shift in our view of "nature" merely as a commodity for human use to a recognition that humans have a stewardship role in taking care of nature. The refinement of society's view regarding its relationship with nature will no doubt inform practices such as elephant spearing.

An exhaustive discussion of the issues raised here is well beyond the scope of this chapter. Nevertheless, even this cursory examination consistently demonstrates that any discussion of what appears to be a very localized phenomenon of Maasai spearing elephants in Amboseli rapidly becomes a global concern with implications of the widest scope. Furthermore, it is clear that as the Maasai, conservationists, and the world community come together as interested parties seeking solutions, the most satisfactory solutions will be those that uphold social justice: meeting the needs of the people that live with wildlife, conserving the Amboseli elephant population, and reflecting the realities of an evolving human society.

Far from recommending practical solutions to conservation problems, I have raised some of the questions that face us as we strive to make space for elephants in human society. Practical solutions will emerge from consultation only when principles guide the discussion and the fundamental issues are explored fully.

Chapter 21 Ethical Approaches to Elephant Conservation

Joyce H. Poole, W. Keith Lindsay, Phyllis C. Lee, and Cynthia J. Moss

Intelligent, social, emotional, personable, imitative, respectful of ancestors, playful, self-aware, compassionate—these are qualities that would gain most of us membership to an exclusive club or at least allow us to be treated decently at its portals. They also describe elephants.

In this chapter, we apply knowledge of elephant behavior and ecology, reproduction and cognition, gained from our studies and those of many others to examine the ethical issues involved in the treatment and management of wild elephants.

The notable capacities of elephants among non-humans for memory, thinking, emotion, awareness, and sociality have been outlined by Byrne and Bates (chapter 11). Elsewhere we have detailed the behavioral and biological requirements for space and stimulus that must be met for elephants' health and well-being (Poole and Moss 2009; Poole and Granli 2009; Lee and Moss 2009) as well as their susceptibility to psychological disturbance that can result from disrupted social and age structures (Bradshaw et al. 2005; Poole and Thomsen 1989).

We contend that in view of the combination of sentient attributes of elephants, they deserve serious moral consideration in our behavior toward them and in framing our approaches to their conservation (see also Poole and Moss 2008).

As we write, the Amboseli Elephant Research Project (AERP) is entering its 38th year. The relationship between elephants and people—in the landscapes they share in Amboseli, in Africa, and globally—is becoming ever more challenging as pressures on land continue to grow. Their future now crucially depends on how elephants are viewed by the people who coexist with them, as discussed by Kangwana and Browne-Nuñez (chapter 3) and Browne-Nuñez (chapter 19) as well as by politicians, planners, and managers working to reach compromises between sustaining biodiversity and meeting ever-increasing human needs (see Croze and Lindsay, chapter 2; Croze, Lindsay, and Moss, chapter 22). At the same time, the ethical dimension to how humans perceive and value non-humans outside the utilitarian domain (e.g., Arluke and Sanders 1996) has become increasingly important.

Since the AERP's inception, much has changed in scientific attitudes toward animals, not to mention those of the general public. When Jane Goodall carried out her pioneering investigations of chimpanzees in the early 1960s, she was criticized by her scientific peers for giving her study animals names instead of numbers (Goodall 1990). The concern was that by naming them, the chimpanzees would acquire anthropomorphic personalities. Recent research has demonstrated that not only do more than 100 species of animals have distinct and statistically validated personalities but also these characteristics have an evolutionary, adaptive function (Dall, Houston, and McNamara 2004). No longer are such perspectives considered subjective; understanding individuality and emotion informs our research on animals and provides the background for an ethical understanding of our relationships with them.

Today, psychology and philosophy departments at major universities offer courses covering animal welfare and theories of animal rights. And an ever-increasing number of scientific journals now cover these issues, too (e.g., *Etica and Animali, Animal Welfare*). Attitudes have gradually shifted, influencing the general public and conservation and management policies for many species of wildlife. Our growing understanding of elephants and the work from Amboseli underpin broader changes in attitudes toward animals (see Poole and Granli 2005).

An example of how far attitudes have shifted during the course of our long-term study can be found in the comparison of two contrasting positions held by elephant management professional bodies. At the 1987 meeting of the African Elephant Specialist Group in the context of southern African countries' desire for a continued ivory trade, members of the AERP asked the group to address the ethics of killing elephants for commercial gain. This appeal was immediately and forcefully rejected, since any discussion of ethics was considered irrelevant at that time. Yet, 12 years later, South Africa's National Parks agency produced a new proposal for management policies for Kruger Park's elephants (Whyte et al. 1999) in which the concept of "ethical considerations" was used throughout. And in 2008, ethical issues were central to the development of the Norms and Standards for the Management of Elephants in South Africa. A thorough discussion of ethical issues surrounding the management of wild elephants specifically in South Africa can be found in Lötter et al. (2008).

The treatment of elephants held in captivity is of equal concern and complexity, and some of us have addressed those issues elsewhere (e.g., Poole and Moss 2008; Lee and Moss 2009; Poole and Granli 2009). Indeed, ethical issues surrounding the treatment of elephants in general have become so topical that in 2008, more than 40 leading elephant biologists signed the online *Elephant Charter* (www.theelephantcharter.info). Written as a consensus on the nature of elephants, the charter is a set of guiding principles for anyone needing to address elephants' interests and is intended to promote scientifically sound and ethical management and care of elephants, whether they be in captivity or in the wild.

Attitudes continue to shift, but are they changing fast enough to make a difference for the lives and very survival of elephants? At the moment, the answer to that question appears to be mixed, and it may be resolved only with the passage of time and future developments in human society. Our position is one of cautious optimism, balanced by pragmatic concern over the enormous challenges (see Croze, Lindsay, and Moss, chapter 22).

In this chapter, we consider the treatment of wild elephants. In doing so, we examine several topics that have significant consequences, and we approach each from an elephant's point of view: How will the consequences affect elephants individually or collectively, physically and psychologically? And what might be the long-term consequences? We consider the exploitation of elephants for commercial gain, including the ivory trade, sport hunting, and the capture of wild elephants; the human and elephant needs for land; and the methods for addressing locally high densities of elephants, such as translocation, birth control, and culling. In all these areas, there may be ethical judgments regarding the trade-off of interests between elephants and people. We recognize that conservation managers in many parts of Africa, after considering the options for addressing the challenges of conflict and coexistence, have adopted approaches that may involve intensive intervention in the lives of elephants. The reasons for these decisions from particular management perspectives have been reviewed thoroughly elsewhere (e.g., Scholes and Mennell 2008), and we will not rehearse them in this chapter. Based on our understanding of elephants, their sociality and their intelligence, our premise is that we should do everything we can to avoid both their immediate distress and suffering and to minimize long-term trauma (e.g., Bradshaw 2004; Bradshaw et al. 2005). We write this chapter in the hope that the lessons we have learned from our own study, as well as those of others, can inform better elephant conservation and management practices.

Commercial Exploitation of Elephants

Elephants in both Africa and Asia have been taken from the wild and trained and exploited by human beings in many ways for at least 4,000 years (Meredith 2001; Sukumar 2003; Hart and Sundar 2000). Elephants have never bred well in captivity (Sukumar 2003), and consequently, a continuous supply of wild elephants has been needed in order to maintain or increase the captive stocks. The offtake of elephants from nature—to fight wars, carry timber, build monuments and canals, and parade in ceremonies—historically has been so great that elephant populations on the subcontinent of India were locally depleted (Sukumar 2003). African elephants were used by Hannibal in his campaigns against Rome, but there was no widespread capture on the scale of their Asian counterparts. In the last century or so, however, both species have been captured in order to perform in circuses, pose in zoos, and more recently, to carry tourists on safari. Throughout this period, elephants have never been domesticated (i.e., bred selectively over generations for adaptation to the conditions of captivity) but have remained wild animals that have been "tamed" to endure confinement and human domination (Csuti 2006).

The greatest exploitation of elephants has been and continues to be the killing of them for their ivory tusks, their skin, and their meat. In addition, they are killed for the amusement of sport hunters and simply because they are in the way of human development.

The Ivory Trade

African elephants have been killed for their ivory for millennia, but such exploitation reached new and devastating peaks during the European colonization of Africa during the 18th and 19th centuries, generally in association with the traffic in human slaves and/or colonial expansion (Moore 1931; Meredith 2001). With the abolition of the slave trade, coupled with the fact that elephants had by then been exterminated from most parts of Africa (Spinage 1973), the ivory trade declined in the early 20th century. Then, in the 1970s and 1980s, it resurfaced, swept across the continent, and significantly impacted elephant populations in much of West, East, and Central Africa (Cobb 1989). The ban on international trade in ivory by the Convention on International Trade in Endangered Species of Wild Fauna and Flora (CITES) in 1989 resulted in a dramatic drop in elephant killing, but with periodic sales of ivory stockpiles in southern Africa and an increase in demand from growing economies in Asia in recent years, there has been a resumption in illegal killing, amounting to the deaths of perhaps as many as 38,000 elephants per year across Africa (Wasser, Clark, and Lurie 2009; Croze, Lindsay, and Moss, chapter 22). These waves of killing for ivory have often been carried out by armed gangs of bandits or combatants in local conflicts, who use the sale of "blood ivory" to fund their expansion of territory. Thus, the parallel suffering of people and elephants has characterized the history of the ivory trade.

Illegal hunting, or poaching, to feed the commercial trade in ivory has always been extremely damaging to the very fabric of elephant society because it removes key segments of the population; killing elephants primarily for their ivory targets older, more experienced individuals (Poole and Thomsen 1989; Poole 1989c; Abe 1995; Gobush, Mutayoba, and Wasser 2008). The tusks of elephants continue to grow throughout life, with the result that older individuals carry the heaviest and most desirable tusks; since males grow to a much larger size than females, their tusks are correspondingly larger. Thus, under poaching pressure, many populations lose the majority of their adult males and many of their older females (e.g., Poole 1989c). In a normally functioning elephant society, old males are the primary breeders (Poole 1989b; Hollister Smith et al. 2007; see Poole et al., chapter 18), while old matriarchs hold the knowledge necessary for finding food and water as well as providing the backbone of social structure (McComb et al. 2001).

During the recent burst of poaching in the 1970s and 1980s, the killing progressed to such an extent that many populations were deprived of almost all of the older adult elephants (Poole 1989c), resulting in devastating consequences for the psychological state and the very survival of remnant families and individuals. In some of the worst hit areas, groups of orphans wandered in bunched, leaderless groups (Poole 1989c). Remaining females and calves formed into small units with low genetic relatedness, low cohesion, and an atypical age structure (Gobush, Mutayoba, and Wasser 2008) or into large aggregations (Abe 1995). This kind of disruption has knock-on effects; calves who are orphaned at under two years of age almost always die, while those who lose their mothers during immaturity (<10 years old) appear to have a shorter lifespan than do their peers with surviving mothers (see Lee and Moss, chapter 17).

How long will it take a population to recover, socially as well as in terms of numbers, from the effects of selective killing for ivory? Even 25 years on from the peak of poaching, some populations still exhibit social and reproductive traces of disruption (e.g., Mikumi-Selous, Tanzania: Gobush et al. 2008) while others have shown some capacity to rebound, restructure, and reorganize socially and in terms of numbers (e.g., Queen Elizabeth National Park, Uganda: Nyakaana et al. 2006). The social flexibility, longevity, and cognitive capacities of elephants, combined with potentially high reproductive rates under the best conditions, suggest that populations, when protected, can recover at least to some extent; the key elements here are protection and leadership. Data from Amboseli suggest that families whose matriarchs survive a period of poaching are likely to have a better chance of recovery than those whose matriarchs are killed (McComb et al. 2001). Yet, without long-term data such as that from Amboseli, it is difficult to assess fully how the loss of knowledge after the death of matriarchs affects long-term success; only long-term data can capture the kinds of rare events (drought, changes in water availability, etc.) that require ecological memory in order for individuals to respond appropriately. Nor can we assess easily the genetic consequences of the loss of diversity associated with the death of primary breeding males. We can speculate that this latter effect could ripple through populations for years, reducing viability and immune function (Gobush, Mutayoba, and Wasser 2008).

The inevitable consequence of local or global promotion of the use of ivory, and the legitimization of its commercial value and trade, is the stimulation of demand and an increase in the killing of elephants, with unavoidable short- and long-term damage to the societies of elephants in which the killing occurs. Some African countries may be able to protect their elephants from illegal killing and may wish to sell the ivory stockpiles that accumulate from elephants that are killed "on control" or from culling operations (see below), or from those who die natural deaths.

Other countries, even without adequate protection in place, may wish to sell their stockpiles in order to benefit from the earnings such sales would bring. It is said that such income from ivory sales could be applied to elephant conservation and benefit the communities that share their land with elephants (Barbier et al. 1990; Blignaut, de Wit, and Barnes 2008). However, the revenue from ivory sales is occasional rather than sustained and contributes small amounts of funds relative to the needs for financing conservation (Bulte and van Kooten 1999). At the same time, as long as there is a demand for ivory, in those countries that are beset with poverty, corruption, war, or simply limited resources, people will continue to supply ivory in order to meet the ever-higher demand, and countries will lose their elephants. It is for these reasons that we are opposed in principle to the commercial trade in ivory in any form.

Trophy Hunting

Trophy or sport hunting has, until recently, been considered a "manly" sport, and in addition to the prize of valuable ivory, the hunting of large and dangerous elephants has long appealed to the male ego and pride (Meredith 2001). As the glory associated with elephant hunting has declined, a variety of other reasons have been put forward to justify the activity, including

- killing the older individuals who allegedly would die anyway or are mistakenly thought to be reproductively senile individuals,
- reducing locally over-abundant populations,
- helping to eliminate crop raiders, and
- bringing much-needed revenue to local communities who suffer from the consequences of having to live with elephants.

These arguments have been summarized and analyzed on utilitarian and ethical grounds (Gunn 2001; LACS 2004). Our concern, and our evidence from experience, is that such selective hunting has a damaging effect on elephants' lives and society.

Traditionally, trophy hunters have targeted male elephants and, by definition, those with the largest tusks. As discussed earlier, these older males are the primary breeders in elephant society. There has been strong selection for longevity; far from being "dead wood," males who live to an old age produce a disproportionate number of offspring (see Poole et al., chapter 18; Hollister-Smith et al. 2007). Females, who must invest so much in each calf (22 months' gestation, 4 years of lactation, and 12 years of parental care), are selective breeders, choosing older males, who have proved they can live a long time, over younger males (Moss 1983; Poole 1989b). In this way, they may produce healthy offspring who can themselves live to an old age and produce many surviving offspring. By selecting older individuals, hunters are negatively influencing the genetic future of populations (see box 18.1).

Using trophy hunting to remove a relatively small number of males (young or old) does not reduce population size since any adult male is capable of mating and with a large number of females. Using trophy hunting to remove problem crop raiders may be seen as "killing two birds with one stone," but it is subject to the same drawbacks as control shooting of problem animals (see below), causing stress and disturbance to the surviving animals.

A final argument used to justify sport hunting is the value it may bring to local communities that would otherwise have only trouble from elephants (e.g., Bond 1994). When managed properly, this source of revenue could indeed assist local communities and promote positive attitudes toward coexisting with elephants. It is often the case, however, that much of the revenue does not make its way to the community level but rather benefits operators of hunting safaris, government officials at higher levels, and corrupt individuals (Muchapondwa 2003, cited in Blignaut, de Wit, and Barnes 2008; LACS 2004). In most cases, non-destructive wildlife-viewing commercial tourism and community-based ecotourism provide both greater direct financial returns to communities and greater benefits in terms of employment and infrastructure development than does trophy hunting (LACS 2004). Creating local economic opportunities from tourism, whether through trophy hunting or other forms, is clearly a governance issue at local, regional, and national levels (Nelson 2004).

Where elephants are regularly hunted, general stress levels are raised, spreading disturbance through the population. Surviving animals become increasingly fearful and/or aggressive toward humans, potentially putting local populations at risk, jeopardizing any potential benefits, and increasing people's antagonism toward the elephants (e.g., Bradshaw 2004). From the elephants' perspective, the killing of mature individuals is likely to have only negative consequences on the lives of the targeted animals and the psychological well-being and genetic health of affected populations. For these reasons, we cannot condone sport hunting of elephants under any circumstances.

The Capture of Calves

The tradition of capturing elephants for human exploitation began more than 4,000 years ago in South Asia, and historically, more Asian than African elephants have been captured and tamed. In recent years, a substantial number of African elephants have been captured for the purpose of selling them to zoos, circuses, and safari parks and to use on elephant-back safaris, generally by private-sector traders who have paid relatively small fees to management authorities or landowners. There have been several highly publicized cases in recent years:

- The so-called Tuli Debacle in 1998 in which 30 calves were abducted from their families in the Tuli Block of Botswana by a South African animal dealer for sale to zoos and circuses (Poole, Moss, and Sheldrick 1999)
- The Selati case in 2006 in which 9 calves were captured for the purpose of sale to elephant-back safari companies in South Africa (van Wyk 2006, cited in Lötter et al. 2008)
- The Shearwater Adventures case in 2006 in which 12 juveniles were captured from Hwange National Park, Zimbabwe
- The Sondelani case in which 10 young elephants were captured in 2008 (ZNSPCA 2009; see also www.elephantvoices.org/news-media-a-reports/97-welfare/689-latest-news-of-the-zimbabwe-9.html).
- The capture of 4 calves in Tanzania in 2009 who were sent to a zoo in Pakistan

This form of exploitation of elephants is particularly repugnant because its practice flies in the face of everything we now know about the social and developmental needs of elephants.

It is clear from years of observations in Amboseli that calves who are orphaned (i.e., those whose mothers have died) become listless and appear depressed (also reported by Daphne Sheldrick, personal communication, 2006). We also know that separating young elephants from their mothers and all other family members can cause lasting trauma and result in abnormal behavior (Bradshaw et al. 2005; Slotow et al. 2000; see Lee and Moss, chapter 14). As mentioned above, our data suggest that some form of physiological or psychological damage results even when such orphaned calves remain with the other relatives in their close families.

Young elephants learn normal behavior in a social context (Lee 1986; Lee and Moss 1999). If removed from a context in which they have an older, experienced individual (a role-model or a "teacher"), they are likely to engage in inappropriate responses to their physical environment, to take foraging risks, or possibly even to starve (Lee 1987). In addition, the trauma of social loss may be even more significant. Young elephants rely on their social companions to learn appropriate behavioral responses to others (Lee and Moss 1999). They are in continual olfactory and vocal contact with mothers and others (Poole, chapter 9) and remain within two meters of their mothers or another caretaker for most of the first five years of their life (Lee 1986). They follow their mothers' social responses and learn who their relatives and friends are and who represents potential threats. In the complex social world of an elephant, the presence of older family members ensures normal, friendly social behavior and reduced levels of aggression. It allows for observation of sexual behavior between adults and the practice of appropriate actions during play—a non-threatening context for learning about size, strength, and the suitable level of physical contact (Lee 1986). Contact with other juveniles during play or caretaking provides vital experience in rearing calves, which is essential to subsequent reproduction and non-abusive rearing of infants (Lee 1987, 1989).

These complex biological and emotional responses of calves (and their mothers) to separation have evolved for a reason. Every elephant calf is biologically extremely important to its mother, representing at least 12 years of rearing investment and protection. As a consequence, elephants have evolved extraordinarily developed behaviors of caring and bonding with their calves (Lee and Moss 1986; Lee 1987; Moss 1988). If a calf is to survive to adulthood, it too must form intense close bonds with its mother and other family members. These bonds involve strong emotional attachment, which if broken cause individuals intense suffering (Moss 2000; Poole 2000b).

In addition, juvenile females act as allomothers, who care for younger calves. The presence of these older sisters and helpers is statistically significant in keeping calves alive (see Moss and Lee, chapter 12). This direct effect on survival is one aspect of the role that allomothers play; they may also free the mother from infant-care duties so that she can spend more time feeding, improving her own nutrition and the quality of the milk that she produces. Young elephants of both sexes act as play partners for their younger siblings, with the benefits for social development noted earlier. Removal of these juvenile animals has a significant impact on the social cohesion and behavioral well-being of the family as a whole.

Due to the long-term effects of trauma, there is the additional problem of a potential increase in aggressive behavior of both the captured animals and their remaining families. Elephants who are "broken" by trauma and physical abuse and trained through fear are perfectly capable of retaliating much, much later on the humans who might be caring for, working with, or riding them. On the other hand, the family members remaining after the removal of juveniles are also likely to be deeply affected. The capture operation is traumatic in itself, involving chasing, noise, and frightening human activity over a period of several hours, and the elephants will not forget it.

For all these reasons, the capture of elephants from the wild for any purpose—whether for circuses, zoos, elephant-back safaris, or other activities—is ethically unjustifiable, and we cannot condone this practice.

The Needs of People and Elephants

Land Use

Discussions about land use and the importance of planning for integrating people and wildlife in the Amboseli ecosystem and in a broader context have been covered in detail in chapters 3, 19, and 22. The expansion of human populations and human land use into areas of elephant and other wildlife habitat is a serious challenge facing wildlife conservation in general but appears particularly acute for species that require sizable areas of open land, such as large predators and elephants. We return to these issues in the final chapter but suggest here that these discussions have an ethical as well as a practical dimension.

On the one hand, a community's economic survival, individual people's livelihoods, and even human lives are at stake. On the other hand, a national heritage, a threatened species, and other beings' lives are in jeopardy. Across Kenya, and indeed across Africa and Asia, this scenario of conflict is increasing as long as human populations continue to grow and convert elephant range for human use (see Croze, Moss, and Lindsay, chapter 22). In each area, elephants are the ultimate losers and will continue to be so unless people and policymakers are willing to develop a new, long-sighted, and indeed ethical perspective that recognizes the limits to human growth and the intrinsic value of elephants and their right to a place in Kenya and elsewhere.

Thoughtful and participatory zoning of Kenya's land into clear areas suitable for forestry, agriculture, livestock, settlement, and urban development or wildlife conservation would make an essential contribution to the development of strategies for improving the livelihoods of the rural poor while promoting the conservation of elephant (and other wildlife) habitat. Difficult decisions must be faced and made regarding the trade-offs that may well involve on the one hand, the removal of small, isolated elephant populations from areas of encircling farmland and on the other, the securing of corridors and elephant range in areas owned by people (Croze, Lindsay, and Moss, chapter 22). However hard these may be to make, such decisions within an equitable land-use planning framework would serve to reduce the likelihood of future conflicts and would be a much more satisfactory, sustainable, and preventive—and at its core, more ethical—approach to the coexistence of elephants and people. Such an approach should reduce the current suffering and preclude the loss of both human and elephant lives and livelihoods.

Human-Elephant Conflict and "Problem Animal" Control

Conflict over land and resources occurs when effective land-use planning has failed or is absent entirely. From the perspective of elephants, the succulent plants of agricultural fields may be a logical feeding alternative at certain times of year (Osborn 2004), notably when wet-season abundance of nutritious vegetation in natural habitats gives way to dry-season shortage (Lindsay, chapter 5). There is not "something wrong" with the elephants but rather "something different" in their local environment. To an individual farmer, elephant-related losses may be dramatic, but in a given area, they may not be as significant as other causes of crop failure and are often exaggerated in media reports or by politicians wishing to score points (Naughton-Treves, Rose, and Treves 1999; Sitati et al. 2003; Lee and Graham 2006). The same attributes apply to conflicts between elephants and livestock interests; incidents of livestock and sometimes human loss of life are certainly momentous—and in the case of humans, tragic—but they are relatively rare when compared with other mortality factors (Malima, Hoare, and Blanc 2005). The alternatives available to farmers and herders and to authorities responsible for their welfare or that of wildlife include preventive methods, the deterrence of animals that have become a "problem," and finally, the killing of the offenders. Clearly, the soundest approach on ethical grounds would be prevention rather than reaction, particularly when it is a fatal reaction, and the reframing of "problem animal" control as rational problem solving. It is our belief that such a preemptive approach is not only more defensible ethically but also more effective.

In the case of crop raiders, Hoare (2000) has clearly shown that killing individual males does nothing to reduce the broader crop-raiding problem. While shooting of "problem animals" is certainly effective as an immediate, short-term solution and may be seen by both conservation authorities and affected communities as a reassuring gesture, crop raiders are simply replaced by others who move in to take advantage of the new opportunities. It is more accurate to identify the heart of the problem as being the presence of the farms, which most likely will not go away, rather than the elephants, which can be encouraged to do so.

In the longer term, methods including early warning systems (Sitati, Walpole, and Leader-Williams 2005; Granli and Kioko 2006a, 2006b; Graham et al. 2009); deterrence with noisemakers, pepper spray, or beehives (Kangwana 1995; Osborn and Parker 2002; Vollrath and Douglas-Hamilton 2005); fencing with electric (Kioko et al. 2008) and chemically treated barriers (Osborn and Parker 2002); and support for the planting and marketing of unpalatable crops (Nelson, Bidwell, and Sillero-Zubiri 2003) can be deployed in order to separate farmlands from elephant incursions. A number of approaches, including barriers and alternative resource provision, may be used to protect livestock watering points and other sites of conflict.

Compensation schemes to replace losses caused by elephants and other wildlife have also been attempted with varying degrees of success (Nelson, Bidwell, and Sillero-Zubiri 2003), most failing because of administrative and financial barriers and the creation of a disincentive to protect crops or livestock properly. Consolation payments have also been applied uniquely in Amboseli (box 19.1). This approach involves a fixed payment as a gesture of goodwill to "console" the people affected for the lost property. Community-based crop/livestock loss insurance schemes, which involve the active investment of resources by the landowners, have been initiated in Namibia (MET 2005) and in Amboseli-Mbirikani (Soila Sayialel, personal communication) and are showing early promise.

It is important to note, however, that all approaches to resolving human-elephant conflict should be accompanied by agreements negotiated with landowners to moderate their expansion of croplands or settled livestock husbandry. In the absence of an acknowledgement that there is a place on the map for elephants, the successful protection of human economic interests will simply result in expansion of land conversion and further reduction of elephant range (Osborn and Parker 2003).

The success or failure of all these actions depends on the consistency of their application, the resources available to maintain them, and their "ownership" and maintenance by the people directly affected, whether farmers or pastoralists. The hoped-for solutions to human-elephant conflict generally require long-term, committed engagement by conservation authorities and non-governmental organization partners with the communities sharing elephant range and the development of realistic incentives, trust, and reciprocal respect for the values held by all parties.

Locally High Elephant Densities

When fences, failure to control incompatible land use, or the danger of hunters (either for ivory or for "problem animal" control) compress or confine elephants within isolated habitat areas, their foraging is likely to have accelerated effects on plant and dependent animal communities. Managers may deem these impacts undesirable, especially if the area contains rare, threatened, or endemic plant and animal species. In such situations, the elephants are described as "overabundant" by local or regional managers (Balfour et al. 2007), and actions may be prescribed to control elephant density in either the entire area or key, localized parts in order to protect the impacted habitat or species. These issues are discussed in other chapters (Croze and Lindsay, chapter 2; Croze, Lindsay, and Moss, chapter 22), but since there are ethical concerns over some of the methods employed by managers to control elephant densities, they are also discussed here. Elephants have value in the important ecological roles they play in shaping habitat structure and heterogeneity as well as in their intrinsic qualities as remarkable animals. Manipulation of their behavior, their physiology, and their very existence should not, we feel, be undertaken lightly without serious consideration of all alternatives.

As noted in chapter 2, management of nature by artificial control of animal density is characteristic of a farming or gardening mindset, which is increasingly seen as misguided. This "command-and-control" approach is being replaced in Africa and elsewhere by the understanding that nature is characterized by constant change (see box 2.1). The "non-equilibrium" viewpoint favors working with ecological processes rather than taking action against individual components of ecosystems. "Overpopulation" is a symptom, as is human-elephant conflict, of the problems caused when humans occupy, dominate, and alienate landscapes from wildlife habitat so that key processes such as dispersal are disrupted. It seems that sustainable solutions to the undesirable effects of elephants on habitats and other species are more likely to be successful when managers treat the underlying causes of the problem rather than just the symptoms.

Alternatives for control of locally high densities include the following (in rough order of increasing artificiality and decreasing ethical soundness):

- increasing the size of the protected area and/or maintaining a "meta-population" landscape with dispersal linkages to other sanctuaries (van Aarde and Jackson 2007);
- manipulation of water supplies in order to alter elephant access to a key resource and thus their effects on nearby ecological communities (Owen-Smith 1996);
- nonlethal deterrence from key habitat areas by means of similar methods as in human-elephant conflict;
- fencing or erecting barriers around the selected habitats to exclude elephants (e.g., moats, ditches, electric fences: Western and Maitumo 2004);

- contraception, generally with drugs or serum administered by projectile darts;
- translocation of adults or family units in order to reduce numbers; and
- culling in order to reduce numbers rapidly.

The issues of concern from an ethical perspective include deterrence shooting (discussed earlier in relation to human-elephant conflict), contraception, translocation, and culling.

Contraception

Delsink et al. (2006) used contraception of female elephants as a non-lethal mechanism to control population growth rate in a small fenced private reserve (Makalali, South Africa). The contraception is based on *Porcine zona pellucida* vaccine, which triggers an immune response that inhibits sperm binding and blocks fertilization in a variety of mammals. This contraceptive has been used, with apparently few long-term negative consequences, in a variety of captive and wild ungulates (Kirkpatrick et al. 1995).

Drawbacks include the potential for ovarian abnormalities, permanent infertility, and the need for a highly invasive initial immobilization followed by subsequent booster administration, which causes stress as well as abscesses at the site of the injections. A major ethical consideration is that the treated females may continue to exhibit estrus every four months, and thus females who have been subjected to contraception measures will be chased, harassed, and mated by males far more frequently than would be expected with a typical 50 percent to 75 percent conception probability at any estrous event (Moss 1983). The energetic, social, and mating stresses on females experiencing constant cycles need careful long-term assessment. A variety of other hormone-based female contraceptives are being explored for small, fenced, and essentially captive populations (see Bertschinger et al. 2008).

For males, while there are some potential hormone treatments, most work to date has been in the area of surgical sterilization (Bertschinger et al. 2008). Since the testes of male elephants are intra-abdominal, such techniques are highly invasive, causing pain and suffering. Mark Stetter, Disney World's Animal Kingdom Head Veterinarian, who is helping to develop a program of surgical sterilization for wild elephants in South Africa, is quoted as saying that an elephant's testicles, which are "the size of cantaloupes, are behind two inches of skin, a foot of muscle and four inches of fat" (www.foxnews.com/story/0,2933,219702,00.html). Yet, despite this highly invasive two-hour surgery, no assessment has been made of how many males would need to be sterilized in order to reduce conception probabilities. No modeling of any genetic consequences has yet been carried out. Indeed, simply looking at the costs to individual males in terms of pain and suffering imposed against minimal benefits suggests that this technique is not ethically acceptable.

Translocation

We have several misgivings about translocation being used as a method to reduce elephant numbers. One is the enormous cost of a translocation operation: the money might be better spent on *in situ* protection or the securing of corridors to link to other elephant ranges. The cost-benefit assessment of returns as a result of removing and moving a small number of individuals suggests that unless the alternative is local extirpation of a population, translocations are difficult to justify.

A major problem associated with translocations is the stress placed on the individuals during the capture, transport, and release process. While overt signs of stress may be mitigated through handling and release procedures (e.g., Dublin and Niskanen 2003), the long-term consequences of subclinical stress for health, welfare, and longevity have not been well considered in most translocation studies (see Teixeira et al. 2007).

A further issue to consider is the fate of the translocated individuals; they are placed into a strange environment, with no knowledge of the locations, distribution, seasonal variation, or constancy of food types or water sources. This loss of ecological information could be life threatening during extreme events such as droughts or floods. Among elephants moved from the Shimba Hills to Tsavo East National Parks in Kenya in 2005, Pinter-Wollman, Isbell, and Hart (2009) found higher mortality rates, sustained reduction in body condition, and differences in habitat preferences compared with the resident Tsavo population. They also suggested that it takes several years before individuals gain sufficient knowledge about their habitat to cope with even seasonal changes. Their results suggest that greater planning and follow-up is needed when assessing whether to translocate individuals as well as where and when they might be translocated.

Culling

The culling (i.e., killing in order to reduce local population density) of elephants has been practiced in Africa over the last century. In colonial days, it was routine to shoot elephants in order to manage their population or presence (see, for example, Laws, Parker, and Johnstone 1975). In Kruger National Park, the practice became an art. Helicopters were used to round up elephant "families" from which adults were darted with scoline, a drug that paralyzed the animals but left them fully cognizant. Ground crews then approached the elephants, shooting older juveniles and adults but sparing calves in order to sell them to zoos and circuses or to stock private reserves and national parks.

Although it is seen by many managers across Africa as the only method by which to reduce elephant densities rapidly (Balfour et al. 2007), culling of elephants is a wasteful action that is inhumane. It is also ineffective beyond the short term and perpetuates a need for constant intervention, since the reduction of numbers removes any potential density-related suppression of reproduction. Artificially maintaining elephant populations at low levels relative to plant resources is likely to prevent natural feedback on birth and survival rates. It is worth noting a recent discussion of the perceived need to cull elephants across the full extent of Kruger National Park (Owen-Smith et al. 2006). After thorough examination of the evidence, a panel of elephant experts concluded that there was no imminent risk of irreversible habitat change or urgent need for widespread population reduction; control of densities in only some localized areas was deemed worthy of consideration.

Culling is disruptive to elephant social structure and peace of mind; elephants are able to detect distress calls over large distances and are fully aware when their fellows are being killed. We value elephants as intelligent, highly social animals and believe that such exceptional, socially complex, and long-lived animals should be treated with respect and empathy and not simply executed for management expediency. In summary, we believe that to casually adopt the destructive practice of culling is unjustified and unethical.

Conclusions

Much of the ethical debate on the treatment of animals revolves around questions of individual suffering and whether individual elephants have any rights in a human-dominated world. Elephants have aesthetic and traditional value in the human domain, and their ecological and biodiversity value as biological entities is often overlooked. They are cognizant creatures. Do we have the right to impose our human value system on elephants, to control and manage them as we see fit? In a broader sense, reflecting upon our attitudes toward elephants leads us to question our approach to "dominion" over the world's resources and all its species. How heavy or light will the human footprint become as we develop an awareness of our collective impact on the planet's vital processes? The answer to these questions essentially determines how we will treat elephants—with either the respect they deserve or continued exploitation.

Elephants are intelligent and highly social and experience a wide variety of emotions and feelings, displaying qualities found only in cetaceans and higher primates. It is past time to accept that our treatment of elephants should be based on careful consideration of alternatives. In developing creative solutions to challenging problems, ethical concerns rather than simple expediency and narrow, self-serving human interests should be paramount.

Chapter 22 The Future of the Amboseli Elephants

Harvey Croze, Cynthia J. Moss, and W. Keith Lindsay

Elephants across Africa face mounting threats to their survival: from poaching for ivory, killing for meat, and competition for space. Although the details of the threats depend on local circumstances, in broad terms Amboseli elephants are little different from those elsewhere in Africa. In this chapter, we discuss the particular threats facing the Amboseli elephants in the early 21st century and the attempts to address those threats.

The future of the Amboseli elephants depends essentially on one thing: assured and secured access to the ecosystem. A protected area the size of Amboseli National Park clearly cannot sustain permanently within its boundaries the current wildlife population, particularly the migratory species (Western 1975). As detailed in Croze and Lindsay (chapter 2), the Park secures only about 5 percent of the wildlife-defined ecosystem (figure 2.5). Adding the relative protection of the various community concession areas (Browne-Nuñez, chapter 19) brings the proportion of ecosystem area that is relatively safe from alienation to approximately 10 percent.

The Group Ranches surrounding the Park comprise key dispersal areas, and through them, there have to be passageways for wildlife to access the rest of the ecosystem (Croze and Moss, chapter 7). The increasing human population, changing land-use practices, and fact that land subdivision is taking place without adequate planning for wildlife corridors are all diminishing the size of the dispersal areas and constricting corridors.

As in most wildlife-dominated ecosystems, sustainability is not entirely a biological problem. Sustainability depends on social and economic drivers, in the midst of which is the biggest "elephant in the room": human population growth. Since it is beyond the scope of the Amboseli Elephant Research Project (AERP) as well as apparently beyond the capacity of most governments to do anything about that particular elephant, all we can do is identify the threats and seek opportunities for mitigation. Thus, in this chapter, apart from the metaphor above, we will hardly discuss elephants at all but rather stray with equal measures of humility and trepidation into current issues impacting future land use, land tenure, and livelihoods.

Threats

Administrative uncertainty

Amboseli has been in a state of uncertain ownership since September 2005, when the Kenyan Minister of Wildlife and Tourism signed a decree to de-gazette (nullify a government enacted directive) Amboseli National Park as a park and turn it over to the Olkejuado County Council (OCC) to be run as a national reserve. In effect, this meant that management of the Park and the key resources of the ecosystem would fall to the OCC, which has no technical or administrative capacity for protected area management, in contrast to the Kenya Wildlife Service (KWS), which manages the entire national parks system. All gate receipts and other revenue would go to the OCC—a serious blow to the finances of the KWS and its ability to operate. A consortium of local and international non-governmental organizations (NGOs) took the matter to the courts, arguing that the de-gazettment was illegal according to the Wildlife Conservation and Management Act, and applied for a stay against change of ownership. The OCC contended that the change was not unlawful because the 1974 gazetting of Amboseli was itself unconstitutional. By giving Amboseli back to the County Council, the government would be redressing a historical wrong.

On the face of it, the transfer of the National Park to the OCC would appear to give the local communities a greater stake in the conservation of the wildlife in their midst, from which they would derive greater benefit. However, the OCC is based in Kajiado, over 160 km by road to the northwest of Amboseli, and its representation is imbalanced: of the 54 councilors, only 8 are from the Ilkisongo Maasai section (Croze, Sayialel, and Sitonic 2006) that makes up most of the Kenya portion of the Amboseli ecosystem. The OCC already receives considerable revenue in fees from lodges based in Ol Tukai, a 162-hectare commercial enclave in the center of the Park. The surrounding community has not shared management of the area or revenue from it apart from limited employment opportunities.

To complicate matters further, Kajiado District has been divided into two new entities, the southern part of which is now Oloitokitok District, which includes Amboseli National Park and its surroundings. It might be one of the few times when an administrative boundary has been made congruent with an ecosystem. The subdivision has strengthened the importance of the government office in Oloitokitok town and should give more locally accountable representation to the people in the region.

At the time of writing in 2009, three and a half years later, the issue of ownership is still in the high court. The status quo prevails, with the KWS in control of Park management. The new administrative structure has not yet come into effect, the Kajiado-based OCC continues to lay claim to the Park, and uncertainty persists.

Local Fronts

In 2004, the then KWS Senior Park Warden met with AERP elephant researchers to define the elements of a campaign to ensure the long-term survival of elephants in the ecosystem. The "fronts" of human-wildlife conflict defined in that exercise pertain to other species as well. The fronts are as follows (see figure 2.1 for place references):

1. *Intensive, irrigated agriculture.* Areas of irrigated and fenced agriculture around key swamp areas (Namalok, Kimana) have two negative impacts. They effectively remove important habitat areas for wildlife (and livestock), and they concentrate attraction at their perimeters, which exacerbates the frequency of negative human-wildlife encounters. Electric fences, initially installed with a grant from the European Community (Browne-Nuñez, chapter 19), could go a long way to reduce incursions and conflict if properly maintained by the community.

2. *Spread of rainfed agriculture.* The march of agriculture from Oloitokitok northward down the lower Kilimanjaro slopes (Browne-Nuñez, chapter 19; see also figure 2.5) represents a moving front of potential human-wildlife conflict. There are no barriers to prevent wildlife from entering these areas.

3. *Closing rangeland corridors.* Subdivision of the group ranches into individually owned plots coupled with unplanned development along potential wildlife corridors, such as east to Kimana or south to Kitenden, is rapidly closing off options for wildlife movements in and out of the Park. The Tanzanian government has set policy to keep a Kitenden Corridor free of agriculture on the southern side of the border, up to Kilimanjaro National Park (see figure 2.1). There is no similar policy on the Kenya side.

4. *Permanent settlements.* The growing number of non-transhumant Maasai settlements (see figure 2.5), including the so-called cultural *bomas* and their attendant infrastructure (see Browne-Nuñez, chapter 19) and a rash of non-eco-friendly lodges close to the Park boundary (see below) are a clear and present threat, both to freely moving wildlife and to people who are put squarely in harm's way.

5. *Maasai warriors.* In the ecosystem, there are several thousand energetic young men who are underemployed. They have great potential for positive action—witnessed by the successful Maasai Game Scouts program (see Browne-Nuñez, chapter 19)—but are also a serious threat to wildlife if their energy is not channeled along constructive lines.

The following sections outline some of the systemic characteristics and causes at work in each and sometimes more than one of the five fronts.

Land-Use Changes

The world's human population is currently increasing by some 5 to 7 million people per month, with much of the increase taking place in the developing world, including elephant range states. The increase in Africa alone is nearly 2 million per month (data from the United Nations Department for Economic and Social Affairs [UNDESA] 2008). In 1965, some 9.5 million people and 165,000 elephants populated Kenya. By 2005, the figures were 35.6 million and 31,600, respectively (UNDESA 2008; Poole et al. 1992;

Blanc et al. 2007). The human population had almost quadrupled in 40 years, while the elephant population plummeted during decades of heavy poaching. During those dark years, Kenya's elephants sought refuge in the protected areas, even though many of those areas were only relatively safe.

Since the ban on international trade in ivory went into effect in January 1990, Kenya's elephant population has slowly increased and has gradually reinhabited some areas of former range. At the same time, land use in Kenya's rangelands has changed dramatically. Agriculturalists, encouraged by government or forced by land pressures, have attempted to farm increasingly marginal land. Seminomadic pastoralists, whose growing populations are beset by periodic drought, have been encouraged to settle and farm. Large tracts of semi-arid rangelands have become incompatible with use by elephants, as group ranch rangelands are subdivided and sometimes fenced.

Land use in the Amboseli ecosystem is also changing rapidly, as discussed in detail by Kangwana and Browne-Nuñez (chapter 3). The human population is steadily increasing, both internally and from immigration of settlers from other parts of Kenya. The group ranch system, itself an imperfect framework imposed on the Maasai traditional communal land-tenure system (see Kangwana and Browne-Nuñez, chapter 3), is breaking down through adjudication and subdivision. As noted above, irrigated agriculture has closed off two of the five main swamps, and rainfed agriculture is inexorably marching down the slopes of Kilimanjaro into wildlife habitat. Water resources are being heavily used, diverted, or polluted in major springs and swamps.

In general, there is no planning for and management oversight of developments in the ecosystem, indeed in most public lands across the country. There is, however, a move afoot in the form of a National Land Policy Formulation Process to rectify the fact that "Kenya has not had a clearly defined or codified National Land Policy since independence" (Government of Kenya 2006). If and when the policy is launched, it will need to address many existing circumstances on the ground that will be difficult to reverse.

The Maasai are fully aware that small plots are not viable for livestock husbandry in semi-arid savannahs with patchy and unpredictable rainfall (Kangwana and Browne-Nuñez, chapter 3) and certainly not for agriculture over most of the ecosystem. The perceived advantage of having a title deed (to secure loans, for example) appears to outweigh the cooperative imperative that was the traditional basis for Maasai society and animal husbandry and the underlying principle of group ranches. But, as noted by Browne-Nuñez (chapter 19), the new holders of title deeds are showing willingness to re-aggregate their holdings under a different management regime.

Stakeholders are aware of the threats to wildlife from rapid land-use change, and over the past half decade, they have gathered on numerous occasions to lay the foundations for an emerging comprehensive ecosystem management plan (Manegene and Bernard 2004). But that plan is seen by landowners as being driven by external groups, including government and wildlife conservation NGOs and generally has not been taken to heart. Yet even if there were a fully participatory process at the local level, the best of plans are doomed to sit on shelves or hard drives in the absence of a "vertically integrated" system of supporting legislation and administration from above.

Maasai Pastoralism

The nature of Maasai pastoralism, past and present with regard to wildlife and elephants in Amboseli, is discussed in chapters 3, 19, and 20. In order for wildlife to have a sustainable future in Amboseli, two issues of potential conflict with pastoralists have to be addressed urgently. One is the issue of distribution of benefits. The Maasai quite reasonably ask, "Why should we tolerate the presence of wildlife on our lands if only a small portion of wildlife-related revenues is going to only a few of us?" The potential benefits range from short-term cash in hand to longer-term development of the region, improvement of ecosystem services, and alleviation of poverty.

The other conflict arena has to do with day-to-day competition for essentials: pasture, water, and living space. The necessarily low density of pastoralists, combined with a cultural propensity to respect wildlife, has traditionally allowed the essentials to be comfortably shared. But that pastoral scene is fading. The absolute numbers of people and livestock are growing. That inexorable growth forces land-use changes, principally subdivision and spread of agriculture, which in turn constrains the movements of all players, crowding them here and there and forcing them into conflict at key sites.

The trade-offs between the two fronts seem quite obvious: if the presence of wildlife is made tolerable through revenue generation, then some pasture, water, and living space may be foregone—given over voluntarily—as a reasonable price to pay, an acceptable opportunity cost, for having wildlife as well as livestock.

Agriculture

Over the past 50 years, on the high-ground fringes of the ecosystem, particularly to the southeast and in the non-Park central swamps, agriculturalists have settled into the Amboseli ecosystem. Whether they are Kikuyu or Kamba agriculturalists or Maasai experimenting with crops, they are agents of land-use change that cannot be ignored (see chapters 3 and 19).

Two issues predominate. One is that contemporary Maasai, particularly those residing near the expanding area of rainfed agriculture (chapters 3 and 19), have become more dependent on buying food from agriculturalists. Diet preferences are changing, apparently driven predominantly by women nurturing children (W. Kiiru, PhD thesis in preparation, personal communication, 2008).

The other issue is that although the agriculturalists suffer opportunity costs from wildlife at least as severe as those experienced by pastoralists, they derive hardly any of the benefits. Unless this imbalance is addressed, the agriculturalists will always be hostile to the presence of wildlife in the ecosystem (Campbell et al. 2003).

The fenced, high-density agriculture areas around Namalok and Kimana swamps (see figure 2.1) have been alienated to wildlife and pastoral livestock and given over entirely to intensive irrigated agriculture. In general, the land-tenure system has shifted from communal stewardship to individual ownership as farmers from other parts of Kenya have moved in and bought or rented plots from the Maasai. The small-plot production of produce such as tomatoes and onions is generating impressive per hectare financial returns, at least two orders of magnitude greater than traditional livestock rearing (D. Campbell, personal communication).

Today's high production levels from the areas of irrigated agriculture have a questionable future. Already the soils are being depleted and the spring water that feeds the swamps is heavily polluted with the effluents of intense cultivation (Githaiga et al. 2003). If productivity falls off over the next few years, either a huge investment in cleaning the polluted waters will be required or the farmers will simply move away, leaving depleted and poisoned soils and water.

Poverty

Contemporary modifications of traditional drought-coping strategies—such as reciprocal grazing arrangements between neighboring clans or setting aside "grass banks" during good rains—seem to be just barely keeping the pastoralists from sliding into poverty. In Kajiado South, basically the Amboseli ecosystem, 48 percent of the population is estimated to be below the poverty line, the level of income required to meet calorific and other basic needs (GoK 2001, Ndegenge 2003).

Poor people have little incentive to conserve wildlife unless they can be shown ways in which they might benefit from the resource, not only monetarily, as Kangwana and Browne-Nunez stress (see chapter 3), but also in terms of a range of ecosystem products and services, such as the health benefits of clean water, the satisfaction of being usefully employed, and the pride and utility of being educated.

Bushmeat

Protein hunger in the urban areas of Kenya and Tanzania is fuelling an informal trade in so-called bushmeat. The Amboseli ecosystem is under siege from the demand, and there are three main routes to the trade: across the porous international border southwest to Arusha, south to Moshi, and north to Nairobi. Tour drivers report that they regularly see pickup trucks plying along the Namanga-Kajiado road (and the dirt tracks that feed into it from the interior of the ecosystem) carrying meat to Kajiado and Nairobi. There is every reason to believe the Amboseli ecosystem is contributing to the estimated 25 percent of urban protein that is currently provided by bushmeat (Born Free 2004).

The bushmeat trade is not yet a threat to elephants in Amboseli as it is elsewhere on the continent (Brashares et al. 2004). The Maasai in fact are repelled by the thought of elephant meat: one elder was observed to have fainted when he saw members of another tribe collecting meat from an elephant shot in the Oloitokitok area "on control" (S. Sayialel, personal communication). But the Maasai are far from alone in the ecosystem (see Kangwana and Browne-Nunez, chapter 3), and hunger for protein is not diminishing.

Development without Planning or Consultation

Just outside the Park boundary to the south and southeast, there are a dozen new enterprises, ranging from small eco-friendly fly camps like Olkanjau (meaning "elephant" in Maa) to large, permanent complexes like that recently constructed by the Mada Hotels group. Mada has erected a fence (unsanctioned, it seems, with no public evidence of planning permission, stakeholder consultation, and the like) around its 60-acre concession, which together with its overdeveloped neighbor, Kibo Safari Camp, effectively blocks east-west elephant movement (see figure 7.2).

These developments have all arisen through direct contracts between concessionaires and individual or small groups of newly title-deeded landowners. A meeting of developers called by the then-District Commissioner of Oloitokitok in August 2007 revealed that none of the individuals or companies had completed an environmental impact assessment as mandated by the Environmental Management and Coordination Act of 1999. However, no action was taken, and in fact one of the larger lodges was subsequently completed without an environmental impact assessment.

The large environmental footprints of such developments disrupt habitats and block wildlife corridors. They also are staging posts for the influx of large numbers of tourists and tour vehicles into the Park. The newest developments have brought the total accommodation in and around the Park to more than 1,100 beds. Conceivably, at peak occupancy, there could be some 200 vehicles driving around the Park, one for every hundred meters or so on the Park's roads. Since all paying visitors spend the night outside its area of control, the Park authority receives reduced revenue that is not commensurate with the increased visitor impact, both in esthetic and ecological terms.

Poaching and the International Ivory Trade

Throughout the 1970s and early 1980s, before the 1989 Convention on International Trade in Endangered Species (CITES) ban on international trade in ivory, poaching was rampant in elephant habitats throughout Asia and Africa. Amboseli was largely spared the slaughter. While many populations across West, Central, and East Africa were drastically reduced (Africa as a whole by 54 percent, Kenya alone by 85 percent), the elephants of Amboseli actually increased (see annex 2.1). Two factors contributed to the relative safety of Amboseli's elephants during this period. One is the intolerance of Maasai to other people poaching on their land. The other is the almost constant presence of researchers providing additional eyes on the ground and early warning of incursion by poachers.

Nonetheless, there has always been some poaching for ivory in Amboseli. It tends to occur in bouts, mainly in the southern part of the ecosystem along the international boundary. Several bull elephants were killed in a period of a few months in 1994 in the Sinya region of the ecosystem in Tanzania (see Croze and Moss, chapter 7), and in April–May 2006 up to 10 more bulls may have been poached in the same region. More worrying is a dramatic increase in poaching incidents in Kenya, starting in late 2008 and extending through 2009, with a minimum of 12 elephants killed for their tusks (AERP records).

It is worth noting that two recent events have accompanied and quite likely encouraged a continent-wide upsurge in poaching for ivory, which has also touched the Amboseli population:

1. In 2007 CITES added China to the countries allowed to purchase legally stockpiled ivory, thus promoting a huge emerging market. Previously, only Japan was allowed to buy ivory from the four southern African countries that were permitted to sell their stockpiles.
2. The 2008 IUCN downlisting of the African elephant from "threatened" to "near threatened" status based primarily on the large Botswana savannah elephant populations that swell the species numbers and mask the ongoing poaching crisis among East African populations and the taxonomically distinct forest elephants of Central Africa (Blake et al. 2007; Hart 2009).

The authors of this volume strongly disagree with both actions.

Human-Wildlife Conflict

Human-wildlife conflict, a general term encompassing a range of interactions between wildlife and human economic interests, is now becoming a minor study discipline in its own right (e.g., Woodroffe et al. 2005; chapters 19 and 20). It is the inevitable consequence of two related deficiencies: (1) no comprehensive land-use policy to define zones for different forms of use and to control access by a growing human population and (2) no management plans to promote and develop the benefits from wildlife and encourage participation in wildlife enterprises. The Amboseli ecosystem is a prime example of both.

The number of wild "culprits" in human-wildlife conflict is legion, ranging from insects and rodents that account for major post-harvest grain losses to hippos that kill more people in Africa than any other species. Yet despite their relatively low ranking in terms of actual losses caused, elephants usually get top billing. For that reason, the *perception* of human-elephant conflict is a major threat to elephants in many parts of Africa and Asia.

Elephant attacks on livestock are discussed in chapter 19 (box 19.1). Although the loss of a cow or bull is undoubtedly a serious economic and spiritual blow to a Maasai, the losses from elephant "predation" are quite small compared with those from drought or disease.

Attacks on humans can also be put into perspective. Unpublished data collected by the KWS and the Amboseli-Tsavo Group Ranch Conservation Association between 1993 and 2005 indicate that over the 12-year period, some 18 people in the greater Amboseli area were killed and the same number injured by elephants (source *KWS Occurrence Book*, Amboseli National Park Headquarters, 2005). As dreadful as any loss of human life may be, an average of some three incidents a year is not a very large number. Being attacked by wildlife, elephants in particular, is a relatively rare event compared with being in a Kenyan road accident, for example. An elephant attack is perhaps more like an airplane crash: relatively infrequent and improbable but spectacular, costly, and well publicized when it does happen.

Spearing of Elephants

The iconic Maasai longspear is a traditional weapon still used today for self-protection as well as self-expression.

Sadly, it is too often used offensively on wildlife, lions and elephants in particular.

The spearing of elephants is not, like the *alamaiyo* ritual hunting of lions, part of Maasai tradition. As with the near-total elimination of rhinos in the 1970s (see Croze and Lindsay, chapter 2), attacks on elephants are motivated by political protest, retaliation, and, undoubtedly, a measure of delinquency. Over the 11-year period from 1997 to 2007, some 91 elephants were speared (see box 19.1), 57 of which died of their wounds. Those that survived—with or without veterinary intervention—are likely to carry psychological scars that make them potentially dangerous to the local human population (Bradshaw et al. 2005).

Opportunities for the Future

The course of Amboseli's conservation history was described in chapter 3. Over the years, despite its small size, Amboseli has come to be one of the top three most visited parks in the Kenya national parks system, along with Nakuru and Tsavo East. There are three iconic elements that continue to give Amboseli its attractive edge over other parks: the most approachable free-living elephants in Africa; a spectacular view most days of Kilimanjaro, the world's tallest free-standing mountain; and an opportunity to meet the Maasai people (see chapter 3 and other citations in this section). The 80,000 to 100,000 visitors a year make Amboseli a major contributor to foreign exchange earnings. In 2005, the Park grossed some $3.5 million in gate takings (KWS, personal communication). Proximity to Nairobi and the open habitat that facilitates viewing of the rich diversity of wildlife undoubtedly contribute to the Park's popularity.

As challenging as the current situation may appear for the moment, there is no reason why Amboseli could not become a model of a well-run conservation area. There are certainly many pitfalls, but on balance, there are real opportunities for a successful, mutually beneficial partnership to be developed between the Maasai landowners and the various conservation and wildlife-related business interests in the region.

Whatever happens over the short and medium term with regard to legal uncertainty and ultimate responsibility for management of the National Park, it would be useful to consider rejuvenating the UNESCO Biosphere Reserve model (e.g., Mburugu 2002) for the Amboseli ecosystem, with a protected and well-managed core area (e.g., the current National Park) surrounded by a buffer zone in which agreements have been negotiated with the community to allow wildlife access. Without this model, Amboseli could go the way of other popular protected areas in Kenya (such as Nairobi and Nakuru National Parks): an isolated, protected island viable only as a glorified zoo, benefiting only a fraction of the surrounding community.

The Maasai tell us clearly, "You no longer have to teach us the benefits of the wildlife resource. We understand that now. What you need to help us achieve is how to tap those benefits and how to make certain that a fair portion of the considerable revenue streams get back to us. And quickly. Else all bets are off." The Maasai landowners should articulate their own community vision of how they want the ecosystem to be in the future. This will help them be both proactive and defensive in the face of corrupt politicians and profiteers from the outside.

Reduction in Elephant Spearing

There is an urgent need to understand the relative importance of the apparent root causes of spearing—protest, retaliation, and delinquency—across the ecosystem. Although, the short-term solution for each cause is probably about the same (namely, appropriate punishment in law as well as by traditional means from the community itself), the long-term solutions are quite different. Political protest, for example, can only be forestalled by good governance and equitable distribution of benefits. Legal deterrents are at perennial risk of being diverted from a just path. Cases of spearing of wildlife brought before local (i.e., non-Nairobi-based) magistrate courts typically are dealt with by imposing fines upon offenders so small as to be without deterrent value. Alternatively, perpetrators may be released with merely a warning into the custody of the community elders to be punished according to traditional fines, such as the slaughter of a prized steer for all to eat. Clearly, the one has little long-term impact and the other can be considered the excuse for a party.

Retaliation spearings arise from an untenable proximity of people and their livestock to wildlife. Seasonal proximity, during which people and wildlife converge on dwindling forage and water resources in the dry season or a drought, is difficult to manage.

The AERP has initiated an important strategy for defusing retaliation spearing. In 1997, after a particularly messy and misinformed eye-for-an-eye killing by the KWS of a tuskless matriarch in exchange for a cow having been tusked to death by an unidentified adult female, the Amboseli Trust for Elephants launched a "consolation scheme" with the three group ranches closest to the National Park (see box 19.1).

It is evident from public statements that the Maasai appreciate the scheme. They recognize and are grateful that even though the AERP has no legal responsibility to provide a compensation payment, the project is concerned about the well-being of the community and the basis of its livelihood.

Often community leaders have stood up in meetings to say how good it is to see that "elephant research" is bringing something back to the community. And, indeed, when payments are suspended pending discussions after a spearing incident, the elders are quick to impose internal discipline.

Other attempts to reduce conflict and improve relations with communities include

- refurbishing earthen dams to extend the period of time that water is available outside the Park into the dry season (under trial by the AERP since 2008);
- promoting incentives such as the strategic locating of schools and boreholes in order to draw settlements away from the Park boundary and known wildlife corridors to places where the maximum population would benefit (dependent on a comprehensive plan for the ecosystem, which is currently lacking); and
- improving defenses for settlements and infrastructure based, for example, on specialized knowledge of elephant behavior, with perhaps subsidy from the wildlife sector (e.g., Granli and Kioko 2006a, 2006b).

Delinquency itself is not a cause of troublemaking but is the result of social and economic discordance. The long-term solution for youth behaving badly is less straightforward and has to do with answering the question, what can be done with a cadre of under- or unemployed young men with lots of energy and an outdated social function? The Maasai Game Scouts program (see Browne-Nunez, chapter 19) is an effective and useful source of employment but can make only a small dent in the ranks of several thousand. Another alternative currently under discussion is a Maasai Games, a kind of regional annual olympiad for the Maasai of Ilkisongo and possibly beyond based on a successful model that has taken the form of the Laikipia Games (K. Gallmann, personal communication) and the Lamu donkey and dhow (traditional sailing boat) race.

Cultural *Bomas*: A Better Model

While there is some concern over the impact of cultural tourism, a few wildlife concessionaires in Amboseli have demonstrated that cultural *bomas* do not have to be a social and economic disruption (Browne-Nunez, chapter 19). The Eselenkei Group Ranch and the management of the Selenkay Conservation Area have designed a "cultural *boma*" experience based on what the tourists truly want to see: not a tacky kiosk with people outside begging for handouts but an honest glimpse into another culture. Each client visiting Porini Camp (the upmarket safari camp in the concession area) is given the opportunity to be taken to the designated *enkang* (traditional Maasai homestead; see chapter 19) after being briefed around a scale model in the camp. They are met by an elder and shown around while people carry on their daily activities. There is no hustling during the visit: handicrafts are sold in a community-stocked and -run shop at the camp. In fact, no money changes hand at all during the visit: the community is paid at the end of the month based on the number of visitors. Typically, the amount is the Kenya shilling equivalent of some $800, several times more than the tour driver–exploited *bomas* of Olgulului/Ololarrashi Group Ranch to the south (see Browne-Nunez, chapter 19). Such a model should be universally adopted in Amboseli. In the meantime, tour companies should be encouraged to discipline their profiteering drivers.

Wildlife Concession Areas

The concept of "renting" wildlife areas for developing enterprises to provide overseas visitors with a "safari experience" and contact with relatively untouched natural Africa is over 100 years old, having begun with colonial "white hunters." Today the practice has a strong experience base and, if well managed, is potentially the single best long-term revenue-generating activity in non-agricultural zones.

Venerable safari companies such as Ker and Downey Ltd have been providing the tented-camp experience for decades. They negotiated a concession southwest of the Park at Nado Soit in 1983. The next significant community-based enterprise was established near the Kimana swamp in 1995 with investment from a Swiss hotelier. Since then, a number of highly successful enterprises have generated significant revenue for group ranch members and provided important centers of conservation in the ecosystem away from the core of Amboseli National Park. As of 2008, there were seven such concessions: Kimana Wildlife Concession Area (Kimana Group Ranch), Maasailand Preservation Trust (Oldonyo Wuas; Mbirikani Group Ranch), Maasai Wilderness Conservation Trust (Campi ya Kanzi; Kuku Group Ranch), Selenkay Conservation Area (Eselenkei Group Ranch), Kitirua Limited Game Concession (Olgulului/Ololarrashi Group Ranch), Elerai Conservation Area (Entonet Group Ranch), and Ol Kanjau (Kimana Group Ranch) (see Croze, Sayialel, and Sitonic 2006 for details).

The conventional wisdom is that such conservancies increase the goodwill of the landowners toward wildlife (and the National Park that sequesters them) by providing direct benefits in the form of fees and employment. There is, however, a measure of mistrust between group ranch members and their own committees concerning the free flow of revenues: one concessionaire took to posting his monthly payments to the group ranch committee on the local schoolhouse door for all members of the ranch to see (L. Belpietro, personal communication). As yet there has been no systematic study to determine in Amboseli (as Bandyopadhyaya et al. [2004] did in Namibia) the answers to three questions: Do conservancies increase household welfare? Are conservancies pro-poor? And, do non-participants in conservancies gain as well as those who choose to participate directly? (The answer in Namibia, by the way, appeared to have been "yes" to all.)

Trust Deed Holdings

The landowners know that having a title deed provides a basis for negotiating agreements with neighbors to form larger areas of common ground that can be better used with economies of scale for whatever kinds of land use the "neighborhood" agrees upon. By the same token, conservationists accept that negotiating contracts for, say, concession areas, could be more straightforward with title deed holders than with a relatively amorphous and ever-changing group ranch leadership.

Against that background, eight landowners on the small Entonet Group Ranch to the southeast of the Park came together with encouragement and guidance from the African Wildlife Foundation to form the Elerai Concession Farm, which included the building of a small, six-bungalow tourist camp.

The Foundation has gone on to negotiate with nearly 150 landowners in the former Kimana Group Ranch to draw up two legally binding trust deeds for two concessions, Kilitome and Osupuko, covering more than 3,600 hectares (AWF 2008; F. Warinwa, personal communication). Together, these trust deed holdings secure nearly 60 percent of the eastern corridor between Amboseli and the Kimana Wildlife Sanctuary. There are plans to negotiate the remaining 2,500 hectares in between the two leaseholds in order to complete the corridor. At the time of writing, there was no secure funding for the vital exercise.

The terms of the agreements provide the landowners with a modest annual rent (equivalent to $3,600 per annum per plot) and in turn ensure that they will not fence large tracts of their land and will allow free movement of wildlife. The landowners can still maintain their livestock herds at traditional levels. There is also provision in the agreements for establishment of one eco-friendly tourist camp in each combined leasehold. The revenues will go directly to the consortium of landowners.

The trust deed model improves greatly on the blanket concessions that group ranch authorities pioneered in the ecosystem (see previous section), since each title deed holder gets his fair share without the sometimes questionable intervention of a group ranch committee.

Community Development Organizations

There has been no effective response by the community surrounding Amboseli National Park to the proposed change in the Park's ownership. One reason is that there is no lead community organization. Concerned NGOs have tried to establish and later resurrect the Amboseli-Tsavo Group Ranch Conservation Association, but the effort appears to have failed for a variety of reasons to do with competing interests between individual group ranches and association leadership. Without such an organization, clever entrepreneurs from outside could overrun the community.

Another advantage of community-based organizations is that they provide employment to local people in their home area. Employment in Amboseli, as in much of the country, is arguably the most important development goal for the individual, once the issues of land ownership and welfare of the cattle have been dealt with.

The Kenya Wildlife Service

As the current manager of the National Park and the authority responsible for wildlife throughout the ecosystem, the KWS has the mandate to serve as a "good contractor," acting fairly and effectively on behalf of the wildlife and the people. It has been hampered by the lack of a comprehensive land-use plan for the region and slowness in the devolution of responsibility and resources to its regions. The KWS is still in a position to seize the opportunity and be a custodian worthy of its mandate and mission.

Toward Better Management

The Olkejuado and Oloitokitok County Councils, for better or worse, are currently the legally constituted representatives of local government in the ecosystem. Leaving aside the uncertainty of their relative jurisdiction and responsibilities, in the best of worlds, the county councils should lead the community in optimizing benefits to the people of the area. Elements of improved management would include definition of responsibilities and the sharing of resources among administrative management and community groups within the partnership. Such a definition is beyond the scope of the current discussion but remains a matter of priority.

Conclusions

The greatest threats to sustainable wildlife conservation in Amboseli are permanent settlements along the National Park boundary, fragmentation of the ecosystem through ongoing subdivision of group ranches, uncontrolled and unplanned development, competition for water and grazing areas (especially in and around the central swamps), the bushmeat trade, the spearing of lions and elephants, and poaching for ivory.

The most urgent conservation goal must be to ensure wildlife access to dispersal areas by means of a lasting compact based on equitable benefit sharing with the surrounding Maasai community. A ground-based patchwork of NGO-driven community concessions and leasehold easements is beginning to work.

It is no longer simple to be an elephant in the human world. Only one or two generations ago—whether in human or elephant terms—there was space enough for people and elephants to coexist with only occasional agonistic encounters. Today the balance is shifting rapidly. As human populations spread across African landscapes in their legitimate quest for a better life, the space left for elephants is diminishing and the potential conflict zones are increasing.

But, in the long run—dare we say it?—people need to decide to limit their population growth, a phrase that epitomizes "easy to say." That oft-thought, seldom-uttered notion should not be seen as being restrictive or confining but as prescriptive and defining of a quality of life that people deserve. Quality is not predicated on constructs of "more" but on designs for "better." Discussion of this assertion is way beyond the scope of this volume. Resolution of such matters lies in the heady realm of human mores and requires the ingredients tabled by Cohen (1995): a bigger pie, fewer forks, and better manners.

We believe there is enough space for both species (and other species of wildlife that make a living in elephant-delineated ecosystems like Amboseli), at least for now. The suggestions we have offered for managing the Amboseli space are neither perfect nor comprehensive, but at least they give us hope that there are immediate concrete steps that a wise and caring leadership would have cause to take, given the will. In the meanwhile, non-governmental organizations and environmentally conscious concessionaires are guiding actions on the ground, bypassing the policy vacuum, and securing corridors. They will hopefully spread and grow and eventually ensure access for elephants to much of the ecosystem.

Whatever it takes to ensure the particular future of the Amboseli elephants and ecosystem, the solutions will have to mix ingredients from four broad arenas—scientific, social, economic, and moral—and are bound to pertain to many other populations in Africa. Throughout this book, we have attempted to illustrate ways in which the unique knowledge of the Amboseli elephants revealed by our studies can inform other scientists and managers. We must now leave it to the next generation of researchers and conservationists to refine the linkages, motivate the decision makers, and further demonstrate the value of this study to understanding elephant biology and behavior and to conserving the elephant's future.

Appendix 1 Methods

Study Area and Animals

The study was conducted in and around Amboseli National Park in southern Kenya from September 1972 to the present and is ongoing. Here we provide an overview of the main study techniques that contributed to the long-term dataset. Chapters that rely on specific data that were not part of the long-term dataset provide information on their collection and analysis techniques in methods boxes.

The park, which was established in 1974, covers an area of 392 km^2, and the Amboseli ecosystem extends over a much wider area of over 10,000 km^2 if both the Kenyan and Tanzanian portions are considered (see below). Further details of the study site are provided throughout chapters as required. By the end of 2002, there were 1,225 living elephants in the population and 791 elephants that had been identified and which subsequently died, making a total of 2,016 individuals in the primary data set. Additional demographic data from 2003 to 2006 were used in some chapters, and sample sizes are given in the relevant chapters. There were 52 family units (Moss and Poole 1983), each made up of related cows and their calves and averaging 17.4 individuals (range 3–48). Adult independent males, of which there were 183, left their natal families at an average age of 14 (Lee and Moss 1999) and moved singly, in loose groupings with other bulls, or in temporary association with family groups. All members of the population were individually known by means of a photographic recognition file (see below) or in the case of young calves, by their association with known mothers.

Individual Recognition

The foundation of all Amboseli Elephant Research Project (AERP) research rests on the ability to recognize individual elephants over time. Starting in September 1972, a photographic recognition file was compiled. Each elephant's head, including the ears and tusks, was photographed from the left and right sides and head-on on occasion. Photographs were pasted on cards dedicated to each individual. Individuals were identified by features including notches, holes, tears, bumps, and vein patterns in the ears and by tusk and body configuration (Moss 1983, 1996).

The ID cards were carried in two boxes, one for the males and one for the females. Male cards were filed by age class; females, by family. In the early days of the study, the association patterns of known individuals gradually provided the information needed to determine family membership. Each family was assigned a letter of the alphabet, and every member (with some exceptions) was given a name beginning with that letter. Thus, the A family members were named Annabel, Alyce, Amy, Alison, Amelia, and Wart Ear (one of the exceptions). For ease of recording and computer database input, the individual names were shortened to codes. When the time came to assign the letter Z to a family, it was necessary to start through the alphabet again, and the A family became the AAs and an AB family was created.

Age Determination

Births have been recorded in the Amboseli population since 1972. Assessment of newborn and young calf age (less than three months old) was based on body size and proportion, skin color, motor coordination, presence of an umbilical cord, and the behavior of both the calf and the mother (Moss 1988, 1996). In the first three years of the study, 1972–75, birth dates were known ±3 months and from 1976 to 1999, they were known, with a few exceptions, to within two weeks. Ages of animals born prior to 1972 were estimated using the following techniques:

1. Length of hind footprint, which has a consistent relationship to growth and age (Lee and Moss 1986, 1995; Western, Moss, and Georgiadis 1983).
2. Tooth eruption and wear sequence, which have been primary methods of estimating ages of dead or immobilized elephants (Laws 1966; Sikes 1971).
3. Tusk eruption and length, which are good age indicators for calves up to six years old (Moss 1988, 1996), and tusk circumference at the lip, which is closely correlated with age throughout life (Pilgram and Western 1986).
4. Visual or photogrammetric assessment of shoulder height and back length (Croze 1972; Laws, Parker, and Johnstone 1975; Lee and Moss 1995; Moss 1996; Shrader et al. 2006).
5. Photographs of recognizable Amboseli elephants taken in the 1960s and early 1970s provided additional information for making age estimates. As the study progressed and known-age individuals became older, the early photos of estimated-age adult and young adult elephants could be compared with the known-age individuals and the age estimates adjusted (Moss 1988).

These methods of assessing age were combined and cross-checked with reference to known-age animals as individuals aged. Animals whose birth dates were unknown were aged using a combination of at least two of these techniques. In this way, all members of the population were assigned a birth year with a degree of accuracy code ranging from 0 (estimate) to 5 (known ±2 weeks). Animals born from 1973 onward were also assigned a birth month, with those born from 1973 to 1975 having a birth accuracy of ±3 months and most of those born after 1976 having an accuracy of ±2 weeks. The birth year of those animals estimated to be born in 1970–72 was considered accurate to ±6 months while the birth year of those born in 1968–70 was considered accurate to ±1 year. The birth year of those animals born before 1968 was estimated at ±2.5 years and that of those born before 1963, to ±5 years. By December 2002, the ages of 1,419 living and dead elephants were known, with a birth accuracy of ±3 months, representing 70 percent of the sample ranging from newborn calves to animals 29 years old.

For the purposes of age structure analysis, eight age classes were created. Individuals were grouped in five 5-year classes until 25 years and then, based on the ability to age the older animals, in one 10-year class and two 15-year classes as follows: 0A = 0–4.99; 0B = 5–9.99; 1A = 10–14.99; 1B = 15–19.99; 2 = 20–24.99; 3 = 25–34.99; 4 = 35–49.99; and 5 = 50+.

Growth Database

We have monitored size of the elephants in the Amboseli population over the past 30 years. The measures used here were hind-foot length from footprints (an excellent estimator of size as noted above), and a non-invasive photographic technique for measuring shoulder height (detailed in Lee and Moss 1995). Hind-foot length and shoulder height are highly correlated within the sexes (Lee and Moss 1995). As well as being indicators of overall stature (Shrader et al. 2006), both shoulder height and hind-foot length are reasonable indicators of body mass, which is difficult to measure non-invasively.

Here, we used a total of 485 measures of height and 709 measures of foot length on 314 males and 297 females ranging in age from birth to 63 years. There were 210 simultaneous measures of height and foot length (97 males and 113 females). Only eight individuals had repeated measures of both foot length and shoulder height while there were 54 repeated measures of height (21 males and 33 females) and 273 repeated measures of foot length (104 males and 63 females). All data, cross-sectional and longitudinal, were used to construct sex-specific asymptotic growth curves and to define residuals from these equations (see Lee and Moss 1995). Size-for-age was calculated for individuals at each measurement age as the residuals from the sex-specific growth curves for foot length and shoulder height. Residuals were used as continuous measures. We were specifically interested in the nature of differences between small and large individuals of the same age. Thus, we constructed categories, and individuals were classed as small-for-age (bottom 25 percent of residuals), average-for-age (26 percent–74 percent), and large-for-age (top 25 percent of residuals) for foot length and shoulder height.

Demographic Data

Over the course of the study, a *census* of each family unit was attempted at least once per month. A census consisted of the identification of all individuals present in the family on that sighting. These census data were used for analyses of individual female-female associations within groups. During a census, absence of any members of a family was noted, especially adult females and males approaching the age of independence; being "away from the family" was a suggestion of estrus sub-grouping, dispersal or death, and such observations were followed up by further searching and sightings.

Censuses were also conducted on the population of adult independent males, with an attempt to find each male once a month. This goal was rarely achieved as males range farther than females, and they were often on their own or in small groups, which made them difficult to locate (see below).

Births. New calves were recorded during a census, but calves were also registered as soon as they were observed, whether or not the family was being censused. As soon as a new calf was found, its age was estimated, sex noted, and mother recorded. Conception dates were estimated by back-dating live births by 22 months (Moss 1983) or 660 days using estrous records as detailed in chapter 18.

Mortalities. Dates of deaths in the Amboseli population were more difficult to pinpoint. Typically, indirect methods were used to determine mortality. Missing animals were recorded during routine monitoring as well as during monthly censuses. If an adult female was absent and her calves were present, it was a good indication that she was dead or seriously ill. That family was then more closely monitored. If the female was not been seen for a week or more, it was assumed she was dead. If a calf less than three years old was missing and its mother was present, the calf was assumed to be dead. If a mother and her calf or calves were absent or a juvenile or adolescent female was missing, they were counted as dead if they were not seen *with or away* from their family for a month or more. Once it was decided that an individual was dead, the date of death was estimated as the midpoint between the last date the animal was seen and the date it was first recorded as missing. Some ill or injured animals were observed and monitored, and their causes and dates of death could be more accurately recorded. In cases of known deaths, lower jaws were collected for confirmation of tooth age.

Attempting to register male mortality was problematic. Until a male calf was seven or eight years old, it was treated in the same way as a female calf, but since males start to spend time away from their natal families as young as six to eight years old (Lee and Moss 1999), absence from the family even for prolonged periods was not necessarily an indication of death. Several months with no sightings either with or away from the family had to pass before an adolescent male was considered potentially dead. An adult male was not considered dead until he had not been sighted for several years. A date of death was assigned based on the last recorded sighting, but often it was only given an accuracy of ±12 months.

Death dates were assigned degrees of accuracy ranging from 0 (estimate) to 4 (±1 week), and death causes were assigned the following degrees of accuracy: 0 = no cause known; 1 = suspected with some reason; 2 = good evidence; 3 = known. Only those causes of mortality that were known or were based on good circumstantial evidence were used in analyses that required finer resolution.

Cause of death. Causes of death again were not easily determined. Surprisingly, few carcasses were found during the course of the study, and the great majority of these were of adult animals. The carcasses or skeletons of calves less than five years old were very rarely discovered. In a few cases, animals showed signs of illness and later were found dead or they disappeared. The only positively known causes of mortality were the result of spearing by Maasai, poisoning or shooting by poachers, control shooting by Park personnel, observed injuries, and starvation of calves whose mothers had died. There was good circumstantial evidence for deaths due to the effects of drought, injuries incurred in fights, complications during births, predation, poisoning, and old age as indicated by the wearing down of the last set of molars. Adult carcasses in Amboseli were observed to be consumed and dispersed by hyenas to the point of almost complete invisibility from the ground and, except for the skull, even from aerial view in less than three weeks.

Sightings Data and Elephant Groups

The Amboseli study was mostly vehicle based. One or two vehicles went out six days a week to gather basic social and ecological data on the elephants. The researchers tried to cover the Park over a period of days, recording all the elephants they saw. At each sighting of a group or a single animal, specific information was recorded on field data sheets and later in the day entered into a computer. The standard definition of a group is given in box 1.1. For some analyses, an absolute threshold distance between individuals or groups of 100 m was used (see, for example, chapter 16). In practice, groups were relatively easy to distinguish in Amboseli, although bulls were often found moving between groups (see also Poole 1987a, 1989b). Elephant groups are socially complex; they can consist of a single family or alternatively several fragments of a number of different families. A group can also consist of individuals of all ages and sexes and contain as many as 20–50 different family units. In analyses of group size, we used the observed size. In some cases, we determined small, medium, and large groups based on frequency distributions. Small and large groups were those in the bottom and top 25 percent of observations.

Each researcher had a personal code to link to his or her observations. The data collected during a sighting were as follows:

Date, Time, Location within a 1 km^2 grid and by global positioning satellite (GPS) recording after 1999.

Number of elephants in the group.

Quality of the count defined as exact, partial count, poor estimate, no estimate.

Type of group: All male group, females and calves only group, or mixed-sex group, which contained females, calves, and independent males.

Presence of estrous female and/or *Presence of musth bull*, with both records linked to an individual identification.

Identity of any family units present and *Cohesiveness of family*—whether all family members were present and if not all present, then whether more than 50 percent or less than 50 percent of the family was present. Each family identification was coded for *the quality of family recognition* as complete, partial, or no identification possible.

Identity of males present and the *Quality of male recognition*, again coded as complete, partial, or no identification possible. A *count* of the number of independent males by age class was made.

Activity of the majority of the group (feeding, walking, feeding while walking, resting, sleeping, drinking, dusting or rubbing, social interaction, or various, where most animals were engaged in diverse activities).

Habitat type (up to two could be listed for a group) coded as wet swamp, swamp edge, *Phoenix* palm plus *Acacia*, young *Acacia xanthophloea* stands, mature *A. xanthophloea* woodland, mature *A. tortilis* woodland, open grasslands, *Suaeda-Salvadora* (salt) bush, mixed bushland, and several specific acacia species bushed woodlands.

For the analyses in this book, a total of 33,420 sightings over the period of 1972 to 2002 or various study-specific subsets have been used. These data were in a Microsoft Access database, with links to enable analysis by individual male or family unit on a day-to-day, month-to-month, yearly, or decade basis. Number of sightings in any month averaged 94 (med = 88) but varied from 1 to 378 in a month (figure A.1a). An average of 1,649 sightings of elephants were made during an average period of 24 days in each month. The total sightings database provides information on relationships among group type, group size, habitat types, locations, and activities. Individual identification of families or males in a sighting contributed to analyses of social structure and dynamics. On average, individual families were recorded as present in sightings 28 times per year, with considerable variation (figure A.2). Those families that were consistently observed in central areas had higher annual sightings than did the more peripheral families during the analysis period (see below).

The 30-plus-year sightings database contains a spatial reference field in the form of a 1 km² grid basemap that was established at the project's outset (see below for further details). AERP researchers generally headed for where elephants were expected to be found in order to maximize data collection on known individuals. Thus, when we aggregate

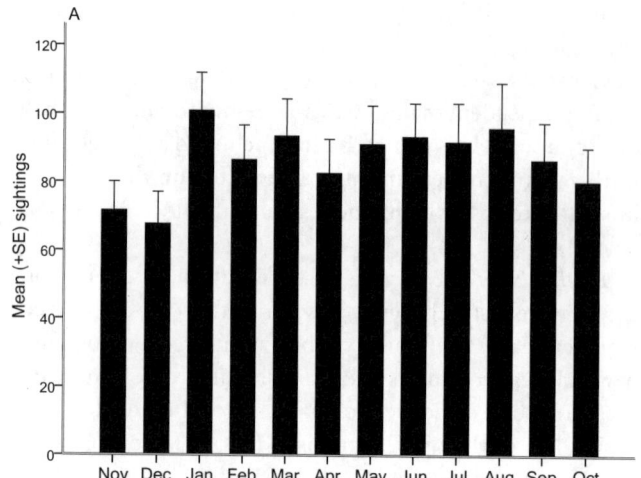

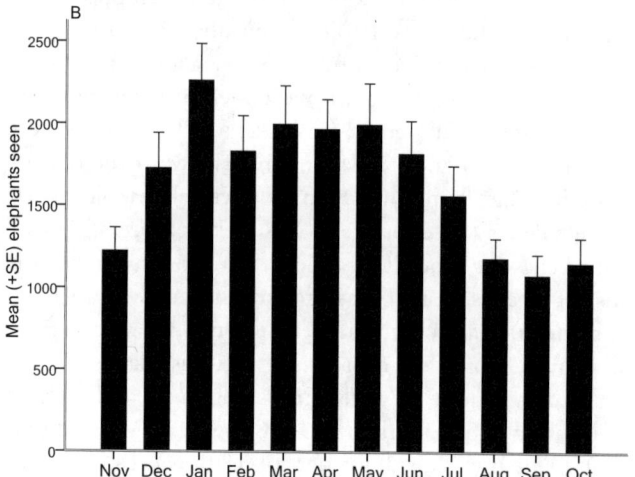

Figure A.1. (A) Mean (+SE) of monthly sightings across 30 years of the study; (B) Mean (+SE) of number of elephants seen per month across 30 years of the study.

observations of elephant groups from the sightings database in order, for example, to illustrate spatial distribution, in fact we get a combined picture of both elephant family distribution and the distribution of observer effort. Thus, the plot of numbers of observations of individual groups over time (figure A.2) illustrates both the magnitude of elephant presence as well as the effort (time, numbers of observers) expended to find the elephants.

In Amboseli, it was not difficult to find elephants, and invariably some could be found within the National Park. The search path was neither random nor systematic but determined by experienced hunches based on a growing knowledge base of where elephants might be at a particular time, constrained by the existing road network. Only after the addition of GPS devices to the researchers' kit in early 2000 were we able to examine search effort in terms of time and tracks followed (figure A.3). Future analyses will be able to take account of spatial expressions of observer effort.

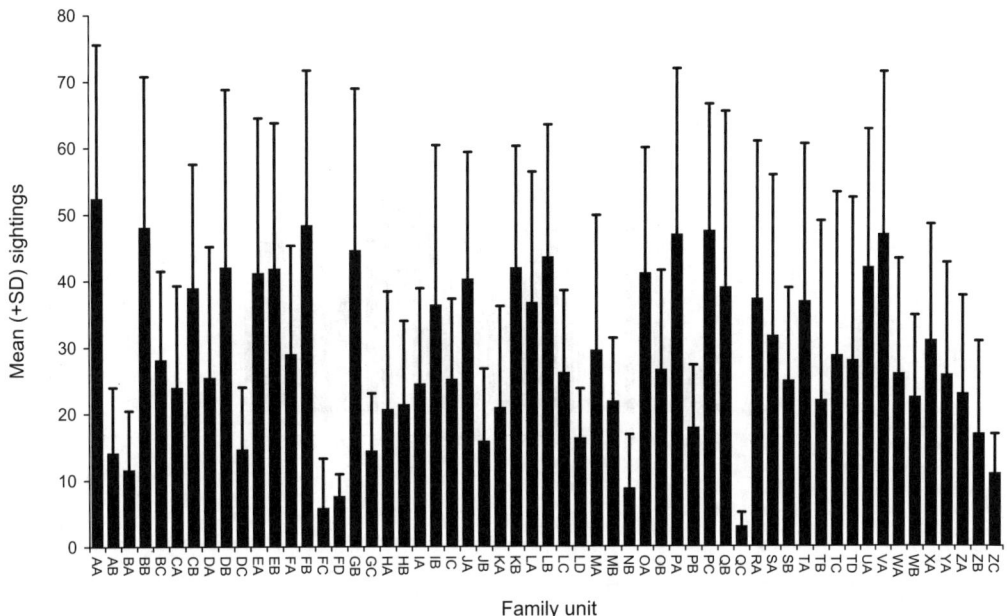

Figure A.2. Mean (+SE) annual sightings for individual family units over the study period.

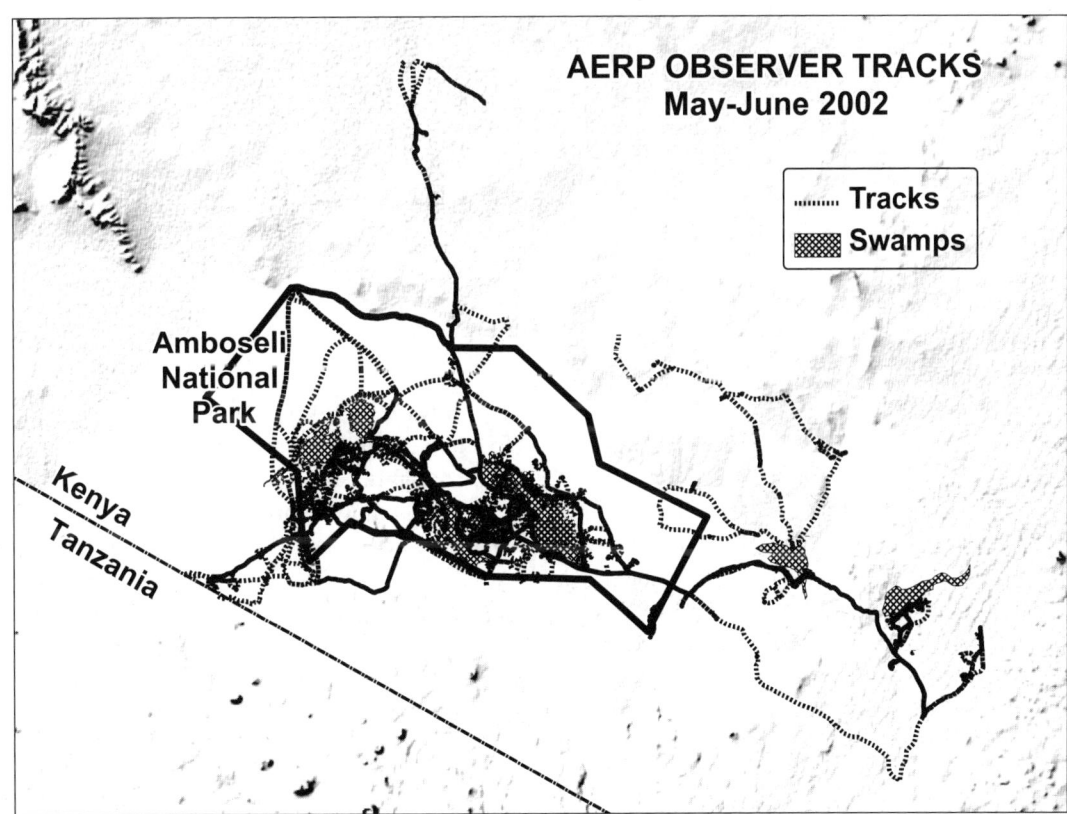

Figure A.3. Search tracks of observers within the basin.

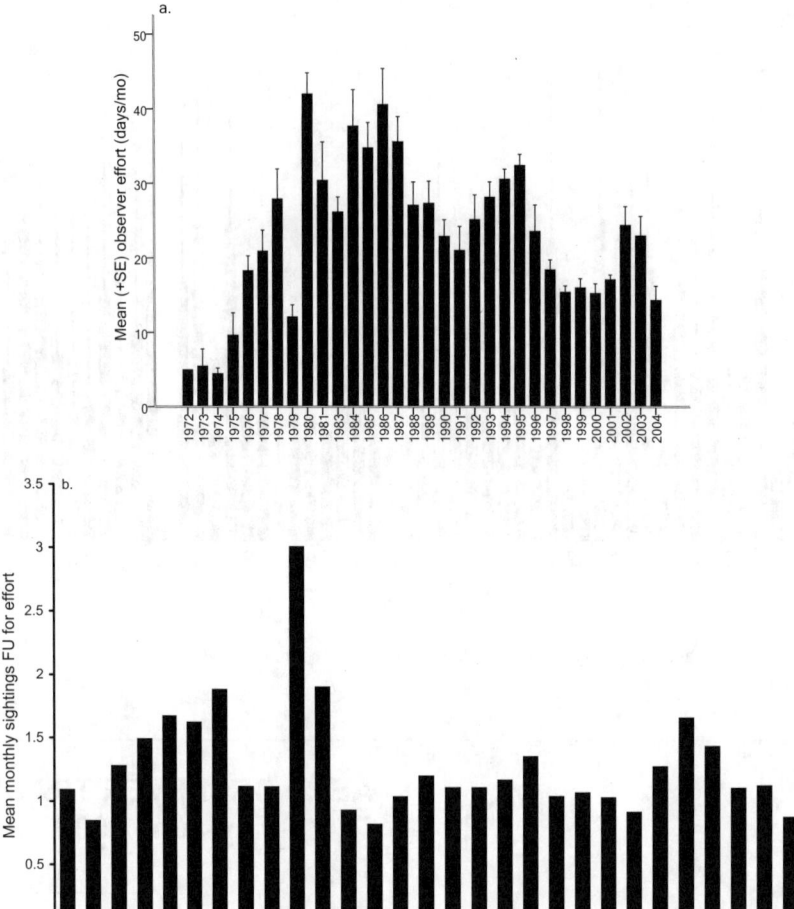

Figure A.4. Observer effort per month in each year (a) and mean number of families seen per month corrected for observer effort (b).

There is, however, a surrogate measurement for observer effort that can be extracted from the long-term database, which is the number of observer days in each month. There was observer presence in the study area for 328 out of the 360 months of the study (all but 5 of 32 missing months were in 1982 or before 1976). Over the course of the study, the number of different observers in any one year ranged from 1 to 9 (mean 3, mode 2), but they were not necessarily simultaneously present. Observer effort was not spread evenly over the year.

The distribution of observer days over the study period (figure A.4a) indicates a peak of observer effort in the 1980s (with the exception of 1982, for which data were lost) and a relatively higher level of effort from 1997 on. The observer effort–corrected sightings (figure A.4b) suggest that between the start of the study in 1972 and 1991, elephants appeared to use the central protected area of the Park very frequently. Most major family units were sighted more frequently ($N = 42/48$) between the decades. However, after 1991 and the end of a period of marked dryness, the probability of sighting any family decreased by an average factor of 3 (43/48 family sightings declined). From 1992 onward, the density of families in the central basin appeared to be far lower, group sizes were smaller, and eastern families were no longer occupying central habitats to the same extent. By 1997, as seen from the figure, despite high observer effort, groups and families were seen more and more rarely, as the population spent more time in areas outside the central protected area. In some analyses, we have used 1992 to distinguish between periods of relatively high density within the central basin and those in which density was increasingly low despite an overall population increase.

As discussed above, some males were rarely seen while some were more regularly contacted. In addition, each male had a unique duration of presence in the sightings, depending on his age at independence and longevity. Thus, the

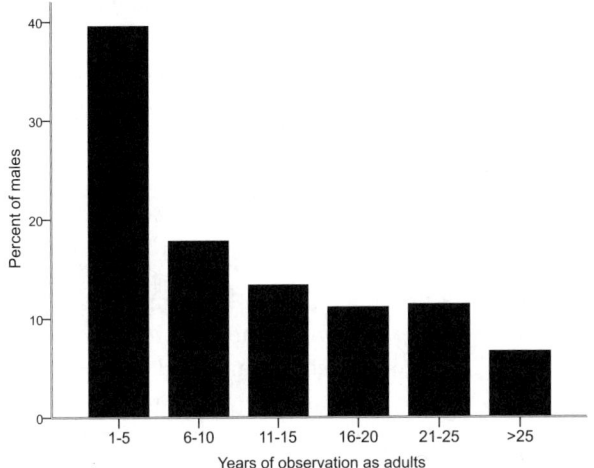

Figure A.5. Distribution of the duration of samples for individual males ($N = 359$) who contributed to sightings of adult males over the period of the study (1–5 years = 142; 6–10 = 64; 11–15 = 48; 16–20 = 40; 21–25=41; >25 = 24).

average annual number of sightings for identified independent males was highly variable, from 1 to 57 (mean = 9, $N = 355$; figure A.5). Males who came into musth were seen more often in a year than were non-musth males, possibly since they associated with larger aggregations and thus were more likely to be detected.

For the sightings of transitional males (males aged 16–18 years), the total group size, number of males, estimated ages of the males, and their identity were recorded whenever possible. Solitary elephants, typically bulls (Poole 1982), were also recorded during sightings. If the contacted group contained females, the family units present in the group were identified, and all members (known from long-term observations) were determined as present or absent. Only one sighting was made for each identified individual male or family in a single day. When all the bulls present in a group could not be individually identified (e.g., groups containing several hundred elephants), their approximate age class was noted.

Sources of Population Numbers and Occupancy

Data on elephant numbers and occupancy in the ecosystem were derived from several sources:

- systematic aerial survey (Croze 1978; Western and Lindsay 1984);
- total elephant counts (e.g., Ojwang', Waragute, and Njiro 2006);
- opportunistic overflights;
- radio tracking (Western and Lindsay 1984) and satellite GPS tracking (Douglas-Hamilton; see box 7.1);
- adventitious data collection from ground observers, in this case, the Maasai Elephant Scouts; and
- the AERP long-term sightings relational database.

Aerial Counts

Systematic reconnaissance flights (SRF). Systematic aerial survey gives an unbiased picture of "instantaneous" distribution in an area (Norton-Griffiths 1978). The SRF originated in the Serengeti Research Institute in Tanzania in the late 1960s, developed by a team of cooperating scientists (H. Croze, H. Kruuk, H. Lamprey, M. Norton-Griffiths, C. Pennycuick, A. Sinclair, and R. Skoog). Subsequent refinements were developed in Kenya by S. Cobb, D. Western, and M. Norton-Griffiths. Typically, crews of four in a high-winged light aircraft fly parallel transects 5–10 km apart at 300–500 feet above the ground, recording animal species and number, habitat features, human activities, and other items of interest. There have been several SRFs over the Amboseli ecosystem during the study period:

- 1973 to the present: H. Croze (until 1975) and subsequently D. Western undertook SRFs over an area of 8,500 km² comprising the southeastern Kajiado District, approximately two flights per year (elephant data unpublished, but see Western 1975; Western and Lindsay 1984).
- From 1973 to 1976, the United Nations Development Programme/Food and Agriculture Organization Wildlife Management Project in Kenya undertook SRFs as part of a Kajiado District–wide survey regime (Croze 1978).
- The Kenyan Department of Resource Survey and Remote Sensing (DRSRS, formerly known as the Kenya Rangeland Ecological Monitoring Unit, or KREMU) conducts SRFs throughout the country's rangelands. The southeastern Kajiado District was surveyed between 1988 and 2000.

Total counts. Counting a low-density, highly aggregated target species requires that the sampling fraction be relatively high in order to achieve acceptable error estimates (Norton-Griffiths 1978). Therefore, throughout Africa, practitioners undertake so-called total counts of elephants (see Douglas-Hamilton 1996). Typically, the survey area is divided into manageable blocks and overflown with parallel overlapping flightlines in a contiguous time period. Considerable flying time and several aircraft are needed in order to minimize missing or double counting groups moving between adjacent blocks. Total counts are expensive and personnel intensive and thus not undertaken lightly or frequently. There was only one total count during the study period, in August 2000. In addition, D. Western (2008, personal communication) has been conducting a series of monthly basin counts since the early 1980s. The publication of these data is in progress.

Opportunistic overflights. Light aircraft were used for transport from Nairobi to the Amboseli study area, with trips in and out of the basin at least once a month. These opportunistic overflights—which included ad hoc reconnaissance forays to look for injured elephants, flights to assist KWS staff in logistical operations, etc.—were an opportunity to gather information on location and numbers of elephants in the ecosystem. Between 1995 and 2002, H. Croze made 65 such flights and recorded location and numbers of elephants seen.

Radio and satellite collars. Elephant tracking using conventional VHF radio beacons was first used in Amboseli in the early 1970s. The technique defined the home ranges of several individuals and allowed them to be followed into inaccessible terrain not normally covered by the monitoring team on the ground. AERP researchers immobilized and put radio-tracking collars on six females between 1974 and 1979. Researchers, together with collaborator Iain Douglas-Hamilton, put collars on four more females and four adult males between 1995 and 1998 (see box 7.1). The reliability and lifetime of the early radio-tracking equipment was such that only a relatively modest amount of data could be collected (Douglas-Hamilton 1971). From the original six females, there were 276 resightings in all, an average of 46 per animal (compared with thousands of data points from contemporary GPS systems). Nonetheless, these data add information to the overall occupancy picture.

Satellite GPS tracking of elephants has been pioneered in Africa by Iain Douglas-Hamilton, and we are grateful to him and Save the Elephants for sharing his Amboseli data with the AERP. The urgent management needed in order to document and understand the movement of Amboseli bulls across the international border into Tanzania inspired the first application of GPS animal tracking in Africa in 1995. The new generation of GPS-linked transponders provides excellent short-term time-series location data and significantly reduces human position recording errors (see Douglas-Hamilton 1998; Douglas-Hamilton, Krink, and Vollrath 2005). In 1996, two Amboseli bulls, Ganesh and Mr. Nick, were immobilized and fitted with Lotek transponders (see below and box 7.1).

Ground Observations

In addition to the ground-based sightings by AERP researchers (see above) were those by the project's Maasai Elephant Scouts. In the late 1990s, the AERP engaged 10 young Maasai warriors to patrol areas of the ecosystem surrounding the National Park. One of the AERP-supported graduate students had trained the scouts to use simple GPS devices to record position and enter elephant observations onto data sheets, including numbers seen and type of group, dung piles, footprints, and any other sign of elephants such as vegetation recently broken by feeding.

Basemap

The sightings Access database contains a spatial reference field in the form of a 1 km^2 grid basemap that was established at the project's outset within the Universal Transverse Mercator (UTM) map projection framework (WGS84 datum, UTM 37S). The grid was established to cover some 3 km beyond the Park boundary (the extent can be seen in occupancy distribution maps such as those in chapter 7). The standard UTM map datum in 1972 was ARC1960 whereas modern GIS work is based on WGS84. In practice, that makes a difference of ca. 200 m on the ground, which is considered small enough to ignore within scale of habitat type changes and the range of positional errors from dead reckoning.

Behavioral Scans and Focal Samples

Detailed definitions and descriptions of elephant behavior are found in the ethogram in chapter 8. Several chapters rely on standard individually based observational sampling methods such as scan samples or focal animal samples as outlined by Altmann (1974), Martin and Bateson (1993), and Lehner (1996). Since each sampling method was unique to the specific study, details of these methods are presented in the relevant chapters. Focal samples tend to record behavior and interactions on a continuous basis from a single target individual while scans are instantaneous observations of either a single animal or all individuals visible in a group.

Over the course of the project, focal samples have been collected on feeding (WKL), calves, mothers and allomothers (CJM, PCL), young male behavior (PCL), musth male behavior (JHP, JH-S), matriarchs and group dynamics (HM), responses to humans (KK, LB), and female-female interactions in relation to kinship and dominance (BA). Other behavior samples have recorded the context of vocalizations (KM, JHP, LB).

Due to the difficulty of locating specific individuals at predetermined times of day or within set periods, sample sizes for individual focal samples tend to be limited compared with those of scans or the sightings data on families or individual males. In most cases, focal samples were used in analyses of hourly rates of behavior or interaction, with all the problems of statistical bias and small sample size that such samples entail.

Scan samples, where specific behaviors were recorded at predetermined intervals, were also widely used to explore diets (WKL), inter-individual proximity and association (PCL, JHP, CJM), activity budgets (HM, WKL), and changing group dynamics over a sampling interval (CJM, PCL, JHP, HM, BA, KM).

As is typical in behavioral studies, all occurrences of specific interactions or behavior were also recorded and analyzed. Such ad libitum data contributed to analyses of rare events or unusual behavior.

Associations and Association Indices

An association is determined by proximity between two individuals or their co-presence in the same group or aggregation. In some analyses, the percentage of observations where two individuals or families were in proximity over a series of scans or sightings was used. Generally, however, an index of association was calculated.

We used two types of association indices (Martin and Bateson 1993). The first is the simple index of association (observations of A + B together/observations of A not with B + observations of B not with A + observations of A + B together). This index was used to explore the strength of an association between two individuals relative to associations with all others (see also Ginsberg and Young 1992). The other index used was the half weight, or Sorensen's index, (observations of A + B together / [observations of A with others + observations of B with others] + ½ · [observations of A alone + observations of B alone]). It was applied to those cases where individuals could be in association with any other individual but could also be alone. Because we were specifically interested in associations between individuals rather than in the probability of being alone, and since the half-weight index biases values toward time in association, we used this index when variation in time alone was known to exist. If all individuals are equally solitary or sociable, then the use of any index will give comparable estimates of the strength of an association. However, if some individuals are relatively solitary, then their associations will appear to be reduced relative to more sociable individuals when using a simple index.

Cluster analysis of associates was done using hierarchical clustering procedures in SPSS Version 14 with either Ward's method (analysis of variance approach to the distance between clusters) or average linkage. While robust and efficient (Wittemeyer, Douglas-Hamilton, and Getz 2005), Ward's method tends to produce small clusters while average linkage may be biased by individuals with low indices.

Statistical Analyses in This Volume

Analyses were carried out using Excel, SPSS Versions 10–15, and Statistica Version 5 along with other statistical packages (detailed in relevant chapters). Specific analysis techniques for genetic data and sound analyses are given in the relevant chapters. Where data were normally distributed, parametric tests were appropriate and means ± SD are presented. In many cases, data were log-normalized for use with parametric statistics. Where data could not be normalized or where sample sizes were very small, non-parametric tests were used and medians plus inter-quartile ranges presented.

A variety of statistical tests were used in relevant comparisons throughout the chapters. These are detailed in each chapter. Comparisons of frequencies used Kruskal-Wallis tests and analysis of variance (ANOVA). When performing analysis of variance with three or more categories in the fixed factors, post hoc analyses were sometimes applied. For equal variances, Tukey's was used; for equal variance not assumed, Dunnett's T3 was used. Student's t-test was used to compare between means.

In analyses of interval data, proportional hazards models were used, either Kaplan-Meier to allow for the entry of covariates or Cox's regression for full-life table statistics. When there were no right censored intervals and large sample sizes (e.g., inter-birth intervals), survival analyses were not used, and standard comparisons were made on means or medians.

The results of statistical tests are presented giving the test statistic, the degrees of freedom (df) or N, and the probability values, which were set to $p < 0.05$ for significance, although some findings on small samples are reported with $p < 0.10$ as trends. Where repeated tests were made on the same dataset, a Bonferroni correction to the p value (P/N tests) was applied to determine significance. The power of the statistical tests varied greatly, as partitioning of data collected on individuals over 30 years tends to produce both extremes of huge and tiny samples. In general, we have not been able to set consistent assumptions about statistical power.

Associations of variables were tested through correlation and regression. In regression analyses, standard least squares models were used, and the unstandardized residual was computed when comparing the deviation of individual points from the mean association. Residuals were also determined for logistic regressions and used in subsequent analyses.

Appendix 2 Large Animal Species Referred to in the Book

African forest elephant (*Loxodonta africana cyclotis*)
African savannah elephant (*Loxodonta africana africana*)
African wild dog (*Lycaon pictus*)
Asian elephant (*Elephas maximus*)
Bohor's reedbuck (*Redunca redunca*)
Burchell's zebra (*Equus burchelli*)
Cape buffalo (*Syncerus caffer*)
Cheetah (*Acinonyx jubatus*)
Common waterbuck (*Kobus ellipsiprymnus*)
Dik-dik (*Madoqua kirkii*)
Domestic goat (*Capra hircus*)
Eland (*Taurotragus oryx*)
Gerenuk (*Litocranius walleri*)
Grant's gazelle (*Eudorcus grantii*)
Hippopotamus (*Hippopotamus amphibius*)
Impala (*Aepyceros melampus*)
Kongoni, Coke's hartebeest (*Alcelaphus buselaphus cokii*)
Lesser kudu (*Tragelaphus imberbis*)
Lion (*Panthera leo*)
Maasai (Zebu) cattle (*Bos indicus*)
Maasai giraffe (*Giraffa camelopardalis*)
Oryx (*Oryx beisa*)
Ostrich (*Struthio camelus*)
Red Maasai sheep (*Ovis aries*)
Rhinoceros, Black rhino (*Diceros bicornis*)
Spotted hyena (*Crocuta crocuta*)
Striped hyena (*Hyaena hyaena*)
Thomson's gazelle (*Eudorcus thomsonii*)
Wildebeest, White-bearded gnu (*Connochaetes taurinus*)

References

Abbot, J. I., D. H. L. Thomas, A. A. Gardner, S. E. Neba, and M. W. Khen. 2001. Understanding the links between conservation and development in the Bamenda Highlands, Cameroon. *World Development* 29:1115–36.

Abe, E. L. 1995. The behavioural ecology of elephant survivors in Queen Elizabeth National Park, Uganda. PhD thesis, University of Cambridge.

Abrahams, M. V. 1986. Patch choice under perceptual constraints: A cause for departures from an Ideal free distribution. *Behavioral Ecology and Sociobiology* 19:409–15.

Adams, J., and J. K. Berg. 1980. Behavior of female African elephants (*Loxodonta africana*) in captivity. *Applied Animal Ethology* 6:267–76.

Adams, J. S., and T. O. McShane. 1992. *The myth of wild Africa: Conservation without illusion.* New York: W. W. Norton & Co.

Adams, W. M., and D. Hulme. 2001. If community conservation is the answer in Africa, what is the question? *Oryx* 35: 193–200.

Adams, W. M., and M. Infield. 2001. Park outreach and gorilla conservation: Mgahinga Gorilla National Park, Uganda. In *African wildlife and livelihoods: The promise and performance of community conservation*, ed. D. Hulme and M. Murphree, 131–47. Oxford: James Currey.

Afolayan, T.A., and M. Fafunsho. 1978. Seasonal variation in the protein content and the grazing of some tropical savannah grasses. *African Journal of Ecology* 16:97–104.

African Wildlife Foundation (AWF). 1999. People, elephants and conservation in the greater Amboseli ecosystem. Unpublished research proposal. Washington, DC: African Wildlife Foundation.

———. 2006. *Annual report.* Nairobi: African Wildlife Foundation.

———. 2008 Lease program established around Amboseli National Park, Kenya. *African Heartland News.* August–December 2008. Washington, DC: African Wildlife Foundation.

Akama, J. S. 1999. Marginalization of the Maasai in Kenya. *Annals of Tourism* 26:716–18.

———. 2002. The creation of the Maasai image and tourism development in Kenya. In *Cultural tourism in Africa: Strategies for the new millennium*, ed. J. Akama and P. Sterry, 43–53. Arnhem: Association for Tourism and Leisure Education.

Akama, J. S., C. L. Lant, and G. W. Burnett. 1995. Conflicting attitudes towards state wildlife conservation programmes in Kenya. *Society and Natural Resources* 8:133–44.

Alberts, S. C., and J. Altmann. 1995. Balancing costs and opportunities—dispersal in male baboons. *American Naturalist* 145:279–306.

Alberts, S. C., H. Watts, and J. Altmann. 2003. Queuing and queue jumping: Long term patterns of dominance rank and mating success in male savannah baboons. *Animal Behaviour* 65:821–40.

Albon, S. D., T. H. Clutton-Brock, and F. E. Guinness. 1987. Early development and population dynamics in red deer. II. Density—independent effects and cohort variation. *Journal of Animal Ecology* 56:69–81.

Alexander, R. D. 1974. The evolution of social behavior. *Annual Review of Ecology and Systematics* 5:324–83.

Ali, I. M. 2006. An anthropocentric approach to saving biodiversity: Kenyan pupils' attitudes towards parks and wildlife. *Applied Environmental Education and Communication* 5:1–32.

Allen, W. R. 2006. Ovulation, pregnancy, placentation and husbandry in the African elephant (*Loxodonta africana*). *Philosophical Transactions of the Royal Society, Series B* 361: 821–34.

Alonso, J. C., J. A. Alonso, L. M. Bautista, and R. Munoz-Palido. 1995. Patch use in cranes: A field test of optimal foraging predictions. *Animal Behaviour* 49:1367–79.

Altmann J. 1974. Observational study of behavior: Sampling methods. *Behaviour* 49:227–65.

———. 1980. *Baboon mothers and infants.* Chicago: University of Chicago Press.

Altmann, J., and P. Muruthi. 1988. Differences in daily life between semi-provisioned and wild-feeding baboons. *American Journal of Primatology* 15:213–22.

Altmann, J., G. Hausfater, and S. A. Altmann. 1988. Determinants of reproductive success in savannah baboons, *Papio cynocephalus*. In *Reproductive success*, ed. T .H. Clutton-Brock, 403–18. Chicago: University of Chicago Press.

Altmann, J., S. Alberts, S. A. Altmann, and S. B. Roy. 2002. Dramatic change in local climate patterns in the Amboseli Basin, Kenya. *African Journal of Ecology* 40:248–51.

Altmann, S. A. 1998. *Foraging for survival: Yearling baboons in Africa*. Chicago: University of Chicago Press.

Alvard, M. 1993. Testing the "ecologically noble savage" hypothesis: Interspecific prey choice by Piro hunters of Amazonian Peru. *Human Ecology* 21:355–87.

———. 1995. Intraspecific prey choice by Amazonian hunters. *Current Anthropology* 36:789.

Amin, M., D. Willetts, and J. Eames. 1987. *The last of the Maasai*. Nairobi: Camerapix Publishers International.

Amos, W., C. Schlotterer, and D. Tautz. 1993. Social structuring of pilot whales revealed by analytical DNA profiling. *Science* 260:670–72.

Amos, W., P. P. Pomeroy, S. D. Twiss, and S. S. Anderson. 1995. Evidence for mate fidelity in the grey seal. *Science* 268: 1897–99.

Ananthasubramanian, C. R. 1980. A note on the nutritional requirements of the Asian elephant (*Elephas maximus indicus*). *Elephant* 1 (supplement): 72–73.

Andelman, S. J. 1985. *Home range utilization and competition in female African elephants*. Unpublished progress report. Nairobi: Kenya Wildlife Service.

Anderson, J. R. 1984. Monkeys with mirrors: Some questions for primate psychology. *International Journal of Primatology* 5: 81–98.

Anderson, R., P. Duncan, and J. D. C. Linnell. 1998. *The European roe deer: The biology of success*. Oslo: Scandinavian University Press.

Anderson, U. S., T. S. Stoinski, M. A. Bloomsmith, and T. L. Maple. 2007. Relative numerousness judgment and summation in young, middle-aged, and older adult orangutans (*Pongo pygmaeus ablii* and *Pongo pygmaeus pygmaeus*). *Journal of Comparative Psychology* 121:1–11.

Anthony, B. 2007. The dual nature of parks: Attitudes of neighboring communities toward Kruger National Park, South Africa. *Environmental Conservation* 34:236–45.

Archabald, K., and L. Naughton-Treves. 2001. Tourism revenue-sharing around national parks in Western Uganda: Early efforts to identify and reward local communities. *Environmental Conservation* 28:135–49.

Archie, E. A. 2005. The relationship between kinship and social behavior in wild African elephants. PhD thesis, Duke University.

Archie, E. A., C. J. Moss, and S. C. Alberts. 2003. Characterization of tetranucleotide microsatellite loci in the African savannah elephant (*Loxodonta africana africana*). *Molecular Ecology Notes* 3:244–46.

———. 2006. The ties that bind: Genetic relatedness predicts the fission and fusion of groups in wild African elephants (*Loxodonta africana*). *Proceedings of the Royal Society Series B* 273:513–22.

Archie, E. A., T. Morrison, C. Foley, C. J. Moss, and S. Alberts. 2006. Dominance rank relationships among wild female African elephants (*Loxodonta africana*). *Animal Behaviour* 71:117–27.

Archie, E. A., J. E. Maldonado, J. A. Hollister-Smith, J. H. Poole, C. J. Moss, R. C. Fleischer, and S. C. Alberts. 2008. Fine-scale population genetic structure in a fission-fusion society. *Molecular Ecology* 17:2666–79.

Archie, E. A., J. A. Hollister-Smith, J. H. Poole, P. C. Lee, C. J. Moss, J. E. Maldonado, R. C. Fleischer, and S. C. Albert. 2007. Behavioural inbreeding avoidance in wild African elephants. *Molecular Ecology* 16:4128–48.

Arctander. P., C. Johansen, and M. A. Coutellec-Vetro. 1999. Phylogeography of three closely related African bovids. *Molecular Biology and Evolution* 16:1724–39.

Arluke, A., and C. Sanders. 1996. *Regarding animals*. Philadelphia: Temple University Press.

Armitage, K. B., and O. A. Schwartz. 2000. Social enhancement of fitness in yellow-bellied marmots. *Proceedings of the National Academy of Sciences USA* 97:12149–52.

Asiema, J. K., and F. D. P. Situma. 1994. Indigenous peoples and the environment: The case of the pastoral Maasai of Kenya. *Colorado Journal of International Environmental Law and Policy* 5:149–71.

Astrom, M. 1994. Travel cost and the ideal distribution. *Oikos* 69:516–19.

Atmoko, S. U., and J. A. R. A. M. van Hooff. 2004. Alternative male reproductive strategies: Male bimaturism in orangutans. In *Sexual selection in primates*, ed. P. M. Kappeler and C. P. van Schaik, 96–207. Cambridge: Cambridge University Press.

Atsalis, S., S. W. Margulis, and P. R. Hof. 2008. *Primate reproductive aging: Cross-taxon perspectives*. Basel: Karger.

Aureli, F., C. M. Schaffner, C. Boesch, S. Bearder, J. Call, C. A. Chapman, R. Connor, A. Di Fiore, R. I. M. Dunbar, P. S. Henzi, K. Holekamp, A. H. Korstjens, R. Layton, P. Lee, J. Lehmann, J. H. Manson, G. Ramos-Fernandez, K. Strier, and C. P. van Schaik. 2008. Fission-fusion dynamics: New research directions. *Current Anthropology* 49:627–54.

Axelrod, R., and W. D. Hamilton. 1981. The evolution of cooperation. *Science* 211:1390–96.

Baird, R. W. 2000. The killer whale: Foraging specializations and group hunting. In *Cetacean societies*, ed. J. Mann, R. C. Connor, P. L. Tyack, and H. Whitehead, 127–53. Chicago: University of Chicago Press.

Balfour, D., and S. Balfour. 1997. *African elephants: A celebration of majesty*. Cape Town: Struik Publishers.

Balfour, D., H. T. Dublin, J. Fennessy, D. Gibson, L. Niskanen, and I. J. Whyte. 2007. *Review of options for managing the impacts of locally overabundant African elephants*. Gland, Switzerland: AESG, IUCN.

Bandyopadhyaya, S. 2004. *Do households gain from community-based natural resource management? An evaluation of community conservancies in Namibia*. Policy Research Working Paper. Washington, DC: World Bank.

Barbier, E. B., J. C. Burgess, T. M. Swanson, and D. W. Pearce. 1990. *Elephants, economics and ivory*. London: Earthscan Publications Ltd.

Barfield, C. H., Z. Tangmartinez, and J. M. Trainer. 1994. Domestic calves (*Bos taurus*) recognize their own mothers by auditory cues. *Ethology* 97:257–64.

Barker, D. J. P. 1994. *Mothers, babies and disease in later life*. London: BMJ Publishing Group.

Barnard C. J. 1984. *Producers and scroungers: Strategies of exploitation and parasitism.* New York: Chapman and Hall.
———. 1991. Kinship and social behaviour: The trouble with relatives. *Trends in Ecology and Evolution* 6:310–12.
Barnard, C. J., and R. M. Sibly. 1981. Producers and scroungers: A general model and its application to captive flocks of house sparrows. *Animal Behaviour* 29:543–50.
Barnes, R. F. W. 1979. Elephant ecology in Ruaha National Park, Tanzania. PhD thesis, University of Cambridge.
———. 1982. Elephant feeding behaviour in Ruaha National Park, Tanzania. *African Journal of Ecology* 20:123–36.
———. 1996. The conflict between humans and elephants in the Central African forests. *Mammal Review* 26:67–80.
Barnes, R. F. W., and K. L. Jensen. 1987. How to count elephants in forests. *Technical Bulletin 1*. Gland, Switzerland: IUCN, African Elephant and Rhino Specialist Group.
Barnes, R. F. W., and E. B. Kapela. 1991. Changes in the Ruaha elephant population caused by poaching. *African Journal of Ecology* 29:89–94.
Barnes, R. F. W., M. P. T. Alers, and A. Blom. 1989. The poor man's guide to counting elephant faeces in forests. Unpublished technical manuscript. Gland, Switzerland: IUCN African Elephant and Rhino Specialist Group.
Barnes, R. F. W., K. L. Barnes, M. P. T. Alers, and A. Blom. 1991. Man determines the distribution of elephants in the rain forests of northeastern Gabon. *African Journal of Ecology* 29:4–63.
Barrett, C. B., and P. Arcese. 1995. Are ICDPs sustainable? On the conservation of large mammals in sub-Saharan Africa. *World Development* 23:1073–85.
Barrett, L., S. P. Henzi, T. Weingrill, J. E. Lycett, and R. A. Hill. 1999. Market forces predict grooming reciprocity in female baboons. *Proceedings of the Royal Society Series B* 266: 665–70.
Barrow, E., H. Gichohi, and M. Infield. 2000. *Rhetoric or reality: A review of community conservation policy and practice in East Africa.* Evaluating Eden Series 5. London: IUCD Biodiversity and Livelihoods Group, World Conservation Union.
Barta, Z., and L. A. Giraldeau. 1998. The effect of dominance hierarchy on the use of alternative foraging tactics: A phenotypic-limited producing-scrounging game. *Behavioral Ecology and Sociobiology* 42:217–23.
Barton, R., A. Purvis, and P. H. Harvey. 1995. Evolutionary radiation of visual and olfactory brain systems in primates, bats and insectivores. *Proceedings of the Royal Society Series B* 348: 381–92.
Bates, L. A., J. H. Poole, and R. W. Byrne. 2008. Elephant cognition. *Current Biology* 18:R544–46.
Bates, L. A., K. Sayialel, N. Njiraini, C. J. Moss, J. H. Poole, and R. W. Byrne. 2007. Elephants classify human ethnic groups by odor and garment color. *Current Biology* 17:1938–42.
Bates, L. A., K. Sayialel, N. Njiraini, J. H. Poole, C. J. Moss, and R. W. Byrne. 2008a. African elephants have expectations about the locations of out-of-sight family members. *Biology Letters* 4:34–36.
Bates, L. A., P. C. Lee, N. Njiraini, J. H. Poole, K. Sayialel, S. Sayialel, C. J. Moss, and R. W. Byrne. 2008b. Do elephants show empathy? *Journal of Consciousness Studies* 15:204–25.
Bates, L. A., R. Handford, P. C. Lee, N. Njiraini, J. H. Poole, K. Sayialel, S. Sayialel, C. J. Moss, and R. W. Byrne. 2010. Why do African elephants (*Loxodonta africana*) simulate oestrus? An analysis of longitudinal data. *PLoS One* 5:e10052.
Bateson, P. P. G. 1981. Ontogeny of behaviour. *British Medical Journal* 37:159–64.
Bauer, H. 2003. Local perceptions of Waza National Park, northern Cameroon. *Environmental Conservation* 30:175–81.
Baum, W. M., and J. R. Kraff. 1998. Group choice, competition, travel and the ideal free distribution. *Journal of Experimental Analysis of Behavior* 69:227–45.
Baumann, C. 1894. *Durch Masailand zur Nilquelle. Reisen und Forschungen der Masai Expedition des deurschen Antisklaverei–Komite in den Jahren 1891–1893.* Berlin: Dietrich Reimer.
Bautista, L. M., J. C. Alonso, and J. A. Alonso. 1995. A field test of ideal free distribution in flocking common cranes. *Journal of Animal Ecology* 64:747–57.
Bayliss, P. 1985. The population dynamics of red and western grey kangaroos in arid New South Wales, Australia. I. Population trends and rainfall. *Journal of Animal Ecology* 64: 111–25.
Beachey, R. W. 1967. The East African ivory trade in the nineteenth century. *Journal of African History* 3:269–90.
Beck, B. B. 1980. *Animal tool behaviour.* New York: Garland Press.
Beer, C. G. 1970. Individual recognition of voice in the social behaviour of birds. *Advances in the Study of Behaviour* 3: 27–74.
Behera, S. K., J.-J. Luo, S. Masson, P. Delecluse, S. Gualdi, A. Navarra, and T. Yamagata. 2005. Paramount impact of the Indian Ocean dipole on the East African short rains: A CGCM study. *Journal of Climate* 18:4514–30.
Behnke, R. 1994. Natural resource management in Africa. *Development Policy Review* 12:5–27.
Behrensmeyer, A. K. 1993. The taphonomic record of ecological change in Amboseli Park, Kenya. *Research and Exploration* 9: 402–21.
Behrensmeyer, A. K., and D. E. Boaz. 1981. Late Pleistocene geology and paleontology of Amboseli National Park, Kenya. *Paleoecology of Africa and the Surrounding Islands* 13:125–88.
Behrensmeyer, A. K., C. T. Stayton, and R. E. Chapman. 1993. Taphonomy and ecology of modern avifaunal remains from Amboseli Park, Kenya. *Paleobiology* 9:52–70.
Bekoff, M., and J. A. Byers. 1998. *Animal play.* Cambridge: Cambridge University Press.
Bell, R. H. V. 1985. Elephants and woodland—a reply. *Pachyderm* 5:17–18.
Bell, W. D. M. 1931. On killing elephants. In *Game animals of the Sudan, their habits and distribution: A handbook for hunters and naturalists*, ed. H. C. Brocklehurst. London: Gurney and Jackson.
Belovsky, G. E. 1978. Diet optimization in a generalist herbivore: the moose. *Theoretical Population Biology* 14:105–34.
———. 1981. Optimal activity times and habitat choice of moose. *Oecologia* 48:22–30.
Benedict, F. G. 1936. *The physiology of the elephant.* Washington DC: Carnegie Institution of Washington.
Bennetts, L. 1996. African dreamer: Leslie Bennetts on Peter Beard. *Vanity Fair* (November). www.vanityfair.com/magazine/archive/1996/11/beard199611.
Ben-Shahar, R. 1999. *On the trail of the wild.* Cologne: Könemann.
Beran, M. 2001. Summation and numerousness judgments of sequentially presented sets of items by chimpanzees (*Pan*

troglodytes). *Journal of Comparative Psychology* 115:181–91.

Bercovitch, F. B. 2000. Behavioral ecology and socioendocrinology of reproductive maturation in cercopithecine monkeys. In *Old world monkeys*, ed. P. Whitehead and C. Jolly, 298–320. Cambridge: Cambridge University Press.

Bercovitch, F. B., A. Widdig, A. Trefilov, M. J. Kessler, J. D. Berard, J. Schmidtke, P. Nurnberg, and M. Krawczak. 2003. A longitudinal study of age-specific reproductive output and body condition among male rhesus macaques. *Macaca Mulatta Naturwissenschaften* 90: 309–12.

Berg, J. K. 1983. Vocalisations and associated behaviours of the African elephant in captivity. *Zeitschrift für Tierpsychologie* 63:63–79.

Berger, D. J. 1993. *Wildlife extension: Participatory conservation by the Maasai of Kenya*. Nairobi: ACT Press.

———. 1996. The challenge of integrating Maasai tradition with tourism. In *People and tourism in fragile environments*, ed. M. F. Price, 175–97. London: John Wiley and Sons Ltd.

Bernsten, J. L. 1976. The Maasai and their neighbors: Variables of interaction. *African Economic History* 2: 1–11.

———. 1980. The enemy is us: Eponymy in the historiography of the Maasai. *History in Africa* 7: 1–19.

Bertschinger, H., A. Delsink, J. J. van Altena, J. Kirkpatrick, H. Killian, A. Ganswindt, R. Solow, and G. Castley. 2008. Reproductive control of elephants. In *Elephant management: A scientific assessment for South Africa*, ed. R. J. Scholes and K. G. Mennell, 255–328. Johannesburg: Wits University Press.

Bérubé, C. H., M. Festa-Bianchet, and J. T. Jorgenson. 1996. Reproductive costs of sons and daughters in Rocky Mountain bighorn sheep. *Behavioral Ecology* 7: 60–68.

Birgersson, B. 1998. Male-biased expenditure and associated costs in fallow deer. *Behavioral Ecology and Sociobiology* 43: 87–93.

Birgersson, B., and K. Ekvall. 1997. Early growth in male and female fallow deer fawns. *Behavioral Ecology* 8:493–99.

Blackburn, R. H. 1982. *Okiek*. London: Evans Brothers Ltd.

———. 1996. Maasai/Dorobo objects: The ethnographic perspective. *African Arts* 29:12–15.

Blake, S., P. Bouche, L. E. L. Rasmussen, A. Orlando, and I. Douglas–Hamilton. 2003. The last Sahelian elephants: Ranging behaviour, population status and recent history of the desert elephants of Mali. Unpublished report. Nairobi: Save the Elephants.

Blake, S., S. Strindberg, P. Boudjan, C. Makombo, I. Bila–Isia, O. Ilambu, F. Grossmann, L. Bene–Bene, B. de Semboli, V. Mbenzo, D. S'hwa, R. Bayogo, L. Williamson, M. Fay, J. Hart, and F. Maisels. 2007. Forest elephant crisis in the Congo basin. *PLoS Biology* 5:e111.

Blanc, J. J., R. F. W. Barnes, G. C. Craig, H. T. Dublin, C. R. Thouless, I. Douglas–Hamilton, and J. A. Hart. 2007. *African elephant status report 2007: An update from the African Elephant Database*. Occasional Paper Series of the IUCN Species Survival Commission 33 Gland, Switzerland: IUCN/SSC African Elephant Specialist Group.

Blaxter, K. L. 1989. *Energy metabolism in animals and man*. Cambridge: Cambridge University Press.

Blaxter, K. L., and W. J. Hamilton. 1980. Reproduction in farmed red deer. 2. Calf growth and mortality. *Journal of Agricultural Science* 95:275–84.

Blignaut, J., M. de Wit, and J. Barnes. 2008. The economic value of elephants. In *Elephant management: A scientific assessment for South Africa*, ed. R. J. Scholes and K. G. Mennell, 446–76. Johannesburg: Wits University Press.

Blumstein, D. T., C. S. Evans, and J. C. Daniel. 1999. An experimental study of behavioral group size effects in tammar wallabies, *Macropus Eugenii*. *Animal Behaviour* 58:351–60.

Boesch, C. 1991. Teaching among wild chimpanzees. *Animal Behaviour* 41:530–532.

———. 1997. Evidence for dominant wild chimpanzees investing more in sons. *Animal Behaviour* 54:811–15.

Boesch, C., and H. Boesch. 1990. Tool use and tool making in wild chimpanzees. *Folia Primatologica* 54:86–99.

Bond, I. 1994. The importance of sport-hunted African elephants to CAMPFIRE in Zimbabwe. *Traffic Bulletin* (WWF/IUCN) 14:117–19.

Born Free. 2004. *A survey on the availability of bushmeat in 202 urban butcheries in Nairobi*. Nairobi: Born Free Foundation UK, Youth for Conservation, Kenya Wildlife Service.

Bouley, D. M., C. N. Alarcon, T. Hildebrandt, and C. E. C. O'Connell-Rodwell. 2007. The distribution, density and three-dimensional histomorphology of pacinian corpuscles in the foot of the African elephant (*Elephas maximus*) and their potential role in seismic communication. *Journal of Anatomy* 211:428–35.

Boulton, T. W., L. L. Tieszen, and S. K. Imbamba. 1988. Seasonal changes in the nutritional content of East African grassland vegetation. *African Journal of Ecology* 26:103–15.

Boysen, S. T., G. G. Bernston, and K. L. Mukobi. 2001. Size matters: Impact of item size and quantity on array choice by chimpanzees (*Pan troglodytes*). *Journal of Comparative Psychology* 115:106–10.

Bradshaw, G. A. 2004. Not by bread alone: Symbolic loss, trauma, and recovery in elephant communities. *Society and Animals* 12:143–58.

Bradshaw, G. A., A. N. Schore, J. L. Brown, J. H. Poole, and C. J. Moss. 2005. Elephant breakdown. *Nature* 433:807.

Brashares, J. S., P. Arcese, M. K. Sam, P. B. Coppolillo, A. R. Sinclair, and A. Balmford. 2004. Bushmeat hunting, wildlife declines, and fish supply in West Africa. *Science* 306:1180–83.

Brody, S. 1945. *Bioenergetics and growth*. New York: Reinhold.

Brommer, J. E. 2000. The evolution of fitness in life-history theory. *Biological Reviews* 75:377–404.

Broquet, T., and E. Petit. 2004. Quantifying genotyping errors in noninvasive population genetics. *Molecular Ecology* 13: 3601–08.

Brown, G. R. 2001. Using proximity measures to describe mother-infant relationships. *Folia Primatologica* 72:80–84.

Brown, G. R., and J. B. Silk. 2002. Reconsidering the null hypothesis: Is maternal rank associated with birth sex ratios in primate groups? *Proceedings of the National Academy of Science USA* 99:11252–55.

Browne-Nuñez, C. 2010. Resident attitudes and behaviors toward elephants around Amboseli National Park, Kenya: Evaluating human-elephant conflict mitigation interventions. PhD diss., University of Florida.

Browne-Nuñez, C., and S. Jonker. 2008. Attitudes toward wildlife and conservation across Africa: A review of survey research. *Human Dimensions of Wildlife* 13:49–72.

Bryant, J. P., P. B. Reichardt, T. P. Clausen, F. D. Provenza, and P. J. Kuropat. 1992. Woody plant–mammal interactions. In

Herbivores: Their interactions with secondary plant metabolites. Volume II. Evolutionary and ecological processes. 2nd ed., ed. G. A. Rosenthal and M. R. Berenbaum, 343–70. San Diego: Academic Press.

Buchan, J. C., E. A. Archie, R. C. Van Horn, and S. C. Alberts. 2005. Locus effects and sources of error in non-invasive genotyping. *Molecular Ecology Notes* 5:680–83.

Buckland, S. T., D. R. Anderson, K. P. Burnham, and J. L. Laake. 1993. *Distance sampling: Estimating abundance of biological populations.* London: Chapman and Hall.

Budde, C., and G. M. Klump. 2003. Vocal repertoire of the black rhino *Diceros bicornis spp.* and possibilities of individual recognition. *Mammalian Biology* 68:42–47.

Bugnyar, T. 2002. Observational learning and the raiding of food caches in ravens, *Corvus corax:* Is it "tactical" deception? *Animal Behaviour* 64:185–95.

Bugnyar, T., and B. Heinrich. 2005. Ravens, *Corvus corax*, differentiate between knowledgeable and ignorant competitors. *Proceedings of the Royal Society Series B* 272:1641–46.

Bulte, E. H., and G. C. van Kooten. 1999. Economics of antipoaching enforcement and the ivory trade ban. *American Journal of Agricultural Economics* 81:453–66.

Bulte, E. H., R. B. Boone, R. Stringer, and P. K. Thornton. 2006. Wildlife conservation in Amboseli, Kenya: Paying for non-use values. Report to Roles of Agriculture Project, Agricultural and Development Economics Division (ESA). Rome: UN FAO.

Burger, J., C. Safina, and M. Gochfield. 2000. Factors affecting vigilance in springbok; importance of vegetative cover, location in herd and herd size. *Acta ethologica* 2:97–104.

BurnSilver, S., and E. Mwangi. 2007. *Beyond group ranch subdivision: Collective action for livestock mobility, ecological viability, and livelihoods.* CAPRi Working Paper 66. Washington DC: International Food Policy Research Institute.

Buss, I. O. 1961. Some observations on food habits and behavior of the African elephant. *Journal of Wildlife Management* 25:131–48.

Buss, I. O., and J. A. Estes. 1971. The functional significance of the movements and position of the pinnae of the African elephant, *Loxodonta africana*. *Journal of Mammalogy* 52:21–27.

Buss, I. O., and N. S. Smith. 1966. Observations on reproduction and breeding behavior of the African elephant. *Journal of Wildlife Management* 30:375–88.

Buss, I. O., L. E. Rasmussen, and G. L. Smuts. 1976. The role of stress and individual recognition in the function of the African elephant's temporal gland. *Mammalia* 40:437–51.

Byers, B. 1996. Understanding and influencing behaviors in conservation and natural resources management. African Biodiversity Series 4, Biodiversity Support Program. Washington, DC: USAID.

Byers, J. A. 1997. *American pronghorn: Social adaptations and the ghosts of predators past.* Chicago: University of Chicago Press.

Byers, J. A., and J. D. Moodie. 1990. Sex-specific maternal investment in pronghorn, and the question of a limit on differential provisioning in ungulates. *Behavioral Ecology and Sociobiology* 26:157–64.

Byrne, R. W. 1979. Memory for urban geography. *Quarterly Journal of Experimental Psychology* 31:147–54.

———. 1996. Relating brain size to intelligence in primates. In *Modelling the early human mind*, ed. P. A. Mellars and K. R. Gibson. Cambridge: Macdonald Institute for Archaeological Research.

———. 1997. The technical intelligence hypothesis: An additional evolutionary stimulus to intelligence? In *Machiavellian intelligence II: Extensions and evaluations*, ed. A. Whiten and R. W. Byrne, 289–311. Cambridge: Cambridge University Press.

———. 2000. How monkeys find their way. Leadership, coordination, and cognitive maps of African baboons. In *On the move: How and why animals travel in groups*, ed. S. Boinski and P. Garber, 491–518. Chicago: University of Chicago Press.

Byrne, R. W., and L. A. Bates. 2006. Why are animals cognitive? *Current Biology* 16:R445–48.

———. 2007. Sociality, evolution, and cognition. *Current Biology* 17:R714–23.

Byrne, R. W., and N. Corp. 2004. Neocortex size predicts deception rate in primates. *Proceedings of the Royal Society of London Series B* 271:1693–99.

Byrne, R. W., and A. Whiten. 1988. *Machiavellian intelligence: Social expertise and the evolution of intellect in monkeys, apes and humans.* Oxford: Clarendon Press.

———. 1992. Cognitive evolution in primates: Evidence from tactical deception. *Man* 27:609–27.

Byrne, R. W., N. Corp, and J. M. Byrne. 2001. Manual dexterity in the gorilla: Bimanual and digit role differentiation in a natural task. *Animal Cognition* 4:347–61.

Caduto, M. J. 1985. A guide on environmental value education. UNESCO–UNEP International Environmental Education Series 13. Paris: UNESCO.

Cameron, E. Z. 1998. Is suckling behavior a useful predictor of milk intake? A review. *Animal Behaviour* 56:521–32.

Campbell, D. J. 1991. The impact of development upon strategies for coping with drought among the Maasai of Kajiado District, Kenya. In *Pastoral economies in Africa and long term responses to drought*, ed. J. Stone. Aberdeen: Aberdeen University African Studies Group.

———. 1993. Land as ours, land as mine: Political and ecological marginalization in Kajiado District. In *Being Maasai*, ed. T. Spear and R. Waller, 258–72. London: James Currey.

Campbell, D. J., H. Gichohi, A. Mwangi, and L. Chege. 2000. Land use conflict in Kajiado District, Kenya. *Land Use Policy* 17:337–48.

Campbell, D. J., H. Gichohi, R. Reid, A. Mwangi, L. Chege, and T. Sawin. 2003. Interactions between people and wildlife in Southeast Kajiado District, Kenya. *Land use change impacts and dynamics (LUCID) project working paper 18.* Nairobi, Kenya: International Livestock Research Institute.

Capone, D. L. 1972. Wildlife, man and competition for land in Kenya. PhD thesis, University of Michigan.

Carey, J. R. 2003. *Longevity: The biology and demography of the lifespan.* Princeton: Princeton University Press.

Carey, J. R., and D. S. Judge. 2001. Lifespan extension in humans is self reinforcing: A general theory of longevity. *Population and Demography Review* 27:411–36.

Caro, T., and P. P. G. Bateson. 1986. Organisation and ontogeny of alternative tactics. *Animal Behaviour* 34:1483–99.

Caro, T. M., and M. D. Hauser. 1992. Is there teaching in non-human animals? *Quarterly Review of Biology* 67:151–74.

Caro, T. M., D. W. Sellen, A. Parish, R. Frank, D. M. Brown, E. Volland, and M. Borgerhoff Mulder. 1995. Termination of reproduction in nonhuman and human primates. *International Journal of Primatology* 16:205–20.

Cartmill, E. A., and R. W. Byrne. 2007. Orangutans modify their gestural signalling according to their audience's comprehension. *Current Biology* 17:1345–48.

Casebeer, R. L. 1975. Wildlife management in Kenya. Unpublished report to the Kenya Game Department and UNDP/FAO Project. April. Project Working Document 8. Nairobi: Kenya Game Department.

Catchpole, C. K., and P. J. B. Slater. 1995. *Bird song: Biological themes and variations*. Cambridge: Cambridge University Press.

Central Bureau of Statistics (CBS). 2005. Who and where are the poor? A constituency level profile. In *Geographic dimensions of well-being in Kenya*. Nairobi: Central Bureau of Statistics, Government of Kenya, World Bank, Swedish International Development Agency, and the Society for International Development.

Chadwick, D. H. 1992. *The fate of the elephant*. San Francisco: Sierra Club Books.

Chambers, R. 1997. *Whose reality counts? Putting the last first*. London: IT Publications.

Chapais, B. 2001. Primate nepotism: What is the explanatory value of kin selection? *International Journal of Primatology* 22:203–29.

Chapais, B., and P. Belisle. 2004. Constraints on kin selection in primate groups. In *Kinship and behavior in primates*, ed. B. Chapais and C. M. Berman, 365–86. Oxford: Oxford University Press.

Chapman, C. A., R. W. Wrangham, and L. J. Chapman. 1995. Ecological constraints on group size: An analysis of spider monkey and chimpanzee subgroups. *Behavioral Ecology and Sociobiology* 36:59–70.

Charif, R. A., R. R. Ramey, W. R. Langbauer, K. B. Payne, R. B. Martin, and L. M. Brown. 2005. Spatial relationships and matrilineal kinship in African savanna elephant (*Loxodonta africana*) clans. *Behavioral Ecology and Sociobiology* 57: 327–38.

Charnov, E. L. 1976. Optimal foraging: The marginal value theorem. *Theoretical Population Biology* 9:129–36.

Cheney, D. L., and R. M. Seyfarth. 1982. Recognition of individuals within and between groups of free-ranging vervet monkeys. *American Zoologist* 22:519–29.

———. 1990. *How monkeys see the world: Inside the mind of another species*. Chicago: University of Chicago Press.

Cheney D., R. M. Seyfarth, and J. B. Silk. 1995. The responses of female baboons to anomalous social interactions: Evidence for causal reasoning? *Journal of Comparative Psychology* 109: 134–41.

Chepko-Sade, B. D., and Z. T. Halpin. 1987. *Mammalian dispersal patterns*. Chicago: University of Chicago Press.

Chesser, R. K. 1991. Gene diversity and female philopatry. *Genetics* 127:437–47.

———. 1998. Relativity of behavioral interactions in socially structured populations. *Journal of Mammalogy* 79:713–24.

Chevalier-Skolnikoff, S., and J. Liska. 1993. Tool use by wild and captive elephants. *Animal Behaviour* 46:209–19.

Clark, C. W., and M. Mangel. 1984. Foraging and flocking strategies: Information in an uncertain environment. *American Naturalist* 123:626–41.

Clauss, M., R. Frey, B. Kiefer, M. Lechner-Doll, W. Loehlein, C. Polster, G. E. Rossner, and W. J. Streich. 2003. The maximum attainable body size of herbivorous mammals: Morphophysiological constraints on foregut and adaptations of hindgut fermenters. *Oceologica* 136:14–27.

Clayton, N. S., and A. Dickinson. 1998. Episodic-like memory during cache recovery by scrub jays. *Nature* 395:272–78.

Clemens, E. T., and G. M. O. Maloiy. 1982. The digestive physiology of three East African herbivores: The elephant, rhinoceros and hippopotamus. *Journal of Zoology* 198:1 41–56.

Clubb, R., and G. J. Mason. 1998. *A review of the welfare of zoo elephants in Europe*. Report to the RSPCA. Oxford: Department of Zoology, University of Oxford.

Clubb, R., M. Rowcliffe, P. Lee, K. U. Mar, C. Moss, and G. J. Mason. 2008. Compromised survivorship in zoo elephants. *Science* 322:1649.

———. 2009. Fecundity, survivorship, fatness and stress: Why are zoo elephant populations not self-sustaining? *Animal Welfare* 18:237–47.

Clutton-Brock, T. H. 1988. *Reproductive success*. Chicago: University of Chicago Press.

———. 1991. *The evolution of parental care*. Princeton: Princeton University Press.

———. 2002. Breeding together: Kin selection and mutualism in cooperative vertebrates. *Science* 296:69–72.

Clutton-Brock, T. H., S. D. Albon, and F. E. Guinness. 1982. *Red deer: The behavior and ecology of two sexes*. Chicago: University of Chicago Press.

———. 1986. Great expectations: Maternal dominance, sex ratios and offspring reproductive success in red deer. *Animal Behaviour* 34:460–71.

———. 1988. Reproductive success in male and female red deer. In *Reproductive success*, ed. T. H. Clutton-Brock, 325–43. Chicago: University of Chicago Press.

Clutton-Brock, T. H., G. R. Iason, and F. E. Guinness. 1987. Sexual segregation and density related changes in habitat use in male and female red deer (*Cervus elaphus*). *Journal of Zoology* 211:275–89.

Clutton-Brock, T. H., M. Major, S. D. Albon, and F. E. Guinness. 1987. Early development and population dynamics in red deer. I. Density-dependent effects on juvenile survival. *Journal of Animal Ecology* 56:53–64.

Clutton-Brock, T. H., I. R. Stevenson, P. Marrow, A. D. MacColl, A. I. Houston, and J. M. McNamara. 1996. Population fluctuations, reproductive costs and life-history tactics in female soay sheep. *Journal of Animal Ecology* 65:675–89.

Clutton-Brock, T. H., P. Brotherton, M. O'Riain, A. Griffin, D. Gaynor, R. Kansky, L. Sharpe, and G. McIlrath. 2001. Contributions to co-operative rearing in meerkats. *Animal Behaviour* 61:705–10.

Cobb, S. 1989. The ivory trade and the future of the African elephant. Volume 1. Summary and conclusions. Volume 2. Technical reports. Unpublished report of the Ivory Trade Review Group. Prepared for the 7th CITES Conference of the Parties, Oxford.

Cohen, J. E. 1995. *How many people can the earth support?* New York: W. W. Norton.

Collett, D. 1987. Pastoralists and wildlife: Image and reality in Kenya Maasailand. In *Conservation in Africa: People, policies, and practice*, ed. D. Anderson and R. Grove, 129–48. Cambridge: Cambridge University Press.

Coltman, D. W. 2005. Differentiation by dispersal: Evolutionary genetics. *Nature* 433:23–24.

Coltman, D. W., D. R. Bancroft, A. Robertson, J. A. Smith, T. H. Clutton-Brock, and J. M. Pemberton. 1999a. Male reproductive success in a promiscuous mammal: Behavioural estimates compared with genetic paternity. *Molecular Ecology* 8:1199–1209.

Coltman, D. W., J. A. Smith, D. R. Bancroft, J. Pilkington, A. D. C. MacColl, T. H. Clutton-Brock, and J. M. Pemberton. 1999b. Density-dependent variation in lifetime breeding success and natural and sexual selection in Soay rams. *American Naturalist* 154:730–46.

Conant, F. 1982. Thorns paired, sharply recurved: Cultural controls and rangeland quality in East Africa. In *Desertification and development: Dryland ecology in social perspective*, ed. B. Spooner and H. Mann, 111–122. New York: Academic Press.

Connor, R. C., M. R. Heithaus, and L. M. Barre. 1999. Superalliance of bottlenose dolphins. *Nature* 397:571–72.

Connor, R. C., A. J. Read, and R. W. Wrangham. 2000. Male reproductive strategies and social bonds. In *Cetacean societies: Field studies of whales and dolphins*, ed. J. Mann, R. C. Connor, P. L. Tyack, and H. Whitehead, 247–69. Chicago: University of Chicago Press.

Connor, R. C., J. Mann, P. L. Tyack, and H. Whitehead. 1998. Social evolution in toothed whales. *Trends in Ecology and Evolution* 13:228–32.

Connor, R. C., R. Wells, J. Mann, and A. Read. 2000. The bottlenose dolphin, *Tursiops* spp.: Social relationships in a fission-fusion society. In *Cetacean societies: Field studies of whales and dolphins*, ed. J. Mann, R. C. Connor, P. L. Tyack, and H. Whitehead, 91–126. Chicago: University of Chicago Press.

Conradt, L., and T. J. Roper. 2000. Activity synchrony and social cohesion: A fission fusion model. *Proceedings of the Royal Society of London Series B* 267:2213–18.

Cowie, M. 1951. *Annual report 1946–1950*. Nairobi: Royal National Parks of Kenya.

———. 1952. *Annual report 1951*. Nairobi: Royal National Parks of Kenya.

———. 1958. *Annual report 1958*. Nairobi: Royal National Parks of Kenya.

Cozzi, B., S. Spagnoli, and L. Bruno. 2001. An overview of the central nervous system of the elephant through a critical appraisal of the literature published in the XIX and XX centuries. *Brain Research Bulletin* 54:219–27.

Crawley, M. J. 1983. *Herbivory: The dynamics of animal-plant interactions*. Oxford: Blackwell Scientific Publications.

Creel, S., and N. M. Creel. 1995. Communal hunting and pack size in African wild dogs, *Lycaon pictus*. *Animal Behaviour* 50:1325–39.

Creel, S. R., and P. M. Waser. 1994. Inclusive fitness and reproductive strategies in dwarf mongooses. *Behavioral Ecology* 5:339–48.

Croze, H. 1972. A modified photogrammetric technique for assessing age structures of elephant populations and its use in Kidepo Valley National Park. *East African Wildlife Journal* 10:91–115.

———. 1974. The Seronera bull problem. 1. The bulls. *East African Wildlife Journal* 12:1–28.

———. 1978. Aerial surveys undertaken by the Kenya Wildlife Management Project: Methodologies and results. UNEP/FAO Wildlife Management Project KEN/71/526: 61. Unpublished report. Nairobi: Wildlife Conservation and Management Department.

Croze, H., A. K. K. Hillman, and E. M. Lang. 1981. Elephants and their habitats: How do they tolerate each other? In *Dynamics of large mammal populations*, ed. C. W. Fowler, 297–316. New York: John Wiley.

Croze, H., J. C. Hillman, E. Migongo, and R. Sinage. 1978. The ecological basis for calculations of wildlife-generated guaranteed minimum returns to landowners in Maasai Mara and Samburu ecosystems. Unpublished report. Nairobi: EcoSystems Ltd.

Croze, H., S Sayialel, and D. Sitonic. 2006. What's on in the ecosystem: Amboseli as a Biosphere Reserve. A Compendium of Conservation and Management Activities in the Amboseli Ecosystem. Nairobi: ATE/AERP, UNESCO/MAB.

Csermely, D., and D. G. M. Wood-Gush. 1986. Agonistic behavior in grouped sows. The influence of feeding. *Biology of Behavior* 11:244–52.

Csuti, B. 2006. Elephants in captivity. In *Biology, medicine, and surgery of elephants*, ed. M. E. Fowler and S. K. Makota, 15–22. Oxford: Blackwell Publishing Professional.

Cutler, R. G. 1979. Evolution of longevity in ungulates and carnivores. *Gerontology* 25:69–86.

Dall, S. R., A. I. Houston, and J. M. McNamara, J.M. 2004. The behavioural ecology of personality: Consistent individual differences from an adaptive perspective. *Ecology Letters* 7:734–39.

Daniel, J. C. 1998. *The Asian elephant: A natural history*. Dehra Dun, India: Natraj Publishers.

Darling, F. F. 1960. An ecological reconnaissance of the Mara Plains in Kenya Colony. *Wildlife Monographs* 5:1–41.

De Boer, W. F., and D. S. Baquete. 1998. Natural resource use, crop damage and attitudes of rural people in the vicinity of the Maputo Elephant Reserve, Mozambique. *Environmental Conservation* 25:208–18.

Deraniyagala, P. E. P. 1955. *Some extinct elephants, their relatives and the two living species*. Ceylon Museums Publication. Ceylon: Government Press.

de Villiers, P. A., and O. B. Kok. 1997. Home range association and related aspects of elephants on the eastern Transvaal lowveldt. *African Journal of Ecology* 35:224–36.

de Waal, F. B. M. 1982. *Chimpanzee politics*. London: Jonathan Cape.

———. 1989. Dominance "style" and primate social organisation. In *Comparative Socioecology*, ed. V. Standen and R. A. Foley, 243–63. Oxford: Blackwell.

Deacon, T. W. 1997. *The symbolic species: The co-evolution of language and the brain*. New York: W. W. Norton and Company.

Deihl, C. 1985. Wildlife and the Maasai: The story of East African parks. *Cultural Survival Quarterly* 9:37–40.

Delsink, A. K., J. J. van Altena, D. Grober, H. Bertschinger, J. F. Kirkpatrick, and R. Slowtow. 2006. Regulation of a small discrete African elephant population through immunocontraception in the Makalali Conservancy, Limpopo, South Africa. *South African Journal of Science* 102:403–05.

Demment, M. W., and P. J. van Soest. 1985. A nutritional explanation for body-size patterns of ruminant and nonruminant herbivores. *American Naturalist* 125:641–72.

Deshmukh, I. K. 1984. A common relationship between precipitation and grassland peak biomass for east and southern Africa. *African Journal of Ecology* 11:181–86.

Distefano, J. A. 1990. Hunters or hunted? Towards a history of the Okiek of Kenya. *History in Africa* 17:41–57.

Dixson, A. 1998. *Primate sexuality*. Oxford: Oxford University Press.

Dneprovskaya, I. A., V. K. Iofe, and F. I. Levitas. 1963. On the attenuation of sound as it propagates through the atmosphere. *Soviet Journal of Physical Acoustics* 8:235–39.

Dolhinow, P., J. McKenna, and J. Laws. 1979. Rank and reproduction among langurs (They're not just getting older, they're getting better). *Aggressive Behavior* 5:19–30.

Doluweera, G., S. Fernando, C. Gunasekera, A. Madanayake, G. Rossel, L. Seneviratne, and S. Yapa. 2003. Infrasound: Can it be more than a means of communication between elephants? In *Endangered elephants, past present and future: Proceedings of the Symposium on Human-Elephant Relationships and Conflicts*, ed. J. Jayewardene, 40. Colombo, Sri Lanka: Biodiversity and Elephant Conservation Trust.

Dobson, A., and J. H. Poole. 1998. Conspecific aggregation and conservation biology. In *Behavioural ecology and conservation biology*, ed. T. M. Caro, 193–208. Oxford: Oxford University Press.

Double, M. C., R. Peakall, N. R. Beck, and A. Cockburn. 2005. Dispersal, philopatry and infidelity: Dissecting local genetic structure in superb fairy-wrens (*Malarus cyaneus*). *Evolution* 59:625–35.

Douglas-Hamilton, I. 1971. Radio-tracking of elephants. *Symposium of Biotelemetry*. Pretoria: CSIR.

———. 1972. *On the ecology and behaviour of the African elephant*. DPhil thesis, University of Oxford.

———. 1973. On the ecology and behaviour of the Lake Manyara elephants. *East African Wildlife Journal* 11:401–3.

———. 1996. Counting elephants from the air—total counts. In *Studying elephants*, ed. K. Kangwana, 28–37. AWF Technical Handbook Series 7. Nairobi: African Wildlife Foundation.

———. 1998. Tracking African elephants with a global positioning system (GPS) radio collar. *Pachyderm* 25:81–92.

Douglas-Hamilton, I., and O. Douglas-Hamilton, O. 1975. *Among the elephants*. London: Penguin.

———. 1992. *Battle for the elephants*. London: Doubleday.

Douglas-Hamilton, I., T. Krink, and F. Vollrath. 2005. Movements and corridors of African elephants in relation to protected areas. *Naturwissenschaften* 92:158–63.

Douglas-Hamilton, I., F. Michelmore, and A. Inamdar. 1992. African Elephant Database. February 1992. Unpublished report. Nairobi: European Commission African Elephant Survey and Conservation Programme, UNEP.

Douglas-Hamilton, I., S. Bhalla, G. Wittemyer, and F. Vollrath. 2006. Behavioural reactions of elephants towards a dying and deceased matriarch. *Applied Animal Behaviour Science* 100: 87–102.

Douglis, C. 2001. Cultural bomas: Business and show biz. *African Wildlife News* 36:3.

Dublin, H. T. 1983. Cooperation and reproductive competition among female African elephants. In *Social behavior of female vertebrates*, ed. S. Wasser, 291–313. New York: Academic Press.

———. 1991. Dynamics of the Serengetiz-Mara woodlands: An historical perspective. *Forest and Conservation History* 35: 169–78.

———. 1994. In the eye of the beholder: Our image of the African elephant. *Endangered Species Technical Bulletin* 19: 5–6.

———. 1996. Elephants of the Masai Mara, Kenya: Seasonal habitat selection and group size patterns. *Pachyderm* 22: 25–35.

Dublin, H. T., and L. S. Niskanan. 2003. *Guidelines for the in situ translocation of the African elephant for conservation purposes*. Gland, Switzerland: African Elephant Specialist Group, IUCN.

Dublin, H. T., T. O. McShane, and J. Newby. 1997. Conserving Africa's elephants: Current issues and priorities for action. Unpublished report. Gland, Switzerland: WWF International.

Dublin, H. T., A. R. E. Sinclair, and J. McGlade. 1990. Elephants and fire as causes of multiple stable states for Serengeti-Mara woodlands. *Journal of Animal Ecology* 59:1157–64.

Dudley, J. P., G. C. Craig, D. Gibson, G. Haynes, and J. Kilmowicz. 2001. Drought mortality of bush elephants in Hwange National Park, Zimbabwe. *African Journal of Ecology* 39: 187–94.

Dugatkin, L. A., and D. S. Wilson. 1992. The prerequisites for strategic behavior in bluegill sunfish, *Lepomis macrochirus*. *Animal Behaviour* 44:223–30.

Dunbar, R. I. M. 1998. The social brain hypothesis. *Evolutionary Anthropology* 6:178–90.

Dunbar, R. I. M., and P. Dunbar. 1975. *Social dynamics of gelada baboons*. Basel: Karger.

Duncan, P. 1985. Determinants of the use of habitat by horses in a Mediterranean wetland. *Journal of Animal Ecology* 52: 93–111.

Duncan, P., T. J. Foose, I. J. Gordon, C. G. Gakahu, and M. Lloyd. 1990. Comparative nutrient extraction from forages by grazing bovids and equids: A test of the nutritional model of equid/bovid competition and coexistence. *Oecologia* 84:411–18.

Dunn, T., and L. B. Leopold. 1978. *Water in environmental planning*. San Francisco: WH Freeman and Company.

Durant, S. M. 2000. Predator avoidance, breeding experience and reproductive success in endangered cheetahs, *Acinonyx jubatus*. *Animal Behaviour* 60:121–30.

Eady, P. E., and I. C. W. Hardy. 2001. Overt versus covert competition in Soay sheep. *Trends in Ecology and Evolution* 16: 279–80.

Economo, E. P., A. Kirkhoff, and B. Enquist. 2005. Allometric growth, life history invariants and population energetics. *Ecology Letters* 8:353–60.

Edwards, A. M. 2008. Using likelihood to test for Lévy flight search patterns and for general power-law distributions in nature. *Journal of Animal Ecology* 77:1212–22.

Eggert, L. S., C. A. Rasner, and D. S. Woodruff. 2002. The evolution and phylogeography of the African elephant inferred from mitochondrial DNA sequence and nuclear microsatellite markers. *Proceedings of the Royal Society of London* 269: 1993–2006.

Eisenberg, J. F. 1981. *The mammalian radiations: An analysis of trends in evolution, adaptation, and behavior*. Chicago: University of Chicago Press.

Eisenberg, J. F., G. M. McKay, and M. R. Jainudeen. 1971. Reproductive behaviour of the Asiatic elephant (*Elephas maximus* L.). *Behaviour* 38:193–225.

Eisenberg, J. F., and M. Lockhart. 1972. An ecological reconnaissance of Wildpattu National Park. *Smithsonian Contributions to Zoology* 101:1–117.

Elgar, M. A. 1989. Predator vigilance and group size in mammals and birds. A critical review of the empirical evidence. *Biological Reviews* 64:13–33.

Eliot, C. 1905. *The East African Protectorate*. London: Edward Arnold.

Ellis, J. E., and D. M. Swift. 1988. Stability of African pastoral ecosystems: Alternative paradigms and implications for development. *Journal of Range Management* 41:450–59.

Ellis, L. 1995. Dominance and reproductive success among non-human animals: A cross-species comparison. *Ethology and Sociobiology* 16:257–33.

Eltringham, S. K. 1982. *Elephants*. Poole, Dorset: Blandford Press.

Emerton, L. 1998. The nature of benefits and the benefits of nature: Why wildlife conservation has not economically benefited communities in Africa. In *Community conservation in Africa: Principles and comparative practice. Discussion paper 7*. Manchester: Institute for Development Policy and Management, University of Manchester.

Emery, N. J., and N. S. Clayton. 2001. Effects of experience and social context on prospective caching strategies by scrub jays. *Nature* 414:443–46.

———. 2004. The mentality of crows: Convergent evolution of intelligence in corvids and apes. *Science* 306:1903–7.

Emlen, S. 1982. The evolution of helping: An ecological constraints model. *American Naturalist* 119:29–39.

———. 1994. Benefits, constraints and the evolution of the family. *Trends in Ecology and Evolution* 9:282–85.

Engh, A. L., K. Esch, L. Smale, and K. E. Holekamp. 2000. Mechanisms of maternal rank "inheritance" in the spotted hyena. *Animal Behaviour* 60:323–32.

Engh, A. L., S. M. Funk, R. C. van Horn, K. T. Scribner, M. W. Bruford, S. Libants, M. Szykman, L. Smale, and K. E. Holekamp. 2002. Reproductive skew among males in a female-dominated mammalian society. *Behavioral Ecology* 13:193–200.

Enghoff, M. 1990. Wildlife conservation, ecological strategies and pastoral communities: A contribution to the understanding of parks and people in East Africa. *Nomadic Peoples* 25:93–107.

Ens, B. J., and J. D. Goss-Custard. 1984. Interference among oystercatchers, *Haematopus ostrabegus*, feeding on mussels, *Mytilus edulis*, on the Exe estuary. *Journal of Animal Ecology* 53: 217–31.

Ensminger, J., and A. Rutten. 1991. The political economy of changing property rights: Dismantling a pastoral commons. *American Ethnologist* 18:683–99.

Estes, R. D. 1991. *The behavior guide to African mammals*. Berkeley: University of California Press.

Evangelous, P. 1984. Livestock development in Kenya's Maasailand: Pastoralists' transition to a market economy. Boulder, CO: Westview Press.

Evans, K. E. 2006. The behavioural ecology and movements of adolescent male African elephant (*Loxodonta africana*) in the Okavango Delta, Botswana. PhD thesis, Bristol University.

Evans, K. E., and S. Harris. 2008. Adolescence in male African elephants (*Loxodonta africana*) and the importance of sociality. *Animal Behaviour* 76:779–87.

Eyring, C. F. 1946. Jungle acoustics. *Journal of the Acoustics Society of America* 18:257–70.

Fagen, R. 1981. *Animal play behavior*. Oxford: Oxford University Press.

Fagen, R., and J. Fagen. 2009. Play behavior and multi-year juvenile survival in free-ranging brown bears, *Ursus arctos*. *Evolutionary Ecology Research* 11:1–15.

Fairbanks, W. S. 1993. Birthdate, birthweight, and survival in pronghorn fawns. *Journal of Mammalogy* 74:129–35.

Faith, J. T., and A. K. Behrensmeyer. 2006. Changing patterns of carnivore modification in a landscape bone assemblage, Amboseli Park, Kenya. *Science* 33:1718–33.

Fant, G. 1960. *Acoustic theory of speech production*. 2nd ed. The Hague: Mouton.

Fay, M. J., and M. Agnagna. 1991. A population survey of forest elephants in Northern Congo. *African Journal of Ecology* 29: 177–87.

Feigenson, L., S. Carey, and M. Hauser. 2002. The representations underlying infants' choice of more: Object files versus analogue magnitudes. *Psychological Science* 13:150–56.

Fernando, P., and R. Lande. 2000. Molecular genetic and behavioral analysis of social organization in the Asian elephant (*Elephas maximus*). *Behavioral Ecology and Sociobiology* 48: 84–91.

Festa-Bianchet, M. 1988. Nursing behavior of bighorn sheep: Correlates of ewe age, parasitism, lamb age, birthdate and sex. *Animal Behaviour* 36:1445–54.

Festa-Bianchet, M., J. T. Jorgenson, and D. Réale. 2000. Early development, adult mass and reproductive success in bighorn sheep. *Behavioral Ecology* 11:633–39.

Festa-Bianchet, M., D. W. Coltman, L. Turelli, and J. T. Jorgenson. 2004. Relative allocation to horn and body growth in bighorn rams varies with resource availability. *Behavioral Ecology* 15:305–12.

Fiedler, P. L., P. S. White, and R. A. Leidy. 1997. The paradigm shift in ecology and its implications for conservation. In *The ecological basis of conservation: Heterogeneity, ecosystems and biodiversity*, ed. S. T. A. Pickett, R. S. Ostfeld, M. Shachak, and G. E. Likens, 83–92. New York: Chapman and Hall.

Field, C. R. 1971. Elephant ecology in the Queen Elizabeth National Park, Uganda. *East African Wildlife Journal* 9:99–123.

Finch, C. E. 1994. *Longevity, senescence, and the genome*. Chicago: University of Chicago Press.

Fischer, M. S. 1990. The unique ear of elephants and manatees (Mammalia): A phylogenetic paradox. *Sciences de la vie* 331: 157–62.

Fishbein, M., and I. Ajzen. 1975. *Belief, attitude, intention and behavior: An introduction to theory and research*. Reading, MA: Addison–Wesley.

Fishlock, V., P. C. Lee, and T. Breuer. 2008. Quantifying forest elephant social structure in Central African bai environments. *Pachyderm* 44:19–28.

Foley, C. A. H. 2002. The effects of poaching on elephant social systems. PhD thesis, Princeton University.

Foley, C. A. H., S. Papageorge, and S. Wasser. 2001. Non-invasive stress and reproductive measures of social and ecological pressures in free-ranging African elephants. *Conservation Biology* 15:1134–42.

Foley, C. A. H., N. Pettorelli, and L. Foley. 2008. Severe drought and calf survival in elephants. *Biology Letters* 4:541–44.

Folland, C. K., T. R. Karl, J. R. Christy, R. A. Clarke, G. V. Gruza, J. Jouzel, M. E. Mann, J. Oerlemans, M. J. Salinger, and S.-W. Wang. 2001. Observed climate variability and change. In *Climate change 2001: The scientific basis. Contribution of Working Group I to the third assessment report of the Intergovernmental Panel on Climate Change (IPCC)*, ed. J. T. Houghton, Y. Ding, D. J. Griggs, M. Noguer, P. J. van der Linden, X. Dai, K. Maskell, and C. A. Johnson, 94. Cambridge: Cambridge University Press.

Foose, T. J. 1982. Trophic strategies of ruminant versus non-ruminant ungulates. PhD thesis, University of Chicago.

Fosbrooke, H. 1972. *Ngorongoro—The eighth wonder.* London: Andre Deutsch Ltd.

Fox, E., A. Sitompul, and C. P. van Schaik. 1999. Intelligent tool use in wild Sumatran orangutans. In *The mentality of gorillas and orangutans*, ed. S. T. Parker, H. L. Miles, and R. W. Mitchell, 99–116. Cambridge: Cambridge University Press.

Frame, L. H., J. R. Malcolm, G. W. Frame, and H. van Lawick. 1979. Social organization of African wild dogs (*Lycaon pictus*) on the Serengeti plains, Tanzania, 1967–1978. *Zeitschrift fur Tierpsycologie* 50:225–49.

Fratkin, E. 1994. *Problems of pastoral land tenure in Kenya: Demographic, economic, and political process among the Maasai, Samburu, Boran, and Rendille, 1950–1990.* Working Paper 177. Boston: Boston University African Studies Center.

Freeman, E. W., E. Weiss, and J. L. Brown. 2004. Examination of the interrelationships of behavior, dominance status and ovarian activity in captive Asian and African elephants. *Zoo Biology* 23: 431–48.

Freeman, E. W., I. Whyte, and J. L. Brown. 2009. Reproductive evaluation of elephants culled in Kruger National Park, South Africa between 1975 and 1995. *African Journal of Ecology* 47: 192–201.

Fretwell, S. D. 1972. *Populations in a seasonal environment.* Princeton, NJ: Princeton University Press.

Fretwell, S. D., and H. L. Lucas. 1970. On territorial behavior and other factors influencing habitat distribution in birds. *Acta Biotheoretica* 19:16–36.

Fritz, H., and M. De Garine-Whichatitsky. 1996. Foraging in a social antelope: Effects of group size on foraging choices and resource perception in impala. *Journal of Animal Ecology* 65: 736–42.

Fritz, H., M. De Garine-Wichatitsky, and G. Letessier. 1996. Habitat use by sympatric wild and domestic herbivores in an African savanna woodland: The influence of cattle on spatial behavior. *Journal of Applied Ecology* 33:589–98.

Fryxell, J. M. 1987. Food limitations and demography of a migratory antelope, the white eared kob. *Oecologia* 72:83–91.

Gadd, M. 2005. Conservation outside of parks: Attitudes of local people in Laikipia, Kenya. *Environmental Conservation* 32: 50–63.

Gaeth, A. P., R. V. Short, and M. B. Renfree. 1999. The developing renal, reproductive, and respiratory systems of the African elephant suggest an aquatic ancestry. *Proceedings of the National Academy of Sciences USA* 96:5555.

Gagneux, P., M. K. Gonder, T. Goldberg, and P. Morin. 2001. Gene flow in wild chimpanzee populations: What genetic data tells us about chimpanzee movement over space and time. *Philosophical Transactions of the Royal Society of London Series B* 356:889–97.

Gaillard, J. M., D. Pontier, D. Allaine, J. D. Lebreton, J. Trouvilliez, and J. Clobert. 1989. An analysis of demographic tactics in birds and mammals. *Oikos* 56:59–76.

Galaty, J. G. 1980. The Maasai Group Ranch: Politics and development in an African pastoral society. In *When nomads settle: Processes of sedentarization as adaptations and response*, ed. J. G. Galaty and P. C. Salzman. Leiden: E. J. Brill.

———. 1981. Land and livestock among Kenyan Maasai: Symbolic perspectives on pastoral exchange, social change and inequality. In *Change and development in nomadic and pastoral societies*, ed. J. G. Galaty and P. C. Salzman. Leiden: E. J. Brill.

———. 1993a. Maasai expansion and the new East African pastoralism. In *Being Maasai*, ed. T. Spear and R. Waller, 61–86. London: James Currey Ltd.

———. 1993b. The eye that wants a person, where can it not see? Inclusion, exclusion and boundary shifters in Maasai identity. In *Being Maasai*, ed. T. Spear and R. Waller, 174–194. London: James Currey Ltd.

Galef, B. G., and L. L. Buckley. 1996. Use of foraging trails by Norway rats. *Animal Behaviour* 51:765–71.

Galef, B. G., and L. Giraldeau. 2000. Social influences on foraging in vertebrates: Causal mechanisms and adaptive functions. *Animal Behaviour* 59:3–5.

Gallup, G. G. 1970. Chimpanzees: Self-recognition. *Science* 167: 86–87.

———. 1982. Self-awareness and the emergence of mind in primates. *American Journal of Primatology* 2:237–48.

———. 1985. Do minds exist in species other than our own? *Neuroscience and Biobehavioural Reviews* 9:631–41.

Ganswindt, A., H. B. Rasmussen, M. Heistermann, and J. K. Hodges. 2005. The sexually active states of free-ranging male African elephants (*Loxodonta africana*): Defining musth and non-musth using endocrinology, physical signals, and behaviour. *Hormones and Behavior* 47:83–91.

Garber, P. 1988. Foraging decisions during nectar feeding by tamarin monkeys (*Saguinus mystax* and *Saguinus fuscicollis*, Callitrichidae, Primates) in Amazonian Peru. *Biotropica* 20: 100–106.

Garstang, M., D. Larom, R. Raspet, and M. Lindeque. 1995. Atmospheric controls on elephant communication. *Journal of Experimental Biology* 198:939–51.

Gasc, J. P. 1967. Squelette hyobranchial. *Traité de Zoologie* 16: 550–83.

Georgiadis, N., L. Bischof, A. Templeton, J. Patton, W. Karesh, and D. Western. 1994. Structure and history of African elephant populations: I. Eastern and Southern Africa. *Journal of Heredity* 85:100–104.

Gese, G. E., and R. L. Ruff. 1997. Scent marking by coyotes *Canis latrans*: The influences of social and ecological factors. *Animal Behaviour* 54:1155–66.

Gesicho, A. 1991. A survey of the Arabuko Sokoke elephant population. Unpublished report, Kenya Wildlife Service Elephant Programme, Nairobi.

Gibson, C. C., and S. A. Marks. 1995. Transforming rural hunters into conservationists: An assessment of community-based wildlife management programs in Africa. *World Development* 23:941–57.

Gillingham, S., and P. C. Lee. 1999. The impact of wildlife-related benefits on the conservation attitudes of local people around the Selous Game Reserve, Tanzania. *Environmental Conservation* 26:218–28.

———. 2003. People and protected areas: A study of local perceptions of wildlife crop-damage conflict in an area bordering the Selous Game Reserve, Tanzania. *Oryx* 37:316–25.

Gillson, L. 2004. Testing non-equilibrium theories in savannas: 1400 years of vegetation change in Tsavo National Park, Kenya. *Ecological Complexity* 1:281–98.

Gilpin, M. E., and M. E. Soulé. 1986. Minimum viable populations: Processes of species extinction. In *Conservation biology: The science of scarcity and diversity*, ed. M. E. Soulé, 19–34. Sunderland, MA: Sinauer.

Ginsberg, J. R., and T. P. Young. 1992. Measuring association between individuals or groups in behavioural studies. *Animal Behaviour* 44:377–79.

Girman, D. J., M. G. L. Mills, E. Geffern, and R. K. Wayne. 1997. A molecular genetic analysis of social structure, dispersal, and interpack relationships of the African wild dog (*Lycaon pictus*). *Behavioral Ecology and Sociobiology* 40:187–98.

Githaiga, J. M., R. Reid, A. Muchiru, and S. van Dijk. 2003. *Survey of water quality changes with land use type, in the Loitokitok Area, Kajiado District, Kenya. Land use change, impacts and dynamics project*. Working Paper 35. Nairobi: International Livestock Research Institute and United Nations Environment Programme/ Division of Global Environment Facility.

Gittleman, J. L. 1986. Carnivore life history patterns: Allometric, phylogenetic and ecological allocations. *American Naturalist* 127:744–71.

Gobush, K., B. Kerr, and S. Wasser. 2009. Genetic relatedness and disrupted social structure in a poached population of African elephants. *Molecular Ecology* 18:722–34.

Gobush, K., B. M. Mutayoba, and S. Wasser. 2008. Long-term impacts of poaching on relatedness, stress physiology, and reproductive output of adult female African elephants. *Conservation Biology* 22:1590–99.

Godard, R. 1991. Long-term memory of individual neighbours in a migratory songbird. *Nature* 350:228–29.

Goldizen, A. 1987. Tamarins and marmosets: Communal care of offspring. In *Primate societies*, ed. B. B. Smuts, D. L. Cheney, R. M. Seyfarth, R. W. Wrangham, and T. T. Struhsaker, 34–43. Chicago: University of Chicago Press.

Goldman, M. 2003. Partitioned nature, privileged knowledge: Community based conservation in Tanzania. *Development and Change* 34:833–62.

Gomendio, M. 1990. The influence of maternal rank and infant sex on maternal investment trends in macaques: Birth sex ratios, inter-birth intervals and suckling patterns. *Behavioral Ecology and Sociobiology* 27:365–75.

Gomendio, M., T. H. Clutton-Brock, S. D. Albon, F. E. Guinness, and M. J. A. Simpson. 1990. Mammalian sex ratios and variation in the costs of rearing sons and daughters. *Nature* 343:261–63.

Gompper, M. E., J. L. Gittleman, and R. K. Wayne. 1998. Dispersal, philopatry, and genetic relatedness in a social carnivore: Comparing males and females. *Molecular Ecology* 7:157–63.

Goodall, J. 1986. *The chimpanzees of Gombe: Patterns of behavior*. Cambridge, MA: Harvard University Press.

———. 1990. *Through a window: 30 years with the chimpanzees of Gombe*. London: Weidenfield and Nicholson.

Gordon, I. J. 1989. Vegetation community selection by ungulates on the Isle of Rhum. III. Determinants of vegetation community selection. *Journal of Applied Ecology* 26:65–79.

Gough, K. F., and G. I. H. Kerley. 2006. Demography and population dynamics in the elephants *Loxodonta africana* of Addo Elephant National Park, South Africa: Is there evidence of density dependent regulation? *Oryx* 40:434–41.

Gouzoules, S., and H. Gouzoules. 1987. Kinship. In *Primate societies*, ed. B. B. Smuts, D. L. Cheney, R. M. Seyfarth, R. W. Wrangham, and T. T. Struhsaker, 299–305. Chicago: University of Chicago Press.

Government of Kenya. 2006. Draft national land policy. *Sunday Nation* (October): i–xvi. Nairobi: Ministry of Lands Kenya Land Alliance.

Graham, M. D., I. Douglas-Hamilton, W. A. Adams, and P. C. Lee. 2009. The movement of African elephants in a human dominated land use mosaic. *Animal Conservation* 12:445–55.

Grandin, B. 1985. Human demography and culture: Factors in range management. In *Wildlife/livestock interfaces on rangelands*, ed. S. MacMillan. Nairobi: Inter-African Bureau for Animal Resources.

Granli, P., and J. Kioko. 2006a. Mitigating human-elephant conflict in the Amboseli ecosystem. Unpublished report to the Amboseli Elephant Research Project.

———. 2006b. Mitigating human-elephant conflict in the Amboseli ecosystem by use of deterrents. Unpublished report to the Amboseli Elephant Project.

Gray, R. D., and M. Kennedy. 1994a. Misconceptions or misreadings? Missing the real issues about the IFD. *Oikos* 71:167–70.

———. 1994b. Perceptual constraints on optimal foraging: A reason for departure from ideal free distribution. *Animal Behaviour* 47:469–71.

Greaves, N. 1996a. *Hwange: Retreat of the elephants*. Halfway House, South Africa: Southern Book Publishers.

———. 1996b. *When elephant was king and other elephant tales from Africa*. Cape Town: Struik Publishers Ltd.

Green, W. C. H., and A. Rothstein. 1991. Sex bias or equal opportunity? Patterns of maternal investment in bison. *Behavioral Ecology and Sociobiology* 27:99–102.

Greenwood, A. D., and S. Paabo. 1999. Nuclear insertion sequences of mitochondrial DNA predominate in hair but not in blood of elephants. *Molecular Ecology* 9:133–37.

Greenwood, P. J. 1980. Mating systems, philopatry and dispersal in birds and mammals. *Animal Behaviour* 28:1140–62.

Grimshaw, J. M., and C. A. H. Foley. 1990. Kilimanjaro elephant project. Unpublished report to Friends of Conservation.

Gunn, A. S. 2001. Environmental ethics and trophy hunting. *Ethics and the Environment* 6:68–95.

Guy, P. R. 1975. The daily food intake of the African elephant, *Loxodonta africana* Blumenbach, in Rhodesia. *Arnoldia* 7:1–8.

Hackel, J. 1990. Conservation attitudes in Southern Africa: A comparison between KwaZulu and Swaziland. *Human Ecology* 18:203–9.

Hakeem, A. Y., P. R. Hof, C. C. Sherwood, R. C. Switzer III, L. E. L. Rasmussen, and J. M. Allman. 2005. Brain of the African elephant: Neuroanatomy from magnetic resonance images. *Anatomical Record* 287A:1117–27.

Hakeem, A. Y., C. C. Sherwood, C. J. Bonar, C. Butti, P. R. Hof, and J. M. Allman. 2009. Von Economo neurons in the elephant brain. *Anatomical Record* 292:242–48.

Hall-Martin, A. J. 1987. Role of musth in the reproductive strategy of the African elephant (*Loxodonta africana*). *South African Journal of Science* 83:616–20.

Hall-Martin, A. J., and L. A. van der Walt. 1984. Plasma testosterone levels in relation to musth in the male African elephant. *Koedoe* 27:147–49.

Hamilton, W. D. 1964a. The genetical evolution of social behaviour. I. *Journal of Theoretical Biology* 7:1–16.

———. 1964a. The genetical evolution of social behaviour. II. *Journal of Theoretical Biology* 7:17–52.

Hand, J. L. 1986. Resolution of social conflicts: Dominance, egalitarianism, spheres of dominance, and game theory. *Quarterly Review of Biology* 61:201–20.

Harcourt, A. H., and F. M. B. de Waal. 1992. *Coalitions and *alliances in humans and other animals*. Oxford: Oxford University Press.

Harcourt, A. H., H. Pennington, and A. W. Weber. 1986. Public attitudes to wildlife conservation in the Third World. *Oryx* 20:152–54.

Hardin, G. 1968. The tragedy of the commons. *Science* 162: 1243–48.

Hare, B., J. Call, and M. Tomasello. 2001. Do chimpanzees know what conspecifics know? *Animal Behaviour* 61:139–51.

Hare, B., M. Brown, C. Williamson, and M. Tomasello. 2003. The domestication of social cognition in dogs. *Science* 298: 1634–36.

Hare, B., J. Call, B. Agnetta, and M. Tomasello. 2000. Chimpanzees know what conspecifics do and do not see. *Animal Behaviour* 59:771–85.

Hart, B. L., L. A. Hart, and N. Pinter-Wollmann. 2007. Large brains and cognition: Where do elephants fit in? *Neuroscience and Biobehavioral Reviews* 32:86–98.

Hart, B. L., L. A. Hart, M. McCoy, and C. R. Sarath. 2001. Cognitive behaviour in Asian elephants: Use and modification of branches for fly switching. *Animal Behaviour* 62:839–47.

Hart, J. T. 2009. How many elephants are left in DR Congo? Unpublished report on www.bonoboincongo.com/2009/02/01.

Hart, L., and Sundar. 2000. Family traditions for mahouts of Asian elephants. *Anthrozoos* 13:34–43.

Hauser, M. D. 1993. Do vervet monkeys cry wolf? *Animal Behaviour* 45:1242–44.

———. 1997a. *The evolution of communication*. Boston: MIT Press.

———. 1997b. Artifactual kinds and functional design features: What a primate understands without language. *Cognition* 64: 285–308.

Hawkes, K., J. F. O'Connell, N. G. Blurton-Jones, H. Alvarez, and E. L. Charnov. 1998. Grandmothering, menopause and the evolution of human life histories. *Proceedings of the National Academy of Science USA* 95:1336–39.

Haynes, G. 1991. *Mammoths, mastodons and elephants*. Cambridge: Cambridge University Press.

Hedges, S. B. 2001. Afrotheria: Plate tectonics meets genomics. *Proceedings of the National Academy of Sciences USA* 98:1–2.

Heffner, R. S., and H. Heffner. 1980. Hearing in the elephant, *Elephas maximus*. *Science* 208:518–20.

———. 1982. Hearing in the elephant: Absolute sensitivity, frequency discrimination and sound localisation. *Journal of Comparative and Physiological Psychology* 96:926–44.

Heffner, R. S., H. E. Heffner, and N. Stichman. 1982. The role of elephant pinna in sound localization. *Animal Behaviour* 30: 628–29.

Held, S., M. Mendl, C. Devereux, and R. W. Byrne. 2000. Social tactics of pigs in a competitive foraging task: The "informed forager" paradigm. *Animal Behaviour* 59:536–76.

Helland, J. 1980. An outline of group ranching in pastoral Maasai areas of Kenya. In *Five essays on the study of pastoralists and the development of pastoralism*. African Savannah Studies Occasional Paper 20. Bergen: Institute of Social Anthropology, University of Bergen.

Hendrichs, H. 1971. Freilandbeobachtungen zum sozial–system des Africanischen elephanten, *Loxodonta africana*. *Dikdik und Elephanten*. Munich: Piper.

Herman, L. M. 1986. Cognition and language competencies of bottlenosed dolphins. In *Dolphin cognition and behaviour: A comparative approach*, ed. R. J. Schusterman, J. A. Thomas, and F. G. Wood, 221–51. Hillsdale, NJ: Lawrence Erlbaum Associates.

Hewison, A. J. M., R. Anderson, J. M. Gaillard, J. D. C. Linnell, and D. Delorme. 1999. Contradictory findings in studies of sex ratio variation in roe deer (*Capreolus capreolus*). *Behavioral Ecology and Sociobiology* 45:339–48.

Hill, C. M. 1998. Conflicting attitudes towards elephants around the Budongo Forest Reserve, Uganda. *Environmental Conservation* 25:244–50.

Hinde, S. L., and H. Hinde. 1901. *The last of the Masai*. London: William Heinemann.

Hiraiwa-Hasegawa, M. 1990. Maternal investment before weaning. In *Chimpanzees of the Mahale Mountains*, ed. T. Nishida, 257–66. Tokyo: University of Tokyo Press.

———. 1993. Skewed birth sex ratios in primates: Should high ranking mothers have daughters? *Trends in Ecology and Evolution* 8:395–400.

Hoare, R. 1995. Options for the control of elephants in conflict with people. *Pachyderm* 19:54–63.

———. 2000. African elephants and humans in conflict: The outlook for co-existence. *Oryx* 34:34–38.

Hodges, J. K. 1998. Endocrinology of the ovarian cycle and pregnancy in the Asian (*Elephas maximus*) and African (*Loxodonta africana*) elephant. *Animal Reproduction Science* 53:3–18.

Hoelzel, A. R., B J. Le Boeuf, J. Reiter, and C. Campagna. 1999. Alpha-male paternity in elephant seals. *Behavioral Ecology and Sociobiology* 46:298–306.

Hogg, J. T., C. C. Hass, and D. A. Jenni. 1992. Sex-biased maternal expenditure in Rocky Mountain bighorn sheep. *Behavioral Ecology and Sociobiology* 31:243–51.

Hogg, R. 1987. Settlement, pastoralism and the commons: The ideology and practice of irrigation development in northern Kenya. In *Conservation in Africa: People, policies and practice*, ed. D. Anderson and R. Grove, 293-306. Cambridge: Cambridge University Press.

Holekamp, K. 2006. Spotted hyenas. *Current Biology* 16: R944–45.

Holekamp, K. E., and L. Smale. 1990. Provisioning and food sharing by lactating spotted hyenas, *Crocuta crocuta* (Mammalia: Hyaenidae). *Ethology* 86:191–202.

Holekamp, K. E., S. M. Cooper, C. I. Katona, N. A. Berry, L. G. Frank, and L. Smale. 1997. Patterns of association among female spotted hyenas (*Crocuta crocuta*). *Journal of Mammalogy* 78:55–64.

Holgrem, N. 1995. The ideal free distribution of unequal competitors: Prediction from a behavior-based functional response. *Journal of Animal Ecology* 64:197–212.

Hollis, A. C. 1905. *The Masai: Their language and folklore*. Oxford: Clarendon Press.

———. 2003. *Masai myths, tales, and riddles*. Mineola, NY: Dover Publications.

Hollister-Smith, J. A. 2005. Reproductive behavior in male African elephants (*Loxodonta africana*) and the role of musth: A genetic and experimental analysis. PhD thesis, Duke University.

Hollister-Smith, J., J. H. Poole, E. A. Archie, E. A. Vance, N. J. Georgiadis, C. J. Moss, and S. C. Alberts. 2007. Age, musth and paternity success in wild male African elephants, *Loxodonta africana*. *Animal Behaviour* 74:287–96.

Holmes, C. M. 2003. The influence of protected area outreach on conservation attitudes and resource use patterns: A case study from western Tanzania. *Oryx* 37:305–15.

Homewood, K., and D. Brockington. 1999. Biodiversity, conservation and development in Mkomazi Game Reserve, Tanzania. *Global Ecology and Biogeography* 8:301–13.

Homewood, K. M., and W. A. Rodgers. 1991. *Maasailand ecology: Pastoralist development and wildlife conservation in Ngorongoro, Tanzania*. Cambridge: Cambridge University Press.

Horgan, J. 1989. The Maasai: These pastoralists are key to the future of Kenya's wildlife. *Scientific American* 261:38–44.

Horner, V., and A. Whiten. 2005. Causal knowledge and imitation/emulation switching in chimpanzees (*Pan troglodytes*) and children (*Homo sapiens*). *Animal Cognition* 8:164–81.

Houston, A. I., J. M. McNamara, and M. Milinski. 1995. The distribution of animals between resources: A compromise between equal numbers and equal intakes rates. *Animal Behaviour* 49:248–51.

Houtsma, A. J. M. 1995. Pitch perception. In *Hearing*, ed. B. C. J. Moore, 267–95. San Diego: Academic Press.

Hrdy, S. B., and D. Hrdy. 1976. Hierarchical relations among female Hanuman langurs. *Science* 193:913–15.

Hughes, C. 1998. Integrating molecular techniques with field methods in studies of social behavior: A revolution results. *Ecology* 79:383–99.

Hulme, D., and M. Murphree. 2001. *African wildlife and livelihoods: The promise and performance of community conservation*. Portsmouth, NH: Heinemann.

Humphrey, N. 1976. The social function of intellect. In *Growing points in animal behaviour*, ed. P. P. G. Bateson and R. A. Hinde, 303–17. Cambridge: Cambridge University Press.

Humphries, S., N. B. Metcalfe, and G. D. Ruxton. 1999. The effects of group size on relative competitive ability. *Oikos* 85:481–86.

Humphries S., G. D. Ruxton, and N. B. Metcalfe. 2000. Group size and relative competitive ability: Geometric progressions as a conceptual tool. *Behavioural Ecology and Sociobiology* 47:113–18.

Hungerford, H. R., and T. L. Volk. 1990. Changing learner behavior through environmental education. *Journal of Environmental Education* 21:8–21.

Hunn, E. 1982. Mobility as a factor limiting resource use in the Columbia Plateau of North America. In *Resource managers: North American and Australian hunter-gatherers*, ed. N. Williams and E. Hunn, 17–43. Boulder: Westview Press.

Hunt, G. R. 1996. Manufacture and use of hook-tools by New Caledonian crows. *Nature* 379:249–51.

———. 2000a. Human-like, population-level specialization in the manufacture of Pandanus tools by New Caledonian crows *Corvus moneduloids*. *Proceedings of the Royal Society Series B* 267:403–13.

———. 2000b. Tool use by the New Caledonian crow *Corvus moneduloides* to obtain *Cerambycidae* from dead wood. *Emu* 100:109–14.

Huntingford, G. W. B. 1953 [1969]. *The southern Nilo-Hamites*. London: International African Institute.

Huxley, J. S. [1924] 1972. *Problems of relative growth*. London: Methuen.

Illius, A. W., and I. J. Gordon. 1987. The allometry of food intake in grazing ruminants. *Journal of Animal Ecology* 56:989–99.

———. 1992. Modelling the nutritional ecology of ungulate herbivores: Evolution of body size and competitive interactions. *Oecologia* 89:428–34.

Illius, A. W., and T. G. O'Connor. 1999. On the relevance of non-equilibrium concepts to arid and semiarid grazing systems. *Ecological Applications* 9:798–813.

Infield, M. 1988. Attitudes of a rural community towards conservation and a local conservation area in Natal, South Africa. *Biological Conservation* 45:21–46.

———. 2001. Cultural values: A forgotten strategy for building community support for protected areas in Africa. *Conservation Biology* 15:800–802.

Infield, M., and A. Namara. 2001. Community attitudes and behaviour towards conservation: An assessment of a community conservation programme around Lake Mburo National Park, Uganda. *Oryx* 35:48–60.

Ingard, U. 1953. A review of the influence of meteorological conditions on sound propagation. *Journal of the Acoustics Society of America* 25:405–11.

Insley, S. J. 2000. Long-term vocal recognition in the fur seal. *Nature* 406:404–5.

Irie-Sugimoto, N., T. Kobayashi, T. Sato, and T. Hasegawa. 2008a. Evidence of means-end behavior in Asian elephants (*Elephas maximus*). *Animal Cognition* 11:359–65.

———. 2008b. Relative quantity judgment by Asian elephants (*Elephas maximus*). *Animal Cognition* 12:193–99.

Isbell, L. A. 1991. Contest and scramble competition: Patterns of female aggression and ranging behavior among primates. *Behavioral Ecology* 2:145–55.

IUCN. 1987. *IUCN directory of Afrotropical protected areas*. Gland, Switzerland: IUCN.

Jachmann, H. 1987. Elephants and woodlands II. *Pachyderm* 8:11–12.

———. 1989. Food selection by elephants in the "miombo" biome, in relation to leaf chemistry. *Biochemical Systematics and Ecology* 17:15–24.

Jachmann, H., and R. H. V. Bell. 1984. Spatial organization of the Kasungu elephants. In Ecology of the elephants in the Kasungu National Park, Malawi. PhD thesis, University of Groningen.

Jacobs, A. H. 1965. The traditional political organization of the pastoral Maasai. DPhil thesis, University of Oxford.

———. 1975. Maasai pastoralism in historical perspective. In *Pastoralism in tropical Africa*, ed. T. Monod, 406–25. Oxford: Oxford University Press.

Jainudeen, M. R., J. F. Eisenberg, and N. Tilakeratne. 1971. Oestrous cycle of the Asiatic elephant, *Elephas maximus*, in captivity. *Journal of Reproduction and Fertility* 27:321–28.

Jainudeen, M. R., G. M. McKay, and J. F. Eisenberg. 1972. Observations on musth in the domesticated Asiatic elephant. *Mammalia* 36:247–61.

Janik, V. M., and P. J. B. Slater. 1997. Vocal learning in mammals. *Advances in the Study of Behavior* 26:59–99.

Janis, C. 1976. The evolutionary strategy of the *Equidae* and the origins of rumen and cecal fermentation. *Evolution* 30:757–74.

Janmaat, K. R. L, R. W. Byrne, and K. Zuberbuhler. 2006. Primates take weather into account when searching for fruits. *Current Biology* 16:1232–37.

Janson, C. H. 2007. Experimental evidence for route integration and strategic planning in wild capuchin monkeys. *Animal Cognition* 10:341–56.

Janson, C. H., and R. Byrne. 2007. What wild primates know about resources: Opening up the black box. *Animal Cognition* 10:357–67.

Jayewardene, J. 1994. *The elephant in Sri Lanka*. Colombo: Wildlife Heritage Trust of Sri Lanka.

Jennions, M. D., and D. W. Macdonald. 1994. Cooperative breeding in mammals. *Trends in Ecology and Evolution* 9:89–93.

Jensen, C. L., and A. J. Belsky. 1989. Grassland homogeneity in Tsavo National Park (West), Kenya. *African Journal of Ecology* 27:3–44.

Jerison, H. J. 1973. *Evolution of the brain and intelligence*. New York: Academic Press.

Jolly, A. 1966. Lemur social behavior and primate intelligence. *Science* 153:501–6.

Kahl, M. P., and B. D. Armstrong. 2000. Visual and tactile displays in African elephants, *Loxodonta africana*: A progress report (1991–1997). *Elephant* 2:19–21.

———. 2002. Visual displays of wild African elephants during musth. *Mammalia* 66:159–71.

Kangwana, K. 1993. Elephants and Maasai: Conflict and conservation in Amboseli, Kenya. PhD thesis, University of Cambridge.

———. 1995. Human-elephant conflict: The challenge ahead. *Pachyderm* 19:11–14.

Kangwana, K., and R. Ole Mako. 2001. Conservation, livelihoods and the intrinsic value of wildlife: Tarangire National Park, Tanzania. In *African wildlife and livelihoods: The promise and performance of community conservation*, ed. D. Hulme and M. Murphree, 148–59. Oxford: James Currey Ltd.

Kappeler, P. M. 1993. Female dominance in primates and other mammals. In *Perspectives in ethology*. Vol. 10. *Behaviour and evolution*, ed. P. P. G. Bateson, P. H. Klopfer, and N. S. Thompson, 143–58. New York: Plenum.

Kapustin, N., J. K. Critser, D. Olsen, and P. V. Malven. 1996. Nonluteal estrous cycles of 3-week duration are initiated by anovulatory luteinizing hormone peaks in African elephants. *Biology of Reproduction* 55:1147–54.

Kasfir, S. L. 1992. Ivory from Zariba Country to the Land of the Zinj. In *Elephant: The animal and its ivory in African culture*, ed. D. H. Ross, 309–27. Berkeley: University of California Press.

———. 1995. Rethinking the Maasai ivories. *African Arts* 28 (2): 12–15.

Keiter, R. B., and M. S. Boyce. 1991. *The greater Yellowstone ecosystem: Redefining America's wilderness*. New Haven: Yale University Press.

Kellert, S. R. 1976. Perceptions of animals in American society. *Transactions of the North American Wildlife and Natural Resource Conference* 41:533–46.

———. 1996. *The value of life*. Washington, DC: Island Press.

Kennedy, M., and R. D. Gray. 1993. Can ecological theory predict the distribution of foraging animals? A critical analysis of experiments on the ideal free distribution. *Oikos* 68:158–66.

Kennedy, M., C. R. Shave, and H. G. Spencer. 1994. Quantifying the effect of predation risk on foraging bullies: No need to assume an IDF. *Ecology* 75:2220–26.

Kenya Wildlife Service. 1990. *Policy framework and development programme 1991–1996*. Nairobi: Kenya Wildlife Service.

———. 1991. *Amboseli National Park Management Plan, 1991–1996*. Nairobi: Kenya Wildlife Service.

Kideghesho, J. R., E. Roskaft, and B. P. Kaltenborn. 2007. Factors influencing conservation attitudes of local people in Western Serengeti, Tanzania. *Biodiversity Conservation* 16: 2213–30.

Kimani, K., and J. Pickard, J. 1998. Recent trends and implications of group ranch sub-division and fragmentation in Kajiado District, Kenya. *Geographical Journal* 164:202–13.

Kinyua, P. I. D., G. C. van Kooten, and E. H. Bulte. 2000. African wildlife policy: Protecting wildlife herbivores on private game ranches. *European Review of Agricultural Economics* 27:227–44.

Kioko, J. 2004. Spatial and temporal distribution of African elephant and their interaction with humans in Kuku-Kimana area of Tsavo-Amboseli ecosystem. MA thesis, University of Greenwich, UK.

Kioko, J., P. Muruthi, P. Omondi, and P. I. Chiyo. 2008. The performance of electric fences as elephant barriers in Amboseli, Kenya. *South African Journal of Wildlife Management* 38: 52–58.

Kipury, N. 1983. *Oral literature of the Maasai*. Nairobi: East African Educational Publishers Ltd.

Kirkpatrick, J. F., W. Zimmermann, L. Kolter, I. K. M. Liu, and J. W. Turner. 1995. Immunocontraception of captive exotic species. I. Przewalski's horse (*Equus caballus*) and banteng (*Bos javancus*). *Zoo Biology* 14:403–13.

Kirshenbaum, A. P., A. D. Szalda-Petree, and N. F. Haddad. 2000. Risk-sensitive foraging in rats: The effects of response-effort and reward-amount manipulations on choice behavior. *Behavior Processes* 50:9–17.

Kiss, A. 1990. *Living with wildlife: Wildlife resource management with local participation in Africa*. Washington, DC: World Bank.

Kituyi, M. 1990. *Becoming Kenyans: Socio-economic transformation of the pastoral Maasai*. Nairobi: African Center for Technology Studies Press.

Kleiber, M. 1961. *The fire of life: An introduction to animal energetics*. New York: Wiley.

Knowles, J. N., and D. P. Collett. 1989. Nature as myth, symbol and action: Notes towards a historical understanding of development and conservation in Kenyan Maasailand. *Africa* 59:433–60.

Koch, P. L., J. Heisinger, C. J. Moss, R. W. Carlson, M. L. Fogel, and A. K. Behrensmeyer. 1995. Isotopic tracking of change in diet and habitat use in African elephants. *Science* 267: 1340–43.

Koenig, A. 2000. Competitive regimes in forest-dwelling Hanuman langur females (*Semnopithecus entellus*). *Behavioral Ecology and Sociobiology* 48:93–109.

Kojola, I. 1993. Early maternal investment and growth in reindeer fawns. *Canadian Journal of Zoology* 71:753–58.

———. 1998. Sex ratio and maternal investment in ungulates. *Oikos* 83:567–73.

Kojola, I., and E. Eloranta. 1989. Influences of maternal body weight, age and parity on sex ratio in semi-domesticated reindeer (*Rangifer t. tarandus*). *Evolution* 43:1331–36.

Korona, R. 1989. Ideal free distribution of unequal competitors can be determined by the form of competition. *Journal of Theoretical Biology* 138:347–52.

Krapf, J. L. 1854. *Vocabulary of the Engutuk Eloikob: Or language of the Wakaufi Nation of the interior of Equatorial Africa*. Tubingen: Ludwig Fried Fues.

———. 1860. *Travels, researches and missionary labours during eighteen years residence in East Africa*. London: Trubner.

Kratz, C. A. 1996. The troublesome Turle collection. *African Arts* 29:187–89.

Krebs, J., and N. B. Davies. 1987. *An introduction to behavioural ecology*. Oxford: Blackwell Scientific.

Krishnan, M. 1972. An ecological survey of the larger mammals of peninsular India. The Indian elephant. *Journal of the Bombay Natural History Society* 69:297–321.

Kruuk, H. 1972. *The spotted hyena: A study of predation and social behaviour*. Chicago: University of Chicago Press.

Kruuk, L. E. B., T. H. Clutton-Brock, S. D. Albon, J. M. Pemberton, and F. E. Guinness. 1999a. Population density affects sex ratio variation in red deer. *Nature* 399:459–61.

Kruuk, L. E. B., T. H. Clutton-Brock, K. E. Rose, and F. E. Guinness. 1999b. Early determinants of lifetime reproductive success differ between the sexes in red deer. *Proceedings of the Royal Society Series B* 266:1655–61.

Kühme, V. W. 1961. Beobachtungen am Africanschen elephanten (*Loxodonta africana* Blumenbach 1797) in gefangenschaft. *Zeitschrift für Tierpsychologie* 18:285–96.

Kühme, V. W. 1962. Ethology of the African elephant. *International Zoo Yearbook* 4:113–21.

———. 1963. Ergänzende beobachtungen an Afrikanischen elephanten (*Loxodonta africana* Blumenbach 1797) im freigehege (2.Teil). *Zeitschrift für Tierpsychologie* 20:66–79.

Kummer H. 1968. *Social organization of hamadryas baboons*. Chicago: University of Chicago Press.

Kuriyan, R. 2002. Linking local perceptions of elephants and conservation: Samburu pastoralists in Northern Kenya. *Society and Natural Resources* 15:949–57.

Kurtis, B. 1994. *Maasai: Secrets of an ancient culture*. Video recording. Chicago: Public Media Education.

Lagendijk, D. D. G., W. F. de Boer, and S. E. Van Wieren. 2005. Can African elephants survive and thrive in monostands of *Colphospermum mopane* woodlands? *Pachyderm* 39: 43–9.

Lahiri-Choudhury, D. K. 1999. *The great Indian elephant book: An anthology of writings on elephants in the Raj*. New Delhi: Oxford University Press.

Lamb, H. H. 1966. Climate in the 1960s: Changes in the world's wind circulation reflected in prevailing temperatures, rainfall patterns and the levels of the African lakes. *Geographical Journal* 132:183–212.

Lambrechts, M. M., and A. A. Dhondt.1995. Individual voice discrimination in birds. *Current Ornithology* 12:115–37.

Lamouse-Smith, B., and J. School. 1998. *Africa Interactive Maps (AFIM)*. Baltimore: University of Maryland.

Lamprey, R. H. 1983. Pastoralism yesterday and today: The overgrazing problem. In *Ecosystems of the world 13, tropical savannas*, ed. F. Bourlière, 643–66. Amsterdam: Elsevier.

Lamprey, R. H., and R. S. Reid. 2004. Expansion of human settlement in Kenya's Maasai Mara: What future for pastoralism and wildlife? *Journal of Biogeography* 31:997–1032.

Lamprey, R. H., and R. D. Waller. 1990. The Loita-Mara region in historical times: Patterns of subsistence, settlement and ecological change. In *Early pastoralists of south-western Kenya*, ed. P. Robertshaw, 16–35. Nairobi: British Institute of East Africa.

Langbauer, W. R. 2000. Elephant communication. *Zoo Biology* 19:425–44.

Langbauer, W. R., K. B. Payne, R. A. Charif, and E. M. Thomas. 1989. Responses of captive African elephants to playback of low-frequency calls. *Canadian Journal of Zoology* 67: 2604–07.

Langbauer, W. R., K. B. Payne, R. A. Charif, L. Rapaport, and F. Osborn. 1991. African elephants respond to distant playbacks of low-frequency conspecific calls. *Journal of Experimental Biology* 157:35–46.

Langman, V. A., T. J. Roberts, J. Black, G. M. Maloiy, N. C. Heglund, J. M. Weber, R. Kram, and C. R. Taylor. 1995. Moving cheaply: Energetics of walking in the African elephant. *Journal of Experimental Biology* 198:629–32.

Larom, D., M. Garstang, K. Payne, R. Raspet, and M. Lindeque. 1997. The influence of surface atmospheric conditions on the range and area reached by animal vocalizations. *Journal of Experimental Biology* 200:421–31.

Laurenson, M. K. 1995. Implications of high offspring mortality for cheetah population dynamics. In *Serengeti II: Dynamics, management and conservation of an ecosystem*, ed. A. R. E. Sinclair and P. Arcese, 385–99. Chicago: University of Chicago Press.

Laws, R. M. 1966. Age criteria for the African elephant, *Loxodonta a. africana*. *East African Wildlife Journal* 4:1–37.

———. 1967. Occurrence of placental scars in the uterus of the African elephant (*Loxodonta africana*). *Journal of Reproduction and Fertility* 14:445–49.

———. 1969. Aspects of reproduction in the African elephant, *Loxodonta africana*. *Journal of Reproduction and Fertility* Supp. 6:193–217.

Laws, R. M., I. S. C. Parker, and R. C. B. Johnstone. 1975. *Elephants and their habitats: The ecology of elephants in North Bunyoro, Uganda*. Oxford: Clarendon Press.

League Against Cruel Sports. 2004. The myth of trophy hunting as conservation. A league against cruel sports submission to Environment Minister, Elliott Morley MP. Unpublished report.

Leakey, R., and V. Morell. 2001. *Wildlife wars: My fight to save Africa's national treasures*. New York: St. Martin's Press.

Le Boeuf, B. J., and J. Reiter. 1988. Lifetime reproductive success in northern elephant seals. In *Reproductive success*, ed. T. H. Clutton-Brock, 344–62. Chicago: University of Chicago Press.

Lee, P. C. 1984. Ecological constraints on the social development of vervet monkeys. *Behaviour* 91:245–62.

———. 1986. Early social development among African elephant calves. *National Geographic Research* 2:388–401.

———. 1987. Allomothering among African elephants. *Animal Behaviour* 35:278–91.

———. 1989. Family structure, communal care and female reproductive effort. In *Comparative socioecology*, ed. V. Standen and R. A. Foley, 323–40. Oxford: Blackwells.

———. 1996. The meanings of weaning: Growth, lactation and life history. *Evolutionary Anthropology* 5:87–96.

———. 1999. Comparative ecology of post-natal growth and weaning among haplorhine primates. In *Comparative primate socioecology*, ed. P. C. Lee, 111–39. Cambridge: Cambridge University Press.

Lee, P. C., and M. Graham. 2006. African elephants and human-elephant interactions: Implications for conservation. *International Zoo Yearbook* 40:9–19.

Lee, P. C., and C. J. Moss. 1986. Early maternal investment in male and female African elephants calves. *Behavioral Ecology and Sociobiology* 18:353–61.

———. 1995. Statural growth in known-age African elephants (*Loxodonta africana*). *Journal of Zoology* 236:29–41.

———. 1999. The social context for learning and behavioural development among wild African elephants. In *Mammalian social learning*, ed. H. O. Box and K. R. Gibson, 102–25. Cambridge: Cambridge University Press.

———. 2009. Welfare and well-being of captive elephants: Perspectives from wild elephant life histories. In *An elephant in the room: The science and well being of elephants in captivity*, ed. D. L. Forthman, L. F. Kane, and P. Waldau, 22–39. North Grafton, MA: Tufts University, Cummings School of Veterinary Medicine's Center for Animals and Public Policy.

Lee, P. C., P. Majluf, and I. J. Gordon. 1991. Growth, weaning and maternal investment from a comparative perspective. *Journal of Zoology* 225:99–114.

Lee, P. C., L. Smith, and C. J. Moss. 2009. Long-term effects of early environmental conditions and growth on African elephant survival and success. *XXXI International Ethological Conference abstracts*: 313.

Leggett, K. E. A. 2006. Home range and seasonal movement of elephants in the Kunene Region, northwestern Namibia. *African Zoology* 41:17–36.

Lehmann, J., and C. Boesch. 2004. To fission or to fusion: Effects of community size on wild chimpanzee *(Pan troglodytes verus)* social organisation. *Behavioral Ecology and Sociobiology* 56: 207–16.

Lehner, P. N. 1996. *Handbook of ethological methods*. Cambridge: Cambridge University Press.

Leighty, K. A., J. Soltis, K. Leong, and A. Savage. 2008. Antiphonal exchanges in African elephants *(Loxodonta africana)*: Collective response to a shared stimulus, social facilitation, or true communicative event? *Behaviour* 145:297–312.

Lembuya, P. 1990. Group ranches. Unpublished report to the African Wildlife Foundation. Nairobi.

Leong, K. M., A. Ortolani, K. D. Burks, J. D. Mellen, and A. Savage. 2003. Quantifying acoustic and temporal characteristics of vocalizations for a group of captive African elephants *(Loxodonta africana)*. *Bioacoustics* 13:213–31.

Lessells, C. M. 1995. Putting resource dynamics into continuous input ideal free distribution models. *Animal Behaviour* 49: 487–94.

Leuthold, W. 1977. Spatial organization and strategy of habitat utilization by elephants in Tsavo National Park, Kenya. *Zoo Savgerterk* 42:358–739.

Lewis, D. M. 1986. Disturbance effects on elephant feeding: Evidence for compression in Luangwa Valley, Zambia. *African Journal of Ecology* 24:227–41.

Lewis, D. M., G. B. Kwaweche, and A. N. Mwenya. 1990. Wildlife conservation outside protected areas: Lessons from an experiment in Zambia. Unpublished report, ADMADE Project. Lusaka: National Parks of Zambia.

Li, X., J. Tao, M. T. Johnson, J. Soltis, A. Savage, K. M. Leong, and J. D. Newman. 2007. Stress and emotion classification using jitter and shimmer features. Presented at the International Conference on Acoustics Speech and Signal Processing 2007. Honolulu, Hawaii.

Limongelli, L., S. T. Boysen, and E. Visalberghi. 1995. Comprehension of cause-effect relations in a tool-using task by chimpanzees *(Pan troglodytes)*. *Journal of Comparative Psychology* 109:18–26.

Lindsay, W. K. 1982. Habitat selection and social group dynamics of the Amboseli elephants. MSc thesis, University of British Columbia.

———. 1987. Integrating parks and pastoralists: Some lessons from Amboseli. In *Conservation in Africa: People, policies, and practice*, ed. D. Anderson and R. Grove, 149–68. Cambridge: Cambridge University Press.

———. 1994. Feeding ecology and population demography of African elephants in Amboseli Kenya. PhD thesis, University of Cambridge.

Lindsay, W. K., and R. Olivier. 1984. Comments on: Why do elephants destroy woodland? *Pachyderm* 4:20.

Lindsey, P. A., J. T. du Toit, and M. G. L. Mills. 2005. Attitudes of ranchers towards African wild dogs *Lycaon pictus*: Conservation implications on private land. *Biological Conservation* 125:113–21.

Little, P. D. 1996. Pastoralism, biodiversity, and the shaping of savanna landscapes in East Africa. *Africa* 66:37–51.

Lopez, P., P. Aragon, and J. Martin. 2003. Responses of female lizards, *Lacertea monticola*, to males' chemical cues reflect their mating preference for older males. *Behavioral Ecology and Sociobiology* 55:73–9.

Lötter, H. P. P., M. Henley, S. Fakir, and M. Pickover. 2008. Ethical considerations in elephant management. In *Elephant management: A scientific assessment for South Africa*, ed. R. J. Scholes and K. G. Mennell, 406–45. Johannesburg: Wits University Press.

Lovatt Smith, D. 1986 [1997]. *Amboseli: Nothing short of a miracle*. Nairobi: East African Publishing House Ltd.

Lugard, F. D. 1893. *The rise of our East African empire: Early efforts in Nyasaland and Uganda, 2 vols*. London: Blackwood.

Lunn, N. J., and J. P. Y. Arnold. 1997. Maternal investment in Antarctic fur seals: Evidence for equality in the sexes? *Behavioral Ecology and Sociobiology* 40:351–62.

Lyon, B. E., and J. M. Eadie. 2000. Family matters: Kin selection and the evolution of conspecific brood parasitism. *Proceedings of the National Academy of Sciences USA* 97:13188–93.

MacDonald, D. W. 1985. The carnivores: Order Carnivora. In *Social odours in mammals*, ed. R. E. Brown and D. W. MacDonald, 619–22. Oxford: Clarendon Press.

Mackenzie, A., J. D. Reynolds, V. J. Brown, and W. J. Sutherland. 1995. Variation in male mating success on leks. *American Naturalist* 145:633–52.

Mackinnon, J. 1978. *The ape within us*. London: Collins.

Makacha, S., M. J. Msingwa, and G. W. Frame. 1982. Threats to the Serengeti herds. *Oryx* 16: 437–44.

Malcolm, J. 1985. Parental care in canids. *American Zoologist* 25:853–59.

Malima, C., R. Hoare, and J. Blanc. 2005. Systematic recording of human–elephant conflict: A case study in south–eastern Tanzania. *Pachyderm* 38:29–38.

Manegene, S., and J. Bernard. 2004. *Scoping report for the development of the Amboseli ecosystem General Management Plan*. Nairobi: AWF, KWS, ACC, ATE, ATGRCA, SFS 15.

Manfredo, M. J., J. J. Vaske, and D. J. Decker. 1995. Human dimensions of wildlife management: Basic concepts. In *Wildlife and recreationists: Coexistence through management and research*, ed. R. L. Knight and K. J. Gutzwiller, 17–31. Washington, DC: Island Press.

Manfredo, M. J., J. J. Vaske, and L. Sikorowski. 1996. Human dimensions of wildlife management. In *Natural resource management: The human dimension*, ed. A. Ewert, 53–72. Boulder: Westview Press.

Mann, J., and B. B. Smuts. 1998. Natal attraction: Allomaternal care and mother–infant separations in wild bottlenose dolphins. *Animal Behaviour* 55:1097–1113.

Mann, J., R. C. Connor, P. L. Tyack, and H. Whitehead, ed. 2000. *Cetacean societies: Field studies of dolphins and whales.* Chicago: University of Chicago Press.

Manser, M., R. M. Seyfarth, and D. L. Cheney. 2002. Suricate alarm calls signal predator class and urgency. *Trends in Cognitive Sciences* 6:55–7.

Marquardt, M., M. Infield, and A. Namara. 1994. Socio–economic survey of communities in the buffer zone of Lake Mburo National Park. Unpublished report for Lake Mburo Community Conservation Project. Kampala: Uganda National Parks.

Marsh, H., and T. Kasuya. 1984. Changes in the ovaries of the short–finned pilot whale, *Globicephala macrorhynchus*, with age and reproductive activity. *Reports of the International Whaling Commission* 6:311–35.

Marshall, T. C., J. Slate, L. E. B. Kruuk, and J. M. Pemberton. 1998. Statistical confidence for likelihood–based paternity inference in natural populations. *Molecular Ecology* 7:639–55.

Martin, A. 1996. *Dolphins and whales.* London: Salamander Books.

Martin, P., and P. Bateson. 1993. *Measuring behaviour.* Cambridge: Cambridge University Press.

Martin, R. B. 1978. Aspects of elephant social organisation. *Rhodesian Science News* 12:184–87.

Martin, R. D. 1990. *Primate origins and evolution.* London: Chapman and Hall.

———. 1996. Scaling of the mammalian brain: The maternal energy hypothesis. *News in Physiological Sciences* 11:149–56.

Martin, R. D., and A. M. MacLarnon. 1985. Gestation period, neonatal size, and maternal investment in placental mammals. *Nature* 313:220–23.

Marx, E. 1977. The tribe as a unit of subsistence: Nomadic pastoralism in the Middle East. *American Anthropologist* 79:343–63.

Mascarenhas, A. 1971. Agricultural vermin in Tanzania. In *Studies in East African geography and development*, ed. S. H. Ominde, 259–67. Berkeley: University of California Press.

Maynard-Smith, J. 1964. Group selection and kin selection. *Nature* 201:1145–47.

———. 1977. Parental investment: A prospective analysis. *Animal Behaviour* 25:1–9.

Maynard-Smith, J., and G. R. Price. 1973. The logic of animal conflict. *Nature* 246:15–18.

Mburugu, J. M. 2002. *The Amboseli biosphere reserve.* AfriMAB technical workshop for anglophone countries, 12–15 September 2000. Nairobi: UNESCO.

Mburur, J., R. Birner, and M. Zeller. 2003. Relative importance and determinants of landowners' transaction costs in collaborative wildlife management in Kenya: An empirical analysis. *Ecological Economics* 45:59–73.

McCabe, J. T. 2003. Sustainability and livelihood diversification among the Maasai of Northern Tanzania. *Human Organization* 62:100–11.

McCabe, J. T., S. Perkin, and C. Schofield. 1992. Can conservation and development be coupled among pastoral people? An examination of the Maasai of the Ngorongoro Conservation Area, Tanzania. *Human Organization* 51:353–66.

McComb, K., L. Baker, and C. Moss. 2006. African elephants show high levels of interest in the skulls and ivory of their own species. *Biology Letters* 2:26–8.

McComb, K., C. Moss, S. Sayialel, and L. Baker. 2000. Unusually extensive networks of vocal recognition in African elephants. *Animal Behaviour* 59:1103–9.

McComb, K., C. J. Moss, S. M. Durant, L. Baker, and S. Sayialel. 2001. Matriarchs as repositories of social knowledge in African elephants. *Science* 292:491–94.

McComb, K., D. Reby, L. Baker, C. Moss, and S. Sayialel. 2003. Long-distance communication of acoustic cues to social identity in African elephants. *Animal Behaviour* 65:317–29.

McCullagh, K. 1969. The growth and nutrition of the African elephant. II. The chemical nature of the diet. *East African Wildlife Journal* 7:91–7.

McElligott, A. G., R. Altwegg, and T. J. Hayden. 2002. Age-specific survival and reproductive probabilities: Evidence for senescence in male fallow deer (*Dama dama*). *Proceedings of the Royal Society of London, Series B* 269:1129–37.

McEwen, E. H., and P. E. Whitehead. 1972. Reproduction in female reindeer and caribou. *Canadian Journal of Zoology* 50:43–6.

McFarland Symington, M. 1987. Sex ratio and maternal rank in wild spider monkeys: When daughters disperse. *Behavioral Ecology and Sociobiology* 20:421–25.

McGregor, P. K. 1992. *Playback and studies of animal communication.* New York: Plenum.

McGregor, P. K., and M. I. Avery. 1986 The unsung songs of great tits (*Parus major*): Learning neighbours songs for discrimination. *Behavioral Ecology and Sociobiology* 18:311–16.

McGrew, W. C. 1989. Why is ape tool use so confusing? In *Comparative socioecology: The behavioural ecology of humans and other mammals*, ed. V. Standen and R. A. Foley, 457–78. Oxford: Blackwell Scientific Publications.

———. 1992. *Chimpanzee material culture: Implications for human evolution.* Cambridge: Cambridge University Press.

McKay, G. M. 1973. Behavior and ecology of the Asiatic elephant in southeastern Ceylon. *Smithsonian Contributions to Zoology* 125:1–113.

McNeill Alexander, R. 1999. One price to run, swim or fly? *Nature* 397:651–52.

Mduma, S., A. E. R. Sinclair, and R. Hilborn. 1999. Food regulates the Serengeti wildebeest, a 40–year record. *Journal of Animal Ecology* 68:1101–22.

Mearns, R. 1997. Livestock and environment: Potential for complementarity. *World Animal Review* 88:2–14.

Meijerink, A. M. J., and W. van Wijngaarden. 1997. Contribution to the groundwater hydrology of the Amboseli ecosystem, Kenya. In *Groundwater / surface water ecotones: Biological and hydrological interactions and management options*, UNESCO International Hydrology Series, ed. J. Gibert, J. Mathieu, and F. Fournier, 111–18. Cambridge: Cambridge University Press.

Meinertzhagen, R. 1983. *Kenya diary (1902–1906).* London: Eland Books.

Meissner, H. H., E. B. Spreeth, P. A. de Villiers, E. W. Pietersen, T. A. Hugo, and B. F. Terblanche. 1990. Quality of food and voluntary intake by elephant as measured by lignin index. *South African Journal of Wildlife Research* 20:104–10.

Melnick, D., M. Pearl, and A. F. Richard. 1984. Male migration and inbreeding avoidance in wild rhesus monkeys. *American Journal of Primatology* 7:229–43.

Menzel, E. W., Jr. 1973. Chimpanzee spatial memory organization. *Science* 182:943–45.

Meredith, M. 2001. *Africa's elephants: A biography*. London: Hodder and Stoughton.

Merker, M. 1910. *The Maasai: Ethnographic monograph of an East African Semite People*. Berlin: Dietrich Reimer.

Mifflin, M. D. 1991. Amboseli hydrology. Unpublished report. Nairobi: Kenya Wildlife Service

Migot–Adholla, S. E. and P. D. Little. 1981. Evolution of policy toward the development of pastoral areas in Kenya. In *The future of pastoral peoples*, ed. J. G. Galaty, D. Aronson, and P. C. Salzman, 144–56. Ottowa: International Development Research Centre.

Miklosi, A., A. Topal, and V. Csanyi. 2004. Comparative social cognition: What can dogs teach us? *Animal Behaviour* 67: 995–1004.

Miller, M. N., and Byers, J. A. 1991. Energetic costs of locomotor play in pronghorn fawns. *Animal Behaviour* 41:1007–13.

Mills, A. J. 2006 The role of salinity and sodicity in the dieback of *Acacia xanthophloea* in Ngorongoro Caldera, Tanzania. *African Journal of Ecology* 44:61–71.

Milner, J. M., S. D. Albon, A. W. Illius, J. M. Pemberton, and T. H. Clutton-Brock. 1999. Repeated selection on morphometric traits in the Soay sheep on St. Kilda. *Journal of Animal Ecology* 68:472–88.

Milton, K. 1981. Distribution patterns of tropical plant foods as a stimulus to primate mental development. *American Anthropologist* 83:534–548.

Ministry of Environment and Tourism. 2005. National workshop on human wildlife conflict management (HWCM) in Namibia. Unpublished report. Windhoek, Namibia: Ministry of Environment and Tourism.

Miquelle, D. G., J. M. Peek, and V. Van Ballenberghe. 1992. Sexual segregation in Alaskan moose. *Wildlife Monographs* 122:1–57.

Mitchell, D. 2005. In the shadow of Mount Kilimanjaro. *Travel News*, September.

Mohr, C. O. 1947. Table of equivalent populations of small North American mammals. *American Midland Naturalist* 37: 223–49.

Mol, F. 1981. The Masai and wildlife. *Swara* 4:24–7.

———. 1996. *Maasai: Language and culture*. Limuru, Kenya: Kolby Press.

Moody, A. L., and A. I. Houston. 1995. Interference and the ideal free distribution. *Animal Behaviour* 49:1065–72.

Moore, E. D. 1931. *Ivory scourge of Africa*. New York; London: Harper & Brothers.

Mordi, R. 1987. Public attitudes toward wildlife in Botswana. PhD thesis, Yale University.

Morgan, B. J. 2001. Ecology of mammalian frugivores in the Reserve de Faune du Petit Loango, Gabon. PhD thesis, University of Cambridge.

Morgan, B., and P. C. Lee. 2007. Forest elephant group composition, frugivory and coastal use in the Réserve de Faune du Petit Loango, Gabon. *African Journal of Ecology* 45:519–26.

Morin, P. A., K. E. Chambers, C. Boesch, and L. Vigilant. 2001. Quantitative polymerase chain reactions analysis of DNA from noninvasive samples for accurate microsatellite genotyping of wild chimpanzees. *Molecular Ecology* 10:1835–44.

Morin, P. A., J. J. Moore, R. Chakraborty, L. Jin, J. Goodall, and D. S. Woodruff. 1994. Kin selection, social structure, gene flow, and the evolution of chimpanzees. *Science* 265: 1193–201.

Morris, M. D. 1986. Large scale deceit: Deception by captive elephants? In *Deception: Perspectives on human and nonhuman deceit*, ed. R. W. Mitchell and N. S. Thompson, 183–91. Albany: State University of New York.

Moss, C. J. 1981. Social circles. *Wildlife News* 16:2–7.

———. 1982. *Portraits in the wild: Behavior studies of East African mammals*, 2nd ed. Chicago: University of Chicago Press.

———. 1983. Oestrous behaviour and female choice in the African elephant. *Behaviour* 86:167–96.

———. 1988. *Elephant memories: Thirteen years in the life of an elephant family*. London: Elm Tree Books.

———. 1992. *Echo of the elephants*. London: BBC Books.

———. 1996. Getting to know a population. In *Studying elephants*, ed. K. Kangwana, 58–74. AWF Technical Handbook Series 7. Nairobi: African Wildlife Foundation.

———. 2000. A passionate devotion. In *The smile of the dolphin: Remarkable accounts of animal emotions*, ed. M. Bekoff, 135–37. New York: Discovery Books.

———. 2001. The demography of an African elephant (*Loxodonta africana*) population in Amboseli, Kenya. *Journal of Zoology* 255:145–56.

Moss, C. J., and J. H. Poole. 1983. Relationships and social structure of African elephants. In *Primate social relationships: An integrated approach*, ed. R. A. Hinde, 314–25. Oxford: Blackwells.

Mosser, A., and C. Packer. 2009. Group territoriality and the benefits of sociality in the African lion, *Panther leo*. *Animal behaviour* 78:359–70.

Mueller, T., and W. F. Fagan. 2008. Search and navigation in dynamic environments—from individual behaviors to population distributions. *Oikos* 117:654–64.

Mugisha, A. 2002. Evaluation of community-based conservation approaches: Management of protected areas in Uganda. PhD thesis, University of Florida.

Mumiukha, P. W. 1976. The vegetation of Kajiado District. Unpublished report, UNEP/FAO Wildlife Management in Kenya Project KEN/71/526. Nairobi: Wildlife Conservation and Management Department.

Murphy, W. J., E. Eizirik, S. O'Brien, O. Madsen, M. Scally, C. J. Douady, E. Teeling, O. A. Ryder, M. J. Stanhope, M. de Jong, and M. S. Springer. 2001. Resolution of the early placental mammal radiation using Bayesian phylogenetics. *Science* 294: 2348–51.

Murray, L. E. 1995. *Personality and individual differences in captive African apes*. PhD thesis, University of Cambridge.

———. 1998. The effects of group structure and rearing strategy on personality in chimpanzees at Chester, London ZSL and Twycross Zoos. *International Zoo Yearbook* 36:97–108.

Muruthi, P., J. Altmann, and S. Altmann. 1991. Resource base, parity, and reproductive condition affect females' feeding time and nutrient intake within and between groups of a baboon population. *Oecologia* 87:467–72.

Mutinda, H. S. 2003. Social determinants of movements and aggregation among free-ranging elephants (*Loxodonta africana* Blumenbach) in Amboseli, Kenya. PhD thesis, University of Nairobi.

Mwangi, E. 2005. The transformation of property rights in Kenya's Maasailand: Triggers and motivations. CAPRi Working Paper no. 35. Washington, DC: International Food Policy Research Institute.

Myers, N. 1973. The relationship of parks and other protected areas to their environs in Masailand, East Africa. PhD thesis, University of California, Berkeley.

Napier, J. R. 1962. The evolution of the hand. *Scientific American* 207:56–62.

Naughton-Treves, L. 1998. Prediction patterns of crop damage by wildlife around Kibale National Park, Uganda. *Conservation Biology* 12:156–68.

———. 2001. Farmers, wildlife and the forest fringe. In *African rain forest ecology and conservation*, ed. W. Weber, L. J. T. White, A. Vedder, and L. Naughton-Treves, 369–84. New Haven: Yale University Press.

Naughton-Treves, L., R. Rose, and A. Treves. 1999. The social dimensions of human–elephant conflict in Africa: A literature review and case studies from Uganda and Cameroon. Unpublished report to the African Elephant Specialist Group, Human–Elephant Conflict Task Force. Gland, Switzerland: IUCN.

Naughton-Treves, L., and A. Treves. 2005. Socioecological factors shaping local support for wildlife in Africa. In *People and wildlife, conflict or coexistence?* ed. R. Woodroffe, S. Thirgood, and A. Rabinowitz, 253–77. Cambridge: Cambridge University Press.

Navidi, W., N. Arnheim, and M. S. Waterman. 1992. A multiple-tubes approach for accurate genotyping of very small DNA samples by using PCR: Statistical considerations. *American Journal of Human Genetics* 50:347–59.

Ndaskoi, N. 2006. The root causes of Maasai predicament. *Fourth World Journal* 7:28–61.

Ndegengy, G. 2003. *Geographic dimensions of human well-being in Kenya: Where are the poor? From districts to locations. vol. 1.* Nairobi: Central Bureau of Statistics.

Nelson, F. 2000. Sustainable development and wildlife conservation in Maasailand. *Environment, Development and Sustainability* 2:107–17.

———. 2004. The evolution and impacts of community-based ecotourism in northern Tanzania. Issue Paper no. 131. London: International Institute for Environment and Development.

Nelson, A., P. Bidwell, and C. Sillero-Zubiri. 2003. *A review of human–elephant conflict management strategies*. People and Wildlife Initiative: Wildlife Conservation Research Unit, Oxford University.

Newmark, W. D., and J. L. Hough. 2000. Conserving wildlife in Africa: Integrated conservation and development projects and beyond. *Bioscience* 50:585–92.

Newmark, W. D., N. L. Leonard, H. I. Sariko, and D. M. Gamassa. 1993. Conservation attitudes of local people living adjacent to five protected areas in Tanzania. *Biological Conservation* 63: 177–83.

Newsome, A. E. 1980. Differences in the diets of male and female red kangaroos in central Australia. *African Journal of Ecology* 18:27–31.

Ngure, N. 1992. Human–elephant interactions: Seeking remedies for conflicts. *Swara* 15:25–6.

Niamir-Fuller, M. 2002. Non–equilibrium theory of African arid ecosystems: Designing for monitoring and evaluation. In *Implementing sustainable development*, ed. H. Abaza and A. Baranzini, 235–53. Cheltenham: Edward Elgar for UNEP.

Nissani, M. 2004. Theory of mind and insight in chimpanzees, elephants and other animals? In *Comparative vertebrate cognition*, ed. L. J. Rogers and G. Kaplan, 227–61. London: Plenum Publishers.

———. 2006. Do Asian elephants (*Elephas maximus*) apply causal reasoning to tool-use tasks? *Journal of Experimental Psychology* 32:91–6.

Nissani, M., and D. Hoefler-Nissani. 2007. Absence of mirror self-referential behavior in two Asian elephants. *Journal of Veterinary Science* 1. www.scientificjournals.org/journals 2007/articles/1043.htm.

Nissani, M., D. Hoefler-Nissani, U. Tin Lay, and U. Wan Htun. 2005. Simultaneous visual discrimination in Asian elephants. *Journal of the Experimental Analysis of Behaviour* 83: 15–29.

Noë, R., and P. Hammerstein. 1994. Biological markets: Supply and demand determine the effect of partner choice in cooperation, mutualism and mating. *Behavioral Ecology and Sociobiology* 35:1–11.

North American Association for Environmental Education. 1993. *Defining environmental education: The NAAEE perspective.* Washington, DC: North American Association for Environmental Education.

Norton-Griffiths, M. 1978. *Counting animals.* AWF Technical Handbook Series 1. Nairobi: African Wildlife Foundation.

———. 1979. The influence of grazing, browsing and fire on the vegetation dynamics of the Serengeti. In *Serengeti: Dynamics of an ecosystem*, ed. A. E. R. Sinclair and M. Norton-Griffiths, 310–52. Chicago: University of Chicago Press.

———. 1998. The economics of wildlife conservation policy in Kenya. In *Conservation of Biological Resources*, ed. E. J. Milner-Gulland and R. Mace, 279–97. Oxford: Blackwells.

Noser, R., and R. W. Byrne. 2007a. Mental maps in chacma baboons (*Papio ursinus*): Using intergroup encounters as a natural experiment. *Animal Cognition* 10:331–40.

———. 2007b. Travel routes and planning of visits to out-of-sight resources in wild chacma baboons, *Papio ursinus*. *Animal Behaviour* 73:257–66.

Ntiati, P. 2002. Group ranches subdivision study in Loitokitok Division of Kajiado District, Kenya. Land Use Change Impacts and Dynamics (LUCID) Project Working Paper 7. Nairobi: International Livestock Research Institute, UNEP/DGEF.

Nunn, C. L., and R. J. Lewis. 2001. Cooperation and collective action in animal behaviour. In *Economics in nature: Social dilemmas, mate choice and biological markets*, ed. R. Noë, J. A. R. A. M. van Hooff, and P. Hammerstein, 42–66. Cambridge: Cambridge University Press.

Nyakaana, S., and P. Arctander. 1998. Isolation and characterization of microsatellite loci in the African elephant, *Loxodonta africana*. *Molecular Ecology* 7:1431–39.

———. 1999. Population genetic structure of the African elephant in Uganda based on variation at mitochondrial and nuclear loci: Evidence for male biased gene flow. *Molecular Ecology* 8: 1105–15.

Nyakaana, S., P. Arctander, and H. R. Siegismund. 2002. Population structure of the African savannah elephant inferred from mitochondrial control region sequences and nuclear microsatellite loci. *Heredity* 89:90-98.

Nyakaana, S., E. L. Abe, P. Arctander, and H. R. Siegismund. 2006. DNA evidence for elephant social behaviour breakdown in Queen Elizabeth National Park, Uganda. *Animal Conservation* 4: 231–37.

O'Connell, C., L. A. Hart, and B. T. Arnason. 1998. Comments on "Elephant hearing." *Journal of the Acoustical Society of America* 105:2051–52.

O'Connell-Rodwell, C., B. T. Arnason, and L. A. Hart. 1997. Seismic transmission of elephant vocalizations and movement. *Journal of the Acoustical Society of America* 105: 2051–52.

———. 2000. Seismic properties of Asian elephant (*Elephas maximus*) vocalizations and locomotion. *Journal of the Acoustical Society of America* 108:3066–72.

O'Connell-Rodwell, C., J. D. Wood, C. Kinzley, T. C. Rodwell, J. Poole, and S. Puria. 2007. Wild African elephants (*Loxodonta africana*) discriminate between familiar and unfamiliar conspecific seismic alarm calls. *Journal of the Acoustical Society of America* 122:823–30.

O'Connell-Rodwell, C., J. D. Wood, T. C. Rodwell, S. Puria, S. R. Partan, R. Keefe, D. Shriver, B. T. Arnason, and L. A. Hart. 2006. Wild elephant (*Loxodonta africana*) breeding herds respond to artificially transmitted seismic stimuli. *Behavioral Ecology and Sociobiology* 59:842–50.

Oftedal, O. 1985. Pregnancy and lactation. In *The bioenergetics of wild herbivores*, ed. R. J. Hudson and R. G. White, 215–38. Boca Raton: CRC Press.

Ogutu, J. O., and N. Owen-Smith. 2003. ENSO, rainfall and temperature influences on extreme population declines among African savannah ungulates. *Ecology Letters* 6:412–19.

Ojwang', G. O., P. Waragute, and L. Njiro. 2006. Trends and spatial distribution of large herbivores in Kajiado District (1978–2000). Unpublished report. Nairobi: Department of Resource Surveys and Remote Sensing (DRSRS).

Okello, M. M. 2005. Land use changes and human-wildlife conflicts in the Amboseli Area, Kenya. *Human Dimensions of Wildlife* 10:19–28.

Oldfield, M. L. 1988. Threatened mammals affected by human exploitation of the female-offspring bond. *Conservation Biology* 2:260–74.

Ole Dapash, M. 2002. Coexisting in Kenya: The human-elephant conflict. *Animal Welfare Institute Quarterly* 51:10–12.

Ole Ndaskoi, N. 2006. The root causes of the Maasai predicament. *Fourth World Journal* 7:28–61.

Ole Parkipuny, M. S., and D. J. Berger. 1993. Maasai rangelands: Links between social justice and wildlife conservation. In *Voices from Africa: Local perspectives on conservation*, ed. D. Lewis and N. Carter, 113–31. Washington, DC: World Wildlife Fund.

Oliveira, R. F., P. K. McGregor, and C. Latruffe. 1998. Know thine enemy: Fighting fish gather information from observing conspecific interactions. *Proceedings of the Royal Society, Series B* 265:1045–49.

Olivier, R. C. D. 1978. On the ecology of the Asian elephant, *Elephas maximus* Linn., with particular reference to Malaya and Sri Lanka. PhD thesis, University of Cambridge.

———. 1982. Ecology and behavior of living elephants: Bases for assumptions concerning the extinct woolly mammoths. In *Paleoecology of Beringia*, ed. D. M. Hopkins, J. V. Matthews, Jr., C. E. Schweger, and S. B. Young, 291–305. New York: Academic Press.

Ollason, J. C., and G. M. Dunnet. 1988. Variation in breeding success in fulmars. In *Reproductive success*, ed. T. H. Clutton-Brock, 263–78. Chicago: University of Chicago Press.

O'Malley, E. 2000. Cattle and cultivation: Changing land use and labor patterns in pastoral Maasai livelihoods, Loliondo Division, Ngorongoro District, Tanzania. PhD thesis, University of Colorado.

Omondi, P. 1994. Wildlife-human conflict in Kenya: Integrating wildlife conservation with human needs in the Maasai Mara Region. PhD thesis, McGill University.

Onetu, L. 1998. The impacts of tourism on the socio-cultural and economic lifestyles of the Maasai. BA thesis, Catholic University of East Africa.

Ono, K. A., and D. J. Boness. 1996. Sexual dimorphism in sea lion pups: Differential maternal investment, or sex-specific differences in energy allocation. *Behavioral Ecology and Sociobiology* 38:31–41.

Organ, G. E., and H. A. Fosbrooke. 1963. *Ngorongoro's first visitor*. Dar es Salaam: East African Literature Bureau.

Orindi, V., and C. Huggins. 2005. The dynamic relationship between property rights, water resource management and poverty in the Lake Victoria Basin. Presented at the International workshop on African Water Laws: Plural legislative frameworks for rural water management in Africa. 26–28 January 2005, Johannesburg, South Africa.

Osborn, F. V. 1998. The ecology of crop-raiding elephants in Zimbabwe. PhD thesis, University of Cambridge.

———. 2004. Seasonal variation of feeding patterns and food selection by crop-raiding elephants in Zimbabwe. *African Journal of Ecology* 42:22–7.

Osborn, F. V., and G. E. Parker. 2002. Community-based methods to reduce crop loss to elephants: Experiments in the communal lands of Zimbabwe. *Pachyderm* 33:32–8.

———. 2003. Linking two elephant refuges with a corridor in the communal lands of Zimbabwe. *African Journal of Ecology* 41:68–74.

Otter, K., P. K. MacGregor, A. M. R. Terry, F. R. L. Burford, T. M. Peake, and T. Dabelsteen. 1999. Do female great tits (*Parus major*) assess males by eavesdropping? A field study using interactive song playbacks. *Proceedings of the Royal Society, Series B* 266:1305–09.

Owaga, M. L. A. 1980. Primary productivity and herbage utilization by herbivores in Kaputei plains, Kenya. *African Journal of Ecology* 18:1–5.

Owen-Smith, N. 1988. *Megaherbivores: The influence of very large body size on ecology*. Cambridge: Cambridge University Press.

———. 1990. Demography of a large herbivore, the greater kudu *Tragelaphus strepsiceros*, in relation to rainfall. *Journal of Animal Ecology* 59:893–913.

———. 1993. Age, size, dominance and reproduction among male kudus: Mating enhancement by attrition of rivals. *Behavioral Ecology and Sociobiology* 32:177–84.

———. 1996. Ecological guidelines for waterpoints in extensive protected areas. *South African Journal of Wildlife Research* 26:107–12.

Owen-Smith, N., and P. Novellie. 1982. What should a clever ungulate eat? *American Naturalist* 119:151–78.

Owen-Smith, N., G. I. H. Kerley, B. Page, R. Slotow, and R. J. van Aarde. 2006. A scientific perspective on the management of elephants in the Kruger National Park and elsewhere. *South African Journal of Science* 102:389–94.

Owens, M. J., and D. Owens. 2009. Early age reproduction in female savanna elephants (*Loxodonta africana*) after severe poaching. *African Journal of Ecology* 47:214–22.

Packer, C. R., S. Lewis, and A. E. Pusey. 1991. A comparative

analysis of non-offspring nursing. *Animal Behaviour* 43: 265–81.
Packer, C. R., M. Tatar, and A. Collins. 1998. Reproductive cessation in female mammals. *Nature* 392:807–11.
Packer, C. R., L. Herbst, A. E. Pusey, J. D. Bygott, J. Hanby, S. J. Cairns, and M. B. Mulder. 1988. Reproductive success of lions. In *Reproductive success*, ed. T. H. Clutton-Brock, 363–83. Chicago: University of Chicago Press.
Parkipuny, M. S. O. 1989. So that the Serengeti shall never die. In *Nature management and sustainable development*, ed. W. D. Verwey, 256–63. Amsterdam; Springfield, VA; Tokyo: IOS.
———. 1997. Pastoralism, conservation, and development in the Greater Serengeti Region. In *Multiple land-use: The experience of the Ngorongoro Conservation Area, Tanzania*, ed. D. M. Thomson, 143–68. Gland, Switzerland: IUCN.
Parry, D., and B. Campbell. 1992. Attitudes of rural communities to animal wildlife and its utilization in Chobe Enclave and Mababe Depression, Botswana. *Environmental Conservation* 19:245–52.
Partridge, L. 1978. Habitat selection. In *Behavioural ecology*, ed. J. R. Krebs and N. B. Davies, 351–76. Oxford: Blackwell Scientific Publications.
Passingham, R. E. 1981. Primate specializations in brain and intelligence. *Symposia of the Zoological Society of London* 46: 361–88.
Passmore, R., and J. B. G. Durnin. 1955. Human energy expenditure. *Physiological Review* 35: 801–35.
Patterson, F. G. B., and R. H. Cohn. 1994. Self-recognition and self-awareness in lowland gorillas. In *Self-awareness in animals and humans: Developmental perspectives*, ed. S. T. Parker, R. W. Mitchell, and M. L. Boccia, 273–90. Cambridge: Cambridge University Press.
Paul, A., and J. Kuester. 1990. Adaptive significance of sex ratio adjustment in semifree-ranging Barbary macaques (*Macaca sylvanus*) at Salem. *Behavioral Ecology and Sociobiology* 27: 287–93.
Payne, K. B. 2000. *Progress report of the elephant listening project.* Cornell. Cornell University.
———. 2003. Sources of social complexity in the three elephant species. In *Animal social complexity: Intelligence, culture, and individualized societies*, ed. F. B. M. de Waal and P. L. Tyack, 57–85. Boston: Harvard University Press.
Payne, K. B., and W. R. Langbauer. 1992. Elephant communication. In *Elephants*, ed. J. Shoshani, 116–23. San Francisco: Weldon Owen.
Payne, K. B., W. R. Langbauer, and E. M. Thomas. 1986. Infrasonic calls of the Asian elephant *Elephas maximas*. *Behavioural Ecology and Sociobiology* 18:297–301.
Payne, K. B., M. Thompson, and L. Kramer. 2003. Elephants calling patterns as indicators of group size and composition: The basis for an acoustic monitoring system. *African Journal of Ecology* 41:99–107.
Payne, R., and D. Webb. 1971. Orientation by means of long range acoustic signaling in baleen whales. *Annals of the New York Academy of Sciences* 188:110–41.
Pélabon, C., J.-M. Gaillard, A. Loison, and C. Portier. 1995. Is sex-biased maternal care limited by total maternal expenditure in polygynous ungulates? *Behavioral Ecology and Sociobiology* 37: 311–19.
Pellew, R. A. 1984. The feeding ecology of a selective browser, the giraffe (*Giraffa camelopardalis tippelskirchi*). *Journal of Zoology* 202:57–81.
Peluso, N. L. 1993. Coercing conservation? The politics of state resource control. *Global Environmental Change* 3:199–217.
Pemberton, J., S. D. Albon, F. E. Guinness, T. H. Clutton-Brock, and G. A. Dover. 1992. Behavioral estimates of male mating success tested by DNA fingerprinting in a polygynous mammal. *Behavioral Ecology* 3:6–75.
Pennington, H. 1983. A living trust: Tanzanian attitudes towards wildlife and conservation. MSc thesis, Yale University.
Pennisi, E. 2001. Behavioural ecology: Elephant matriarchs tell friends from foe. *Science* 292:417–18.
Pennycuick, C. J., J. B. Sale, M. Stanley-Price, and G. M. Jolly. 1977. Aerial systematic sampling applied to censuses of large mammals in Kenya. *African Journal of Ecology* 15:139–46.
Pereira, M. E., and L. A. Fairbanks. 1993. *Juvenile primates: Life history, development and behavior.* New York: Oxford University Press.
Pereira, M. E., and S. R. Leigh. 2003. Models of primate development. In *Primate life history and socioecology*, ed. P. M. Kappeler and M. E. Pereira, 149–76. Chicago: University of Chicago Press.
Peters, R. H. 1983. *The ecological implications of body size.* Cambridge: Cambridge University Press.
Petrides, G. A., and R. G. Swank. 1966. Estimating the productivity and energy relations of an African elephant population. *Proceedings of the IXth International Grasslands Congress*, 831–42. Sao Paulo, Brazil.
Petrie, M., A. Krupa, and T. Burke. 1999. Peacocks lek with relatives even in the absence of social and environmental cues. *Nature* 401:155–57.
Pettorelli, N., J. O. Vik, A. Mysterude, J.-M. Gaillard, C. J. Tucker, and N. C. Stenseth. 2005. Using the satellite derived NDVI to assess ecological responses to environmental change. *Trends in Ecology and Evolution* 20:503–10.
Phillips, P. K., and J. E. Heath. 1992. Heat exchange by the pinna of the African elephant (*Loxodonta africana*). *Comparative Physiology* 101:693–9.
Pido, D. K. 1994. Review of *The art of the Maasai: 300 newly discovered objects and works of art. African Arts* 27:15–19.
———. 1995. Donna Klump Pido responds. *African Arts* 29:105.
Pilgram, T., and D. Western. 1986. Inferring the sex and age of African elephants from tusk measurements. *Biological Conservation* 36:39–52.
Pinter-Wollman, N., L. A. Isbell, and L. A. Hart. 2009. Assessing translocation outcome: Comparing behavioral and physiological aspects of translocated and resident African elephants (*Loxodonta africana*). *Biological Conservation* 142: 1116–24.
Plotnik, J. M., F. B. M. de Waal, and D. Reiss. 2006. Self-recognition in an Asian elephant. *Proceedings of the National Academy of Sciences USA* 103:17053–7.
Poole, J. H. 1982. Musth and male-male competition in the African elephant. PhD thesis, University of Cambridge.
———. 1987a. Rutting behaviour in African elephants: The phenomenon of musth. *Behaviour* 102: 283–316.
———. 1987b. Elephants in musth, lust. *Natural History* 96: 46–55.
———. 1987c. Raging bulls. *Animal Kingdom* 90:18–25
———. 1989a. Announcing intent: The aggressive state of musth in African elephants. *Animal Behaviour* 37:140–52.

———. 1989b. Mate guarding, reproductive success and female choice in African elephants. *Animal Behaviour* 37:842–49.

———. 1989c. The effects of poaching on the age structures and social and reproductive patterns of selected East African elephant populations. In *The ivory trade and the future of the African elephant. Volume 2: Technical reports*, ed. S. Cobb. Oxford: Unpublished report of the Ivory Trade Review Group, Prepared for the 7th CITES Conference of the Parties.

———. 1994a. Sex differences in the behavior of African elephants. In *The differences between the sexes*, ed. R. Short and E. Balaban, 331–46. Cambridge: Cambridge University Press.

———. 1994b. Logistical and ethical considerations in the management of elephant populations through fertility regulation. In *Proceedings of the 2nd international conference on advances in reproductive research in man and animals*, ed. C. Singh Bambra, 278–83. Nairobi: Institute of Primate Research, National Museums of Kenya.

———. 1996. *Coming of age with elephants*. New York: Hyperion.

———. 1997. *Elephants*. Grantown-on-Spey: Colin Baxter Photography.

———. 1998a. Communication and social structure of African elephants. In *Elephants*. Care for the Wild International, UK.

———. 1998b. An exploration of a commonality between ourselves and elephants. *Etica & Animali* 9:85–110.

———. 1999a. Signals and assessment in African elephants: Evidence from playback experiments. *Animal Behaviour* 58: 185–93.

———. 1999b. Voices of elephants. *Sotokoto* 8:14–6.

———. 1999c. Ella's Easter baby. *Care for the Wild News* 15: 24–5.

———. 2000a. Family reunions. In *The smile of the dolphin: Remarkable accounts of animal emotions*, ed. M. Bekoff, 22–3. New York: Discovery Books.

———. 2000b. When bonds are broken. In *The smile of the dolphin: Remarkable accounts of animal emotions*, ed. M. Bekoff, 142–43. New York: Discovery Books.

Poole, J. H., and P. K. Granli. 2003. Elephant visual and tactile signals database. Available at: www.ElephantVoices.org.

———. 2004. The visual, tactile and acoustic signals of play in African savannah elephants. In *Endangered elephants, past present and future*, ed. J. Jayewardene, 44–50. Proceedings of the Symposium on Human Elephant Relationships and Conflicts, Sri Lanka, September 2003. Colombo: Biodiversity and Elephant Conservation Trust.

———. 2005. The ethical management of elephants and the value of long-term field research. *American Anti–Vivisection Society* 63:2–5.

———. 2009. Mind and movement: Meeting the interests of elephants. In *An elephant in the room: The science and well being of elephants in captivity*, ed. D. L. Forthman, L. F. Kane, and P. Waldau, 2–21. North Grafton, MA: Tufts University, Cummings School of Veterinary Medicine's Center for Animals and Public Policy.

Poole, J. H., and C. J. Moss. 1981. Musth in the African elephant, *Loxodonta africana*. *Nature* 292: 830–31.

———. 1989. Elephant mate searching: Group dynamics and vocal and olfactory communication. In *The biology of large African mammals in their environment*, ed. P. A. Jewell and G. M. O. Maloiy, 111–25. Oxford: Clarendon Press.

———. 2008. Elephant sociality and complexity: The scientific evidence. In *Never forgetting: Elephants and ethics*, ed. C. M. Wemmer and C. A. Christen, 69–98. Boston: Johns Hopkins University Press.

Poole, J. H., and J. B. Thomsen. 1989. Elephants are not beetles: Implications of the ivory trade for the survival of the African elephant. *Oryx* 23:188–98.

Poole, J. H., C. J. Moss, and D. Sheldrick. 1999. The sad plight of the Tuli elephants. The last word. *Swara* 22:32–3.

Poole, J. H., L. H. Kasman, E. C. Ramsay, and B. L. Lasley. 1984. Musth and urinary testosterone concentrations in the African elephant, *Loxodonta africana*. *Journal of Reproduction and Fertility* 70:255–60.

Poole, J. H., K. B. Payne, W. R. Langbauer, Jr., and C. J. Moss. 1988. The social contexts of some very low frequency calls of African elephants. *Behavioral Ecology and Sociobiology* 22: 385–92.

Poole, J. H., P. L. Tyack, A. S. Stoeger-Horwath, and S. Watwood. 2005. Elephants are capable of vocal learning. *Nature* 434:455–56.

Poole, J. H., N. Aggarwal, R. Sinange, S. Nganga, and M. Broten. 1992. The status of Kenya's elephants 1992. Unpublished report. Nairobi: Kenya Wildlife Service and Department of Resource Surveys and Remote Sensing.

Pope, T. R. 1998. Effects of demographic change on group kin structure and gene dynamics of populations of red howling monkeys. *Journal of Mammalogy* 79:692–712.

Portier, C., M. Festa-Bianchet, J.-M. Gaillard, J. T. Jorgenson, and N. G. Yoccoz. 1998. Effects of density and weather on survival of bighorn sheep lambs (*Ovis canadensis*). *Journal of Zoology* 245:271–78.

Povinelli, D. J. 1989. Failure to find self-recognition in Asian elephants (*Elephus maximus*) in contrast to their use of mirror cues to discover hidden food. *Journal of Comparative Psychology* 103:122–31.

———. 1993. Reconstructing the evolution of mind. *American Psychologist* 48:493–509.

Povinelli, D. J., and T. J. Eddy. 1996. What young chimpanzees know about seeing. *Monographs of the Society for Research in Child Development* 61:1–189.

Povinelli, D. J., and J. Vonk. 2003. Chimpanzee minds: Suspiciously human? *Trends in Cognitive Sciences* 7:157–60.

Pratt, D. J., P. J. Greenway, and M. D. Gwynne. 1966. A classification of East African rangeland, with an appendix on terminology. *Journal of Applied Ecology* 3:369–82.

Prentice, A. M., and R. G. Whitehead. 1987. The energetics of human reproduction. *Symposia of the Zoological Society of London* 75:275–304.

Preston, B. T., I. R. Stevenson, J. M. Pemberton, and K. Wilson. 2001. Dominant rams lose out by sperm depletion—A waning success in siring counters a ram's high score in competition for ewes. *Nature* 409: 681–82.

Prins, H. H. T. 1989. Condition changes and choice of social environment in African buffalo bulls. *Behaviour* 108:298–324.

———. 1992. The pastoral road to extinction: Competition between wildlife and traditional pastoralism in East Africa. *Environmental Conservation* 19:117–23.

Prins, H. H. T., and H. P. v. d. Jeugd. 1993. Herbivore population crashes and woodland structure in East Africa. *Journal of Ecology* 81:305–14.

Promislow, D. E. L., and P. H. Harvey. 1990. Living fast and dying young: A comparative analysis of life history variation among mammals. *Journal of Zoology* 220:417–37.

Proops, L., K. McComb, and D. Reby. 2009. Cross-modal

individual recognition in domestic horses. *Proceedings of the National Academy of Sciences* 106:947–51.
Pusey, A. E., and C. Packer. 1987. Dispersal and philopatry. In *Primate societies*, ed. B. B. Smuts, D. L. Cheney, R. M. Seyfarth, and R. W. Wrangham, 250–66. Chicago: University of Chicago Press.
Queller, D. C., and K. F. Goodnight. 1989. Estimating relatedness using genetic markers. *Evolution* 43:258–75.
Ranta, E., N. Peuhkuri, R. Lauroa, H. Rita, and N. B. Metcalfe. 1996. Producers, scroungers and foraging group structure. *Animal Behaviour* 51:171–75.
Rasmussen, H. B. 2005. Reproductive tactics in male African savannah elephants (*Loxodonta africana*). DPhil thesis, University of Oxford.
Rasmussen, H. B., G. Wittemyer, and I. Douglas-Hamilton. 2006. Predicting time-specific changes in demographic processes using remote sensing data. *Journal of Applied Ecology* 43: 366–76.
Rasmussen, L. E. L. 1988. Chemosensory responses in two species of elephants to constituents of temporal gland secretion and musth urine. *Journal of Chemical Ecology* 16:687–711.
———. 1998. Chemical communication: An integral part of functional Asian elephant *(Elephas maximus)* society. *Ecoscience* 5:410–26.
Rasmussen, L. E. L, and V. Krishnamurthy. 2000. How chemical signals integrate Asian elephant society: The known and the unknown. *Zoo Biology* 19:405–23.
Rasmussen, L. E. L., and B. Munger. 1996. The sensorimotor specializations of the trunk tip of the Asian elephant, *Elephas maximus*. *The Anatomical Record* 246:127–34.
Rasmussen, L. E. L., and B. A. Schulte. 1998. Chemical signals in the reproduction of Asian (*Elephas maximus*) and African (*Loxodonta africana*) elephants. *Animal Reproduction Science* 53:19–34.
Rasmussen, L. E. L., and G. Wittemyer. 2002. Chemosignaling of musth by individual wild African elephants *(Loxodonta africana)*: Implications for conservation and management. *Proceedings of the Royal Society, Series B* 269:853–60.
Rasmussen, L. E. L., A. Hall-Martin, and D. L. Hess. 1996. Chemical profiles of African bull elephants (*Loxodonta africana*); physiological and ecological implications. *Journal of Mammalogy* 77:422–39.
Rasmussen, L. E. L., M. J. Schmidt, and G. D. Daves. 1986. Chemical communication among Asian elephants. In *Chemical signals in vertebrates: Evolutionary, ecological, and comparative aspects*, ed. D. Duvall, M. Silverstein, and D. Muller-Schwarze, 627–46. New York: Plenum Press.
Rasmussen, L. E. L., T. D. Lee, G. D. Daves, and M. J. Schmidt. 1993. Female–male pheromones of low volatility in the Asian elephant, *Elephas maximus*. *Journal of Chemical Ecology* 19: 2115–28.
Rasmussen, L. E. L., M. J. Schmidt, R. Henneous, D. Groves, and G. D. Daves. 1982. Asian bull elephants: Flehmen-like responses to extractable components in female elephant estrous urine. *Science* 217:159–62.
Ratcliffe, N., R. W. Furness, and K. C. Hamer. 1998. The interactive effects of age and food supply on the breeding ecology of great skuas. *Journal of Animal Ecology* 67:853–62.
Reby, D., and K. McComb. 2003. Anatomical constraints generate honesty: Acoustic cues to age and weight in the roars of red deer stags. *Animal Behaviour* 65:519–30.
Reed, J. D. 1986. Relationships among soluble phenolics, insoluble proanthocyanidins, and fiber in East African browse species. *Journal of Range Management* 39:5–7.
Reid, R. S., L. N. Gachimbi, J. Worden, E. E. Wangui, S. Mathai, S. M. Mugatha, D. Campbell, B. Butt, J. Maitima, H. Gichohi, and E. Ogol. 2004. Linkages among changes in land use, biodiversity and land degradation in the Loitokitok area of Kenya. LUCID Working Paper no. 29. Nairobi: ILRI/LUCID, Michigan State University, KARI, NMK, AWF, ACC.
Reid, R. S., M. Rainy, J. Ogutu, R. L. Kruska, M. McCartney, M. Nyabenge, K. Kimani, M. Kshatriya, J. Worden, L. Ng'ang'a, J. Owuor, J. Kinoti, E. Njuguna, C. J. Wilson, and R. Lamprey. 2003. People, wildlife and livestock in the Mara ecosystem: The Mara count 2002. Unpublished report. Nairobi: International Livestock Research Institute.
Reiss, D., and L. Marino. 2001. Mirror self-recognition in the bottlenose dolphin: A case of cognitive convergence. *Proceedings of the National Academy of Sciences, USA* 98:5937–42.
Rendall, D., M. J. Owren, and P. S. Rodman. 1998. The role of vocal tract filtering in identity cueing in rhesus monkey (*Macaca mulatta*) vocalizations. *Journal of the Acoustical Society of America* 103:602–14.
Rendall, D., P. S. Rodman, and R. E. Emond. 1996. Vocal recognition of individuals and kin in free-ranging rhesus monkeys. *Animal Behaviour* 51:1007–15.
Rendell, L., and H. Whitehead. 2001. Culture in whales and dolphins. *Behavioral and Brain Sciences* 24: 309–82.
Rensch, B. 1957. The intelligence of elephants. *Scientific American* 196:44–49.
Reuling, M. A. 1991. Elephant use of the Marang Forest Reserve in Northern Tanzania. MSc thesis, University of Washington.
Reuter, T., S. Nummela, and S. Hemila. 1998. Elephant hearing. *Journal of the Acoustical Society of America* 104:1122–23.
Ritsma, N., and S. Ongaro. 2002. The commoditization and commercialization of the Maasai culture: Will cultural manyattas withstand the 21st century? In *Cultural tourism in Africa: Strategies for the new millennium*, ed. J. Akama and P. Sterry, 127–35. Arnhem: Association for Tourism and Leisure Education.
Robbins, C. T. 1983. *Wildlife feeding and nutrition*. New York: Academic Press.
Roca, A. L., N. Georgiadis, J. Pecon-Slattery, and S. J. O'Brien. 2001. Genetic evidence for two species of elephant in Africa. *Science* 293:1473–77.
Roed, K. H., O. Holand, M. E. Smith, H. Gjostein, J. Kumpula, and M. Nieminen. 2002. Reproductive success in reindeer males in a herd with varying sex ratio. *Molecular Ecology* 11: 1239–43.
Rogers, K. H. 1997. Operationalizing ecology under a new paradigm: An African perspective. In *The ecological basis of conservation: Heterogeneity, ecosystems, and biodiversity*, ed. S. T. A Pickett, R. S. Ostfeld, M. Shachak, and G. E. Likens, 60–77. New York: Chapman and Hall.
———. 2003. Adopting a heterogeneity paradigm: Implications for management of protected savannas. In *The Kruger experience: Ecology and management of savanna heterogeneity*, ed. J. du Toit, K. H. Rogers, and H. C. Biggs, 41–58. Washington, DC: Island Press.
Rolston, H. 1988. Human values and natural systems. *Society and Natural Resources* 1: 271–83.
Roth, G., and U. Dicke. 2005. Evolution of the brain and intelligence. *Trends in Cognitive Sciences* 9:250–57.
Ruggiero, R. G. 1989. The ecology and conservation of the

African elephant *(Loxodonta africana oxyotis* Matschie, 1900*)* in Gounda–St. Floris N.P., Central African Republic. PhD thesis, Rutgers University.

———. 1992. Seasonal forage utilization by elephants in Central Africa. *African Journal of Ecology* 30:137–48.

Ruiz, A., J. C. Gómez, J. J. Roeder, and R. W. Byrne. 2009. Gaze following and gaze priming in lemurs. *Animal Cognition* 12 (3): 427–34.

Ruttan, L. M., and M. Borgerhoff Mulder. 1999. Are East African pastoralists truly conservationists? *Current Anthropology* 40:621–52.

Rutten, M. 1992. *Selling wealth to buy poverty: The process of individualization of landownership among the Maasai pastoralists of Kajiado District, Kenya, 1890–1990*. Saaarbrucken: Verlag Breitenbach Publishers.

———.2005. *Shallow wells: A sustainable and inexpensive alternative for boreholes in Kenya*. ASC Working Paper 66. Leiden: African Studies Centre.

Sadleir, R. M. F. S. 1969. *The ecology of reproduction in wild and domestic mammals*. London: Methuen & Co. Ltd.

Sæther, B.-E. 1997. Environmental stochasticity and population dynamics of large herbivores: A search for mechanisms. *Trends in Ecology and Evolution* 12:143–49.

Saitoti, T.O. 1986. *The worlds of a Maasai warrior*. New York: Dorset Publishing.

Saitoti, T. O., and C. Beckwith. 1980. *Maasai*. New York: Harry N. Abrams, Inc.

Sanderson, G. P. 1907. *Thirteen years among the wild beasts of India*, 6th ed. Edinburgh: John Grant.

Santos, L. R., C. T. Miller, and M. D. Hauser. 2003. Representing tools: How two non-human primate species distinguish between the functionally relevant and irrelevant features of a tool. *Animal Cognition* 6:269–81.

Sanz, C. M., and D. B. Morgan. 2007. Chimpanzee tool technology in the Goualougo Triangle, Republic of Congo. *Journal of Human Evolution* 52:420–33.

Schaller, G. B. 1972. *The Serengeti lion*. Chicago: University of Chicago Press.

Schildkrout, E. 1996. Bone picking. *African Arts* 29:15–7.

Schmidt, K. T., and H. Hoi. 1999. Feeding of low ranking red deer stags at supplementary feeding site. *Ethology* 105:349–60.

Schmidt-Neilsen, K. 1984. *Scaling: Why is animal size so important?* Cambridge: Cambridge University Press.

Scholes, R. J., and K. G. Mennell. 2008. *Elephant management: A scientific assessment for South Africa*. Johannesburg: Wits University Press.

Seed, A. M., S. Tebbich, N. Emery, and N. S. Clayton. 2006. Investigating physical cognition in rooks, *Corvus frugilegus*. *Current Biology* 16:697–701.

Seno, S. K., and W. W. Shaw. 2002. Land tenure policies, Maasai traditions, and wildlife conservation in Kenya. *Society and Natural Resources* 15:79–88.

Setchell, J. M., and P. C. Lee. 2004. Development and sexual selection in primates. In *Sexual selection in primates: Causes, mechanisms and consequences*, ed. P. M. Kappeler and C. P. van Schaik, 175–95. Cambridge: Cambridge University Press.

Setchell, J. M., M. Charpentier, and E. J. Wickings. 2005. Mate-guarding and paternity in mandrills (*Mandrillus sphinx*): Factors influencing monopolisation of females by the alpha male. *Animal Behaviour* 70:1105–20.

Setchell, J. M., P. C. Lee, E. J. Wickings, and A. F. Dixson. 2001. Growth and ontogeny of sexual size dimorphism in the mandrill (*Mandrillus sphinx*). *American Journal of Physical Anthropology* 115:349–60.

Seyfarth, R. M., D. L. Cheney, and P. Marler. 1980. Monkey responses to three different alarm calls: Evidence of predator classification and semantic communication. *Science* 210: 801–3.

Shannon, G., B. R. Page, K. J. Duffy, and R. Slotow. 2006. The role of foraging behaviour in the sexual selection of the African elephant. *Oecologia* 150:344–54.

Shoshani, J. 1991. Anatomy and physiology. In *The illustrated encyclopedia of elephants*, ed. S. K. Eltringham, 30–47. London: Salamander Books.

———. 1997. It's a nose! It's a hand! It's an elephant's trunk! *Natural History* 106:37–43.

———. 1998. Understanding Proboscidean evolution: A formidable task. *Trends in Ecology and Evolution* 13:480–87.

Shoshani, J., W. J. Kupsky, and G. H. Marchant. 2001. Elephant brain structures and possible functions inferred from human brain anatomy. Abstracts of papers and posters, *Eighth International Theriological Congress*, Sun City (South Africa), 12–17 August 2001: 127–28.

———. 2006. Elephant brains. Part I: Gross morphology, functions, comparative anatomy, and evolution. *Brain Research Bulletin* 70:124–57.

Shrader, A. M., S. M. Ferreira, M. E. McElveen, P. C. Lee, C. J. Moss, and R. J. van Aarde. 2006. Growth and age determination of African savanna elephants. *Journal of Zoology* 270: 40–8.

Shultz, S., and R. I. M. Dunbar. 2006. Both ecological and social factors predict ungulate brain size. *Proceedings of the Royal Society, Series B* 273:207–15.

Sikes, S. K. 1971. *The natural history of the African elephant*. London: Weidenfeld and Nicolson.

Silk, J. B. 1987. Social behavior in evolutionary perspective. In *Primate societies*, ed. B. B. Smuts, D. L. Cheney, R. M. Seyfarth, R. W. Wrangham, and T. T. Struhsaker, 318–29. Chicago: University of Chicago Press.

Silverman, B. W. 1986. *Density estimation for statistics and data analysis*. New York: Chapman and Hall.

Simon, N. 1963. *Between the sunlight and the thunder: The wildlife of Kenya*. Boston: Houghton Mifflin Co.

Simpson, M. J. A., A. E. Simpson, J. Hooley, and M. Zunz. 1981. Infant-related influences on birth intervals in rhesus monkeys. *Nature* 290:49–51.

Sinclair, A. R. E., and P. Arcese, eds. 1995. *Serengeti II: Dynamics, management and conservation of an ecosystem*. Chicago: University of Chicago Press.

Sinclair, A. R. E., S. A. R. Mduma, and P. Arcese. 2000. What determines phenology and synchrony of ungulate breeding in Serengeti? *Ecology* 81:2100–11.

Sindinga, I. 1994. Employment and tourism training in Kenya. *The Journal of Tourism Studies* 5:45–52.

Sitati, N. 2003. Human–elephant conflict in the Maasai Mara dispersal areas of Transmara District. PhD thesis, University of Kent.

Sitati, N. W., M. J. Walpole, and N. Leader-Williams. 2005. Factors affecting susceptibility of farms to crop-raiding by African elephants: Using a predictive model to mitigate conflict. *Journal of Applied Ecology* 42:1175–82.

Sitati, N. W., M. J. Walpole, R. J. Smith, and N. Leader-Williams. 2003. Predicting spatial aspects of human–elephant conflict. *Journal of Applied Ecology* 40:667–77.

Slotow, R., G. van Dyke, B. Page, J. H. Poole, and A. Klocke. 2000. Older bull elephants control young males. *Nature* 408:425–26.

Smit, I. J. P., C. C. Grant, and I. J. Whyte. 2007a. Landscape-scale sexual segregation in the dry season distribution and resource utilization of elephants in Kruger National Park, South Africa. *Diversity and Distributions* 13:225–36.

———. 2007b. Elephants and water provision: What are the management links? *Diversity and Distributions* 13:666–69.

Smith, E. A., and M. Wishnie. 2000. Conservation and subsistence in small-scale societies. *Annual Review of Anthropology* 29:493–524.

Smith, G. E. 1907. From the Victoria Nyanza to Kilimanjaro. *The Geographical Journal* 29:249–69.

Sombroek, W. G., H. M. H. Braun, and B. J. A. V. Pouw. 1982. *Exploratory soil map and agro-climatic zones of Kenya, 1980*. Nairobi: Ministry of Agriculture, Government of Kenya.

Songorwa, A. N. 1999. Community-based wildlife management (CWM) in Tanzania: Are the communities interested? *World Development* 27:2061–79.

Songorwa, A., N. Alexander, T. Buhr, and F. Hughley. 2000. Community based wildlife management in Africa: A critical assessment of the literature. *Natural Resource Journal* 40:603–43.

Soltis, J. 2009. Vocal communication in African elephants (*Loxodonta africana*). *Zoo Biology* 28:1–18.

Soltis, J., K. Leong, and A. Savage. 2005a. African elephant vocal communication I: Antiphonal calling among affiliated females. *Animal Behaviour* 70:579–87.

———. 2005b. African elephant vocal communication II: Rumble variation reflects the individual identity and emotional state of callers. *Animal Behaviour* 70:589–99.

Soltis, J., K. A. Leighty, C. M. Wesolek, and J. Savage. 2009. The expression of affect in African elephant (*Loxodonta africana*) rumble vocalizations. *Journal of Comparative Psychology* 123: 222–25.

Sorenson, A. E. 1984. Nutrition, energy and passage time. Experiments with fruit preference in European blackbirds (*Turdus nurela*). *Journal of Animal Ecology* 53:545–57.

Spear, T. 1993. Being "Maasai" but not "People of Cattle": Arusha agricultural Maasai in the nineteenth century. In *Being Maasai: Ethnicity and identity in East Africa*, ed. T. Spear and R. Waller, 120–36. London: James Currey Ltd.

Spear, T., and R. Waller. 1993. *Being Maasai: Ethnicity and identity in East Africa*. London: James Currey Ltd.

Spencer, H. G., M. Kennedy, and R. D. Gray. 1995. Patch choice with competitive asymmetries of perceptual limits: The importance of history. *Animal Behaviour* 50:497–508.

Spencer, P. 1988. *The Maasai of Matapato*. Bloomington, IN: University Press for the International African Institute.

Spinage, C. A. 1973. A review of ivory exploitation and elephant population trends in Africa. *East African Wildlife Journal* 11:281–89.

———. 1994. *Elephants*. London: Poyser.

Spong, C., J. Stone, S. Creel, and M. Björklund. 2002. Genetic structure of lions (*Panthera leo* L.) in the Selous Game Reserve: Implications for the evolution of sociality. *Journal of Evolutionary Biology* 15:945–53.

Stacey, P. B., and D. J. Ligon. 1991. The benefits of philopatry hypothesis for the evolution of cooperative breeding: Variation in territory quality and group size effects. *American Naturalist* 137:831–46.

Stearns, S. C. 1992. *The evolution of life histories*. Oxford: Oxford University Press.

Stephenson, P. J. 2005. *African elephant update No. 5*. Gland, Switzerland: WWF.

Stern, P. C. 2000. Toward a coherent theory of environmentally significant behavior. *Journal of Social Issues* 56:407–24.

Stevenson-Hinde, J., and M. Zunz. 1978. Subjective assessment of individual rhesus monkeys. *Primates* 19:473–82.

Stevick, P. T. 1999. Age-length relationships in humpback whales: A comparison of strandings in the western north Atlantic with commercial catches. *Marine Mammal Science* 15:725–37.

Stillman, R. A., J. D. Goss-Custard, and R. W. Caldow. 1997. Modeling interference from basic behaviour. *Journal of Animal Ecology* 66:692–703.

Stillman, R. A., J. D. Goss-Custard, R. T. Clarke, and E. E. A. Durrell. 1996. Shape of the interference function in the foraging vertebrate. *Journal of Animal Ecology* 65:813–24.

Stoeger-Horwath, A. S., S. Stoeger, H. M. Schwammer, and H. Kratochvil. 2007. Call repertoire of infant African elephants: First insights into the early vocal ontogeny. *Journal of the Acoustical Society of America* 121:3922–31.

Stokke, S., and J. du Toit. 2002. Sexual segregation in habitat use by elephants in Chobe National Park, Botswana. *African Journal of Ecology* 40:360–71.

Struhsaker, T. T. 1976. A further decline in numbers of Amboseli vervet monkeys. *Biotropica* 8: 211–14.

Suarez, R. K., C. A. Darveau, and J. J. Childress. 2004. Metabolic scaling: A many-splendoured thing. *Comparative Biochemistry and Physiology B: Biochemistry and Molecular Biology* 139: 531–41.

Sugg, D. W. C. R., F. S. Dobson, and J. L. Hoogland. 1996. Population genetics meets behavioral ecology. *Trends in Ecology and Evolution* 11:338–42.

Sukumar, R. 1989. *The Asian elephant: Ecology and management*. Cambridge: Cambridge University Press.

———. 1990. Ecology of the Asian elephant in Southern India: Feeding habits and crop raiding patterns. *Journal of Tropical Ecology* 6:33–53.

———. 1991. The management of large mammals in relation to male strategies and conflict with people. *Biological Conservation* 55:93–102.

———. 1994. *Elephant days and nights: Ten years with the Indian elephant*. Oxford: Oxford University Press.

———. 2003. *The living elephants*. Oxford: Oxford University Press.

Sukumar, R., and M. Gadgil. 1988. Male–female differences in foraging on crops by Asian elephants. *Animal Behaviour* 36:1233–55.

Sutherland, W. J. 1983. Group and "ideal free" distributions. *Journal of Animal Ecology* 52:821–28.

———. 1992. Games theory models of functional and aggressive responses. *Oecologia* 90:150–52.

Sutherland, W. J., and P. M. Dolman. 1994. Combining behaviour and population dynamics with applications for predicting the consequences of habitat loss. *Proceedings of the Royal Society, Series B* 255:133–38.

Sutherland, W. J., and G. A. Parker. 1992. The relationship

between conditions inputs and interference models of ideal free distribution with unequal competitors. *Animal Behaviour* 44: 345–255.

Sutton, J. E. G. 1993. Becoming Maasailand. In *Being Maasai: Ethnicity and identity in East Africa*, ed. T. Spear and R. Waller, 38–86. London: James Currey Ltd.

Taberlet, P., S. Griffin, B. Goossens, S. Questiau, V. Manceau, N. Escaravage, L. P. Waits, and J. Bouvet. 1996. Reliable genotyping of samples with very low DNA quantities using PCR. *Nucleic Acids Research* 24:3189-94.

Talbot, L. M. 1972. Ecological consequences of rangeland development in Masailand, East Africa. In *The careless technology: Ecology and international development*, ed. M. T. Farvar and J. P. Milton, 694–711. Garden City, NY: The Natural History Press.

Tchamba, M. N. 1995. The problem elephants of Kaele: A challenge for elephant conservation in Northern Cameroon. *Pachyderm* 19:26–32.

Tchamba, M. N., H. Bauer, A. Hunia, H. H. de Langh, and H. Planton. 1994. Some observations on the movements and home range of elephants in Waza National Park, Cameroon. *Mammalia* 58: 527–33.

Teixeira, C. P., C. S. Azevedo, M. De, Mendl, C. F. Cipreste, and R. J. Young. 2007 Revisiting translocation and reintroduction programmes: The importance of considering stress. *Animal Behaviour* 73:1–13.

Tennent, J. E. 1867. *The wild elephant, and the method of capturing and taming it in Ceylon*. London: Longmans, Green.

Theuerkauf, J., H. Ellenberg, and Y. Guiro. 2000. Group structure of forest elephants in the Bossematié Forest Reserve, Ivory Coast. *African Journal of Ecology* 38:262–64.

Thompson, M., and K. Homewood. 2002. Entrepreneurs, elites, and exclusion in Maasailand: Trends in wildlife conservation and pastoralist development. *Human Ecology* 30:107–38.

Thompson, L. G., E. Mosley-Thompson, M. E. Davis, K. A. Henderson, H. H. Brecher, V. S. Zagorodnov, T. A. Mashiotta, P. N. Lin, V. N. Mikhalenko, D. R. Hardy, and J. Beer. 2002. Kilimanjaro ice core records: Evidence of Holocene climate change in tropical Africa. *Science* 298:589–93.

Thomson, J. [1885] 2006. *Through Masai Land: A journey of exploration among the snowclad volcanic mountains and strange tribes of Eastern Equatorial Africa*. East Sussex: Rediscovery Books Ltd.

Thornton, A., and N. Raihani. 2008. The evolution of teaching. *Animal Behaviour* 75:1823–36.

Thornton, P. K., S. B. BurnSilver, R. B. Boone, and K. A. Galvin. 2006. Modeling the impacts of group ranch subdivision on agro-pastoral households in Kajiado, Kenya. *Agricultural Systems* 87:331–56.

Thouless, C. R. 1996. Home ranges and social organisation of female elephants in northern Kenya. *African Journal of Ecology* 34:284–97.

Thouless, C. R., and J. Sakwa. 1995. Shocking elephants: Fences and crop raiders in Laikipia District, Kenya. *Biological Conservation* 72:99–107.

Titze, I. R. 1994. *Principles of voice production*. New Jersey: Prentice-Hall Inc.

Tomasello, M., and J. Call. 1997. *Primate cognition*. Oxford: Oxford University Press.

Tomasello, M., J. Call, and B. Hare. 2003. Chimpanzees understand psychological states— the question is which ones and to what extent. *Trends in Cognitive Sciences* 7:153–56.

Tomasello, M., B. Hare, and B. Agnetta. 1999. Chimpanzees, *Pan troglodytes*, follow gaze direction geometrically. *Animal Behaviour* 58:769–77.

Tregenza, T., and D. J. Thompson. 1998. Unequal competitor ideal free distribution in fish? *Evolutionary Ecology* 12: 655–66.

Tregenza, T., G. A. Parker, and D. J. Thompson. 1996. Interference and the ideal free distribution: Models and tests. *Behavioral Ecology* 7:379–86.

Treves, A., R. B. Wallace, L. Naughton-Treves, and A. Morales. 2006. Co-managing human–wildlife conflicts: A review. *Human Dimensions of Wildlife* 11:1–14.

Trillmich, F. 1996. Parental investment in pinnipeds. *Advances in the Study of Behavior* 25:533–77.

Trivers, R. L. 1971. The evolution of reciprocal altruism. *Quarterly Review of Biology* 46:35–57.

———. 1972. Parental investment and sexual selection. In *Sexual selection and the descent of man*, ed. B. Campbell, 139–179. Chicago: Aldine.

Trivers, R. L., and D. E. Willard. 1973. Natural selection of parental ability to vary the sex ratio. *Science* 179:90–2.

Turkalo, A. 1996. Studying forest elephants by direct observation in the Dzanga clearing: An update. *Pachyderm* 22:59–60.

Turkalo, A., and J. M. Fay. 1995. Studying forest elephants by direct observation: Preliminary results from Dzanga clearing, Central African Republic. *Pachyderm* 20:45–54.

Turle, G. 1992. *The art of the Maasai: 300 newly discovered works of art*. New York: Alfred A. Knopf, Inc.

UNDESA. 2008. *World population prospects: The 2006 revision—comprehensive tables: 1 population studies*. Population Division; United Nations Department for Economic and Social Affairs. New York: UNDESA.

———. 2009. *2008 revision: 21st projection for world population 2050*. Population Division; United Nations Department for Economic and Social Affairs. New York: UNDESA.

UNDP/FAO (United Nations Development Program/Food and Agricultural Organization). 1980. *Wildlife management in Kenya: Project findings and recommendations*. Nairobi: Ministry of Tourism and Wildlife.

UNEP. 1992. *The El Niño phenomenon*. GEMS Environmental Library no. 8. Nairobi: United Nations Environment Program.

———. 2000. *The atmospheric brown cloud: Climate and other environmental impacts*. Bangkok: UNEP Regional Office for Asia and the Pacific.

Van Aarde, R. T., and T. Jackson. 2007. Megaparks for metapopulations: Addressing the causes of locally high elephant numbers in South Africa. *Biological Conservation* 134: 289–97.

Vancuylenberg, B. W. B. 1977. Feeding behaviour of the Asiatic elephant in south-east Sri Lanka in relation to conservation. *Biological Conservation* 12:33–54.

Van der Merwe, N. J., J. A. Lee-Thorp, and R. H. V. Bell. 1988. Carbon isotopes as indicators of elephant diets and African environments. *African Journal of Ecology* 26:163–72.

Van Horn, R. C., A. L. Engh, K. T. Scribner, S. M. Funk, and K. E. Holekamp. 2004. Behavioral structuring of relatedness in the spotted hyena (*Crocuta crocuta*) suggests direct fitness benefits of clan-level cooperation. *Molecular Ecology* 13: 449–58.

Van Hoven, W., R. A. Prins, and A. Lankhorst. 1981. Fermentative digestion in the African elephant. *South African Journal of Wildlife Research* 11:78–86.

Vanleeuwé, H., and A. Gautier-Hion. 1998. Forest elephant paths and movement at the Odzala NP in Congo: The role of clearings (salt-licks) and Marantaceae on elephant distribution and movement. *African Journal of Ecology* 36:174–82.

van Noordwijk, M. A., and C. P. van Schaik. 2005. Development of ecological competence in Sumatran orangutans. *American Journal of Physical Anthropology* 127:79–94.

van Schaik, C. P. 1989. The ecology of social relationships amongst female primates. In *Comparative socioecology*, ed. V. Standen and R. A. Foley, 195–218. Oxford: Backwells.

van Schaik, C. P., and S. B. Hrdy. 1991. Intensity of local resource competition shapes the relationship between maternal rank and sex ratios at birth in cercopithecine primates. *American Naturalist* 138:1555–61.

Van Soest, P. J. 1982. *Nutritional ecology of the ruminant*. Corvallis: O & B Books.

Vehrencamp, S. L. 1983. A model for the evolution of despotic versus egalitarian societies. *Animal Behaviour* 31:667–82.

Vickery, W. L., L. R. Giraldeau, J. J. Templeton, D. L. Kramer, and C. A. Chapman. 1991. Producers, scroungers and group foraging. *American Naturalist* 137:847–63.

Vidya, T. N. C., and R. Sukumar. 2005. Social organization in the Asian elephant (*Elephaus maximus*) in southern India inferred from microsatellite DNA. *Journal of Ethology* 23:205–10.

Viljoen, P. J. 1989. Spatial distribution and movements of elephants (*Loxodonta africana*) in the northern Namib desert region of Kaokoveld, South West Africa–Namibia. *Journal of Zoology* 219:1–19.

Voeten, M. M., and H. H. T. Prins. 1999. Resource partitioning between sympatric wild and domestic herbivores in the Tarangire region of Tanzania. *Oecologia* 120:287–97.

Vollrath, F., and I. Douglas-Hamilton. 2005. African bees to control African elephants. *Naturwissenschaften* 92:508–11.

Von Hoehnel, L. 1894. *Discovery of lakes Rudolf and Stefanie: A narrative of Count Samuel Teleki's exploring and hunting expedition in Eastern Equatorial Africa in 1887 and 1888*. London: Longmans, Green & Co.

Waithaka, J. 1993. The elephant menace. *Wildlife Conservation* 96:62–3.

Walker, B. H., and D. A. Salt. 2006. *Resilience thinking: Sustaining ecosystems and people in a changing world*. Washington, DC: Island Press.

Wall, J., I. Douglas-Hamilton, and F. Vollrath. 2006. Elephants avoid costly mountaineering. *Current Biology* 16:527–29.

Waller, R. 1976. The Maasai and the British 1895–1905: The origins of an alliance. *Journal of African History* 17:529–53.

Waser, P. M. 1977. Individual recognition, intragroup cohesion and intergroup spacing: Evidence from sound playback to forest monkeys. *Behaviour* 60:28–74.

Wasser, S. K., B. Clark, and C. Laurie. 2009. The ivory trail. *Scientific American* 22: 68-76.

Wauters, L. A., S. A. Crombrugghe, A. de Nour, and E. Matthysen. 1995. Do female roe deer in good condition produce more sons than daughters? *Behavioral Ecology and Sociobiology* 37: 189–93.

Wayne, R. K., A. Meyer, N. Lehman, B. Van Valkenburgh, P. W. Kat, T. K. Fuller, D. Girman, and S. J. O'Brien, 1990. Large sequence divergence among mitochondrial DNA genotypes within populations of eastern African black-backed jackals. *Proceedings of the National Academy of Sciences* 87: 1772–76.

Webber, A. D., C. M. Hill, and V. Reynolds. 2007. Assessing the failure of a community-based human–wildlife conflict mitigation project in Budongo Forest Reserve, Uganda. *Oryx* 41: 177–84.

Weber, A. W. 1987. Socioecologic factors in the conservation of the Afromontane forest reserves. In *Primate conservation in the tropical rain forest*, ed. J. S. Gartlan, C. W. Marsh, and R. A. Mittermeier, 205–29. New York: Alan R. Liss.

Weiss, A., J. E. King, and W. D. Hopkins. 2007. A cross-setting study of chimpanzee (*Pan troglodytes*) personality structure and development: Zoological parks and Yerkes National Primate Research Center. *American Journal of Primatology* 69: 1264–77.

Wells, M., and K. Brandon. 1992. *People and parks: Linking protected area management with local communities*. Washington, DC: World Bank.

Wemmer, C. M., and C. A. Christen. 2009. *Never forgetting: Elephants and ethics*. Boston: Johns Hopkins University Press.

Wemmer, C. M., and H. R. Mishra. 1982. Observational learning by an Asian elephant of an unusual sound production method. *Mammalia* 46:557.

Wemmer, C. M., H. R. Mishra, and E. Dinerstein. 1985. Unusual use of the trunk for sound production in a captive Asian elephant: A second case. *Journal of the Bombay Natural History Society* 82:187.

Wesolek, C. M., J. Soltis, K. A. Leighty, and A. Savage. 2009. Infant African elephant rumble vocalizations vary according to social interactions with adult females. *Bioacoustics* 18: 227–39.

West, G. B., J. H. Brown, and B. J. Enquist. 1997. A general model for the origin of allometric scaling laws in biology. *Science* 276:122–26.

———. 1999. The fourth dimension of life: Fractal geometry and allometric scaling of organisms. *Science* 284:1677–79.

West-Eberhard, M. J. 1975. The evolution of social behavior by kin selection. *Quarterly Review of Biology* 50:1–33.

Western, D. 1973. The structure, dynamics and changes of the Amboseli ecosystem. PhD thesis, University of Nairobi.

———. 1975. Water availability and its influence on the structure and dynamics of a savannah large mammal community. *East African Wildlife Journal* 13:265–86.

———. 1982a. Amboseli National Park. Enlisting landowners to conserve migratory wildlife. *Ambio* 11:302–8.

———. 1982b. Patterns of depletion in a Kenya black rhino population and the conservation implications. *Biological Conservation* 24:147–56.

———. 1982c. Amboseli. *Swara* 5:8–14.

———. 1994. Ecosystem conservation and rural development: The case of Amboseli. In *Natural connections: Perspectives in community-based conservation*, ed. D. Western and R. M. Wright, 15–53. Washington, DC: Island Press.

———. 1997. *In the dust of Kilimanjaro*. Washington, DC: Island Press.

———. 2005. The ecology and changes in the Amboseli ecosystem: Recommendations for planning and conservation. Unpublished report. Nairobi: African Conservation Centre (ACC).

———. 2006. A half a century of habitat change in Amboseli National Park, Kenya. *African Journal of Ecology* 45:302–10.

Western, D., and W. K. Lindsay. 1984. Seasonal herd dynamics of a savanna elephant population. *African Journal of Ecology* 22:229–44.

Western, D., and D. Maitumo. 2004. Woodland loss and

restoration in a savanna park: A 20-year experiment. *African Journal of Ecology* 42:111–21.

Western, D., and D. L. Manzolillo-Nightingale. 2003. Environmental change and the vulnerability of pastoralists to drought: A case study of the Maasai in Amboseli, Kenya. GEO: African Environment Outlook, no. 35. Nairobi: UNEP/GEF.

Western, D., C. Moss, and N. Georgiadis. 1983. Age estimation and population age structure of elephants from footprint dimensions. *Journal of Wildlife Management* 47:1192–97.

Western, D., and P. Thresher. 1973. Development plans for Amboseli. Unpublished report. Washington: International Bank for Reconstruction and Development.

Western, D., and C. van Praet. 1973. Cyclical changes in the habitat and climate of an East African ecosystem. *Nature* 241: 104–06.

Western, D., and R. M. Wright. 1994. *Natural connections: Perspectives in community-based conservation*. Washington, DC: Island Press.

Westoby, M. 1978. What are the biological bases of varied diets? *American Naturalist* 112:627–31.

Whitehead, H. P. 1995. Investigating structure and temporal scale in social organization using identified individuals. *Behavioral Ecology* 6:199–208.

———. 1996. Babysitting, dive synchrony, and indications of alloparental care in sperm whales. *Behavioral Ecology and Sociobiology* 38:237–44.

———. 1997. Analyzing social structure. *Animal Behaviour* 53: 1057–67.

———. 2003. *Social evolution in the ocean*. Chicago: University of Chicago Press.

Whitehead, H., and J. Mann. 2000. Female reproductive strategies of cetaceans: Life histories and calf care. In *Cetacean societies: Field studies of dolphins and whales*, ed. J. Mann, R. C. Connor, P. L. Tyack, and H. Whitehead, 219–46. Chicago: University of Chicago Press.

Whitehead, H., and L. Weilgart. 2000. The sperm whale: Social females and roving males. In *Cetacean societies: Field studies of dolphins and whales*, ed. J. Mann, R. C. Connor, P. L. Tyack, and H. Whitehead, 154–72. Chicago: University of Chicago Press.

Whitehouse, A. M., and E. H. Harley. 2001. Post-bottleneck genetic diversity of elephant populations in South Africa, revealed using microsatellite analysis. *Molecular Ecology* 10: 2139–49.

Whyte, I. J., H. C. Biggs, A. Gaylard, and L. E. O. Braack. 1999. A new policy for the management of Kruger National Park's elephant population. *Koedoe* 42:111.

Wickler, W., and U. Seibt. 1997. Aimed object-throwing by a wild African elephant in an interspecific encounter. *Ethology* 103: 365–68.

Wiley, R. H., and D. G. Richards. 1978. Physical constraints on acoustic communication in the atmosphere: Implications for the evolution of animal vocalizations. *Behavioral Ecology and Sociobiology* 3:69–94.

Williams, G. C. 1979. The question of adaptive sex ratio in outcrossed vertebrates. *Proceedings of the Royal Society of London, Series B* 205:567–80.

Wing, L. D., and I. O. Buss. 1970. Elephants and forests. *Wildlife Monographs* 19:1–92.

Wittemyer, G., and W. M. Getz. 2007. Hierarchical dominance structure and social organization in African elephants, *Loxodonta africana*. *Animal Behaviour* 73:671–81.

Wittemyer, G., I. Douglas-Hamilton, and W. M. Getz. 2005. The socioecology of elephants: Analysis of the processes creating multitiered social structures. *Animal Behaviour* 69:1357–71.

Wittemyer, G., W. M. Getz, F. Vollrath, and I. Douglas-Hamilton. 2007. Social dominance, seasonal movements, and spatial segregation in African elephants: A contribution to conservation behaviour. *Behavioural Ecology and Sociobiology* 61: 1919–31.

Wittemyer, G., D. Daballen, H. Rasmussen, O. Kahindi, and I. Douglas-Hamilton. 2005. Demographic status of elephants in the Samburu and Buffalo Springs National Reserves, Kenya. *African Journal of Ecology* 43:44–7.

Wittemyer, G., J. B. A. Okello, H. B. Rasmussen, P. Arctander, S. Nyakaana, I. Douglas-Hamilton, and H. R. Siegismund. 2009. Where sociality and relatedness diverge: The genetic basis for hierarchical social organization in African elephants. *Proceedings of the Royal Society Series B* 276: 1–9.

Wood, D. S., and D. W. Wood. 1990. *How to plan a conservation education program*. Washington, DC: Center for International Development, Environment of the World Resources Institute, and U.S. Fish and Wildlife Service.

Wood, J. D., B. McCowan, W. R. Langbauer, J. J. Viloen, and L. A. Hart. 2005. Classification of African elephant *Loxodonta africana* rumbles using acoustic parameters and cluster analysis. *Bioacoustics* 15:143–61.

Wood, J. R. 1994. *Dynamics of human reproduction*. New York: Aldine.

Wrangham, R. W. 1980. An ecological model of female-bonded primate groups. *Behaviour* 75: 262–300.

———. 1987. Evolution of social behavior. In *Primate societies*, ed. B. B. Smuts, D. L. Cheney, R. M. Seyfarth, R. W. Wrangham, and T. T. Struhsaker, 282–96. Chicago: University of Chicago Press.

Wyatt, J. R., and S. K. Eltringham. 1974. The daily activity of the elephant in the Rwenzori National Park, Uganda. *East African Wildlife Journal* 12:273–89.

Xu, F., and E. Spelke. 2000. Large number discrimination in 6-month-old infants. *Cognition* 74:B1–B11.

Yokayama, S., N. Takenaka, D. W. Agnew, and J. Shoshani. 2005. Elephants and human color-blind deuteranopes have identical sets of visual pigments. *Genetics* 170:335–44.

Young, T. P. and W. K. Lindsay. 1988. Role of even-age population structure in the disappearance of *Acacia xanthophloea* woodlands. *African Journal of Ecology* 26: 69-72.

Young, T. P., T. M. Palmer, and M. E. Gadd. 2005. Competition and compensation among cattle, zebras, and elephants in a semi-arid savanna in Laikipia, Kenya. *Biological Conservation* 122: 351–59.

Zimbabwe National Society for the Prevention of Cruelty to Animals. 2009. A summary of wild elephants captured for commercial use housed on Sondelani Ranch. Unpublished report to the Zimbabwe National Society for the Prevention of Cruelty to Animals.

Zuberbuhler, K. 2000. Referential labelling in Diana monkeys. *Animal Behaviour* 59:917–27.

Contributors

Susan C. Alberts
Department of Biology
Duke University
Durham, NC 27708
USA

Elizabeth A. Archie
Department of Biological Sciences
University of Notre Dame
Notre Dame, IN 46556
USA

Lucy A. Bates
Centre for Social Learning and Cognitive Evolution and Scottish
Primate Research Group
School of Psychology
The University of St Andrews
Scotland
UK

Christine Browne-Nuñez
Wildlife Ecology and Conservation
University of Florida
Gainesville, FL 32611-0430
USA

Richard W. Byrne
Centre for Social Learning and Cognitive Evolution and Scottish
Primate Research Group
School of Psychology
The University of St Andrews
Scotland
UK

Harvey Croze
Amboseli Trust for Elephants
Nairobi
Kenya

Iain Douglas-Hamilton
Save the Elephants
Nairobi
Kenya

Courtney L. Fitzpatrick
Department of Biology
Duke University
Durham, NC 27708
USA

Petter Granli
ElephantVoices
Sandefjord
Norway

Seif Mutinda Hamisi
Wildlife-Quest Kenya
Nairobi
Kenya

Julie Hollister-Smith
Division of Neuroscience
Oregon National Primate Research Center
Oregon Health and Science University
Beaverton, OR 97006
USA

Kadzo Kangwana
5 Chestnut Hill Road
West Hartford, CT 06107
USA

Phyllis C. Lee
Amboseli Trust for Elephants and Behaviour and Evolution
Research Group
Department of Psychology
University of Stirling
Scotland
UK

W. Keith Lindsay
Amboseli Trust for Elephants and the Environment and
Development Group
Oxford
UK

Karen McComb
Centre for Mammal Vocal Communication Research
School of Psychology
University of Sussex
Brighton
UK

Cynthia J. Moss
Amboseli Trust for Elephants
Nairobi
Kenya

Norah Njiriani
Amboseli Trust for Elephants
Nairobi
Kenya

Joyce H. Poole
Amboseli Trust for Elephants and ElephantVoices
Sandefjord
Norway

David Reby
Centre for Mammal Vocal Communication Research
School of Psychology
University of Sussex
Brighton
United Kingdom

Catherine Sayialel
Amboseli Trust for Elephants
Nairobi
Kenya

Soila Sayialel
Amboseli Trust for Elephants
Nairobi
Kenya

Index

Page numbers followed by *f* and *t* indicate figures and tables, respectively.

AA family: affiliation patterns, 206; associates, 221*t*; ecological influences on, 219*t*; formation of, 337; gregariousness of, 218; identification of, 206; index of association, 212, 212*t*; life histories, 207*t*

Acacia spp.: *A. senegalia,* 16; *A. tortilis,* 16, 20, 22*f,* 22*t,* 23–25, 56, 57, 340; *A. xanthophloea,* 16, 20, 21*f,* 22*f,* 22*t,* 23–25, 56, 57, 95, 340; in the Amboseli ecosystem, 16; energy and protein levels, 63; grass biomass in, 57

acoustic communication: behavioral contexts, 125–61; call nomenclature, 132*t*; comparison of calls, 133*f*; frequencies, 133*f*; parameters measured, 130*t*; rumbles, 131–33. *See also* calls; *specific* calls

activities, defined, 3

adaptation, intelligence and, 4

Addo National Park elephants, 44, 46

adolescence, male, 260

adult females, defined, 2

adult males, defined, 2

Advance-Toward threat, 110, 111

advertisement/attraction behaviors, 117–18

aerial counts, 343–44

affiliations: AA family, 206; calf interactions, 229–30; DB family, 206; FB family, 206; female sociality and, 240; kinship and, 243, 244*f*

African elephants: daily food intake, 72*f*; ethogram of, 109; pick rates, 70*t*–71*t*; signals and gestures, 109–24

African Elephant Specialist Group, 319

African Wildlife Foundation (AWF), 295, 302

age/aging: biological, 4; calf mortality trends and, 196; call characteristics and, 137; classes, 128, 196*t*; determination of, 337–38; discrimination of calls in, 172–73; experience and, 258; at first birth, 188; at first conception, 188; at first musth, 278, 278*f*; last observed musth, 282*f*; longevity and offspring production, 203*f*; male dominance and, 282; male reproductive strategies and, 273; matriarch leadership and, 250; of matriarchs, 175; movement initiation and, 249; patterns of musth and, 276–79; reproductive success and, 274, 275*f,* 284*t*; at sexual maturity, 192; success in competition and, 251*t*; of weaning, 189*f*

age-power index, 254

age-sex classes, 2–3, 121, 131, 338

aggregations: association dynamics and, 254; benefits of, 257–58; definition of, 2, 206; formation of, 247–50; joining, 247; leaving, 247; matriarch age and, 250*f*; positions within, 257*t*; size of, 247*f,* 248*f*; subgroups within, 257

aggression: aggregation size and, 258; calf interactions, 229–30; calls-related to, 149; within family, 265–66; of male elephants, 260–61; during musth, 273; postures, 110–13; toward transitional males, 265

aging, 4. *See also* longevity

agriculture: in the Amboseli ecosystem, 31; government encouragement of, 329; high-intensity, 14; impacts of, 328; irrigated fields, 13; Maasai, 29; productivity of, 330; swamp-edge, 96; trends in, 301, 316, 328–30

alarm-trumpets, 147

alert behaviors, 141

allometry, defined, 52

allomothers: calf play and, 231, 232, 232*f*; calf proximity to, 159, 235, 235*f*; calf social development and, 236–37, 260; calf survival and, 79, 88, 197–201, 198*f,* 243; numbers of, 190, 203–4, 217, 228; roles of, 136, 147, 151, 152, 187, 234, 322

altruistic behaviors, 238–39

AMB1 haplotype, 41

AMB2 haplotype, 41

AMB3 haplotype, 41, 46

AMB4 haplotype, 46

ambivalent behaviors, 113–14

Amboseli basin ecosystem: agriculture in, 31–32; boreholes, 16; context of, 11–28; core spatial distribution, 97*f*; dams, 16; geology, 13; group ranches, 31–32; habitat changes, 20–24; Ilkisongo section topography, 12*f*; kernel distribution, 98*f*; location, 12–13, 337; mutability of, 11–12; non-equilibrium, 19; occupancy by elephants, 95*f*; peoples of, 16–17; pipelines, 16; rainfall, 14–16; soils, 13; total occupancy, 97*f*; vegetation, 16, 20–24, 21*f*; water resources, 13–14, 13*f*; wells, 16; wildlife of, 17–20

Amboseli Elephant Research Project (AERP), 294–95, 302; in conflict resolution, 296; duration of, 318; historical perspective, 1; origin of, 2; reduction of spearing and, 332–33; scope of, 327, 337

Amboseli Game Reserve, 32–33

Amboseli National Park: administrative insecurity, 327–28; boundaries of, 327; history of, 33–34. *See also* Amboseli basin ecosystem

Amboseli Trust for Elephants (ATE), 306, 332

Amboseli-Tsavo Group Ranch Association, 295, 331, 334

Amboseli-Tsavo Group Ranch Conservation Association, 295

anovulatory period, 276

anti-predator behaviors, 160*t*

apparent digestible protein (ADP), 60–67; daily requirements, 62–67, 62*t*; drought years and, 64*f,* 65*f,* 66–67; by food types, 63*f*; from forages, 60

apprehension, 113–14

approach, defined, 254

Approach behaviors, 113–24, 114*f,* 145, 151, 164*f,* 250

ArcGIS tools, 96

Art of the Maasai (Turle), 293

as-touched rumbles, 141*t,* 147, 148*f,* 149, 159

Asian elephants: daily food intake, 72*f*; pick rates, 70*t*

Assembly behaviors, 141

association indices, 211–13, 212*t,* 345

association patterns, 345; age classes and, 265–66; female elephants, 211–17; independent males, 267–68; juvenile males, 268; older males, 267–71, 271*f*; younger mature males, 269, 270*f*

Athi-Kapiti Plains, 12

attentive behaviors, 122–23

attitudes, measurement of, 301

"August Vacation" food choices, 95

avoid, defined, 254

avoidance behaviors, 112, 169–70, 190, 313, 315–16

Back-Bowing, 114
back length, 338
Back-Towards behavior, 112, 115, 119
barks, 125, 133*f*, 136, 145
Baroo-rumble, 129, 140, 141, 148*f*, 149, 151, 160*t*
basemaps, 344
BB family: associates, 212*t*, 219*t*, 221*t*; ecological influences on, 219*t*; range, 103*f*
begging rumbles, 128, 141*t*, 145, 146*f*, 147*f*, 149
behavioral scans, 344–45
births: data collection, 339; first, age at, 192; intervals between, 188
birth weights, 188*f*
black rhinos, 18
Body-Guarding, 124
body sizes: early drought experience and, 85–86; of males, 278–79; measurements for elephants, 53*t*; size-for-age data, 338–39
bomas, 30, 33, 94, 293, 303, 333
bond groups: associates, 221*t*; between-group recognition, 170–73; definition of, 206; distinguishing members of, 163; distributions, 101–3, 103*f*; family formation and, 210; family unit flexibility and, 222; social recognition beyond, 170
bonding, calls related to, 151, 152*f*
bonding ceremonies, 115
bone, sound conduction, 126
boreholes, 16
Botswana, 322
Bow-Neck charge, 111
brain size, 52, 174, 188–89
Brody's regression, 52
browse biomass, 58
buffalo, 18
bull areas: description of, 27, 266–67; musth *versus* non-musth, 266–67, 267*f*; use of, 104, 121*f*, 251, 260, 262, 268–71
bull elephants: fertilization probability scores, 282*f*; poaching of, 331; ranges of, 92–93, 92*f*; sightings of, 267*f*, 337, 339, 340, 343, 344. *See also* ivory trade; male elephants
bull groups: behavioral changes, 103–4; daytime diets, 61*f*; definition of, 2; food choices, 58, 60*f*; non-musth retirement, 104; nutrient intake, 68*f*, 69; response to cattle sounds, 314–16, 315*f*
Bunching behaviors: description, 117; in potentially dangerous situations, 106, 141, 173; response to calls, 164*f*, 172, 172*f*
Bush-Bashing, 111, 112, 113, 122
bushmeat, 18, 330, 335

CA family, 99, 99*f*, 101*f*
cadenced-rumble, 154–57, 154*f*
Calandre, 209
calf-calf interactions, 229, 230–32
calls: acoustic quality of, 139–41; behavioral contexts of, 139–41, 160*t*–161*t*; chorused, 139; decision making, 155–57; departure signals, 155; duration of, 130, 131*f*; food-related, 142–43; frequency attenuation, 168–70; group defense, 141–42; laryngeal types, 131–38, 136*t*; learned, 137*t*; logistical, 154–57, 154*f*; long-term memory of, 163–65; play-related, 157, 158*f*; range in frequency, 130, 130*t*, 131*f*; related to bonding, 152*f*; related to sexual behaviors, 143–45; social identity coded in, 165–68; trunk types, 137*t*, 138–39; types of, 130–39. *See also* cries; vocal communication; *specific* calls
calves (0-4.9 years), 128; allomothers and survival of, 197–99, 198*f*; analysis of rumbles, 141*t*; calls of, 129–30; capture of, 321–23; causes of death, 199; cries, 135–36; development of, 224–37; developmental milestones, 225, 226*t*; distress calls, 149; early experiences of, 85–86; effect of inter-birth intervals, 199–200, 200*f*; exploration by, 228–32; focal hours in study of, 225*t*; growth of, 85–86; interactions, 229–30; lost, 146–47; maternal experience and, 196*f*; maternal investment and, 196–97, 225; metabolic energy requirements and, 233; mortality rates, 78*f*, 79–80, 80*f*, 188–89; orphaned, 322; rates of, 84–85, 85*f*; sex ratios at birth, 196; social interactions of, 258; survival of, 195–96
Camilla, 209
Caress behavior, 117, 119
carrying capacities, 19, 31, 298
Casual-Walk behavior, 119
cattle: compensation for, 33, 293; cultural value of, 292, 297, 298; elephant responses to, 310, 313–16; in the park, 17, 34, 294, 308, 311; quotas, 31; uses of dung from, 17
cattle bells, 315–16, 315*t*
causality, understanding of, 176–77
CB family: associates, 221*t*; ecological influences on, 219*t*; fission-plus-fusion, 101; fusion, 100–101, 101*f*; life histories, 207*t*, 209
Celeste, 209
census data analysis, 210–11, 338–39. *See also* demographics
Central Hills, Kajiado District Kenya, 12
Cerise, 209
Charlotte, 209
Chase behavior, 111, 119, 120, 122, 149, 151
cheetahs, 20
chemical markers, 259
Chin-In posture, 117
Chloe, 101, 209
chorused calls, 139
Chyulu Hills, 12, 13*f*, 14, 16, 18, 92, 295
clan areas, 251
clans: defined, 206; home ranges and, 38
classification abilities, of elephants, 177
Climb-On-play, 122
coalitions, 111, 125, 149, 179, 240, 243, 258, 266
cognition, 6, 174–84; physical, 175–78
cognitive mapping skills, 175
cold stress, 52
colonial period: elephant populations, 26–27; Maasailand, 32
comfort activities, 3
"command and control" approaches, 324
commercial exploitation, 319–23
communication, manipulation and, 179
community development organizations, 334

competition: avoidance of, 265; within family units, 246; habitat use and, 69; of male elephants, 260–61; reproductive, 282–83; success in, 251*t*
conception: annual rates of, 77–79; calculating dates of, 75; fertilization probability scores, 282*f*; monthly rates of, 75–79, 75*f*; musth and, 275*f*; peak times of, 280, 281*f*; priority-of-access models, 282–85; rainfall and rates of, 76, 76*f*, 77*f*, 79*f*; temporal patterns of, 276; within-family synchrony, 77
conflict, calls related to, 149–51, 150*f*, 160*t*
Conflict Resolution Committee, 295
conservation: challenges of, 291–306; costs of, 292; ethical approaches, 318–26; ethical issues, 317; goals of, 335; impact of poverty on, 330; longitudinal studies, 291–92; Maasai relationship with, 299–300
consolation scheme, 294–97, 325, 332
Consorting behaviors, 120
Consorting pairs, 120*f*
contact calls, 154–57; characteristics, 154*f*; formant frequencies, 165–66; identity characteristics in, 167–68, 167*f*; playback studies, 127; responses to, 164*f*; spectrograms, 171, 171*f*; structure of, 166, 166*f*
contraception, 325
Convention on International Trade in Endangered Species of Flora and Fauna (CITES), 28, 320, 331
coo-rumbles, 129, 148*f*, 149, 157, 159
cooperation: kinship and, 243; problem solving and, 179
courtship behaviors, 119–20
cow-calf groups: calf development, 224–37; definition of, 2; food choices, 58, 59*f*; group size, 218, 218*f*, 312, 313*f*; habitat choices, 57; Njiri transect, 313*t*. *See also* mother-offspring behaviors; parent-offspring behaviors
cries: analysis of, 138*t*; characteristics, 135–36; of infants, 146. *See also* calls
croaks, 134*f*, 135*f*, 139
crop-raiding, 323–24
Croze, Harvey, 2
culling, 325–27
Cynodon spp., 56, 57, 95
Cyperus spp., 56, 57

d-loop data, 239, 240
daily food intake, 72*f*
dams, 13*f*, 16, 333
data collection: aerial counts, 343–44; assumption of death, 339; births, 339; causes of death, 339; coding, 339–40; demographics, 338–39; growth data, 338–39; mortality rates, 339; population numbers, 343–44; sightings, 339–43
DB/DC family, 207*t*, 209
DB family: affiliation patterns, 206; associates, 221*t*; ecological influences on, 219*t*; fission-plus-fusion, 101; life histories, 207*f*
death: behavior related to, 123–24; causes of, 339; census data, 339; seasonality of, 79–85
Deborah, 209

decision making, 155–57, 246–59
Defecation behavior, 115, 153
defense: affiliative behavior and, 258; group size and, 244; role of allomothers, 235–36; social dynamics and, 223
Delia, 101, 209
demographics: data collection, 338–39; data used in, 191; family unit flexibility and, 222t. *See also* census data analysis
Department of Resource Survey and Remote Sensing, Kenya, 28
departure signals, 155
deterrents, use of, 324
development: milestones for calves, 226t; physical, 236; social, 237
development planning, 330–31
diet choices, 68, 69. *See also* food
dietary crude protein, 60
discrimination learning, 177
dispersal: of female elephants, 263; of male elephants, 261–63
displacement behaviors, 113
Displacement-Feeding, 113
Displacement-Grooming, 113
Distant-Frontal-Attitude behavior, 121
DNA extraction, 40, 239, 274
dogs, wild, 20
dominance: age-related male, 282; in family units, 244–45; in female elephants, 190–91; inter-family ranks, 250; musth and, 273; procession order and, 259; trail use and, 250–57
Dorobo people, 3, 7, 26, 34–35, 292, 293
Douglas-Hamilton, Iain, 1
drinking activities, defined, 3
Driving behavior, 119
drought-coping strategies, 330
dry matter (DM), 61, 64f, 65–67, 65f
dry matter digestibility (DMD), 60
Dry Season Intensity (DSI), 15–16, 15f, 75
dry seasons, biomass levels, 63
Dueling, 111, 112f, 113
dung: burning of, 296; counts, 308–10; DNA extraction from, 4, 239, 274

EA family, 103, 212t, 219t, 221t
Ear-Flap-Slide behavior, 116
Ear-Flattening behavior, 111, 112
Ear-Folding threat, 110, 110f, 111
Ear-Lifting behavior, 113
Ear-Slap behavior, 116
Ear-Spreading behavior, 110, 111
Ear-Stiffening behavior, 114
Ear-Wave behavior, 117, 117f
Ears-Lifted behavior, 116
Ears-Tense posture, 117
Earthworks QTC1 omni-directional microphones, 128
EB family, 207t, 210, 212t, 214t, 214t, 216f
ecological factors: demographic rates and, 74–88; family cohesiveness and, 218–19, 219t; long-term effects of, 85–88; in musth times, 262
ecosystems: corridors, 96, 334; elephant range, 89–93
Ed, 103, 104f
elatia, 30
electric fencing project, 302

Elephant Charter, 319
elephants: Amboseli populations, 18; commercial exploitation of, 319–23; deterrents, 324; early attitudes toward, 300–301; future of, 304–5; gestation length, 188f; life span, 188; the Maasai and, 34–35, 308–10; in myth, 292–98; population genetics, 37–47; sound detection, 125–26; sound production, 125–26; traditional uses of, 296; use of water resources, 310–13, 311f
elephant studies: cognition, 174–84; historical perspective, 1–2; Njiri transect, 312, 312f; population sizes over time, 25–28, 28f
ElephantVoices.org, 109
Elephas iolensis, 45
Elerai Conservation Area, 333, 334
emotions, understanding of, 180, 182
Empaash-Oldepe area, 97f, 251–53, 255, 257
employment issues, 334
encephalization quotients, 174
energetics, body size and, 52–53
energy balance, intake and, 53
energy intake, diet choice and, 69
Enkai (God), 292
enkang, 30
Enkong'u Narok, 7, 13f, 14, 16, 18, 23, 93, 100
enkutoto, 30
environmental spaces, knowledge of, 175–78
Escalated-Contest, 111, 112
escalation postures, 111–12
Eselenkei Group Ranch, 12f, 306, 333
Eselenkei-Kiboko river, 13–14
Eselenkei River, 295
Esme, 171f
estrous: anovulatory period of, 276; characteristics of, 119; definition, 3; determinants of, 195; guarding by males, 285f; seasonal trends, 282; suckling time and, 233
estrous-roars, 143, 144f
Estrous-Walk, 119
estrus-rumbles, 127
ethical issues, conservation-related, 318–26
Exaggerated-Walk, 112
experience, calf rearing and, 195–96, 233–34
Explore-Touch behavior, 123
Eyes-Blinking behavior, 123
Eyes-Open behavior, 123
Eyes-Wide behavior, 114, 122

FA family, 207t, 209, 212t, 219t, 221t
family units: characteristics of, 212t; cohesion within, 212, 213f; competition within, 246; at the core of the range, 96; data analysis, 210–11, 312–13, 339–43; definition, 3, 205–6; dispersal of male elephants, 261; distinguishing members of, 163; dominance relationships in, 244–45; ecological influences on, 218–19, 219t; female elephant associations, 211; fission of, 222t; flexibility within, 216–17; "floaters," 206; formation of, 210; fully independent males and, 268–69; fusion and fission, 100, 240–43; fusion of, 222t; genetic relatedness, 240–41, 242f; histories of, 207t–208t; integration of calves into, 225; kernel distribution, 98f; linkage between, 220f; matriarch age and, 247–50, 248f; mother-calf interactions and, 234–36; permanent fission of, 213–16; procession order in, 256; proximity issues, 228, 229f; range shifts, 100f; response to cattle sounds, 314, 314t; seasonal movement, 99f; short-term fission, 216; sightings, 94f, 341f; size of, 211, 212f, 220f; size *versus* gregariousness of, 218–19; social dynamics, 211–13; temporary divisions of, 38–39; tracking position of members, 175–76; trail use and, 253f. *See also specific* family units
FB family, 206, 207f, 219t, 221t
fecal samples, 40, 274. *See also* DNA extraction
fecundity: determinants of, 195; ecological factors, 76; impact of calf death on, 78, 78f
feeding: daytime activity, 55, 55f; focal samples during, 344; stationary, 3; while moving, 3. *See also* foraging
feet, mechanoreceptors, 126
Feigned-Fear behavior, 122
Female-Chorus display, 117, 143
female elephants: adult, 2; age at first birth, 192; age of reproductive maturity, 224; age of sexual maturity, 192; association patterns, 39, 211–17; chorus calls, 143; conception by week, 277f; diets, 61f; dispersal of, 263; dominance in, 190–91, 244–45; in estrous, 143; fecundity, 76; habitat choices, 56, 57f; hazard rates, 87f; intake in good and drought years, 64f; investment in offspring, 201–3; life expectancy, 192; life histories, 191–93; life tables, 82t–84t; mate choice, 200–201; ranging patterns, 38; relatedness, 241f; reproductive lifespan, 187; reproductive strategies, 187–204; reproductive success, 38; rumbles, 129, 140; social dynamics, 205–23; social relationships, 238–45; status among, 194
Female-Male-Test-Genitals, 119
fences, 324
fertilization probability scores, 282f
Fifi, 209
Fiona, 209
fission: fusion issues, 4
fission-fusion sociality, 210
fitness, 246
Flehmen response, 119
Floppy-Running behavior, 122
focal samples, 344–45
food: in the "August Vacation," 95; behavioral ecology, 51–55; calls related to, 142–43; competition and, 69; context for learning, 229; cow-calf groups, 59f; daily intake, 72f; family unit flexibility and, 222; measures of availability, 75; seasonal variation, 55, 58; study of, 56–60. *See also* diet choices; nutrient intake
Foot-Swing, 113–14
foraging: apparent digestible protein (ADP), 60; gender differences, 71; nutrient intake, 60–62, 68–69; optimal, 246; social environment and, 55; trail use and, 252. *See also* feeding
forests, thinning of, 296

Forward-Trunk-Swing, 110
fragments, definition of, 2
Freezing behavior, 117, 123, 141
friendships. *See* association patterns
Full-Retreat behavior, 117
future, focus on, 4

Ganesh, 92f, 93
Gathering behavior, 115, 117
gaze priming, 180
gender: calf development and, 224, 236f; calf mortality trends and, 196, 203; developmental milestones and, 226t, 227–28, 227f; inter-birth intervals and, 234; metabolic energy requirements and, 233; play partners and, 230; sizes for age, 236f; time spent suckling and, 233–34. *See also* sex differences
gene flow, 38, 39
general terminology, 3
genetic diversity, habitat loss and, 37
genetic sampling, non-invasive, 239. *See also* DNA extraction
genotyping, 40–41, 274
geology, Amboseli ecosystem, 13
gestation length, 188f, 276, 276f
giraffes, 18
Goodall, Jane, 318
grandmothers, 199, 201–3, 235
grasses: biomass in *Acacia* woodlands, 57; fires, 16; as forage, 54
grasslands: in the Amboseli basin, 56; habitat choices, 57f; population changes, 22t
Great Rift Valley, 12
Greeting-Ceremony behavior, 115, 151–53
gregariousness: definition of, 211; determinants of, 218–19, 222; of families, 264; of male elephants, 263, 270–71
Group-Charge, 111, 117
group defense behaviors, 117
group defense calls, 141–42
Group Ranch Conservation Association, 295
Group Ranches, 31–32, 327
Group Representatives Act of 1978, 31
groups: coordinated behavior, 206; definition of, 2, 206; formation of, 248–49; fragmentation of, 248–49; males in, 269; short-term fission, 249–50; short-term fusion, 249–50; size of, 251t, 252–53, 252t, 253f, 312, 312f
group types, defined, 2
growth: age at first reproduction and, 192–93; assessment of, 225; database, 338–39; drought experience and, 85–86, 87f; gender and rates of, 233–34; gender differences, 236f; male elephants, 279f; measures of, 192–93, 193f, 337–38; size-for-age data, 338–39
grumbling-rumbles, 145, 149, 150f
grunts, 133f, 136
Guarding behavior, 120, 266, 272, 274, 283, 284

habitat: loss of, 37; types of, 57–58, 68; use trends, 67–68
habitat choices: behavioral ecology, 51–55; competition and, 69; high-density areas, 324; seasonal variation, 55; sex differences, 57f; social environment and, 55; study of, 56; trail use and, 255
Hardy-Weinberg equilibrium, 41
harmonic-play-trumpets, 157
Head-High posture, 111, 112, 114, 117
Head-Low posture, 112
Head-Nod behavior, 111
Head-Raising behavior, 113, 114–15
Head-Raising posture, 110
Head-Shaking behavior, 110, 122
Head-Toss behavior, 117
Head-Waggle behavior, 121
height: calf growth and, 236, 236f; at shoulder, 338
herbivores: Amboseli populations, 17–19; hindgut fermentation, 54
herbs/forbs, biomass of, 57
Herding-Push behavior, 111, 115, 115f
hind foot growth, 193f, 225, 279f
hindgut fermentation, 54
hippos, 18
historical perspective, 1–2
Hoehnel, Ludwig von, 26
home ranges, clans and, 38
Horizontal-Sniffs, 123
Hovering-Sniffs, 123
human-elephant interactions: behavioral responses to the Maasai, 307–17; conflicts, 323–24, 331; exploitation of elephants, 319–23; human population growth and, 327; influence on distribution, 95–96; land-use issues, 323; land-use planning and, 328–29; local fronts of, 328; Maasai settlements and, 316–17; negative, 294
human interventions: conservation efforts and, 291–306; elephant social relationships and, 243
human society, defined, 3
hunting, illegal, 320
husky-cries, 133f, 136–37, 138t, 146
hyenas, 17, 20

Ideal Free Distribution theory, 258
Ilkisongo section, Kenya Kajiado District, 7, 12–13, 12f, 30, 295, 328, 333
ilmoran (Maasai warriors), 3, 293, 294, 298, 302, 328
impala, 17
inbreeding, minimization of, 262–63
independence: age of, 264f, 265f; transitions to, 263–65, 263f
individuals, recognition of, 1–2, 337
infants (less than 6 months), 128; cries, 135–36, 146, 148f; focal hours in study of, 225t; survival of, 232
infrasonic frequencies, 168–70, 169f
intelligence, adaptive edge and, 4
interactions, defined, 3
inter-birth intervals, 199–200, 200f, 203–4, 234
intimidation, in group defense, 141–42
ivory trade, 293, 320–21, 331

JA family, 99f, 100f, 207t, 209, 219t, 221t
Jaw-Tilted-Upward, 114, 122
Joan, 209
J-Sniff, 122f, 123
juveniles (5-9.9 years), 128, 263

KA family, 212t, 221t
Kajiado District, Kenya, 12, 12f, 301t, 328
Kenya Wildlife Service (KWS), 28, 34, 294, 302; in conflict resolution, 296; operation of, 327–28; park management by, 334
Ker and Downey, Ltd., 333
Kernel Density Analysis (KDA), 96
Kibo Safari Camp, 330
Kick-Back behavior, 115
Kick-Dust, 113
Kidepo Valley, Uganda, 46
KIL3 haplotype, 46
Kilimanjaro, eruption of, 13
Kilimanjaro elephants, 44; gene flow, 46–47; haplotypes, 44; microsatellite allele frequencies, 45f; origin of, 44; phylogeography, 44–46; population genetics, 37–47; relationship to Amboseli elephants, 44
Kimana Group Ranches, 295, 333
Kimana swamp area, 328
Kimana Wildlife Concession Area, 333
Kimana Wildlife Sanctuary, 93, 334
kinship, 238–45. *See also* allomothers; family units
Kitiura Limited Game Concession, 333
Kleiber equation for metabolic rates, 52
Kneel-Down behavior, 121
Kruger National Park, South Africa, 19, 44, 46, 62, 258–59, 319, 325
kudu, 17

LA family, 212t, 219t, 221t
lactation: anestrous related to, 77, 78, 188–89, 195; costs of, 53, 69, 76, 78, 188, 232–33; duration of, 224; peak, 62t, 63, 66f; trade-offs, 193, 197, 200, 204
laibons (diviners, medicine men), 293
Lake Amboseli, formation of, 13
Lake Manyara National Park, Tanzania, 1–2, 20, 23, 77f, 191
landowners, conservancies and, 333–34
land-use planning, 306, 328–29
leadership: decision making and, 246–59; movement initiation and, 116; trail use and, 250–57
leaf biomass, 58f, 63
Leakey, Richard, 299
"let's-go" rumble, 154f, 155, 156t
Let's-Go-Stance behavior, 116, 116f
life expectancy. *See* longevity
life histories: data used in, 191, 337–45; for family units, 207t–208t; female elephants, 187–204; patterns of females, 191–93; reproductive, 192f
life span. *See* longevity
life tables, 81t–82t, 82t–84t
lions, 17, 19–20, 35
Listening behaviors, 123, 123f, 141, 164f
little-greeting-rumbles, 147, 151–53, 152t, 159
livestock, 20, 294–97. *See also* cattle
llmarn, 294
localization skills, 126
logistical calls, 154–57, 154f
Lone-Play, 121
longevity: challenges to study of, 4; male elephants, 280; offspring production and, 203f; reproduction and, 192; reproductive

potential and, 262; reproductive senescence and, 189–90; reproductive success and, 232; size and, 188; timing of musth and, 280–81
Longido Game Controlled Area, 12, 13f, 93
long-term memory, 163–65
Lonkinya spring, 7, 13f, 14, 99
Look-Back, 112, 117
lost-calls, 146
Loxodonta adaurora, 45
Loxodonta africana (African savanna elephants), 44–46, 126
Loxodonta cyclotis (African forest elephants), 126

M07, 267f, 285f
M13, 267f
M22, 267f
M28, 279
M103, 278
M126, 277f
Maa place-names, 6–7
Maasailand, 32–34, 299, 333
Maasai *moran*. See *morans*
Maasai peoples: age-set system, 30–31; animal husbandry, 329; conservation and, 299–300; cultural identity, 298; current attitudes, 303–4; early attitudes of, 300–310; economic opportunities, 305–6; elephant approaches to settlements, 310; elephant responses to, 307–17; Elephant Scouts, 344; elephants in tradition, 34–35, 292–98; fatalism in culture of, 297–98; future of, 304–5; herder competition for water, 16; hunting by, 293; occupation of settlements, 23–25; pastoralism, 31, 293, 298–300, 329; permanent homesteads, 308–9, 328; relationship with elephants, 291–306; settlements, 308, 308f; spearing of elephants, 331–32; temporary settlements, 309–10; use of water resources, 310–13; wildlife conservation and, 29–31
Maasai warriors. See *ilmoran*
Maasai Wilderness Conservation Trust, 306
Mada Hotels group, 330
male elephants: age classes, 265f; age of musth phase, 260; age of reproductive maturity, 224; associations, 261–62, 265–66, 268–71, 268f; daytime diets, 61f; definition of adult, 2; gene flow and, 39; growth curves, 279f; habitat choices, 57f; habitat distribution, 56; hazard rates, 87f; intake in good and drought years, 65f; life expectancy, 280; life stages, 260, 261f; life tables, 81t–82t; musth distributions, 267f; non-musth distributions, 267f; reproductive strategies, 272–88; reproductive success, 38, 201, 279–80; social dynamics, 260–71; sociality, 266–67; sterilization of, 325; transitional, 265f. See also bull elephants; bull groups
Male-Female-Test-Genitals, 118
male-male associations, 269f
Male-Male behaviors, 120
Male-Male Chase, 120
Male-Male Greeting, 115, 119
Male-Male-Mounting, 117, 120
Male-Male-Reach-Over, 120

Male-Male-Test-Genitals, 118
manyatta, 30
Marking behaviors, 117
"mark test" paradigm, 181
marshes, population changes, 22t
mate choice, 200–201, 285–86
maternal condition, 195–96, 232
mating, calls related to, 160t
Mating-Pandemonium behavior, 120
mating-pandemonium calls, 144f, 145
matriarchs: call discrimination by, 172–73; definition, 3; drought response and age of, 175; experience of poaching, 320; family size and, 247; genetic relatedness, 242f; group aggregation and, 250f, 254–56; influence over decision making, 246, 258; leadership by, 250; loss of, 101, 194, 207t–208t, 209, 211, 213; status of, 194; success in competition and, 251t
MB family, 98–99, 99f
Mbirikani region, 295
mechanoreceptors, 126
metabolized energy, 61; daily requirements, 62–67, 62t; by food types, 63f; intake in good and drought years, 64f, 65f, 66–67
methods, 337–45
microphones, omni-directional, 128
microsatellite allele frequencies, 45f
microsatellite amplification, 40
migrant species, 17
mirror self-recognition, 181
mitochondria, d-loop data, 239
mitochondrial DNA (mtDNA), 242
mixed groups, defined, 2
mixed roars, 133f, 134–35
Mobbing behaviors, 141–42
Mock-Charge, 111
Monitoring behavior, 123
morans, 3, 30, 35, 178, 295, 302
mortality rates: age-based trends, 196; birth order and, 197f; causes of death and, 339; data collection, 339; precipitation and, 197–98; sex-based trends, 196
Moss, Cynthia, 2
Mother is Matriarch, 194
mother-offspring behaviors, 116–17; begging calls, 145, 146f; calf proximity and, 225–28; calls eliciting care, 145–47, 147f, 148f; family unit size and, 234–36; maternal investment and, 232–34, 322; proximity issues, 228; rearing strategies, 232–34. See also cow-calf groups; parent-offspring behaviors
mother-offspring pairs, 206, 241
Mounting behavior, 119–20, 120f, 123
Mouth-Opening, 115, 120
movement: association dynamics and, 254; initiation of, 116, 248, 249f, 254
movement initiation index, 254
movement initiator, 254
moving activities, 3
Mr. Nick, 92–93, 92f, 93
mtDNA amplification techniques, 40
mtDNA haplotypes: Amboseli, 40–44; comparison across Africa, 41; control region distributions, 42t–43t; control region frequencies, 43t; minimum spanning network, 44t

musth: age at first, 278f; age at last, 282f; age of, 260; characteristics of, 273; conception timing and, 281f; context of, 272–73; cortisol levels during, 285; definition, 3; distributions, 267f; duration, 277, 278, 279f; female guarding behaviors, 285f; longevity and timing of, 280–81; number of males in, 280f; reproductive success and, 274, 275–76; temporal patterns of, 276–79, 277f
musth-rumbles, 127, 143, 144f
Musth-Temporal-Gland-Secretion, 117
Musth-Walk behavior, 117
myths, elephants in, 292–98

Namalok swamp area, 328, 330
nasal trumpets, 131, 132t, 134f, 137t, 138, 138t, 157, 161t
natal dispersal process, 262
National Land Policy Formulation Process, Kenya, 329
nepotistic interactions, 191, 244
neurons in elephant brains, 175
Njiri transect, 311–13, 312f, 313t
noisy roars, 133f, 134–35, 148f
non-mother caretakers, 235. See also allomothers
non-musth core ranges, 104, 105f
Norms and Standards for the Management of Elephants in South Africa, 319
nutrient intake: estimates of, 60–62; implications, 68–73; observations, 68–73; observed by season, 62–67; observed by sex, 62–67. See also food

Object-Play, 121, 121f
objects, knowledge of, 175–78
occupancy: ecosystem level, 89–93; patterns of, 89–106
Oldono Orok, length of, 13
Olekjuado County Council, 12
Olgulului Trust Group Ranch, 295, 303
Ol Kanjau, 330, 333
Olkejuado Country Council (OCC), 12, 16, 327–28, 334
Oloitokitok County Council, Kenya, 334
optimal foraging theory, 246
organ scaling, 52, 53t
orphans, 322

PA family, 101, 129t, 212t, 213, 221t
Pacinian corpuscles, 126
Panic-Running, 114
PA-PC family, 102f, 209
Parallel Walk behavior, 111
parent-offspring behaviors, 160t, 201–3. See also mother-offspring behaviors
Parsitau, 92, 92f
partiesm, defined, 2
pastoralism, 31, 293, 298–99, 329
paternity analysis, 274–75
Penelope, 209
Pennisetum straminium, 56
Periscope-Sniff, 121, 122f
personalities, 214–16
pharyneal pouch, 126
Phoebe, 209
Phoenix reclinata (wild date palm), 57, 95, 340

photogrammetric assessment, 337, 338
phylogeography, 44–46
physical abuse, long-term effects of, 322–23
pick rates: African elephants, 70t–71t; of Asian elephants, 70t; food types and, 64; by gender, 70t; seasonality, 64–65
pipelines, Amboseli, 16
Pirouette-Run, 122
playback experiments: frequency attenuation studies, 168–69; response to, 127, 141, 142f, 172f, 313–15; use of, 162, 165, 170–71, 178; volume of, 168, 169t
play behaviors, 121–22; allomothering and, 231, 232f; bouts per hour by gender, 231f; calf-calf interactions, 229, 230–32; calls related to, 157, 158f, 161t; categories of, 230; Lone-Play, 121; Object-Play, 121; partners for, 231f; social development and, 237; social play, 122; Solicit-Play, 121
Play-Bush-Bashing, 122
Play-Climb-Upon, 121f
Play-Kick-Back behavior, 122
Play-Kneel-Down play, 121f
Play-Mounting, 122
Play-Pursuit behavior, 122
Play-Social-Rub behavior, 122
Play-Throw-Debris, 122
Play-Trumpets, 122
Play-Tusk-Ground, 121f, 122
poaching, 320, 329, 331
population genetics, 37–47
populations: age-structure pulses in, 86; definition of, 206; density of, 75, 343–44; growth of, 3; rate of increase, 88f; size over time, 25–28
porcine zona pellucida vaccine, 325
Porini Camp, 333
post-conflict displays, 112–13
post-copulatory behavior, 120
post-copulatory rumbles, 127, 145
Post-Copulatory Stance, 120
postures: aggressive, 110–13; threat, 110–11. *See also specific* postures
poverty, conservation issues and, 330
precipitation: Amboseli ecosystem, 14–16; drought-coping strategies, 330; dry season, 15f; monthly conception rates by, 76. *See also* rainfall
predator species: Amboseli populations, 20; calf mortality and, 199; calls in response to, 142f, 160t; confrontation with, 141
primiparous females, 76
priority-of-access models, 282–85
"problem animal" control, 323–24
problem solving, cooperative, 179
Prolonged-Pursuit, 111
protection behaviors, 117
protein: dietary, 55, 60; digestible, 60, 64f, 65f, 66–69; female intake, 66; male intake, 66; requirements, 62t, 161
protein hunger, 330
protests, calls related to, 149–51
"pull" factors, defined, 246
pulsated-play-trumpets, 157
Pulsated-Trumpets, 122
Pursuit, 111
"push" factors, defined, 246

Pushing, 111
pyramidal neurons, 175

QB-DB family, 102f, 209
QB family, 129t, 221t
quantity judgments, 178
Quilla, 101, 209

RA family, 129t, 212t, 221t
radio-tracking data, 89–93, 90f, 344
Raggedy, 209
rainfall: aggregation size and, 247; Amboseli ecosystem, 14–16; conception rates and, 77f, 79f; family cohesiveness and, 218–19; growth rates and, 193; influences on sociality, 217–18; monthly conception rates by, 76, 76f. *See also* precipitation
Ramming, 111, 112f
ranges: changes over time, 94–96; of elephants, 89–93; long-term core shifts, 99–100, 99f, 100f; non-musth, 104; observed, 90–93; occupancy at the core of, 96–104; patterns of female elephants, 38; radio-tracking data, 89–90
Rapid-Ear-Flapping posture, 110, 115, 120
Rayleigh waves, 126
Reaching-High, 113
Reach-Over behavior, 119
Real-Charge, 111
Rear-guarding behavior, 117
rearing strategies: family size and, 234–36; maternal, 232–34
reassurance behaviors, 117, 149
Reconciliation, 113, 151
Redirected-Aggression, 113
Reject-Suck behavior, 116
relatedness, coefficients of, 239
reproduction: decline with age, 189; female strategies, 187–204; influence of growth on, 192–93; influence of size on, 192–93; life expectancy and, 192; life histories in females, 192f; monthly conception rates, 75–79, 75f; size and, 4; social knowledge and, 173
reproductive senescence, 189–90
reproductive success: age and, 275f; age of male elephants, 201; components of, 232; density-dependent, 74; experience and, 258; factors influencing, 38–39; growth curves as calves and, 236; life span and, 203; male, 279–80, 283–85, 284t; for males in musth, 273; musth and, 275–76; of older males, 266; priority-of-access models, 282–85; trade-offs for older females, 201, 202f
resources, partitioning of, 251
resting, defined, 3
retreat behaviors, 112
Retreat-From, 112
revs, frequencies, 133f
rhinos, 17
roars, 134–35, 141
Rombo region, 295
rubbing behavior, 243
rumbles: of adult females, 129; analysis of, 140t, 141t; call duration, 137f; calves', 141t; characteristics, 131–33, 157t; context-types, 140; frequencies, 133f; in group defense, 141; seismic wave produced, 127; sub-types, 126, 127; types of, 140
Rump-Present behavior, 117, 119
Run-After behavior, 111

SA family, 129t, 212t, 221t
safaris, 322, 333–34
Salvadora persica, 21, 22f, 22t, 23–25, 56
Samburu National Reserve, Kenya, 46
Samburu people, 293
sample collection, 40, 239
satellite tracking, 92–93, 344
scramble competition, 257
seasonality: of breeding, 75–76; of death, 79–85; nutrient intake observed, 62–67; range changes, 98–99
sedge swamps (SWPs), 57–59
Selati case, 322
Selenkay Conservation Area, 333
self-recognition, mirror studies, 181
separated-rumbles, 146, 148f
Setter, Mark, 325
sex differences: daily food intake, 72f; in food choice, 51; habitat choices, 57f; habitat distribution, 56; in habitat use, 51; nutrient intake and, 62–67; responses to cattle sounds, 315–16. *See also* gender
sex ratios, at birth, 196, 233
sexual behaviors: advertisement/attraction, 117–18; calls related to, 143–45; courtship, 119–20; Male-Male, 120; post-copulatory, 120; sexual monitoring, 118–19; sexual solicitation, 119
sexual maturation: age for female elephants, 192, 224; age for male elephants, 224; age of, 188; reproductive success and, 232
sexual monitoring, 118–19
sexual solicitation behaviors, 119
Shearwater Adventures, 322
Shepherding behaviors, 117, 123
sightings: of adult males, 343, 343f; of bull elephants, 267f, 337, 339, 340, 343, 344; data analysis, 210–11; data collection, 339–43; definition, 3, 339–434; of family units, 94f, 341f; ground observations, 344; methods, 339–43; observer efforts, 341–42, 342f
SIGNAL RTSD analysis, 128
sizes: age and, 338–39; age at first reproduction and, 192–93; energetics and, 52–53; plasticity of, 53; study challenges, 4
Skirt-Around, 112
Slapping, 111
Smelling behavior, 164f
Sniff-Toward, 122f
snorts, 126, 134f, 138–39, 141
social clusters, 96, 98, 206
social cognition, 178–80
social complexity hypothesis, 240
social dynamics: within families, 211–13; family unit flexibility and, 222t; female elephants, 205–23; life stages, 260; male, 260–71; temporal changes in, 217; trail use and, 255
social environment, foraging and, 55
social identity, calls coding, 165–68

social integrative behaviors, 114–16, 151–53, 161t
social intelligence hypothesis, 174
sociality: in African elephants, 239–40; of calves, 225–28; dynamic nature of, 4; environmental influences on, 217–18; male elephants, 265–67; proximity issues, 228
social learning, 179–80
social manipulation, 179
social play, 122
social recognition, 168–70
social relationships: evolution of, 238–39; female elephants, 238–45
social-roars, 145
social roles, 193–95
Social-Rub behavior, 114
social terms, 3
social-trumpets, 145
soils, Amboseli ecosystem, 13
Solanga, 92, 92f
Soliciting-Guarding behavior, 120
Solicit-Play behavior, 121, 121f
Solicit-Suckling behavior, 116
Sondelani case, 322
sound, detection/production of, 125–26
Sparring, 122
sparring play, 230
spatial proximity behaviors, 115
spearing, 302; conservation policies, 317; elephant responses to, 313; of elephants, 297f; as political protest, 307; reduction of, 332–33; ritual, 307; tradition of, 331–32
Spinning behavior, 115
Split Matriarch, 194
Sporobolus spp., 56
squelches, 134f, 135f, 139
Standing-Tall threat, 110, 110f, 114, 122
Startle behavior, 142
statistical analyses, 345
status: age at first birth and, 193–95; assessment of, 230; definition of, 194; of matriarchs, 250; personality traits and, 215t, 216f; procession order and, 255, 256–57
sterilization, male elephant, 325
stress: cold-related, 52; competition and, 246; hunting and, 321
study area, 337
Suaeda monoica, 21, 22f, 22t, 23–26, 56
sub-groups, defined, 2
sub-populations, defined, 206
submission behaviors, 112
subunits, defined, 2
Suckle-Rumble call, 116
Suckle-Stance behavior, 116, 116f
suckling: behaviors related to, 116–17; delay of estrus and, 233; by grandmother elephants, 201; time spent by calves, 233–34
sustainability: drivers of, 327; threats to, 334–35
swamp-edge grasslands (SEGs), 56, 57–60
swamp-edge woodlands, 23–24, 23f, 24f, 56, 57f
swamps: agriculture on the edge of, 96; in the Amboseli basin, 56; grass biomass in, 57–58; habitat choices, 57f; map of, 341f
systematic reconnaissance flights (SRFs), 343

TA family, 221t
Tail-Raising behavior, 111, 114, 122
Tail-Swat behavior, 115
tall grasslands (TGRs), 57–58
Tanzania: bull elephant ranges, 93; calf capture in, 322; park regions in, 337
teeth, eruption of, 338
Teleki, Samuel, 26
temporin secretion, 115, 117
Test-Dung behavior, 119
Test-Genitals behavior, 118
Test-Mouth behavior, 114
Test-Penis behavior, 120
Test-Semen behavior, 120
Test-Temporal-Glands, 119
Test-Urine behavior, 119
theory of mind, 180–81
thermoregulatory costs, 52
threat postures, 110–11
threats, coalition, 111
Through Maasailand (Joseph), 26
Throw-Debris, 111, 113
Tilted-Ear-Spread, 114
Tinbergen, Niko, 1
tissue samples, 40
toes, mechanoreceptors, 126
tonal roars, 133f, 138t; characteristics, 134–35
tools: construction of, 176; use of, 176–77
Touch-Face, 113–14, 113f
tourism: cultural *bomas* and, 333; development planning and, 330–31; economic opportunities, 305–6; impact of, 292; revenues, 327–28; wildlife concession areas, 333–34
Tracking behavior, 123
Tracking musth, 119
tracks, observer, 341f
trail use: for ceremonies, 296; dominance relations and, 250–57; frequency of, 256f, 256t; habitat selection and, 255; map of, 252f; processions, 254–57; seasonal variability in, 253–54; sources of information, 252; trail types and, 259
Trample-Ground behavior, 124
Trans Mara District, Kenya, 296
translocation, 325
"trap tube" test, 176
trauma, long-term effects of, 322–23
travel, by elephants, 175–76. *See also* trail use
trophy hunting, 321
truck-like calls, 134f, 135f, 139
Trumpet-Blast escalation, 111
Trumpets, 122; analysis of, 138t; by calves, 230; characteristics, 138, 150f; frequencies, 134f; in group defense, 141; production of, 126; social integration and, 153; sub-types, 158f
trunk, vocal tract and, 125–26

Trunk-Bounce behavior, 117
Trunk-Curl behavior, 117
Trunk-Curled-Under behavior, 114
Trunk-Drag behavior, 117
trunkful weights, 64
Trunk-Grasping behavior, 115
Trunk-Out-Stretched behavior, 111, 113
Trunk-Squelching behavior, 121
Trunk-Sucking behavior, 114
Trunk-Twining behavior, 115, 122
Trunk-Twisting behavior, 113–14
trust deed holdings, 334
Tsavo East National Park, Kenya, 325
Tsavo West National Park, 89–90
Tuli Debacle, 322
Turn-Away, 112
Turn-Toward threat, 110
Tusk-Clicking behavior, 111, 115
Tusk-Ground behavior, 111
Tusking behavior, 111, 112f
Tuskless (matriarch), death of, 294
twig biomass, 58f, 63
twinning, 188
Typha spp., 56, 57

UNESCO Biosphere Reserve model, 332
Universal Transverse Mercator (UTM) maps, 344
Urination behavior, 115
Urine-Dribbling behavior, 117
urine-moving experiments, 179

V-8 rumbles, 149, 150f, 159
Valechia. See *Acacia*
vegetation, Amboseli, 16, 56
vocal communication, 162–73
vocal discrimination, 170–73
vocalizations: by calves, 229; definition, 3; sound acquisition, 128; sound measurement, 128; statistical analysis, 128–29; study methodology, 128–30
vocal learning, 127, 131, 139
"Von Economo" neurons, 175

WA family, 221t
Waiting behavior, 116
Wariness behavior, 119
water resources: Amboseli ecosystem, 13–14, 13f; control of, 324; diversion of, 329; elephant use of, 310–13, 311f; irrigated fields, 13; Maasai use of, 310–13, 311f; refurbishing of dams, 333
WB family range, 98
weaning: age of, 189f; definition of, 225
wells, Amboseli ecosystem, 16
Western, David, 2
wild date palm (*Phoenix reclinata*), 57
wildebeest playbacks, 313, 314, 314t, 315t
wildlife concession areas, 333–34
wildlife corridors, risks to, 328, 334
woodlands: in the Amboseli basin, 56; grass biomass in, 57; habitat choices, 57f

zebras, 18–19